Reference Series in Phytochemistry

Series Editors
Jean-Michel Mérillon
Faculty of Pharmaceutical Sciences
University of Bordeaux
Bordeaux, France

Kishan Gopal Ramawat
Botany Department
M.L. Sukhadia University
Udaipur, India

This reference works series provides a platform for all information on plant metabolites and phytochemicals, their chemistry, properties, applications, and methods. By the strictest definition, phytochemicals are chemicals derived from plants. However, the term is often used to describe the large number of secondary metabolic compounds found in and derived from plants. These metabolites exhibit a number of nutritional and protective functions for human welfare such as colorants, fragrances and flavorings, amino acids, pharmaceuticals, hormones, vitamins and agrochemicals. Besides food, fibers, fuel, cloth and shelter, a vast number of wild plants can hence provide important sources for medicines, especially in developing countries for their traditional health systems. Natural products have inspired and provided the foundation to the bulk of FDA-approved compounds and there is tremendous increase in natural products and natural products derived compounds that have been registered against many prevailing diseases. Natural product industry has shown tremendous growth and is expected to continue to do so in the near future. The present series compiles reference information on various topics and aspects about phytochemicals, including their potential as natural medicine, their role as chemo-preventers, in plant defense, their ecological role, their role in plants as well as for pathogen adaptation, and disease resistance. Volumes in the series also contain information on methods such as metabolomics, genetic engineering of pathways, molecular farming, and obtaining metabolites from lower organisms and marine organisms besides higher plants. The books in the series are hence of relevance in various fields, from chemistry, biology, biotechnology, to pharmacognosy, pharmacology, botany, or medicine. Each volume is edited by leading experts and contains authoritative contributions by renowned authors.

More information about this series at: http://www.springer.com/series/13872

Jean-Michel Mérillon
Kishan Gopal Ramawat
Editors

Sweeteners

Pharmacology, Biotechnology, and Applications

With 108 Figures and 57 Tables

Springer

Editors
Jean-Michel Mérillon
Faculty of Pharmaceutical Sciences
University of Bordeaux
Bordeaux, France

Kishan Gopal Ramawat
Botany Department
M.L. Sukhadia University
Udaipur, India

ISBN 978-3-319-27026-5 ISBN 978-3-319-27027-2 (eBook)
ISBN 978-3-319-27028-9 (print and electronic bundle)
https://doi.org/10.1007/978-3-319-27027-2

Library of Congress Control Number: 2017951226

Printed on acid-free paper

This Springer imprint is published by Springer Nature
The registered company is Springer International Publishing AG
The registered company address is: Gewerbestrasse 11, 6330 Cham, Switzerland

Preface

Sweeteners, artificial sweeteners, or high intensity sweeteners are natural or synthetic products imparting sweetness to the food when used as food additives. These nonsugar compounds are hardly metabolized and have little or no calorific value. Thus, demand for these sweeteners is increasing day by day for health cautious people and for diabetics. In America, average consumption of added sugar is 165 pounds (~75 kg) per person per year. Due to increase in obesity-related diseases, certain governments have taken measures to reduce sugar consumption, which causes indirectly financial burden on government exchequer in the form of medical treatment and healthcare. Though some of these sweeteners were known since long time, recent surge in demand of these products has initiated research for search of novel compounds and their derivatives, genetic modifications leading to production of these compounds in other systems, their biological effects, and finding their industrial applications. Many confectionary and soft drink products as sugar free or "diet" are available in the market. This type of product development requires evaluation on several parameters for their industrial use. This treatise is a timely compilation of all facts about the sweeteners. The chapters are written by leading scientists of the world from all the continents.

The book contains 27 chapters which are assembled in five sections: Section I: Biology and Occurrence, Section II: Biotechnology and Genetic Modifications, Section III: Biological Effects, Section IV: Methods of Analysis, and Section V: Industrial Applications. The book covers biology of sweeteners including physical and chemical properties; biotransformation and production through transgenes; health effects of sweeteners such as on gastrointestinal tract, in diabetes, in carcinogenesis, as antioxidants, and on signaling receptors; methods of their analysis; industrial applications of sweeteners; sweeteners regulations and safety aspects.

This complete book on sweeteners will be useful for all those concerned with artificial sweeteners from students to researchers in the field of botany, biotechnology, agriculture, recombinant DNA technology and transgenics, production of sweeteners, and entrepreneur for developing industrial applications of sweeteners.

We are thankful to our contributors for their support in contributing the comprehensive chapters and patience during the publication of the book. We are also grateful to Springer team, namely Tobias Wasserman, Sofia Costa, Sylvia Blago, Heike Rossel, and Clifford Nwaeburu, for the editorial expertise and continued support during the project.

July 2017

Professor J.-M. Mérillon
Professor K. G. Ramawat

Contents

1 **An Introduction to Sweeteners** 1
Arpita Das and Runu Chakraborty

Part I Biology and Occurrence 15

2 **Brazzein: A Natural Sweetener** 17
V. Rajan and J. A. Howard

3 **Cultivation of *Stevia rebaudiana* Bertoni and Associated Challenges** 35
Luciana G. Angelini, Andrea Martini, Barbara Passera, and Silvia Tavarini

4 ***Glycyrrhiza glabra*: Chemistry and Pharmacological Activity** 87
Varsha Sharma, Akshay Katiyar, and R. C. Agrawal

5 **Steviol Glycosides: Natural Noncaloric Sweeteners** 101
Indra Prakash and Venkata Sai Prakash Chaturvedula

6 **Xylitol as Sweetener** 129
Salim-ur Rehman, Mian Anjum Murtaza, and Zarina Mushtaq

Part II Biotechnology and Genetic Modifications 151

7 **Biotransformation of Mogrosides** 153
Reuben Wang, Chun-Hui Chiu, Ting-Jang Lu, and Yi-Chen Lo

8 **Mass Production of the Taste-Modifying Protein Miraculin in Transgenic Plants** 167
Hiroshi Ezura and Kyoko Hiwasa-Tanase

9 **Transgenic Plants as Producers of Supersweet Protein Thaumatin II** 185
A. P. Firsov, A. S. Pushin, and S. V. Dolgov

10 **Transgenics with Monellin** ... 211
Tanushri Kaul, Chinreddy Subramanyam Reddy, and Saurabh Pandey

Part III Biological Effects ... **223**

11 **The Role of Dietary Sugars and Sweeteners in Metabolic Disorders and Diabetes** ... 225
Motahar Heidari-Beni and Roya Kelishadi

12 **The Pharmacological Activities of Glycyrrhizinic Acid ("Glycyrrhizin") and Glycyrrhetinic Acid** ... 245
Cedric Stephan Graebin

13 **Beneficial Effects of *Stevia rebaudiana* Bertoni and Steviol-Related Compounds on Health** ... 263
Alan Talevi

14 **Health Implications of Fructose Consumption in Humans** ... 285
Luc Tappy

15 **Potential Carcinogenic Risks of Aspartame** ... 311
Morando Soffritti, Angela Guaragna, and Marco Manservigi

16 ***Stevia Rebaudiana*'s Antioxidant Properties** ... 349
Cecilia Bender

17 **Nonnutritive Sweeteners and Their Role in the Gastrointestinal Tract** ... 377
Allison C. Sylvetsky, Jenna Abdelhadi, Najy Issa, and Kristina I. Rother

18 **Sweet-Taste Receptor Signaling Network and Low-Calorie Sweeteners** ... 393
Menizibeya O. Welcome, Nikos E. Mastorakis, and Vladimir A. Pereverzev

19 **Xylitol: One Name, Numerous Benefits** ... 409
Chika Ifeanyi Chukwuma and Md. Shahidul Islam

Part IV Methods of Analysis ... **437**

20 **Analytical Strategies to Determine Artificial Sweeteners by Liquid Chromatography-Mass Spectrometry** ... 439
Rosa A. Lorenzo Ferreira, María Asunción Lage Yusty, and Antonia María Carro Díaz

21 **Characterization of Artificial Sweeteners Using Raman Spectroscopy** ... 479
Anna Grazia Mignani, Leonardo Ciaccheri, Gabriele Manca, Andrea Azelio Mencaglia, Tom Verschooten, Heidi Ottevaere, and Hugo Thienpont

22 **Sweet-Tasting Protein Thaumatin: Physical and Chemical Properties** ... 493
Tetsuya Masuda

Part V Industrial Applications ... **525**

23 **The Recent Development of a Sweet-Tasting Brazzein and its Potential Industrial Applications** ... 527
Fabrice Neiers, Christian Naumer, Michael Krohn, and Loïc Briand

24 **Sugar Alcohols as Sugar Substitutes in Food Industry** ... 547
Małgorzata Grembecka

25 **Tagatose Stability Issues in Food Systems** ... 575
Leonard N. Bell

26 **Low Calorie Intense Sweeteners: Safety Aspects** ... 591
Qurrat ul-Ain, Madiha Sikander, Sohaib Ahmed Khan, Muhammad Sikander Ghayas Khan, and Rabia Ghayas

27 **Sweeteners: Regulatory Aspects** ... 613
Jose Manuel García-Almeida, Isabel M. Cornejo-Pareja, Araceli Muñoz-Garach, Ana Gómez-Pérez, and Jorge García-Alemán

Index ... 643

About the Editors

Jean-Michel Mérillon
Faculty of Pharmaceutical Sciences
University of Bordeaux
Bordeaux, France

Prof. Dr. Jean-Michel Mérillon received his Pharm.D. (1979), Ph.D. (1984), and HDR (1992) from the University of Tours in France. He joined this same university as Assistant Professor in 1981 and became Associate Professor in 1987. In 1993, he moved to the Faculty of Pharmacy, University of Bordeaux, France, accepting a position as Full Professor. He is currently leading the "study group on biologically active plant substances" at the Institute of Vine and Wine Sciences, which comprises 25 scientists and research students. The group has been working on phenolic compounds from vine and wine for many years, mainly complex stilbenes and their involvement in health. Prof. Mérillon has supervised the doctoral theses of 20 students. He is involved in developing teaching on plant biology, natural bioactive compounds, and biotechnology.

Prof. Mérillon has published more than 150 research papers in internationally recognized journals, resulting in an H-index of 38 (documents published between 1996 and 2016). He has coedited books and reference works on secondary metabolites and biotechnology.

Throughout his career, Prof. Mérillon has traveled widely as a Senior Professor. Scientists from several countries have been and are working in his laboratory, and his research is supported by funding from the Aquitaine Regional Government, the Ministry of Higher Education and Research, and various private companies. In 2004, he founded the technology transfer unit "Polyphenols Biotech," providing support for R&D programs for SMEs and major groups from the cosmetic, pharmaceutical, agricultural, and health-nutrition sectors.

Kishan Gopal Ramawat
Botany Department
M.L. Sukhadia University
Udaipur, India

Prof. Dr. Kishan Gopal Ramawat is Former Professor and Head of the Botany Department, M.L. Sukhadia University, Udaipur, India, and can look back on longstanding research experience. He received his Ph. D. in Plant Biotechnology in 1978 from the University of Jodhpur, India, and afterwards joined the university as a faculty member. In 1991, he moved to the M.L. Sukhadia University in Udaipur as Associate Professor and became Professor in 2001. He served as the Head of the Department of Botany (2001–2004, 2010–2012); was in charge of the Department of Biotechnology (2003–2004); was a member of the task force on medicinal and aromatic plants of the Department of Biotechnology, Government of India, New Delhi (2002–2005); and coordinated UGC-DRS and DST-FIST programs (2002–2012).

Prof. Ramawat had done his postdoctoral studies at the University of Tours, France, from 1983 to 1985, and later returned to Tours as Visiting Professor (1991). He also visited the University of Bordeaux 2, France, several times as Visiting Professor (1995, 1999, 2003, 2006, 2010), and in 2005 Poland in an academic exchange program (2005). Through these visits in France, Prof. Ramawat and Prof. Mérillon established a strong connection, which has resulted in productive collaborations and several book and reference work publications.

Prof. Ramawat has published more than 170 well-cited peer-reviewed papers and articles, and edited several books and reference works on topics such as the biotechnology of medicinal plants, secondary metabolites, bioactive molecules, herbal drugs, and many other topics. His research was funded by several funding agencies.

In his research group, Prof. Ramawat has supervised doctoral theses of 25 students. He is an active member of several academic bodies, associations, and editorial boards of journals.

Contributors

Jenna Abdelhadi Department of Exercise and Nutrition Sciences, The George Washington University, Washington, DC, USA

R. C. Agrawal Department of Research, Priyamvada Birla Cancer Research Institute, M.P. Birla Hospital, Satna, MP, India

Luciana G. Angelini Department of Agriculture, Food and Environment (DAFE), University of Pisa, Pisa, Italy

Leonard N. Bell Department of Poultry Science, Food Science Program, Auburn University, Auburn, AL, USA

Cecilia Bender Institut Kurz GmbH, Köln, Germany

Istituto Kurz Italia SRL, Parma, Italy

Loïc Briand Centre des Sciences du Goût et de l'Alimentation, INRA UMR1324, CNRS UMR6265, Université de Bourgogne – Franche-Comté, Dijon, France

Antonia María Carro Díaz Department of Analytical Chemistry, Nutrition and Bromatology, Faculty of Chemistry and Health Research Institute of Santiago de Compostela (IDIS), Campus Vida, University of Santiago de Compostela, Santiago de Compostela, Spain

Runu Chakraborty Department of Food Technology and Biochemical Engineering, Jadavpur University, Kolkata, India

Venkata Sai Prakash Chaturvedula Wisdom Natural Brands, Gilbert, AZ, USA

Chun-Hui Chiu Chang Gung University of Science and Technology, Taoyuan City, Taiwan (R.O.C.)

Chika Ifeanyi Chukwuma Department of Biochemistry, School of Life Sciences, University of KwaZulu-Natal, Durban, South Africa

Leonardo Ciaccheri CNR-Istituto di Fisica Applicata "Nello Carrara", Sesto Fiorentino (FI), Italy

Isabel M. Cornejo-Pareja Endocrinology and Nutrition Department, Universitary Hospital Virgen de la Victoria, Malaga, Andalucía, Spain

Arpita Das Department of Pharmaceutical Technology, Jadavpur University, Kolkata, India

S. V. Dolgov Branch of the Shemyakin-Ovchinnikov Institute of Bioorganic Chemistry of the RAS, Russian Federation, Pushchino, Moscow Region, Russian Federation

Hiroshi Ezura Faculty of Life and Environmental Sciences, Graduate School of Life, University of Tsukuba, Tsukuba, Japan

A. P. Firsov Branch of the Shemyakin-Ovchinnikov Institute of Bioorganic Chemistry of the RAS, Russian Federation, Pushchino, Moscow Region, Russian Federation

Jorge García-Alemán Endocrinology and Nutrition Department, Universitary Hospital Virgen de la Victoria, Malaga, Andalucía, Spain

Jose Manuel García-Almeida Endocrinology and Nutrition Department, Universitary Hospital Virgen de la Victoria, Malaga, Andalucía, Spain

Rabia Ghayas Arif Memorial Teaching Hospital, Lahore, Pakistan

Ana Gómez-Pérez Endocrinology and Nutrition Department, Universitary Hospital Virgen de la Victoria, Malaga, Andalucía, Spain

Cedric Stephan Graebin Chemistry Department, Federal Rural University of Rio de Janeiro, Seropédica, RJ, Brazil

Małgorzata Grembecka Department of Food Sciences, Faculty of Pharmacy with Subfaculty of Laboratory Medicine, Medical University of Gdansk, Gdansk, Poland

Angela Guaragna Clinical Center of Cancer Prevention, Bologna, Italy

Motahar Heidari-Beni Child Growth and Development Research Center, Research Institute for Primordial Prevention of Non Communicable Disease, Isfahan University of Medical Sciences, Isfahan, Iran

Kyoko Hiwasa-Tanase Faculty of Life and Environmental Sciences, Graduate School of Life, University of Tsukuba, Tsukuba, Japan

J. A. Howard Applied Biotechnology Institute, San Luis Obispo, CA, USA

Md. Shahidul Islam Department of Biochemistry, School of Life Sciences, University of KwaZulu-Natal, Durban, South Africa

Najy Issa Department of Exercise and Nutrition Sciences, The George Washington University, Washington, DC, USA

Akshay Katiyar Plant Virology Unit, Division of Plant Pathology, Indian Agriculture Research Institute, New Delhi, India

Tanushri Kaul Nutritional Improvement of Crops Group, Plant Biology, International Centre for Genetic Engineering and Biotechnology (ICGEB), New Delhi, India

Roya Kelishadi Child Growth and Development Research Center, Research Institute for Primordial Prevention of Non Communicable Disease, Isfahan University of Medical Sciences, Isfahan, Iran

Muhammad Sikander Ghayas Khan Riphah International University, Lahore campus, Lahore, Pakistan

Sohaib Ahmed Khan Civil Hospital, Karachi, Pakistan

Michael Krohn BRAIN AG, Zwingenberg, Germany

María Asunción Lage Yusty Department of Analytical Chemistry, Nutrition and Bromatology, Faculty of Pharmacy, Campus Vida, University of Santiago de Compostela, Santiago de Compostela, Spain

Yi-Chen Lo Institute of Food Science and Technology, National Taiwan University, Taipei, Taiwan (R.O.C.)

Rosa A. Lorenzo Ferreira Department of Analytical Chemistry, Nutrition and Bromatology, Faculty of Chemistry and Health Research Institute of Santiago de Compostela (IDIS), Campus Vida, University of Santiago de Compostela, Santiago de Compostela, Spain

Ting-Jang Lu Institute of Food Science and Technology, National Taiwan University, Taipei, Taiwan (R.O.C.)

Gabriele Manca CNR-Istituto di Chimica dei Composti Organometallici and CNR-Centro Ricerca&Impresa, Sesto Fiorentino (FI), Italy

Marco Manservigi Cesare Maltoni Cancer Research Center, Bologna, Italy

Andrea Martini Department of Agriculture, Food and Environment (DAFE), University of Pisa, Pisa, Italy

Nikos E. Mastorakis World Scientific and Engineering Society and Academy, Athens, Greece

Technical University of Sofia, Sofia, Bulgaria

Tetsuya Masuda Division of Food Science and Biotechnology, Graduate School of Agriculture, Kyoto University, Uji, Japan

Andrea Azelio Mencaglia CNR-Istituto di Fisica Applicata "Nello Carrara", Sesto Fiorentino (FI), Italy

Anna Grazia Mignani CNR-Istituto di Fisica Applicata "Nello Carrara", Sesto Fiorentino (FI), Italy

Araceli Muñoz-Garach Endocrinology and Nutrition Department, Universitary Hospital Virgen de la Victoria, Malaga, Andalucía, Spain

Mian Anjum Murtaza Institute of Food Science and Nutrition, University of Sargodha, Sargodha, Pakistan

Zarina Mushtaq Institute of Food Science and Nutrition, GC University, Faisalabad, Pakistan

Christian Naumer BRAIN AG, Zwingenberg, Germany

Fabrice Neiers Centre des Sciences du Goût et de l'Alimentation, INRA UMR1324, CNRS UMR6265, Université de Bourgogne – Franche-Comté, Dijon, France

Heidi Ottevaere Department of Applied Physics and Photonics, Brussels Photonics Team, Vrije Universiteit Brussel (VUB), Brussel, Belgium

Saurabh Pandey Nutritional Improvement of Crops Group, Plant Biology, International Centre for Genetic Engineering and Biotechnology (ICGEB), New Delhi, India

Barbara Passera University of Pisa, Pisa, Italy

Vladimir A. Pereverzev Department of Normal Physiology, Belarusian State Medical University, Minsk, Belarus

Indra Prakash The Coca-Cola Company, Atlanta, GA, USA

A. S. Pushin Branch of the Shemyakin-Ovchinnikov Institute of Bioorganic Chemistry of the RAS, Russian Federation, Pushchino, Moscow Region, Russian Federation

V. Rajan Applied Biotechnology Institute, San Luis Obispo, CA, USA

Chinreddy Subramanyam Reddy Nutritional Improvement of Crops Group, Plant Biology, International Centre for Genetic Engineering and Biotechnology (ICGEB), New Delhi, India

Salim-ur Rehman National Institute of Food Science and Technology, University of Agriculture, Faisalabad, Pakistan

Kristina I. Rother Section on Pediatric Diabetes and Metabolism, National Institute of Diabetes, Digestive, and Kidney Diseases, Bethesda, MD, USA

Varsha Sharma Department of Research, Priyamvada Birla Cancer Research Institute, M.P. Birla Hospital, Satna, MP, India

Madiha Sikander University of Lahore, Lahore, Pakistan

Morando Soffritti Clinical Center of Cancer Prevention, Ramazzini Institute, Bologna, Italy

Allison C. Sylvetsky Section on Pediatric Diabetes and Metabolism, National Institute of Diabetes, Digestive, and Kidney Diseases, Bethesda, MD, USA

Department of Exercise and Nutrition Sciences, The George Washington University, Washington, DC, USA

Alan Talevi Medicinal Chemistry, Department of Biological Sciences, Faculty of Exact Sciences, National University of la Plata (UNLP) - Argentinean National Council of Scientific and Technical Research (CONICET), La Plata, Buenos Aires, Argentina

Luc Tappy Physiology Department, Faculty of Biology and Medicine, University of Lausanne UNIL, Lausanne, Switzerland

Silvia Tavarini Department of Agriculture, Food and Environment (DAFE), University of Pisa, Pisa, Italy

Hugo Thienpont Department of Applied Physics and Photonics, Brussels Photonics Team, Vrije Universiteit Brussel (VUB), Brussel, Belgium

Qurrat ul-Ain Shalamar Hospital, Lahore, Pakistan

Dow Medical College DUHS, Civil Hospital Karachi, Karachi, Pakistan

Tom Verschooten Department of Applied Physics and Photonics, Brussels Photonics Team, Vrije Universiteit Brussel (VUB), Brussel, Belgium

Reuben Wang Institute of Food Science and Technology, National Taiwan University, Taipei, Taiwan (R.O.C.)

Menizibeya O. Welcome Department of Physiology, Faculty of Basic Medical Sciences, Madonna University, Elele, Rivers State, Nigeria

World Scientific and Engineering Society and Academy, Athens, Greece

An Introduction to Sweeteners

1

Arpita Das and Runu Chakraborty

Abstract

A preference for sweet taste is innate and sweeteners can increase the pleasure of eating. Nutritive sweeteners contain carbohydrate and provide energy. They occur naturally in food or may be added in food processing or by consumers before consumption. Higher intake of added sugars is associated with higher energy intake and lower diet quality, which can increase the risk for obesity, prediabetes, type 2 diabetes, and cardiovascular disease. On average, adults in the United States consume 14.6% of energy from added sugars. Polyols (also referred to as sugar alcohols) add sweetness with less energy and may reduce risk for dental caries. The body does not differentiate between naturally occurring sugars and those added to food, but those that are added to food are most often associated with low nutrient-dense food. Consumers should limit these empty sources of energy to help achieve or maintain a healthy weight. Consumers who want a sweet taste without added energy can choose from five FDA-approved non-nutritive sweeteners based on their personal taste preference and the intended use (e.g., for cooking or for tabletop use). Non-nutritive sweeteners, when substituted for nutritive sweeteners, may help consumers limit carbohydrate and energy intake as a strategy to manage blood glucose or weight.

Keywords

Nutritive sweetener • Non-nutritive sweetener • Obesity • Sucralose • Acceptable daily intake

A. Das (✉)
Department of Pharmaceutical Technology, Jadavpur University, Kolkata, India
e-mail: arpita_84das@yahoo.co.in

R. Chakraborty (✉)
Department of Food Technology and Biochemical Engineering, Jadavpur University, Kolkata, India
e-mail: rchakraborty@ftbe.jdvu.ac.in; crunu@hotmail.com

J.-M. Mérillon, K.G. Ramawat (eds.), *Sweeteners*, Reference Series in Phytochemistry,
https://doi.org/10.1007/978-3-319-27027-2_1

Contents

1 Introduction 2
2 Classification of Sweeteners 3
2.1 Low Potency Sweetener and High Potency Sweeteners 3
2.2 Natural and Artificial Sweetener 3
2.3 Nutritive and Non-nutritive Sweeteners 4
3 Health Aspects of Sweeteners 8
3.1 During Pregnancy 8
3.2 During Childhood 9
3.3 Dental Caries 9
3.4 Obesity 9
3.5 Diabetes and Glycemic Response 10
4 Conclusions 10
References 11

1 Introduction

Everyone is interested to crave a sweet treat. Maybe it is sometime a morning macchiato, a mid afternoon cola, or to wind things down, a bowl of extreme chocolate fudge brownie overload with caramel swirl ice cream. Regardless of your favorite confectionary indulgences, we are actually born with a sweet tooth. Sweet-tasting food is more calorically dense than less sweet food, making sweetness a valid predictor of subsequent energy availability. The fondness of humans for sweet food is inborn: studies have shown a preference for sweet-tasting nutrition in newborns. Therefore, mankind has always added sweet substances to their food. Recognizing our desire for sweet flavors, the food industry has developed and supplied sugar free alternatives designed to satisfy our cravings, referred to as sugar substitutes, artificial sweeteners, or non-nutritive sweeteners. Uses of sweeteners have largely increased in the past 20 years. Nowadays they occupy a large portion of commercial space on supermarket shelves worldwide. These products are available mainly for people who are diabetic or who are looking for low-calorie materials. Alternative sweeteners are produced to be used in several products ranging from cookies to soft drinks, in order to satisfy the consumers. Where conventional sugars such as glucose, fructose, and sucrose are to be replaced by highly sweet alternatives, the fact that such alternatives often do not have the bulk that the conventional carbohydrates do, on account of their higher sweetening intensity per unit weight, must be borne in mind.

Sweet substances are compounds with diverse chemical structures and sizes, for example, sugars (sucrose), sugar alcohols (xylitol), sulfonyl amides (saccharin), peptides (aspartame), D-amino acids (D-tryptophan), and proteins (thaumatin) [1]. The worldwide demand for high-potency sweeteners is expected to increase, especially with the new practice of combining different sweeteners. Amongst naturally derived sweeteners, stevia and liquorice root are likely to become prevalent sources of high-potency sweeteners for the growing natural food market in the future. Rebaudioside A deriving from stevia is the least astringent, the least bitter, has the least persistent aftertaste, and was judged to have the most favorable sensory attributes of the four major steviol glycosides [2]. Ideally, sweeteners should be of low-caloric value, able to mask the test at lower concentration, and it should be free from harmful side effects and suitable for long-term use. It should remain stable at wide range of temperature and pH condition. It should have quick onset of action and no lingering after taste. Sweetener should be water soluble with high dissolution rate. In addition, it should be nonhygroscopic and should give synergistic effect with other sweeteners. Therefore, in addition to other factors, commercialization of sweetener needs to qualify most of these parameters [3].

2 Classification of Sweeteners

Sweeteners can be grouped in various ways. Sweetener potency is defined as the number of times that a sweetener is sweeter than sucrose. The potency of a sweetener is compared with sucrose mainly in the threshold levels of the sweetener and sucrose.

2.1 Low Potency Sweetener and High Potency Sweeteners

Sugars and sugar alcohols, such as sucrose and xylitol, are low-potency sweeteners, whose sweetener potencies are about 1 and under. On the other hand, sweeteners which have a sweetener potency exceeding 10 are called high-potency sweeteners, such as saccharin and aspartame. Interestingly, low-potency sweeteners, such as sucrose, exhibit higher sweetness intensity than high-potency sweeteners at very high concentrations. That is why low-potency sweeteners are also called high-intensity sweeteners [4–6].

2.2 Natural and Artificial Sweetener

Sweeteners are functional food additives that impart sweetness in food [7]. Sweetener can be broadly divided into two categories, natural and artificial or synthetic sweetener. Natural sweetener can be further divided into saccharide and nonsaccharide sweeteners [3]. Many other natural alternatives to sugar are available, though not widely used, despite the fact that natural nonrefined sugar alternatives potentially contain beneficial bioactive compounds, especially polyphenolic compounds, known and appreciated for their antioxidant properties. Some of these natural sugar alternatives include plant saps/syrups (e.g., maple syrup, agave nectar), syrups made from raw sugar and grains (e.g., molasses, barley malt, and brown rice syrup), honey, and fruit or vegetable sugars (e.g., date sugar, carrot). Among natural alternatives to sucrose that can be used as sweeteners and are favored due to their low glycemic index are also lucuma (*Pouteria obovata*) and yacon (*Smallanthus sonchifolius*), which do not undergo any refining process and may therefore provide a substantial content of other beneficial nutrients and bioactives. Likewise, the use of Stevia and liquorice (*Glycyrrhiza glabra*) are also well known in confectionery industry, but they still do not have a wider application due to the aftertaste that often occurs in the products. Black locust (*Robinia pseudoacacia*) is a tree native and widely spread in the southeast European region, appreciated for its medicinal properties (prepared and consumed as tea) and culinary uses (the flower nectar is used for production of honey, flowers are fried, added to dishes, or used for preparation of beverages), mostly due to its specific sweet taste and mild, flowery aroma [8]. Artificial sweeteners are being used as sugar substitutes in considerable and increasing amounts in food and beverages, especially for those who are diabetic and/or obese. They have also been used in other personal care and pharmaceutical products such as toothpastes [9]. Although, from the beginning of their use, there has been a controversy over their risk as potential carcinogens [10], these sweetener compounds are generally considered to be safe for use in food stuffs [11–13]. Some of the low-calorie sweeteners at present approved by different international authorities as direct food additives include acesulfame, aspartame, cyclamate, saccharin, and sucralose [14, 15]. Other flavorings are continually being developed and are gradually more commonly used in foodstuffs, especially because they confer longer shelf life. Just as these compounds are metabolically inert in the human body, so scientists are finding, they are also inert in the environment. Concern is shifting from health concerns to ecosystem concerns. In terms of environmental degradation, among the five most commonly used artificial sweeteners named above, only aspartame decomposes under normal usage conditions, and safety clearance was given to the intake of even its breakdown derivatives [16].

2.3 Nutritive and Non-nutritive Sweeteners

Artificial sweeteners have been classified as nutritive and non-nutritive depending on whether they are a source of calories. The nutritive sweeteners include the monosaccharide polyols (e.g., sorbitol, mannitol, and xylitol) and the disaccharide polyols (e.g., maltitol and lactitol). They are approximately equivalent to sucrose in sweetness [17]. The non-nutritive sweeteners, better known as artificial sweeteners, include substances from several different chemical classes that interact with taste receptors and typically exceed the sweetness of sucrose by a factor of 30–13,000 times. Nutritive sweeteners (e.g., sucrose, fructose) are generally recognized as safe (GRAS) by the Food and Drug Administration (FDA), yet concern exist about increasing sweetener intakes relative to optimal nutrition and health. In the United States, estimated intakes of nutritive sweeteners fall below this, although one in four children (ages 9–18 years) can surpass this level. Polyols (sugar alcohols), GRAS affirmed or petitions filed for GRAS, add sweetness with reduced energy and functional properties to foods/beverages and promote dental health. In addition to their sensory qualities, nutritive sweeteners add functional properties to foods through their effects on physical (e.g., crystallization, viscosity), microbial (e.g., preservation, fermentation), and chemical (e.g., caramelization, antioxidation) characteristics [18]. Nutritive sweeteners are easily digestible except in the cases of rare genetic abnormalities of carbohydrate metabolism (e.g., galactosemia, inherited fructose intolerance) [19]. Some unrefined nutritive sweeteners provide minerals (e.g., molasses contains calcium, iron, magnesium, and potassium), the amount per teaspoon of these minerals is practically negligible compared with the dietary reference intakes [20]. Thus, consumers should base their selection of nutritive sweeteners on sensory or functional properties, not on misconceptions of differences in nutrient value. The use of non-nutritive sweeteners began with the need for cost reduction and continued on with the need for calorie reduction. It is interesting that artificial sweeteners were actually chemicals being developed for another purpose when the researcher tasted it and found that it was sweet. Since the 1950s, non-nutritive sweeteners have become a weight-loss wonder that allowed us to have our sweets without the calories and cavities [21]. Since the discovery of saccharin in the late 1800s, non-nutritive sweeteners have been used by consumers to achieve a sweet taste, for reasons of economics, blood glucose control, or energy control. Non-nutritive sweeteners approved for use in the United States have been tested and determined to be safe at levels that are within the acceptable daily intake (level that a person can safely consume everyday over a lifetime without risk, ADI) [22]. Non-nutritive sweeteners have also seen increased use in European countries (due to the growing interest in health and an aging population) as well as in developing countries (with interest in making limited diets more palatable) [23]. One group of such sweeteners consists of substances with a very intense sweet taste and is used in small amount to replace the sweetness of a much higher amount of sugar. The sweeteners of this type currently approved for use in the United States are- aspartame, acesulfane-K, neotame, saccharin, sucralose, cyclamate, and alitame [24]. Five non-nutritive sweeteners (Table 1) with intense sweetening power have FDA approval (acesulfame-K, aspartame, neotame, saccharin, sucralose) and estimated intakes below the ADI. By increasing palatability of nutrient-dense foods/beverages, sweeteners can promote diet healthfulness.

2.3.1 Acesulfame-K

Acesulfame-K (potassium) was discovered in 1967 by chemist Karl Clauss. He noticed a sweet taste when he licked his finger while working in the laboratory. Acesulfame-K was approved in the United States in 1988 for specific uses, including a tabletop sweetener. In 1998, the FDA-approved acesulfame-K for use in beverages. In particular, it has been used to decrease the bitter aftertaste of aspartame and can be found in NutraSweet-containing products. In 2003, it was approved for general use in food, excluding meat or poultry [25]. Synthesis of acesulfame-K involves the treatment of acetoacetamide with at least two equivalents of sulfur trioxide. This results in formation of *N*-sulfoacetoacetamide, which is then dehydrated by sulfur trioxide to form oxathiaazinone dioxide. Neutralization with potassium hydroxide gives acesulfame-K [26]. Acesulfame-K is 200 times sweeter than sugar and has no calories. Brand names include Sunett and Sweet One. It can be found in baked

Table 1 Characteristics of FDA-approved artificial sweeteners [25]

Sl. no	Common name	Brand names	FDA approval	Number of times sweeter than sucrose	kcal/g	Commercial uses
1.	Acesulfame-K	Sunett, Sweet One	1988 – tabletop 1993 – beverages 2003 – general use, but not in meat or poultry	200	0	Baked goods, frozen desserts, candies, beverages, cough drops, breath mints
2.	Aspartame	NutraSweet, Equal	1981 – tabletop 1996 – general purpose	200	4	General-purpose food
3.	Neotame		2002	7000–13000	0	Baked goods, soft drinks, chewing gum, frosting, frozen desserts, jams, jellies, gelatins, puddings, processed fruit and fruit juices, toppings, syrups
4.	Saccharin	Sweet'N Low, Sweet Twin, Necta Sweet	1970	200–700	0	Tabletop sweetener, baked goods, soft drinks, jams, chewing gum
5.	Sucralose	Splenda	1998 – in 15 food categories 1999 – general-purpose sweetener	~600	0	Tabletop sweetener, beverages, chewing gum, frozen desserts, fruit juices, gelatins

goods, frozen desserts, candies, beverages, cough drops, and breath mints. Acesulfame-k is heat stable, so can be used in cooking and baking [27]. It may have a bitter after taste when used alone to sweeten food or beverage [28] Ace-k is often blended with other sweetener (usually sucralose or aspartame) whereby each sweetener masks the other's after taste and exhibit a synergistic effects by which the blend is sweeter than its components.

2.3.2 Aspartame

Aspartame was discovered in 1965 by James Schlatter a chemist [29]. It is an artificial, nonsaccharide sweetener, L-aspertyl-L phenylalanine methyl ester that is a methyl ester of the dipeptide of the amino acids aspartic acid and phenylalanine. Under strongly acidic or alkaline conditions, aspartame may generate methanol by hydrolysis. Under more severe conditions, the peptide bonds are also hydrolyzed, resulting in the free amino acids. It is slightly soluble in water. The solubility increases with higher or lower pH as well as with increased temperature. In aqueous solution the relationship between pH and stability of aspartame is a bell-shaped curve with the maximum stability at pH 4.3 [30]. The two major processes are known as the Z- and F-processes named after the protecting group used on the aspartyl group. Both of these processes produce some β-coupled products together with the desired α-aspartame. The Z-process mainly involves the dehydration of the benzyloxycarbonyl-L-aspartic acid with acetic anhydride [31]. The F-process involves the protection of the amino group of aspartic acid with a formyl group and concomitant dehydration to form anhydride [32]. This sweetener is marketed under a number of trademark names including Equal, Nutrasweet, and Candere and has a good clean sweet taste but its time-intensity profile differs from sucrose.

2.3.3 Neotame

Neotame is the newest artificial sweetener, a derivative of aspartame. Another similar compound, alitame, is pending approval before the FDA. Neotame is 7,000–13,000 times sweeter than sugar and has no calories. Synthesis method of neotame involves the hydrogenation of L-α-aspartyl–L-phenylalanine I methyl ester and 3–3 dimethyl-butyraldehyde produced in situ by the hydrolysis or cleavage of a 3-3-dimethylbutyr-aldehyde precursor [33]. Neotame is an odorless white to gray-white powder with a strong sweetness and is readily soluble in alcohols and slightly soluble in water. The 0.5% aqueous solution of neotame is weakly acidic (pH 5.8) [34]. The FDA approved neotame in 2002 as a general-purpose sweetener, excluding in meat and poultry. It can be found in baked goods, soft drinks, chewing gum, frosting, frozen desserts, jams, jellies, gelatins, puddings, processed fruits, toppings, and syrups.

2.3.4 Saccharin

Saccharin, the first artificial sweetener, was discovered serendipitously, as were most artificial sweeteners. In 1879, Constantine Fahlberg was researching the oxidation mechanisms of toluenesulfonamide while working at Johns Hopkins University in the laboratory of Ira Remsen. During his research, a substance accidentally splashed on his finger; he later licked his finger and noticed the substance had a sweet taste, which he traced back to saccharin [35]. Since that time, a number of compounds have been discovered and used as food additives for their sweetener properties. Saccharin has been in use since 1900 and obtained FDA approval in 1970. Saccharin has no calories and is 300 times sweeter than sugar [25]. Synthesis involves diazotization of methyl anthranilate and then treatment of the diazonium salt with sulfur dioxide and chloride gas to give the sulfonyl chloride which is then treated with ammonia to give saccharin [36]. It is marketed as Sweet'N Low and sweetens various products, including soft drinks, baked goods, jams, chewing gum, canned fruit, candy, dessert toppings, and salad dressings. Saccharin is also used in cosmetic products (e.g., toothpaste, mouthwash, and lip gloss), vitamins, and medications.

2.3.5 Sucralose

Sucralose was accidentally discovered in 1976 when Tate & Lyle, a British sugar company, was looking for ways to use sucrose as a chemical intermediate. This non-nutritive sweetener is made from sucrose by a process that substitutes 3 chloride atoms for 3 hydroxyl groups on the sucrose molecule [25]. Sucralose is 600 times sweeter than sugar and contains no calories. Sucralose was approved by the FDA in 1998 for use in 15 food categories, including a tabletop sweetener under the brand name Splenda. It is used in beverages, chewing gum, frozen desserts, fruit juices, and gelatins. In 1999, the FDA expanded its use as a general-purpose sweetener in all food. Sucralose is very much soluble in water and is stable over a wide range of pH and temperature [37].

3 Health Aspects of Sweeteners

Both nutritive and non-nutritive sweeteners have generated health concerns among health care providers and the public for many years [38, 39]. Concerns related to safety of non-nutritive sweeteners are addressed primarily in animal studies. Artificial sweeteners are present in many food consumed by whole world. Their use is beneficial in that they provide sweetness, increasing the palatability of food without the added sugar and resulting calories, an important adjunct to weight loss and diet regimens. Most artificial sweeteners are not metabolized by the body and are therefore considered safe. However, scientists disagree about safety because the metabolites of the “nonmetabolized” compounds have been shown to produce deleterious effects in mice, rats, and dogs.

3.1 During Pregnancy

Pregnancy is a time of special concern because the focus is on maternal and fetal health. All FDA-approved nutritive sweeteners and non-nutritive sweeteners are approved for use by the general public, which includes pregnant and lactating women. The position of the Academy is that use of nutritive sweeteners is acceptable during pregnancy [40]. The safety of acesulfame-K, aspartame, sucralose, and neotame in pregnancy has been determined with rat studies; the scientific community accepts rats and some other animals as appropriate models for reproductive toxicology testing that is applicable to human beings. At high doses, there was no change observed in fertility, size of litter, body weight, growth, or mortality for acesulfame-K, sucralose, or neotame [41–43]. Use of aspartame within the FDA guidelines appears safe for pregnant women. Thus, consumption of these sweeteners within the acceptable daily intakes appears safe during pregnancy.

3.2 During Childhood

Because of their size and relatively high food and fluid intakes compared with adults, children will have the highest intake of nutritive and non-nutritive sweeteners as calculated by milligram intake/kg bw/day. Children can safely consume nutritive sweeteners. However, healthy young children (6–18 months) can exhibit malabsorption because of incomplete digestion of fructose found naturally in fruit juices or added to fruit drinks and carbonated sodas. Therefore, children exhibiting nonspecific diarrhea may benefit from a reduction in fruit juice or drinks containing fructose and polyols [19]. The estimated intakes of non-nutritive sweeteners in children are below the established acceptable daily intakes for all approved sweeteners. As a percentage of ADI, they are as low as 10.4% for aspartame to as high as 60% for acesulfame-K. It has been suggested that caregivers may want to limit intake of saccharin by young children because of the limited amount of data available for its use in children [44]. The wide range of nutritive and non-nutritive sweeteners available in the food supply, as well as blending these sweeteners in food and beverage systems, should continue to keep estimated intakes of non-nutritive sweeteners in children well below the acceptable daily intakes.

3.3 Dental Caries

Risk of dental caries increases with intake of nutritive sweeteners; this risk, however, does not work independently of factors such as oral hygiene and fluoridation. Dental caries are the localized destruction of dental hard tissue by acidic material from bacterial fermentation of dietary carbohydrate [45]. Factors that influence the development of dental caries include microbiological shifts in the biofilm, salivary flow, buffering capacity of saliva, frequency and kind of dietary sugars consumed, length of time oral bacteria have to ferment the fermentable carbohydrate and make organic acids, tooth susceptibility, preventive behaviors such as cleaning of teeth [46], and exposure to fluoride [47]. Use of polyol-based gum can reduce the risk of dental caries in children, with the greatest benefit in xylitol-based gums [48]. The FDA authorizes use of the health claim in food labeling that sugar alcohols and some novel sugars (xylitol, sorbitol, erythritol, tagatose, mannitol, maltitol, isomalt, lactitol, hydrogenated starch hydrolysates, hydrogenated glucose syrups, or a combination of these) do not promote tooth decay. Non-nutritive sweeteners do not promote dental caries.

3.4 Obesity

The causes of overweight and obesity are multifactorial, and the focus on any single factor no doubt oversimplifies the issue. Nevertheless, with regard to recent and rapid increases in the prevalence of obesity, scientific evidence has implicated a number of dietary factors as likely contributors. Most recently, special attention has been focused on the extremely high levels of consumption of sugars in general and sugar-sweetened beverages in particular. For example, in the United States overall consumption of sugar-sweetened soft drinks in 2001 was roughly 37 gallons per capita [49]. In 2012 over 70% of adults reported that they consumed sugar-sweetened beverages (SSB; soft drinks or fruit drinks with added sugar) [50], with over 25% reporting daily intake. A recent meta-analysis also showed strong links between SSB consumption and increased body weight [51]. The prevalence of obesity has increased substantially at the same time as the consumption of non-nutritive sweeteners has increased. The question is, do these sweeteners maintain a highly sweet food environment to increase risk of obesity through appetite, intake, and energy regulation mechanisms? Some evidence primarily from studies with animals suggests that high intakes of sweets (nutritive sweeteners alone or in mixtures with fat) promotes weight gain through changes in neuropeptide control of appetite, intake, and energy expenditure [52]. Thus, the rise in prevalence clearly relates to all factors that cause an energy imbalance. Individuals who wish to lose weight may choose to use non-nutritive sweeteners but should do so within the context of a sensible weight management program including a balanced diet and exercise.

3.5 Diabetes and Glycemic Response

It is well recognized that sweeteners do not cause diabetes. Increasing intakes of sugars are not associated with increasing risk of diabetes [53, 54], with the latest affirmation from a prospective study of over 39,000 women [55]. Intakes as high as 60 g fructose or sucrose per day may not adversely affect glycemic or lipid response in persons with type 2 diabetes [56]. However, because there exists concern for increased blood lipid levels with high intakes of fructose, addition of fructose as a sweetening agent is not recommended for people with diabetes [57]. Polyols, including trehalose, produce a lower glycemic response than fructose, glucose, or sucrose, most likely because of their incomplete absorption [58]. Therefore, these substances can be used safely in the diets of people with diabetes; however, because of its laxative effect, the amount of polyols consumed may need to be limited (especially in children). The non-nutritive sweeteners do not affect glycemic response and can be safely used by those with diabetes.

4 Conclusions

Sweeteners elicit pleasurable sensations with (nutritive) or without (non-nutritive) energy. Consumers can enjoy a wide range of sweeteners in a wide variety of food and beverages. The range of nutritive and non-nutritive sweeteners allows choice in the type and amount of sweeteners to include in the diet. Non-nutritive sweeteners are safe for use within the approved regulations. Non-nutritive sweeteners are those that sweeten with minimal or no carbohydrate or energy. They are regulated by the

Food and Drug Administration as food additives or generally recognized as safe. The Food and Drug Administration approval process includes determination of probable intake, cumulative effect from all uses, and toxicology studies in animals. Five non-nutritive sweeteners are approved by FDA; those are acesulfame-K, aspartame, neotame, saccharin, and sucralose. They have different functional properties that may affect perceived taste or use in different food applications. Consumers must be aware of science-based information about sweeteners and supportive research on the use of sweeteners to promote eating enjoyment, optimal nutrition, and health.

References

1. Shallenberger RS, Acree TE (1967) Molecular theory of sweet taste. Nature 216:480–482
2. Brandle JE, Starratt AN, Gijzen M (1998) Stevia rebaudiana: its agricultural, biological, and chemical properties. Can J Plant Sci 78:527–536
3. Surana SJ, Gokhala SB, Rajmane RA, Jadhav RB (2006) Non- saccharide natural intense sweeteners – an overview of current status. Nat Prod Radiance 5(4):270–278
4. Hayes JE (2008) Trans disciplinary perspectives on sweetness. Chemosens Percept 1:48–57
5. Cardello HMAB, Da Silva MAPA, Da Masio MH (1999) Measurement of the relative sweetness of stevia extract, aspartame and cyclamate/saccharin blend as com-pared to sucrose at different concentrations. Plant Foods Hum Nutr 54:119–130
6. Farkas A, Hid J (2011) The black agonist-receptor model of high potency sweeteners, and its implication to sweetness taste and sweetener design. J Food Sci 76:S465–S468
7. DuBois GE, Prakash I (2012) Non-caloric sweeteners, sweetness modulators, and sweetener enhancers. Annu Rev Food Sci Technol 3:353–380
8. Belšcak-Cvitanovic A, Komes D, Dujmović M, Karlović S, Biškic M et al (2015) Physical, bioactive and sensory quality parameters of reduced sugar chocolates formulated with natural sweeteners as sucrose alternatives. Food Chem 167:61–70
9. Zygler A, Wasik A, Namiesnik J (2009) Analytical methodologies for determination of artificial sweeteners in food stuffs. TRAC Trends Anal Chem 28(9):1082–1102
10. Weihrauch MR, Diehl V (2004) Artificial sweeteners-do they bear a carcinogenic risk? Ann Oncol 15(10):1460–1465
11. Kroger M, Meister K, Kava R (2006) Low calorie sweeteners and other sugar substitutes: a review of the safety issues. Compr Rev Food Sci Food Saf 5(2):35–47
12. Ahmed FE, Thomas DB (1992) Assessment of the carcinogenicity of the non-nutritive sweetener cyclamate. Crit Rev Toxicol 229(2):81–118
13. Cohen SM, Arnold LL, Emerson JL (2008) Safety of saccharin. Agro Food Ind Hi-tech 19(6):24
14. US FDA (2006) Artificial sweeteners: no calories. sweet! FDA Consum Mag 40(4):27–28
15. EU (2003) European Parliament and Council Directive 2003/115/EC of 22 December 2003 Amending Directive 94/35/EC on Sweeteners for Use in Food stuffs
16. US FDA (1983) Food additives permitted for direct addition to food for human consumption; aspartame. Final rule. Fed Regist 48:31376–31382
17. Dills WL (1989) Sugar alcohols as bulk sweeteners. Annu Rev Nutr 9:161–186
18. Davis E (1995) Functionality of sugars: physicochemical interactions in foods. Am J Clin Nutr 62:170S–177S
19. Rumessen J (1992) Fructose and related food carbohydrates: sources, intake, absorption, and colinical implications. Scand J Gastroenterol 27:819–828
20. National Academy of Sciences and Institute of Medicine (1997) Dietary reference intake for calcium, phosphorus, magnesium, vitamin D and fluoride. N.A. Press, Washington, DC

21. Kovacs Betty Artificial sweeteners (2011) http://www.medicinenet.com/script/main/art.asp?articlekey=81475
22. Renwick AG (2006) The intake of intense sweeteners – an updated review. Food Addit Contam 23(4):327–338
23. Bright G (1999) Low-calorie sweeteners- from molecules to mass markets. In: Corti A (ed) Lowcalorie sweeteners: present and future, vol 85, World Review of Nutrition and Dietetics. S. Karger AG, Basel, pp 3–8
24. Godshall MA (2007) The expanding world of nutritive and non- nutritive sweeteners. Sugar J 69:12–20
25. Food and Drug Administration (2006) Artificial sweeteners: no calories. . .sweet! Retrieved 2 Aug from www.fda.gov/fdac/features/2006/406_sweeteners.html
26. Clauss J, Schmidt RK, Spiess HW (1993) Determination of domain sizes in heterogeneous polymers by solid-state NMR. Acta Polym 44:1–17
27. Nabors LO (2002) Sweet choices: sugar replacements for foods and beverages. Food Technol 56:28–32
28. Horne J, Lawless HT, Speirs W, Sposato D (2002) Bitter taste of saccharin and acesulfame-K. Chem Senses 31:8–27
29. Mazur RH, Goldkamp AH, James PA, Schlatter JM (1970) Structuretaste relationships of aspartic acid amides. J Med Chem 6:1217–1221
30. Mazur RH, Ripper A (1979) Peptide-based sweeteners. Applied Science Publishers, London
31. Ager DJ, Pantaleone DP, Henderson SA, Katritzky AR, Prakash I, Walters DE (1998) Commercial, synthetic nonnutritive sweeteners. Angew Chem Int Edit 37:1802–1817
32. Hill JB, Gelman Y, Dryden Jr HL, Erickson R, Hsu K, Johnson MR (1991) One-pot process for the preparation of α-L-aspartyl-L-phenylalanine methyl ester hydrochloride. JustiaPatents US Patent 5053532. http://patents.justia.com/1991/05053532.html. Accessed 15 June 2010
33. Prakash I (2007) Synthesis of *N*-[*N*-(3,3-dimethylbutyl)-L-α-aspartyl]-L-phenylalanine 1-methyl ester using 3,3-dimethylbutyraldehyde precursors. Ip.com US Patent 7288670. http://ip.com/patent/US7288670. Accessed 2 June 2010
34. Prakash I, Corliss G, Ponakala R, Ishikawa G (2002) Neotame: the next-generation sweetener. Food Technol 56:36–40
35. Arnold DL (1983) Two-generation saccharin bioassays. Environ Health Perspect 50:27–36
36. Tarbell DS, Tarbell AT (1978) The discovery of saccharin. J Chem Educ 55:161–162
37. Beyts PK, Lillard DW, Batterman CK (1995) A sucralose compositon. PatentstromUS patent 5380541. http://www.patentstorm.us/patents/5380541/fulltext.html. Accessed 21 Aug 2009
38. Bray GA, Nielsen SJ, Popkin BM (2004) Consumption of high-fructose corn syrup in beverages may play a role in the epidemic of obesity. Am J Clin Nutr 79(4):537–543
39. Whitehouse CR, Boullata J, McCauley LA (2008) The potential toxicity of artificial sweeteners. AAOHN J 56(6):251–259
40. Position of the American Dietetic Association (2008) Nutrition and lifestyle for a healthy pregnancy. J Am Diet Assoc 108(3):553–561
41. Toxicological Evaluation of Certain Food Additives (1980) WHO Food Additives Series.16:11
42. Food and Drug Administration (1998) Food additives permitted for direct addition to food for human consumption; Sucralose. 21CFR172.64
43. Food and Drug Administration (1998) Food additives permitted for direct addition to food for human consumption: Neotame. Washington, DC: Federal Register; April 3, 1998 2002. 21 CFR 172.829.
44. Council on Scientific Affairs (1985) American Medical Association. Saccharin: review of safety issues. JAMA 254:2622
45. American Academy of Pediatric Dentistry (2011) Guideline on infant oral health care. http://aapd.org/media/Policies_Guidelines/G_InfantOralHealthCare. pdf. Accessed 12 Dec 2011
46. Selwitz RH, Ismail AI, Pitts NB (2007) Dental caries. Lancet 369(9555):51–59

47. National Institutes of Health (2001) Diagnosis and management of dental caries throughout life. NIH Consensus Statement. http://consensus.nih.gov/2001/2001DentalCaries115html.htm. Accessed 30 Apr 2011
48. Makinen K, Bennett C, Hujoel P, Isokangas P, Isotupa K, Pape HJ (1995) Xylitol chewing gums and caries rates: a 40-month cohort study. J Dent Res 74:1904–1913
49. USDA, Economic Research Service (2008) Beverages. Per capita consumption. http://www.ers.usda.gov/data/foodconsumption/spreadsheets/beverage.xls. Last accessed 26 Oct 2008
50. Kumar GS, Pan L, Park S, Lee-Kwan SH, Onufrak S, Blanck HM (2014) Sugar-sweetened beverage consumption among adults. 18 states, 2012. MMWR Morb Mortal Wkly Rep 63(32):686–690
51. Malik VS, Pan A, Willett WC, Hu FB (2013) Sugar-sweetened beverages and weight gain in children and adults. A systematic review and meta-analysis. Am J Clin Nutr 98(4):1084–1102. doi:10.3945/ajcn.113.058362
52. Levine A, Kotz C, Gosnell B (2003) Sugars: hedonic aspects, neuroregulation and energy balance. Am J Clin Nutr 78:834S–842S
53. Meyer KA, Kushi LH, Jacobs DR Jr, Slavin J, Sellers TA, Folsom AR (2000) Carbohydrates, dietary fiber, and incident type 2 diabetes in older women. Am J Clin Nutr 71:921–930
54. Colditz GA, Manson JE, Stampfer MJ, Rosner B, Willett WC, Speizer FE (1992) Diet and risk of clinical diabetes in women. Am J Clin Nutr 55:1018–1023
55. Janket SJ, Manson JE, Sesso H, Buring JE, Liu S (2003) A prospective study of sugar intake and risk of type 2 diabetes in women. Diabetes Care 26:1008–1015
56. Malerbi D, Paiva E, Duarte A, Wajchenberg B (1996) Metabolic effects of dietary sucrose and fructose in type II diabetic subjects. Diabetes Care 19:1249–1256
57. Association AD (2003) Evidence-based nutrition principles and recommendations for the treatment and prevention of diabetes and related complications. Diabetes Care 26:51S–61S
58. Jentjens RL, Jeukendrup AE (2003) Effects of pre-exercise ingestion of trehalose, galactose and glucose on subsequent metabolism and cycling performance. Eur J Appl Physiol 88:459–465

Part I

Biology and Occurrence

Brazzein: A Natural Sweetener

2

V. Rajan and J. A. Howard

Abstract

Brazzein is a 54-amino acid, heat-stable sweetener derived from the ripe fruit of the West African plant *Pentadiplandra brazzeana* Baillon. Its historical use as a sweetener testifies to its safety, and it has been extensively studied, both in regard to its structure and interaction with taste receptors. It has also been investigated as a possible commercial product to fulfill the need for a natural, low-calorie, heat-stable sweetener. Here, we review brazzein in the context of other sweeteners and examine attempts to produce brazzein in recombinant systems.

Keywords

Brazzein • Sweet protein • Natural sweetener • Recombinant sweetener • Low-calorie sweetener • pyrE brazzein • Type 1 brazzein • Des-pyrE brazzein • Type 2 brazzein • E1 brazzein • Type 3 brazzein • T1R2/T1R3 receptors

Contents

1 Introduction 18
  1.1 Evolutionary Importance of Glucose 18
  1.2 Small Carbohydrate Sweeteners 18
  1.3 Synthetic, Noncarbohydrate Sweeteners 19
  1.4 Plant-Derived Sweeteners 20
2 Discovery of Protein Sweeteners 20
  2.1 Structure and Action of Sweet Proteins 21
  2.2 Protein Sweeteners and Their Interactions with the Taste Receptor T1R2/T1R3 22

V. Rajan (✉) • J.A. Howard (✉)
Applied Biotechnology Institute, San Luis Obispo, CA, USA
e-mail: vidyarajan@hotmail.com; jhoward@appliedbiotech.org

J.-M. Mérillon, K.G. Ramawat (eds.), *Sweeteners*, Reference Series in Phytochemistry,
https://doi.org/10.1007/978-3-319-27027-2_3

3 Brazzein: A High-Intensity Natural Protein Sweetener 23
3.1 Structure of Brazzein 24
3.2 Recombinant Production of Brazzein 26
3.3 Considerations for Brazzein as a Commercial Product 29
4 Conclusions 30
References 31

1 Introduction

1.1 Evolutionary Importance of Glucose

The monosaccharide glucose is the primary substrate for the operation of the glycolytic pathway, an ancient and primitive catabolic pathway that releases energy from stored covalent bonds. Unsurprisingly the presence of the metabolite glucose can be sensed by heterotrophs which are highly affected by variations in nutrient concentrations in their environment. In bacteria, many sugar chemosensors promote positive chemotaxis of bacteria toward sugars [1]. In yeast, two unrelated glucose sensors [2] stimulate the expression of hexose transporters that bind and take up the sugar. Photoautotrophs, such as plants, also contain sensors for sugars, but these have intracellular effects such as activation or repression of glucose catabolic pathways [3]. Animals, too, have metabolite sensors called "taste receptors" that can guide them to sweet foods with high nutritive value, although the types of receptors vary between groups of animals.

Taste and odor are primitive functions that discriminate between nutritive and nonnutritive food materials [4]. Generally speaking, sweeteners impart sweet taste by binding to G-protein-linked taste receptors TIR1, TIR2, and TIR3. There is, however, some variation in the receptors that respond to different sweeteners in different phyla. Insects contain the G-protein-linked GR5α receptor [5] and [6], whereas mammals have a dimeric G-protein-linked T1R2/T1R3 [5, 7, 8]. Ligands which activate the mammalian T1R2/T1R3 sweetness receptor range from small carbohydrates such as glucose (180 Da) to large proteins such as the 207-amino acid thaumatin (2220 Da) [9].

The evolutionary association of sweet with nutrition has been exploited by both animals and plants, one for obtaining nutrition and the other for dispersal of pollen or seeds by luring animals to collect nectar or fruit. Thus, sweet molecules are often found associated with products for dispersal such as pollen or seed, with animals acting as willing dispersal agents. Sweet molecules are also important for metabolism in the plants themselves. Glucose is converted to the disaccharide sucrose in mesophyll cells before loading into phloem for transport to sink regions of roots, stems, and fruit [10].

1.2 Small Carbohydrate Sweeteners

The abundant production of sucrose by sugarcane, sugar beet, and maple and its storage as honey by honeybees were exploited by humans craving sweeteners, leading to the industrial production of all these products, especially sugarcane and beet that yielded highly transportable, crystallized sucrose. Over time, high cost and low production were replaced by low cost and high production such that a kilogram of sucrose in the United States in February 2016 cost only around \$0.13/lb or approximately \$0.26/kg (http://www.indexmundi.com/commodities/?commodity=sugar). The cost of the other, most widely used sweetener used in the prepared food industry, high fructose corn syrup, has more than doubled in price from \$0.16 in 2000 to \$0.41/lb in 2016 (http://www.ers.usda.gov/datafiles/Sugar_and_Sweeteners_Yearbook_Tables/World_and_US_Sugar_and_Corn_Sweetener_Prices/TABLE09.XLS), probably due to diversion of cornstarch to ethanol production. Still, the relatively low price of sweeteners makes them a valuable commodity as raw material costs relative to the cost of end products, such as carbonated beverages, are high. The comparatively low cost for the finished product makes them a financially accessible item, making them proportionately high in the diet profile of low-income households. The proliferation and easy access to these sugar-rich foodstuffs have provided calorie-rich, but nutrient poor, foods at low monetary, but high public health, costs [11].

1.3 Synthetic, Noncarbohydrate Sweeteners

Burgeoning demand by a health-aware public for non-caloric, non-glycemic sweeteners resulted in the production of many artificial, non-metabolized sweeteners such as saccharin (from coal tar) and cyclamate (from cyclohexamine) to sucralose (Splenda), which fulfilled the need for a time. Early artificially synthesized sweeteners such as saccharin (synthesized in 1879 from coal tar), cyclamate (synthesized in 1937 from cyclohexylamine), and aspartame (synthesized in 1965) were small molecules with molar masses of about 183, 201, and 294 g/mol, which are similar to the masses of glucose (180 g/mol) and sucrose (360 g/mol) (see Fig. 1). Because of concerns about aftertaste, natural and artificial sweeteners were blended to improve taste and potency [12]. Artificial sweeteners do not trigger insulin production but had other problems such as potential carcinogenicity or metabolism to potentially toxic chemicals [13]; animal studies linked some artificial sweeteners to cancer [14] and alteration of gut microbiota ([15]. Note that sugar, glucose, fructose, and other naturally occurring sweet carbohydrates are D-sugars, but L-sugars also have sweetness without being metabolized by the body.

Public health concerns about the potential for toxicity of artificial sweeteners are somewhat exaggerated (http://www.cancer.gov/about-cancer/causes-prevention/risk/diet/artificial-sweeteners-fact-sheet). Nonetheless, negative public perception

Fig. 1 Structures of the synthetic sweeteners saccharin and cyclamate

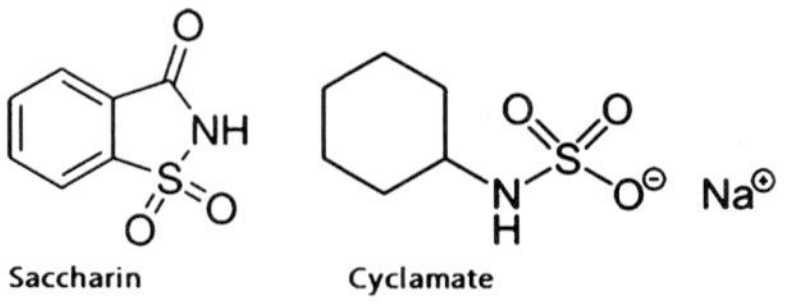

about nonfood origin synthetic sweeteners spurred the search for natural sweeteners with low glycemic indices [13]. Attention turned to the identification of natural sweeteners from plants used historically by indigenous populations. Their safe use over many generations indicates they may pass regulatory muster by the FDA and be approved for commercial use more rapidly than synthetic compounds.

1.4 Plant-Derived Sweeteners

It was recognized that many fruits are sweet and fulfilled an ecological and evolutionary role for dispersal of their seeds. It was rationalized that to successfully recruit animal dispersal, they should generally be nontoxic. A slew of sweet molecules chemically classified as phenylpropanoids, amino acids, glycosides, terpenoids, flavonoids, and proteins were consequently identified from green plants. Some of these molecules were modified for better sweetener outcomes and developed commercially. By mid-2002, there were approximately 20 families of compounds obtained from vascular green plants with about 100 discrete compounds that exhibit natural sweetness [16–18].

Glycoside sweeteners from plants include the no-calorie, non-glycemic stevioside, rebaudioside A, from *Stevia rebaudiana* (Bertoni) [19], and the cucurbitane glycosides mogroside V and 11-oxo-mogroside V from *Siraitia* (previously *Momordica) grosvenorii* [20]. Kinghorn and Soejarto [21] describe a large number of terpenoid and phenolic sweeteners that could be further modified for commercial purposes. Of the saponins, one of the most widely used sweeteners is glycyrrhizin, obtained from the root of the licorice plant *Glycyrrhiza glabra*. This compound is often used in naturotherapy, but glycyrrhizin has been shown to inactivate the cortisol-inactivating enzyme, 11β-hydroxysteroid dehydrogenase, in the kidney [22] and leads to cortisol-mediated stimulation of aldosterone receptors causing hypertension and hypokalemia [23]. Thus, although it has been used over several centuries, has many apparently therapeutic effects, and has GRAS status, conferred by the FDA, the intake of more than 200 mg/day of licorice is not recommended [24].

Plant-derived sweet proteins are non-glycemic and do not trigger the release of insulin. They also do not cause dental caries, which is a problem with carbohydrate sweeteners which are metabolized in the mouth by bacteria, producing acids that can erode enamel. The amino acids they are composed of, while not contributing nutritive benefit due to the small quantities consumed, have no deleterious side effects.

2 Discovery of Protein Sweeteners

The US Department of Health and Human Services in their 2010 guidelines asked Americans to cut back on the amount of added salt and sugar in their foods and to eat "dense" foods high in nutritional value (http://www.cnpp.usda.gov/sites/default/files/dietary_guidelines_for_americans/DG2010Brochure.pdf). This publication advises consumers to "Check for added sugars using the ingredients list. When a

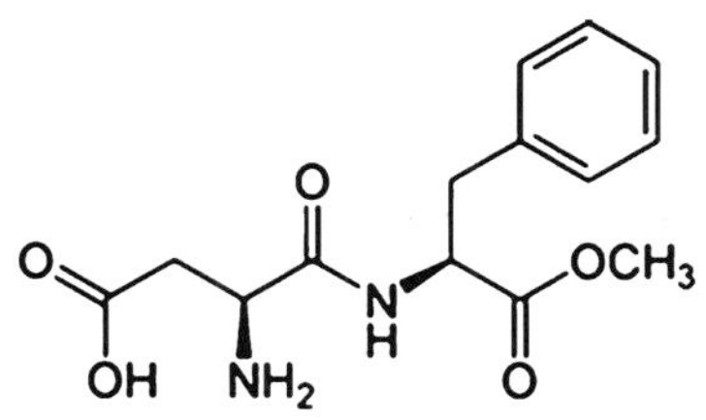

Aspartame

Fig. 2 Structure of the aspartate-phenylalanine dipeptide sweetener, aspartame

sugar is close to first on the ingredients list, the food is high in added sugars. Some names for added sugars include sucrose, glucose, high fructose corn syrup, corn syrup, maple syrup, and fructose." This advice stems from public health concerns over the consumption of large amounts of carbohydrate sweeteners which represents a greater than 15% intake of calories daily in sweeteners alone (http://www.cdc.gov/nchs/data/databriefs/db87.htm#x2013;2008</a>). Increases in levels of obesity and type II diabetes across the age spectrum underlined concerns that these glycemic sweeteners were detrimental to health. The main advantage of protein sweeteners over carbohydrate sweeteners is that they have an extremely high sweetness intensity thereby requiring an insignificant amount of calories in comparison to sucrose.

The first amino acid sweetener, aspartame, a synthetic dipeptide, was discovered serendipitously in 1965. It is composed of a methylated L-aspartate-L-phenylalanine dipeptide. Aspartate is negatively charged at neutral pH and phenylalanine is aromatic in nature. Aspartame contains 50% phenylalanine and is not recommended for consumption for individuals with the genetic condition phenylketonuria. Other protein sweeteners have varying phenylalanine quantities but none as high as aspartame (Fig. 2).

2.1 Structure and Action of Sweet Proteins

Before examining protein sweeteners, it is worthwhile to examine the factors that contribute to sucrose being the benchmark against which all other sweeteners are measured (https://www.ncsu.edu/project/sweeteners/sweetener_comparisons.html). Sucrose has excellent stability properties at high temperatures as well as humectant properties and mouthfeel that enhance the taste and texture of cooked products. It is also highly soluble in water. Disadvantages of sucrose are its high glycemic index and high calorie value. Therefore, an ideal sweetener would be heat stable for long shelf life and use in cooking, be soluble in water, and be easy to produce large amounts at a low cost, either by recombinant methods or purification.

Currently, there are eight known sweet proteins ranging in size from 54 residues and 6 kDa to 207 residues and 22 kDa (Table 1; modified from [9]). The most

Table 1 Comparison of characteristics of protein sweeteners (Modified from ([9]). Licensed under an Open Access Creative Commons attribution license)

Protein	Size	Amino acids	Sweetness (relative to sucrose)	Source
Brazzein	6.5 kDa	54	500–2000	*Pentadiplandra brazzeana* Baillon
Thaumatin	22.2 kDa	207	3000	*Thaumatococcus daniellii* Benth
Monellin	10.7 kDa	45 (A chain) 50 (B chain)	3000	*Dioscoreophyllum cumminsii* Diels
Curculin	24.9 kDa	114 (active form is dimer)	500	*Curculigo latifolia*
Mabinlin	12.4 kDa	33 (A chain) 72 (B chain)	100	*Capparis masaikai* Levi
Miraculin	98.4 kDa	191 (active form is tetramer)	n/a	*Richadella dulcifica*
Pentadin	12.0 kDa	Not available	500	*Pentadiplandra brazzeana* Baillon
Aspartame	294.31 g/mol	2	200	Synthetic dipeptide

important features of a successful sweetener are related to stability, solubility, and taste. Amino acid sequences of protein sweeteners can be directly targeted for genetic modification to improve these features and also provide enhanced expression for commercial benefit.

2.2 Protein Sweeteners and Their Interactions with the Taste Receptor T1R2/T1R3

Taste is an important sensory pathway to distinguish desired nutritive foodstuffs from undesired and toxic materials. Sweet and umami (amino acid) taste receptors were found to be clustered at the distal end of chromosome 4 and are mediated by the T1R family of receptors in heteromeric form. In mice, sweet taste is mediated by the T1R2/T1R3 combination in all circumvallate, foliate, and palate taste buds [8] and correlates to S fibers in the marmoset [25], whereas umami (a savory taste) is mediated by the T1R1/T1R3 combination in fungiform and palate taste buds [26]. Bitter taste, including bacterial secretions, is sensed by T2R receptors which correlate to Q fibers in marmoset [25]. Activation of T2R receptors also induces the secretion of antimicrobial peptides in the upper respiratory pathway [27]. T1R2/T1R3 receptors thus antagonize the T2R pathway, downregulating the innate immune response [27]. Sour and salty tastes are either encoded elsewhere, probably through ion transport pathways, and do not share cross-talk between T1R and T2R receptors [28].

The T1R2/T1R3 G-protein-coupled receptors were initially isolated using degenerate PCR based on the T1R1 sequence [29, 30]. In functional assays, transfected human T1R2/T1R3 receptors were shown to respond to sucrose as well as monellin and aspartame in HEK-293 T cells. This interaction was blocked by lactisole, a sweet taste inhibitor, showing that the response was specific to taste perception. Transfected rat T1R2/T1R3 showed a lack of sweet perception of aspartame, thaumatin, cyclamate, neotame, and monellin and was not affected by lactisole, which does not block sweet perception in rats [30]. Both rat and human receptors were stimulated by a variety of natural and synthetic sweeteners, including acesulfame-K, dulcin, glycine, lactose, maltose, fructose, sucrose, saccharin, sucralose, and D-tryptophan but not L-tryptophan, quinine, NaCl, or L-glutamate. Sucrose stimulated both murine and human receptors equivalently. These results indicate species-specific differences in taste perception [30]. Further refinement showed that a cysteine-rich region of receptor T1R3 was responsible for the perception of sweet tastes of the proteins brazzein and thaumatin in humans, apes, and old-world monkeys [31]. Electrophysiological responses in monkeys and sweetness tests in mice and humans have indicated that sweetness is perceived in the anterior areas of the tongue [32] rather than the back of the tongue, which is where histochemical data show co-expression of these receptors in mice [8]. According to [32], this might indicate the presence of other sweet receptors or receptor interactions.

Temussi and colleagues have hypothesized that small carbohydrate and protein sweeteners such as aspartame and larger protein sweeteners stimulate the same T1R2/T1R3 receptors in distinct ways. According to their hypothesis, the heterodimeric receptor exists in two equilibrium states. Low molecular weight sweeteners bind to the resting state and shift it to its active state where it stimulates a bound G-protein cascade, whereas large proteins bind a negatively charged extracellular domain and stabilize the receptor in its active state [7]. Note that the onset of sweetness is variable with either a rapid sweetness onset (sucrose, aspartame) or slow onset (thaumatin, brazzein) [32]. This hypothesis is consistent with protein sweeteners generally having a positive charged surface, such as for monellin.

3 Brazzein: A High-Intensity Natural Protein Sweetener

Brazzein is one of two sweet proteins, the other being pentadin, produced by the West African plant, *Pentadiplandra brazzeana* Baillon, the ripe fruit of which contains up to 0.2% brazzein on a weight basis [33–35]. Indigenous West African populations have used the plant for sweetening beverages; thus, it is culturally considered to be safe to consume although it has not been formally recognized with GRAS status by the FDA. Brazzein joins a family of large, thermostable proteins which include monellin, thaumatin, curculin, mabinlin, and pentadin (see [33] and references therein).

3.1 Structure of Brazzein

Brazzein contains 54-amino acid residues and weighs in at 6,473 Da. There are two naturally occurring forms: 80% of brazzein in the seed is pyrE-brazzein which has a pyroglutamic acid residue at the N-terminus, and 20% is of the des-pyrE-brazzein form, which lacks the N-terminal pyroglutamic acid residue. Brazzein is highly water soluble (50 mg/mL or >7.7 M) [33]. Due to the presence of four disulfide cross-links between the eight cysteine residues within the small protein, it is highly thermostable, persisting for up to 4 h at 80 °C. It is about 2000 times sweeter than a 2% sucrose solution on a weight basis and is heat stable [33]. Brazzein shows significantly higher sweetness than sucrose – EC_{50} for brazzein forms is 16 μM and its sweetness potency is 800, while that of sucrose is 10 mM at the reference potency of 1 [36].

Brazzein is made of all L-amino acids; a mirror-image conformer of brazzein synthesized from all D-amino acids did not show sweet taste [37]. The amino acid sequence of brazzein is shown in Fig. 3. 80% of the protein from the fruit contains a pyroglutamic acid residue, pyrE, at the N-terminus of the protein (pyrE-bra or type 1), and the remaining 20% lacks this residue (des-pyrE-bra or type 2). Heterologous synthesis in the yeast *Pichia pastoris* also produced a form with a glutamine residue at this position (Q-bra). Both des-pyrE-bra and Q-bra were sweeter than the major natural pyrE form [38]. New-world monkeys, mice, and rats do not perceive brazzein as tasting sweet; thus brazzein stimulates the receptor differently from small sweet carbohydrates [31].

The organoleptic properties of brazzein are highly suitable for use as a sweetener, with no sourness, saltiness, or bitterness. The onset of sweetness with brazzein is slower than sucrose, making it better suited for use in conjunction with other sweeteners rather than on its own. Tests showed that brazzein combines well with synthetic sweeteners such as acesulfame-K, aspartame, and stevioside [34]. A third form of brazzein, containing the N-terminal Glu residue, has been made and is also sweet. This form has been used in the production of recombinant brazzein and will be discussed later in Sect. 3.3.

Disulfide bridges are formed from Cys4-Cys52, Cys16-Cys37, Cys22-Cys47, and Cys26-Cys49 (highlighted in yellow). Red highlights residues important for loops for sweet taste, flanking a beta-sheet structure (boxed residues). The single phenylalanine residue in brazzein is highlighted in green (Fig. 3).

The solution structure of brazzein obtained by NMR spectroscopy at pH 5.2 and 22 °C showed one alpha-helix and three antiparallel beta-sheets (Fig. 4) similar to the structure of a newly identified class of serine protein inhibitors that include brazzein, plant gamma-thionins and defensins, and arthropod toxins [39]. Further, its crystal structure at 1.8A [40] and NMR-soluble secondary structure at pH 3.5 by

pyrE-D-K-C4-K-K-V-Y-E-N-Y-P-V-S-K-C16-Q-L-A-N-Q-C22-N-Y-D-C26-K-L-D-K-H-A-R-S-G-E-C37-F-Y-D-E-K-R-N-L-Q-C47-I-C49-D-Y-C52-E-Y

Fig. 3 54-amino acid single-letter abbreviation sequence of brazzein (Modified; reproduced with permission from [34])

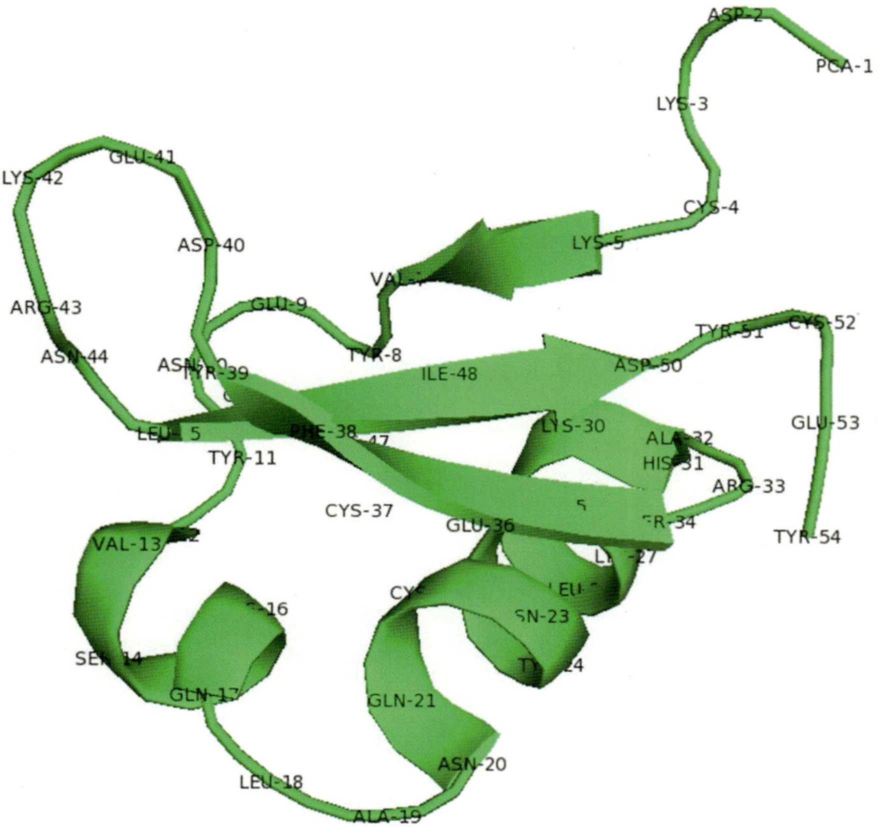

Fig. 4 Structure of brazzein. Ribbon structure created with PyMOL (pymol.org) using brazzein protein sequence PBD ID 4HE7 [40] from the RCSB protein database at www.rcsb.org [53]. Amino acid positions on the peptide are indicated

two-dimensional NMR techniques showed that brazzein shared "cysteine-stabilized alpha-beta" (CS$\alpha\beta$) motifs in common with scorpion neurotoxins, insect, and plant defensins ([41] and see http://www.rcsb.org/pdb/explore/jmol.do?structureId=4he7&bionumber=1&jmolMode=HTML5 for crystal structure).

Brazzein's secondary structure of alpha-helix and beta-sheets contributes significantly to the perceptions of sweetness. It is known that hydrogen bonds between the *i* and *i*-3 or *i*-4 residues occur in alpha helices, and intra-strand hydrogen bonds occur in beta -pleat structures [42]. Finally, covalent S-S bonds between Cys4-Cys52, Cys16-Cys37, Cys22-Cys47, and Cys26-Cys49 stabilize the three-dimensional shape of the molecule [43]. Site-directed mutation studies have implicated residues 29–33, 36, and 39–43 and the C-terminus of the protein in the sweet taste, presumably through interactions within the native protein or between the protein and the receptor.

Mutagenesis, followed by NMR folding studies and gustatory studies, indicated that the N- and C-termini and a fold containing Arg43 were responsible for sweetness in brazzein [44]. Mutations at position 29 (Asp29Ala, Lys, or Asn) increased the sweet response in electrophysiological stimulation of S-taste fibers from monkeys and in human taste profiles [32]. However, changes in amino acids in the same region (Lys30Asp or Arg33Ala) abolished sweetness [32, 34]. Similar studies on N- and C-termini showed they were important for sweetness of brazzein [34]. Other locations showing significant impact on sweetness were Glu41Lys and Asp29Ala/Asn/Lys (increase in sweetness); Gln17Ala, Asp2Asn/Ala, and Ala2ins (no change in sweetness); Lys15Ala, Lys6Ala, Asp50Ala, Arg33Asp, and His31Ala (slight decrease in sweetness); Lys6Asp, Lys30Asp, Tyr54del, Glu36Ala/Lys/Gln, Arg19Ile20ins, Arg33Ala, Arg19ins, and Arg43Ala (abolishment of sweetness) [7, 32, 44] (see Table 2).

3.2 Recombinant Production of Brazzein

The potential for brazzein as a heat-stable, nutritive sweetener is well recognized. As the plant is not a cultivated crop, it is not well suited for large-scale production. Consequently, recombinant expression of brazzein has been examined on several platforms with the goal of making a low-cost version that could be commercialized. One of the key factors for this is the ability to accumulate large amounts of the product in the host tissue. This latter feature has turned out to be more of a challenge than initially anticipated in order to make the product at a cost competitive with sucrose.

3.2.1 Recombinant Expression of Brazzein in Different Hosts

A variety of platforms have been used for the expression of brazzein ranging from bacteria to higher animals. The key difference in the platform is the use of GRAS vs. non-GRAS systems. The benefits of a GRAS system are potentially more rapid approval by regulatory agencies, but additionally provide the benefit of the use of brazzein directly in food products made using the host without the need for purification of brazzein.

[45] made several synthetic brazzein constructs by fusing *Staphylococcus* nuclease incorporating a cyanogen bromide cleavage site and expressed high levels of recombinant brazzein in *Escherichia coli* bacteria. Following purification, [44] used NMR to correlate the structure of the mutated protein to the contributions of different protein regions to sweetness. While this was very useful and highly successful, the purpose was to generate adequate protein for functional studies, not to optimize expression levels. Similarly, [38] expressed pyr-GluE bra/type 1 and des-Pyr-GluE bra/type 2 as well as a synthetic Q1-bra (containing glutamate at the N-terminus) in the yeast *Pichia pastoris*. Purification showed that the proteins were correctly folded and that they were sweet to the taste for human volunteers. The key item of note is the relative ease of making substantial quantities of recombinant brazzein.

[46] used three GRAS *Lactobacillus* species and transformed them with brazzein gene constructs along with a nisin-controlled expression (NICE) system. In this

Table 2 Summary of changes in brazzein sweetness related to mutagenesis of specific residues from work by Jin et al. [41] and Assadi-Porter et. al. [44]

	[32]	[44] Fruit = 500; sucrose = 1	Comments
Increase sweetness	pyrGludel Asp29Ala Asp29Lys Asp29Asn Glu41Lys	DespGlu1 (1000; WT) Ala2ins (2000) His31Ala (2000)	Des-pGlu1 is the same as pyrGludel. Increased sweetness of pyrGlu deletion maybe attributed to removal of the bulky cyclic pyrGlu residue or presence of positive charge at N-terminus upon pyrGlu removal Change of negatively charged Asp 29 to neutral (Ala) or positively charged residues (Lys, Asn) increases sweetness
Decrease sweetness	Lys6Ala Lys15Ala Arg33Asp His31Ala Asp50Ala	Asp2Asn (100) Cys4Ala (50) Lys15Ala (500) Ala19Ins (500) Arg33Ala (100) Arg43Ala (<33) Asp50Ala (200) Tyr54del (<33) Arg55Arg56ins (<33)	His31Ala increased sweetness at low concentrations [44]
Abolish sweetness	Lys6Asp Arg19Ile Arg20ins Lys30Asp Arg33Ala Glu36Ala Glu36Lys Glu36Gln Arg43Ala Tyr54del	Tyr51Ala (0)	Arg33 hydrogen bonded to Asp50 bringing it close to aromatic residues Tyr51 and Tyr54, at the C-terminus Change of negatively charged Glu36 to neutral (Ala), positively charged (Lys) or polar (Gln) residues abolishes sweetness Tyr51Ala did not fold correctly
No change in sweetness	Ala2ins Asp2Ala Asp2Asn Gln17Ala	Tyr8Ala Gln17Ala	

inducible system, the biochemical nisin causes activation of the *nis*A promoter to which heterologous genes of interest can be fused [47]. Fermentation parameters for brazzein expression using this system optimized by [48] and showed that levels could match *E. coli* expression [49]. Obviously, the use of a GRAS platform is recognized as a valuable economic parameter. Yogurt is made using *Lactobacillus* fermentation and commands a big sector of the dairy market. To address the dairy market, [50] expressed des-pGlu/type 2 brazzein in secreted milk of recombinant mice using a goat beta-casein promoter which directs expression specifically in mammary glands. Additional changes were codon optimization for animals and changes to the 29th and 41st amino acids to lysine (K) for optimized sweetness.

Baby mice consuming the transgenic milk grew normally, and human volunteers rated boiled milk as being sweet with no untoward taste. While the specter of milking enough mice for human consumption boggles the imagination, the proof of concept for the production of sweetened milk which could be subjected to heat treatment was established by this experiment.

3.2.2 Expression of Brazzein in Corn

Corn was selected for the production of brazzein as it has several advantages for the production of recombinant proteins that are targeted for consumption. Corn is edible, generally recognized as safe (GRAS) product, tolerates high level of recombinant protein expression, and is a convenient long-term storage system for product, as corn seed can be dehydrated.

For expression, open reading frames for type 2 (des-pyrE) and type 3 (additional N-terminal E) brazzein were synthesized using corn-optimized sequences, taking care to eliminate internal restriction sites, polyadenylation sequences, mRNA destabilizing sequences, and splice sites. The brazzein gene was then cloned into several expression vectors with combinations of promoters and targeting sequences. The promoters included constitutive polyubiquitin-like promoter, 22 kDa alpha-zein endosperm-preferred promoter, and the globulin-1 embryo-preferred promoter targeting cytoplasm or the cell wall. For the constructs targeting the cell wall, UTRs of plant viral origin were additionally inserted upstream of the gene. Agrobacterium was used to transform the constructs into corn. The offspring showed normal phenotype. Screening for brazzein production by Western blot analysis showed that levels of approximately 4% TSP were obtained from constructs containing cell wall-targeted protein expressed from the embryo-preferred promoter for both type 2 and 3 brazzein, whereas cytoplasmic targeting generally has very low levels of protein accumulation [51] (Fig. 5).

The germ from embryo-driven, cell wall-targeted protein was fractionated to obtain a fivefold increase in concentration in the germ fraction compared to whole grain. The protein was purified from constructs to yield the type 2 and type 3 protein. N-terminal sequencing of the purified protein showed that the type 2 version (lacking glutamate (E) at the N-terminal sequence) contained an additional serine at the N-terminal end which originated from an improperly processed cell wall-targeting sequence. The type 3 version (containing glutamate/E at the N-terminus) was normally processed [51].

This type 3, 54-amino acid brazzein, containing the N-terminal E residue showed a sweetness intensity that was between 940–1200 times the sweetness intensity of sucrose, compared to type 2 brazzein purified from corn, which was 900 times as sweet as sucrose under the same conditions. This matches the sweetness of recombinant brazzein expressed in bacteria [44]. The sweetness of the purified proteins further showed that brazzein in corn could readily withstand storage, fractionation, defatting, and milling procedures. Due to the unique platform of corn as a GRAS product, it was also possible to use germ flour directly as a sweet product, bypassing the purification process, which distinguished the recombinant corn product from other expression systems [51].

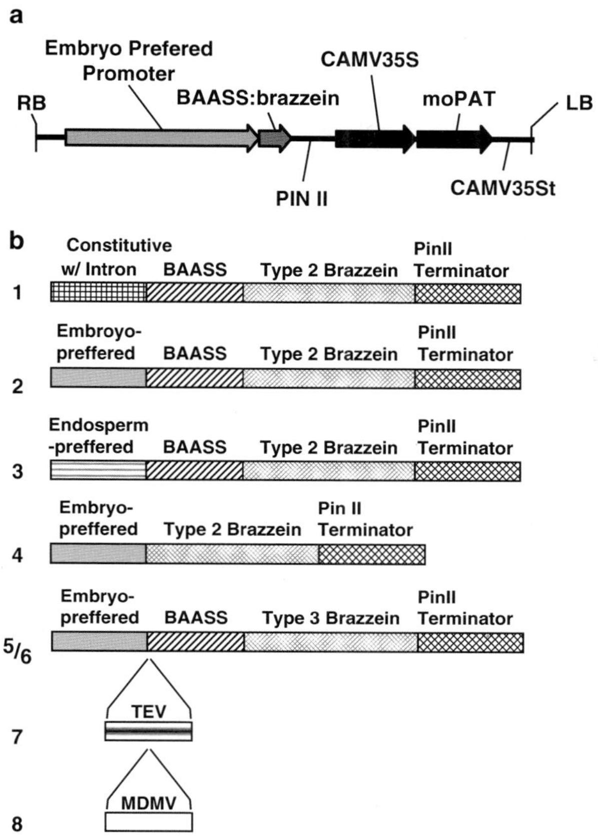

Fig. 5 Eight constructs generated using various combinations of promoter, brazzein, and subcellular targeting sequence [51]. Reprinted under Open Access license

Importantly, using cost models, accumulation in corn had the potential to make a purified commercial product competitive with sucrose. In addition, since corn flour is used in many food applications, there is the potential to use the flour directly without purification thereby creating a huge savings while adding the nutritive value and/or the bulking agent of the corn flour [52]. The brazzein-expressing corn flour has been shown to retain its sweetness following heat treatment on several prototype products. For example, muffins containing germ flour cooked for 35 min at 325°F, as well as corn bread made from meal and germ flour and cooked for 25 min at 400 °F retained their sweetness [52]. Other forms of heat treatment, including brief boiling, did not affect sweetness [52].

3.3 Considerations for Brazzein as a Commercial Product

As a protein with a long history of human consumption, brazzein may be considered GRAS in its natural host. In the natural host, most (80%) of brazzein is of pyrE-bra/type 1 with smaller amounts (20%) of des-pyrE-bra/type 2. Most of the studies conducted with heterologously expressed brazzein use the des-pyrE-bra/type 2 version, the E1-bra/type 3 version with a glutamine (Q) at the N-terminus, or the Q1-bra version with an N-terminal glutamate (E). The E1-bra/type 3 brazzein is significantly sweeter than both type 1 and type 2.

Mutagenesis studies have established the significance of the N- and C-terminal domains and some residues, particularly changes in residues 29 and 41 to lysine significantly increased sweetness. Increased sweetness of brazzein will obviously be of commercial value as smaller amounts of the product will be required to achieve the same final sweetness. Many of the modified proteins were taste-tested by human volunteers with institutional ethical oversight. Since structural studies indicated that the protein's conformation was retained, it may expedite regulatory approval for the modified versions of the protein.

The second consideration for commercial development is economics. Health concerns about sugar consumption have propelled the introduction of low-calorie and non-caloric sweeteners which have themselves come under scrutiny. Consensus opinion has moved to the use of natural rather than synthetic sweeteners. Protein sweeteners have been identified and developed from plant sources and provide nutritive material in addition to sweetness. Brazzein is well poised to be a successful sweetener for its intrinsic heat stability and high sweetness, but also because recombinant studies have successfully expressed high levels for purification and use in situ. In the latter regard, expression in cereal and grains can provide a valuable source of both nutrition as well as heat stability for use of the flour in baked goods. It is hoped brazzein will soon join the legion of sweet molecules that are available in the grocery aisle.

4 Conclusions

The push to replace carbohydrate sweeteners with healthier alternatives has received a boost from the identification of naturally occurring protein sweeteners from plants. Besides the obvious advantages of not triggering an insulin response or supporting dental caries, protein sweeteners are more nutritive, are safe for consumption, and used at lower concentrations in food. Other benefits derive from their nature – their sequences may be modified for optimal sweetness, and they can be expressed at high levels in heterologous hosts.

Brazzein is the poster child for the advantages of protein sweeteners due to its small size, stability due to four disulfide bridges, high solubility, and amenability to heat treatment without denaturation. It can be used in purified form or expressed in heterologous hosts which are directly used in food preparation. Other work has established the regions that contribute to the interactions of brazzein with the taste receptor, which helps to identify regions that may be modified for best outcomes.

Our work has dealt with the expression of brazzein in corn and highlights the feasibility of production of brazzein for commercial use. In addition to expressing brazzein in corn, we have used both flour and purified material in taste tests and determined the conditions for its use in each for best outcomes. In our analysis, brazzein is a suitable product for commercial development for use as a sweetener.

References

1. Adler J (1975) Chemotaxis in bacteria. Annu Rev Biochem 44:341–356. doi:10.1146/annurev.bi.44.070175.002013
2. Özcan S, Dover J, Johnston M (1998) Glucose sensing and signaling by two glucose receptors in the yeast *Saccharomyces cerevisiae*. EMBO J 17:2566–2573
3. Rolland F, Baena-Gonzalez E, Sheen J (2006) Sugar sensing and signaling in plants: conserved and novel mechanisms. Annu Rev Plant Biol 57:675–709
4. Lindemann B (2001) Receptors and transduction in taste. Nature 413:219–225
5. Scott K (2005) Taste recognition: food for thought. Neuron 48:455–464
6. Amrein H, Thorne N (2005) Gustatory perception and behavior in *Drosophila melanogaster*. Curr Biol 15:R673–R684
7. Temussi PA (2002) Why are sweet proteins sweet? Interaction of brazzein, monellin and thaumatin with the T1R2-T1R3 receptor. Febs Lett 526:1–4
8. Nelson G, Hoon MA, Chandrashekar J, Zhang Y, Ryba NJ, Zuker CS (2001) Mammalian sweet taste receptors. Cell 106:381–390
9. Kant R (2005) Sweet proteins–potential replacement for artificial low calorie sweeteners. Nutr J 4:1–6
10. Doidy J, Grace E, Kühn C, Simon-Plas F, Casieri L, Wipf D (2012) Sugar transporters in plants and in their interactions with fungi. Trends Plant Sci 17:413–422
11. Bray GA, Popkin BM (2014) Dietary sugar and body weight: have we reached a crisis in the epidemic of obesity and diabetes? Health be damned! Pour on the sugar. Diabetes Care 37:950–956
12. Nabors L, Gelardi R (2001) Alternative sweeteners: an overview. In: Nabors, LO (ed) Alternative sweeteners. CRC Press, Boca Rotan, FL. pp 1–10
13. Weihrauch M, Diehl V (2004) Artificial sweeteners – do they bear a carcinogenic risk? Ann Oncol 15:1460–1465
14. Gallus S, Scotti L, Negri E, Talamini R, Franceschi S, Montella M, Giacosa A, Dal ML, La Vecchia C (2007) Artificial sweeteners and cancer risk in a network of case–control studies. Ann Oncol 18:40–44
15. Suez J, Korem T, Zeevi D, Zilberman-Schapira G, Thaiss CA, Maza O, Israeli D, Zmora N, Gilad S, Weinberger A (2014) Artificial sweeteners induce glucose intolerance by altering the gut microbiota. Nature 514:181–186
16. Kim N-C, Kinghorn AD (2002) Highly sweet compounds of plant origin. Arch Pharm Res 25:725–746
17. Kinghorn A, Chin Y, Pan L, Jia Z (eds) (2010) Natural products as sweeteners and sweetness modifiers. In: Townsend CA, Ebizuka Y (eds) Comprehensive natural products II: chemistry and biology. Elsevier Publishers, London, UK. pp 269–315
18. DuBois GE, Prakash I (2012) Non-caloric sweeteners, sweetness modulators, and sweetener enhancers. Annu Rev Food Sci Technol 3:353–380
19. Prakash I, DuBois G, Clos J, Wilkens K, Fosdick L (2008) Development of rebiana, a natural, non-caloric sweetener. Food Chem Toxicol 46:S75–S82
20. Takasaki M, Konoshima T, Murata Y, Sugiura M, Nishino H, Tokuda H, Matsumoto K, Kasai R, Yamasaki K (2003) Anticarcinogenic activity of natural sweeteners, cucurbitane glycosides, from *Momordica grosvenorii*. Cancer Lett 198:37–42
21. Kinghorn AD, Soejarto DD (2002) Discovery of terpenoid and phenolic sweeteners from plants. Pure Appl Chem 74:1169–1179

22. Kageyama Y, Suzuki H, Saruta T (1992) Glycyrrhizin induces mineralocorticoid activity through alterations in cortisol metabolism in the human kidney. J Endocrinol 135:147–152
23. Walker BR, Edwards C (1994) Licorice-induced hypertension and syndromes of apparent mineralocorticoid excess. Endocrinol Metab Clin North Am 23:359–377
24. Fenwick G, Lutomski J, Nieman C (1990) Liquorice, Glycyrrhiza glabra L. – composition, uses and analysis. Food Chem 38:119–143
25. Danilova V, Hellekant G (2004) Sense of taste in a New World monkey, the common marmoset. II. Link between behavior and nerve activity. J Neurophysiol 92:1067–1076
26. Nelson G, Chandrashekar J, Hoon MA, Feng L, Zhao G, Ryba NJ, Zuker CS (2002) An amino-acid taste receptor. Nature 416:199–202
27. Lee RJ, Kofonow JM, Rosen PL, Siebert AP, Chen B, Doghramji L, Xiong G, Adappa ND, Palmer JN, Kennedy DW (2014) Bitter and sweet taste receptors regulate human upper respiratory innate immunity. J Clin Invest 124:1393
28. Zhang Y, Hoon MA, Chandrashekar J, Mueller KL, Cook B, Wu D, Zuker CS, Ryba NJ (2003) Coding of sweet, bitter, and umami tastes: different receptor cells sharing similar signaling pathways. Cell 112:293–301
29. Hoon MA, Adler E, Lindemeier J, Battey JF, Ryba NJ, Zuker CS (1999) Putative mammalian taste receptors: a class of taste-specific GPCRs with distinct topographic selectivity. Cell 96:541–551
30. Li X, Staszewski L, Xu H, Durick K, Zoller M, Adler E (2002) Human receptors for sweet and umami taste. Proc Natl Acad Sci 99:4692–4696
31. Jiang P, Ji Q, Liu Z, Snyder LA, Benard LM, Margolskee RF, Max M (2004) The cysteine-rich region of T1R3 determines responses to intensely sweet proteins. J Biol Chem 279:45068–45075
32. Jin Z, Danilova V, Assadi-Porter FM, Markley JL, Hellekant G (2003) Monkey electrophysiological and human psychophysical responses to mutants of the sweet protein brazzein: delineating brazzein sweetness. Chem Senses 28:491–498
33. Ming D, Hellekant G (1994) Brazzein, a new high-potency thermostable sweet protein from Pentadiplandra brazzeana B. FEBS Lett 355:106–108
34. Hellekant G, Danilova V (2005) Brazzein a small, sweet protein: discovery and physiological overview. Chem Senses 30:i88
35. Van der Wel H, Larson G, Hladik A, Hladik CM, Hellekant G, Glaser D (1989) Isolation and characterization of pentadin, the sweet principle of Pentadiplandra brazzeana Baillon. Chem Senses 14:75–79
36. Venkitakrishnan RP, Choudhury M (2015) Importance of method validation: implications of non-correlated observables in sweet taste receptor studies. arXiv preprint arXiv:1511.07977
37. Izawa H, Ota M, Kohmura M, Ariyoshi Y (1996) Synthesis and characterization of the sweet protein brazzein. Biopolymers 39:95–101
38. Poirier N, Roudnitzky N, Brockhoff A, Belloir C, Maison M, Thomas-Danguin T, Meyerhof W, Briand L (2012) Efficient production and characterization of the sweet-tasting brazzein secreted by the yeast Pichia pastoris. J Agric Food Chem 60:9807–9814
39. Caldwell JE, Abildgaard F, Džakula Ž, Ming D, Hellekant G, Markley JL (1998) Solution structure of the thermostable sweet-tasting protein brazzein. Nat Struct Mol Biol 5:427–431
40. Nagata K, Hongo N, Kameda Y, Yamamura A, Sasaki H, Lee WC, Ishikawa K, Suzuki E-i, Tanokura M (2013) The structure of brazzein, a sweet-tasting protein from the wild African plant Pentadiplandra brazzeana. Acta Crystallogr D Biol Crystallogr 69:642–647
41. Gao G, Dai J, Ding M, Hellekant G, Wang J, Wang D (1999) Studies on solution NMR structure of brazzein. Sci China C Life Sci 42:409–419
42. Assadi-Porter FM, Abildgaard F, Blad H, Markley JL (2003) Correlation of the sweetness of variants of the protein brazzein with patterns of hydrogen bonds detected by NMR spectroscopy. J Biol Chem 278:31331–31339
43. Kohmura M, Ota M, Izawa H, Hellekant M, Göran D, Ariyoshi Y (1996) Assignment of the disulfide bonds in the sweet protein brazzein. Biopolymers 38:553–556

44. Assadi-Porter FM, Aceti DJ, Markley JL (2000) Sweetness determinant sites of brazzein, a small, heat-stable, sweet-tasting protein. Arch Biochem Biophys 376(2):259–265
45. Assadi-Porter FM, Aceti DJ, Cheng H, Markley JL (2000) Efficient production of recombinant brazzein, a small, heat-stable, sweet-tasting protein of plant origin. Arch Biochem Biophys 376:252–258
46. Lee Y-W, Kim K-Y, Han S-H, Kang C-H, So J-S (2012) Expression of the sweet-tasting protein brazzein in *Lactobacillus* spp. Food Sci Biotechnol 21:895–898
47. Zhou XX, Li WF, Ma GX, Pan YJ (2006) The nisin-controlled gene expression system: construction, application and improvements. Biotechnol Adv 24:285–295
48. Berlec A, Tompa G, Slapar N, Fonović UP, Rogelj I, Štrukelj B (2008) Optimization of fermentation conditions for the expression of sweet-tasting protein brazzein in *Lactococcus lactis*. Lett Appl Microbiol 46:227–231
49. Berlec A, Štrukelj B (2009) Large increase in brazzein expression achieved by changing the plasmid/strain combination of the NICE system in *Lactococcus lactis*. Lett Appl Microbiol 48:750–755
50. Yan S, Song H, Pang D, Zou Q, Li L, Yan Q, Fan N, Zhao X, Yu H, Li Z (2013) Expression of plant sweet protein brazzein in the milk of transgenic mice. Plos One 8:e76769
51. Lamphear B, Barker D, Brooks C, Delaney D, Lane J, Beifuss K, Love R, Thompson K, Mayor J, Clough R (2005) Expression of the sweet protein brazzein in maize for production of a new commercial sweetener. Plant Biotechnol J 3:103–114
52. Fake G, Howard J (2014) Brazzein: a high-intensity natural sweetener. In: Howard JA, Hood EE (eds) Commercial plant-produced recombinant protein products. Springer, Berlin, pp 247–257
53. Berman HM, Westbrook J, Feng Z, Gilliland G, Bhat T, Weissig H, Shindyalov IN, Bourne PE (2000) The protein data bank. Nucleic Acids Res 28:235–242

Cultivation of *Stevia rebaudiana* Bertoni and Associated Challenges

3

Luciana G. Angelini, Andrea Martini, Barbara Passera, and Silvia Tavarini

Abstract

The rising concern about the spread of obesity and diabetes, and a growing awareness about healthy foods in western societies have stimulated, in the last years, a strong interest toward stevia sweeteners as alternative of sucrose and artificial intensive sweeteners. The worldwide demand for purified steviol glycosides from stevia is steadily increasing, and it is expected that in the future the agricultural production capacity will be lower than the market demand. This provides a strong incentive to explore the possibilities to cultivate stevia and to produce leaves and extracts, beyond the traditional production zones. The cultivation of stevia might represent a formidable opportunity for the growers, in order to diversify the cropping systems and to meet the increasing market demand for high-quality and traceable raw material. In addition, several legislative initiatives, such as the steviol glycosides approval as food additive in several countries, represent favorable factors for the development of a stevia-based agro-industry. To foster the introduction of this novel species, a significant improvement of its cultivation should be achieved and a modern agronomical blueprint defined. It is, in fact, imperative to develop economically viable and environmentally sustainable crop production systems through the integration of site-specific agronomic techniques and efficient mechanization technologies for the production and processing of a higher quality product. Several agronomic aspects require still to be optimized (e.g., choice of the cultivar, propagation and transplanting, sustainable weed management, nutrition, irrigation, harvesting), in order to improve, not only the leaf yield, but also its quality in terms of steviol glycosides

L.G. Angelini (✉) • A. Martini (✉) • S. Tavarini (✉)
Department of Agriculture, Food and Environment (DAFE), University of Pisa, Pisa, Italy
e-mail: luciana.angelini@unipi.it; andrea.martini87@gmail.com; silvia.tavarini@for.unipi.it

B. Passera (✉)
University of Pisa, Pisa, Italy
e-mail: passera.barbara@gmail.com

J.-M. Mérillon, K.G. Ramawat (eds.), *Sweeteners*, Reference Series in Phytochemistry,
https://doi.org/10.1007/978-3-319-27027-2_8

and bioactive compounds. Traceability and crop quality control along the production chain are the strongest points, in order to obtain a certification that will provide to the farmers preferential market access and to sell the raw material at a differential price.

The present chapter, therefore, aims to provide updated scientific information regarding the most important agronomic factors in order to foster stevia cultivation and attain maximum yield and quality.

Keywords

Agronomy • Cultivation and sustainability • Preharvest factors • *Stevia rebaudiana* • Sustainable crop production

Abbreviations

2,4-D	2,4-Dichlorophenoxyacetic acid
ADI	Acceptable daily intake
AMF	Arbuscular mycorrhizal fungi
B	Boron
BA	6-Benzyladenine
BW	Body weight
CCC	Chlorocholine chloride
Chl	Chlorophyll
Cu	Cupper
DPPH	2,2-Diphenyl-1-picrylhydrazyl
ET_0	Reference (or potential) evapotranspiration
ETc	Crop evapotranspiration
FC	Field capacity
Fe	Iron
FYM	Farmyard manure
GACP	Good agricultural and collection practices
GAs	Gibberellins
IBA	Indole-3-butyric acid
IPM	Integrated pest management
IWM	Integrated weed management
K	Potassium
Kc	Crop coefficient
Kn	Kinetin
LAI	Leaf area index
LDs	Long-day conditions
MEP	Methyl-D-erythritol 4-phosphate
Mg	Magnesium
Mn	Manganese
MS	Murashige and Skoog medium
N	Nitrogen
NAA	α-Naphthaleneacetic acid
P	Phosphorus

PBZ	Paclobutrazol
PGPRs	Plant growth promoting rizhobacteria
PNUE	Photosynthetic nitrogen use efficiency
PSB	Phosphorous solubilizing bacteria
Reb A	Rebaudioside A
S	Sulfur
SDs	Short-day conditions
SLW	Specific leaf weight
Stev	Stevioside
SVglys	Steviol glycosides
TDZ	Thidiazuron
UPOV	Union for the Protection of New Varieties of Plants
Zn	Zinc

Contents

1 Introduction 37
2 Origin and Botany 38
3 Cultivation and Sustainability 41
3.1 Interest in Europe 43
3.2 Stevia – Guidelines for Good Agricultural Practices 44
4 Optimal Cultivation Conditions 45
4.1 Climate and Soil Requirements 45
4.2 Light and Photoperiod Requirements 46
4.3 Varieties 47
4.4 Rotation 49
4.5 Tillage and Soil Preparation 49
4.6 Propagation Methods 51
4.7 Planting 55
4.8 Crop Nutrition and Fertilization 56
4.9 Mycorrhizal Association and Plant Growth Promoting Rhizobacteria 58
4.10 Water Requirement, Irrigation, and Salinity Tolerance 60
4.11 Plant Protection (Weed Control, Fungal Diseases, and Insects) 62
5 Harvesting, Storage, and Primary Processing 66
6 Preharvest Factors Affecting Biomass Yield, Steviol Glycosides, and Antioxidants Compounds 69
6.1 Case Study 1: Nitrogen Fertilization 69
6.2 Case Study 2: Drought Stress 70
6.3 Case Study 3: Plant Density 71
6.4 Case Study 4: Growth Regulators 71
6.5 Case Study 5: Harvest Time 72
7 Conclusion 73
References 74

1 Introduction

Stevia rebaudiana Bertoni is a member of the Asteraceae family and is one of 154 members of the genus Stevia. This perennial herbaceous plant, native to Northeast Paraguay, produces, in its leaves, *ent*-kaurene diterpenoid glycosides, namely, steviol glycosides, which are no-calorie, intensive sweeteners more powerful than sucrose [1–3]. To date, 34 steviol glycosides (SVglys) are known to occur within leaf stevia extracts [4, 5]; the most abundant are stevioside and rebaudioside A, followed by rebaudioside B–E, dulcoside A, and steviolbioside [6]. Stevia leaves have been used as natural sweeteners for centuries by indigenous Guaraní people, and actually the purified SVglys are used as sweeteners, sugar substitutes, and dietary supplements in the global food market [7], following the rising concern about the spread of obesity and diabetes and a growing awareness about healthy foods in Western societies [8]. In the European market the introduction of steviol glycosides as food additives, with a purity of more than 95% was given at the end of 2011, after EFSA issued a positive opinion on their safety and raised the acceptable daily intake (ADI) for steviol glycosides, expressed as steviol equivalents, to 4 mg/kg BW/day [9]. It has been reported that stevioside, along with related compounds (e.g., rebaudioside A, steviol, isosteviol), has antihyperglycemic, antihypertensive, anti-inflammatory, antitumor, antidiarrheal, diuretic, and immunomodulatory actions [10–12]. Moreover, *Stevia rebaudiana* elaborates a complex mixture of sterols, triterpenoids, essential oils, and flavonoids that can explicate an antioxidant activity [6, 13, 14].

Demand for both stevia leaves and SVglys is expected to grow more and more in the coming years [15], in particular in Europe, following the new Regulation on novel foods approved on November 2015 [16]. Consequently, there is a formidable opportunity for the growers of Western countries to produce, in a sustainable way, traceable raw material in order to meet the growing market demands. Currently, in Europe this species was grown by smallholders or at experimental level by several public or private institutions, within research programs. The lack of suitable adapted and available varieties, the high input costs (in particular for planting and establishment), and the lack of leaf processing facilities for the production of extracts represent all critical points to be overcome. Therefore, the agronomic research should be of high priority to develop modern and sustainable agro-techniques in order to obtain a raw material designed to meet consumer needs and bio-industry requirements for high-quality, safe, and environment-friendly products. At the same time, the optimization of cultural practices and the increase of sustainability, through modern agricultural production systems, may ensure high yield and reduce production costs, with consequent economic benefits for the farmers.

2 Origin and Botany

Stevia rebaudiana Bert. (2n = 22) is native to the highlands of Northeastern Paraguay, in particular to the valley of Rio Monday between 25 and 26°S latitude and 55° to 56° W longitude. The leaves of this sweet plant, known as Kaáhe'é, were historically used by the indigenous Guaranì people for centuries as sweetener for beverages (such as maté, a typical South American drink) and as herbal remedy [2, 17]. *S. rebaudiana* was brought to the attention of Europeans in 1888, by the

Swiss botanist, Dr Moisés Santiago Bertoni, who learned about the species and its sweetening properties from herbalists and indigenous people in Northeastern Paraguay [18]. He described and classified stevia, giving it its scientific name.

This plant is typically considered a perennial herbaceous shrub although, under some environmental conditions, it behaves as annual plant [17]. *S. rebaudiana* grows up to 65 cm tall with green, erect, slender stems with internodes 2–4 cm long covered with very fine, short, and whitish hairs; the plant from its base easily generates secondary shoots. Under cultivation management, with optimal environmental conditions, the plant can grow taller and wider than normally. When pruned, usually produces lateral branches, forming more or less circular stem crown [1, 12, 17]. *S. rebaudiana* roots are fibrous, filiform, and perennial; the roots form abundant stocks that hardly ramified and does not deepen in the soil, consequently distributing on to the soil surface [17] (Fig. 1). Leaves are relatively small, simple, opposite, subsessile, and serrated above the middle, with a large variability both in shape and size, ranging from oblanceolate to lanceolate or ovate, often toothed, and they showed three primary veins and secondary reticulate venation. *S. rebaudiana* has trichome structures on the leaf epidermal surface, which are of two distinct sizes: one large (4–5 μm) and one small (2.5 μm) [1, 12, 19].

Stevia is a short-day photoperiod sensitive plant [20–23] with a critical day length of 12–13 h. Flowering starts from January to March in the southern hemisphere and

Fig. 1 *Stevia rebaudiana* roots: on the *left*, stevia plant cultivated in floating system; on the *right*, roots of stevia cultivated in open-field conditions at the Experimental Centre of the Department of Agriculture, Food and Environment, of the University of Pisa

from September to December in the northern hemisphere [3]. When the plants were subjected to the short day light period (12 h), flowering commenced early, after a minimum of four true leaves have been formed. Flowers are pentamerous, perfect (hermaphrodite), small (7–15 mm), and are arranged in corymbs of capitulum inflorescences (Fig. 2); the corolla of flowers is white and has glandular external hairs [1, 12, 17, 24–29]. The ovary of the flowers has a single egg topped by the stylus, which ends with a bifid stigma protruding from perianth. The androecium consists of five stamens, which wrap the stylus [1, 19]. Stevia takes more of a month in producing all its flowers [17]. Stevia is an allogamous species, with entomophylous pollination. The flowers, in fact, show a sporophytic self-incompatibility.

Seeds have very little endosperm and they are contained in slender achenes (3 mm in length). Stevia achene has a pappus composed of 13–15 persistent appendices which, in addition to its small size (weight of 1,000 achenes usually is 0.15–0.30 g), facilitate seed wind-borne dispersion [17, 19]. Two types of achenes are produced by stevia plants, dark and pale colored (Fig. 3) [30]. The development of these two types of achenes seems to be controlled by compatibility factor. The black type is produced by cross-pollination, while the tan one is originated from self-pollination [29].

Fig. 2 Stevia corymbs of capitulum inflorescences – from the beginning of flowering to ripe seed formation (Source: experimental trials carried out at the Department of Agriculture, Food and Environment, of the University of Pisa)

Fig. 3 Stevia seeds: on the *left*, black (obtained from open-pollinated plants) and tan seed (obtained from plants grown in greenhouse); on the *right*, a particular of stevia capitulum with ripe seeds (Source: experimental trials carried out at the Department of Agriculture, Food and Environment, of the University of Pisa)

3 Cultivation and Sustainability

The commercial use of stevia, mainly as steviol glycosides, began in the early 1970s when Japanese introduced stevia from Paraguay after the intensive sweeteners, such as cyclamate and saccharin, became suspected of being carcinogens [31]. A large effort aimed at establishing stevia as a crop in Japan was begun by Sumida [32], and subsequently, the Japanese company Morita Kagaku Kogyo Co., Ltd. became the first to produce a commercial sweetener from stevia [33]. Over the following years, the SVglys, extracted from stevia leaves, have been used extensively in a number of countries as a nonsucrose and no-calorie sweeteners, as a result of growing awareness about healthy foods in Western societies. At the same time, the health controversy about artificial sweeteners and the lifting of restrictions on their application, encouraged the use of steviol glycosides in a wide range of beverages and food products, including cereals, teas, juices, flavored milks, yogurts, and carbonated soft drinks [34].

The worldwide demand for stevia is steadily increasing, since worldwide main regularity authorities (European Food Safety Authority, The US Food and Drug Administration, The Joint FAO/WHO Expert Committee on Food Additives, Food Standards Australia New Zealand) have approved the use of SVglys as dietary supplement and food additive [9, 35–37]. In 2009, the World Health Organization (WHO) estimated that steviol glycosides have the potential to replace 20–30% of all dietary sweeteners in the coming years [15]. The expected revenue from food and beverages containing steviol glycosides as sweeteners is expected to be US$ 8–11 billion in 2015 [38]. Mintel also gives figures for the growing market for steviol glycosides themselves, estimating that this will more than double between 2013 and 2017, jumping from US$ 110 million to US$ 275 million [39]. Actually eight multinational corporations are controlling the market with several patents – an increasing number of patents focus on ways of producing steviol glycosides [40] – and are now successfully marketing steviol glycosides, benefitting from different rules and regulations applying to the import and use of stevia leaves, which prohibit the direct use of stevia leaves as a sweetener [41]. For instance, stevia leaves are not authorized for sale in US, European, or Swiss market [41]. In Europe, stevia leaves would need authorization as a novel food. A novel food is in fact a food that does not have a significant history of consumption within the European Union before 15 May 1997.

It is estimated that, in 2011, a total of 30,000 ha of stevia plants was globally cultivated for SVglys production [41]. Eighty percent of global cultivation was situated in China, 5% in Paraguay, 3% in Argentina, 3% in Brazil, and 3% in Colombia. It was also grown in Peru, India, Japan, Kenya, South Korea, Taiwan, Vietnam, and the USA [42, 43]. In Paraguay – similarly to other southern American countries – stevia is cultivated mainly by smallholders, most of them are mestizos, who deliver the dry leaves to the collection centers. Farmers can start harvesting in the first year, with up to four harvests per year possible [44]. Even if the cultivation done on small areas is labor intensive, it gives a more profitable income than traditional crops. China not only holds the largest area of *S. rebaudiana* plantation but also is the biggest supplier of stevia extracts in the world. The main plantation area of *S. rebaudiana* in China covers the provinces of Jiangsu, Shandong, Anhui, Helan, Zhejiang, and Xinjiang [41, 45]. Growers first started cultivating stevia around 1977. Similarly to Paraguay, in China, stevia is typically produced by contracted smallholders on plots of 1 mu, i.e., 667 m^2 [46, 47]. In 2005, the cultivation reached around 4,450 ha. In 2010, it rapidly expanded to around 24,700 ha with a total annual output of about 96,000 T. This equals 80% of the entire global stevia output, which was fueled by the worldwide growing trend of sugar-free beverages and supported by giants like The Coca-Cola Company. Even after the significant increase in stevia usage, the Chinese production capacity was greater than the market demand. As a consequence, the cultivation started to decrease around 2011, and by the end of 2013, the total Chinese cultivation remained at around 2,670 ha. Recently it has been also reported that salinization occurred in many irrigated areas with consequent production problems [45].

Stevia is a profitable crop for smallholder farmers in Asia, South America, and Africa, because it requires little land and gives high yield of leaves for 3–5 years. It is cultivated as a cash crop on smaller plots of farmlands in addition to food crops for added income. Several reasons for stevia's environmental advantages are associated with its lower inputs of land, water, and energy than other traditional sweetener crops such as sugar beet and sugar cane. In a 2013 study carried out by Global Stevia Institute [48], the carbon footprint of stevia was shown to be 55% lower than sugar beet and 29% lower than sugar cane, based on industry production standards. In the same study, the water footprint of stevia leaf extract – both from rain and irrigation or process – was shown to be 92% lower than sugar beet and 96% lower than sugar cane, according to available benchmarks in water consumption, when comparing these sweeteners at the same sweetness equivalence.

Ramesh et al. [17] listed a series of challenges that the cultivation of stevia can provide in the temperate climates, where it is generally grown over the spring-summer period as a transplanted annual or perennial crop. In these conditions the economic feasibility of growing stevia, taking into account the high production costs (in particular for planting and establishing), should be carefully evaluated. It is therefore of crucial importance to develop a science-based information about stevia production and processing practices, in order to reduce production costs and to increase the yield and quality, that result in acceptable financial returns to growers.

3.1 Interest in Europe

The growing consumer demand for stevia-based products have been strongly increased in the Europe [43], in particular after the SVglys approval as food additive within the EU since December 2011. Furthermore, the recent EU Regulation on novel food, approved on November 2015, suggests a change of course and opens new perspectives for the possibility to produce and sell stevia leaves as novel food [16].

In the 2009–2014 period, Europe had the highest annualized growth rate (+149%), leading the way as food and beverage companies strive to use plant-based ingredients in new products and reformulations [49]. At the same time multinational companies involved in the refining and commercialization of steviol glycosides are more and more interested in diversifying their stevia leaf sources because of rising production costs in China and the uncertainties about raw material production and quality from traditional exporting countries [44]. All these facts make stevia cultivation an interesting possibility for European farmers, more and more interested to explore the possibility to grow stevia as new alternative crop. Furthermore, in the last years, several research studies have highlighted that stevia goes beyond its sweetening power and may also offer therapeutic benefits, thanks to several bioactive metabolites which are involved in insulin regulated glucose metabolism [50, 51]. This opens new perspectives toward the traditional use of whole stevia leaves, instead of chemically purified or synthetically produced SVglys, with consequent benefits for smallholder farmers and rural development. In Europe, it may become a profitable, sustainable cash crop, providing an opportunity to diversify cropping systems in accordance with the greening measures of the new EU CAP (Common Agricultural Policy) (2014–2020). According to the EU-funded project "Go4Stevia" (*Stevia rebaudiana* as a diversification alternative for European tobacco farmers to strengthen the European competitiveness), the yearly demand of artificial sweeteners at European level, which was estimated about 18,000 t, can be substituted with 36,000 ha of stevia. For the farms which will decide for the stevia diversification alternative, the income can be maintained like tobacco, making possible the conversion of 860 ha of tobacco to stevia [52].

In the European Union, despite rising consumption of stevia products, its share in global cultivated area is still negligible, even if an increasing and strong interest in the growing of the plant and processing by a range of individuals and groups, including farmers and farm organizations, has been expressed. A particular interest is coming from organic producers. The potential of stevia to be grown with organic cultivation methods represents an aspect of pivotal importance in obtaining, not only a sustainable and safe final leaf products, but also in increasing the economic return and market competitiveness.

3.2 Stevia – Guidelines for Good Agricultural Practices

In many parts of the world, stevia is grown on small scale, and large-scale mechanized production has not yet been established. To meet the burgeoning demand of stevia around the world, it is imperative to develop economically viable and environmentally sustainable crop production systems, through the integration of site-specific agronomic techniques and efficient mechanization technologies for the production and processing of a higher quality product. Thus, major emphasis has been given to the knowledge of basic and applied aspects of the agronomy of the crop, addressing the need to increase the yield production while minimizing pressure on the environment. The guidelines for the *Good Agricultural and Collection Practices* of Medicinal and Aromatic Plants (GACP-MAP) [53] are intended to apply to the cultivation and primary processing practices of stevia using all methods of agricultural production, including organic production ones, in accordance with the different regulations (Reg. EC 834/2007 in Europe, NOP-USDA in USA, JAS in Japan). GACP provide a detailed description of the techniques and measures required for the appropriate cultivation and collection of stevia and for the recording and documentation of necessary data and information during its processing. Quality control directly impacts the safety and efficacy of final products. According to GACP, any agrochemicals used to promote the growth of or to protect stevia should be kept to a minimum and applied only when no alternative measures are available. Integrated pest management (IPM) and sustainable weed management systems (SWMS) should be followed where appropriate. When necessary, only approved pesticides and herbicides should be applied at the minimum effective level, in accordance with the labeling and/or package insert instructions of the individual product and the regulatory requirements that apply for the grower and the end-user countries. GACP are voluntaristic guidelines, and all the operations should comply within force regional and/or national regulations. GACP for stevia is only the first step in quality assurance, on which the safety and efficacy of the products directly depend upon, and will also play an important role in the protection of natural resources for sustainable use.

Stevia can be cultivated according to organic certified production systems, also within GACP rules. Organic production is an environmentally friendly form of agricultural production that is based on cultivation and animal husbandry practices that are in tune with natural cycles, excluding the use of synthetic fertilizers, pesticides, and growth regulators and applying a number of modern preventive methods to maintain the natural soil fertility (cover crops, intercropping, conservative tillage, etc.), to control diseases (stimulating useful insects and pathogens) and weeds (stimulating biodiversity, integrated weed management methods, etc.). Organic production is an overall system of farm management and food production that combines best environmental practices, a high level of biodiversity, the preservation of natural resources, the application of high animal welfare standards, and a production method in line with the preference of certain consumers for products produced using natural substances and processes. The use of genetically modified organisms is prohibited and that of synthetic chemicals kept as low as possible.

4 Optimal Cultivation Conditions

4.1 Climate and Soil Requirements

Although originated from Paraguayan highland of Amambay, between 22° and 25° S latitude and 55° to 56° W longitude with average annual temperature of 25 °C and average rainfall of 1,375 mm year^{-1} [54–59], stevia can be grown over a wide climatic range, from semi-humid, subtropical to temperate zones [17]. While tolerant of mild frost, hard frosts will kill the roots of the plant. This lack of winter hardiness means that stevia can be grown as perennial in subtropical latitudes. Stevia is generally considered an herbaceous tender or half-hardy perennial. Under mid to high latitude regions, it is not hardy enough to survive winters. It behaves as annual crop in Northern and Central Europe, with hardiness zones below 9 (average annual minimum of −7 °C), and as perennial, at lower latitudes of Southern Europe, with hardiness zones from 9 to 11 (average annual minimum temperatures at or above 4.4 °C), that are essentially frost free.

Ideal growing areas will have long periods when day temperatures are over 25 °C but under 35 °C. Stevia is highly sensitive to the day length and requires long-day conditions (above 14 h day length) to remain in vegetative stage. Therefore, under the long-day conditions of the temperate zones of Central and South Europe, the vegetative growth (shoot length, leaf area, dry weight, leaf-stem ratio) is enhanced [17, 60, 61], similarly to the SVglys content [17, 22, 56]. Stevia prefers full sun or partial shade, in particular in hot, sunny climates.

Regarding the soil, stevia in its native place is found to be grown in sandy soils and marshy land, being well adapted to acid, infertile soils with ample supply of water. When cultivated, stevia prefers lightly textured and well-drained soil rich in organic matter. Stevia is in fact sensitive to water logging, and any soil that retains the moisture for very long period of time is unsuitable for stevia cultivation and should be avoided. It may be grown on acid soils but can also be cultivated on more neutral ones (pH 6.5–7.5). Even if stevia can be considered moderately tolerant to salt stress, saline soils should be avoided in order to prevent yield losses and low SVglys accumulation in the leaves [45]. Soils should not be contaminated by sludge, heavy metals, and residues of plant protection products, as well as other not naturally occurring chemicals.

4.2 Light and Photoperiod Requirements

Light quality/quantity greatly influences plant architecture development, morphogenetic responses, crop productivity, and synthesis of valuable bioactive compounds [62]. However, the physiological and morphological responses of plants toward light are greatly diverse depending upon the plant species. In stevia, the vegetative growth and the accumulation of SVglys are strongly influenced by photoperiodism [22] and irradiation levels, the latter becoming an increasingly important determinant when days grow longer [63, 64]. Stevia is an obligate short-day plant with a critical day length of about 12–13 h, and this prompted many investigators to examine the effect of length of day and night and temperature variation on the cultivation and the resultant SVglys levels [1, 17, 65–69]. When the plants were subjected to the short day light period, flowering commenced early. The early commencement of flowering decreased the vegetative growth of the plant. Therefore, the total biomass of stevia decreased significantly, irrespective of variety. Under long-day conditions (LDs), the vegetative growth phase is retained for a long time by prohibiting precocious flowering. Metivier and Viana [22] observed that plants maintained under long-day conditions were characterized by long internodes and a single main stem, bearing large, horizontally held ovate leaves. In addition, an increase in leaf biomass as well as in the stevioside content in the leaves was also noted under LDs [67, 68]. The first studies on this topic assumed that the synthesis of stevioside was linked with total carbon fixation and the allocation of photosynthates [22, 23]. So, under LD conditions or high irradiance, leaves with a greater dry weight were produced, able to allocate relatively more carbon into stevioside, as a possible defense mechanism, compared to SDs or lower irradiance. Recently, Ceunen and Geuns [68] investigated the effect of photoperiodism (16 h or 8 h photoperiod) and ontogenesis on dry matter and SVglys accumulation dynamics. These authors observed that LDs prolonged vegetative growth, thereby significantly increasing leaf biomass and SVglys amounts. In both photoperiods, SVglys declined during reproductive development, occurring mainly in mature leaves under LDs or young leaves under short-day conditions (SD). At the same time, also the Reb A/Stev ratio was influenced by day length, with higher ratios during vegetative stages of plants under SDs. Ceunen et al. [67] reported that phytochrome activation during SDs using midnight interruptions by red LED light provides an easy way to simulate LD conditions. After 7 weeks of treatment, total SVglys content was already twofold higher in LED-treated plants than in the SD control group. These findings conclude that the stevia plants should be grown under LD conditions in order to obtain greater leaf biomass with higher SVglys content [70].

Also shade affects plant growth and development. Shade not only influences the amount of photosynthetically active radiation (PAR) and alters spectral quality – thus affecting plant photosynthesis, dry matter production, and crop yield [71] – but also changes other environmental conditions, such as air and ground temperature, humidity, carbon dioxide (CO_2) concentrations, and so on, which are also important for plant growth [72]. In stevia, the negative effect of shade was reported for the first time by Slamet and Tahardi [73]. They confirmed that shade reduced growth and rate of flowering. Furthermore, about 60% reduction in light delayed the flowering, decreased plant biomass production, significantly decreased the percentage of flowering plants, and also reduced rate of flowering.

Interception and utilization of light can be directly affected by plant geometry and architecture. Kumar et al. [74], in a 2-year field experiment, assessed the effect of different levels of shade (no shade, 25% shade, 50% shade, 75% shade) and plant geometries [30 × 15 cm (north–south, N–S), 45 × 10 cm (N–S), 30 × 15 cm (east–west, E–W), and 45 × 10 cm (E–W)] on the growth, yield, and quality of stevia. These authors observed as plants grown at minimum light (75% shade) were characterized by slower development, delayed bud formation and flowering, and lower dry leaf yield, if compared with plants grown under 25% shade. Plants grown under 25% shade showed highest LAI values and produced significantly taller plants than the other conditions, which may be due to long internodal lengths and thinner stems. On the contrary, total SVglys decreased with increasing shade levels. All shade levels resulted in the same leaf-stem ratio, which was significantly higher than that obtained under normal irradiation. Also plant geometry, modifying the microclimate conditions of plant growth, significantly affected the number of leaves and leaf dry weight. Generally, the leaves in the middle stem portion present the highest SVglys content [75].

Finally, it is reported as light sources can directly stimulate the production of important secondary metabolites including anthocyanins, polyphenols, phenolic acids, and flavonoids [76]. In fact, the application of colored lights is one of the important elicitation strategies to enhance biomass accumulation and production of bioactive compounds. For this reason, Ahmad et al. [77] studied the effect of various spectral lights on biomass accumulation, secondary metabolite production, and antioxidant activity in stevia callus cultures. The results indicated as white light (used as control) was more effective for callogenesis and biomass accumulation than colored lights (yellow, blue, and green lights). However, the blue light improved phenolic and flavonoid contents than control, suggesting that the application of colored lights may be a promising approach for enhanced production of stevia antioxidant secondary metabolites.

4.3 Varieties

There are about 90 varieties of *Stevia rebaudiana* developed throughout the world [17]. These varieties were developed for different climatic requirements. As the varietal diversity in stevia is reportedly quite high, selection of appropriate varieties for specific areas is essential. Brandle and Rosa [78] found that there is high heritability for leaf yield (75%) and leaf-stem ratio (83%) and noted that leaf yield or the leaf-stem ratio were not related to Stev concentration in the landrace cultivar (imported from China) grown in Delhi Research Station, Ontario, Canada. They also found that heritable variation in the Stev concentration was also high. Being an introduced species, the number of stevia accessions available is still small. The most known and studied stevia varieties are Criolla and Morita II, the first seems to be the original stevia variety from Paraguay, while the second one has been selected for

Fig. 4 Phenotypic variability between two *Stevia rebaudiana* genotypes, in trials carried out at the Experimental Centre of the Department of Agriculture, Food and Environment, of the University of Pisa

high content of Reb A. Other known and cultivated varieties are Eirete, a hybrid developed in Paraguay for intensive cultivation; Morita III, obtained from Morita II and characterized by low water requirements; and Katupyry, a recent selected variety in Paraguay for growing in arid soils and characterized by high sweetening power.

Studying phenotypic and genotypic diversity is a key step to develop a plant breeding program for stevia [79]. For this reason, researchers have been attempted to enhance knowledge about stevia variability and to develop a breeding strategy, studying the genotypic and phenotypic variability of several genotypes under different environmental and agroclimatic conditions. In Fig. 4, an example of phenotypic variability between two genotypes is reported. At the same time, a better understanding of the genotypic control of SVglys composition would be of interest for plant breeding. Plant breeding efforts in stevia have been largely focused on improving leaf yield and Reb A concentration in the leaves [75, 78, 80–88].

By the end of the 1990s, several patent applications have been filed for new stevia varieties, in particular with improved sweetening power and taste [89–94]. Conventional plant breeding approaches, such as selection and intercrossing among various desirable genotypes along with chemical profiling for high Reb A content, are considered one of the best methods for improving quality traits in a highly cross-pollinated crop like stevia.

The International Union for the Protection of New Varieties of Plants (UPOV) database showed that there are some 40 applications worldwide for Plant Breeders Rights (n. 31) or Plant Patents (n. 9) concerning stevia [95], of which only ten applications for Plant Breeders Rights in Paraguay [41]. Currently, at the European Union agency CPVO (Community Plant Varieties Office) only one stevia variety is in the list of entrusted examination by the GEVES (Group d'Etude et de contrôle des Variétés Et des Semences) office. In order to discriminate this as new variety, the DUS (Distinctness Uniformity and Stability) criteria should be respected.

Patent applications concerning new stevia varieties have recently raised several concerns about the vexed question of how to ensure the fair and equitable sharing of the benefits arising from the utilization of genetic resources and the associated traditional knowledge according to Nagoya Protocol, which was adopted in 2010, and came into force in October 2014. According to the spirit of the Convention on Biological Diversity (CBD) and the Nagoya protocol, signed by different countries, the knowledge and the share of benefits from any commercialization have to be taken into account. In fact, since global demand for natural and sugar-free products is rapidly expanding, stevia plants are being grown and processed commercially in many countries outside Paraguay. However, the Guaraní people's right to benefit from its use, as established under the Convention on Biological Diversity's Nagoya Protocol, is being ignored, determining, in this way, a clear case of biopiracy [41].

4.4 Rotation

Both as annual and perennial crop, stevia does not tolerate to be grown in monoculture (i.e., growing the same crop year after year on the same piece of land), but it should be grown in short- or long-term rotation with other crops. It is important to be followed or preceded by plants belonging to a different family, which do not share some common disease problems. Green-manure legumes, such as fava bean, winter pea, or hairy vetch may come before stevia as spring-summer crop. For instance, before the planting of stevia in late spring, it is possible to incorporate a hairy vetch cover crop into the soil to add organic matter and nitrogen to the soil. Hairy vetch, with its low carbon to nitrogen ratio and rapid decomposition rate, can fulfill the stevia need of nitrogen during the early season. At present most stevia crops are sole cropped, but intercropping is possible during the initial growing period of stevia, in particular when grown as perennial crop. Due to the cessation of growth during winter, a legume (e.g., lentil) or clover (e.g., white clover) – stevia intercrops – can provide robust and resilient crop stands, as showed by Ramesh et al. [96], which determined the suitability of *S. rebaudiana* for food crops–based intercropping, testing wheat, barley, lentil, and rapeseed, as intercrops, during the winter season.

4.5 Tillage and Soil Preparation

Different tillage options can be used for stevia depending on type of soil, weed competition, incorporation of crop residues and/or fertilizers, planting type, crop duration, costs, and benefits, etc. To minimize the negative consequences of tillage on soil structure and organic matter, the timing of tillage, equipment operation, and soil conditions should be taken into careful consideration. In stevia, conventional tillage by moldboard, disk, and/or chisel plowing, followed by secondary tillage operations with disks and/or field cultivators were generally used to prepare a fairly smooth, firm-planting surface. However, problems may occur on very poorly drained clay soils or when it is necessary to incorporate organic fertilizers into the soil. When stevia is grown as perennial crop, it is recommended to use moldboard or chisel plow instead of disk and field cultivators to reduce the risks of soil compaction and, consequently, to reduce water infiltration and drainage from the compacted layer. In poorly drained silt loam or silty clay loam soils, or when a frost action is required to create a desirable seedbed or to improve water retention under semiarid climate, fall plowing (20–30 cm depth) is better than spring one. In addition, it is important to have the soil as weed free as possible prior planting. Fall plowing and disking are done to reduce weed populations. When stevia is grown as annual spring crop, disks and field cultivators are often used for primary tillage, though the depth of tillage is less than with moldboard and chisel plows. Conservation tillage, leaving all or part of the previous crop's residue on the soil surface, may be helpful in reducing erosion and surface runoff and in increasing water infiltration in particular on sloping land. The practices vary (e.g., strip-till, mulch-till, ridge-till) and, even if their beneficial environmental impacts are generally recognized [97], no information are available for stevia.

Stevia can be grown on plastic mulch film, under greenhouse and open field with several advantages (e.g., reduced weed problems, reduced soil evaporation, increased soil temperature, increased growth, cleaner production). When stevia is grown as perennial crop, the mulch film can help the crop to reduce the competition from weeds in the first period of growth, before being established. The soil should be fertilized prior to preparing the beds for transplanting. With some bedding machines, the soil is lifted and then bedded in one operation. Drip irrigation is recommended for use with plastic mulches. The plastic mulch is generally 150 cm wide, and with a double/quadruple rows crop, the drip tube should be placed directly on the center of the bed (Fig. 5). In organic systems, biodegradable bioplastic mulch films were used

Fig. 5 Stevia cultivated with biodegradable mulch film at the Experimental Centre of the Department of Agriculture, Food and Environment of the University of Pisa

with a duration that depends on environmental factors and film thickness. When the crop is perennial, biodegradable mulch films are more favorable because they do not obstacle the crop rationing in the second year. The methods used for working and preparing the land (plowing, etc.) are largely the same as those used with traditional plastics. The film can be laid mechanically and perforated with the same machinery used for traditional plastic film. It is advisable to lay the film and transplant cuttings at the same time. It does not have to be removed or disposed of at the end of the crop cycle, thanks to its capacity to biodegrade when incorporated into the soil.

4.6 Propagation Methods

Stevia is propagated by seeds (gamic reproduction) and from cuttings or tissue culture (agamic reproduction) [3, 98, 99].

Stevia seeds are used to produce low-cost plantlets, useful especially for northern farmers of temperate zones where stevia grows as an annual crop. Seed propagation is generally considered the most feasible and cheap method for reproducing stevia [99] and could be used as easy way for a commercial scale propagation [100]. Unfortunately, stevia seeds are smaller in size with very low germination percentage, and these negative characteristics cause serious obstacles for the establishment of gamic propagation [75, 101–103]. Consequently, stevia is conventionally propagated by stem cuttings, which root easily, but this traditional method is expensive and requires high labor inputs [17, 102], and it is necessary to collect a relative high number of healthy mother plants from which vegetative portions are taken in the proper time. Furthermore, in the temperate climates the mother plants should be grown under greenhouse to anticipate the regrowth and make possible to take the vegetative portions in time for the spring planting.

Micropropagation is a rapid technique widely used to produce large quantity of new plantlets from a single mother plant in a short span of time [102, 104, 105] and, at the same time, it represents an opportunity for germplasm conservation. This technique requires very high and qualified labor inputs and advanced technology laboratories. Both stem cuttings and micropropagated plants could provide genetically uniform plant populations with homogenous SVglys profile [106]. On the other hand, plants obtained by seed are not uniform, and they could be useful for breeding programs.

4.6.1 Seeds and Propagation Material

According to GACP, seeds are to be identified botanically, indicating plant variety, cultivar, chemotype (if available), and origin. The material used should be traceable. The same applies to vegetatively propagated starting material. Starting materials used in organic production have to be certified as "organic" according to Reg. 834/2007. Seed should meet the requirements/standards concerning purity and germination (wherever available, certified seed/propagation material should be used). The starting material should be as free as possible, of pests and diseases in order to guarantee healthy plant growth. Plant material or seeds derived from or comprising genetically modified organisms have to be in accordance with national and European regulations. Starting materials used in organic production should be OGM free.

4.6.2 Seed Production and Quality

Stevia is characterized by a complex mechanism of seed multiplication. The species is hermaphroditic, highly cross-pollinated, and photoperiodically sensitive [3, 24, 101, 103]. Stevia flowers have sporophytic self-incompatibility, and the pollinators are needed to produce viable seeds, whereby three to four hives per hectare are recommended [25–29, 101, 107]. In support of this argument, stevia produces two types of achenes (that contain seed and embryo), black and tan colored [29, 30, 99, 101]. Black type is produced by cross-pollination, while tan achenes, which do not have the seed embryo, are originated from self-pollination [29]. Late flowering may adversely affect seed production, especially if it occurs during an unfavorable season for pollination [101]. Seed production and yield depend on latitude and climate. For example, yields of 8.1 kg ha^{-1} were obtained in Brazil [19]. Under temperate climate and mid latitudes of Italy, earlier flowering genotypes produced higher seed quantity, due to favorable weather conditions during the reproductive phase [107] and, at the same time, cross-pollination and the insect abundance (i.e., Apidae and Syrphidae) played a crucial role for enhancing seed yield and quality [107]. Seed production and viability are linked to genetic characteristics as well as to plant development during pollination, seed filling, and ripening phase [2, 107]. The presence of pappus, the small achene size, and the seed production dispersed along the time [18, 108] are all problematic traits for seed management and mechanical harvest. At the same time, the harvest of immature seeds may also contribute to stevia poor germination [100]. Several studies analyzed different kinds and levels of light and temperature to improve seed germination, showing as, the best results were possible adopting a temperature of 20–25 °C under light conditions, with positive effects of red light [101, 109–111]. Furthermore, the minimum germination temperature was found at 5 °C, showing a good adaptability to cooler areas than the native ones [109].

Several authors found that the viability of stevia seeds is pretty low, in particular when stored under not suitable conditions (high humidity and temperatures) [19, 30, 112]. Germination tests showed that, in order to extend seed viability from a few months to more than 3 years, the optimal preservation method is to store seeds in insulating containers maintained at low temperature (−10 °C, −18 °C) [112].

Recently, Martini et al. [107] investigated the influence of insect pollinators and harvesting time on the qualitative and quantitative characteristics of stevia ripe seeds, through a pot trial carried out under open air conditions in central Italy. In this study, 36 F1 open-pollinated genotypes were compared. A clear effect of pollinator visits and harvest time on ripe seed quality, in terms of germination percentage, mean germination time, and thousand seed weight (TSW), was assessed. On the basis of the achieved results, the authors confirmed the complex interactions among the investigated parameters in defining the yield and quality of stevia seeds. Results indicated that earlier flowering genotypes produced higher seed quantity, due to favorable weather conditions during the reproductive phase.

Fig. 6 *Stevia rebaudiana* insect pollinators (**A** = *Celastina argiolus*, **B** = *Pieris brassicae*, **C** = *Vanessa cardui*, **D** = *Eristalinus taeniops*, **E** = *Apis mellifera*, **F** = *Xylocopa violacea*) visiting stevia in the experimental trials carried out at the Department of Agriculture, Food and Environment, of the University of Pisa

Moreover, cross-pollination and the insect abundance (Fig. 6) played a crucial role for enhancing the stevia seed yield and quality. Therefore, all the conditions that allow open pollination and insect visit are necessary because of the self-incompatibility of this plant species. Proper seed harvesting techniques may help to prevent the losses of early ripe seeds by wind dispersal. At the same time, the selection of early flowering populations with short reproductive stage may help to increase seed production and raise profits for seed producers.

4.6.3 Agamic Propagation

Stem cuttings propagation. To produce stem cuttings, healthy mother plants to take vegetative portion are needed. Mother plants should be growing in vegetative condition, avoiding short day length that cause flowering induction. In fact, the cuttings generated from induced mother plants (or already flowering mother plants) grow slowly [113]. In order to obtain well-formed plants, the vegetative portions, from which cuttings are taken, are extremely important. Best results have been achieved taking stem portions from the top part of the main shoots with two or three well-formed nodes, each with a couple of expanded leaves [3, 113]. Also current year's shoots from the leaf axils give high performances [114]. The best period to take cuttings is late winter [115]. Stem-cutting rooting percentage and relative growth are affected by cuttings length. Chalapathi et al. [116] found that with 15 cm long cuttings, both rooting percentage and growth were higher, compared with 7.5 cm long cuttings, since food reserves was double. Similarly to other plants, cutting rooting was stimulated by using growth regulators [3, 102, 113, 117]. Interesting results were achieved using solutions of indole-3-butyric acid (IBA) and α-naphthalene acetic acid (NAA). Highest sprouting survival rate was observed when dipped in IBA solution (1,000 ppm) of 33% and NAA solution (1,000 ppm) of 22.2% [102]. On the other hand, Smitha and Umesha [118] reported a very high survival rate (76.8%) using IBA both 500 and 1,000 ppm; these data are not in agreement with those of other authors who reported that, cuttings were not influenced by NAA or IBA treatment at 500 ppm. Furthermore IBA or NAA at 1,000 ppm caused complete inhibition of cutting rooting [3]. Ingle and Venugopal [117] showed the highest rooting percentage both for IBA 500 and 400 ppm (92% and 90%, respectively). Moreover, with these treatments the highest number of roots and leaves, and root and cutting length, has been registered, according to Bagoury et al. [119]. After growth regulators treatment, it is necessary to place stem cuttings in planting trays filled with sterile growth media, because direct planting in field showed limited success [17, 113].

In vitro propagation. Suitable alternative method for large-scale plant production within a short period is the use of in vitro culture technology [105, 120]. To obtain micropropagated plants it is necessary to follow five phases: mother plants choice, culture initiation, multiplication, rooting, and acclimatization. Mother plants should be healthy and virus free, kept in long-day photoperiod avoiding flowering phase. For micropropagation several plant parts could be used, such as leaves, auxiliary shoots, root-neck sprouts, shoot primordial, and internodal explants [3]. For stevia micropropagation, it was observed that MS medium [121] gave the best results [104, 122–130]. Usually, a 1–3% (v/v) of sucrose is used as carbohydrate source and 0.7–0.8% of agar as gelling agent [102, 120, 123, 126, 128–131]. During the years several researchers studied the best performance for shoot induction, callogenesis, and rooting, evaluating different growth regulators. Indole-3-acetic acid (IAA), 6-benzyladenine (BA), kinetin (Kn), 2,4-dichlorophenoxyacetic acid (2,4-D), naphthalene acetic acid (NAA), indole-3-butyric acid (IBA), and thidiazuron (TDZ) were tested on several kind of explants (i.e., flower, nodal segments, apical and nodal meristems, axillary buds, leaf), in different medium and concentrations with interesting and promising results [104, 124, 125, 127–129]. The last micropropagation step is the acclimatization of the plantlets, which represents a decisive phase in order to obtain plants ready for the market. Several factors are important for a successful acclimatization, such as substrate, light, temperature, and humidity [132]. Different mixtures (i.e., sand, soil, and peat – 1:1:1; sand, soil, and vermicompost 1:1:1; soil, sand, and manure 2:1:1) and arrangements (i.e., rooted plantlets in plastic cups covered with polythene bags to ensure high humidity) were successfully tested [102, 120, 124, 125].

Finally, also the effects of micropropagation methods on SVglys content and antioxidant properties have been evaluated. Bondarev et al. [133] showed that the nutrition medium composition had no effects on the ratio of individual glycosides, and Hwang [122] found that the Stev content, in vitro propagated plants and relative mother plants, was similar. Zayova et al. [134] observed that the leaf extracts of plants propagated in vitro (and well adapted to field conditions) were characterized by high levels of water-soluble antioxidant capacity, phenols, flavonoids, and high total antioxidant potential, expressed as DPPH radical scavenging activity.

4.7 Planting

Planting represents also an important agronomic factor that plays a significant role in improving the crop yield, the plant's phenological development, and total biomass production, along with efficient conversion of biomass into economic yield. Planting (planting material and field operations) is one of the major cost items in stevia production, mainly when it is managed as annual crop [135]. The pedoclimatic conditions of the cultivation site should be taken into account for planning the optimal planting time. Change in planting time leads to significant changes in weather parameters and, consequently, in the crop performance [100]. Taleie et al. [136], in a 2-year field experiment carried out in Guilan (northern Iran), observed that transplanting date significantly affected plant height, leaf and stem fresh and dry biomass, and also Stev, phenol, and flavonoid yield. These authors, comparing three transplanting dates (15 March, 30 March, and 15 April) and four spacing arrangements in culture-derived stevia plantlets, found that the maximum plant development, leaf yield, and secondary metabolite levels were obtained in the 50 × 20 cm spacing (10 plants m^{-2}), when plants were transplanted on March 15th. Delayed transplanting date decreased the product quantity. In the northern hemisphere, such as Canada, stevia planting is done during mid-May [1], while under the Indian climate conditions, the crop could be transplanted from January to March in order to obtain maximum plant height, total dry matter production, and fresh and dry leaf yield [137, 138]. In the climate conditions of the Mediterranean basin, the optimal planting time is during the spring season from March to May according to the site-specific climatic conditions. Serfaty et al. [139] found that in Israel the optimal time for planting stevia was during the spring season (April and May); plantings in January–March and in October were found to decrease crop biomass, due to delayed acclimatization and development. In such pedoclimatic conditions, early planting of stevia in autumn was found unfavorable, due to the short-day light conditions that stimulate flowering. Consequently, much of the growing period had no additive developmental value with regard to biomass accumulation parameters. Moreover, the winter harsh climatic conditions negatively affected survival rate, conversely to spring planting. Spring-planted plants had a higher maximal biomass yield than autumn-planted ones [139]. Similarly, in central Italy, where 14 h photoperiod prevailing in the spring-summer growing season, the optimal planting time for stevia is during spring (April–May) [140].

Also the choice of optimal plant density plays an important role in the establishment of a successful stevia cultivation, influencing canopy development, radiation interception, evaporation of water from soil, dry matter production, weed competition, and the development of fungal and viral diseases. Initial trials indicated that higher growth and biomass yield were reached when a low plant density was adopted (8.3 plants m^{-2} with 60 × 20 cm), while dry leaf yield was higher in denser planting (16.6 plants m^{-2} with 60 × 10 cm) [141]. In contrast, Lee et al. [142] reported that plant height, number of branches, and number of nodes were unaffected by planting density, while dry leaf yield per plant decreased with increasing plant density. Accordingly, Donalisio et al. [143] recommended a plant population of

80,000–100,000 plants ha^{-1}. Rank and Midmore [144] reported that higher plant densities can promote weed control, but, at the same time, increase production costs per hectare [144]. Taleie et al. [136] observed that, under Iranian conditions of Guilan province, spacing significantly affected plant height, leaf and stem fresh and dry weights, and also stevioside, phenol, and flavonoid yields. The maximum plant height (80 cm), total fresh (2,017.21 g m^{-2}) and dry (588.69 g m^{-2}) weight, and the highest stevioside (34.51 g m^{-2}), phenol (1.5 g m^{-2}), and flavonoid (1.97 g m^{-2}) yields were obtained with 10 plants m^{-2} (50 $\times$ 20 cm row spacing). Recently, Serfaty et al. [139] and Rashid et al. [145] in studies carried out in Israel and India, respectively, concluded that with a stand of about 10 plants m^{-2}, it was possible to obtain higher biomass levels, dry leaf weight, and higher SVglys yields than those obtained with lower stands. Anyway, plant density is strongly dependent on the pedoclimatic conditions of the cultivation site in which stevia is cultivated, varying according to different geographical areas. Few information are available about plant density and the productive performances along the years when stevia is grown as perennial crop.

4.8 Crop Nutrition and Fertilization

In stevia, similarly to other crops, the standardization/optimization of nutritional doses, particularly N, P, and K, for different agroclimatic conditions, is essential for increasing the biomass yield and its quality. The type, timing, and quantity of fertilization should be determined by the effective nutritional requirements of the crop and the supporting capacity of the soil. A soil test should be obtained prior to planting to determine the soil pH and levels of available macronutrients.

The nutrient requirement for this crop is low to moderate, as it is adapted to poor quality soils in its natural habitat at Paraguay [17]. It varies according to the environment, soil type, expected yield, previous crop, and waste management. Generally, stevia plants respond well to mineral fertilizers. In integrated production systems, particular attention should be paid to mineral nitrogen fertilization by adopting management strategies able to reduce the risk of N losses to air and water, and improve N use efficiency. Stevia grown in organic systems rely on several management practices (such as biological N fixation, amendments, cover crops, perennial crops, and legumes) that enhance soil fertility.

The actual rates of fertilizer application vary according to soil type and environment of the growing region and, consequently, need to be optimized for each specific situation. First reports from Japan demonstrated that, at the time of maximum dry matter accumulation, dry stevia shoots consisted of 1.4% N, 0.3% P, and 2.4% K [55]. Kawatani et al. [146–148] found that nitrogen and potassium determined an increase in growth, stem thickness, and number of branches. Under Canadian conditions, the crop required approximately 105 kg N, 23 kg P, and 180 kg K for a biomass dry yield of 7.5 t ha^{-1}; thus to produce 1 t of total aboveground biomass, the crop needed 14 kg, 3 kg, and 24 kg of N, P, and K, respectively [1].

Recently, Angelini and Tavarini [140], in a field experiment carried out in the northeastern Italy, showed that the average amount of each macronutrient extracted by the crop and required to produce 1 t of total aboveground biomass was 16.4 kg, 1.9 kg, and 22.4 kg of N, P, and K in the 1-year-old crop and 14.9 kg, 2.8 kg, and 28.5 kg of N, P, and K in the 2-year-old crop. These data suggested that, in the tested environment, in the second year of growth, higher doses of phosphorus and potassium are needed to produce 1 t of total aboveground biomass. These authors determined also the contribution to the overall nutrient uptake by the different plant organs, showing that the leaves absorbed higher amounts of total nitrogen, phosphorous, and potassium than the stems (80%, 65%, and 59% of total N, P, and K were taken up by the leaves, as mean values over the 2 years of the trial). Pal et al. [70], through a field experiment comparing three levels of nitrogen, two levels of phosphorus, and three levels of potassium at three locations of northern India, established sole and interaction effects of N, P, and K on yield, and SVglys accumulation in leaves. Principal component analysis (PCA) indicated that the applications of 90 kg N, 40 kg P_2O_5, and 40 kg K_2O ha^{-1} were the best nutritional conditions in terms of dry leaf yield, in two of the three pedoclimatic conditions tested here. As general trend, in this study, the dry leaf yield was increased by increasing the N level at all three locations, due to the fact that higher dose of N increased the availability of N in soil, and subsequently induced cytokinin synthesis in root tips and maintained desirable cytokinin and auxin ratio [70].

On the other hand, reducing the amount of mineral fertilizers and increasing the use of amendments is an important objective in the sustainable production of many crops, including stevia. Liu et al. [149] conducted a field experiment to study the effect of organic manure and mineral fertilizers on growth and development of stevia in China. Plants grown with organic fertilization, sampled 80 days after transplanting, showed higher root activity, stem diameter and length, and aboveground dry biomass than those grown with mineral fertilizers, even if in the organic treatment, the plant growth was lower at the earlier stages. Dube [150] showed that organic manures, biofertilizers (PSB, Phosphate Solubilizing Bacteria), and plant growth regulators (GA, IAA, IBA) induced an increase in all qualitative parameters related to morphological characters, growth, yield and yield components, over the control. Among the organic fertilizers, farmyard manure (FYM) and vermicompost allowed the best crop performances. Kumar et al. [151] reported that farmyard manure (at doses up to 30 t ha^{-1}) enhanced the bioactive compounds than mineral fertilizers, even if achieving lower values than the unfertilized control. On the other hand, the mineral fertilization of 50-60-50 kg N-P-K ha^{-1} determined to highest plant height and dry leaf yield per hectare compared with other treatments. Dry leaf yield, number of leaves/plant, leaf area index, and dry matter accumulation/plant were highest in plants grown at the highest level of FYM (45 t ha^{-1}) and with 60 kg N ha^{-1} mineral fertilization [152].

Lima Filho and Malavolta [153] described the foliar symptoms of macronutrient deficiency, and B and Zn toxicity in stevia grown in hydroponics. Biomass and chlorophyll content decreased with increasing level of B concentration when applied as foliar spray. Utumi et al. [154] described the macronutrient deficiency in stevia in relation to plant growth, chemical composition, and SVglys content. They found that total aboveground biomass decreased in all the macronutrient deficiencies; however, the percentage of reduction was higher in treatments without N, P, and Mg; Stev concentration decreased with the deficiency of all macronutrients except for P. Regarding micronutrients, Utumi et al. [154] reported that deficiencies of Ca, Mg, and S induced some morphological changes such as apical necrosis, chlorosis, inverted "V" shaped necrosis (leaves drying from the tip), and small pale green leaves. Nutritional interactions have been studied showing synergistic effects on foliar content between N and P, P and Cu, and P and Fe; antagonistic effects between N and K, N and Zn, K and Mg, and K and S; and synergistic and/or antagonistic effects between Zn and B, and Mn and Mg [155].

4.9 Mycorrhizal Association and Plant Growth Promoting Rhizobacteria

Arbuscular mycorrhizal fungi (AMF) are soilborne mutualistic fungi belonging to the phylum of Glomeromycota that can improve plants' growth rate, pathogen resistance, water and mineral nutrient uptake (macro- and micronutrients), and photosynthesis rate. This effect is also mediated through production of enzymes by AMF, such as phosphatases [156–158]. Thus, these soil-dwelling symbiotic microorganisms are useful for agriculture and for cultivation of medicinal plants, stevia included [159, 160]. Several studies showed the potential of AMF to augment the secondary metabolite production in plants [161–165]. Portugal et al. [166] have made a selection of the better arbuscular mycorrhizal fungi for stevia, founding that the best AMF is *Glomus intraradices*, which enhanced the plant biometric parameters, dry biomass and leaf yield, and Stev content. Mandal et al. [167] demonstrated that AMF association increased Stev and Reb A production. For these authors, the increase in yield of these two steviol glycosides in mycorrhizal plants was mainly due to their increased concentration and enhanced biomass of the tissue in which they were found. In turn, the increase in biomass was directly related to improved uptake of nutrients (N, K, Mg, Cu, Fe, Mn, and Zn), chlorophylls, and carbohydrates. These authors also observed an higher jasmonic acid content and glandular trichome density in mycorrhizal plants in comparison to the control. Greater concentrations of total carbohydrates and jasmonic acid in mycorrhizal plants contributed to increased biosynthesis of SVglys. In fact, the increase in jasmonic acid concentration is associated with the presence of methyl jasmonates that has been shown to alter the expression of genes in the methyl-D-erythritol 4-phosphate (MEP) pathway (MEP is the pathway used for the SVglys biosynthesis) and the concentration of products derived from these genes [168–170]. So, the presence of AMF stimulated the induction of MEP pathway through enhanced jasmonic acid levels which, consequently, lead to an increased SVglys content. Accordingly, Mandal et al. [171] studied the interactions between mycorrhizal fungi and SVglys production at molecular level, investigating the transcription profiles of 11 key genes involved in the MEP pathway. They found that the concentration of Stev and Reb

A increased 1.68- and 3.4-fold, respectively, in mycorrhizal (M) plants with respect to nonmycorrhizal (NM) ones, as well as the Reb A/Stev ratio. Furthermore, they observed a better nutrition uptake in M plants with respect to NM plants, underlying an improved plant's nutritional status. Studying the effects of AMF on MEP pathway in stevia, Mandal et al. [171] concluded that AMF symbiosis upregulates the transcription of all 11 genes involved in the SVglys biosynthesis. This might be a result of improved nutrition and enhanced sugar concentration due to increased photosynthesis in M plants. Angelini et al. (unpublished data) are studying the effect of AMF association and phosphorus nutrition on growth, yield, and photosynthetic parameters as well as on phytochemical properties and antioxidant capacity of stevia plants grown in pots under Central Italy climatic conditions. In Fig. 7, mycorrhizal infection is reported, in comparison with a control (plant without AMF symbiosis).

Another field of research in stevia is about the use of plant growth promoting rhizobacteria (PGPRs) with AMF. PGPRs have different benefits for plants: (i) to increase the availability of nutrients in the soil (phosphate solubilization, siderophore production); (ii) to increase root length, resulting in a more nutrient uptake by plant; (iii) to stimulate plant growth (through phytohormone production), (iv) to control or inhibit the activity of plant pathogens (by enhancing the amount of plant enzymes); and (v) to improve the chemical, microbial and structural properties of soil. For these characteristics, they are used as biofertilizers and can be considered a valuable strategy to help sustainable and environmentally friendly agricultural practices

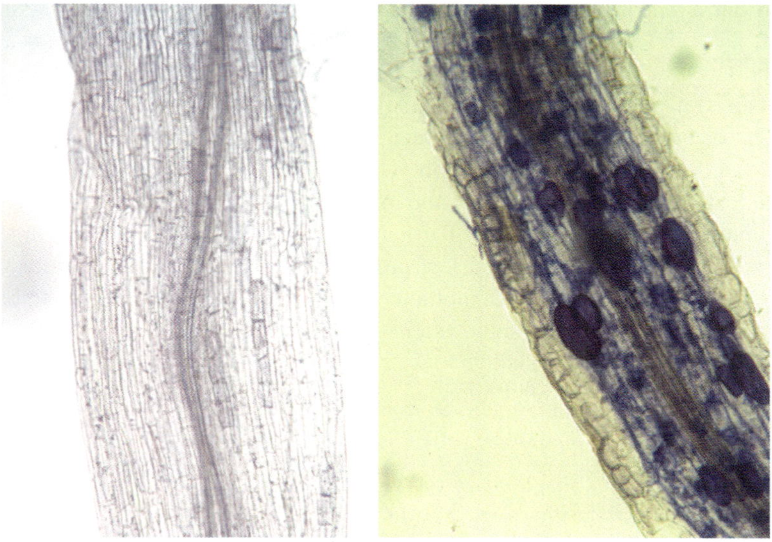

Fig. 7 *Stevia rebaudiana* Bert. secondary radicles: on the *left*, radicle without mycorrhizal symbiosis; on the *right*, radicle with mycorrhizal infection by *Glomus intraradices*. Mycorrhizal portions are colored in *blue* by trypan blue (notice the small ifae penetrating the radicle) (Courtesy of M. Giovannetti and L. Avio, Department of Agriculture, Food and Environment, University of Pisa)

[172]. In stevia, several works have been carried out about the effects of association between AMF and PGPRs. Several studies [173–181] showed the beneficial consortium between AMF, phosphorous solubilizing bacteria (PSB), and *Azospirillum* or *Azotobacter* or *Rhizobium*. In stevia, these triple associations improved biometric parameters, uptake of nutrients and water from soil, SVglys content, and protein, mineral, and phenol contents. These latter compounds in turn improved the antioxidant activity of leaf extracts [182]. Zare Hoseini et al. [183] observed an increased biomass production and microelements content in plants biofertilized with *Piriformaspora indica* and *Glomus mosseae*. Furthermore, PSB and *Azotobacter* can be also used in the treatment of plantlets before transplanting, for improving the biometric and qualitative parameters in stevia plants at harvest [184–188]. At the same time, the soil can benefit from the use of biofertilizers, as reported by Ramakrishnaiah and Vijaya [189]. These authors evaluated the effect of *Glomus mosseae* associated with *Azotobacter* sp. on soil phosphatase activity and soil nutrients in comparison with a control, without inoculation. The treated soils showed higher content of nitrogen, phosphorous, potassium, microelements (Zn, Fe, Cu, Mn), and alkaline and acid phosphatases, than the control.

4.10 Water Requirement, Irrigation, and Salinity Tolerance

Stevia can be cultivated profitably wherever irrigation facilities are available, taking into account the needs of the plant. Irrigation and drainage should be controlled and carried out in accordance with the needs of the plants during its various stages of growth and environmental conditions. Water used for irrigation purposes should comply with local, regional, and/or national quality standards and should be as free as possible of contaminants, such as any organic or inorganic solids, including domestic or industrial wastes, heavy metals, pesticides, herbicides, and toxicologically hazardous substances. Irrigation water can be optimized by using efficient irrigation systems, such as sprinkler and microirrigation systems whenever possible, or by improving the scheduling of irrigation. Normally, stevia requires frequent, shallow irrigation for commercial leaf production, especially under Mediterranean zones. It is reported that stevia cannot tolerate drought [190] even if once established it can tolerate water stress to some extent as reported by Tonello et al. [191]. At the same time, it is extremely sensitive to water logging. Water logging can rot the plant and results in poor crop stand. Prolonged water logging induces senescence, which is harmful to the crop. Thus, it is necessary to have proper water disposal system to ward off adverse effects of sudden and heavy downpour and subsequent water logging. The excess moisture and water logging conditions during rainy season creates unfavorable conditions for growth, reduced aeration, hampered root development, reduced nutrient uptake, and favorable environment for disease incidence, finally leading to reduced crop stand [151, 192].

Several studies have been conducted with the aim to define proper irrigation scheduling in order to achieve good results in terms of plant growth, dry matter accumulation, and crop yield as well as SVglys production. Goenadi [193] reported that 2.33 mm/plant/day on latosoils could be considered a correct water supply. Under these conditions, the growth of stevia was optimum when the soil content was 47%, while below 30% the plant reached permanent wilting point (−1.5 MPa). A field experiment carried out by Fronza and Folegatti [194] on a sandy loam soil of the low Arno river plain in Pisa (central Italy 43°N lat) using microlysimeters in a 2-year-old crop. They reported a total water consumption of 464.55 mm over 80 days dry season from June to August, with an average ETc (crop evapotranspiration) 5.44 mm/day, double that found by Goenadi [193]. The crop coefficient value (Kc), calculated as ratio of the ETc to the ETo, was 1.45 (from 0 to 25 days), 1.14 (from 26 to 50 days), and 1.16 (from 51 to 80 days). These results are higher than those found by Gonzalez [195] in Paraguay and justified by higher leaf dry yield (about 4,370 kg ha^{-1} with 6,600 plants ha^{-1}) and LAI values (4.83) achieved under LDs (15.9 h), high solar radiation, and fertile soil in Central Italy. Lavini et al. [196] in a field plot trial on sandy clay soil of the Volturno river plain (South Italy, 40°N lat) compared three irrigation regimes (100%, 66%, and 33% of soil water consumption) and estimated stevia ETc using the soil water balance method and soil water content measure gravimetrically. Leaf yield increased by up to 40% when soil moisture was increased from 33% to 100% of soil water consumption (based upon the soil water balance method), suggesting 100% soil water consumption is best for stevia cultivation. Midmore et al. [197] in a series of trials carried out in central Queensland, Australia, observed high leaf yield in plants grown under field capacity and 80% of FC (field capacity). The SVglys concentration on waterlogged conditions was similar to that of the plants grown under 100% and 80% of the FC. Decreased SVglys content in drought stressed stevia plants was more due to the reduced yield of the leaf biomass than the SVglys concentration.

Guzman [198], in a study carried out in greenhouse conditions, observed that moderate water-deficit stress (8-days irrigation period) did not significantly affect the SVglys content [198]. Aladakatti et al. [199] found that an irrigation based on the full replenishment of crop evapotranspiration resulted in a maximum leaf yield. These authors suggested that irrigation supply at 5-days interval was suitable for summer cultivation of stevia in the semiarid regions of India. Shi and Ren [200] studied the effects of drought stress on the photosynthesis and dry leaf yield in different stevia accessions, observing that there were no significant changes in net photosynthesis, transpiration rate, and leaf dry matter in short-term drought-stress treatment (5-day irrigation interval). Significant decrease of these parameters was observed with a prolonged drought stress (10-day irrigation intervals). Under in vitro culture conditions and using polyethylene glycol to stimulate drought stress on stevia, it has been reported that fresh and dry weight, water content, chlorophylls, carotenoids, anthocyanins, water-soluble carbohydrates, and reducing sugars were negatively affected by drought stress [201]. On the contrary, antioxidant activity and enzymatic defense systems (catalase, ascorbate peroxidase, polyphenol oxidase, and peroxidase) remarkably increased [202].

The salinity of the irrigation water (ECw) significantly reduced plant growth and yield, as well as SVglys content of the leaves. So, it is important to understand the mechanisms of salt adaptation and tolerance of stevia, since, as the global climate changes, water resources become more scarce in many regions of the world and exploiting saline water and/or soil becomes more urgent. Stevia shows some variability in what concerns the sensitivity or tolerance to salt stress. First reports [203] suggested that stevia is a moderately salt-tolerant plant. This preliminary assessment was recently confirmed by Zeng et al. [45]. These authors evaluated the effects of different NaCl concentrations on the growth, physiological responses, and main SVglys composition, in plants grown in a controlled chamber under three different NaCl concentrations (60, 90, and 120 mM). The biomass production did not show significant reduction with lower and moderate salt concentration (60 and 90 mM NaCl, respectively), while drastically decreased by 40% with high salt treatment (120 mM NaCl), suggesting that stevia is a moderately salt-tolerant plant. In addition, these authors observed leaf chlorosis, wilting, and necrosis in plants treated with 90 and 120 mM NaCl for 28 days. Disruption of the thylakoid and stromal membranes may result in leaf chlorosis and necrosis, which indicated that salt concentration above 90 mM NaCl could impair the chloroplast structure of stevia. To confirm this statement, the contents of Chl a, Chl b, and Chl tot decreased notably with increasing NaCl concentration. On the contrary, the increments of the antioxidant enzyme activities (superoxide dismutase, peroxidase, and catalase) were observed with increasing NaCl treatment, and the proline contents in salt-treated plants were 17–42 times higher than that in the control, indicating that high salt stress resulted in a high level of osmotic stress. Finally, Zeng et al. [45] found that, under all salt treatments, Reb A content, Reb A/Stev, and Reb A/(Reb A + Stev) increased, whereas Stev content and Stev/(Reb A + Set) decreased, indicating NaCl could modulate the composition of SVglys for its function of promoting the transformation of Stev to Reb A. In this regard, Mohamed et al. [66] discovered that UGT76G1 played the key role in the transformation process. It was suggested that NaCl stress could enhance the transcription of UGT76G1. Thus, the relationship between salt stress and the transcription level of UGT76G1 is interesting and needs to be further studied. Reis et al. [204] studied the growth development and the yield response of stevia to the salinity of the irrigation water in a Mediterranean region (Algarve, Portugal), testing six irrigation water salinity levels and the possibility to realize two harvests. These authors concluded that stevia is suitable to be grown in semiarid and saline regions, if only one harvest is accomplished. In order to obtain two or more harvests, the authors suggested that only fresh water with low electrical conductivity should be used. In fact, salinity dramatically reduced plant regrowth after the first harvest, with an important effect in plant productivity and survival. Plants irrigated with higher salinity water did not regrow or even died before the second harvest.

4.11 Plant Protection (Weed Control, Fungal Diseases, and Insects)

Stevia has a poor capacity to compete with weeds, in particular during the initial growth period. Apart from reducing the overall biomass, weed competition resulted in reduced branching and delayed flowering [205]. Up to a 40% yield loss may be realized as a result of weed pressure by pigweed species (*Amaranthus* spp.), lambsquarters (*Chenopodium album*), foxtail species, and other annual grasses. Consequently, an efficient weed control is an important issue in order to establish the cultivation of stevia at large scale.

Currently, most strategies aimed at reducing the abundance of agricultural weeds involve herbicide applications and/or mechanical disturbance. The lack of herbicides registered for stevia often leads farmers to control weeds by mechanical control (between rows) and by hand (within rows) and/or by the integration of biological, cultural, mechanical, and chemical practices.

In a sustainable stevia cultivation system, the adoption of integrated weed management (IWM) strategies is strongly recommended. Weed management goes beyond control of existing weed problems and focuses on reducing weed invasion and emergence, preventing weed reproduction, and minimizing weed competition with crops. Weed management emphasizes understanding the causes of weed problems with the goal of preventing weeds from becoming problematic in the first place. Integrated weed management combines the use of biological, cultural, mechanical, and chemical practices to manage weeds, so that reliance on any one weed management technique is reduced. The main goals are to (1) use preventive tools to maintain weed density at a level that does not harm the crop; (2) prevent shifts toward more difficult to control weeds; and (3) develop agricultural systems that maintain or improve crop productivity and environmental quality. Alternative methods – such as the use of allelopathy phenomenon, cover crops and living mulches, competitive crop cultivars, suitable nutrient management, appropriate rotations, intercropping, planting arrangement, and biological, physical, and mechanical control – are all effective and eco-friendly practices for a sustainable weed management in stevia cultivation.

When the crop is grown in raised beds, manual operations are easier. Plastic mulch can be an effective means of controlling weeds. Basuki [206] indicated that establishing high plant densities and using black plastic mulch help to control weeds in stevia.

Few studies about IWM strategies in stevia are available, despite their importance, while most of the scientific information are dealing with chemical weed control in order to achieve the immediate elimination of an existing weed population through the use of herbicides.

Several studies evaluated the possibility of a weed chemical control in stevia, by testing a number of commercially available herbicides. Andolfi et al. [207] and Midmore and Rank [61] indicated that stevia was tolerant to trifluralin. Covarelli et al. [208], in a 2-year field experiment under the pedoclimatic conditions of central Italy, tested three preemergence and postemergence herbicides [Stomp – active ingredient (a.i.): pendimethalin; Challenge – a.i. aclonifen; and Targa Flo – a.i. quizalofop-ethyl isomer D]. The authors found that a preemergence treatment with pendimethalin allowed the best weed control and the greater selectivity. Regarding postemergence herbicides, a good selectivity of aclonifen was observed, while slight symptoms of phytotoxicity were noted with quizalofop-ethyl isomer D treatment. Angelini et al. [75], in a field trial conducted in northeastern Italy, tested four pretransplant herbicides (Goal 480 SC – a.i. oxyfluorfen; Stomp

330 E – a.i. pendimethalin; Ronstar FL – a.i. oxadiazon; and Sipcam Flow – a.i. linuron) and five posttransplant herbicides (Goal 480 SL – a.i oxyfluorfen; Stomp 330 E – a.i. pendimethalin; Sipcam Flow – a.i. linuron; Ronstar FL – a.i. oxadiazon; Whip S – a.i. fenoxaprop-p-ethyl; Challenge – a.i. aclonifen) in order to assess their selectivity and efficacy in stevia weed control. This experiment showed that, in the pre-transplanting treatments, both pendimethalin and oxadiazon showed high selectivity, while oxyfluorfen and linuron caused phytotoxicity symptoms to stevia plants. Despite this initial negative effect, the best herbicide efficacy was obtained by using oxyfluorfen and oxadiazon, with a reduction of more than 70% of weeds with respect to control. In the post-transplanting treatments, fenoxaprop-p-ethyl and pendimethalin showed the best selectivity. On the other hand, oxyfluorfen and linuron initially caused phytotoxicity in stevia plants (overcome in short time). Regarding the efficacy of the tested herbicides, the best results, after an initial phytotoxic effect, were observed for oxyfluorfen and linuron. Harrington et al. [205] carried out a series of trials in New Zealand, to evaluate the safety of 25 herbicides (many of which have been used for selective control in other Asteraceae crops), both in pre- and postemergence treatments. A number of herbicides appeared to be safe for stevia. The preemergence herbicides that can be used in transplanted stevia included trifluralin, pendimethalin, oryzalin, bromacil, terbacil, linuron, methabenzthiazuron, and alachlor. Among the postemergence treatments, stevia plants tolerated many of the herbicides listed above, plus clethodim, haloxyfop, propyzamide, thifensulfuron, flumetsulam, and pyridate. In particular, terbacil and bromacil resulted as the most effective herbicides for preemergence weed control, although some crop damages occurred. The most effective postemergence herbicides were clethodim, haloxyfop, and propyzamide, active mainly on grasses, and thifensulfuron and pyridate, for broad-leaved weed control. Recently, a study conducted by Zachokostas [209] in southern Balkans, underlined promising results with pendimethalin, napronamide, and dimethanamid herbicides, whereas prometryn, floumeturon, and clopiralid proved to be too toxic for stevia plants. Hopkins and Midmore [210], within an intense research activity carried out in Australia, assessed the efficacy of 16 herbicides on stevia and associated weeds. Among the tested herbicides, the post-planting, pre-weed emergence, oryzalin-containing herbicide Prolan 500 (at rate of 6.8 L ha^{-1}) provided the best weed control without affecting the biomass of stevia crop. Despite these results, further investigations are required to determine the optimal herbicide (or herbicide combination) for weeds management on stevia. Furthermore, the integration of chemical control with biological, cultural, and mechanical practices – shifting the focus from chemical weed control to weed management – can help to reduce weed colonization, establishment, and impact.

Production of high-quality stevia requires careful attention to pests and diseases, as well as agricultural practices that may exacerbate or suppress these pests. Stevia is not known to have serious insect pest problems, and it is often reported as exhibiting insect-repellant qualities. Similarly, documented disease problems are few and seemingly insignificant. However, insect and plant pathogens have been documented in several geographic areas. The damage by these organisms may cause ranges from insignificant to significant loss due to direct reduction in quantity of yield or diminished yield quality that can render stevia unsalable. Young plantlets, especially in the first 3 weeks after germination, are susceptible, in greenhouse conditions, to insect damage (aphids, white flies, mites, etc.). Once established, insect damage is not common, although an incidence of grasshoppers has been seen in trials carried out in North Queensland, Australia [191]. Minimal attack by aphids, spider mites, white flies has also been reported [211]. Slags and snails can attack stevia, causing direct (leaves consumption) and indirect effects (top necrosis, stunting, and reduced growth). Leafhoppers (Cicadellidae, subfamily Deltocephalinae), leaf beetles (Coleoptera: Chryssomelidae), and shield or stink bugs (Hemiptera: Pentatomidae) can be possible causes of discoloration symptom observed in some trials.

Fungal diseases have been recorded mainly in warm and moist conditions and occasionally in wet, cool climates. In particular, two fungal diseases caused by *Septoria steviae* and *Sclerotinia sclerotiorum* have been reported in stevia grown in Canada [1], India [212], and Italy [213]. Septoria disease was characterized by depressed, angular, shiny olive-gray lesions, sometimes surrounded by a chlorotic halos that rapidly coalesce. Sclerotinia disease was characterized by brown lesions on the stem, near the soil line, followed by wilting and eventually by the complete collapse of affected [1]. During DIVAS project [214], three *Fusarium* species (*F. oxysporum*, *F. solani*, and *F. semitectum*) were isolated from few plants with vascular diseases, while *Verticilium* sp. and *Alternaria alternata* were isolated from one plant; however, they were only of minor importance. These diseases can be minimized by promoting air circulation in the canopy and timing irrigations to reduce periods of wetness on the plant. According to Yadav et al. [3], further work is needed to develop and identify resistant sources to develop varieties resistant or tolerant to these diseases. Cultural practices reduce the duration of wetness on plants, and the soil surface can reduce these disease incidences. These cultural practices may include limiting nitrogen fertilization, removing excess basal shoots and leaves, and timing irrigations to allow the top 5 cm of the soil to dry completely between irrigations. Formulations for biological control of *Sclerotinia sclerotiorum* are also available, even if their efficacy in stevia has not yet been investigated. The adoption of integrated pest management (IPM) is recommended for a sustainable stevia production. IPM emphasizes the management of agricultural systems (i.e., crop rotations, tillage, irrigation, fertilization, variety selection), rather than individual pests, so as to prevent or reduce the number and severity of pest outbreaks. Included in prevention is the conscious selection of all agronomic procedures that optimize plant production and reduce plant susceptibility to pests. In IPM, the concepts of acceptable pest levels, economic injury levels, and economic thresholds are defined, which imply a need to monitor for levels of pests or pest damage in relation to these levels. Pesticides (mainly synthetic insecticides and fungicides) are essential tools in IPM when other management tactics fail to control pests at acceptable levels. Chemical controls include synthetic and natural pesticides used to reduce pest populations. Insecticides derived from naturally occurring microorganisms, such as *Bacillus thuringiensis*, entomopathogenic fungi, and entomopathogenic nematodes, and natural insecticides, such as nicotine, pyrethrin, and spynosins, are important tools when stevia is cultivated in organic and nonorganic integrated systems. Pesticides should be used with care, applying them at the dose, and on the specific pest and diseases listed on the label. Unfortunately, until now, few pesticides are registered for stevia.

5 Harvesting, Storage, and Primary Processing

The individuation of optimal harvest time mainly depends on the cultivar, growing season, and the type of climate of the geographical area in which the crop is cultivated [100]. The main phenological factor affecting the SVglys content in stevia plants is flowering, which is induced by day length and temperature [78], and the optimal time to harvest the leaves is at the onset of flowering, when the accumulation of steviol glycosides reaches its peak [13, 17, 139, 215]. At the same time, harvest frequency results of particular importance in affecting the quantitative and qualitative characteristics of stevia. A single harvesting at flowering stage may lead to lower leaf yield compared with multicut management system, due to defoliation of lower and old leaves [216]. Therefore, harvest management of stevia grown both as annual and perennial crop requires a negotiation between quality and quantity of biomass. The regrowth of stevia after every harvest depends upon energy reserves stored in the roots. In case of the perennial stevia, substantial energy reserve is also needed for the development of cold hardiness, which allows the plant to survive during winter and to regrowth during spring [140, 216]. Consequently, the harvesting time of stevia should be decided based on a sound understanding of growth behavior and survival mechanism [216].

Serfaty et al. [139], in a study carried out under the agroclimatic conditions of Israel, with the aim to evaluate the performances of single or two consecutive harvests, found that performing a single harvest in late summer (September) was preferable than two harvests during the whole growth season. These authors observed the maximal biomass produced by performing two harvests was lower than that produced in a single harvest. They highlighted that the date of the first harvest (accomplished in June) had a considerable effect on the total biomass produced, since it affected the biomass yield obtained in the second harvest, due to the relatively slow recovery rate of plants. This was mainly due to the particular climatic characteristics of the cultivation site; in fact, the Jordan Valley, where the experiment was conducted, is characterized by extremely hot temperatures during summer, which seemed to suppress the plants ability to recover after the first harvest. These authors also concluded that the highest Stev yield was obtained with the harvest, carried out in September, since the Stev/dry weight leaves ratio increased from June to September. Moraes et al. [217] evaluated stevia's cold hardiness in North Mississippi during four growing seasons and determined the effects of different harvest timing on leaf production and yield of diterpene glycosides. The harvesting regimes tested here were: three harvests at 60-day intervals; two harvests at a 90-day interval, and a single harvest 180 days after transplanting, at the onset of flowering. This study revealed that leaf production from plants harvested once a year was higher than those obtained from multiple harvests (two and three harvests per growing season). Similarly, the SVglys productivity obtained from a single harvest, in terms of Reb A and Stev, was also greater than the yields of multiple harvests (two and three harvests). Even if the traditional recommendation for stevia harvest is at the onset of flowering stage in relation to higher SVglys content in leaves, Pal et al. [216] believed that the harvesting time of stevia should be decided based on the type of cropping system (crop duration, rotation, etc.) and on a sound understanding of growth behavior and survival mechanism in the site-specific growing conditions. For this reason, these authors evaluated the efficiency of annual versus perennial stevia cultivation, in the subtemperate region of western Himalaya in India, with the aim to standardize the harvesting regime in the specific cropping systems. Two life cycles (annual and perennial) and three harvest regimes (single cut, double cut and triple cut) were tested. Stevia when grown as perennial produced significantly higher dry leaf and stem yield compared with the annual crop at all harvesting stages. This was probably due to extensive living root systems of the perennial crop and to the benefits of rationing. In fact, rhizome-derived shoots of stevia grow more rapidly compared with newly transplanted small plantlets. Consequently, the maximum leaf yield was obtained with the perennial stevia as a result of higher LAI, which enhanced light interception capacity. In particular, highest dry leaf yield was recorded with the perennial stevia under three-cut and two-cut management systems, while the maximum SVglys was produced by perennial crop with two-cut regimes. According to principle component analysis (PCA), Pal et al. [216] concluded that, in the conditions of western Himalaya region, perennial stevia cultivation with two- or three-cut management appeared to be the most effective and suitable management practices for both dry leaf and SVglys yield. These authors also observed that, in the case of stevia grown as annual crop, total SVglys content was remarkably reduced during second or third cut (depending on the management system). In the case of the perennial, total SVglys content was reduced during the latest cut in the three-cut management system, probably because SVglys biosynthesis in regrowing leaves was adversely affected by short sunshine hours.

Regarding crop yields, experimental trials suggested that it is possible to reach, under optimal conditions, leaf dry yields ranging from 2 to 4 t ha^{-1}. Midmore and Rank [61] reported that systems with multiple harvests per year could give higher yields, and, where ratoon crops are grown, harvests in the second and third years are likely to be greater than for crops requiring replanting each year.

A range of yields, achieved under different pedoclimatic conditions are listed below. It is important to point out that commercial large-scale growers should expect only 50–70% of experimental yields [61].

In the natural habitat of Paraguay, the dry leaf yield was between 1.5 and 2.5 t ha^{-1} per year under dry land conditions and around 4.3 t ha^{-1} per year with irrigation [17, 218]. Leaf yields of 3.0 t ha^{-1} per year was obtained in Canada [78], while in Japan, with one or two harvests per year, it is possible to achieve 3–3.5 t ha^{-1} dry leaf yield in the first year, 4.0–4.5 t ha^{-1} in the second, 4.0–6.0 t ha^{-1} in third, diminishing in the fourth year [219]. Megeji et al. [190], in India, observed a production of 3.5 t ha^{-1} of dry stems and leaves. An average dried leaf yield of 1.7, 2.0, 2.3, and 2.5 t ha^{-1} could be produced in the first, second, third, and fourth year after planting, respectively, under agro-climatic conditions of Palampur, India [220]. In the latitudes of central and southern Italy, where stevia is grown as a perennial with an economically profitable cultivation until the sixth year, dry leaf yield are equal to 3.6 t ha^{-1} per year, as 1–6 year mean value, with a peak of 6.1 t ha^{-1} of dry leaves in the sixth year of cultivation [75, 109, 196, 221]. Similar leaf yields were obtained in the north part of the State of Mississippi [217]. In the northeastern Italy conditions, where the stevia production has to rely on annual cultivation, a mean leaf dry yield of 3.1 t ha^{-1} was recorded [75, 140].

Without proper machinery, the cultivation areas and the quantity of the harvested raw materials remain small and costly, and labor intensive. Nevertheless, the grower must ensure that the machinery employed does not negatively affect the quality of the raw material. The machinery employed were quite variable and depends to some extent on local availability, cultivation area, and type of farms. Generally, at least in Europe, machinery specific to stevia does not exist; therefore the machinery available should be tested and adapted for the requirements of stevia. For example, in central Italy a modified spinach harvester with a cutter bar width of 180 cm has been used to harvest a hectare of stevia. At large scale a specially designed harvester, similar to that used for mint or lemon balm, that cuts the crop at ground level (~10 cm above soil) might be used, while small plots are often harvested by hand.

Drying is the important activity in postharvest handling of stevia. Freshly harvested leaves contain high moisture content (80–90%) and drying should be completed immediately after harvesting. The method and temperature used for drying may have a considerable impact on the quality of the resulting medicinal plant materials. Artificial drying within forced air circulating ovens/rooms is better than natural drying in the open air. The plants should be placed in thin layers on drying frames at low temperature (30–40 °C) and good air circulation for 24–48 h. Temperature and humidity should be controlled to avoid damage to the active chemical constituents. Dried plants are then threshed to separate the sweet leaves from the bitter stems. A specially designed thresher/separator is necessary to separate dry stevia leaves from the stem. The yield of stem and leaf is similar (approximately 3,000 kg ha^{-1} each).

Once dried, leaves can be stored for long periods in airtight containers and can be easily transported for commercialization or further processing (for steviol glycosides extraction). Processing generally involves either crushing the leaves or making a crude extract using a water extraction process. Commercial processing to isolate pure rebaudioside A for the various sugar substitutes on the market involves a much more complicated procedure. Some of these extraction processes are patented.

Quality and safety are the basic requirements that stevia as dry leaves material should ensure in terms of the identity, purity, and content of desired compounds. A combination of both physical and chemical tests for quality standards is described and established in national, European, or other international monographs, similarly to other herbs.

6 Preharvest Factors Affecting Biomass Yield, Steviol Glycosides, and Antioxidants Compounds

The growth and development of stevia plants, as well as the biosynthesis of its secondary metabolites, are dependent on several important preharvest factors, including genotype, climatic conditions, and agronomic practices. It is known that genetic characteristics, as well as environmental (drought, salinity, temperatures, light conditions, atmospheric CO_2 concentrations, soil type, soil mineral deficiency, geographical area) and agronomic conditions (i.e., fertilization, harvest time, plant density, growth regulators), affect growth, photosynthesis, crop productivity, and other parts of the primary metabolism, as well as secondary metabolic pathways. Biotic and abiotic stresses act as inductors of different responses in plants. Stress response is initiated when plants recognize a disorder at a cellular level, activating signal transduction pathways that transmit information within individual cell and throughout plant, leading to changes in expressing many gene networks [222, 223]. Specific studies have been conducted in order to evaluate the crop responses to both growth conditions and agricultural management and the identification of their best combinations.

6.1 Case Study 1: Nitrogen Fertilization

The knowledge of the changes in leaf yield and accumulation patterns of SVglys and other bioactive compounds in response to nutritional variations could provide useful information for developing appropriate strategies in nutrient management to improve the stevia productivity and secondary metabolite profiles under different agroclimatic conditions. The optimization of the N fertilization is gaining greater attention to avoid irrational fertilization management, minimizing the potential N losses, and to combine high yields with relatively low environmental impacts and product safety. Tavarini et al. [224] evaluated, through a pot trial in open air conditions carried out in central Italy, the physiological, agronomic, and phytochemical responses of stevia plants grown under four mineral N rates (0, 50, 150, and 300 kg N ha^{-1}). The behavior of stevia regarding N availability is similar to other crops showing the highest specific leaf weight (SLW) in plants grown under maximum N rate (300 kg N ha^{-1}) suggesting that leaf thickness was altered. However, in plants subjected to the highest N dose, the photosynthetic nitrogen use efficiency (PNUE) was the lowest, to indicate a poor N utilization in relation to the carbon assimilation by the plant. These authors observed that the application of 150 kg N ha^{-1} seemed to be the most effective dose to improve Reb A content, Reb A/Stev ratio, photosynthetic CO_2 assimilation, stomatal conductance, N use efficiency, Rubisco activity, and PSII efficiency. Conversely, crop quality was not improved by greater N supplies. These findings underlined the possibility, by using an appropriate N rate, to modulate the SVglys biosynthesis, with a significant increase in the Reb A content and Reb A/Stev ratio. Thus, the possibility to shift, through N fertilization, the SVglys biosynthesis toward Reb A represents an important finding in order to obtain plants characterized by high levels of this compound. These findings were confirmed by Tavarini et al. [225] that demonstrated as stevia bioactive compounds (SVglys, polyphenols, total flavonoids and composition) and their related antioxidant activity increased by optimizing the N supply and decreased with increasing or decreasing N rates compared to the optimal N dose (150 kg N ha^{-1}, applied as mineral fertilizer). At high supply rates, N might inhibit the synthesis of phenols and flavonoids via enhancing the channeling of L-phenylalanine toward proteins. Regarding the N form (mineral and organic), Tavarini et al. [225] observed both lower phytochemical content and antioxidant capacity in plants treated with organic N in comparison with plants grown with mineral N. This fact is attributable to a distinct temporal N availability that characterized organic and inorganic N sources.

6.2 Case Study 2: Drought Stress

Drought is the most severe abiotic stress factor limiting plant growth and crop production, and it represents one of the major environmental factors affecting many aspects of plant physiology and biochemistry. It is known that plant response to drought stress is associated with the time and intensity of stress, plant species, genotypes, and environmental conditions [226, 227]. In stevia, proper irrigation scheduling is fundamental to achieve good results, in terms of plant growth, dry matter accumulation, and crop yield, as reported by many authors. At the same time, drought stress significantly affected the level of bioactive compounds. Even if the increase of secondary metabolites under drought stress has been frequently reported for many plant species, this phenomenon has not been sufficiently elucidated in stevia. Karimi et al. [228], through a greenhouse experiment, determined threshold values of soil moisture content for stevia and evaluated the effects of drought stress on plant growth and the main metabolites. In this research, four soil moisture levels were tested scheduling irrigation at 3, 6, 9, and 12 day irrigation intervals based on soil moisture content at 90%, 75%, 60%, and 45% of field capacity (FC), respectively. The results showed that soil water depletion up to 60% FC had no negative effect on plant growth and leaf dry weight, whereas a significant growth reduction occurred at 45% FC. Similarly, the total SVglys content increased when soil moisture was depleted to 60% FC, but these metabolites' contents decreased by 45% FC treatment. Although plant growth and SVglys content significantly decreased under severe drought stress, the total antioxidant capacity and soluble sugars increased in this condition. The increase in soluble sugars (in particular glucose) can be considered an osmotic adjustment in response to drought. The stimulation of antioxidant systems and the increase in the biosynthesis of antioxidant metabolites, similar to that observed in vitro condition [229], can be considered as physiological and biochemical response to a progressive drought stress in stevia. On the basis of these results, the authors concluded that stevia well tolerated mild water stress, while severe drought can cause serious damages during its vegetative growth. This finding is noteworthy, especially in water limited regions. The improvement of antioxidant capacity and soluble sugar content by soil water stress conditions could be considered as physiological and biochemical responses to a progressive drought stress in stevia and maybe an acclimation response to drought stress. Karimi et al. [230] observed the functional and positive role of SVglys compounds in stevia plants affected by drought stress. In this experiment, liquid blend of SVglys (200 ppm) was sprayed on stevia plants grown in well-watered (90% field capacity) and drought-stress conditions (45% field capacity), evaluating the effects on the morphological characteristics and the main stevia metabolites. It was observed that leaf losses caused by drought stress were stopped through external application of SVglys and consequently the harvest index of stevia was increased. Metabolite analysis of stevia leaves showed that the total SVglys content was significantly decreased due to drought stress but was compensated by external application of SVglys. Among the SVglys, Reb A responded more to external SVglys. A slight promotion in total antioxidant activity was also detected.

6.3 Case Study 3: Plant Density

The adoption of twin rows arrangement, as alternative cropping design for innovation of the cultural techniques, could be an interesting choice in stevia in order to improve the productivity, enhance the mechanization of the cultural practices, decrease plant-to-plant and weed competition, and alleviate crop crowding stress. Angelini and Tavarini [231], in a field trial carried out in northeastern Italy, evaluated the effect of two plant arrangements (single vs. twin rows) and three plant densities (7, 10, and 14 plants m^{-2}) on the biometric, productive, and phytochemical characteristics of stevia. These authors observed that the twin rows plant arrangement, although reduced total yield, increased SVglys content. In fact, with a plant density of about 15 plants m^{-2}, in single row arrangement (45 × 15 cm), improved plant biometric characteristics and crop yield, while the highest SVglys contents were obtained with 10 plants m^{-2}, in a twin rows arrangement (50 × 30 × 25 cm).

6.4 Case Study 4: Growth Regulators

SVglys are diterpenoids whose biosynthetic pathways share four steps in common with gibberellic acid formation. Gibberellins (GAs) and SVglys biosynthetic pathways have many common stages and intermediate metabolites, although it has been reported that they are present in stevia leaf in different amounts [232]. The SVglys and GAs belong to diterpenes and are produced by methylerythritol 4-phosphate (MEP) pathway but diverged from (−)-kaurenoic acid [233]. In one of the divergences, steviol is produced by hydroxylation of (−)-kaurenoic acid at C-13 position [234], while in another divergence, GAs is generated by hydroxylation at C-7 position [235]. Regarding the biosynthesis pathway, SVglys and GAs could be produced in a competitive process, especially for substrate obtaining at the shared biosynthesis stages. Since, in stevia, SVglys and gibberellins have a common precursor in their biosynthesis pathway, it is important to study the role of gibberellin inhibitors in order to better understand their effects on stevia growth responses, SVglys and antioxidant systems induction. Recently, Karimi et al. [236] evaluated, through two greenhouse experiments, the effect of two plant growth retardants, chlorocholine chloride (CCC) and paclobutrazol (PBZ), on growth, SVglys content, and antioxidant capacity. They found that CCC reduced plant height, but improved leaf and stem dry weight, especially with 750 ppm concentration. SVglys content and yield were significantly reduced by CCC application. PBZ had no effect on plant height, while it significantly enhanced stem and dry weight at 12 ppm concentration. Moreover, PBZ remarkably increased total SVglys contents, SVglys yield, and Reb A/Stev ratio. Total antioxidant capacity significantly varied with CCC and PBZ, and the highest activity was obtained with 1,000 and 12 ppm of CCC and PBZ, respectively. Karimi et al. [236] concluded that, although CCC and PBZ are plant growth retardants and act as antigibberellins, only CCC reduced plant height and SVglys production in stevia. On the contrary, PBZ at 12 ppm concentration improved stevia growth, SVglys production, and antioxidant capacity.

Karimi et al. [237] studied, in two greenhouse experiments, the opposing effects of external gibberellin and daminozide (a gibberellin inhibitor) on stevia growth and metabolites. This study highlighted a positive response of stevia to external gibberellin in growth aspects, meaning the high precise-regulated gibberellins biosynthesis is at a specific level inside the stevia tissues. Growth induction in stevia by external gibberellin has been implicated in segregated regulation of gibberellins biosynthesis and preservation of gibberellins in metabolic demand threshold. Gibberellin inhibitor (daminozide) application revealed that the gibberellins could not compete with the SVglys production. Karimi et al. [237] found that SVglys production was affected by exceeding gibberellin, but not by daminozide, demonstrating the separation in regulation among the main stages of biosynthesis pathway. On the contrary, daminozide, at 30 ppm concentration, was able to significantly increase the soluble sugar production in stevia leaves. Definitively, this study demonstrated that, since daminozide had no effects on SVglys production, gibberellins biosynthesis could not act as a competitive factor for SVglys production in stevia leaves.

6.5 Case Study 5: Harvest Time

Tavarini and Angelini [14], in a 2-year study carried out in two locations of northeastern Italy aimed to compare three time of harvest (beginning of July, end of July, and mid-September), showed that harvest time was certainly one of the major factors determining the most important qualitative traits of stevia in both years of growth, independently of site and year. These authors showed the possibility of obtaining very high SVglys yields thanks to the long-day conditions (>14 h photoperiod) during the spring/summer season of the tested environments. Choosing the optimal harvest, it was possible to improve the stevia quality, mainly influencing the Reb A/Stev ratio, that increased from July to September, in both years of cultivation. On the other hand, Tavarini and Angelini [14] observed a decrease of the antioxidant activity and polyphenol content at leaf level, in the last harvest of September, when the vegetative growth declined and the physiological state moved toward flowering. These findings have been confirmed by Tavarini et al. [225] in a pot study under the climatic conditions of central Italy. In this study, a significant increase in the RebA/Stev ratio was observed in the leaves from July to September. At the same time, leaves collected at the beginning of July generally showed higher phenolic and flavonoid contents, as well as antioxidant activity, compared to leaves collected in the other two harvests (end of July and beginning of September). In particular, luteolin-7-*O*-glucoside, the most abundant flavonoid detected in stevia leaves in this study, appeared to be strongly affected by the harvest time, with a twofold higher value in the first harvest, than in the other ones. Photoperiod regimes, similarly to that observed for SVglys (i.e., long-day conditions significantly increase leaf biomass and Stev content), can influence phenolic acids and flavonoids by increasing the total contents of these compounds under long-day conditions, suggesting a role for these compounds in protection against enhanced light exposure.

7 Conclusion

The agronomic researches conducted around the world show that stevia can be grown in many geographic areas, also very different each other. The possibility to introduce a modern and fully mechanized stevia cultivation chain in the temperate climates of Europe gives a new opportunity to the farming community and to scientists. The possibility to create a full production chain of stevia might represent an interesting and economically feasible opportunity for farmers addressing the needs of bio-industry for high-quality, safe, and fully traceable products. However, to develop environmentally and economically sustainable production scheme and to address the value chain for high-quality products, an integrated approach by a team of multidisciplinary scientists is required. The overall objective is the development of stevia agricultural standard based on best practices as a core component of the strategy, making the stevia cultivation socially acceptable, cheaper, and economically viable. Traceability and crop management are the strongest points in the stevia production chain, in order to obtain a certification that will provide to the farmers preferential market access and to sell raw material at a differential price. In particular, for the European agriculture, the introduction of stevia in the intensive cropping systems can be a viable alternative in order to increase the agro-environmental sustainability and cropping system diversification, as well as to contribute to diversify and increase the income from agriculture. Research should be directed toward the achievement of optimal biomass and SVglys, improving plant yield per production unit in a sustainable manner and reducing water and the amount of needed agrochemicals. Research is of pivotal importance to address these challenges on any and all parts of the agricultural supply chain by encompassing innovations in science, technology, and economics, as well as on the means to effectively disseminate and extend the knowledge to the relevant clientele groups comprising the chains.

References

1. Brandle JE, Starratt AN, Gijzen M (1998) *Stevia rebaudiana*: its agricultural, biological, and chemical properties. Can J Plant Sci 78:527–536
2. Madan S, Ahmad S, Singh GN, Kohli K, Kumar Y, Singh R, Garg M (2010) *Stevia rebaudiana* (Bert.) Bertoni – a review. Indian J Nat Prod Resour 1:267–286
3. Yadav AK, Singh S, Dhyani D, Ahuja PS (2011) A review on the improvement of stevia [*Stevia rebaudiana* (Bertoni)]. Can J Plant Sci 91:1–27
4. Otha M, Sasa S, Inoue A, Tamai T, Fujita I, Morita K, Matsuura F (2010) Characterization of novel steviol glycosides from leaves of *Stevia rebaudiana* Morita. J Appl Glycosci 57:199–209
5. Chaturvedula VSP, Clos JF, Rhea J, Milanowski D, Mocek U, DuBois GE, Prakash I (2011) Minor diterpenoids glycosides from the leaves of *Stevia rebaudiana*. Phytochem Lett 4:209–212
6. De S, Mondal S, Banerjee S (2013) Enhancement of stevioside recovery by diafiltration. In: Stevioside: technology, applications and health. Wiley, Oxford. doi:10.1002/9781118350720.ch9
7. OECD/FAO (2013) Agricultural outlook 2013–2022. http://groupedebruges.eu/sites/default/files/publications/downloads/agricultural_outlook_2013-2022_-_oecd-fao.pdf. Accessed 27 Oct 2015
8. Lemus-Mondaca R, Vega-Gálvez A, Zura-Bravo L, Ah-Hen K (2012) *Stevia rebaudiana* Bertoni, source of a high-potency natural sweetener: a comprehensive review on the biochemical, nutritional and functional aspects. Food Chem 132:1121–1132
9. EU (2011) Commission Regulation (EU) No. 1131/2011. Off J Eur Union L 295:205–211
10. Jeppesen PB, Gregersen S, Poulsen CR, Hermansen K (2000) Stevioside acts directly on pancreatic β cells to secrete insulin: actions independent of cyclic adenosine monophosphate and adenosine triphosphate – sensitivie K^+-channel activity. Metabolism 49:208–214
11. Jeppesen PB, Gregersen S, Rolfsen SED, Jepsen M, Colombo M, Agger A, Xiao J, Kruhøffer M, Ørntoft T, Hermansen K (2003) Antihyperglycemic and blood pressure-reducing effects of stevioside in the diabetic Goto-Kakizaki rat. Metabolism 52:372–378
12. Chatsudthipong V, Muanprasat C (2009) Stevioside and related compounds: therapeutic benefits beyond sweetness. Pharmacol Ther 121:41–54
13. Kinghorn AD (2002) Overview. In: Kinghorn AD (ed) *Stevia*, the genus of Stevia, Medicinal and aromatic plants-industrial profiles. Taylor and Francis, London/New York, pp 1–17
14. Tavarini S, Angelini LG (2013) *Stevia rebaudiana* Bertoni as a source of bioactive compounds: the effect of harvest time, experimental site and crop age on steviol glycoside content and antioxidant properties. J Sci Food Agric 93:2121–2129
15. WHO (2009) Evaluation of certain food additives – sixty-ninth report of the Joint FAO/WHO Expert Committee on Food Additives. WHO technical report series 952. http://apps.who.int/iris/bitstream/10665/44062/1/WHO_TRS_952_eng.pdf?ua=1. Accessed 8 Jan 2016
16. EU (2015) Regulation (EU) No. 2015/2283. Off J Eur Union L 327:1–22
17. Ramesh K, Singh V, Megeji NW (2006) Cultivation of stevia [*Stevia rebaudiana* (Bert.) Bertoni]: a comprehensive review. Adv Agron 89:137–177
18. Bertoni MS (1918) La *Stevia rebaudiana* Bertoni. An Cient Parag 2:129–134
19. Carneiro JWP (1990) *Stevia rebaudiana* (Bert) Bertoni: Produção de Sementes. Paraná State University at Maringa, Maringá
20. Kudo M, Koga Y (1977) Photoperiodic response and its variation in *Stevia rebaudiana* Bertoni. Jpn J Trop Agric 20:211–217
21. Valio IFM, Rocha RF (1977) Effect of photoperiod and growth regulators on growth and flowering of *Stevia rebaudiana* Bertoni. Jpn J Crop Sci 46:243–248
22. Metivier J, Viana AM (1979) The effect of long and short day length upon the growth of whole plants and the level of soluble proteins, sugars, and stevioside in leaves of *Stevia rebaudiana* Bert. J Exp Bot 30:1211–1222

23. Zaidan LPB, Dietrich SMC, Felippe GM (1980) Effect of photoperiod on flowering and stevioside content in plants of *Stevia rebaudiana* Bertoni. Jpn J Crop Sci 49:569–574
24. Miyagawa H, Fujikowa N, Kohda H, Yamasaki K, Taniguchi K, Tanaka R (1986) Studies on the tissue culture of *Stevia rebaudiana* and its components: (II). Induction of shoot primordia. Planta Med 4:321–324
25. Proctor M, Yeo P, Lack A (1996) Natural history of pollination. Timber Press, Portland
26. Chalapathi MV, Thimmegowda S, Rama Krishna Prama VR, Prasad TG (1997) Natural non-calorie sweetener Stevia (*Stevia rebaudiana* Bertoni): a future crop of India. Crop Res 14:347–350
27. Oddone B (1997) How to grow stevia. Technical manual. Guarani Botanicals, Pawtucket
28. Maiti RK, Purohit SS (2008) Stevia: a miracle plant for human health Agrobios, Jodhpur. ISBN 9788177543384
29. Raina R, Bhandari S, Chand R, Sharma Y (2013) Strategies to improve poor seed germination in *Stevia rebaudiana*, a low calorie sweetener. J Med Plant Res 7:1793–1799
30. Goettemoeller J, Ching A (1999) Seed germination in *Stevia rebaudiana*. In: Janick J (ed) Perspectives on new crops and new uses. ASHS Press, Alexandria, pp 510–511
31. Crammer B, Ikan R (1986) Sweet glycosides from the Stevia plant. Chem Br 22:915–918
32. Sumida T (1968) Reports on stevia introduced from Brazil as a new sweetness resource in Japan (English summary). J Cent Agric Exp Stn 31:1–71
33. Morita Kagaku Kogyo Co., Ltd. (2007) Greeting from the President. www.morita-kagaku-kogyo.co.jp/e/about_us.html. Accessed 21 July 2015
34. Evolva (2014) Zero-calorie, natural sweetener. www.evolva.com/products/stevia. Accessed 19 Jan 2016
35. JECFA (1999) Safety evaluation of certain food additives. WHO food additive series: 42. In: Prepared by the fifty-first meeting of the Joint FAO/WHO Expert Committee on Food Additives (JECFA). International Programme on Chemical Safety, World Health Organization, Geneva
36. Food and Drug Administration (FDA) (2007) Letter Department of Health and Human Services. Food and Drug Administration to Hain Celestial Group, Washington, DC, www.fda.gov/foi/warning_letters/s6500c.htm
37. EU (2012) Commission Regulation (EU) No. 231/2012. Off J Eur Union L 83:1–295
38. Industry ARC (2014) Global Stevia/Stevia Rebaudiana Market (2013–2018). http://industryarc.com/Report/103/Stevia-Rebaudiana-Food-Beverages-Market-Report.html. Accessed 16 Sept 2015
39. Mintel (2014) Stevia set to steal intense sweetener market share by 2017, reports Mintel and Leatherhead Food Research, 13 January 2014. http://www.mintel.com/press-centre/food-and-drink/stevia-set-to-steal-intense-sweetener-market-share-by-2017-reports-mintel-and-leatherhead-food-research. Accessed 16 Jan 2016
40. Lens (2015) Abstract search: steviol glycosides. www.lens.org/lens/search?q=abstract%3A%28steviol+glycosides%29&l=en. Accessed 22 July 2015
41. Meienberg F, Sommer L, Lebrecht T, Lovera M, Gonzalez S, Luig B, Bremen V, Steiner K, Glauser M, Kienle U (2015) The bitter sweet taste of stevia. Berne Declaration, CEIRAD, Misereor, Pro Stevia Switzerland, SUNU, University of Hohenheim. https://www.bernedeclaration.ch/fileadmin/files/documents/Biodiversitaet/FINAL_EN_STEVIA_REPORT_20151112_letzte_Korr.pdf. Accessed 15 Jan 2016
42. SteviaOne (2012) La experiencia peruana. Presented at the VIth international symposium of *Stevia rebaudiana*, Asuncion
43. Gmuer H (2015) Zuckersüss und Kalorienfrei. www.nzz.ch/finanzen/devisen-und-rohstoffe/rohstoffe/zuckersuess-und-kalorienfrei-1.18524753. Accessed 1 July 2015
44. Nikkei Asian Review (2015) Appetite for Stevia brings Japan back to Paraguay, 20 August 2015. http://asia.nikkei.com/magazine/20150820-STICKER-SHOCK/Business/Appetite-for-stevia-brings-Japan-back-to-Paraguay. Accessed 1 Oct 2015
45. Zeng J, Chen A, Li D, Yi B, Wu W (2013) Effects of salt stress on the growth, physiological responses, and glycoside contents of *Stevia rebaudiana* Bertoni. J Agric Food Chem 61:5720–5726

46. Bamber P, Fernandez-Stark K (2012) Strengthening the competitiveness of the stevia value chain in Paraguay. Duke Centre on Globalization, Governance & Competitiveness (CGGC), Durham, http://www.cggc.duke.edu/pdfs/STEVIA_final_2012ENG.pdf. Accessed 19 Jan 2016
47. Kienle U (2011) *Stevia rebaudiana*: der Zucker des 21. Jahrhunderts. Spurbuchverlag, Baunach
48. GSI (Global Stevia Institute) (2015) Stevia farming. http://globalsteviainstitute.com/consumers/stevia-story/stevia-farming/. Accessed 12 Oct 2015
49. GSI (Global Stevia Institute) (2015) Stevia product growth accelerates throughout the world. http://globalsteviainstitute.com/stevia-leaf-extract/stevia-product-growth-accelerates-throughout-world/. Accessed 12 Oct 2015
50. Jeppesen PB, Barriocanal L, Meyer MT, Palacios M, Canete F, Benitez S, Logwin S, Schupmann Y, Benitez G, Jimenez JT (2006) Efficacy and tolerability of oral stevioside in patients with type 2 diabetes: a long-term, randomized, double-blinded, placebo-controlled study. Diabetol Suppl 49:511–512
51. Rizzo B, Zambonin L, Angeloni C, Leoncini E, Viecelle Dalla Sega F, Prata C, Fiorentini D, Hrelia S (2013) Steviol glycosides modulate glucose transport in different cell types. Oxid Med Cell Longev 2013:1–11
52. GO4STEVIA Report Summary, CORDIS (Community Research and Development Information Service). http://cordis.europa.eu/result/rcn/155604_en.html. Accessed 4 Feb 2016
53. WHO (2003) WHO guidelines on good agricultural and collection practices (GACP) for medicinal plants. WHO, Geneva
54. Sumida T (1973) Reports on *Stevia rebaudiana* Bertoni. Introduced from Brazil as a newsweetness resource in Japan. Misc Publ Hokkaido Natl Exp Stn 2:79–83
55. Katayama O, Sumida T, Hayashi H, Mitsuhashi H (1976) The practical application of Stevia and research and development data. I.S.U. Company, Tokyo, p 747
56. Shock CC (1982) Experimental cultivation of Rebaudis Stevia in California. Agron Prog Rep 122, Davis, CA: University of California
57. Alvarez M (1984) *Stevia rebaudiana* Bert. estado atual do conhecimento. Universitadade Estaudual de Maringa, Maringa, p 118
58. Bertonha A, Muniz AS, Carneiro JWP, Martins EN, Jabur IC, Thomaz SI (1984) Estudo de cultivo, reproducao e selecao das variedades mais productivas de *Stevia rebaudiana*, en solos do norte de Prana. In: Mimeo (ed) 2nd Maringa, UEM, p 103
59. Monteiro R (1986) Taxonomia e bologia da reproducao de *Stevia rebaudiana*. Campinas. Dissertacan (Mestrado), Instituto de Biologia, Universitadade Estaudual de Maringa, p 104
60. Lankes C, Pude R (2008) Possibilities for growth of Stevia in Europe temperate zones. In: Geuns JMC (ed) Steviol glycosides: technical and pharmacological aspects. Proceedings of the 2th Stevia symposium 2008 of European Society of Stevia (EUSTAS), 27 June 2008, KU Leven, Leuven, pp 103–116. ISBN: 9789074253031
61. Midmore DJ, Rank AH (2002) A new rural industry – stevia – to replace imported chemical sweeteners. A report for the Rural Industries Research and Development Corporation. RIRDC Publication No W02/022
62. Tariq U, Ali M, Abbasi BH (2014) Morphogenic and biochemical variations under different spectral lights in callus cultures of *Artemisia absinthium* L. J Photochem Photobiol B Biol 130:264–271
63. Ermakov EI, Kochetov AA (1994) Growth and productivity of Stevia under regulated conditions, depending on the photoperiod and light intensity. Russ Agric Sci 11:11–14
64. Ermakov EI, Kochetov AA (1996) Specificities of the growth and development of Stevia plants under different controllable light conditions. Russ Agric Sci 1:9–11
65. Lester T (1999) *Stevia rebaudiana* sweet leaf. Aust New Crop Newsl 11:1
66. Mohamed AAA, Ceunen S, Geuns JMC, Van Den Ende W, De Ley M (2011) UDP-dependent glycosyltransferases involved in the biosynthesis of steviol glycosides. J Plant Physiol 168:1136–1141

67. Ceunen S, Werbrouck S, Geuns JMC (2012) Stimulation of steviol glycoside accumulation in *Stevia rebaudiana* by red LED light. J Plant Physiol 169:749–752
68. Ceunen S, Geuns JMC (2013) Influence of photoperiodism on the spatio-temporal accumulation of steviol glycosides in *Stevia rebaudiana* (Bertoni). Plant Sci 198:72–82
69. Ceunen S, Geuns JMC (2013) Spatio-temporal variation of the diterpene steviol in *Stevia rebaudiana* grown under different photoperiods. Phytochemistry 89:32–38
70. Pal PK, Kumar R, Guleria V, Mahajan M, Prasad R, Pathania V, Gill BS, Singh D, Chand G, Singh B, Singh RD, Ahuja PS (2015) Crop-ecology and nutritional variability influence growth and secondary metabolites of *Stevia rebaudiana* Bertoni. BMC Plant Biol 15:67
71. Mu H, Jiang D, Wollenweber B, Dai T, Jing Q, Cao W (2010) Long-term low radiation decreases leaf photosynthesis, photochemical efficiency and grain yield in winter wheat. J Agron Crop Sci 196:38–47
72. Song R, Kelman D, Johns KL, Wright AD (2012) Correlation between leaf age, shade levels, and characteristic beneficial natural constituents of tea (*Camellia sinensis*) grown in Hawaii. Food Chem 133:707–714
73. Slamet IH, Tahardi S (1988) The effect of shading and nitrogen fertilisation on the flowering of *Stevia rebaudiana*. Menara Perkeb 56:34–37
74. Kumar R, Sharma S, Ramesh K, Singh B (2013) Effects of shade regimes and planting geometry on growth, yield and quality of the natural sweetener plant stevia (*Stevia rebaudiana* Bertoni) in north-western Himalaya. Arch Agron Soil Sci 59:963–979
75. Angelini LG, Leone G, Russo C, Tavarini S (2013) Sustainable production chain of stevia in Italy: agronomy, phytochemical assessment and downstream processing. In: Geuns JMC (ed) Knowledge on tour in Europe. Proceedings of the 7th Stevia symposium 2013 of European Society of Stevia (EUSTAS), 24–26 June 2013, Toulouse, pp 93–96. ISBN: 9789074253277
76. Ramakrishna A, Ravishankar GA (2011) Influence of abiotic stress signals on secondary metabolites in plants. Plant Signal Behav 6:1720–1731
77. Ahmad N, Abdur R, Nisar A (2016) Light-induced biochemical variations in secondary metabolite production and antioxidant activity in callus cultures of *Stevia rebaudiana* (Bert). J Photochem Photobiol B 154:51–56
78. Brandle JE, Rosa N (1992) Heritability for yield, leaf: stem ratio and stevioside content estimated from a landrace cultivar of *Stevia rebaudiana*. Can J Plant Sci 72:1263–1266
79. Othman HS, Osman M, Zainuddin Z (2015) Morphological assessment of *Stevia rebaudiana* Bertoni accessions in IIUM's germplasm as initial material for stevia breeding. Aust J Basic Appl Sci 9:1–9
80. Brandle J (1999) Genetic control of rebaudioside A and C concentration in leaves of the sweet herb, *Stevia rebaudiana*. Can J Plant Sci 79:85–92
81. Lee JI, Kang KH, Park HW, Ham YS (1982) New high rebaudioside-A stevia variety "Suweon 11" (English abstr.). Res Rep ORD 24:186–188
82. Shizhen S (1995) A study on good variety selection in *Stevia rebaudiana*. Sci Agric Sin 28:37–41
83. Morita T (1987) Dried leaves. Japanes Patent 62-96025 [English abstract]
84. Anami ET, Poletine JP, Gonçalves-Vidigal MC, Vidigal Filho PS, Lacanallo GF, Kvitschal MV, Gonela A (2010) Characterization and genetic divergence in *Stevia rebaudiana* (Bert.) Bertoni clones based in agronomical and morphological characteristics. J Food Agric Environ 8:463–469
85. Lankes C, Zabala UM, Müller V (2010) Performance of *Stevia rebaudiana* Bertoni genotypes under European temperate zone conditions. In: XXVIII International Horticultural Congress on science and horticulture for people (IHC2010): international symposium, vol 936. pp 273–276
86. Yadav AK, Singh S, Singh B, Sharma RK, Vijaylata S, Sood A, Ahuja PS (2014) Release of *Stevia rebaudiana* cultivar HIM Stevia (CSIR-IHBT-ST-01). www.ihbt.res.in

87. Barbet-Massin C, Giuliano S, Alletto L, Daydé J, Berger M (2015) Towards a semi-perennial culture of *Stevia rebaudiana* (Bertoni) Bertoni under temperate climate: effects of genotype, environment and plant age on steviol glycoside content and composition. Genet Resour Crop Evol. doi:10.1007/s10722-015-0276-9
88. Hastoy C, Cosson P, Sclaunich E, Jannot P, Boutié P, Rolin D, Schurdi-Levraud V (2015) Towards breeding of new *Stevia rebaudiana* cultivars adapted to organic production in the South West of France: first results on genetic and phenotypic evaluation. In: Ceunen S, Geuns JMC (eds) Stevia: growth in knowledge and taste. Proceedings of the 8th Stevia symposium 2015 of European Society of Stevia (EUSTAS), 27–29 Jan 2015, Bonn, pp 225–232. ISBN: 9789074253291
89. Brandle J (1998) *Stevia rebaudiana* with altered steviol glycoside composition. US Patent 6255557 B1
90. Marsolais AA, Brandle J, Sys EA (1998) Stevia plant named "RSIT 94-751". US Patent PP10564 P
91. Wang Q (2012) High rebaudioside-a plant and methods of producing the same and uses thereof. Patent WO 2012088612 A1
92. Alvarez Britos ER (2012) Stevia plant named "AKH L1". US Patent PP23,164 P3
93. Alvarez Britos ER (2012) Stevia plant named "AKH L4". US Patent 20120090063 P1
94. Yadav AK, Singh S, Rajeevr R (2014) Self-incompatibility evidenced through scanning electron microscopy and pollination behaviour in *Stevia rebaudiana*. Indian J Agric Sci 84:93–100
95. UPOV (2015) PLUTO – plant variety database. www3.wipo.int/pluto/en/. Accessed 23 June 2015
96. Ramesh K, Singh V, Singh Ahuja P (2007) Production potential of *Stevia rebaudiana* (Bert.) Bertoni. under intercropping systems. Arch Agron Soil Sci 53:443–458
97. Schertz D, Towery D (2006) In: Lal R (ed) Encyclopedia of soil science, vol 1, 2nd edn. Taylor and Francis, New York, pp 351–353. doi:10.1081/E-ESS-120006662
98. Southward RC, Kitchen KL, Fountain DW (2004) Flowering and seed production in a model pot-grown specimen of the sweet herb *Stevia rebaudiana* Bertoni (Asteraceae). Agron N Z 34:183–190
99. Kumar R (2013) Seed germination of *Stevia rebaudiana* influenced by various potting media. Octa J Biosci 1:143–146
100. Sharma S, Walia S, Singh B, Kumar R (2015) Comprehensive review on agro technologies of low-calorie natural sweetener stevia (*Stevia rebaudiana* Bertoni): a boon to diabetic patients. J Sci Food Agric. doi:10.1002/jsfa.7500
101. Macchia M, Andolfi L, Ceccarini L, Angelini LG (2007) Effects of temperature, light and pre-chilling on seed germination of *Stevia rebaudiana* (Bertoni) Bertoni accessions. Ital J Agron 2:55–62
102. Khalil SA, Zamir R, Ahmad N (2014) Selection of suitable propagation method for consistent plantlets production in *Stevia rebaudiana* (Bertoni). Saudi J Biol Sci 21:566–573
103. Abdullateef RA, Osman MB, Zainuddin ZB (2015) Acclimatized apparatus enhanced seed germination in *Stevia rebaudiana* Bertoni. Int J Biol 7:28–34
104. Ahmed MB, Salahin M, Karim R, Razvy MA, Hannan MM, Sultana R, Hossain M, Islam R (2007) An efficient method for in vitro clonal propagation of a newly introduced sweetener plant (*Stevia rebaudiana* Bertoni.) in Bangladesh. Am Eurasian J Sci Res 2:121–125
105. Knafla K (2015) In vitro culture of *Stevia rebaudiana*: principles and possibilities. In: Ceunen S, Geuns JMC (eds) Stevia: growth in knowledge and taste. Proceedings of the 8th Stevia symposium 2015 of European Society of Stevia (EUSTAS), 27–29 Jan 2015, Bonn, pp 117–182. ISBN: 9789074253291
106. Tamura Y, Nakamura S, Fukui H, Tabata M (1984) Comparison of Stevia plants grown from seeds, cuttings and stem-tip cultures for growth and sweet diterpene glucosides. Plant Cell Rep 3:180–182
107. Martini A, Tavarini S, Macchia M, Benelli G, Romano D, Canale A, Angelini LG (2015) Floral phenology, insect pollinators and seed quality of 36 genotypes of *Stevia rebaudiana*

Bert. cultivated in Italy. In: Ceunen S, Geuns JMC (eds) Stevia: growth in knowledge and taste. Proceedings of the 8th Stevia symposium 2015 of European Society of Stevia (EUSTAS), 27–29 Jan 2015, Bonn, pp 13–26. ISBN: 9789074253291

108. Carneiro JWP (2007) *Stevia rebaudiana* (Bert) Bertoni: stages of plant development. Can J Plant Sci 87:861–865
109. Macchia M, Morelli I, Angelini LG, Flamini G (1999) Agronomic characteristics and quantitative analysis of stevioside in *Stevia rebaudiana* Bert. a new source of sweet compounds. In: Proceeding fourth European symposium on industrial crops and products, 23–25 Mar, Bonn, Edited by FNR Band 14, Gulzow, pp 331–332. ISBN 3-7843-3019-3
110. Kumar R, Sharma S (2012) Effect of light and temperature on seed germination of important medicinal and aromatic plants in north western Himalayas. Int J Med Aromat Plants 2:468–475
111. Abdullateef RA, Osman MB (2011) Effects of visible light wavelengths on seed germinability in *Stevia rebaudiana* Bertoni. Int J Biol 3:83–91
112. Shuping C, Shizhen S (1995) Study on storage technique of *Stevia rebaudiana* seed. Acta Agron Sin 21:102–105
113. Shock CC (2015) Propagation of *Stevia rebaudiana* by rooted cuttings, Sustainable Agriculture Techniques. Oregon State University, Department of Crop and Soil Science Ext/CrS 154
114. Gvasaliya VP, Kovalenko NV, Garguliya MC (1990) Studies on the possibility of growing honey grass in Abkhazia conditions. Subtropicheskie Kul'tury 5:149–156
115. Carvalho MAM, Zaidan LBP (1995) Propagation of *Stevia rebaudiana* from stem cuttings. Pesq Agrop Bras 30:201–206
116. Chalapathi MV, Thimmegowda S, Kumar ND, Rao GGE, Mallikarjuna K (2001) Influence of length of cutting and growth regulators on vegetative propagation of stevia (*Stevia rebaudiana* Bert.). Crop Res 21:53–56
117. Ingle MR, Venugopal CK (2009) Effect of different growth regulators on rooting of stevia (*Stevia rebaudiana* Bertoni) cuttings. Karnataka J Agric Sci 22:455–456
118. Smitha GR, Umesha K (2012) Vegetative propagation of stevia [*Stevia rebaudiana* (Bertoni) Hems.] through cuttings. J Trop Agric 50:72–75
119. Bagoury EL, Olfat H, Allam AI, Abdelghany AM, Attia AE (2006) IBA application to stevia stem cutting: its productivity and quality of plant and ratoon crops. Egypt J Agric Res 84:299
120. Thiyagarajan M, Venkatachalam P (2012) Large scale in vitro propagation of *Stevia rebaudiana* (bert) for commercial application: pharmaceutically important and antidiabetic medicinal herb. Ind Crop Prod 37:111–117
121. Murashige T, Skoog F (1962) A revised medium for rapid growth and bioassay with tobacco tissue cultures. Physiol Plant 15:473–495
122. Hwang SJ (2006) Rapid in vitro propagation and enhanced stevioside accumulation in *Stevia rebaudiana* Bert. J Plant Biol 49:267–270
123. Kalpana M, Anbazhagan M, Natarajan V (2009) Utilization of liquid medium for rapid micropropagation of *Stevia rebaudiana* Bertoni. J Ecobiotechnol 1:16–20
124. Ali A, Gull I, Naz S, Afghan S (2010) Biochemical investigation during different stages of in vitro propagation of *Stevia rebaudiana*. Pak J Bot 42:2827–2837
125. Anbazhagan M, Kalpana M, Rajendran R, Natarajan V, Dhanavel D (2010) In vitro production of *Stevia rebaudiana* Bertoni. Emir J Food Agric 22:216–222
126. Ojha A, Sharma VN, Sharma V (2010) An efficient protocol for in vitro clonal propagation of natural sweetener plant (*Stevia rebaudiana* Bertoni). Afr J Plant Sci 4:319–321
127. Ahmad N, Fazal H, Zamir R, Khalil AS, Abbasi BH (2011) Callogenesis and shoot organogenesis from flowers of *Stevia rebaudiana* (Bert.). Sugar Tech 13:174–177
128. Jitendra M, Monika S, Ratan SD, Priyanka G, Priyanka S, Kiran DJ (2012) Micropropagation of an anti-diabetic plant-*Stevia rebaudiana* Bertoni, (natural sweetener) in Hadoti region of South-East Rajasthan, India. Int J Biol Sci 1:37–42
129. Lata H, Chandra S, Wang YH, Raman V, Khan IA (2013) TDZ-induced high frequency plant regeneration through direct shoot organogenesis in *Stevia rebaudiana* Bertoni: an important medicinal plant and a natural sweetener. Am J Plant Sci 4:117–128

130. Rathore S, Yadav K, Singh N, Singh SK (2014) Comparative study on callus induction, proliferation and plantlets regeneration in two cultivars of *Stevia rebaudiana* Bertoni-the only non caloric natural sweetener. Pertanika J Trop Agric Sci 37:499–508
131. Modi A, Patil G, Kumar N, Singh A, Subhash N (2012) A simple and efficient in vitro mass multiplication procedure for *Stevia rebaudiana* Bertoni and analysis of genetic fidelity of in vitro raised plants through RAPD. Sugar Tech 4:391–397
132. Gantait S, Das A, Mandal N (2015) Stevia: a comprehensive review on ethnopharmacological properties and in vitro regeneration. Sugar Tech 17:95–106
133. Bondarev N, Reshetnyak O, Nosov A (2003) Effects of nutrient medium composition on development of *Stevia rebaudiana* shoots cultivated in the roller bioreactor and their production of steviol glycosides. Plant Sci 165:845–850
134. Zayova E, Stancheva I, Geneva M, Petrova M, Dimitrova L (2013) Antioxidant activity of in vitro propagated *Stevia rebaudiana* Bertoni plants of different origins. Turk J Biol 37:106–113
135. Rebeaux D, Perret J (2015) Sourcing stevia from Europe, economical key questions. In: Ceunen S, Geuns JMC (eds) Stevia: growth in knowledge and taste. Proceedings of the 8th Stevia symposium 2015 of European Society of Stevia (EUSTAS), 27–29 Jan 2015, Bonn. ISBN: 9789074253291
136. Taleie N, Hamidoghli Y, Rabiei B, Hamidoghli S (2012) Effects of plant density and transplanting date on herbage, stevioside, phenol and flavonoid yield of *Stevia rebaudiana* Bertoni. Int J Agric Crop Sci 4:298–302
137. Maheshwar HM (2005) Effect of different levels of nitrogen and dates of planting on growth and yield of Stevia (*Stevia rebaudiana* Bert.). MSc thesis, Department of Horticulture College of Agriculture/University of Agricultural Sciences, Dharwad
138. Singh A, Singh K, Singh P, Singh MP (2015) Medicinal prospective and floral biology of candy leaf (*Stevia rebaudiana* Bertoni). Int J Adv Res 3:628–636
139. Serfaty M, Ibdah M, Fischer R, Chaimovitsh D, Saranga Y, Dudai N (2013) Dynamics of yield components and stevioside production in *Stevia rebaudiana* grown under different planting times, plant stands and harvest regime. Ind Crop Prod 50:731–736
140. Angelini LG, Tavarini S (2014) Crop productivity, steviol glycoside yield, nutrient concentration and uptake of *Stevia rebaudiana* Bert. under Mediterranean field conditions. Commun Soil Sci Plant Anal 45:2577–2592
141. Murayama S, Kayano R, Miyazato K, Nose A (1980) Studies on the cultivation of *Stevia rebaudiana*. II. Effects of fertilizer rates, planting density and seedling clones on growth and yield. Sci Bull Coll Agric 27:1–8
142. Lee JI, Kang KH, Park HW, Ham YS, Park CH (1980) Studies on the new sweetening source plant, *Stevia rebaudiana* in Korea. II. Effects of fertilizer rates and planting density on dry leaf yields and various agronomic characteristics of *Stevia rebaudiana*. Res Rep Off Rural Dev (Crop Suwon) 22:138–144
143. Donalisio MGR, Duarte FR, Pinto AJDA, Souza CJ (1982) *Stevia rebaudiana*. Agronomico 34:65–68
144. Rank AH, Midmore DJ (2006) Stevia, an intense, natural sweetener. Australian Government: Rural Industries Research and Development Corporation. Publication No 06/020. ISBN 1741512832
145. Rashid Z, Inamullah S, Peer QJA, Rashid M, Souliha R (2015) Influence of crop geometry on yield, yield attributes and glycoside yield of *Stevia rebaudiana* Bertoni. J Appl Nat Sci 7:339–343
146. Kawatani T, Kaneki Y, Tanabe T, Takahashi T (1978) On the cultivation of Kaa He-e (*Stevia rebaudiana* Bertoni). III. Response of Kaa He-e to fertilizer application amount and to nitrogen fertilization rates. Jpn J Trop Agric 21:165–172
147. Kawatani T, Kaneki Y, Tanabe T, Sakamoto I, Murakami K, Tanaka O (1978) On the cultivation of Kaa He-e (*Stevia rebaudiana* Bertoni). IV. Response of Kaa He-e to nitrogen fertilization rates and to the three major elements of fertilizer. Jpn J Trop Agric 21:173–178

148. Kawatani T, Kaneki Y, Tanabe T, Takahashi T (1980) On the cultivation of Kaa He-e (*Stevia rebaudiana* Bertoni). V. Response of Kaa He-e to phosphorus fertilization rates and to the three major elements of fertilizer. Jpn J Trop Agric 24:54–61
149. Liu X, Ren G, Shi Y (2011) The effect of organic manure and chemical fertilizer on growth and development of *Stevia rebaudiana* Bertoni. Energy Procedia 5:1200–1204
150. Dube K (2011) Organic production of Stevia. Effect of organic manures, biofertilizers and growth regulators on the productivity of *Stevia rebaudiana* Bertoni. Asiat J Biotechnol Resours 2:403–413
151. Kumar R, Sharma S, Ramesh K, Prasad R, Pathania VL, Singh B, Singh RD (2012) Effect of agro-techniques on the performance of natural sweetener plant-stevia (*Stevia rebaudiana*) under western Himalayan conditions. Indian J Agron 57:74–81
152. Rashid Z, Rashid M, Inamullah S, Rasool S, Bahar FA (2013) Effect of different levels of farmyard manure and nitrogen on the yield and nitrogen uptake by stevia (*Stevia rebaudiana* Bertoni). Afr J Agric Res 8:3941–3945
153. Lima Filho OF, Malavolta E (1997) Symptoms of nutritional disorders in stevia *Stevia rebaudiana* (Bert.) Bertoni. Sci Agric 54:53–61
154. Utumi MM, Monnerat PH, Pereira PRG, Fontes PCR, Godinho VdPC (1999) Macronutrient deficiencies in Stevia: visual symptoms and effects on growth, chemical composition, and stevioside production. Pesq Agrop Bras 34:1039–1043
155. Lima Filho OF, Malavolta E (1997) Nutritional interactions in stevia (*Stevia rebaudiana* (Bert.) Bertoni). Arq Biol Tecnol 40:351–357
156. Gosling P, Hodge A, Goodlass G, Bending GD (2006) Arbuscular mycorrhizal fungi and organic farming. Agric Ecosyst Environ 113:117–135
157. Smith SE, Read DJ (2008) Mycorrhizal symbiosis. Academic, London
158. Miransari M (2011) Arbuscular mycorrhizal fungi and nitrogen uptake. Arch Microbiol 193:77–81
159. Gianinazzi S, Gollotte A, Binet MN, Van Tuinen D, Redecker D, Wipf D (2010) Agroecology: the key role of arbuscular mycorrhizas in ecosystem services. Mycorrhiza 20:519–530
160. Zeng Y, Guo LP, Chen BD, Hao ZP, Wang JY, Huang LQ, Yang G, Cui XM, Yang L, Wu ZX (2013) Arbuscular mycorrhizal symbiosis for sustainable cultivation of Chinese medicinal plants: a promising research direction. Am J Chin Med 41:1199–1221
161. Kapoor R, Giri B, Mukerji KG (2002) *Glomus macrocarpum*: a potential bioinoculant to improve essential oil quality and concentration in Dill (*Anethum graveolens* L.) and Carum (*Trachyspermum ammi* (Linn.) Sprague). World J Microbiol Biotechnol 18:459–463
162. Kapoor R, Giri B, Mukerji KG (2002) Mycorrhization of coriander (*Coriandrum sativum* L.) to enhance the concentration and quality of essential oil. J Sci Food Agric 82:339–342
163. Kapoor R, Giri B, Mukerji KG (2004) Improved growth and essential oil yield and quality in *Foeniculum vulgare* mill on mycorrhizal inoculation supplemented with P-fertilizer. Bioresour Technol 93:307–311
164. Kapoor R, Chaudhary V, Bhatnagar AK (2007) Effects of arbuscular mycorrhiza and phosphorus application on artemisinin concentration in *Artemisia annua* L. Mycorrhiza 17:581–587
165. Copetta A, Lingua G, Berta G (2006) Effects of three AM fungi on growth, distribution of glandular hairs, and essential oil production in *Ocimum basilicum* L. var. Genovese. Micorrhiza 16:485–494
166. Portugal EP, Mercuri Quitério GC, Honório SL (2006) Seleção de fungos micorrízicos arbusculares para estévia, *Stevia Rebaudiana* (bert.) Bertoni. Construindo a história dos produtos naturais
167. Mandal S, Evelin H, Giri B, Singh VP, Kapoor R (2013) Arbuscular mycorrhiza enhances the production of stevioside and rebaudioside-A in *Stevia rebaudiana* via nutritional and non-nutritional mechanisms. Appl Soil Ecol 72:187–194
168. Zeneli G, Krokene P, Christiansen E, Krekling T, Gershenzon J (2006) Methyl jasmonate treatment of mature Norway spruce (*Picea abies*) trees increases the accumulation of terpenoid

resin components and protects against infection by *Ceratocystis polonica*, a bark beetle-associated fungus. Tree Physiol 26:977–988

169. Okada A, Shimizu T, Okada K, Kuzuyama T, Koga J, Shibuya N, Nojiri H, Yamane H (2007) Elicitor induced activation of the methylerythritol phosphate pathway toward phytoalexins biosynthesis in rice. Plant Mol Biol 65:177–187
170. Goklany S, Loring RH, Glick J, Lee-Parsons CW (2009) Assessing the limitations to terpenoid indole alkaloid biosynthesis in *Catharanthus roseus* hairy root cultures through gene expression profiling and precursor feeding. Biotechnol Prog 25:1289–1296
171. Mandal S, Upadhyay S, Singh VP, Kapoor R (2015) Enhanced production of steviol glycosides in mycorrhizal plants: a concerted effect of arbuscular mycorrhizal symbiosis on transcription of biosynthetic genes. Plant Physiol Biochem 89:100–106
172. Kennedy AC, Smith KL (1995) Soil microbial diversity and the sustainability of agricultural soils. Plant Soil 170:75–86
173. Das K, Dang R, Shivananda TN, Sekeroglu N (2007) Influence of bio-fertilizers on the biomass yield and nutrient content in *Stevia rebaudiana* Bert. grown in Indian subtropics. J Med Plant Res 1:5–8
174. Das K, Dang R, Shivananda TN, Sekeroglu N (2007) Comparative efficiency of bio-and chemical fertilizers on nutrient contents and biomass yield in medicinal plant *Stevia rebaudiana* Bert. IJNES 1:35–39
175. Das K, Dang R, Shivananda TN, Hegde L (2007) Effect of bio-fertilizers on biomass yield of *Stevia rebaudiana*. Biomed 2:278–282
176. Das K, Dang R, Shivananda TN (2008) Influence of bio-fertilizers on the availability of nutrients (N, P and K) in soil in relation to growth and yield of *Stevia rebaudiana* grown in South India. Int J Appl Res Nat Prod 1:20–24
177. Das K, Dang R, Shivananda TN (2009) Effect of biofertilizers on the nutrient availability in soil in relation to growth, yield and yield attributes of *Stevia rebaudiana*. Arch Agron Soil Sci 55:359–366
178. Das K, Dang R (2010) Influence of biofertilizers on stevioside content in *Stevia rebaudiana* grown in acidic soil condition. Arch Appl Sci Res 2:44–49
179. Eranna N (2007) Response of *Stevia rebaudiana* to biofertilizers. Karnataka J Agric Sci 20:616–617
180. Umesha K, Smitha GR, Sreeramu BS, Waman AA (2011) Organic manures and bio-fertilizers effectively improve yield and quality of stevia (*Stevia rebaudiana*). J Appl Hortic 13:157–162
181. Vafadar F, Amooaghaie R, Otroshy M (2014) Effects of plant-growth-promoting rhizobacteria and arbuscular mycorrhizal fungus on plant growth, stevioside, NPK, and chlorophyll content of *Stevia rebaudiana*. J Plant Interact 9:128–136
182. Das K, Dang R (2014) Influence of biofertilizers on protein, moisture and ash content in relation to swelling property, water absorption capacity, mineral elements, total phenolic level of Stevia (*Stevia rebaudiana* Bert.) plant grown under acidic soil zone of Shout India. Int Lett Nat Sci 1:78–97
183. Zare Hoseini R, Mohammadi E, Kalatejari S (2015) Effect of bio-fertilizer on growth, development and nutrient content (leaf and soil) of *Stevia rebaudiana* Bertoni. J Crop Prod 4:691–704
184. Mamta RP, Pathani V, Gulati A, Singh B, Bhanwra RK, Tewari R (2010) Stimulatory effect of phosphate solubilizing bacteria on plant growth, stevioside and rebaudioside A contents of *Stevia rebaudiana* Bertoni. Appl Soil Ecol 46:222–229
185. Patil NM (2010) Biofertilizer effect on growth, protein and carbohydrate content in *Stevia rebaudiana* var Bertoni. Recent Res Sci Technol 2:42–44
186. Borda-Molina D, Pardo-García JM, Montaña-Lara JS, Martínez-Salgado MM (2011) Influencia de la materia orgánica y Azotobacter nigricans en un cultivo de *Stevia rebaudiana* B. Univ Sci (Bogota) 16:282–293
187. Gupta M, Bisht S, Singh B, Gulati A, Tewari R (2011) Enhanced biomass and steviol glycosides in *Stevia rebaudiana* treated with phosphate-solubilizing bacteria and rock phosphate. Plant Growth Regul 65:449–457

188. Kumar R, Sharma S, Sood P, Dubey YP (2015) Bioorganic nutrient source effect on growth, biomass and quality of natural sweetener plant stevia and soil fertility in the Western Himalayas. Commun Soil Sci Plant Anal 46:1170–1186
189. Ramakrishnaiah G, Vijaya T (2013) Influence of VAM fungi, *Azotobacter* sp. and PSB on soil phosphatase activity and nutrients (N, P, K, Cu, Zn, Fe and Mn) status in the rhizosphere of *Stevia rebaudiana* (Bert.) plants. Am J Plant Sci 4:1443–1447
190. Megeji NW, Kumar JK, Singh V, Kaul VK, Ahuja PS (2005) Introducing *Stevia rebaudiana*, a natural zero-calorie sweetener. Curr Sci 88:801–804
191. Tonello PE, DeFaveri C, Weeden BR (2006) Agronomic assessment of Stevia for the development of a non-caloric natural sweetener industry in north Queensland. A final report to plant science. Queensland Department of Primary Industries and Fisheries, Department of Primary Industries and Fisheries, Queensland Government, Brisbane
192. Parihar CM, Rana KS, Kantwa SR (2010) Nutrient management in pearlmillet (*Pennisetum glaucum*) – mustard (*Brassica juncea*) cropping system as affected by land configuration under limited irrigation. Indian J Agron 55:191–196
193. Goenadi DH (1983) Water tension and fertilization of *Stevia rebaudiana* Bertoni on Oxic Tropudalf (English abstr.). Menara Perkeb 51:85–90
194. Fronza D, Folegatti MV (2003) Water consumption of the estevia (*Stevia rebaudiana* (Bert.) Bertoni) crop estimated through microlysimeter. Sci Agric 60:595–599
195. González RE (2000) Necesidad de agua para el cultivo de KA'A HE'E (*Stevia rebaudiana* Bert) bajo riego por goteo, calculado sobre la base de lectura de microlisimetro. Universidad Nacional de Asunción, Faculdad de Ciencias Agrarias, San Lorenzo, 37 p
196. Lavini A, Riccardi M, Pulvento C, De Luca S, Scamosci M, d'Andria R (2008) Yield, quality and water consumption of *Stevia rebaudiana* Bertoni grown under different irrigation regimes in Southern Italy. Ital J Agron 3:135–143
197. Midmore DJ, Rank AH, Walsh KB, Reyes R, Gautam G, Hopkins KC (2012) Further development of the stevia natural sweetener industry. Rural Industries Research and Development Corporation. RIRDC Publication No 12/051, pp 107
198. Guzman RD (2010) Autoecological role of steviol glycosides in *Stevia rebaudiana* Bertoni. Central Queensland University Australia, Rockhampton
199. Aladakatti Y, Palled Y, Chetti M, Halikatti S, Alagundag S, Patil P, Patil V, Janawade A (2012) Effect of irrigation schedule and planting geometry on growth and yield of stevia (*Stevia rebaudiana* Bertoni.). Karnataka J Agric Sci 25:30–35
200. Shi Y, Ren GX (2012) The effects of drought stress on the photosynthetic parameters and dry leaf yield of *Stevia rebaudiana* Bertoni. Adv Mater Res 518:4786–4789
201. Hajihashemi S, Ehsanpour AA (2013) Influence of exogenously applied paclobutrazol on some physiological traits and growth of *Stevia rebaudiana* under in vitro drought stress. Biologia 68:414–420
202. Hajihashemi S, Ehsanpour AA (2014) Antioxidant response of *Stevia rebaudiana* B. to polyethylene glycol and paclobutrazol treatments under in vitro culture. Appl Biochem Biotechnol 172:4038–4052
203. Maas EV, Hoffman GJ (1977) Crop salt tolerance – current assessment. ASCE J Irrig Drain Div 103:115–134
204. Reis M, Coelho L, Santos G, Kienle U, Beltrão J (2015) Yield response of stevia (*Stevia rebaudiana* Bertoni) to the salinity of irrigation water. Agric Water Manag 152:217–221
205. Harrington KC, Southward RC, Kitchen KL, He XZ (2011) Investigation of herbicides tolerated by *Stevia rebaudiana* crops. N Z J Crop Hortic Sci 39:21–33
206. Basuki S (1990) Effects of black plastic mulch and plant density on the growth of weeds and stevia. BIOTROP Spec Publ 38:107–113
207. Andolfi L, Ceccarini L, Macchia M (2002) Bioagronomic characteristics of *Stevia rebaudiana*. Inf Agrar 58:48–51
208. Covarelli G, Peccetti G, Pannacci E, Graziani F (2008) Stevia (*Stevia rebaudiana* Bertoni), produttività, concimazione e diserbo in Umbria. Progetto Co.Al.Ta. II, pp 105–107

209. Zachokostas P (2015) Effects of chemicals weed management on the quantitative and qualitative characteristics of *Stevia rebaudiana* Bertoni, under Mediterranean conditions (Karditsa, Thessaly, Greece). In: Ceunen S, Geuns JMC (eds) Stevia: growth in knowledge and taste. Proceedings of the 8th Stevia symposium 2015 of European Society of Stevia (EUSTAS), 27–29 Jan 2015, Bonn, pp 5–12. ISBN: 9789074253291
210. Hopkins KC, Midmore DJ (2015) Assessment of herbicides for use on the sweet herb, *Stevia rebaudiana* in the tropics. J Hortic Sci Biotechnol 90:530–536
211. Li TSC (2000) Medicinal plants: culture, utilization and phytopharmacology. Technomic Publishing Company, Lancaster
212. Bhandari S, Harsh NSK (2006) First report of *Septoria steviae* on *Stevia rebaudiana* in India. Indian Forester 132:385
213. Angelini LG, Tavarini S (2010) Studio sulla ottimizzazione della coltivazione di *Stevia rebaudiana* Bertoni. Final report Eridania Sadam project contract, p 80
214. DIVAS EU financed project (FP7-SME) Diversification for tobacco growing farms by the alternative crop *Stevia rebaudiana* Bertoni. http://cordis.europa.eu/result/rcn/58290_en.html. Accessed 12 Oct 2015
215. Kang KH, Lee EW (1981) Physio-ecological studies on stevia (*Stevia rebaudiana* Bertoni). Korean J Crop Sci 26:69–89
216. Pal PK, Mahajan M, Prasad R, Pathania V, Singh B, Ahujac PS (2015) Harvesting regimes to optimize yield and quality in annual and perennial *Stevia rebaudiana* under sub-temperate conditions. Ind Crop Prod 65:556–564
217. Moraes RM, Donega MA, Catrell CL, Mello S, McChesney JD (2013) Effect of harvest timing on leaf production and yield of diterpeneglycosides in *Stevia rebaudiana* Bert: a specialty perennial crop for Mississippi. Ind Crop Prod 51:385–389
218. Jordan Molero F (1984) La propagación de ka'a he'e, *Stevia rebaudiana* Bertoni. Primer Simposio Nacional de la Stevia (kaa hee) Julio 1983, Asunción, p 29
219. Taiariol DR (2004) Characterization of the *Stevia rebaudiana* Bert. http://www.monografias.com/trabajos13/Stevia/stevia.html
220. Singh V, Kaul VK (2005) *Stevia rebaudiana* for income generation. Vigyan Pragati 54:10–15
221. Andolfi L, Macchia M, Ceccarini L (2006) Agronomic productive characteristics of two genotype of *Stevia rebaudiana* in central Italy. Ital J Agron 2:257–262
222. Gorovits R, Czoznek H (2007) Biotic and abiotic stress responses in tomato breeding lines resistant and susceptible to tomato yellow leaf curl virus. In: Czosnek H (ed) Tomato yellow leaf curl virus disease. Springer, Chapter 6, pp 223–237 Dordrecht, The Netherlands
223. Jimenez-Garcia SN, Vazquez-Cruz MA, Guevara-Gonzalez RG, Torres-Pacheco I, Cruz-Hernandez A, Feregrino-Perez AA (2013) Current approaches for enhanced expression of secondary metabolites as bioactive compounds in plants for agronomic and human health purposes – a review. Pol J Food Nutr Sci 63:67–78
224. Tavarini S, Pagano I, Guidi L, Angelini LG (2015) Impact of nitrogen supply on growth, steviol glycosides and photosynthesis in *Stevia rebaudiana* Bertoni. Plant Biosyst. doi:10.1080/11263504.2014.993743
225. Tavarini S, Sgherri C, Ranieri AM, Angelini LG (2015) Effect of nitrogen fertilization and harvest time on steviol glycosides, flavonoid composition, and antioxidant properties in *Stevia rebaudiana* Bertoni. J Agric Food Chem 63:7041–7050
226. Chaves MM, Pereira JS, Maroco J, Rodrigues M, Ricardo CPP, Osorio M, Carvalho I, Faria T, Pinheiro C (2002) How plants cope with water stress in the field? Photosynthesis and growth. Ann Bot 89:907–916
227. Chaves MM, Maroco JP, Pereira JS (2003) Understanding plant responses to drought – from genes to the whole plant. Funct Plant Biol 30:239–264
228. Karimi M, Ahmadi A, Hashemi J, Abbasi A, Tavarini S, Guglielminetti L, Angelini LG (2015) The effect of soil moisture depletion on Stevia (*Stevia rebaudiana* Bertoni) grown in greenhouse conditions: growth, steviol glycosides content, soluble sugars and total antioxidant capacity. Sci Hortic 183:93–99

229. Hajihashemi S, Geuns JMC, Ehsanpour AA (2012) Molecular and physiologicalanalysis of *Stevia rebaudiana* after treatment with polyethylene glycol, paclobu-trazol and gibberellic acid. In: Geuns JMC (ed) Stevia: 6 months beyond authorisation. Proceedings of the 6th Stevia symposium 2012 of European Society of Stevia (EUSTAS), 3–4 July 2012, KU Leven, Leuven, pp 71–95. ISBN: 9789074253291
230. Karimi M, Ahmadi A, Hashemi J, Abbasi A, Tavarini S, Pompeiano A, Guglielminetti L, Angelini LG (2015) The positive role of steviol glycosides in stevia (*Stevia rebaudiana* Bertoni) under drought stress condition. Plant Biosyst. doi:10.1080/11263504.2015.1056857
231. Angelini LG, Tavarini S (2011) Effetto della densità d'impianto sulle caratteristiche produttive e sul contenuto di steviol glicosidi in *Stevia rebaudiana* Bert. In: Pisante M, Stagnari F (eds) Proceedings XL Convegno Nazionale della Società Italiana di Agronomia, 7–10 settembre 2011, Teramo, pp 54–55
232. Alves LM, Ruddat M (1979) The presence of gibberellin A20 in *Stevia rebaudiana* and its significance for the biological activity of steviol. Plant Cell Physiol 20:123–130
233. Brandle JE, Telmer PG (2007) Steviol glycoside biosynthesis. Phytochemistry 68:1855–1863
234. Kim KK, Sawa Y, Shibata H (1996) Hydroxylation of ent-kaurenoic acid to steviol in *Stevia rebaudiana* Bertoni – purification and partial characterization of the enzyme. Arch Biochem Biophys 332:223–230
235. Hedden P, Kamiya Y (1997) Annu Rev Plant Physiol Plant Mol Biol 48:431–460
236. Karimi M, Ahmadi A, Hashemi J, Abbasi A, Angelini LG (2014) Effect of two plant growth retardants on steviol glycosides content and antioxidant capacity in Stevia (*Stevia rebaudiana* Bertoni). Acta Physiol Plant 36:1211–1219
237. Karimi M, Hashemi J, Ahmadi A, Abbasi A, Pompeiano A, Tavarini S, Guglielminetti L, Angelini LG (2015) Opposing effects of external gibberellin and daminozide on stevia growth and metabolites. Appl Biochem Biotechnol 175:780–791

Glycyrrhiza glabra: Chemistry and Pharmacological Activity

4

Varsha Sharma, Akshay Katiyar, and R. C. Agrawal

Abstract

Nature is an attractive source of new therapeutic candidate compounds as a tremendous chemical diversity is found in millions of species of plants, animals, marine organisms, and microorganisms as potential medicinal agents. This chapter of research is an effort to highlight the phytochemical/chemical constituents of an ancient medicinal plant *G. glabra* and their pharmacological importance. *G. glabra* is an old age medicinal plant that belongs to Leguminosae/Fabaceae/Papilionaceae family and commonly known as mulaithi in north India. The chemical composition of *G. glabra* is glycyrrhizin, glycyrrhetic acid, isoliquiritin, isoflavones, etc., and their derivatives have been reported for several pharmacological activities like, expectorant, antidemulcent, antiulcer, anticancer, anti-inflammatory, antidiabetic, etc. These phytochemicals hold strong promise for designing new herbal drugs, and derivatives of these compounds are being generated to evaluate their pharmacological purposes for future drug use. Natural products have been a prime source for the treatment of many forms of ailments, many of which are consumed daily with the diet. They provide significant protection against various diseases and disorders.

Keywords

Chemistry • *Glycyrrhiza glabra* • Licorice • Medicinal plant • Pharmacology

V. Sharma (✉) • R.C. Agrawal
Department of Research, Priyamvada Birla Cancer Research Institute, M.P. Birla Hospital, Satna, MP, India
e-mail: drvarsha.research81@gmail.com; rcagrawal60@yahoo.com

A. Katiyar
Plant Virology Unit, Division of Plant Pathology, Indian Agriculture Research Institute, New Delhi, India
e-mail: akshay.katiyar@gmail.com

J.-M. Mérillon, K.G. Ramawat (eds.), *Sweeteners*, Reference Series in Phytochemistry,
https://doi.org/10.1007/978-3-319-27027-2_21

Abbreviations

AP-1	Activator protein-1
Bcl2	B cell lymphoma-2
CCl_4	Carbon tetrachloride
CYP1A1	Cytochrome P1450A1
DGL	Deglycyrrhizinated licorice
DNA	Deoxyribonucleic acid
GA	Glycyrrhetinic acid
GG/*G*	*glabra Glycyrrhiza glabra*
GST	Glutathione-S-transferase
H1N1	Hemagglutinin type-1 and neuraminidase type-1
HIV	Human immunodeficiency virus
KK-Ay	Knockout diabetic mice
MAP	Mitogen activated protein
NADH	Nicotinamide adenine dinucleotide (reduced)
PAF	Platelet aggregating factor
SARS	Severe acute respiratory syndrome
SNMC	Stronger neo-minophagen-C
TPA	12-*O*-tetradecanoylphorbal-13-acetate

Contents

1 Introduction 89
1.1 Scientific Classification 90
1.2 Binomial Name 90
1.3 Synonyms 90
1.4 Vernacular Names [4] 90
1.5 Origin 91
1.6 Ecology 91
1.7 Morphology 91
1.8 Medicinal Parts Used 91
2 Phytochemistry/Chemistry 91
3 Pharmacological Activity 93
3.1 Antitussive and Expectorant 93
3.2 Antimicrobial 93
3.3 Anticoagulant and Memory Enhancing Activity 94
3.4 Antiviral 94
3.5 Antioxidant and Anti-inflammatory 94
3.6 Antiulcer Activity 94
3.7 Anticarcinogenic and Antimutagenic Activity 95
3.8 Hepatoprotective Activity 95
3.9 Antidiabetic 96
3.10 Immunomodulator Activity 96
4 Side Effects and Toxicity 97
5 Future Aspects 97
6 Conclusion 97
References 98

1 Introduction

Plants have been one of the important sources of medicines since the beginning of human cultivation. There is a growing demand for plant-based medicines, health products, pharmaceuticals, food supplements, etc. Medicinal plants are of great importance to the health of individuals and communities. The medicinal value of these plants lies in some chemical substances that produce a definite physiological action on the human body. The most important of these bioactive constituents of plants are triterpenoid, saponin, flavonoids, tannins, alkaloids, and phenolic compounds [1]. Many of these indigenous medicinal plants are used as spices and food plants. They are also sometimes added to foods meant for pregnant and nursing mothers for medicinal purposes [2, 3].

Glycyrrhiza glabra (Fig. 1) is one of the useful medicinal plants. *Glycyrrhiza* is derived from the ancient Greek term *glykos*, meaning sweet, and *rhiza*, meaning root. *Glycyrrhiza glabra* is known as mulaithi in north India. *Glycyrrhiza glabra*, also known as licorice and sweet wood, is native to the Mediterranean and certain areas of

Fig. 1 *Glycyrrhiza glabra* plant

Asia. A number of traditional healers have claimed the efficacy of *Glycyrrhiza* species for a variety of pathological conditions as a diuretic, choleretic, used as insecticide, and indicated in traditional medicine for coughs, colds, and painful swellings [4, 5].

1.1 Scientific Classification

Kingdom: Plantae
Division: Angiospermae
Class: Dicotyledoneae
Order: Rosales
Family: Leguminosae
Genus: *Glycyrrhiza*
Species: *glabra* Linn

1.2 Binomial Name

Glycyrrhiza glabra Linn.

1.3 Synonyms

Glycyrrhiza glandulifera

1.4 Vernacular Names [4]

Sanskrit: Yashti-madhu, madhuka
Bengali: Jashtimadhu, jaishbomodhu
Gujarat: Jethimadhu
Hindi: Jothi-madh, mulaithi
Kannada: Yastimadhuka, atimaddhura
Malayalam: Iratimadhuram
Marathi: Jeshtamadha
Oriya: Jatimadhu
Tamil: Atimaduram
Telugu: Atimadhuranu, yashtimadhukam
English: Licorice, liquorices, sweet wood
Arab: Aslussiesa
Persia: Ausareha mahaka
France: Boisdoux
Germany: Sussholz

1.5 Origin

The roots are unearthed in the autumn of the fourth season. It is grown in India, Spain, Iran, Russia, China and Italy.

1.6 Ecology

Glycyrrhiza glabra enjoys fertile, sandy, and clay soil near a river or stream where enough water is available for the plant to flourish in the wild, or under cultivation where it can be irrigated.

1.7 Morphology

Glycyrrhiza glabra is herbaceous perennial, growing to 1 m in height, with pinnate leaves about 7–15 cm long, with 9–17 leaflets. The flowers are 0.8–1.2 cm long, purple to pale whitish blue, produced in a loose inflorescence. The fruit is an oblong pod, 2–3 cm long, containing several seeds [6]. The *Glycyrrhiza* shrub is a member of the pea family and grows in subtropical climates in rich soil. Below ground, the *Glycyrrhiza glabra* plant has an extensive root system with a main taproot and numerous runners. The main taproot, which is harvested for medicinal use, is soft, fibrous, and has a bright yellow interior [7].

1.8 Medicinal Parts Used

Roots and Rhizome (powder, teas, tonic, extracts, tinctures, decoction).

2 Phytochemistry/Chemistry

A number of components have been isolated from the roots of *Glycyrrhiza glabra*, including a water-soluble, biologically active complex that accounts for 40–50% of total dry material weight. This complex is composed of triterpene, saponin, flavonoids, polysaccharides, pectins, simple sugars, amino acids, mineral salts, asparagines, bitters, essential oil, fat, female hormone estrogen, gums, mucilage (rhizome), protein, resins, starches, sterols, volatile oils, tannins, glycosides, and various other substances [8, 9]. Glycyrrhizin (Fig. 2), a triterpenoid compound, accounts for the sweet taste of licorice root. This compound represents a mixture of potassium-calcium-magnesium salts of glycyrrhizic acid that varies within a 2–25% range. Among the natural saponin, glycyrrhizic acid is a molecule composed of a

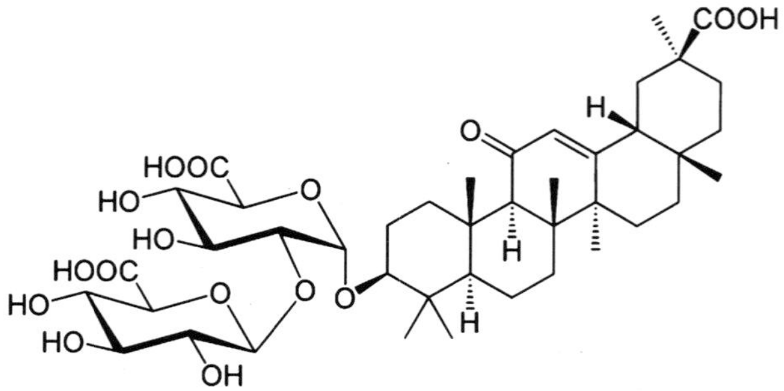

Fig. 2 Glycyrrhizin

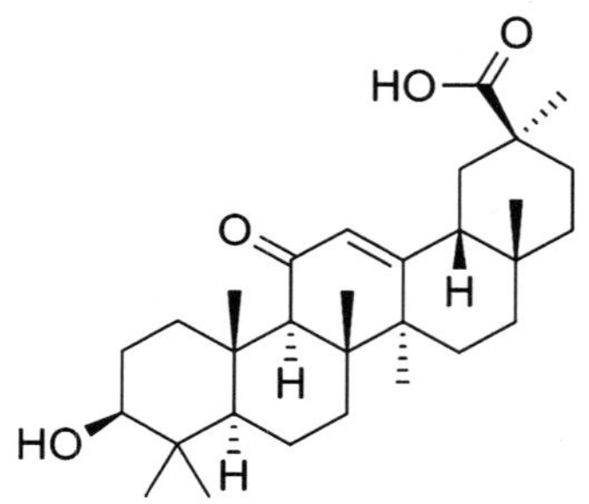

Fig. 3 Glycyrrhetic acid

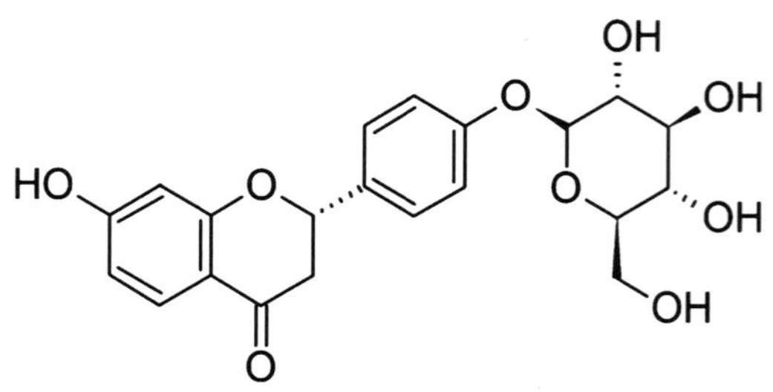

Fig. 4 Liquiritin

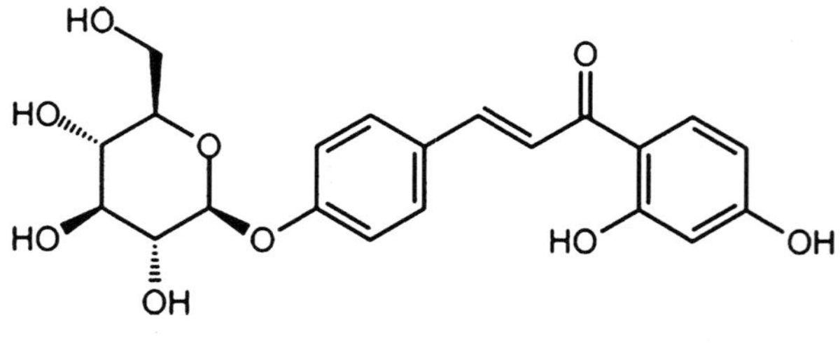

Fig. 5 Isoliquiritin

hydrophilic part, two molecules of glucuronic acid, and a hydrophobic fragment, glycyrrhetic acid (Fig. 3) [10]. The yellow color of licorice is due to the flavonoid content of the plant, which includes liquiritin (Fig. 4), isoliquiritin (Fig. 5) (a chalcone) and other compounds [11]. The isoflavones, glabridin (Fig. 6) and

Fig. 6 Glabridin

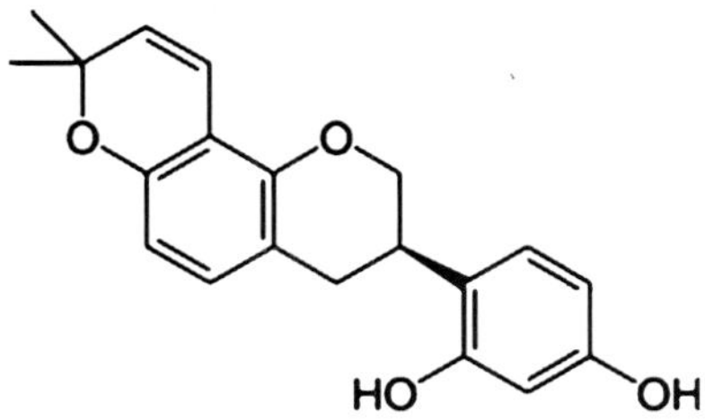

hispaglabridins A and B have significant antioxidant activity [12], and both glabridin and glabrene possess estrogen-like activity [13].

3 Pharmacological Activity

The various studies carried out by ethnobotanists, phytochemists, and experimental pharmacologists on its bioactivities revealed that the plant may be a source of new drugs, and therapeutic agents for the treatment of a variety of diseases and ailments could be manufactured. Report of various activities is given here.

3.1 Antitussive and Expectorant

The licorice powder and extract was found to be useful for the treatment of sore throat, cough, and bronchial catarrh. It has antitussive, demulcent, and expectorant loosening activities which may attribute due to presence of glycyrrhizin and helping to expel congestion in the upper respiratory tract as it accelerates tracheal mucus secretion [14]. It has been recently found that liquiritin apioside is an active compound present in the methanolic extract of liquorice. The compound inhibits capsaicin-induced cough [15].

3.2 Antimicrobial

Multidrug-resistant microorganisms pose a serious infestation in clinical medicine today due to the rapid spread as well as chronic infections caused by them. Each species of the genus *Glycyrrhiza* Linn is characterized by isoprenoid phenols, which have selective antimicrobial activity. Recent research has shown antibacterial effects of hydromethanolic root extract of *G. glabra* against some gram-positive and negative pathogens [16]. A number of components isolated from *Glycyrrhiza* include glabridin, gabrin, glabrol, glabrene, hispaglabridin A, hispaglabridin B, 40-methylglabridin, and 3-hydroxyglabrol have exhibited potential in vitro antimicrobial activity [17, 18]. Glycyrrhizinic acids have been used to cure atopic dermatitis, pruritis, and cysts due to parasitic infestations of skin [19, 20].

3.3 Anticoagulant and Memory Enhancing Activity

Glycyrrhizin, an already known anti-inflammatory compound, has also been found as the first plant-based inhibitor of thrombin. It prolonged the thrombin and fibrinogen clotting time and increased plasma recalcification duration. The thrombin-induced platelet aggregation was found to be inhibited by the action of glycyrrhizin, but PAF (platelet aggregating factor)- or collagen-induced agglutination was not affected by glycyrrhizin [21, 22]. One of the laboratory-based research has shown memory enhancing activity of *G. glabra* in experimental animals [23].

3.4 Antiviral

Glycyrrhizin has a prominent antiviral activity, as it does not allow the virus cell binding. Recently antiviral activities of ribavirin, 6-azauridine, pyraziofurin, mycophenolic acid, and glycyrrhizin against two clinical isolates of SARS (severe acute respiratory syndrome) virus, i.e., FFM-1 and FFM-2 were evaluated. It was observed that glycyrrhizin was the most effective in controlling viral replication and could be used as a prophylactic measure. Glycyrrhizin has been previously used to treat patients suffering from HIV-1 and chronic hepatitis C virus [24–26].

3.5 Antioxidant and Anti-inflammatory

Hydromethanolic root extract of *Glycyrrhiza glabra* exhibited marked antioxidant activity in a test tube system [16]. *Glycyrrhiza* (root) have a plenty of polyphenolic components as a potential source of antioxidants. Licochalcones B and D exhibit a potential activity by inhibiting the microsomal lipid peroxidation. Retrochalcones exhibit mitochondrial lipid peroxidation and prevent red blood corpuscles from oxidative hemolysis. Isoflavones like glabridin, hispaglabridin A and 3′hydroxy-4-*O*-methylglabridin present in *Glycyrrhiza glabra* were found to have potential antioxidant activity. More recently, dehydrostilbene derivatives like α-dihydro-3,5,4-trihydroxy-4,5-diiodopentenylstilbene have been isolated and reported as free radical scavengers [27–29]. Research shows that on being broken down in the gut, glycyrrhizin exerts an anti-inflammatory action similar to hydrocorticosone and other corticosteroid hormones.

3.6 Antiulcer Activity

Licorice has been used as an antiulcer agent since early 1970s. The extracted glycyrrhizin, DGL (deglycyrrhizinated licorice) is generally employed for the effective treatment of ulcers. Carbenoxolone from liquorice roots produce the antiulcerogenic effect by inhibiting the secretion of gastrin [28]. Liquorice can raise the concentration of prostaglandins in the digestive system that promote mucus secretion from the stomach. It was also reported that liquorice prolongs the life span of surface cells in the stomach and has an antipepsin effect [29].

3.7 Anticarcinogenic and Antimutagenic Activity

The aqueous extract of *G. glabra* inhibits in vivo and in vitro proliferation of Ehrlich ascites tumor cells and inhibits angiogenesis in in vivo assay, peritoneal and choreoallantonic membrane assay [30]. On the other hand, there are many reports about the anticancer effects of several derivatives of its components both in in vivo and in vitro studies. Glycyrrhetic acid could trigger the proapoptotic pathway by inducing mitochondrial permeability transition, and this property may be useful for inducing apoptosis of tumor cells [31, 32]. The licorice extract induced the Bcl2 phosphorylation and G2/M cycle arrest in tumor cell lines, and 70% methanol soluble fraction of licorice acetone extract was found to induce apoptosis in human monoblastic leukemia U937 cells. The compound was identified to be licocoumarone also responsible for antioxidant and antimicrobial activity [33]. The hydromethanolic root extract of *G. glabra* also exhibited antimutagenic potential by suppressing micronucleus formation and chromosomal aberration in bone marrow cells of albino mice [34, 35]. It was found that glycyrrhizin induced AP-1 (activator protein-1) activity in untreated cells whereas inhibited TPA (12-*O*-tetradecanoylphorbal-13-acetate) induced AP-1 activity in TPA treated cells. This mechanism could serve as a model for development of new chemoprotective agents [36]. Recently licochalcone E, a new retrochalcone from the roots of *G. inflate*, exhibited the most potent cytotoxic effect compared with the known antitumor agents, lichochalcone A and isoliquiritigenin [37].

3.8 Hepatoprotective Activity

Chronic hepatitis (viral as well as nonviral) is a slowly progressive liver disease that may evolve into cirrhosis with its potential complications of liver failure or hepatocellular carcinoma. In Japan, glycyrrhizin has been used for more than 60 years as treatment for chronic hepatitis under the name of SNMC (stronger neo–minophagen-C) clinically as an antiallergic and antihepatitis agent [38]. Glycyrrhizin induced significant reduction in serum aminotransferases and improved the liver histology when compared with the placebo. It has also been implicated that long-term usage of glycyrrhizin prevents development of hepatocellular carcinoma in chronic hepatitis C. In vitro studies have indicated that glycyrrhizin modifies the intracellular transport and uppresses hepatitis B virus surface antigen [39, 40]. It has been found that 18β-glycyrrhetinic acid (GA), an aglycone of glycyrrhizin decreases the expression of P450 E1 thereby protecting the liver [41]. GA also prevents the oxidative and hepatic damage caused by aflatoxins by increasing the CYP1A1 and GST (glutathione-S-transferase) activities and may also contribute to anticarcinogenic activity by metabolic deactivation of the hepatotoxin [42]. One of the researches indicated that the hydromethanolic root extract of *G. glabra* exhibited significant protection from hepatotoxicity induced by CCl_4 (carbon tetrachloride) in liver tissue of experimental mice [43]. It has also been experimentally investigated that glycyrrhizin and its analogues have a mitogenic effect via epidermal growth factor receptors subsequently stimulating the MAP (mitogen activated protein) kinase pathway to induce hepatocyte DNA synthesis and proliferation [44].

3.9 Antidiabetic

Type 2 (noninsulin dependent) diabetes mellitus, an insulin resistant syndrome, is a growing health concern in the modern society. PPAR's (peroxisome proliferation activated receptors) are ligand-dependent transcriptional factors regulating the expression of a group of genes that play an important role in glucose and lipid metabolism. The PPAR receptors are classified as PPAR-α, PPAR-γ, and PPAR-δ. The PPAR-α is found in liver, muscle, and kidney. PPAR-γ is associated with adipose tissue, adrenals, and small intestine whereas PPAR-δ is expressed ubiquitously. PPAR-γ serves as a predominant target for insulin sensitizing drugs like pioglitazone and roziglitazone. Ethyl acetate extract of licorice using GAL-4-PPAR-γ chimera assay exhibited a significant PPAR-γ binding activity which was attributed to seven phenolic compounds, viz., dehydroglyasperin, glyasperin B, glyasperin D, glycycoumarin, glycyrin, glycyrol, and isolglycyrol. Pioglitazone and glycyrin were found to suppress the increased blood glucose level in mice after sucrose loading during the oral sucrose tolerance test. Pioglitazone, a potent PPAR-γ agonist ameliorated the insulin resistance and type 2 diabetes mellitus. Similarly, glycyrin also exhibited a potent PPAR-γ ligand binding activity and therefore reduces the blood glucose level in KK-Ay (knockout diabetic mice). This finding is of much significance as licorice has also been traditionally used as an artificial sweetening agent and could be helpful in insulin resistance syndrome prevalent in the modern society [45].

3.10 Immunomodulator Activity

Swine flu is a highly contagious respiratory disease of pigs with low mortality (1–4%), is species-specific in nature, and outbreak usually occurs once in a year with an upsurge in autumn and winter in temperate zones. One such virus, namely, influenza A H1N1 virus has evolved the capacity to cross species barrier (i.e., pig to humans) and has spread widely among humans. Polysaccharide fractions obtained from *Glycyrrhiza glabra* stimulate macrophages and hence elevate and assist immune stimulation [46]. *N*-acetylmuramoyl peptide is glycyrrhizin analogue having potential in vitro immune-stimulating properties, [47] also animal studies have revealed its efficacy against the influenza virus that is mediated by stopping the virus replication. Glycyrrhizic acid present in the plant inhibits virus growth and inactivates virus particles is a potential source of immunomodulator [48].

4 Side Effects and Toxicity

Everything on the earth has a purpose whether they are plants, animals, or microbes, but there is a limit for consumption because it is well known that anything can be harmful or poisonous after a limit. The most common reported side effect with licorice supplementation is elevated blood pressure. This is thought to be due to the effect of licorice on the rennin-angiotensin-aldosterone system. It is suggested licorice saponins are capable of potentiating aldosterone action while binding to mineral-corticoid receptors in the kidneys. In addition to hypertension, patients may experience hypokalemia and sodium retention, resulting in edema. All symptoms usually disappear with discontinuation of therapy. Many studies report no side effects during the course of treatment [27, 28]. Generally the onset and severity of symptoms depend on the dose and duration of licorice intake, as well as individual susceptibility. The amount of licorice ingested daily by patients with mineral-corticoid excess syndromes appears to vary over a wide range, from as little as 1.5 g daily to as much as 250 g daily [5, 29].

5 Future Aspects

In ancient Indian literature, it is mentioned that every plant on this earth is useful for human beings, animals, and other plants. *G. glabra* is an herbal plant which has lots of medicinal properties such as antimicrobial, antioxidant, anti-inflammatory, antitussive, antidiabetic, antiviral, anticancer, antimutagenic, antiulcer, and hepatoprotective. The phytochemicals present in *G. glabra* have been of immense importance in phytotherapeutics. Thus there is an immense need to modify the natural *Glycyrrhiza* constituents to reduce these side effects thereby generating the advanced versions of the bioactive compounds to be used as drugs in future. There are so many bioactive compounds that have been isolated from *G. glabra* like glycyrrhizin, glycyrrhetic acid, glabridin, liquiritin, isoliquiritigenin, isoflavones, etc. and have been assessed for medicinal potential which showed that these compounds hold a strong promise in designing future drugs. Derivatives of these compounds are being generated to evaluate their pharmacological purposes for future drug use. There are ample chances of arriving to pharmacophores with least toxic side effects using combinatorial chemistry.

6 Conclusion

Glycyrrhiza glabra (GG) is a plant with strong ethnobotanical history. The root and rhizome parts of this plant are used as a folk medicine both in Europe and eastern countries. *G. glabra* extracts have been shown to possess lot of medicinal properties like antitussive, antimicrobial, antioxidant, anti-inflammatory, antiulcer, anticancer, etc. due to presence of so many bioactive components, triterpene, saponins, flavonoids, alkaloids, glycyrrhizin, glycyrrhetic acid, glabridin, liquiritin, etc. *G. glabra* is one of those ancient plants, which have been used in the traditional pharmacopoeias for its multifaceted activities against variety of systematic and nonsystematic ailments. The chemical foundations of *G. glabra* have been discovered in the last era. The chemical constituents of *G. glabra* hold a strong promise for providing new molecules, which could be of immense medicinal applications in the drug discovery process for the development of new drugs present era.

References

1. Hill AF (1952) Economic botany: a textbook of useful plants and plant products, 2nd edn. McGarw Hill, New York
2. Okwu DE (1999) Flavouring properties of spices on cassava Fufu. Afr J Roots Tuber Crops 3(2):19–21
3. Okwu DE (2001) Evaluation of the chemical composition of indigenous spices and flavouring agents. Global J Pure Appl Sci 7(3):455–459
4. Chopra RN, Nayar SL, Chopra IC (2002) Glossary of Indian medicinal plants. NISCAIR, CSIR, New Delhi
5. Sharma V, Agrawal RC (2013) *Glycyrrhiza glabra*: a plant for the future. Mintage J Pharm Med Sci 2(3):15–20
6. Huxley A (ed) (1992) New RHS dictionary of gardening. ISBN 0-333-47494-5
7. Olukoga A, Donaldson D (1998) Historical perspectives on health. The history of liquorice: the plant, its extract, cultivation, and commercialization and etymology. J R Soc Health 118:300–304
8. Bradley PR (ed) (1992) British herbal compendium, vol 1. BHMA, Bournemouth
9. Hoffmann D (1990) The new holistic herbal, 2nd edn. Element, Shaftesbury
10. Obolentseva GV, Litvinenko VI, Ammosov AS, Popova TP, Sampiev AM (1999) Pharmacological and therapeutic properties of licorice preparations (a review). Pharm Chem J 33:24–31
11. Yamamura Y, Kawakami J, Santa T, Kotaki H, Uchino K, Sawada Y, Tanaka N, Iga T (1992) Pharmacokinetic profile of glycerrhizin in healthy volunteers by a new high-performance liquid chromatographic method. J Pharm Sci 81(10):1042–1046
12. Vaya J, Belinky PA, Aviram M (1997) Antioxidant constituents from licorice roots: isolation, structure elucidation and antioxidative capacity toward LDL oxidation. Free Radic Biol Med 23:302–313
13. Tamir S, Eizenberg M, Somjen D, Izrael S, Vaya J (2001) Estrogen like activity of glabrene and other constituents isolated from licorice root. J Steroid Biochem Mol Biol 78:291–298
14. Hikino H, Wagner H, Farnsworth NR (eds) (1985) Recent research on oriental medicinal plants. Economic and medicinal plant research, vol 1. Academic, London, pp 53–85
15. Kamei J, Nakamura R, Ichiki H, Kubo M (2003) Anti-tussive principles of *Glycyrrhiza radix*, a main component of Kampo preparations Bakumondo-to. E J Pharm 69:159–163
16. Sharma V, Agrawal RC, Pandey S (2013) Phytochemical screening and determination of anti-bacterial and anti-oxidant potential of *Glycyrrhiza glabra* root extracts. J Environ Res Dev 7 (4A):1552–1558
17. Haraguchi H, Tanimato K, Tamura Y, Kinoshita T (1998) Antioxidative and Superoxide scavenging activities of retrochalcones in *Glycyrrhiza inflate*. Phytochemistry 48:125–129
18. Mitscher LA, Park YH, Clark D, Beal JL (1980) Antimicrobial agents from higher plants: antimicrobial isoflavanoids and related substances from *Glycyrrhiza glabra* L. var typical. J Nat Prod 43:259–262
19. Saeedi M, Morteza-Semnani K, Ghoreishi MR (2003) The treatment of atopic dermatitis with licorice gel. J Dermatol Treat 14:153–157

20. Indian Herbal Pharmacopoeia (1998) Indian drug manufacturer's association and Regional Research Laboratory, vol 1. Regional Research Laboratory, Jammu Tawi/Mumbai, pp 89–97
21. Mauricio I, Francischett B, Monterio RQ, Guimaraeas JA (1997) Identification of glycyrrhizin as thrombin inhibitor. Biochem Biophys Res Commun 235:259–263
22. Mendes-Silva W, Assafim M, Ruta B, Monteiro RQ, Guimaraes JA, Zingali RB (2003) Antithrombotic effect of glycyrrhizin, a plant-derived thrombin inhibitor. Thromb Res 112:93–98
23. Dhingra D, Parle M, Kulkarni SK (2004) Memory enhancing activity of *Glycyrrhiza glabra* in mice. J Ethnopharmacol 91(2–3):361–365
24. De Clercq E (2000) Current lead natural products for the chemotherapy of human immunodeficiency virus (HIV) infection. Med Res Rev 20:323–349
25. Badam L (1997) In vitro antiviral activity of indigenous glycyrrhizin, licorice and glycyrrhizic acid (Sigma) on Japanese encephalitis virus. J Commun Dis 29:91–99
26. Badam L (1994) In vitro studies on the effect of glycyrrhizin from *Glycyrrhiza glabra* on some RNA and DNA viruses. Indian J Pharmacol 26:194–199
27. Biondi DM, Rocco C, Ruberto G (2003) New dihydrostilbene derivatives from the leaves of *Glycycrrhiza glabra* and evaluation of their anti-oxidant activity. J Nat Prod 66:477–480
28. Masoomeh MJ, Kiarash G (2007) In vitro susceptibility of Helicobacter pylori to licorice extract. Iran J Pharm Res 6:69–72
29. Adel M, Alousi LA, Salem HA (2005) Licorice: a possible anti-inflammatory and anti-ulcer drug. AAPS Pharm Sci Tech 6:74–82
30. Sheela ML, Ramakrishna MK, Salimath BP (2006) Angiogenic and proliferative effects of the cytokine VEGF in Ehrlich ascites tumor cells is inhibited by *Glycyrrhiza glabra*. Int Immunopharmacol 6:494–498
31. Salvi M, Fiore C, Armanini D, Toninello A (2003) Glycyrrhetinic acid induced permeability transition in rat liver mitochondria. Biochem Pharmacol 66:2375–2379
32. Fiore C, Salvi M, Palermo M, Sinigagliab G, Armaninia D, Toninello A (2004) On the mechanism of mitochondrial permeability transition induction by glycyrrhetinic acid. Biochim Biophys Acta 1658:195–201
33. Watanabe M, Hayakawa S, Isemura M, Kumazawa S, Nakayama T, Mori C, Kawakami T (2002) Identification of licocoumarone as an apoptosis inducing component in licorice. Biol Pharm Bull 25:1388–1390
34. Sharma V, Agrawal RC, Shrivastava VK (2014) Assessment of median lethal dose and antimutagenic effects of *Glycyrrhiza glabra* root extract against chemically induced micronucleus formation in *swiss albino* mice. Int J Basic Clin Pharmacol 3:292–297
35. Sharma V, Agrawal RC (2015) Evaluation of Anticlastogenic effects of *Glycyrrhiza glabra* root extract against cyclophosphamide induced chromosomal aberration in *swiss albino* Mice. J Appl Pharm Sci 5(6):127–132
36. Hsiang CY, Lai IL, Chao DC, Ho TY (2002) Differential regulation of activator protein-1 activity by glycyrrhizin. Life Sci 70:1643–1656
37. Yoon G, Jung YD, Cheon SH (2005) Cytotoxic allyl retrochalcone from the roots of *Glycyrrhiza inflate*. Chem Pharm Bull 53:694–695
38. Acharaya SK, Dasarathy S, Tandon A, Joshi YK, Tandon BN (1993) Preliminary open trial on interferon stimulator (SNMC) derived from *Glycyrrhiza glabra* in the treatment of subacute hepatic failure. Indian J Med Res 98:69–74
39. Sato H, Goto W, Yamamura J, Kurokawa M, Kageyama S, Takahara T, Watanabe A, Shiraki K (1996) Therapeutic basis of glycyrrhizin on chronic hepatitis B. Antivir Res 30:171–177
40. Van Rossum TG, Vulto AG, De Man RA, Brouwer JT, Schalm SW (1998) Glycyrrhizin as a potential treatment of chronic hepatitis C. Aliment Pharmacol Ther 12:199–205
41. Jeong HG, You HJ, Park SJ, Moon AR, Chung YC, Kang SK, Chun HK (2002) Hepatoprotective effects of 18β-glycyrrhetinic acids on carbon tetrachloride-induced liver injury, inhibition of cytochrome P450 2E1 expression. Pharmacol Res 46(3):221–227

42. Chan HT, Chan C, Ho JW (2003) Inhibition of glycyrrhizic acid on alfatoxin B1-induced cytotoxicity of hepatoma cells. Toxicology 188:211–217
43. Sharma V, Agrawal RC (2014) In vivo antioxidant and hepatoprotective potential of *Glycyrrhiza glabra* extract on carbon tetra chloride (CCl_4) induced oxidative stress mediated hepatotoxicity. Int J Res Med Sci 2:314–320
44. Kimura M, Inoue H, Hirabayahi K, Natsume H, Ogihara M (2001) Glycyrrhizin and some analogues induce growth of primary cultured adult rat hepatocytes via epidermal growth factor receptors. Eur J Pharmacol 431:151–161
45. Takii H, Kometani T, Nishimura T, Nakae T, Okada S, Fushiki T (2000) Anti-diabetic effect of glycyrrhizin in genetically diabetic KK-Ay mice. Biol Pharm Bull 24:484–487
46. Wagner H, Jurcic K (2002) Immunological studies of Revitonil: a phytopharmaceutical containing *Echinacea purpurea* and *Glycyrrhiza glabra* root extract. Phytomedicine 9(5):390–397
47. Blatina LA (2003) Chemical modification of glycyrrhizic acid as a route to bioactive compounds for medicine. Curr Med Chem 10:155–171
48. Arora R, Chawla R, Marwah R, Arora P, Sharma RK, Kaushik V, Goel R, Kaur A, Silambarasan M, Tripathi RP, Bhardwaj JR (2011) Potential of complementary and alternative medicine in preventive management of novel H1N1 Flu (Swine Flu) pandemic: thwarting potential disasters in the bud. Evid Based Complement Alternat Med 2011:1–16

Steviol Glycosides: Natural Noncaloric Sweeteners

5

Indra Prakash and Venkata Sai Prakash Chaturvedula

Abstract

Steviol glycosides are a group of highly sweet diterpene glycosides isolated in only a few plant species of the Paraguayan shrub *Stevia rebaudiana*. Stevioside and rebaudioside A are the most abundant steviol glycosides which are responsible for its sweet taste and have commercial value all over the world as sugar substitute in foods, beverages, or medicines. More than 30 additional steviol glycosides have been described in the scientific literature to date implies a significant metabolic investment and poses questions regarding the benefits *S. rebaudiana* might gain from their accumulation. Due to the increase in demand for the major constituents in the leaves of *S. rebaudiana*, it is now grown commercially in a number of countries, particularly in China, Japan, Taiwan, Korea, Thailand, and Indonesia. It is a magical plant that offers sweetness with fewer calories and do not show any side effects after consumption on human health. They are thermostable even at higher temperatures making them suitable for use in cooked foods. Stevia cultivation and production would further help those who have to restrict carbohydrate intake in their diet, to enjoy the sweet taste with minimal or no-calories. During the past few decades, the nutritional and pharmacological benefits of these secondary metabolites have become increasingly apparent. While these properties are now widely recognized, many aspects related to their in vivo biochemistry and metabolism and their relationship to the overall plant physiology of *S. rebaudiana* are not yet understood.

I. Prakash (✉)
The Coca-Cola Company, Atlanta, GA, USA
e-mail: iprakash@coca-cola.com

V.S.P. Chaturvedula
Wisdom Natural Brands, Gilbert, AZ, USA
e-mail: sprakash@wisdomnaturalbrands.com

J.-M. Mérillon, K.G. Ramawat (eds.), *Sweeteners*, Reference Series in Phytochemistry,
https://doi.org/10.1007/978-3-319-27027-2_9

Keywords
Stevia rebaudiana Bertoni • Diterpene Glycosides • History • Chemistry • Stability • Sensory • Metabolism • Analytical methods • Regulatory

Contents

1 Introduction ... 102
  1.1 Stevia Leaf Extract and Steviol Glycosides ... 102
2 History/Progression of Steviol Glycoside Approval ... 103
  2.1 Molecules Covered by Current Regulation and Ninety Five Percent Purity ... 103
3 Chemistry and Grouping of Steviol Glycosides ... 103
4 Production ... 107
5 Stability of Steviol Glycosides ... 108
6 Sensory Properties ... 113
  6.1 Sweetness Potency ... 113
  6.2 Flavor Profile ... 115
  6.3 Sweetness Temporal Profile ... 115
7 Food Applications ... 115
8 Blending ... 117
9 Metabolism ... 118
10 Analytical Methods ... 120
11 Regulatory Status ... 123
12 Conclusion ... 125
References ... 125

1 Introduction

1.1 Stevia Leaf Extract and Steviol Glycosides

Stevia rebaudiana (Bertoni) Bertoni is a perennial shrub in the sunflower family of the Asteraceae (Compositae) found originally by the Guarani natives of Paraguay growing along the edges of the rainforest. It is often referred to as "The sweet herb of Paraguay" [1–3]. Dr. Moises Santiago Bertoni discovered this plant in 1888 and the plant was scientifically named as *Stevia rebaudiana* after a Paraguayan chemist Dr. Rebaudi in 1905. The Guarani and Mato-Grosso Indians believe *S. rebaudiana*, also known as stevia, was first used by their ancestors more than 1,500 years ago as a sweetener in foods and beverages and medicinal benefits. The leaves of stevia are about 30 times sweeter than sugar, whereas the compounds called steviol glycosides isolated from the leaves of stevia are the potently sweet diterpenoid glycosides stevioside and rebaudioside A [4, 5] are 200–300 times sweeter than sucrose; these compounds are also known as stevia sweeteners. Further, reports indicated that the extract of stevia has been used for centuries to sweeten food and beverages in other part of the world like Japan and China. Due to the increase in demand for the major constituents in the leaves of stevia which are the potently sweet diterpenoid glycosides, it is now grown commercially in a number of countries, particularly in China, Japan, Taiwan, Korea, Thailand, and Indonesia.

2 History/Progression of Steviol Glycoside Approval

2.1 Molecules Covered by Current Regulation and Ninety Five Percent Purity

Crude extracts of stevia often sold as dietary supplements in some countries, but it is important to note that only purified stevia leaf extract has been evaluated and approved for use as an ingredient in foods and beverages by the leading regulatory agencies. Purified stevia leaf extract (also known as high-purity stevia) generally describes stevia that has 95% or greater steviol glycoside content. This purity specification was set as part of a thorough safety review by the Joint FAO/WHO Expert Committee on Food Additives (JECFA) in 2008 and is supported by several regulatory authorities including the US Food and Drug Administration (FDA) and the European Commission.

JECFA has finalized specifications for steviol glycosides (95% purity) including nine steviol glycosides, namely, stevioside; rebaudiosides A, B, C, D, and F; steviolbioside; rubusoside; and dulcoside A, on a dried basis (JECFA 2010) [6]. The European Food Safety Authority (EFSA) evaluated the safety of steviol glycosides as a food additive (sweetener) and expressed its opinion in 2010 which has been designated with E number. Steviol glycosides, E 960 is currently an authorized food additive in the European Union for use in several food categories and specifications; presently those specifications stipulate that steviol glycoside preparations contain not less than 95% of the ten named steviol glycosides: stevioside; rebaudiosides A, B, C, D, E, and F; steviolbioside; rubusoside; and dulcoside A, on a dried basis. The specifications further define the preparations as comprising mainly (at least 75%) of stevioside and/or rebaudioside A. The use of steviol glycosides as a food sweetener is regulated under the European Parliament and Council Regulation (EC) No 1333/2008 on food additives. Several other assessments were made for steviol glycosides as a sweetener by the Scientific Committee for Food (SCF in 1984–1999), the Joint FAO/WHO Expert Committee on Food Additives (JECFA in 2000–2010 time frame), and EFSA (2010–2015). JECFA and EFSA have established an Acceptable Daily Intake (ADI) for steviol glycosides as 4 mg/kg bw (body weight) per day, expressed as steviol equivalents. The difference between JECFA and EFSA is in the presence of an additional steviol glycoside rebaudioside E in EFSA. Further, High-Purity Rebaudioside M (minimum purity 95%) has been approved GRAS status by FDA in 2014. Structures of various steviol glycosides covered by JECFA and EFSA along with their structures, molecular description, and approximate sweetness potency compared to sugar are given in Fig. 1.

3 Chemistry and Grouping of Steviol Glycosides

The two major steviol glycosides isolated from the leaves of *S. rebaudiana* are stevioside and rebaudioside A, and their structures were fully determined between 1955 and 1965 [7, 8]. Besides the major sweetening compounds, stevia also contains

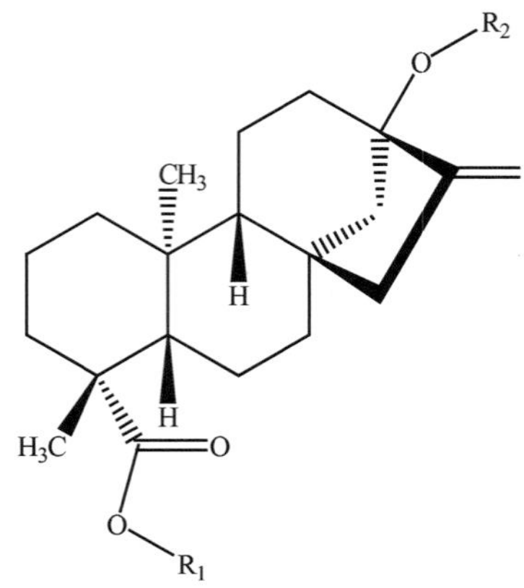

Sweetener	R_1	R_2	Molecular Formula	Molecular Weight	Sweetness Potency
Rebaudioside A	Glcβ1-	Glcβ(1-2)[Glcβ(1-3)]Glcβ1-	$C_{44}H_{70}O_{23}$	967.01	200
Rebaudioside B	H	Glcβ(1-2)[Glcβ(1-3)]Glcβ1-	$C_{38}H_{60}O_{18}$	804.88	150
Rebaudioside C	Glcβ1-	Rhaα(1-2)[Glcβ(1-3)]Glcβ1-	$C_{44}H_{70}O_{22}$	951.01	30
Rebaudioside D	Glcβ(1-2)Glcβ1-	Glcβ(1-2)[Glcβ(1-3)]Glcβ1-	$C_{50}H_{80}O_{28}$	1129.15	221
Rebaudioside E	Glcβ(1-2)Glcβ1-	Glcβ(1-2)Glcβ1-	$C_{44}H_{70}O_{23}$	967.01	174
Rebaudioside F	Glcβ1-	Xylβ(1-2)[Glcβ(1-3)]Glcβ1-	$C_{43}H_{68}O_{22}$	936.99	200
Stevioside	Glcβ1-	Glcβ(1-2)Glcβ1-	$C_{38}H_{60}O_{18}$	804.88	210
Steviolbioside	H	Glcβ(1-2)Glcβ1-	$C_{32}H_{50}O_{13}$	642.73	90
Rubusoside	Glcβ1-	Glcβ1-	$C_{32}H_{50}O_{13}$	642.73	114
Dulcoside A	Glcβ1-	Rhaα(1-2)Glcβ1-	$C_{38}H_{60}O_{17}$	788.87	30

glc = glucose rha = rhamnose xyl = xylose

Fig. 1 Molecular formulae, molecular weights, structures and potencies of the *Stevia* sweeteners

other minor diterpene glycosides like rebaudiosides B, C, E, steviobioside, dulcoside A, isosteviol, and dihydroisosteviol. The most common composition of the wild variety of *S. reabaudiana* is often reported as follows: stevioside (5–10% w/w), rebaudiosides A (2–5%), and C (1%), dulcoside A (0.5%), rebaudiosides D, E, and F (0.2%), rebaudioside M (0.05%), and steviolbioside (0.1%), on dry basis. Based on the genotype of the plant or due to the cultivation conditions, it was observed as a large difference in total steviol glycoside content as well as percentage steviol glycoside compositions. For example, the yield of stevioside from dried leaves varying from 5% to 22% whereas that of rebaudioside A content varies from 25% to 54%. A yield of 9.2% stevioside and of 61.6% rebaudioside A was described in the special species *S. rebaudiana* Morita, which was produced by selection and breeding of *S. rebaudiana* Bertoni [9]. Most of the glycosides isolated from this plant have the same diterpenoid backbone of steviol but differ in the content of saccharides. Many of these glycosides are natural sweeteners, which are not metabolized by humans and therefore do not provide energy in the diet, hence noncaloric. Though JECFA analytical method lists nine different steviol glycosides (Fig. 1), there were several more detected in recent years across the world present in trace quantities in dried leaf extracts originating from different cultivars. So far, over 40 steviol glycosides have been identified in *S. rebaudiana*, but the organoleptic properties of majority of these newly described steviol glycosides have not been reported due to lack of quantity for taste testing. Ohta et al. [9] found 10 new steviol glycosides in the special species *S. rebaudiana* Morita which was produced by selection and breeding of *S. rebaudiana* Bertoni and their structures were confirmed using liquid chromatography electrospray ionization mass spectrometry (LC-ESI-MS) in negative ion mode, ^{1}H- and ^{13}C NMR spectroscopy, and several chemical studies. Chaturvedula and Prakash isolated a series of new steviol glycosides from commercially available stevia leaf extracts belonging to the rebaudioside A and E family, anomers of dulcoside A, and rebaudioside C, with an α-glycosyl linkage, and six diterpene glycosides with modifications in the *ent*-kaurene body [10–26]. Structural confirmations were performed using a quadrupole time-of-flight (Q-TOF) mass spectrometer equipped with an electrospray ionization (ESI) source operating in positive ionization mode, 1D and 2D NMR spectroscopic data as well as chemical studies. Ibrahim et al. [27] reported the presence of a new diterpene glycoside isolated from a commercial extract of the leaves of *S. rebaudiana*, namely, rebaudioside KA that was shown to be 13-[(*O*-β-D-glucopyranosyl)oxy]ent-kaur-16-en- 19-oic acid 2-*O*-β-D-glucopyranosyl-β-D-glucopyranosyl ester. Zimmermann succeeded in confirming 5 of the 10 new steviol glycosides evaluated in leaves produced in Greece as well as in a commercially available stevia extract from China, certified to contain 95% steviol glycosides.

However, some evidence exists that rebaudioside B and steviolbioside are not native constituents of *S. rebaudiana* but are formed by partial hydrolysis during extraction [28], being thus artifacts of the extraction procedure. Hence, it is a debatable situation whether all of these new compounds, mainly those detected in purified leaf extracts, are genuine or artifacts due to purification procedures.

Majority of the steviol glycosides isolated so far are having mainly β-D-glucosyl moieties connected to the C-13 and C-19 position on steviol with varying attachments of sugars attached at individual positions. The structures of various steviol glycosides isolated along with their grouping, sugars attached, and corresponding references are provided below in Fig. 2:

In addition, three novel diterpene glycosides were isolated for the first time from the commercial extract of the leaves of *S. rebaudiana* and were identified as 13-[(2-*O*-β-D-glucopyranosyl-3-*O*-β-D-glucopyranosyl-β-Dglucopyranosyl) oxy] *ent*-kaur-15-en-19-oic acid, 13-[(2-*O*-β-D-glucopyranosyl-3-*O*-β-D-glucopyranosyl-β-D-glucopyranosyl) oxy]-16β-hydroxy-*ent*-kauran-19-oic acid, and 13-methyl-16-oxo-17-nor-*ent*-kauran-19-oic acid-β-D-glucopyranosyl ester on the basis of extensive 2D NMR and MS spectroscopic data as well as chemical studies [26].

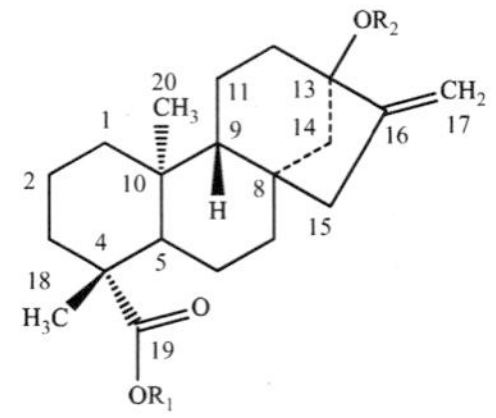

S. No	Common Name	R1	R2	Reference
1) Steviol + Glucose				
1.1	Steviolmonoside	H	Glcβ1-	[9]
1.2	Steviol-19-O-β-D-glucoside	Glcβ1-	H	[29]
1.3	Rubusoside	Glcβ1-	Glcβ1-	[9]
1.4	Steviolbioside	H	Glcβ(1-2)Glcβ1-	[30]
1.5	Stevioside	Glcβ1-	Glcβ(1-2)Glcβ1-	[2]
1.6	Rebaudioside KA	Glcβ(1-2)Glcβ1-	Glcβ1-	[27]
1.7	Rebaudioside B	H	Glcβ(1-2)[Glcβ(1-3)]Glcβ1-	[30]
1.8	Rebaudioside G	Glcβ1-	Glcβ(1-3)Glcβ1-	[9]
1.9		Glcβ(1-3)Glcβ1-	Glcβ1-	[31]
1.10	Rebaudioside E	Glcβ(1-2)Glcβ1-	Glcβ(1-2)Glcβ1-	[23]
1.11	Rebaudioside A	Glcβ1-	Glcβ(1-2)[Glcβ(1-3)]Glcβ1-	[30]
1.12		Glcβ1-	Glcβ(1-6)Glcβ(1-2)Glcβ1-	[11]
1.13	Rebaudioside D	Glcβ(1-2)Glcβ1-	Glcβ(1-2)[Glcβ(1-3)]Glcβ1-	[32]
1.14	Rebaudioside I	Glcβ(1-3)Glcβ1-	Glcβ(1-2)[Glcβ(1-3)]Glcβ1-	[9]
1.15	Rebaudioside L	Glcβ1-	Glcβ(1-6)Glcβ(1-2)[Glcβ(1-3)]Glcβ1-	[9]
1.16		Glcα(1-2)Glcα(1-4)Glcβ1-	Glcβ(1-2)Glcβ1-	[23]
1.17		Glcβ1-	Glcα(1-4)Glcβ(1-2)[Glcβ(1-3)]Glcβ1-	[11]
1.18		Glcβ1-	Glcα(1-3)Glcβ(1-2)[Glcβ(1-3)]Glcβ1-	[12]
1.19		Glcβ1-	Glcα(1-4)Glcβ(1-3)[Glcβ(1-2)]Glcβ1-	[13]
1.20		Glcα(1-2)Glcβ1-	Glcβ(1-2)[Glcβ(1-3)]Glcβ1-	[14]
1.21	Rebaudioside M	Glcβ(1-2)[Glcβ (1-3)]Glcβ1-	Glcβ(1-2)[Glcβ(1-3)]Glcβ1-	[9, 25]
1.22		Glcβ(1-2)[Glcβ(1-6)]Glcβ1-	Glcβ(1-2)Glcβ1-	[31]
1.23		Glcβ(1-2)Glcβ1-	[Glcβ(1-3)Glcβ(1-6)]Glcβ1-	[24]
2) Steviol + Rhamnose + Glucose				
2.1	Dulcoside A	Glcβ1-	Rhaα(1-2)Glcβ1-	[34]
2.2	Dulcoside B	H	Rhaα(1-2)[Glcβ(1-3)]Glcβ1-	[9]
2.3	Rebaudioside C	Glcβ1-	Rhaα(1-2)[Glcβ(1-3)]Glcβ1-	[24]
2.4		Rhaα(1-2)Glcβ1-	Glcβ(1-3)Glcβ1-	[35]
2.5	Rebaudioside H	Glcβ1-	Glcβ(1-3)Rhaα(1-2)[Glcβ(1-3)]Glcβ1-	[34]
2.6	Rebaudioside K	Glcβ(1-2)Glcβ1-	Rhaα(1-2)[Glcβ(1-3)]Glcβ1-	[9, 18]
2.7	Rebaudioside J	Rhaα(1-2)Glcβ1-	Glcβ(1-2)[Glcβ(1-3)]Glcβ1-	[35]
2.8	Rebaudioside N	Rhaα(1-2)[Glcβ(1-3)]Glcβ1-	Glcβ(1-2)[Glcβ(1-3)]Glcβ1-	[9, 20]
2.9	Rebaudioside O	Glcβ(1-3)Rhaα(1-2)[Glcβ(1-3)]Glcβ1-	Glcβ(1-2)[Glcβ(1-3)]Glcβ1-	[21]
3) Steviol + Xylose + Glucose				
3.1		Glcβ1-	Xylβ(1-2)Glcβ1-	[15]
3.2	Rebaudioside F	Glcβ1-	Xylβ(1-2)[Glcβ(1-3)]Glcβ1-	[36]
3.3		Glcβ1-	Glcβ(1-2)[Xylβ(1-3)]Glcβ1-	[15]
3.4		Xylβ(1-6)Glcβ1-	Glcβ(1-2)Glcβ1-	[16]
3.5		Glcβ(1-2)[Glcβ (1-3)]Glcβ1-	Xylβ(1-2)[Glcβ(1-3)]Glcβ1-	[22]
4) Steviol + Fructose + Glucose				
4.1		Glcβ1-	Glcβ(1-2)[Fruβ(1-3)]Glcβ1-	[10]
5) Steviol + deoxyGlucose + Glucose				
5.1		Glcβ1-	6-deoxyGlcβ(1-2)Glcβ1-	[16]
5.2		Glcβ1-	6-deoxyGlcβ(1-2)[Glcβ(1-3)]Glcβ1-	[17]
5.3		6-deoxyGlcβ1-	Glcβ(1-2)[Glcβ(1-3)]Glcβ1-	[19]

Fig. 2 Structures of various identified Steviol glycosides grouped on sugars attached to steviol backbone

4 Production

Stevia rebaudiana (Bertoni) Bertoni is a member of a New World genus of 150–300 species that belong to the tribe Eupatorieae of the family Asteraceace (sunflower family). The plant may reach a height of 80 cm when fully grown and is native to Paraguay in the Department of Amambay on the border of Paraguay with Brazil. Stevia plants require long daylight and water and can be grown all around the world, although most commercial stevia agriculture is done in mainland China. Steviol glycosides from dried leaves of *S. rebaudiana* are usually obtained after extraction with water or water-organic solvent mixture or organic solvent, precipitation of high molecular weight substances (sometimes combined with a defatting step), decolorization, purification, concentration, and drying. Pressurized hot water and microwave-assisted water extractions were observed to have higher or comparable efficiencies than heating under reflux. Ultrasonically assisted extraction is said to increase the yield by a factor of 1.5 at a lower temperature (68 °C) and to shorter extraction time (32 min) as compared to classical extraction. When compared different extraction techniques, it was found that conventional cold extraction was performed at 25 °C for 12 h, ultrasound extraction at 35 ± 5 °C for 30 min, and using microwave-assisted extraction, the extraction time could be reduced to 1 min at 50 °C. Methods using chloroform and ethanol for extraction or even supercritical fluid extraction (SFE) have also been published. The extraction step is sometimes followed by solid phase extraction (SPE), mostly with C18 cartridges, to remove disturbing matrixes. Stevioside and rebaudioside A are obtained from *S. rebaudiana* leaves in two stages. First step involved the extraction of steviol glycosides from the leaves with hot water or alcohol/water and the extract dried. In the second stage, further purification via alcohol/water crystallization yields very pure stevioside and/or rebaudioside A. In modern processes, hot-water extraction of stevia leaves gives a "primary extract" from which other plant components are then removed by flocculation. The cleared solution is passed through adsorption resins to concentrate the steviol glycosides, which are then eluted with alcohol. The dried eluate, comprising mixed steviol glycosides, may be stored and transported in this form before final purification. In this last step, the dried eluate is redissolved in a lower alcohol (pure or an aqueous solution) and recrystallized. Refluxing the dried solid in a methanol solution followed by cooling enables the isolation of stevioside. Rebaudioside A is obtained either from the filtrate after stevioside is removed or by directly crystallizing the solid with alcohol or aqueous alcohol. Traditionally, methanol has been used as the solvent, but ethanol has the advantage of selectively increasing the yield of rebaudioside A. Finally, the crystallized product is filtered and dried.

Refining methods of stevioside into solvent partition extraction mainly methanol or water extraction and solvent partition extraction, incorporating mainly in situ precipitation with calcium hydroxide–carbon dioxide to remove impurities, similar to the purification process in the sugar industry. They also reported different methods of purification, such as adsorption, chromatography, ion-exchange, plasmid gel, or adsorption by activated carbon. JECFA also cited that the typical manufacture starts with extracting leaves with hot water, and the aqueous extract is then passed through an adsorption resin to trap and concentrate the component steviol glycosides. The resin is washed with methanol to release the glycosides, and the product is recrystallized with methanol. Ion exchange resins may be used in the purification process. The final product is commonly spray dried. There are several extraction patents for the isolation of steviol glycosides. Kinghorn et al. [5] has categorized the extraction patents into those based on solvent, solvent plus a decolorizing agent, adsorption and column chromatography, ion exchange resin, and selective precipitation of individual glycosides. In recent patents, methods such as ultrafiltration, metallic ions, supercritical fluid extraction with CO_2, and extract clarification with zeolite have been employed.

5 Stability of Steviol Glycosides

In order to be an effective sweetener in foods, beverages, and functional foods, stability of steviol glycosides in different matrices is one of the most important characters. Under controlled humidity conditions, dry powder of stevioside is stable for 2 years whereas rebaudioside A is stable for 3 years. Stevioside showed stability when incubated at a high temperature (120 °C) for one hour, although it degraded at temperatures above 140 °C, whereas rebaudioside A and D demonstrated higher stability at elevated temperatures to stevioside. At 200 °C, stevioside showed completed degradation, 63% and 32% rebaudioside A and rebaudioside D remain unchanged, respectively. Several investigations have reported in the literature about the stability of stevioside, rebaudiosides A, and D, and steviol glycoside mixture under room temperature, neutral pH, and sunlight conditions. These sweeteners can be used under pH conditions ranging from 2 to 10 and at temperatures of up to 120 °C.

The stability of a stevioside based sweetener in solutions formulated with different organic acids (citric, tartaric and phosphoric acid) and vitamins (thiamine, riboflavin, niacin and pyridoxine) under different storage conditions was evaluated and its application as a sweetener in coffee and tea was also assessed. An aqueous solution of the stevioside remained stable over a pH range of 2–10 at 80 °C. Stevioside content did not exhibit significant decomposition when incubated for up to 4 h with water-soluble vitamins at 80 °C. When subjected to extremely acid conditions, however, a significant decrease in stevioside concentration was observed, whereas ascorbic acid had a protective effect against the degradation of the stevioside. Moreover, both the major steviol glycosides of *S. rebaudiana* stevioside and rebaudioside A were slightly less stable in phosphoric acid compared to citric acid.

The major reaction pathways during degradation of steviol glycosides could be: 1. Isomerization of the C-16 olefin double bond to form the corresponding C-15 isomer (exocyclic to endo cyclic), 2. Hydration of the C-16 olefin to form an alcohol, and 3. Hydrolysis of the glycosidic ester at C-19 position to form a carboxylic acid (Fig. 3).

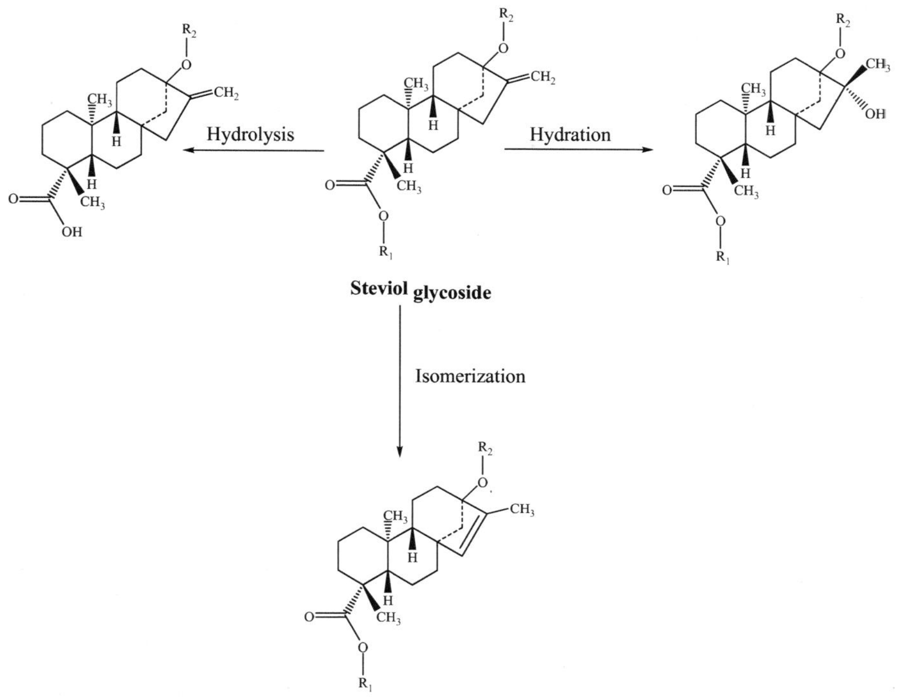

Fig. 3 Pathways of degradation of steviol glycosides

Photostability of a mixture of steviol glycosides was studied under fluorescent light exposure at 25 °C in mock beverages at pH 3.8 using International Conference on Harmonization (ICH) technical requirements for 2 weeks yielded three minor compounds which were identified as 13-[(2-*O*-β-D-glucopyranosyl*O*-β-D-glucopyranosyl)oxy]-17-hydroxy-*ent*-kaur-15-en-19-oic acid β-D-glucopyranosyl ester (**a**), 13-[(2-*O*-α-L-rhamnopyranosyl-3-*O*-β-D-glucopyranosyl-β-D-glucopyranosyl)oxy] *ent*-kaur-16-en-19-oic acid-(2-*O*-β-D-glucopyranosyl-β-D-glucopyranosyl) ester (**b**), and 13-[(2-*O*-β-D-glucopyranosyl-3-*O*-β-D-glucopyranosyl-β-D-glucopyranosyl)oxy]-16-hydroxy-*ent*-kauran-19-oic acid β-D-glucopyranosyl ester (**c**) (Fig. 4). The mass balance was calculated against their controls and was found as 98.3% suggested that there was not any appreciable amount of undetected degradation products were formed under the conditions of the study [36].

Similarly, the photostability of rebaudioside A was studied under fluorescent light exposure at 25 °C in mock beverages at pH 3.8 using ICH technical requirements indicated that rebaudioside A did not undergo any major decomposition with fluorescent light exposure for 2 weeks. Identification of the degradation products furnished two minor compounds which were identified as 13-[(2-*O*-β-D-glucopyranosyl-3-*O*-β-D-glucopyranosyl-β-Dglucopyranosyl) oxy]-17-hydroxy-*ent*-kaur-15-en-19-oic acid β-D-glucopyranosyl ester (**d**) and 13-[(2-*O*-β-Dglucopyranosyl-3-*O*-β-D-glucopyranosyl-β-D-glucopyranosyl)oxy]-16β-hydroxy *ent*-kauran-19-oic acid

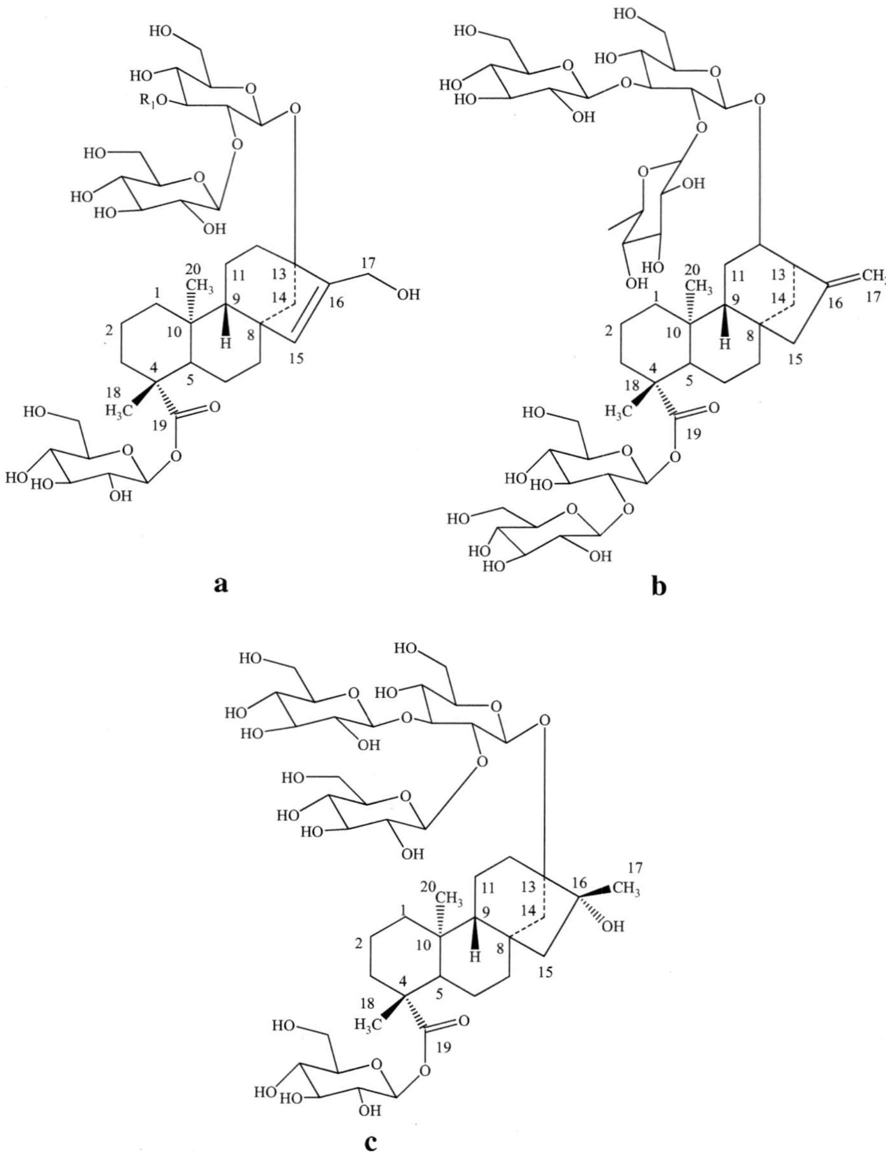

Fig. 4 Degradation compounds of steviol glycosides under fluorescent light exposure

β- D-glucopyranosyl ester (**e**) (Fig. 5). The mass balance for rebaudioside A was calculated against their control samples was found 99.1%, suggesting any appreciable amount of undetected degradation products were formed under the conditions of the study [37].

This suggested that steviol glycoside mixture and purified rebaudioside A are considered relatively stable using the conditions of this study when exposed to fluorescent light.

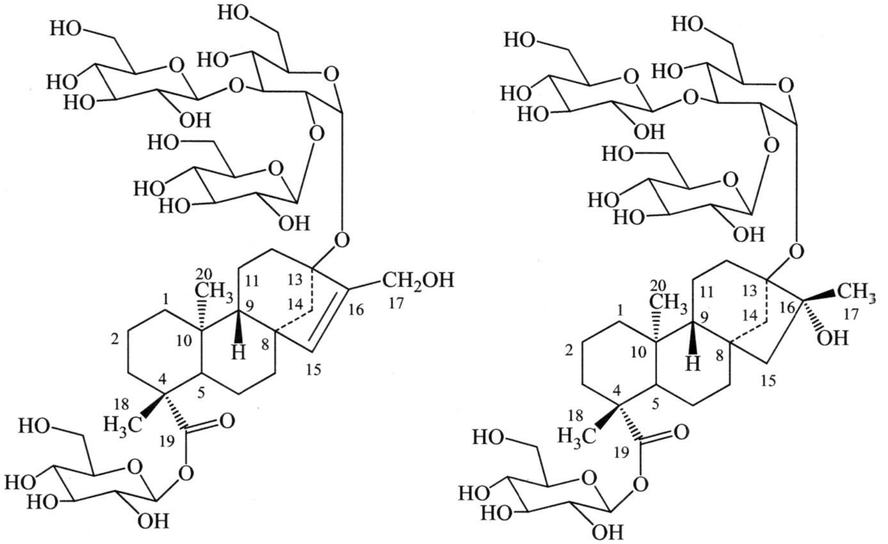

d e

Fig. 5 Degradation compounds of Rebaudioside A under fluorescent light exposure

In another report, stability of a mixture of steviol glycosides was studied under a typical pH range and various temperatures that simulated both relevant and extreme beverage storage conditions. Thus, steviol glycosides were evaluated in mock beverage solutions by simulating formulations used in commercial cola soft drinks (pH 2.8 and pH 3.2), lemon-lime soft drinks (pH 3.8), and root beer soft drinks (pH 4.2) but lacking corresponding flavor components. Experimental results indicated that steviol glycoside mixture did not undergo any major decomposition but yielded two minor compounds which were identified as 13-[(2-*O*-β-D-glucopyranosyl-β-D-glucopyranosyl) oxy] *ent*-kaur-15-en-19-oic acid β-D-glucopyranosyl (**f**) and 13-[(2-*O*-β-D-glucopyranosyl-β-D-glucopyranosyl)oxy]-16β-hydroxy-*ent*-kauran-19-oic acid β-D-glucopyranosyl ester (**g**) (Fig. 6). The stability of steviol glycosides in mock beverage solutions is pH, temperature, and time dependent; the rate and extent of degradation product formation is increased under acidic conditions (lower pH) and at higher temperatures with the majority of degradation product formation occurring after extended period of storage [38].

Likewise, stability of rebaudioside A was studied in mock beverage solutions by simulating formulations used in commercial cola soft drinks (pH 2.8 and pH 3.2), lemon–lime soft drinks (pH 3.8), and root beer soft drinks (pH 4.2) but lacking the flavor components. Samples were analyzed at scheduled intervals throughout the 26-week period for rebaudioside A, its known impurities, and degradation products. Experimental results indicated that rebaudioside A yielded six minor degradation compounds (**h-m**) (Fig. 7) whose structural characterization was performed on the basis of 1D (^{1}H, ^{13}C) and 2D (COSY, HSQC, HMBC) NMR, HRMS, MS/MS

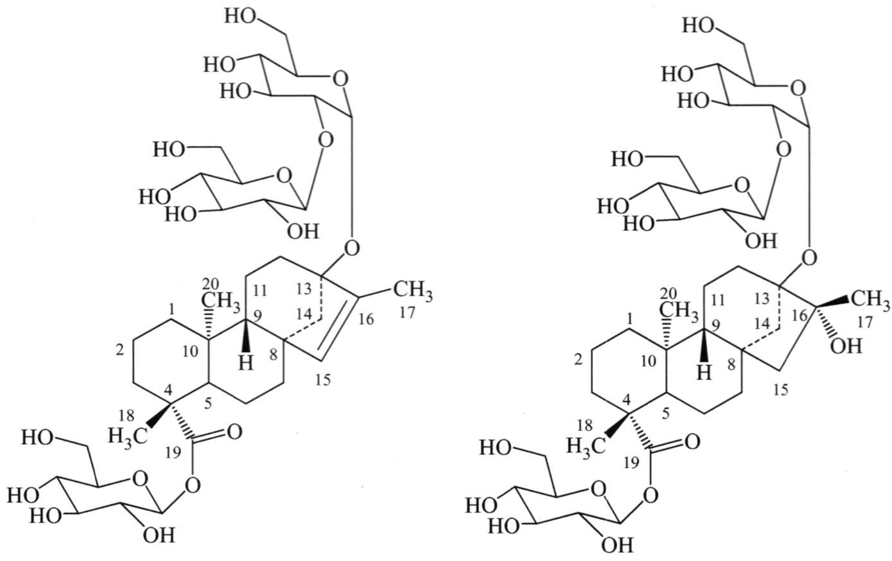

f g

Fig. 6 Degradation compounds of steviol glycosides under acidic conditions

spectral data. Results indicated that the stability of rebaudioside A in mock beverage solutions is pH, temperature, and time dependent. The rate and extent of degradation product formation is increased under acidic conditions (lower pH) and at higher temperatures with the formation of degradation products mainly occurring after extended period of storage. But in each case, excellent mass balance was achieved at all conditions by the identification of known impurities and degradation products suggested that rebaudioside A is considered stable [39].

Degradation of rebaudioside M, a minor sweet component of *S. rebaudiana* Bertoni, under simulated extreme pH and temperature conditions has been studied. Thus, rebaudioside M was treated with 0.1 M phosphoric acid solution (pH 2.0) and 80 °C temperature for 24 h, and experimental results indicated that rebaudioside M under low pH and higher temperature yielded three minor degradation compounds (**n-p**)(Fig. 8), whose structural characterization was performed on the basis of 1D (1H-, 13C-) & 2D (COSY, HSQC, HMBC) NMR, HRMS, MS/MS spectral data as well as enzymatic and acid hydrolysis studies [40].

Moreover, the stability of steviol glycosides in diverse food products semi skimmed milk, soy drinks, fermented milk drinks, ice cream, yogurt (made with both full cream and skimmed milk), dry biscuits, and jam was assessed. Storage conditions were those recommended for each type of food (−18 °C for ice cream, 6 °C for yogurt and 20 °C for skimmed and fermented milk). Stevioside recovery was between 96% and 103%, thus demonstrating that the steviol glycosides had not decomposed in any of the samples tested [41].

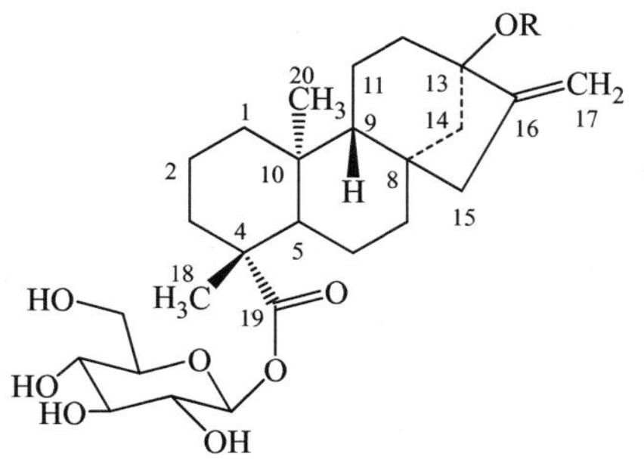

h; R = Glcβ1-3Glcβ1-
i; R = H

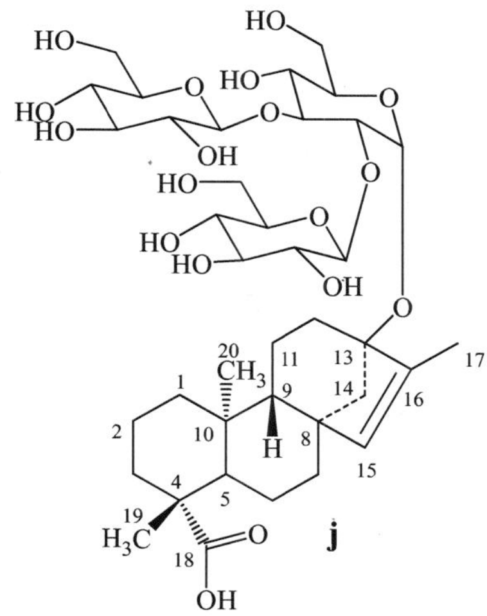

j

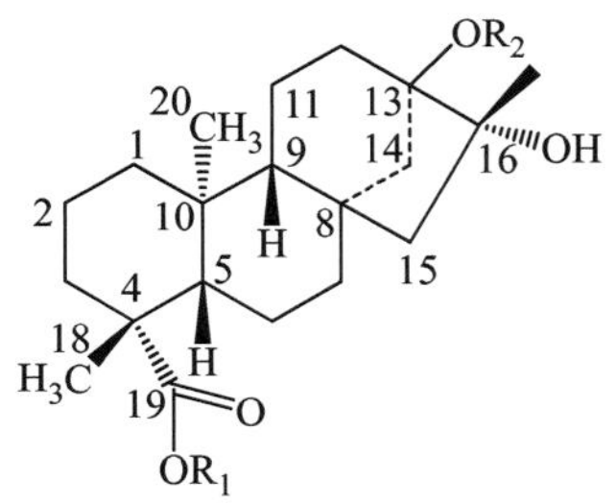

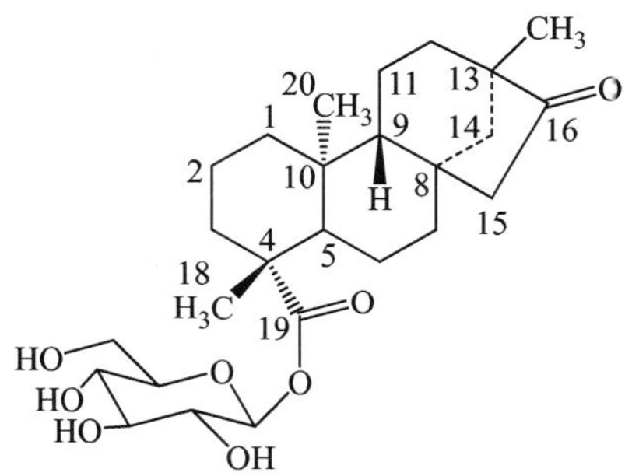

m

Compound	R_1	R_2
k	H	Glcβ1-2(Glcβ1-3)Glcβ1-
l	Glcβ1	Glcβ1-2(Glcβ1-3)Glcβ1-

Fig. 7 Degradation compounds of rebaudioside A under acidic conditions

6 Sensory Properties

6.1 Sweetness Potency

In the literature, stevioside, rebaudioside A, reabudioside D, and rebaudioside M are often reported as 150–250, 200–300, 200–350, and 200–450 times sweeter than sucrose, respectively. However, sweetness potency is strongly dependent on sucrose equivalency (SE) level for all high-potency sweeteners (HPS). Therefore, it is important to state the SE level at which sweetness potency has been determined.

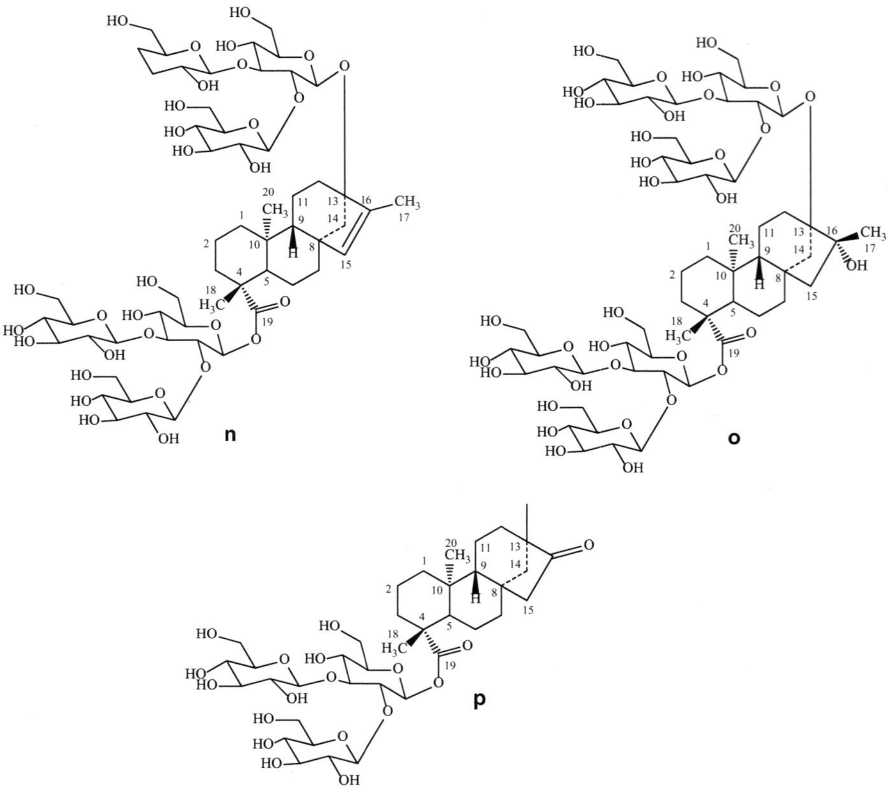

Fig. 8 Degradation compounds of rebaudioside M under acidic conditions

Sweetness potency is also system dependent and, therefore, it is important to also define the medium (e.g., water or phosphoric acid at pH 2.5). For comparison of different HPS, the most common medium is water (in the scientific literature where the medium is not specified, it is assumed to be water). While the choice of test medium is important, SE level is the major determinant of sweetness potency. There is, however, no industry-wide agreement on a common SE at which to report sweetness potency.

Typical use-levels of HPS in beverages are generally in the range of 4–10% SE. Consequently, 6% SE represents a reasonable average value at which to compare the sweetness potency of HPS in a plain water vehicle. At 6% SE, the sweetness potency of rebaudioside A is 200 times [28], whereas rebaudioside M is 310 times [25] to that of sucrose. This is similar to the sweetness potency of aspartame, a HPS that is widely marketed and approved for addition to numerous foods and beverages in many countries, which is 180 times as potent at 6% SE. Although the concentration-response functions for rebaudioside A ($R = 8.2C/(194 + C)$), rebaudioside M ($R = 14.2 \times C/(265 + C)$), and aspartame ($R = 25.5C/(1{,}160 + C)$) differ slightly in water, the three sweeteners are similar in sweetness over the range of SE levels. Rebaudioside M can be used as a single sweetener or blends to make zero calorie beverages.

6.2 Flavor Profile

As with most high-potency sweeteners, stevioside, rebaudioside A, rebaudioside D, and rebaudioside M exhibit clean sweetness at low SE levels but have other negative taste attributes (e.g., bitterness and black licorice) at higher SE levels. Stevioside exhibits much more bitterness than rebaudioside A, rebaudioside D, and rebaudioside M. Rebaudioside M also exhibits a clean sweet taste without any bitter or licorice aftertaste but is present at relatively low levels in currently available stevia plants [25, 28].

6.3 Sweetness Temporal Profile

Sweetness temporal profiles demonstrate changes in perception of sweetness over time. This property is key to the utility of a sweetener in foods and beverages and is complementary to its flavor profile. Every sweetener exhibits a characteristic Appearance Time (AT) and Extinction Time (ET). All high-potency sweeteners, in contrast to carbohydrate sweeteners, display prolonged ETs. This can be beneficial in some products such as chewing gum, where prolonged sweetness is desirable. Among all the steviol glycosides known today, rebaudioside M exhibits the quick onset (Fig. 9).

7 Food Applications

Based on the growing number of stevia-based products on the world market, such as drinks, table-top sweeteners, candy and other processed foods, personal hygiene products, and various delicacies, it is clear that the addition of steviol glycosides can increase the palatability and enjoyment of food by improving flavor and smell [42, 43]. The key factors affecting the stability of steviol glycosides are pH, humidity/moisture, and temperature, based on known results from aspartame and neotame. Products comprised carbonated and still nonalcoholic beverages, table-top sweetener formulations, chewing gum, yogurt, and cake were packed, stored (mostly at 25 °C and 60% relative humidity), and evaluated at intervals using both chemical (HPLC) and sensory analyses. Sweetness was assessed using panels consisting of 35–50 persons. Samples were evaluated using a five-point scale of categories ranging from 5 (much too sweet) to 1 (not at all sweet). Samples were considered satisfactory if at least 80% of the panelists rated the sweetness in category 3 (just about right) or above. Key findings were by Prakash et al. [28] on soft drinks (100–600 ppm): rebaudioside A remained acceptably sweet throughout 26 weeks

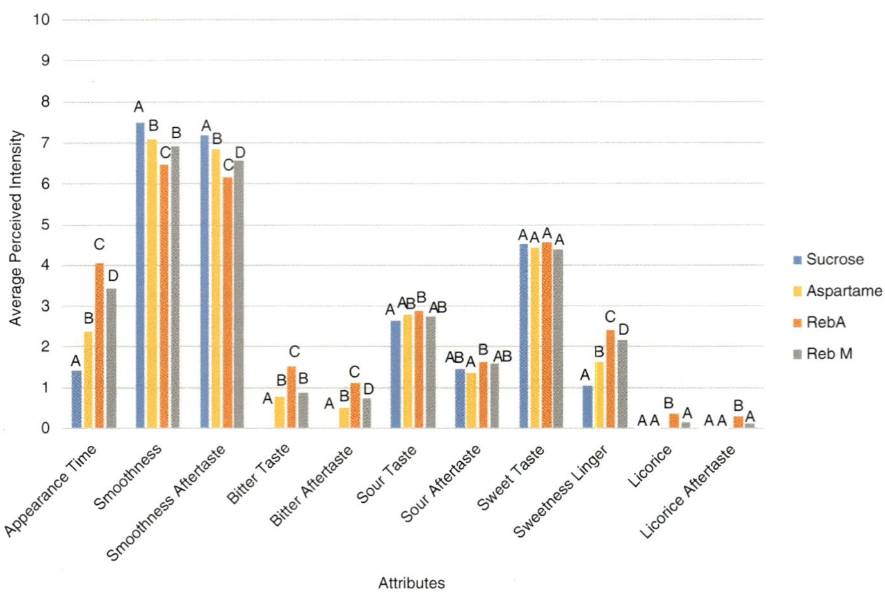

Fig. 9 Descriptive Analysis results of specific attributes for sucrose, aspartame, rebaudioside A, and rebaudioside M sweeteners at 400 ppm at room temperature in water

storage (cola and lemon-lime); for comparison, most soft drinks are consumed within 16 weeks of production.

1. Tabletop sweetener (200–2,000 ppm): rebaudioside A was tested in a number of formulations and sweetness remained stable for at least 52 weeks in all of them.
2. Chewing gum (300–10,000 ppm): rebaudioside A was considered stable and functional in chewing gum for 26 weeks.
3. Plain yogurt (100–1,000 ppm): No significant loss was measured during pasteurization (190 °F for 5 min) and fermentation. Rebaudioside A sweetness was stable throughout a 6-week storage period (40 °F).
4. White cake (200–1,000 ppm): No significant loss of sweetness was measured during the baking process (350 °F for 20–25 min) or during 5 days subsequent storage (25 °C and 60% RH).
5. Other products (100–1,000 ppm): rebaudioside A has been successfully formulated into cereals and cereal-based foods, dietary supplements, pharmaceuticals, edible gels, and confectionery products.

Specifically, stevia extracts have been used to sweeten low-calorie soft drinks, soy sauce, dried seafood, candy, ice cream, chewing gum, and yogurt in several countries, particularly in Japan, Korea, and Brazil. There have been several investigations undertaken aimed at evaluating the potential of stevia as a sweetener of specific products. Reports concluded that low-calorie yogurts could be manufactured using commercial sweeteners including stevia without modifying standard technological procedures [44]. A similar study, but applied to cake, assessed the rheological and microstructural properties and the final quality of cakes made by replacing sugar with stevioside and liquid sorbitol. The results showed that the addition of stevioside did not change the amylographic viscosity of wheat flour batter during heating and cooling, unlike sucrose, which increases this property which concluded that it is possible to replace sugar with stevioside using this cake recipe while maintaining the rheological properties of the final product.

The stevia glycosides which are natural noncaloric and low-energy sweeteners can be used as functional ingredients in nutrition, by diabetic and phenylketonouric patients as well as by obese people due to the ability to reduce the craving for sweet and fatty foods. Stevioside is very stable at high temperatures up to 200 °C and in a wide range of pH values, is not fermented, and does not support formation of plaque on the teeth. Accordingly, stevioside has a high degree of stability and can find application as natural stabilizer in a variety of dairy products, beverages, confectionery, and other foods. According to reported literature, healthy humans who consumed stevioside did not differ from controls in blood pressure and blood biochemical parameters, while no gastrointestinal uptake was detected for stevioside or steviol. Also, no genetic toxicity or mutagenicity was found for steviol glycosides. Additionally, rebaudioside A is stable in a wide variety of foods and beverages such as flavored ice-tea, juices, flavored milk, and "live" yogurt. Since these products are usually consumed cold, Fry et al. [45] found that rebaudioside A is significantly sweeter under these conditions. According to Brusick [46], human consumption of rebaudioside A does not pose a risk for genetic damages.

8 Blending

Blending of stevia sweeteners with other high-potency sweeteners may reduce bitterness and improve temporal profile and often result sweetness synergy. Lingering sweetness of steviol glycosides is able to be masked by addition of certain substances with floral and sweet note. Further, combinations with other high-potency sweeteners result in reduction of the sweetener cost as well as improvement in stability. The bitter after taste and lingering properties of steviol glycosides can be complemented to flavors of coffee, chocolate, spicy sauces, and to the stringency and bitterness of tea. Overall blending of sweeteners result in suppression or addition or synergy.

At higher concentrations of steviol glycosides, mainly rebaudiosde A and stevioside (>6% sugar equivalence) exhibit "off-taste" and this could be the reason these sweeteners are unlikely to be used as a sole sweetener. This limitation can be addressed by blending with other nutritive and nonnutritive sweeteners, especially for stevia sweetener rebaudioside A. Examples for nutritive sweeteners those go well with rebaudioside A include sorbitol and xylitol belongs to polyol family, and carbohydrates such as glycerol, fructose, glucose, sucrose, and HFCS. Similarly, the nonnutritive sweeteners suitable to be partnered with reabudioside A are Luo

Han Guo extract; monatin; brazzein; erythritol; sweet tasting amino acids such as glycine, alanine, and serine; and the artificial sweeteners like aspartame, acesulfame potassium, cyclamate, sucralose, and saccharin. These blending formulations may create a good tasting, improved temporal profile and reduced lingering.

9 Metabolism

Studies have been conducted in humans to document the metabolism of steviol glycosides. Steviol glycosides are not absorbed in the small intestine and once steviol glycosides reach the colon, gut bacteria hydrolyze steviol glycosides into steviol by snipping off their glucose units. Steviol is then absorbed via the portal vein and primarily metabolized by the liver forming steviol glucoronide and then excreted in the urine. Research has shown that there is no accumulation of stevia (or any by-product of stevia) in the body during metabolism. It is a result of this essentially poor absorption in the digestive tract which ultimately contributes to the fact that stevia has zero calories and does not raise blood glucose or insulin levels when digested.

Studies in rats [47] and other animal models, including chickens [3], hamsters [48], and pigs [49], indicated that stevioside is not readily absorbed from the GI tract. Available evidence from in vitro metabolism studies suggests that bacteria in the colon of rats and humans can transform various stevia glycosides into steviol [29]. Slow absorption of steviol was indicated by detection in the plasma of rats given oral stevioside [50]; however, did not detect plasma steviol following oral administration of steviosides to rats. In studies with human and rat liver extracts, Koyama et al. [47] demonstrated that steviol can be converted to various glucuronides. Excretion of metabolites of stevioside after oral doses has been shown in urine and feces in rats and hamsters [48]. Oral doses in pigs led to the detection of steviol glycoside metabolites in feces but not in urine [49]. Steviol was shown to be more readily transported with in vitro intestinal preparations than various steviosides [47, 49].

Renwick et al. [51] reviewed studies on microbial hydrolysis of steviol glycoside concluding that stevioside and reabudioside A are not absorbed directly and both are converted to steviol by gut microbiota in rats and in humans. This hydrolysis occurs more slowly for reabudioside A than for stevioside may be associated with the molecular structure. Studies have shown that steviol-16,17 epoxide is not a microbial metabolite. The authors concluded that there is a single hydrolysis product and that toxicological studies on stevioside are relevant to the safety assessment for reabudioside A. In a human study with 10 healthy subjects, Geuns et al. [52] measured blood, urine, and fecal metabolites in subjects that received 3 doses of 250 mg of purified stevioside (>97%) three times a day for 3 days. Urine was collected for 24 h on day 3, and blood and fecal samples were also taken on day 3. Free steviol was detected in feces but not in blood or urine. Steviol glucuronide was detected in blood, urine, and feces. Approximately 76% of the total steviol equivalents dosed were recovered in urine and feces. Based on these measurements, the authors concluded that there was complete conversion of stevioside in the colon to steviol, which was absorbed and rapidly converted to the glucuronide.

Metabolism of steviol glycosides in humans and rats is the same, but the pattern of metabolite excretion is different [53]. In humans, as in rats, both stevioside and rebaudioside A are metabolized by bacteria in the lower gut to steviol, which is absorbed into the portal blood system and transported to the liver where it is glucuronidated, the same process observed in the rat. However, in humans most steviol glucuronide appears in the plasma instead of the bile. Peak plasma concentrations of steviol glucuronide in humans occur approximately 8 and 12 h post-dosing for stevioside and rebaudioside A, respectively. Like the rat, peak plasma metabolite concentrations were lower after rebaudioside A ingestion than after stevioside ingestion. The half-life of steviol glucuronide in human plasma is approximately 14 h. Steviol glucuronide is the major excretion form of absorbed steviol in humans, and excretion occurs primarily via the urine rather than the feces. Only a small amount of steviol excretion occurs via the feces in humans. The metabolism of orally ingested steviol is interesting but not relevant for the understanding of steviol glycoside metabolism or safety. Humans do not ingest steviol, and orally ingested steviol kinetics are quite different compared to those of steviol glycosides.

For comparative purposes to determine whether toxicological studies conducted previously with stevioside would be applicable to the structurally related glycoside, rebaudioside A, toxicokinetics and metabolism of rebaudioside A, stevioside, and steviol were examined in rats [54]. Orally administered single doses of the radiolabeled compounds were extensively and rapidly absorbed with plasma concentration-time profiles following similar patterns for stevioside and rebaudioside A. Radioactivity from orally administered steviol glycosides slowly increased in plasma over a period of hours and was excreted primarily via the feces within 48 h of oral dosing. While the half-life of plasma radioactivity was 5 h in male rats and 10 h in female rats, other kinetic parameters were similar in males and females. Both steviol and steviol glucuronide were identified in plasma. Peak plasma levels of radioactivity were slightly lower for rebaudioside A compared to stevioside. As indicated above, this is not unexpected given the time required to remove an additional glucose moiety present in rebaudioside A. Less than 2% of the radioactivity was found in the urine and virtually no residual radioactivity was observed in any organ 96 h after dosing. The predominant compound observed in bile from cannulated rats was steviol glucuronide, while steviol was the predominant compound found in rat feces. Radioactivity in the feces accounted for 97–98% of the administered dose of both stevioside and rebaudioside A demonstrating that excretion of steviol glucuronide via bile is the major excretory route of steviol in the rat. The authors concluded that the overall data on toxicokinetics and metabolism indicate that rebaudioside A and stevioside are handled in an almost identical manner in the rat after oral dosing. In a randomized, double blind, cross-over study in healthy male subjects, Wheeler et al. [53] assessed the comparative pharmacokinetics of steviol and steviol glucuronide following single oral doses of rebaudioside A and stevioside. Steviol glucuronide was eliminated from the plasma, with similar t1/2 values of approximately 14 h for both compounds. Administration of rebaudioside A resulted in a significantly (approximately 22%) lower steviol glucuronide geometric mean Cmax value (1,472 ng/mL) than administration of stevioside (1,886 ng/mL). Steviol glucuronide was excreted primarily in the urine of the subjects during the 72-h collection period, accounting for 59% and 62% of the rebaudioside A and stevioside doses, respectively.

Pharmacokinetic analysis indicated that both rebaudioside A and stevioside were hydrolyzed to steviol in the gastrointestinal tract prior to absorption. The majority of circulatory steviol was in the form of steviol glucuronide indicating rapid first-pass conjugation prior to urinary excretion. Only a small amount of steviol was detected in urine (rebaudioside A: 0.04%; stevioside: 0.02%). The investigators concluded that rebaudioside A and stevioside underwent similar metabolic and elimination pathways in humans with steviol glucuronide excreted primarily in the urine and steviol in the feces. No safety concerns were noted as determined by reporting of adverse events, laboratory assessments of safety, or vital signs [53]. Another pharmacokinetic investigation was done as a toxicokinetic (TK) phase of a dietary study to determine the potential of rebaudioside A toxicity in rats at levels up to 2,000 mg/kg bw/day. Rebaudioside A and total steviol were detected in peripheral blood of rats during daily administration of 2,000 mg/kg bw/day of rebaudioside A at extremely low levels, with mean plasma concentrations of approximately 0.6 and 12 ug/mL, respectively. Estimates of absorbed dose for rebaudioside A and total steviol were approximately 0.02% and 0.06%, respectively, based on the amounts measured in urine collected over 24 h in comparison to daily administered dietary dose to rats. Mean fecal rebaudioside A and measured hydrolysis products expressed as total rebaudioside A equivalents compared to daily administered dose results in an estimate of percent of dose recovered is 84%.

10 Analytical Methods

Early analysis of sweet-tasting steviol glycosides involved Thin Layer Chromatography (TLC) coupled with densitometry, while other protocols like colorimetric detection has been used for the specific detection of isosteviol as its methyl ester. Several methods are known for determining the quantitative content of glycosides in plant material like gas chromatography or infrared spectroscopy; however, the simplest and most reliable method is HPLC, which has been used to determine the composition of *S. rebaudiana* growing in various geographical areas due to the easiness of sample preparation and the more satisfactory separation of stevioside, rebaudioside A, and other minor steviol glycosides, compared to methodologies based on TLC.

The initial HPLC analysis results were produced on amino-based or reversed-phase columns (C18) in combination with UV-detection. All JECFA methods before 2010 proposed the amino column as well. Amino-based stationary phases have a high selectivity for all steviol glycosides and provide good separation of the most abundant isomer pair rebaudioside B/stevioside and rebaudioside A/rebaudioside E. The separation order predominately depends on the glucose units attached to the *ent*-kaurene backbone, the higher is their retention time on the column. Accordingly, stevioside (3 glucose moieties) elutes before rebaudioside A (4 glucose moieties); both are well separated. Unfortunately, amino-based columns suffer from poor reproducibility and long equilibration times, and they cannot be used in combination with MS detection due to their strong bleeding. Moreover, they are not suitable for the determination of the aglycon steviol, which is poorly retained on these columns, and co-elution occurs with some nonspecific matrix peaks. An improved method has been developed by changing the HPLC conditions and including the use of an octadecylsilyl column instead of an amino-bonded column to enable the rapid and reliable determination of the nine steviol glycosides reported in JECFA by an isocratic HPLC-UV method. With the developed method, the nine steviol glycosides can be separately determined and identified using individual reference chemicals as standards, unlike the previous identification method, which was based on the relative retention times. In addition, the single stevioside quantification standard was replaced with both reference standards of stevioside and rebaudioside A. Importantly, the validation of the developed method was successful with the limits of quantification for the nine steviol glycosides were between 0.2% and 0.6%. The developed assay method for the nine steviol glycosides was proposed to JECFA and adopted as the revised assay method for the steviol glycosides specifications at its 73rd meeting in 2010 [6].

Eventually, highly specific HPLC-based analytical methods have evolved for the separation and quantitation of the different steviol glycosides with ever higher resolution and sensitivity, using a variety of different HPLC techniques like RP-HPLC, 110 2D-HPLC, 111 ultra-HPLC, 112 and 2D-ultra-HPLC48, using a wide variety of columns, namely, reversed-phase, amino, and HILIC with mobile phases AcCN and MeOH using either gradient or isocratic elution, under various detection systems (e.g., UV, DAD, and amperometry) [55]. As regulatory constraints require sensitive methods to analyze the sweet-tasting steviol glycosides in foods and beverages, a HILIC-MS/MS method was developed enabling the accurate and reliable quantitation of the major steviol glycosides stevioside, rebaudiosides A–F, steviolbioside, rubusoside, and dulcoside A by using the corresponding deuterated 16,17-dihydrosteviol glycosides as suitable internal standards. This quantitation not only enables the analysis of the individual steviol glycosides in foods and beverages but also can support the optimization of breeding and postharvest downstream processing of Stevia plants to produce preferentially sweet and least bitter tasting Stevia extracts. The determination of stevioside, rebaudioside A, and steviol was carefully pursued through different methods as indicated in the scientific literature, including enzymatic hydrolysis and chemical detection, high-performance TLC, over-pressured layer chromatography, capillary electrophoresis, high-speed counter-current chromatography, 2D-GC, quantitative NMR, near-infrared reflectance spectroscopy, and square-wave polarography. Recently, ambient ionization MS techniques such as desorption electrospray ionization (DESI) have been applied successfully for the direct analysis of steviol glycosides in *S. rebaudiana* leaves with minimal sample preparation [56]. HPLC technology and a near infrared (NIR) spectroscopy model was established to directly measure the stevioside glycosides (rebaudioside A and stevioside) content in the leaves of *S. rebaudiana* Bertoni. This model can be applied directly to measure the content of rebaudioside A and stevioside and resolved the problem of high cost and complex operation in using the current chemical laboratory methods. Though most HPLC-based methodologies use an external standard as reference for quantitation like rebaudioside A or stevioside, the major steviol glycosides from the plant with >99% purity; recently, an internal standard has been developed for steviol glycoside analysis, namely, the 19-*O*-β-D-galactopyranosyl ester of steviolmonoside. Use of an internal standard allows for the correction of losses due to sample cleanup and is independent of errors in injection volume or detector sensitivity. A qualitative LC-TOF method was also proposed to evaluate steviol glycosides together with a validated HPTLC procedure with densitometric detection and a NIR procedure for the quantification of steviol glycosides. Recently, a semi-quantitative determination of steviol glycosides was also performed by desorption electrospray ionization mass spectrometry. As for steviol quantification, a validated an RP-LC method with fluorometric detection after derivatization by a coumarin byproduct has been reported. Recently, a study has concluded that ultra-HPLC methods with electrospray ionization mass spectrometry (UHPLC-MS) can be used for the routine evaluations of steviol glycosides in crude extracts [57].

Mass spectrometry is one of the most sensitive detection methods for steviol glycosides. Frequently, these detectors operate in the electrospray ionization (ESI) negative ion mode 19, 24, 32, 33, 46 and are linked to HPLC. It has been stated that negative ion mode is 10 times more sensitive than positive ion mode. Using MS detection, poor resolution for some critical pairs of steviol glycosides can be acceptable because of the high selectivity of the MS detector. Mobile phases contain acetonitrile–water mixtures and additives such as ammonium formate or dichloromethane for ionization enhancement. In ESI MS/MS fragmentation, the steviol glycosides were readily confirmed through subsequent glycosidic losses of fragments of 162 Daltons. This makes it rather difficult to distinguish between isomers such as rubusoside/steviolbioside or stevioside/rebaudioside B, especially when LC resolution is not sufficient. Some authors confirmed that a distinction is possible when applying low, intermediate, and high -collision energies (20, 40, and 60 V, respectively) in MS detection. They were able to prove that the ester bond between the glucose moiety and the carboxyl group at C4 of the kaurene backbone fragments quite easily, even at low collision energies. The corresponding steviol glycosides (e.g., stevioside, rubusoside) could only be confirmed by their fragment ions. The bonds with and between the sugar chains at C19 are more stable, and the resulting steviol glycosides (i.e., steviolbioside, steviol monoside, rebaudioside B) have stable $[M - H] +$ ions even at higher voltage settings. Desorption electrospray ionization (DESI) mass spectrometry was found as a rapid, qualitative, and semi-quantitative method that does not require sample preparation for steviol glycosides estimation in stevia leaves [58].

Quantification by ^{1}H NMR spectroscopy is possible for the major components stevioside and rebaudioside A – C [47]. The solvent mixture pyridine-d5-DMSO-d6 (6:1) enables satisfactory separation of various steviol glycosides. Similar results were obtained after comparing the quantitative results with those obtained using the JECFA method. The advantage of this method is that NMR analysis does not require reference compounds and it is significantly faster than HPLC analysis. For quantification, the internal standard anthracene was used.

As regulatory constraints require sensitive methods to analyze the sweet-tasting steviol glycosides in foods and beverages, a HILIC-MS/MS method was developed enabling the accurate and reliable quantitation of the major steviol glycosides stevioside, rebaudiosides A – F, steviolbioside, rubusoside, and dulcoside A by using the corresponding deuterated 16,17-dihydrosteviol glycosides as suitable internal standards. This quantitation not only enables the analysis of the individual steviol glycosides in foods and beverages but also can support the optimization of breeding and postharvest downstream processing of stevia plants to produce preferentially sweet and least bitter tasting Stevia extracts.

11 Regulatory Status

Since 1995, the extract of the leaves of the plant *S. rebaudiana* with a mixture of steviol glycosides have been used as a dietary supplement in the US [3]. Based on the available information, no New Dietary Ingredient Notification for dietary supplement use of purified rebaudioside A has been made to the US-FDA. Since 1989 and prior to 2008, at least two GRAS petitions seeking authorization for the addition of stevioside or steviol glycosides to foods had been submitted to FDA. However, no authorizations had been issued by FDA in response to these filings, and subsequently these petitions were withdrawn. It appears that the previously available safety data including purity considerations for stevia, stevioside, or steviol glycosides were inadequate.

In December 2008, in response to Generally Recognized As Safe (GRAS) notifications submitted to the US Food and Drug Administration (FDA), the FDA stated it has no questions regarding the conclusion of expert panels that rebaudioside A is GRAS for use as a general purpose sweetener in foods and beverages, excluding meat and poultry. Rebaudioside A is a stevia sweetener isolated and purified from the leaves of the stevia plant. In June 2009, FDA stated it has no questions regarding the conclusion of an expert panel on the GRAS status of steviol glycoside extract with high rebaudioside A content for use as a tabletop sweetener. In 2010, the European Food Safety Authority (EFSA) assessed the safety of steviol glycosides from stevia and established an Acceptable Daily Intake (ADI) for their safe use. In November 2011, the European Commission authorized the use of steviol glycosides as a sweetener in foods and beverages [59]; it is also approved as a dietary supplement in the EU. Stevia and steviol glycosides have a long history of use in several countries, including Japan and Paraguay. In South America and in several countries in Asia, including China, Japan, and Korea, stevia derived-sweeteners are permitted as a food additive. The Food Standards Australia New Zealand (FSANZ) has completed its evaluation of an application for use of steviol glycosides in foods in 2008. FSANZ recommended that the Australia and New Zealand Food Regulation

Ministerial Council (Ministerial Council) amend the Australia New Zealand Food Standards Code to allow the use of steviol glycosides in food (FSANZ, 2008). In 2008, Switzerland's Federal Office for Public Health approved the use of stevia as a sweetener citing the favorable actions of JECFA. Subsequently, France published its approval for the food uses of rebaudioside A with a purity of 97% (AFSSA, 2009). Based on a review of the international regulation of *Stevia rebaudiana* and the clinical evidence for safety and efficacy, the Natural Health Products Directorate, Health Canada (2009), has adopted the following guidelines for the use of stevia and steviol glycosides in Natural Health Products (NHPs) on September 18, 2009. Stevia sweeteners are approved for use in many other countries, including Korea, Mexico, Taiwan, China, Russia, Australia, Argentina, New Zealand, Colombia, Peru, Uruguay, Brazil, Malaysia, and Switzerland. Since 2008, the US FDA has issued "no questions" letters in response to the multiple GRAS notifications filed on Reb A and steviol glycosides.

In 2008 and 2009, the Food and Agriculture Organization/World Health Organization's Joint Expert Committee on Food Additives (JECFA), a global panel of food ingredient safety experts, and the United States FDA stated the use of high-purity steviol glycosides (≥95%) is safe for human consumption, with an ADI expressed in steviol equivalents of up to 4 mg per kilogram of body weight per day. The European Food Safety Authority (EFSA) in 2010 assessed the safety of steviol glycosides from Stevia and established an ADI for their safe use. Daily Intake (ADI) of 4 mg/kg body weight is expressed as steviol equivalents [60]. The ADI is listed in units of mg per kg of body weight. The European Commission on 11th November 2011 allowed the usage of steviol glycosides as a food additive which will probably lead to wide-scale use in Europe. In 2011, the European Commission authorized the use of high-purity steviol glycosides (≥95%) in foods and beverages across the European Union. Globally, scientists have concluded that Stevia sweeteners are safe for people of all ages. The Dietary Supplement Health and Education Act (DSHEA) passed in the United States in 1994 were also approved steviol glycosides to be used as a functional ingredient in dietary supplements [61]. Recently the Food Safety and Standards Authority of India (FSSAI) has allowed the use of natural sweetener stevia (steviol glycoside) in selected products, including soft drinks, dairy-based beverages, and desserts.

Though not exhaustive, below are links to key multiple global regulatory agencies that have approved the use of high-purity stevia extracts:

- Agence Française de Sécurité Sanitaire des Aliments (AFSSA, or French Agency for Food Safety)
- Codex Alimentarius Commission (CAC)
- European Food Safety Authority (EFSA)
- European Commission
- Food Standards Australia New Zealand (FSANZ)
- Health Canada
- Joint FAO/WHO Expert Committee on Food Additives (JECFA)
- US Food and Drug Administration (FDA)

12 Conclusion

The sweet herb *S. rebaudiana* (Bertoni) has a valuable future and is extensively used in various areas of the world. Stevia, steviol glycosides, and their metabolites have commercial value in number of countries as sugar substitutes in foods, beverages, and medicines. Judging by the published literature, the use of steviol glycosides as sweeteners is an area that requires much further research, due to both their high appeal and commercial potential. This is very important, not only for manufacturers but also in general for various food applications. In this context, much research is still needed, not only to develop and optimize steviol glycoside extraction but also to improve the taste of products sweetened with these compounds. Thus, the continued evaluation of these ingredients as regards aspects such as the intensity, persistence of sweet taste, and absence of other residual flavors is necessary in order to meet the demands of today's consumers and ensure their acceptance, preference, and choice by the general public. Furthermore, and given that there is currently a sizeable and growing market for the commercialization of Stevia-containing products, optimization of production and processing should be undertaken concurrently in order to avoid limitations in the supply of steviol glycosides, an aspect that could restrict their extensive use in the demanding future that lies ahead. Finally, the patterns of use of stevia and its approval by the major international regulatory organizations, who have confirmed both the safety of this product for human consumption and its stability over time, point towards its development as an ubiquitous sugar alternative. It is expected that steviol glycosides will be used mainly in the manufacture of beverages, along with other traditional foods (such as dairy products, bread and cakes, confectionery, etc.), table-top sweeteners, functional food and beverages, and nutritional supplements, in addition to their use in personal care products (such as toothpaste) and as an active pharmaceutical ingredient or excipient. Steviol glycosides, with rebaudioside A in particular, are widely used commercially as a healthy, noncariogenic, zerocalorie alternative for sucrose. In addition, many studies with both animal models and human volunteers have shown clearly the beneficial pharmacological effects of stevioside (37) against type-II diabetes, hypertension, metabolic syndrome, and atherosclerosis. Besides steviol glycosides, the leaves of S. rebaudiana are also known for their important natural antioxidants such as flavonoids and various phenolics, tannins, essential oils, and other compounds. These substances have antioxidant, antimicrobial, immunostimulatory, and sweetening activities. According to recent research, stevia can be used as a novel functional ingredient in the food, feed, and medical industry, although further research is needed to be continued.

References

1. Brandle JE, Starrratt AN, Gijen M (1998) *Stevia rebaudiana*: its agricultural, biological and chemical properties. Can J Plant Sci 78:527–536
2. Bridel M, Lavielle R (1931) The sweet principle in Kaa-he-e (*Stevia rebaudiana* Bertoni). II. Hydrolysis of stevioside by enzymes. III. Steviol by enzymic hydrolysis and isosteviol by acid hydrolysis. Bull Soc Chim Biol 13:781–796

3. Geuns JMC (2003) Stevioside. Phytochemistry 64:913–921
4. Lewis WH (1992) Early uses of *Stevia rebaudiana* (Asteraceae) leaves as a sweetener in Paraguay. Econ Bot 46:336–337
5. Kinghorn AD (2002) Overview [of Stevia]. Med Aromat Plant Ind Prof 19:1–17
6. JEFCA (2010) Steviol glycosides. In: Combined compendium of food additive specifications [Online edition]. Food and Agriculture Organization of the United Nations (FAO), Joint FAO/-WHO Expert Committee on Food Additives (JECFA), Rome. Available from http://www.fao.org/ag/agn/jecfa-additives/specs/monograph10/additive-442-m10.pdf
7. Mosettig E, Nes WR (1955) Stevioside. II. The structure of the aglucon. J Org Chem 20:884–899
8. Mosettig E, Beglinger U, Dolder F, Lichiti H, Quitt P, Waters JA (1963) The absolute configuration of steviol and isosteviol. J Am Chem Soc 85:2305
9. Ohta M, Sasa S, Inoue A, Tamai T, Fujita I, Morita K, Matsuura F (2010) Characterization of novel steviol glycosides from leaves of *Stevia rebaudiana* Morita. J Appl Glycosci 57:199–209
10. Chaturvedula VSP, Rhea J, Milanowski D, Mocek U, Prakash I (2011) Two minor diterpene glycosides from the leaves of *Stevia rebaudiana*. Nat Prod Commun 6:175–178
11. Chaturvedula VSP, Prakash I (2011) Utilization of RP-HPLC fingerprinting analysis for the identification of diterpene glycosides from *Stevia rebaudiana*. Int J Res Phytochem Pharmacol 1:88–92
12. Chaturvedula VSP, Prakash I (2011) Diterpene glycosides from *Stevia rebaudiana*. J Med Plant Res 5:4838–4842
13. Chaturvedula VSP, Prakash I (2011) Structures of the novel diterpene glycosides from *Stevia rebaudiana*. Carbohydr Res 346:1057–1060
14. Chaturvedula VSP, Mani U, Prakash I (2011) Diterpene glycosides from *Stevia rebaudiana*. Molecules 16:3552–3562
15. Chaturvedula VSP, Prakash I (2011) Additional minor diterpene glycosides from *Stevia rebaudiana*. Nat Prod Commun 6:1059–1062
16. Chaturvedula VSP, Clos JF, Rhea J, Milanowski D, Mocek U, DuBois GE, Prakash I (2011) Minor diterpenoid glycosides from the leaves of *Stevia rebaudiana*. Phytochem Lett 4:209–212
17. Chaturvedula VSP, Mani U, Prakash I (2011) Structures of the novel α-glucosyl linked diterpene glycosides from *Stevia rebaudiana*. Carbohydr Res 346:2034–2038
18. Chaturvedula VSP, Prakash I (2011) A new diterpenoid glycoside from *Stevia rebaudiana*. Molecules 16:2937–2943
19. Chaturvedula VSP, Rhea J, Milanowski D, Mocek U, Prakash I (2012) Isolation and structure elucidation of two new minor diterpene glycosides from *Stevia rebaudiana*. Org Chem Curr Res 1:102
20. Chaturvedula VSP, Chen S, Yu O, Mao G (2013) Isolation, NMR spectral analysis and hydrolysis studies of rebaudioside N, a Minor Steviol glycoside of *Stevia rebaudiana* Bertoni. Food Nutr Sci 4:1004–1008
21. Chaturvedula VSP, Chen S, Yu O, Mao G (2013) Isolation, NMR spectral analysis and hydrolysis studies of a hepta Pyranosyl diterpene glycoside from *Stevia rebaudiana* Bertoni. Biomolecules 3:733–740
22. Chaturvedula VSP, Prakash I (2013) Additional minor diterpene glycosides from *Stevia rebaudiana* Bertoni. Molecules 18:13510–13519
23. Chaturvedula VSP (2014) Isolation and structural characterization of a new minor penta β-D-glucopyranosyl diterpene from *Stevia rebaudiana* Bertoni. Am J Plant Sci 5:3519–3525
24. Chaturvedula VSP, Meneni SR (2016) Isolation and structural characterization of a new minor Steviol glycoside of *Stevia rebaudiana*. IOSR J Pharm 6:1–5
25. Prakash I, Chaturvedula VSP, Markosyan A (2013) Isolation, characterization and sensory evaluation of a hexa β-D-glucopyranosyl diterpene from *Stevia rebaudiana*. Nat Prod Commun 11:1523–1526
26. Prakash I, Chaturvedula VSP (2013) Additional minor diterpene glycosides from *Stevia rebaudiana* Bertoni. Molecules 18:13510–13519

27. Ibrahim MA, Rodenburg DL, Alves K, Fronczek FR, McChesney JD, Wu C, Nettles BJ, Venkataraman SK, Jaksch F (2014) Minor diterpene glycosides from the leaves of *Stevia rebaudiana*. J Nat Prod 77:1231–1235
28. Prakash I, DuBois GE, King GA, San Miguel RI (2008) Sweetness enhancers, sweetness enhanced sweetener compositions, methods for their formulation, and uses. US 20080292765 A1
29. Gardana C, Simonettin P, Canzi E, Zanchi R, Pietta P (2003) Metabolism of stevioside and rebaudioside A from *Stevia rebaudiana* extracts by human microflora. J Agric Food Chem 51:6618–6622
30. Kohda H, Kasai R, Yamsaki K, Murakami K, Tanaka O (1976) New sweet diterpene glucosides from *Stevia rebaudiana*. Phytochemistry 15:981–983
31. Chaturvedula VSP, Zamora J (2014) Isolation and structural characterization of a new minor diterpene glycoside from *Stevia rebaudiana*. Nat Prod Commun 9:1677–1679
32. Sakamoto I, Yamasaki K, Tanaka O (1977) Application of ^{13}C NMR spectroscopy to the chemistry of natural glycosides: rebaudioside C, a new sweet diterpene glycoside of *Stevia rebaudiana*. Chem Pharm Bull 25:844–846
33. Kobayashi M, Horikawa S, Degrandi IH, Ueno J, Mitsuhashi H (1977) Dulcosides A and B, new diterpene glycosides from *Stevia rebaudiana*. Phytochemistry 16:1405–1408
34. Sakamoto I, Yamasaki K, Tanaka O (1977) Application of ^{13}C NMR spectroscopy to chemistry of plant glycosides: rebaudiosides D and E, new sweet diterpene glucosides of *Stevia rebaudiana* Bertoni. Chem Pharm Bull 25:3437–3439
35. Starratt AN, Kirby CW, Pocs R, Brandle JE (2002) Rebaudioside F, a diterpene glycoside from *Stevia rebaudiana*. Phytochemistry 59:367–370
36. Chaturvedula VSP, Clos JF, Prakash I (2011) Stability study of steviol glycosides in mock beverages using fluorescent light exposure under ICH guidelines. Int J Pharm Pharm Sci 3:316–323
37. Chaturvedula VSP, Clos JF, Prakash I (2012) Fluorescent light exposure of rebaudioside A in mock beverages under international conference on Harmonization (ICH) guidelines. Int J Chem 4:1–9
38. Prakash I, Clos JF, Chaturvedula VSP (2011) Stability of steviol glycosides in mock beverages under acidic conditions. Int J Pharm Pharm Sci 3(Suppl 5):421–425
39. Prakash I, Clos JF, Chaturvedula VSP (2012) Stability of rebaudioside A under acidic conditions and its degradation products. Food Res Int 48:65–75
40. Prakash I, Chaturvedula VSP, Markosyan A (2014) Structural characterization of the degradation products of a minor natural sweet diterpene glycoside rebaudioside M under acidic conditions. Int J Mol Sci 15:1014–1025
41. Jooken E, Amery R, Struyf T, Duquenne B, Geuns J, Meesschaert B (2012) Stability of steviol glycosides in several food matrices. J Agric Food Chem 42:10606–10612
42. Christaki E, Bonos E, Giannenas I, Karatzia MA, Paneri PF (2013) *Stevia rebaudiana* as a novel source of food additives. J Food Nutr Res 52:195–202, and references cited therein
43. Gonzalez C, Tapia M, Perez E, Pallet D, Dornier M (2013) Main properties of steviol glycosides and their potential in the food industry: a review. Fruits 69:127–141, and references cited therein
44. Gupta E, Purwar S, Sundaram S, Rai GK (2013) Nutritional and therapeutic values of *Stevia rebaudiana*: a review. J Med Plant Res 7:3343–3353
45. Fry JC, Yurttas N, Biermann KL (2011) Sweetness concentration-response behavior of rebiana at room and refrigerator temperatures. J Food Sci 76:S545–S548
46. Brusick DJ (2008) A critical review of the genetic toxicity of steviol and steviol glycosides. Food Chem Toxicol 46(7S):S83–S91
47. Koyama E, Kitazawa K, Ohori Y, Izawa O, Kakegawa K, Fujino A, Ui M (2003) In vitro metabolism of the glycosidic sweeteners, stevia mixture and enzymatically modified stevia in human intestinal microflora. Food Chem Toxicol 41:359–374
48. Hutapea AM, Tolskulkao C, Wilairat P, Buddhasukh D (1999) High-performance liquid chromatographic separation and quantitation of stevioside and its metabolites. J Liq Chromatogr Relat Technol 22:1161–1170

49. Geuns JMC, Malheiros RD, Moraes VMB, Decuypere EMP, Compernolle F, Buyse JG (2003) Metabolism of stevioside by chickens. J Agric Food Chem 51:1095–1101
50. Wang LZ, Goh BC, Fan L, Lee HS (2004) Sensitive high-performance liquid chromatography/mass spectrometry method for determination *of* steviol in rat plasma. Rapid Commun Mass Spectrom 18:83–86
51. Renwick AG, Tarka SM (2008) Microbial hydrolysis of steviol glycosides. Food Chem Toxicol 46:S70–S574
52. Geuns JM, Buyse J, Vankeirsbilck A, Temme EH, Compernolle F, Toppet S (2006) Identification of steviol glucuronide in human urine. J Agric Food Chem 5:2794–2798
53. Wheeler A, Boileau AC, Winkler PC, Compton JC, Prakash I, Jiang X, Mandarino DA (2008) Pharmacokinetics of rebaudioside A and stevioside after single oral doses in healthy men. Food Chem Toxicol 46(Suppl 7):S54–S60
54. Roberts A, Renwick AG (2008) Comparative toxicokinetics and metabolism of rebaudioside A, stevioside, and steviol in rats. Food Chem Toxicol 46(Suppl 7):S31–S39
55. Espinoza MI, Jean-Paul Vincken J–P, Sanders M, Castro C, Stieger M, Agosin E (2014) Identification, quantification, and sensory characterization of Steviol glycosides from differently processed *Stevia rebaudiana* commercial extracts. J Agric Food Chem 62:11797–11804
56. Ceunen S, Geuns JMC (2013) Steviol glycosides: chemical diversity, metabolism, and function. J Nat Prod 76:1201–1228
57. Wölwer-Rieck U (2012) The leaves of *Stevia rebaudiana* (Bertoni), their constituents and the analyses thereof: a review. J Agric Food Chem 60:886–895, and references cited therin
58. Tada A, Ishizuki K, Iwamura J, Mikami H, Hirao Y, Fujita I, Yamazaki T, Akiyama H, Kawamura Y (2013) Improvement of the assay method for steviol glycosides in the JECFA specifications. Am J Anal Chem 4:190–196
59. EFSA Panel on Food Additives and Nutrient Sources added to Food (2015) Scientific opinion on the safety of the proposed amendment of the specifications for steviol glycosides (E 960) as a food additive. EFSA J 13:4316
60. Carakostas MC, Curry LL, Boileau AC, Brusick DJ (2008) The history, technical function and safety of rebaudioside A, a naturally occurring steviol glycoside, for use in food and beverages. Food Chem Toxicol 46(7S):S1–S10
61. Williams LD, Burdock GA (2009) Genotoxicity studies on a high-purity rebaudioside A preparation. Food Chem Toxicol 47:1831–1836

Xylitol as Sweetener

6

Salim-ur Rehman, Mian Anjum Murtaza, and Zarina Mushtaq

Abstract

Xylitol is a low-calorie, crystalline sweetener which is naturally present in fibrous plant foods and hardwood trees. It can be produced by hydrolysis from different plant sources having polysaccharides. Hemicellulose (**xylan**) is transformed to xylose that is further converted into xylitol through hydrogenation. In the human body, more than half of the ingested xylitol is not absorbed in the small intestine. It moves to the gut and serves as a substrate for growth of intestinal flora. Blood glucose and insulin responses to xylitol are very low as compared to sucrose. Its energy value is calculated as 2.4 kcal/g. It is the only sugar alcohol having sweetness intensity equivalent to sucrose. Although, its solubility is comparable to sucrose at ambient temperatures and greater at higher temperatures but being a monosaccharide sugar alcohol, it shows lower viscosity than sucrose in a solution of similar concentration. Due to a low-calorific-value sweetener, it is being used as a food additive in confectionery, bakery, drinks, and dairy products, as well as in pharmaceutical industry. Xylitol imparts numerous potential health benefits, being low in calories; having insulin-independent metabolism, prebiotic nature, and anabolic effects; and being safe to use.

S.-u. Rehman (✉)
National Institute of Food Science and Technology, University of Agriculture, Faisalabad, Pakistan
e-mail: drsalim_rehman@yahoo.com

M.A. Murtaza (✉)
Institute of Food Science and Nutrition, University of Sargodha, Sargodha, Pakistan
e-mail: mian.anjum@uos.edu.pk

Z. Mushtaq (✉)
Institute of Food Science and Nutrition, GC University, Faisalabad, Pakistan
e-mail: zee_ft@hotmail.com

J.-M. Mérillon, K.G. Ramawat (eds.), *Sweeteners*, Reference Series in Phytochemistry,
https://doi.org/10.1007/978-3-319-27027-2_30

Keywords
Sweetener • Sugar alcohol • Calories • Hemicellulose • Xylan • Hydrolysis • Food • Pharmaceutical • Health benefits • Safety

Contents

1 Xylitol: A Reduced Calorie Bulk Sweetener ... 130
2 Chemistry and Calorific Value ... 131
3 Natural Sources ... 131
4 Xylitol Production ... 132
4.1 Preparation of the Raw Materials ... 133
4.2 Detoxification Practices ... 134
4.3 Chemical Production ... 135
4.4 Microbial Production ... 136
5 Applications of Xylitol ... 138
5.1 Food Industry ... 138
5.2 Pharmaceutical Industry ... 139
6 Digestion, Absorption, and Metabolism of Xylitol ... 140
7 Xylitol and Health ... 140
7.1 Impact on Blood Glucose Levels ... 141
7.2 Impact on Blood Lipids ... 142
7.3 Impact on Dental Health ... 142
7.4 Other Health Impacts ... 142
8 Safety Perspectives of Xylitol Uses ... 143
9 Conclusion ... 143
References ... 144

1 Xylitol: A Reduced Calorie Bulk Sweetener

The artificial sweeteners or chemically instigated substitutes for sugar are commonly used in overweights, obese, diabetics, and people with metabolic syndromes owing to the sweetening power as well as noncalorific values. Xylitol was discovered nearly concurrently by German and French scientists around the year 1891, but there was no substantial accomplishment in the next 50 years. The enormous shortage of sugar observed during the Second World War provoked the researchers for finding innovative sweeteners [1]. Primarily, it was propagated as a safe sweetener with no anticipated effects on glucose or insulin levels of diabetics in Europe. It is used as sweetener in food as well as in pharmaceutical industries. It is utilized in food products for the prevention of dental caries and as sugar substitute for insulin-dependent diabetics. It is just about as sweet as sucrose but has about 33% less calories per unit weight. In contrast to the other sweeteners, it is dynamically advantageous in reducing dental cavities and supports the remineralization on regular use [2]. When utilized through chewing gum, tablets, and/or nasal spray, it reduces the prevalence of infection of middle ear in the children [3]. The sugar alcohols, mainly xylitol, are acknowledged as healthier sweetener being instigated naturally, being low in calories and carcinogenicity, as well as having numerous persuasive health effects [4].

2 Chemistry and Calorific Value

Xylitol ($C_5H_{12}O_5$) is a natural pentose polyol sweetener [5]. It is a white, crystalline powder with no odor and has a molecular weight of 152.15. Its solubility in water is 169 g/100 g with 5–7 pH. The melting and boiling points (at 760 mmHg) and bulk density are 93–94.5 °C, 216 °C, and 1.50 g/L, respectively. It is optically inactive with a viscosity (20 °C) of 10% as 1.23 cP and 50% as 8.04 cP [1].

The chemical structure of xylitol is given in Fig. 1.

The sweetening power of xylitol and sucrose is almost the same; however, other polyols are relatively less sweet. As the caloric value of xylitol is one third of conventional sugar, so, low caloric food products could be prepared by replacing sucrose with it.

The laboratory trials carried out with xylitol and some other alcohols proved that almost 25% of the ingested xylitol is absorbed in GIT. It is efficiently digested by the pentose phosphate and delivers almost 4 kcal/g energy. The remaining 75% dose is totally metabolized by the microflora in the large intestine. Hence, based on comprehensive evaluation, the fermentable energy values are ranged from 2.8 to 2.9 kcal/g. The xylitol has been allocated to have 2.4 kcal/g of disposable calorific value. This calorific value is described as the energy dumped or metabolized in a body to perform metabolic, physical, and mental works [6]. The equal sweetness but lower calorific value (2.5 kcal) and glycemic index (13) than sucrose containing 4 kcal and 65, respectively, made it a recognized sugar substitute. It is popular in diabetic, obese, and overweight people with the intention of reducing calorie intake from sucrose [4].

3 Natural Sources

In humans and animals, xylitol is considered as an intermediate product of carbohydrate metabolism. An adult human produces about 5–15 g xylitol daily [7]. Fruits and vegetables are natural sources of xylitol; however, it could be traced in minute quantities [8]. It can efficiently be extracted from other sources such as oats, berries, and mushrooms besides the fibrous foods like corn husks and sugarcane bagasse [9].

Fig. 1 Xylitol (Structure)

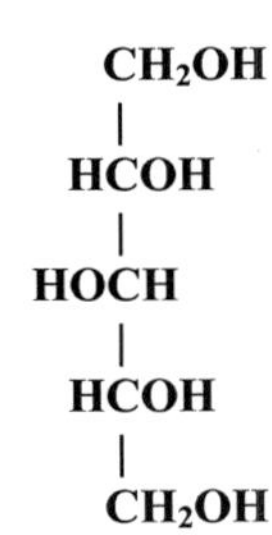

Industrial and commercial scale productions begin from a hemicellulose (xylan) extracted from corncobs and hardwoods. The xylan is first hydrolyzed into xylose and then hydrogenated with the help of catalysts into the xylitol by adjusting the process conditions [10].

4 Xylitol Production

It can be prepared by hydrogenation of xylose that converts the sugars into alcohol or extracted from natural sources. Another practice is the use of microbial processes comprising the fermentation and biocatalyst applications through bacteria, fungi, and yeast [11]. These employ xylose-fermenting yeasts that consume xylose reductase to reduce xylose into xylitol. The *Candida tropicalishas* and *Candida guilliermondii* are the most widely used yeasts for production purpose [12, 13]. Biotechnological methods of xylitol production are safe and ecologically approachable as well as result in better yield and productivity. Its huge demands by the food and pharmaceutical sectors forced to develop and optimize the approaches for biological production of xylitol using yeasts [14, 15].

Production of xylitol initiates with hydrolysis of lignocellulosic materials. Multiple detoxifying techniques are adopted to purify xylitol from the hydrolysates. Microorganisms are utilized to remove furfurals, acetic acid, and phenolic compounds from acid hydrolysates in a series of steps [16]. It is followed by xylitol hydrogenation by a chemical process. The hydrogenation is facilitated with Raney nickel catalyst or by biological methods (microbes as catalyst) and then hydrogenated xylitol is further purified [8].

Lignocellulosic materials are the widely spread, excessively available, lower cost, and economical sources of polysaccharides. A wide range of biotechnological products including xylitol could be prepared from lignocellulosic materials. These materials could be produced from residues of agriculture, forestry, and agro-industrial wastes. Their residues are present as organic matter which are composed of 34–50% cellulose, 19–34% hemicellulose, and 11–30% lignin and reduced quantities of pectins, protein, and inorganic matters [17]. The most plentiful polymer of lignocellulosic material is hemicelluloses (xylan) that contains pentoses, hexoses, and acids. The xylans are the ample hemicelluloses which are heteropolysaccharides containing arabinose, ether, and many acids [18]. Xylans from various sources including cereals, grasses, and soft- and hardwoods differ in their compositions as detailed in Table 1.

Important rheological characteristics of powdered xylan taken out from the corncobs are tap density (0.2256 g/ml), bulk density (0.1336 g/ml), Hausner ratio (1.68), compressibility index (40.77%), angle of repose (40.70°), and compactability (32.6 mL) [23]. About 80% of the xylan chain is greatly surrogated with monomeric side chains of glucuronic acid/arabinose attached to O-2 and/or O-3 of xylose residues and also to oligomeric side chains having xylose, galactose, and arabinose [24]. The diferulic cross-bridges intensely affect the heteroxylans which are intersected to form a complex and where the cellulose microfibrils are fixed. The

Table 1 Composition of xylan from various sources

Sources	Xylose (%)	Arabinose (%)	Glucose (%)	Galactose (%)	Anhydrouronic acid (%)
Birchwood xylan [19]	89.3	1.0	1.4	–	8.3
Rice bran xylan [20]	46	44.9	1.9	6.1	1.1
Wheat arabinoxylan [21]	65.8	33.5	0.3	0.1	–
Corn fiber xylan [22]	51.0	34.0	–	8.0	4.5

structural wall proteins form an insoluble network by cross-linking mutually by isodityrosine bridges and with feruloylated heteroxylans [25]. A range of other plant-based biomasses are potential sources of raw materials like corn [9], sugarcane bagasse [26], eucalyptus [27], spent brewing grain [28], olive tree and soy hull [29], and rice straw [30].

4.1 Preparation of the Raw Materials

Lignocellulosic materials remain stable against chemical and biological compounds. At normal conditions, they could not be converted into simple sugars. Some pretreatments are practiced to break its structural integrity. In this regard, lignin is removed from lignocellulosic materials to enhance its surface area and to ensure its availability as fermentable sugars. Pretreatment performance is associated with harvesting nature of selected material, composition of lignin and other components, nature of chemicals, time, and temperature [31]. These materials are fractionated, solubilized, and hydrolyzed into cellulose, hemicellulose, and lignin by applying of a variety of pretreatments like addition of concentrated acid [32], dilute acid [33], alkaline solution [34], SO_2 [35], hydrogen peroxide [36], steam explosion (autohydrolysis) [37], ammonia fiber explosion [38], wet oxidation [39], lime [40], liquid hot water [41], and CO_2 explosions [24]. In each practice, lignocellulosic material is gone to decrease in molecular size [42].

Chemical hydrolysis is considered as an easy and quick procedure for hemicellulosic materials. However, treatment environment may differ by means of materials and nature of chemical agents, their quantities, time, and temperature of incubation [43]. Many mineral acids like phosphoric, sulfuric, nitric, and hydrochloric acids are used for acid hydrolysis [44, 45]. During pretreatment, dilute acid is used at high temperature. It facilitates the fractionation of hemicellulose to water-soluble sugar component like xylose, arabinose, and other sugars. Cellulose and lignin are also obtained as a hydrolysate. Formic acid or ethanol is used to separate the lignin [46].

Hydrolysis of lignocellulosic material yields many compounds. Further degradations of pentoses and hexoses result in furfural and hydroxymethylfurfural, respectively. Formic acid is produced by further degradation of furfurals. In hemicelluloses, less quantity of hexoses are present, so, hydroxymethylfurfural is produced in lesser amount as compared to furfural. Breakdown of hemicellulosic acetyl yields acetic acid. A series of aromatic compounds are liberated by hydrolysis of lignin with dilute acids [47]. Chemical hydrolysis results in terpenes acidic resins, vanillic, tannic, caprylic, syringic, caproic, pelargonic, and palmitic acids [48]. It also induces the production of monosaccharide derivatives like furan, hydroxymethylfurfural, and phenolic toxic materials which reduces the availability of monosaccharides. Monosaccharide derivatives act as microbial growth inhibitors and further biotransformation is ceased. Microbial fermentative inhibitory compounds concentrations are associated with type of raw materials and operational conditions. However, microbial toxicity depends upon fermentation variables including growth conditions of microorganisms [49]. During chemical hydrolysis, pentose sugar-derived furfurals are considered as microbial growth inhibitor for production of xylitol. 25–99% microbial growth is inhibited in regard with 0.5–2 g/L furfural concentration. Respiration processes are interrupted due to cell mass ATP. The inhibitory effects against *Saccharomyces cerevisiae* and *Pichia stipitis* reach at 100% when hydroxymethylfurfural is consumed as a growth medium (1.5 and 1 g/L). However, inhibitory effect varies from strain to strain [50]. Less quantity of inhibitory compounds in the medium of fermentation leads to improve the growth of microflora. The inhibitory effects of furfural and hydroxymethylfurfural on microbial growth during xylose fermentation are proven [51].

Microbial membrane permeability may be affected by lignin degradation products including polyaromatic, aromatic, aldehydic, and phenolic compounds of hydrolysate [52]. Toxic effect of acetic acid is generally determined with the pKa level. Cell internal pH is reduced due to the death of the cells. Bioconversion of xylose to xylitol may be modified at lower concentrations of acetic acid [53]. It may be happening as a result of enhanced rate of diffusion of total xylitol during the metabolism of xylose. At this stage, inadequate acetic acid outcome is originated on cell membrane. Moreover, heavy metals such as copper, nickel, iron, and chromium are turned out during hydrolysis, which are generally created from corrosion of the utensils and caused the toxicity of the cells by restraining metabolic pathway of enzyme system [54].

4.2 Detoxification Practices

Physical, chemical, and biological techniques may be applied to eliminate inhibitors of microbial growth and to improve the fermentation efficiency of the hydrolysates. However, detoxification requirements depend upon the strains and composition of metabolites. The detoxification procedure varies depending upon raw materials, hydrolysis type, process type, and microorganism employed [9]. An inhibitory effect of hydrolysate may be minimized by adopting various processes including hydrolysis process, prior fermentation, detoxification, use of inhibitor-resistant microorganism, and altering the nature of toxic compounds into nontoxic. The product cost increases too much by applying detoxification techniques, so, during selection of the technique, the choice should be cost-effectiveness and efficient detoxification. Another suitable biological method for removing toxic compounds may be the use of inhibitor-tolerant microbial strains [55].

Physical method of detoxification such as a vacuum evaporation has limited benefits. It could only be helpful to eliminate volatile toxins. More than 90% inhibitory compounds could be eliminated from rice straw, sugarcane bagasse, and wood hemicellulosic hydrolysates through vacuum evaporation technique [49]. Nevertheless, this process results in increased quantities of nonvolatile inhibitors and reduced amounts of metabolites [56]. Ionization of inhibitory compounds could be minimized by using different techniques. However, the best practiced, reliable, and cost-effective method of detoxification is pH adjustment [57].

The absorbing capacity of activated charcoal is of much importance [9]. Different process variables including temperature, pH, time, and solid to liquid ratio govern over activated charcoal efficiency. During activated charcoal treatment, lower pH facilitates the elimination of phenolic compounds, whereas higher pH facilitates the elimination of organic sources. Clarification process requires increased contact time. During charcoal treatment, the absorption process may be enhanced by increasing temperatures [58]. Comparative assessment of various detoxification procedures showed that anion exchange resins omitted greater quantities of inhibitors as compared to cation exchange resins [9, 49].

Biological detoxification can be facilitated either by the use of particular microorganism or enzyme. Laccases and peroxides are enzymes which are commonly applied for detoxification purpose [49]. Microbial hydrolysate detoxification implicates the use of toxics for growth of microorganisms or version of particular microbes [9]. It was reported that *S. cerevisiae* has a capacity to remove more than 90% acetic acid available in hydrolysate [59]. Production of xylitol from corncob and rice straw by using *C. tropicalis* is the most efficient and economical method. A study established that combined application of one or more chemical and biological techniques poses the effective results on xylitol fabrication from corncob and bagasse. The modification of pH followed by charcoal activation and resin application facilitates the adaptation of certain microbial strains [9].

4.3 Chemical Production

Xylose is subjected to hydrogenation at hydrogen pressures up to 50 atm and 80–140 °C in the presence of Raney nickel catalyst of hemicellulosic detoxified hydrolysate. The fractionation through chromatography and crystallization are further applied to obtain pure xylitol [60]. Thus, more than half of the xylose is converted into xylitol by applying this procedure [11].

4.4 Microbial Production

Yeast is considered as a substitute for chemical process in respect of xylitol production [61]. First the facilitated passive diffusion is adopted for transporting the sugars from a membrane into the cell during xylose metabolism. The passive diffusion may be performed through produced gradient in a substrate concentration and molecular factors. Though facilitated diffusion also happened by the concentration gradient using carrier transport protein substrate, the process is influenced by temperature and pH during fermentation of the substrate [62]. D-Xylose is consumed during cellular breathing and biomass production in aerobic environments; however, in the presence of limited oxygen, xylitol is produced by yeast. It is happened because the concentration of intracellular NADH is greater owing to lower respiration rate. Thus, NADH acts as a xylitol dehydrogenase inhibitor and redirects conversion of xylose to xylitol [61, 63].

The pathway for xylose deployment by microorganisms is shown in the Fig. 2.

Several yeasts and fungi possess the enzyme xylose reductases that yield xylitol such as *Candida* species [64]. These species applied for the production of xylitol have established phosphate pathways and can grow only in xylose substrate. During phase of phosphate pathway, oxidative pentose phosphate produces hexose phosphates provided that NADPH required for their biosynthesis is available. The non-oxidative stage yields triglycerides pentose phosphates and hexose phosphates. Among microflora, *Candida tropicalis* is one of the most thriving microbes used for

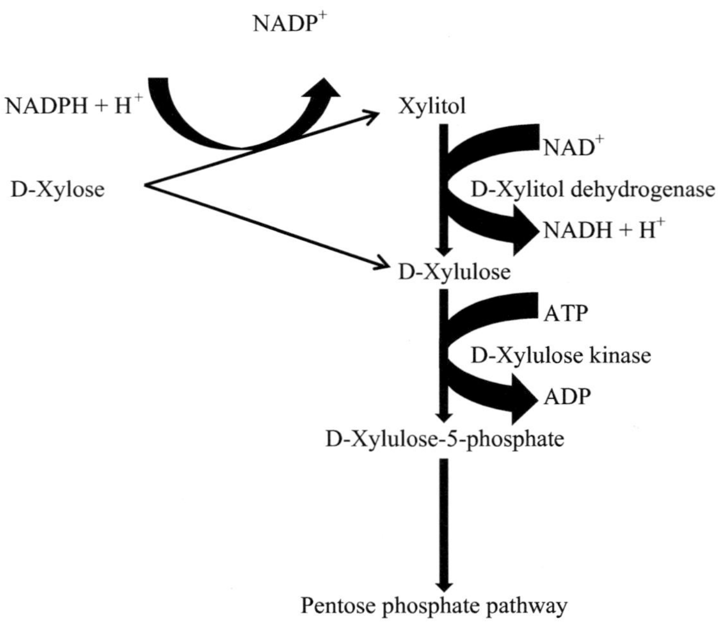

Fig. 2 Xylitol utilization leading to the pentose phosphate pathway

the preparation of xylitol from xylose. Its large-scale commercial importance is owing to the more absorption of xylose, capacity for xylitol fabrication and fatty acids, and their degradation peroxisome [9].

In a study, the pH and acetic acid were optimized by using *Candida tropicalis* for the conversion of hemicellulose into xylitol from corncobs. Xylitol was produced from corncob successively by hydrolysis, detoxification during liming, and extraction through solvents. The glucose produced in the hydrolysate gets better growth of *Candida tropicalis*, whereas acetic acid exhibits inhibitory impact. The acetic acid inhibition could be removed by maintaining pH 6 prior to fermentation. In this optimal environment, a maximum quantity of xylitol (68.4 g/L) was obtained after 72 h of fermentation, providing a concentration of 0.7 g xylitol/L of xylose and the 0.95 g/L/h productivity [65].

4.4.1 Impact of Process Variables on Xylitol Production

The inhibition of microbial growth occurs because of the undissociated form of acetate. However, on increasing the pH, undissociated type of acetic acid is found to decrease. It has been observed that rise in pH from 4.5 to 6.5 results in an augment in the productivity of xylitol. However, the highest production is obtained at a pH 6.0 [66]. Its production was found to decrease at a pH 3.0 as compared to a pH 6.0 [67]. It was noted that the toxic outcome of acetic acid was augmented at low-pH environment owing to an entry of acid in its undissociated type in the cells. However, acetic acid may lead to acidification of the cytoplasm of the cells [68].

Correct temperature for the production of the xylitol is noted as 30 °C. However, yield of xylitol is found to be independent of the temperatures, if yeast is grown in a range of 30–37 °C. The temperature beyond 37 °C is found to considerably decrease its yield. It was perceived that a variation of 3 °C influenced about 27–30% yield of xylitol in the presence of *Candida tropicalis*. Nevertheless, transformation to xylitol by *Candida* sp. is gone to constantly improve between 35 °C and 40 °C, but at 45 °C and greater temperatures, production is found to significantly decrease [69]. It might be traced due to the thrashing of activity of NADH- and NADPH-dependent enzyme xylose reductase associated with rise in temperature [70]. The microbial production of xylitol may also be influenced by the nature and concentration of nitrogenous compounds. It has been found that the xylose consumption increases as a source of organic nitrogen present in the media [71]. A maximum production (100 g/L) of xylitol may be achieved by using 114g/L xylose and adding 3 g/L urea as a nitrogen source. Yeast extract is a vibrant component for xylitol manufacturing, since it delivers all the vitamins essential for microorganisms. The rate of production increased when media contains 20 g/L yeast extract mixed with *Candida tropicalis* [72].

4.4.2 Recovery of Xylitol

The scums existing in the fermentation broth have residual nutrients coming from fermentation including the polypeptides from yeast extract, the salts, and colorants. Reclamation of weak concentration of xylitol from a composite is a great confront. The yeast extract, an impurity in the melt, may contain amino acids, polypeptides, peptides, and proteins. Activated charcoal chromatography and ion exchange resin techniques are mostly applied for recovery of xylitol. Both anionic and cationic exchange resins are used to purify xylitol from hydrolysate of sugarcane bagasse [73]. However, 40–55% loss of the product is observed in both cation and anion exchange resin techniques.

In a study, xylitol was produced with *C. guilliermondii* from a hydrolysate of sugarcane bagasse. Various tests were performed so as to evaluate the best state for clarifying the fermented broth by using activated carbon. Crystallization of xylitol was carried out after exposing the media with ion exchange resins. Though optimum environment for clarification was maintained by the addition of 25 g charcoal into 100 mL of broth maintained at pH 6 and 80 °C for one hour, the ion exchange resins were observed inefficient for the present situation. Although crystallization method exhibited improvement in the results, the crystals were found in the form of viscous-colored solution [74].

5 Applications of Xylitol

Xylitol is being successfully utilized by food industries as a low-calorie sweetener in different types of food products like candies, chewing gum, bakery items, soft and energy drinks, dental care, and other medicated products [75].

5.1 Food Industry

Food processors are adding xylitol in food products during preparation in order to enhance appearance, color, and flavor. However, it does not obscure or decrease the nutritive value of proteins being inert in Maillard's reaction. It may be used alone or in a blend with other sugars for the preparation of chewing gum, unsweetened chocolate, candies, filled wafers, chocolate, dragées, and foods for diabetic patients [76]. Pectin jellies without sugar can be prepared by blending it and hydrogenating starch hydrolysates. Xylitol's crystals are an exceptional material for blending with pectin in confectionery items. The leading xylitol world application is in the sugarless chewing gums. It can be used to sweeten the two forms of wood and pellet gum, since it provides a quick sweetness, flavor, as well as the cooling effect [77]. Owing to the fast drying properties and crystallization, it is often used to form layers and granules in sugarless gum. To ensure a satisfactory sweetness, the profile of sugarless chewing gum is prepared by the assortment of polyol and the deep sweeteners. However, only some blends contain the xylitol and maximum foods have less than half of the xylitol [78]. It is commercially utilized in numerous countries, independently or as a blend with sugar confectionery. Though it is utilized alone in producing acacia tablets, chocolate, candy, ice cream, etc., some sugar alternatives can be utilized to optimize taste and texture as well as shelf stability of the food products [77]. It can be utilized as a sweetening agent in bakery products for babies and adults. Addition of xylitol in bakery products imparts distinctive color and flavor. Some golden color anticipated from reducing sugars in wheat flour might be due to Maillard's reaction. In the case of a cake, xylitol turned out as a worthy replacement of sucrose. Its cakes very much looked like those containing sucrose with respect to texture and color. Some cookies manufactured with xylitol possess brown striped lines, possibly owing to low xylitol solubility in a dough containing fat. Nevertheless, the mouthfeel of the cookies containing xylitol is better than the usual cookies [79].

In a study, the prospective application of xylitol as a modest energy sweetener in bakery foods was observed. The cookies were manufactured fully using xylitol and its features were matched with those containing sucrose. The shelf life of 7–14 days exhibited no substantial influence on the texture and flavor of the product. Moreover, after 90 days, a considerable impact on textural and flavor characteristics was exhibited as compared to the cookies made with sucrose having more freshness and tenderness [80]. However, the perpetually increased mandate of the food industry for foods sweetened with xylitol like sweeteners has gained the need to apprehend the sensory perception particularly profiled by high concentration of sweetener frequently ingested in large quantities [81].

5.2 Pharmaceutical Industry

Xylitol may be utilized as a sweetener in many pharmaceutical preparations due to its many benefits such as the appropriateness for diabetic patients and its anticariogenic and non-fermentative properties. Tonics, cough syrups, and vitamin products prepared with it cannot be fermented/attacked by molds. Since it is a chemically inert ingredient, it does not experience Millard's reaction or interaction with other excipients and active pharmaceutical ingredients. Medicines sweetened with xylitol can be taken by the children after brushing their teeth at night without any anticipated damage to them [82]. Those chewing gums holding it has been shown to protect against ear infection in the children [83]. Xylitol-coated pharmaceuticals, confectionery, and the dietary supplements impart a pleasant cooling similar to vaporization in oral and nasal cavities. Moreover, it may be used as a stabilizer in protein extraction in order to control protein denaturation [84] and anticariogenic and plaque formation-preventing agent [7].

In tablets, it may be utilized both as a carrier and sweetener [85]. In addition to sweetness, it has microbial stability, no reactivity, and the benefits of elevated solubility at a body temperature and offers a pleasant cooling outcome. Coating of tablets with xylitol alone or in combination of sorbitol (up to 20%) and xylitol mixtures either prepared by sintering of surface of the tablets in a current of warm air or conventional pan coating has proved better choices. In toothpaste, it can replace sorbitol completely or partially as a humectant. However, owing to its better sweetness, it ameliorates the taste of toothpaste, and in the manufacture of transparent gels, it exhibits slightly superior properties than sorbitol [77].

An inhibitory effect of xylitol on enamel demineralization has also been reported. It has been observed that the use of a dentifrice having xylitol can lead to a substantial decline in *Streptococcus* mutants in saliva. Due to its favorable effects on dental health, it may also be used in many oral care preparations such as toothpastes and mouthwashes [86, 87].

6 Digestion, Absorption, and Metabolism of Xylitol

Many investigations have been carried out on the digestion of xylitol by oral bacteria that lead to resistant anticaries properties. It was testified that xylitol absorption in a stomach and an upper gastrointestinal tract is 20–30% than that of glucose. Moreover, absorbed xylitol is not actively transported across the intestinal tract. A study revealed that its absorption may vary from 49% to 95% in 3–4 h after intake either to its first or constant feeding. However, other investigators suggested that the adjustment may occur over a time, but it may happen owing to the variation in the intestinal microflora. Another factor affecting the absorption of xylitol is gastric emptying [88]. In another study regarding the influence of xylitol on gastric emptying and absorption, it has been observed that intake of xylitol has improved the gastric emptying rate. Thus, some modifications are possible, if it is incorporated into a diet. After its uptake in the bloodstream, its liver uptake becomes independent of insulin and increases the blood glucose, insulin, and glucagon levels. The liver is a main organ which eliminates it from the bloodstream and metabolizes 50–80%, and the remaining 20–50% of it can be metabolized by the fat deposits, kidneys, lungs, and erythrocytes. It is not absorbed through active transport and does not need insulin uptake by a liver for conversion into glucose, glycogen, and L-lactic acid under definite pathological circumstances. These metabolites, thus, may be converted into water and carbon dioxide through normal physiological conditions of carbohydrate metabolism [88, 89].

Two types of enzymes are involved in the metabolism of xylitol: one is linked with NAD-specific polyol dehydrogenase and the other is xylitol dehydrogenase enzyme that is linked with specific NADP. After its metabolism, 4.06 kcal/g of energy and 35 mol of ATP are produced, as compared to 38 mol of glucose. It may mostly be absorbed in the distal intestine by the intestinal microflora [88].

7 Xylitol and Health

Xylitol is metabolized independently of insulin and may replace sucrose on weight basis [90] and make it an appropriate sweetener for diabetic patients [91]. It is recommended for diabetics to take in daily because a very small increase in glucose and insulin levels happens in the blood [92]. In addition, it can be used in postoperative or post-traumatic conditions, when consumption of glucose is subdued. Certain metabolic disorders can be corrected due to its anabolic effects. Moreover, when it is consumed on a regular basis in the diet, obesity may be controlled. It can be utilized for parenteral nutrition infusion therapy because it is found to be inert to amino acids. It can also be used in the treatment of disorders of lipid metabolism. There are numerous functional impacts of xylitol due to a low-calorie sweetener and its prebiotic character [93–95].

7.1 Impact on Blood Glucose Levels

It may not be passed actively from the intestine, nor required insulin for its absorption in the liver, and thus, it provides a low index of sugar in the blood. Its incomplete and slow absorption by the upper digestive tract escorts the modest influence on a level of sugar in the blood. It was found that the diabetic rats consuming it reinstated the glycolytic function and boosted up glycogen production. Mostly, its consumption reduces the level of blood glucose and releases the free fatty acids during severe sepsis and trauma stages. It is described that the augment in the rate of oxidation of carbohydrates on xylitol consumption has been found 25% of that resulted through the intake of glucose. A study concluded that metabolism of xylitol marked a slighter boost in insulin and glucose levels and thermogenic response when matched with the metabolism of glucose [96, 97].

The controlling of levels of blood sugar, fat, and weight is the one mainly significant perspective in diabetes managing goals. Minimum impact on blood sugar is the provision of fewer calories and numerous other health benefits. It has been observed that the addition of 30–60 g xylitol per day in a diet has exhibited no deleterious effect, especially in interactive metabolism of fat and carbohydrates [98]. It was primarily anticipated that absence of insulin reaction with xylitol might be lacking in case of complex meal consumption. A study conducted in this regard, using xylitol and sucrose in equal quantities to surrogate the starch in a standard serving of food of a diabetic patient professed noticeably that xylitol has performed even more in lowering the blood glucose and insulin levels than sucrose in perspective of multifaceted meals [92].

Xylitol provides almost 33% less calories and has been proved to be a lower-calorie substitute of sucrose. Since being sluggishly absorbed as compared to sucrose, it does not subsidize the higher sugar levels in blood causing hyperglycemia owing to inadequate insulin response. The distinctive role is also favorable for the persons distressed by metabolic syndromes: a collective condition comprising the weak insulin response, hypercholesterolemia, hypertension, and more threats of blood clotting [99].

The xylitol serves as a dietary fiber serving in maintaining persuaded facets, since it is not absorbed in human gut. The unabsorbed portion of xylitol leads to bacterial fermentation, mainly in the large intestine, partially converting it into short-chain fatty acids which are absorbed in the gut as a source of energy during oxidative metabolic pathways. It is also beneficial in retrieval after heavyweight workout, since our body transforms the absorbed xylitol into glucose-6-phosphate and glycogen. As this conversion is very slow, it acts as a low gastrointestinal energy resource [100, 101].

7.2 Impact on Blood Lipids

Replacement of sucrose with xylitol in foods might have an impact on reducing the triglycerides and cholesterol levels in blood. The physiology and mechanism working behind it is the viscous nature of fibers binding dietary or biliary cholesterol in the colon resulting in increased excretion of bile acids through feces. The motivation of liver cholesterogenesis occurred by an enhanced action of the 3-hydroxy-3-methyl-glutaryl CoA reductase (HMG-CoA). The cholesterol decreasing trend also stimulates acceptance of lipoprotein cholesterol of the liver through the regulation activity of low-density lipoprotein receptor. An investigation completed on rats fed on diet containing 10–15% oligosaccharides resulted in the decrease of body fat, phospholipids, and triglycerides [102].

7.3 Impact on Dental Health

Xylitol being a non-fermentable sugar alcohol shows more dental health assistance than other alcohols. Its structure has a tridentate that reshuffles with polyvalent cations swallowing them to convey through gut wall and saliva; the remineralizing dental coatings has taken place before the development of the cavities. The cavity-causing microorganisms ferment 6-carbon sugars but xylitol cannot be fermented or used as a source of energy. It is taken up into the cells due to similarity in shape imparting no opportunity to the disaccharides to interfere with microbial growth. Moreover, mouth remineralizes the injured teeth without any hindrance, because the detrimental microorganisms become ravenous in the presence of xylitol. About six grams of dose of xylitol per day from a slow-releasing source may be recommended for efficient dental care. Moreover, the secretion of saliva owing to the perceived sweetness from xylitol serves as a buffer in declining pH during dental decay by microorganisms [103–105]. However, in spite of such auspicious assumptions, it was not found through the clinical trials that xylitol is certainly more effective or alike the polyols or topical fluorides in dental care. It is an FDA-recognized food additive, and hence, xylitol-sweetened products are permitted to be marked as these do not stimulate dental cavities [106, 107].

7.4 Other Health Impacts

Acute otitis media is a common disease related to middle ear infection in childhood that compels for the frequent visits to the pediatricians. This problem has been only cured by the application of antibiotics; however, effective and harmless protective treatments are still unavailable. Numerous studies in children validated the preemptive impact of xylitol treatment on acute otitis media. It has shown no toxicity in the middle ear as well as inner ear cells, and hence, it might be safely applied to preclude and treat the acute otitis media [108]. Chewing gum containing xylitol seems to decline the prevalence of acute otitis media in the children attending day-care by about 25% [3]. The nasal sprays based on xylitol have also been revealed to control the prevalence of acute otitis media along with supporting and prompting the natural nasopharyngeal washings of a body and decreasing both bacterial growth and an allergic contamination and their related complications [109]. The rats fed on xylitol-based diets are found to have augmented bone volume resulting in developing the curiosity in using this sweetener to prevent from and treat human osteoporosis [110].

8 Safety Perspectives of Xylitol Uses

Injurious side effects of synthetic sweeteners enforced the production and applications of naturally instigated sucrose alternates or the sweeteners with no or very low side effects. There are several natural sweeteners accessible in the market like glucose, fructose, corn and maple syrups, sugarcane molasses, and honey having sweetness comparable to sucrose. However, prolonged and/or greater uses of these have been found to result in many side effects including hypertension, obesity, increased cholesterol and glucose levels, insulin resistance, diabetes, and metabolic syndromes [111].

In the year 1986, FDA had approved xylitol administration as a food additive and professed it as a safe sweetener for human utilization. The Experts Committees of WHO and FAO regarding food additives have endorsed the indefinite ADI for xylitol and advocated that no further toxicological studies are needed for it. It is a safe additive for globally registered food products, medicines, and dental care commodities [112]. It has no identified toxicity in human beings. Like many other sugar alcohols, it has purgative effects, because such alcohols are not fully metabolized during the process of digestion. However, its impact varies from person to person. In a study on xylitol effects in children, only 4 out of 13 suffered from diarrhea on consuming more than 65 g per day. Like other sugar alcohols, the excessive intake of xylitol beyond the threshold level can cause the momentary gastrointestinal problems. Regular consumption results in the adaptation by increasing the laxation edge. It has proven to have lesser threshold as compared to some other sugars and it can easily be tolerable by the digestive system of a body [88].

9 Conclusion

Xylitol, a low-calorie sugar alcohol, is grabbing worldwide interest owing to numerous probable food and pharmaceutical applications. Xylose is a raw material which is utilized for its manufacturing through chemical-based hydrogenation or biodegradation by microorganisms. The chemical methods are expensive as well as energy consuming. Hemicellulosic xylan is transformed into xylose through hydrolysis by chemicals or enzymes depending upon process parameters. Agricultural biomass yields the growth inhibitors for microorganisms through hydrolysis that require detoxification. The detoxification of these metabolites is conceded through physical, chemical, and/or biological techniques. The permutation of these methods is the most appropriate and economical approaches. The *Candida* species of yeast have been observed as the best for microbial preparation of xylitol, through intensive research on xylose-consuming species. Such xylitol production method is affected by certain process parameters like temperature and pH. Xylitol is effectively employed in foods like confectionery, drinks, bakery, and milk products. Moreover, it has a proven prospective in pharmaceutical preparations like syrups, tonics, and tablets for diabetic patients and sugar-avoiding people. It may efficiently be absorbed and consumed by the microflora of intestine because of its prebiotic nature resulting in the lowering of the blood sugar, triglycerides, and cholesterol levels.

There is a need to explore indigenous native sources, metabolic reactions, detoxification mechanisms, and process controls for xylitol production. In addition, the xylitol applications and consumption in daily life should be increased as well as its more health and toxic effects should be explored through laboratory and clinical trials.

References

1. Rehman SU, Mushtaq Z, Zahoor T, Jamil A, Murtaza MA (2015) Xylitol: a review on bioproduction, application, health benefits, and related safety issues. Crit Rev Food Sci Nutr 55:1514–1528
2. Steinberg LM, Odusola F, Mandel ID (1992) Remineralizing potential, antiplaque and antigingivitis effects of xylitol and sorbitol sweetened chewing gum. Clin Prev Dent 14:31–34
3. Azarpazhooh A, Limeback H, Lawrence HP, Shah PS (2011) Xylitol for preventing acute otitis media in children up to 12 years of age. Cochrane Database Syst Rev 11:CD007095
4. Rahman M, Islam M (2014) Xylitol improves pancreatic islets morphology to ameliorate type 2 diabetes in rats: a dose response study. J Food Sci 79:436–1442
5. Soderling E, ElSalhy M, Honkala E, Fontana M, Flannagan S, Eckert G, Kokaras A, Paster B, Tolvanen M, Honkala S (2015) Effects of short-term xylitol gum chewing on the oral microbiome. Clin Oral Invest 19:237–244
6. FASEB (1994) The evaluation of the energy of certain sugar alcohols used as food ingredients. Federation of American Societies for Experimental Biology, Bethesda
7. Pepper T, Olinger PM (1988) Xylitol in sugar-free confections. Food Technol 10:98–106
8. Parajo JC, Domınguez H, Domınguez JM (1998) Biotechnological production of xylitol. Part 2: operation in culture media made with commercial sugars. Bioresour Technol 65:201–212
9. Rao RS, Pavana Jyothi C, Prakasham RS, Sharma PN, Venkateswar Rao L (2006) Xylitol production from corn fiber and sugarcane bagasse hydrolysates by *Candida tropicalis*. Bioresour Technol 97:1974–1978
10. Liu S, Okuyama Y, Tamura M, Nakagawa Y, Imai A, Tomishige K (2016) Selective transformation of hemicellulose (xylan) into n-pentane, pentanols or xylitol over a rhenium-modified iridium catalyst combined with acids. Green Chem 18:165–175
11. Nigam P, Singh D (1995) Processes for fermentative production of xylitol – a sugar substitute. Process Biochem 30:117–124
12. Kang HY, Kim YS, Kim GJ, Seo JH, Ryu YW (2005) Screening and characterization of flocculent yeast, *Candida* sp. HY200, for the production of xylitol from D-xylose. J Microbiol Biotechnol 15:362–367
13. Kwon SG, Park SW, Oh DK (2006) Increase of xylitol productivity by cell-recycle fermentation of *Candida tropicalis* using submerged membrane bioreactor. J Biosci Bioeng 101:13–18

14. Sampaio FC, Chaves-Alves VM, Converti A, Lopes Passos FM, CavalcanteCohelo JL (2008) Influences of cultivation conditions on xylose-to-xylitol bioconversion by a new isolate of *Debaryomyces hansenii*. Bioresour Technol 99:502–508
15. Silva CJSM, Mussatto SI, Roberto IC (2006) Study of xylitol production by *Candida guilliermondii* a bench reactor. J Food Eng 75:115–119
16. Du Preez JC (1994) Process parameters and environmental factors affecting D-xylose fermentation by yeasts. Enzyme Microb Technol 16:944–956
17. Bobleter O (1994) Hydrothermal degradation of polymers derived from plants. Prog Polym Sci 19:797–841
18. Eda SA, Ohnishi A, Kato K (1976) Xylan isolated from the stalk of *Nicoliana labacul*1l. Agric Biol Chem 40:359–364
19. Kormelink FJ, Voragen AG (1993) Degradation of different glucurono arabinoxylans by a combination of purified xylan-degrading enzymes. Appl Microbiol Biotechnol 38:688–695
20. Shibuya N, Iwasaki T (1985) Structural features of rice bran hemicellulose. Phytochem 24:285–289
21. Gruppen H, Hamer RJ, Voragen AGJ (1992) Water-unextractable cell wall material from wheat flour. 2. Fractionation of alkali-extracted polymers and comparison with water extractable arabinoxylans. J Cereal Sci 16:53–67
22. Doner LW, Hicks KB (1997) Isolation of hemicellulose from corn fiber by alkaline hydrogen peroxide extraction. Cereal Chem 74:176–181
23. da Silva AE, Marcelino HR, Gomes MCS, Oliveira EE, Egito EST, Toshiyuki N Jr (2012) Xylan, a promising hemicellulose for pharmaceutical use. In: Verbeek J (ed) Products and applications of biopolymers. In Tech, University Campus, Slavka Krautzeka, Rijeka, pp 61–84. ISBN 978-953-51-0226-7
24. Saulnier L, Vilarot C, Chanliaud E, Thibault JF (1995) Cell wall polysaccharide interactions in maize bran. Carbohydr Polym 26:279–287
25. Hood EE, Hood KR, Fritz SE (1991) Hydroxyproline-rich glycoproteins in cell walls of pericarp from maize. Plant Sci 79:13–22
26. Carvalho W, Santos JC, Canilha L, Silva SS (2005) Xylitol production from sugarcane bagasse hydrolysate; Metabolic behaviour of *Candida guilliermondii* cells entrapped in Ca-alginate. Biochem Eng J 25:25–31
27. Villarreal MLM, Prata AMR, Felipe MGA, Almeida E, Silva JB (2006) Detoxification procedures of eucalyptus hemicellulose hydrolysate for xylitol production by *Candida guilliermondii*. Enzyme Microb Technol 40:17–24
28. Carvalheiro F, Duarte LC, Lopes S, Parajo JC, Pereira H, Giyrio FM (2005) Evaluation of the detoxification of brewery's spent grain hydrolysate for xylitol production by *Debaryomyces hansenii* CCMI 941. Process Biochem 40:1215–1223
29. Schirmer-Michel AC, Flores SH, Hertz PF, Matos GS, Ayub MAZ (2008) Production of ethanol from soybean hull hydrolysate by osmotolerant *Candida guilliermondii* NRRL Y-2075. Bioresour Technol 99:2898–2904
30. Liaw WC, Chen CS, Chang WS, Chen KP (2008) Xylitol production from rice straw hemicellulose hydrolysate by polyacrylic hydrogel thin films with immobilized *Candida subtropicalis* WF79. J Biosci Bioeng 105:97–105
31. Iranmahboob J, Nadim F, Monemi S (2002) Optimizing acid hydrolysis: a critical step for production of ethanol from mixed wood chips. Biomass Bioenerg 22:401–404
32. Goldstein IS, Easter JM (1992) An improved process for converting cellulose to ethanol. Tappi 75:135–140
33. Saha BC, Bothast RJ (1999) Pretreatment and enzymatic saccharification of corn fiber. Appl Biochem Biotechnol 76:65–77
34. Koullas DP, Christakopoulos PF, Kekos D, Koukios EG, Macris BG (1993) Effect of alkali delignification on wheat straw saccharification by *Fusarium oxysporum* cellulases. Biomass Bioenerg 4:9–13
35. Clark DP, Mackie KL (1987) Steam explosion of the softwood Pinus radial Cl with sulphur dioxide addition I process optimization. J Wood Chem Technol 7:373–403

36. Gould JM (1984) Alkaline peroxide delignification of agricultural residues to enhance enzymatic saccharification. Biotechnol Bioeng 26:46–52
37. Fernandez-Bolanos J, Felizon B, Heredia A, Rodriquez R, Guillen R, Jimenez A (2001) Steam-explosion of olive-stones: hemicellulose solubilization and enhancement of enzymatic hydrolysis of cellulose. Bioresour Technol 79:53–61
38. Dale BE, Leong CK, Pham TK, Esquivel VK, Rios L, Latimur VM (1996) Hydrolysis at low enzyme levels: application of the AFEX process. Bioresour Technol 56:111–116
39. Schimidt AS, Thomsen AB (1998) Optimization of wet oxidation pretreatment of wheat straw. Bioresour Technol 64:139–151
40. Kaar WE, Holtzaple MT (2000) Using lime pretreatment to facilitate the enzymatic hydrolysis of corn stover. Biomass Bioenerg 18:189–199
41. Laser M, Schulman D, Allen SG, Lichwa J, Antal MJ, Lynd LR (2002) A comparison of liquid hot water and steam pretreatments of sugar cane bagasse for bioconversion to ethanol. Bioresour Technol 81:33–44
42. Chum HL, Johnsoon DK, Black S (1988) Organosolv pretreatment for enzymatic hydrolysis of poplars: 1. enzyme hydrolysis of cellulosic residues. Biotechnol Bioeng 31: 643–649
43. Sun Y, Cheng J (2002) Hydrolysis of lignocellulosic materials for ethanol production: a review. Bioresour Technol 83:1–11
44. Vinals-Verde MI, Maciel-de-Mancilha J, Batista-de-Almeida ES, NapolesSolenzar AL (2006) Métodos de purificación de hidrolizados de bagazo de caña de azúcarpara la obtención de xilitol. Cienc Tecnol Aliment 5:129–134
45. Herrera A, Tellez-Luis SJ, Gonzalez-Cabriales JJ, Ramírez JA, Vazquez M (2004) Effect of the hydrochloric acid concentration and time on the hydrolysis of sorghum straw at atmospheric pressure. J Food Eng 63:103–109
46. Bungay H (1992) Product opportunities for biomass refining. Enzyme Microb Technol 14:501–507
47. Palmqvist E, Hahn-Hagerdal B (2000) Fermentation of lignocellulosic hydrolysates I: inhibition and detoxification. Bioresour Technol 74:17–24
48. Bower S, Wickramasinghe SR, Nagle NJ, Schell DJ (2008) Modeling sucrose hydrolysis in dilute sulfuric acid solutions at pretreatment conditions for lignocellulosic biomass. Bioresour Technol 99:7354–7362
49. Mussatto SI, Roberto IC (2004) Alternatives for detoxification of diluted acid lignocellulosic hydrolysates for use in fermentative processes: a review. Bioresour Technol 93:1–10
50. Martinez A, Rodriguez ME, Wells ML, York SW, Preston JF, Ingram LO (2001) Detoxification of dilute acid hydrolysates of lignocellulose with lime. Biotechnol Prog 17:287–293
51. Vogel-Lowmeier EM, Sopher CR, Lee H (1998) Intracellular acidification as a mechanism for the inhibitors of xylose fermentation by yeast. J Indus Microbiol Biotechnol 20:75–81
52. Villa P, Felipe MGA, Rodriguez RCL, Vitolo M, Luis dos Reis E, Silva SS, Napoles AI, Mancilha IM (1998) Influence of phenolic compounds on the bioprocess of xylitol production by *Candida guilliermondii*. In: Esbes-2 European symposium on biochemical engineering science, Porto
53. Felipe MGA, Vieira DC, Vitolo M, Silva SS, Roberto IC, Mancilha IM (1995) Effect of acetic-acid on xylose fermentation to xylitol by *Candida guilliermondii*. J Basic Microbiol 35:171–177
54. Prakasham RS, Merrie JS, Sheela R, Saswathi N, Ramakrishna SV (1999) Biosorption of chromium VI by free and immobilized *Rhizopus arrhizus*. Environ Pollut 104:421–427
55. Taherzadeh MJ, Niklasson C, Liden G (2000) On-line control of fed-batch production of xylose. Bioresour Technol 98:554–559
56. Larsson S, Reimann A, Nilvebrant N, Jonsson LJ (1999) Comparison of different methods for the detoxification of lignocellulose hydrolysates of spruce. Appl Biochem Biotechnol 77–79:91–103
57. Nilvebrant NO, Reimann A, Larsson S, Jonsson LJ (2001) A detoxification of lignocellulose hydrolysates with ion exchange resins. Appl Biochem Biotechnol 91(93):35–49

58. Moreira RFPM, Jose HJ, Soares JL (2000) Isotermas De Adsorção de Corantes Reativos Sobre Carvão Ativado. In: Florianopolis SC, Pinto LT (eds) 2 Encontro Brasileiro Sobreadsorçao, Brasil, pp 85–91
59. Schneider H (1996) Selective removal of acetic acid from hardwood-spent sulfite liquor using mutant yeast. Enzyme Microb Technol 19:94–98
60. Hyvonen L, Koivistoinen P, Voirol F (1982) Food technological evaluation of xylitol. In: Chichester CO, Mrak EM, Stewart G (eds) Advances in food research. Academic, New York, pp 373–403
61. Granstrom T, Ojamo H, Leisola M (2001) Chemostat study of xylitol production by *Candida guilliermondii*. Appl Microbiol Biotechnol 55:36–42
62. Nicklin I, Graeme-Cook K, Paget T, Killington RA (1999) Notes in microbiology. Springer, New York
63. Granstrom T, Wu X, Airaksinen U, Leisola M (2002) *Candida guilliermondii* grows on rare pentoses – implications on production of pure xylitol. Biotechnol Lett 24:507–510
64. Saha BC, Bothast RJ (1997) Fuels and chemicals from biomass, vol. 666. ACS symposium series. American Chemical Society, Washington, DC, pp 307–319
65. Cheng KK, Zhang JA, Ling HZ, Ping WX, Huang W, Ge JP, Xu JM (2009) Optimization of pH and acetic acid concentration for bioconversion of hemicellulose from corncobs to xylitol by *Candida tropicalis*. Biochem Eng J 43:203–207
66. Kim JH, Ryuo YW, Seol JH (1999) Analysis and optimization of two-substrate fermentation for xylitol production using *Candida tropicalis*. J Ind Microbiol Biotechnol 22:181–186
67. Fond O, Jansen NB, Tsao GT (1985) A model of acetic acid and 2, 3-butanediol inhibition of the growth and metabolism of Klebsiella oxytoca. Biotechnol Lett 7:727–732
68. El-Batal A, Khalaf S (2004) Xylitol production from corn cobs hemicellulosic hydrolysate by *Candida tropicalis* immobilized cells in hydrogel copolymer carrier. Int J Agri Biol 6:1066–1073
69. Rodrigues DCGA, Silva SS, Prata MR, Felipe MGA (1998) Biotechnological production of xylitol from agricultural residues evaluation bioprocess. Appl Biochem Biotechnol 70–72:869–875
70. Sreenivas RR, Prakasham RS, Prasad KK, Rajesham S, Sharma PN, Rao LV (2004) Xylitol production by *Candida* sp.: parameter optimization using Taguchi appraoach. Process Biochem 39:951–956
71. Slininger PJ, Bolen PL, Kurtzman CP (1987) *Pachysolen tannophilus*: properties and process consideration for ethanol production from D-xylose. Enzyme Microb Technol 9:5–15
72. Palnitkar S, Lachke A (1992) Effect of nitrogen sources on oxidoreductive enzymes and ethanol production during xylose fermentation by *Candida shehatae*. Can J Microbiol 38:258–260
73. Winkelhausen E, Kuzmanova S (1998) Microbial conversion of D-xylose to xylitol. J Ferment Bioeng 861:1–14
74. Gurgel PV, Mancilha IM, Pecanha RP, Siqueira JFM (1995) Xylitol recovery from fermented sugarcane bagasse hydrolysate. Bioresour Technol 52:219–223
75. Siripurkpong P, Prajan S, Kongkhum S (2014) Xylitol does not directly affect adiponectin production and adipogenesis in 3T3-L1 cells. Songklanakarin J Sci Technol 36:425–432
76. Bar A (1991) Xylitol. In: Nabors LO, Gelardi RC (eds) Alternative sweetener, 2nd edn. Marcel Dekker Inc, Hong-Kong, pp 349–379
77. Olinger PM, Pepper T (2001) Xylitol. In: Nabors LO (ed) Alternative sweetners. Marcel Dekker, New York, pp 335–365
78. Edgar E (1998) Sugar substitutes, chewing gum and dental caries-a review. Br Dent J 184:29–31
79. Hyvonen L, Espo A (1981) Replacement of sucrose in bakery products. In Cakes and cookies, EKT-Ser. 569. University of Helsinki, Finland
80. Winkelhausen E, Malinovska RJ, Velickova E, Kuzmanova S (2007) Sensory and microbiological quality of a baked product containing xylitol as an alternative sweetener. Int J Food Prop 10:639–649
81. Withers C, Barnagaud C, Mehring P, Ferris S, Thomson DMH (2016) Adapting and enhancing sequential profiling to understand the effects of successive ingestion, using the sensory characteristics of high intensity sweeteners as a case study. Food Qual Prefer 47:139–147

82. Feigal RJ, Jensen ME, Mensing CA (1981) Dental caries potential of liquid medications. Pediatrics 68:416–419
83. Uhari M, Kontiokari T, Koskela M, Niemela M (1996) Xylitol chewing gum in prevention of acute otitis media: double blind randomized trial. BMJ 313:1180–1184
84. Maloney PC, Amburdkar SV (1989) Functional reconstitution of prokaryote and eukaryote membrane-proteins. Arch Biochem Biophys 269:1–10
85. Laakso R, Sneck K, Kristoffersson E (1982) Xylitol and Avicel PH-102 as excipients in tablets made by direct compression and from granulate. Acta Pharm Fenn 91:47–54
86. Smits MT, Arends J (1988) Influence of extraoral xylitol and sucrose dippings on enamel demineralization in vivo. Caries Res 22:160–165
87. Featherstone JDB, Cutress TW, Rodgers BE, Dennison PJ (1982) Remineralization of artificial caries-like lesions in vivo by a self-administered mouthrinse or paste. Caries Res 16:235–242
88. Wang YM, Eys V (1981) Nutritional significance of fructose and sugar alcohols. Ann Rev Nutr 1:437–475
89. Salminen E, Prokka L, Koivistoinen P (1984) The effects of xylitol on gastric emptying and secretion of gastric inhibitory polypeptide in the rat. J Nutr 114:2201–2203
90. Cao NJ, Gong RCS, Chen LF (1994) The effect of cell density on the production of xylitol from D-xylose by yeast. Appl Biochem Biotechnol 46:515–519
91. Emodi A (1978) Xylitol: its properties and food applications. Food Technol 32:20–32
92. Hassinger W, Auer G, Cordes U (1981) The effects of equal caloric amounts of xylitol, sucrose and starch on insulin requirements and blood glucose levels in insulin-dependent diabetics. Diabetologia 21:37–40
93. Beutler HO (1984) Xylitol. In: Hans Ulrich BW (ed) Methods of enzymatic analysis, vol 6, 3rd edn. Verlag Chemie, Deerfield Beach, pp 484–490
94. Aguirre-Zero O, Zero DT, Proskin HM (1993) Effect of chewing xylitol chewing gum on salivary flow rate and the acidogenic potential of dental plaque. Caries Res 27:55–59
95. Manz U, Vanninen E, Voirol F (1973) Xylitol- its properties and use as a sugar. In: Food RA Symposium sugar and sugar replacements, London
96. Schricker T, Gross G, Wolfel R, Georgieff M (1995) Enhancement of fatty acid mobilization and oxidation by glucose-xylitol compared to glucose alone in posttraumatic and septic patients. Nutr Hosp 10:13–18
97. Natah SS, Hussien KR, Tuominen JA, Koivisto VA (1997) Metabolic response to lactitol and xylitol in healthy men. Am J Clin Nutr 65:947–950
98. Hamber O, Almdal TP (1996) Effects of xylitol on urea synthesis in normal humans: relation to glucagon. J Parentr Enteral Nutr 20:139–144
99. Marti N, Funes LL, Saura D, Micol V (2008) An update on alternative sweeteners. Int Sugar J 110:425–429
100. Vasilescu R, Ionescu AM, Mihai A, Carniciu S, Ionescu-Tirgoviste C (2011) Sweeteners and metabolic diseases: xylitol as a new player. Proc Rom Acad B 2:125–128
101. Chen X, Jiang ZH, Chen S, Qin W (2010) Microbial and bioconversion production of D-xylitol and its detection and application. Int J Biol Sci 6:834–844
102. Mazur A, Remedy C, Gurus E, Leveret AM, Demigne C (1990) Effects of diet rich in fermentable carbohydrates on plasma lipoprotein levels and on lipoprotein catabolism in rats. J Nutr 120:1037–1045
103. Maguire A, Rugg-Gunn AJ (2003) Xylitol and caries prevention – is it a magic bullet? Br Dent J 194:429–436
104. Reusens B (2004) Functional foods, ageing and degenerative disease. Woodhead Publishing, Cambridge, UK, p 202
105. Scheinin A (1993) Dental caries, sugars and xylitol. Ann Med 25:519–521
106. Mickenautsch S, Yengopal V (2012) Effect of xylitol versus sorbitol: a quantitative systematic review of clinical trials. Int Dent J 62:175–188
107. Ritter AV, Bader JD, Leo MC, Preisser JS, Shugars DA, Vollmer WM, Amaechi BT, Holland JC (2013) Tooth-surface-specific effects of xylitol: randomized trial results. J Dent Res 92:512–517

108. Lee BD, Park MK, Djeric DR, Folic MM, Blazic SR, Djoric IB, Czbay C, Dundar R, Kulduk E, Soy KF, Aslan M (2014) Effects and safety of xylitol on middle ear epithelial cells. Int Adv Otol 10:19–24
109. Jones A (2001) Intranasal xylitol, recurrent otitis media, and asthma: report of three cases. Clin Pract Alt Med 2:112–117
110. Sato H, Ide Y, Nasu M, Numabe Y (2011) The effects of oral xylitol administration on bone density in rat femur. Odontology 99:28–33
111. Stanhope KL, Bremer AA, Medici V, Nakajima K, Ito Y, Nakano T, Chen G, Fong TH, Lee V, Menorca RI, Keim NL (2011) Consumption of fructose and high fructose corn syrup increase postprandial triglycerides, LDL-cholesterol, and apolipoprotein-B in young men and women. J Clin Endocrinol Metabol 96:E1596–E1605
112. Grizard D, Barthomeuf C (1999) Non-digestible oligosaccharides used as prebiotic agents: mode of production and beneficial effects on animal and human health. Reprod Nutr Dev 39:563–588

Part II

Biotechnology and Genetic Modifications

Biotransformation of Mogrosides

7

Reuben Wang, Chun-Hui Chiu, Ting-Jang Lu, and Yi-Chen Lo

Abstract

Mogrosides are natural sweeteners. Previous studies have shown that mogroside extracts exert numerous functions including antioxidation, anti-inflammatory, and blood glucose modulation effects. Chemical structure of mogrcsides consists of mogrol and glycosylated sugar moieties linked by β-linkage to the mogrol. The complexities of the mogroside structures hinder the purification or synthesis of mogrosides and result in difficulties for further producing specific mogrosides. In this review, we discuss the available methods, particularly biotransformation, to convert mogrosides. We anticipate providing the whole pictures of mogroside biotransformation.

Keywords

Mogrosides • Biotransformation • Chemical modification • Enzymatic conversion

Contents

1 Introduction 154
2 Mogroside Structure 154
3 Mogroside Extraction and Their Contents 155
4 Sweetness 157
5 Functional Properties 159
6 Mogrosides Conversion 159
6.1 Chemical Conversion 160
6.2 In Vitro Enzymatic Conversion 160

R. Wang • T.-J. Lu • Y.-C. Lo (✉)
Institute of Food Science and Technology, National Taiwan University, Taipei, Taiwan (R.O.C.)
e-mail: d99641004@ntu.edu.tw; tjlu@ntu.edu.tw; loyichen@ntu.edu.tw

C.-H. Chiu
Chang Gung University of Science and Technology, Taoyuan City, Taiwan (R.O.C.)
e-mail: chchiu@mail.cgust.edu.tw

J.-M. Mérillon, K.G. Ramawat (eds.), *Sweeteners*, Reference Series in Phytochemistry,
https://doi.org/10.1007/978-3-319-27027-2_13

6.3 Bioconversion ... 161
6.4 Mogrosides Bioconversion in Yeast ... 162
7 Conclusions and Future Perspectives ... 162
References ... 163

1 Introduction

Siraitia grosvenori, also called Luo-Han-Kuo (LHK) or monk fruit, cultivated in restricted areas of southern China, is a perennial vine of the Cucurbitaceae (cucumber and melon) family [31]. It has been widely used in traditional Chinese medicine. Because LHK contains a strong sweetness and flavor [22] known as mogrosides, LHK has now become a popular natural sweetener due to the global concerns of the occurrence of obesity and diabetes [25, 28]. In 2010, US Food and Drug Administration (FDA) has certified LHK extracts and mogrosides as generally recognized as safe substances (GRAS) [8].

However, different structures of mogrosides exhibit distinct intensities of sweetness and show divergent potent activities in biological functions. Thus, the identification or production of individual mogroside is a key issue to understand the chemical and biological properties of mogrosides. In this chapter, chemical structures of mogrosides are reviewed, and the recent advances in mogrosides biotransformation will be discussed.

2 Mogroside Structure

Siraitia grosvenori fruit was originated in the mountain areas of Thailand and China, such as Guilin, Guangxi, province of China [37], where the specific climate conditions are particularly suitable for the growth of LHK. In 1975, the sweetness of LHK fruit has been found to come from a mixture of triterpenoid glycosides [15]. Takemoto et al. further isolated and characterized series of cucurbitane-type triterpene glycosides, including mogroside IV (MG IV), mogroside V (MG V), and mogroside VI (MG VI) from LHK fruit [33–35]. Chemical structure of mogrosides consists of mogrol, also called aglycone, and glycosylated sugar moieties linked by β-linkage to carbon 3 and carbon 24 of the mogrol (Fig. 1). Other minor aglycones exist based on the structure of mogrol with several modifications, such as having hydroxyl group side chains or double bonds in the ring structure. For example, 5α, 6α-epoxymogrol has an epoxy group between its carbon 5 and carbon 6. 20-Hydrooxy-11-oxomogrol contains each of a hydroxyl group linked to carbon 20 and carbon 11. In addition, 11-oxomogrol and 7-oxomogrol are two other mogrols with their carbon 11 and carbon 7 being oxidized to carbonyl group, respectively. 11-Dehydroxymogrol is another kind of mogrol with its carbon 11 reduced to hydroxyl group (Table 1).

Depending on the numbers of glycosides (glucose) attached to a mogrol, mogrosides are classified as triterpene glucosides, designated as the diglucoside,

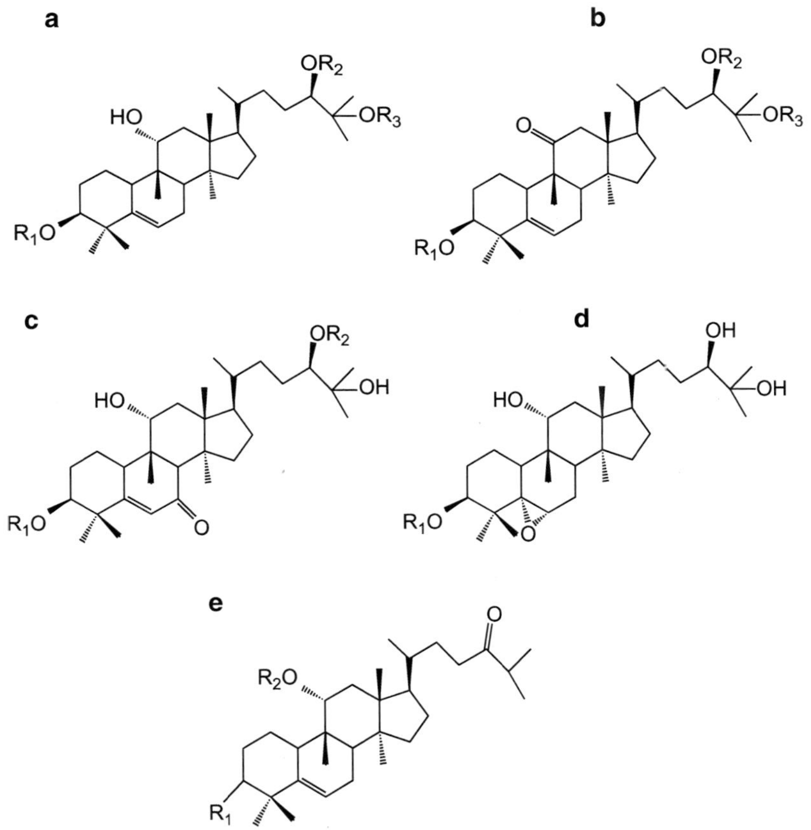

Fig. 1 Structure of mogroside aglycones. **(a)** The original mogroside aglycone structure, mogrol; **(b)** 11-oxo-mogrol; **(c)** 7-oxo-mogrol; **(d)** 5α, 6α-epoxymogrol; **(e)** 25-dehydroxy-24-oxo-mogrol

triglucoside, tetraglumside, pentaglucoside, and hexaglucoside [19, 35]. For example, mogroside II has a total of two glucose molecules with one glucose residue attached to each of carbons 3 and 24. Base on the frame of mogrosides II, mogroside III has an additional glucose residue linked to carbon 24, while mogroside IV has 2-unit glucose side chains at both carbon 3 and 24. Mogroside V and 11-oxomogroside V have five glucose molecules attached to their specific aglycones. To date, the glycosidic bonds of all the identified mogrosides are in beta configuration.

3 Mogroside Extraction and Their Contents

The most common and traditional way to consume LHK is to make water extract of dried LHK fruit. Several investigators analyzed the contents of mogrosides in LHK from different cultivation areas in China. Although all these analyses are based on

Table 1 Structure of mogrosides

Compound name	Aglycone	R1	R2	R3	References
Mogrol	A	H	H	H	[38, 47]
25-methoxy-Mogrol	A	H	H	OCH3	[4]
Mogroside I A1	A	H	Glc	H	[38]
Mogroside I E1	A	Glc	H	H	[38, 44]
Mogroside II A	A	H	Glc2-Glc	H	[20]
Mogroside II A1	A	H	Glc6-Glc	H	[1]
Mogroside II A2	A	Glc6-Glc	H	H	[35]
Mogroside II E	A	Glc	Glc	H	[19, 35]
Mogroside II B	A	Glc	H	Glc	[1]
Mogroside III A1	A	H	Glc6-Glc2-Glc	H	[35]
Mogroside III A2	A	Glc6-Glc	Glc	H	[1, 35]
Mogroside III E	A	Glc	Glc2-Glc	H	[19]
Mogroside III	A	Glc	Glc6-Glc	H	[19]
Mogroside IV	A	Glc6-Glc	Glc2-Glc	H	[19]
Mogroside IV A	A	Glc6-Glc	Glc6-Glc	H	[47]
Siamenoside I	A	Glc	Glc6-Glc2-Glc	H	[19]
Mogroside V	A	Glc6-Glc	Glc6-Glc2-Glc	H	[19]
Mogroside VI	A	Glc6-Glc2-Glc	Glc6-Glc2-Glc	H	[41]
11-oxo-Mogrol	B	H	H	H	[38]
25-methoxy-11-oxo-Mogrol	B	H	H	OCH3	[4]
11-oxo- Mogroside I A1	B	H	Glc	H	[38]
11-oxo- Mogroside I E1	B	Glc	H	H	[38]
11-oxo- Mogroside II A1	B	H	Glc6-Glc	H	[1]
11-oxo- Mogroside II E	B	Glc	Glc	H	[41]
11-oxo- Mogroside III	B	Glc	Glc6-Glc	H	[41]
11-oxo- Mogroside IVA	B	Glc6-Glc	Glc6-Glc	H	[1]
11-oxo- Mogroside V	B	Glc6-Glc	Glc6-Glc2-Glc	H	[19]
7-oxo- Mogroside IIE	C	Glc	Glc	–	[1]
7-oxo- Mogroside V	C	Glc6-Glc	Glc6-Glc2-Glc	–	[1]
5α, 6α-epoxymogroside IE	D	Glc	–	–	[38]
25-dehydroxy-24-oxo-Mogrol	E	β-OH	H	–	[4]
3α-hydroxy-25-dehydroxy-24-oxo-Mogrol	E	α-OH	H	–	[4]
Bryogenin	E	OH	–	–	[4]

the dried fruit, they employed different solvents and various extraction procedures, which make it difficult to compare the mogroside contents among studies from the same basis. For example, researchers extracted mogrosides using methanol combined with water from dried LHK whole fruits or endocarp [18]. They found that mogroside V contents of two tested samples from the dried fruits were 0.80% and 1.29% (w/w), respectively. In the extracts from endocarp of the fruit, the average contents of mogroside V were higher (~1.5%) compared to those in the dried fruit extracts [18]. However, Matsumoto et al. [19] measured the mogroside contents using the similar extraction methods, but only detected 0.45% (w/w) of mogroside V and 0.034% (w/w) of mogroside IV. Additionally, they identified the presence of 7 11-oxo-mogroside V (~0.18%), siamenoside I (~0.044%), mogroside III E (~0.029%), and mogroside II E (~0.025%) and mogroside III (~0.008%) [19].

To compare with the extraction efficiency among various solvents, Ukiya et al. [38] performed a series of extractions from 2 kg of dried and powdered LHK using water, methanol, or butanol as solvent. Their results demonstrated that mogroside V is the most abundant mogrosides in the water-soluble fraction, while mogrosides with fewer glucose molecules, such as mogroside IIE or mogrol I AI and mogrol, tend to be extracted with less polar reagents, butanol or methanol (Table 2) [38]. Recently, Xia et al. applied subcritical water extraction method in order to maximize the mogroside V concentration from LHK extracts. Using subcritical water extraction technique, mogroside V with high solubility in water can be extracted with most efficiency at high water temperature. Elevated pressure compared to other methods such as Soxhlet (hexane: ethanol, 1: 4 v/v) extraction or supercritical ethanol: carbon dioxide extraction was found to be the most efficient method for mogroside V extraction [40]. Others identified several minor compounds, such as mogroside II B, 11-deoxymogroside III, 7-oxo-mogroside II E, 7-oxo-mogroside V, 11-oxomogroside II A1, and 11-oxomogroside IV, from ethanol extracts of dried LHK powder [1]. Interestingly, unripen LHK may contain mogrosides with less glucose moieties and other oxomogrosides [16, 17]. This also indicated that mogroside conversion occurs during the ripening process which leads to the change of the taste. Indeed, Chen et al. [3] found that mogroside content varies with the maturity of the fruit, and the content of mogroside V increases rapidly after 50 days and remains stable after about 80 days growth of the fruit body [3]. As a result, ripened LHK contains the most mogroside V that brings the sweet flavor of the LHK extracts.

4 Sweetness

Although the contents of mogrosides fluctuate during the ripening process, mogroside V is the major sweet compound in LHK extract from ripened LHK fruits. Other mogrosides, including mogroside III, and mogroside II E also exist at various stages of growth [17], but decline to non-detectable levels [23] in the ripened fruit. Accordingly, dried LHK fruits contain about 0.5–1% of mogrosides, particularly, mogroside V consists of half of the mogrosides in mature fruits. Comparing the relative sweetness intensity of mogrosides to sucrose, the relative sweetness of mogroside V, mogroside VI, 11-oxo-mogroside V, mogroside IV, siamenoside I, and mogroside III were 378, 125,

Table 2 Content of mogrosides (mg) in every 100 g of fruit

Compound name	[38]	[16]	[17]	[1]
Mogroside IV		0.188[c]		
Mogroside V	135.7[a]	0.504[c]		
Siamenoside I	4.5[a]			
11-oxomogroside V	18.3[a]			
Mogroside II E	45.85[b]	8.112[c]		
Mogroside III	20.25[b]	2.56[c]		
Mogrol	3.05[c]			
5α,6α-epoxymogroside I E1	0.165[c]			
11-oxomogroside I A1	1.55[c]	0.64[c]		
11-oxomogroside I E1	0.325[c]			
Mogroside I A1	10.9[c]			
Mogroside I E1	8.6[c]			
Mogroside IV A	20.4[a]			
Mogroside IV E	17.6[a]			
20-hydroxy-11-oxo mogroside I A1		0.202[c]		
11-oxomogroside II E		6.84[c]		
11-oxomogroside III			1.832[c]	
11-dehydroxymogroside III			1.338[c]	
11-oxomogroside IV A			0.188[c]	0.28[d]
Mogroside II A1				1.255[d]
Mogroside II B				0.11[d]
Mogroside III A2				0.235[d]
11-deoxymogroside III				2.078[d]
7-oxomogroside II E				0.345[d]
7-oxomogroside V				0.36[d]
11-oxomogroside II A1				0.28[d]

[a]Water extracts
[b]Butanol extracts
[c]Methanol extracts
[d]Ethanol extracts

68, 300, 465, and 195 times as sucrose, respectively [22, 33], but mogroside II and mogroside I were tasteless (Table 3) [22]. In addition, a recent study identifies a novel iso-mogroside V, 3-[(4-O-β-D-glucopyranosyl- β-D-glucopyranosyl)oxy]-mogrol-24- O-β-D- glucopyranosyl- (1->2)-O- [β-D-glucopyranosyl-(1->6)]-β-D-glucopyranoside, as a high-potency sweetener and flavor enhancer with estimated 500 times sweeter than sucrose [13]. Further investigation of the sweetness of mogrosides found that the configuration of hydroxyl group on the carbon 11 of mogrol is the key factor to determine the sweetness of mogrosides. The hydroxyl group with alpha configuration contributes to the sweet taste, while beta configuration turns into bitter taste.

Table 3 Sweetness of mogrosides (Data source: [19])

Compound name	The fold of sweetness to that of sucrose (sucrose = 1)
Mogroside IV	125
Mogroside V	256–378
11-oxo-mogroside V	68
Iso- mogroside V	500
Mogroside IV	300
Siamenoside I	465
Mogroside III	195
Mogroside II	<1
Mogroside I	<1

5 Functional Properties

Mogrosides are known to function as antioxidants, anticarcinogens, and anti-inflammatory substances. Several studies have demonstrated that mogrosides may prevent diabetic complications via their antioxidant properties in diabetic mouse [26, 29, 30]. The anti-inflammatory activities of mogrosides may also contribute to the anti-diabetic effects of mogrosides in both murine macrophage RAW264.7 cells and a murine ear edema model [7]. A number of studies also reported that mogrosides had potent inhibitory effects on early Epstein-Barr virus antigens and the carcinogenesis of mouse skin tumors [1, 14, 32, 38]. Interestingly, a recent study reveals the profiles of mogroside metabolites in rats with oral administration of triterpenoid fractions purified from LHK extracts. The results indicate that the pharmaceutical activities of mogrosides are likely due to the presence of mogrol and monoglucosides in portal blood as sulfates/or glucuronide conjugates [21]. More recently, the distribution of mogroside V metabolites was explored in rats. Seven bioactive metabolites of mogroside V were identified, among which mogroside II E was abundant in the heart, liver, spleen, and lung in rats [42]. So far, the effective forms of mogrosides and the bioavailability of specific mogrosides remain unclear. One of the obstacles is the difficulty to obtain sufficient quantities of specific mogrosides. Methods that can generate specific mogrosides may overcome this restriction in the future.

6 Mogrosides Conversion

The sweetness properties, specific functionalities, and health-promoting effects of individual mogrosides have raised the interest of mogroside conversion to get purified mogrosides. Mogrosides conversion can be achieved by chemical modification, enzyme hydrolysis, or microbial fermentation. Each conversion method provides unique feathers, advantages, and drawbacks as discussed below.

6.1 Chemical Conversion

Chemical conversion, particularly acid hydrolysis, of mogrosides is a common method. The application is straightforward due to the glucose linkages between molecules which can be hydrolyzed by various acids, such as sulfuric acid (H_2SO_4), hydrochloric acid (HCl), phosphoric acid, and trifluoroacetic acid (TFA). Each acid has its specific hydrolyzing properties, and several parameters, such as acid concentration, acid-to-substrate ratio, reaction temperature, and reaction period, are needed to be carefully considered for hydrolysis of mogrosides. Typically, mogrosides can be treated with 2–4 N HCl or 5–10% of H_2SO_4 at 40–90 °C for 30 min to 12 h [48]. Previous study showed that LHK extracts treated with HCl (4 N) at 80 °C for 2 h completely degraded mogrosides and generated eight different cucurbitane triterpenoids including five new aglycones which were not naturally existing [4]. The results suggest that the formation of artificial mogrosides or mogrol compounds at the presence of acids may be due to the combination effects of protonation, dehydration, oxidation, and the shift or rearrangement of double bonds. In addition, the use of concentrated H_2SO_4 can lead to dehydrated byproducts, such as hydroxymethylfurfural (HMF) and levulinic acid [12]. In contrast, TFA, a relative weak acid, could result in mild glucose hydrolysis and obtain broad spectrum of mogrosides containing various numbers of glucose residues (our unpublished data). Although the establishment of acid hydrolysis of mogrosides is easily accessible, several potential drawbacks of chemical conversion need to be taken into account including poor specificity of hydrolysis, low yields, and quality of mogrosides. Chemical conversion also brings concerns of causing safety hazards, environment risks, and toxicological effects [24].

6.2 In Vitro Enzymatic Conversion

Although chemical hydrolysis is favorable due to its low cost, alternative methods need to be considered to provide efficient conversion of mogrosides. Currently, in vitro enzymatic hydrolysis has been extensively used in producing various mogrosides and offers substantial benefits for mogroside hydrolysis. Particularly, glycoside hydrolases, such as β-glucosidase and glucanase, are good candidates to hydrolyze glucose moieties from mogrosides. Cellulase, a 1,4-(1,3;1,4)-β-D-glucan-4-glucanohydrolase, is able to convert mogroside V into mogroside III and mogroside II E [21]. Maltase, an α-glucosidase, converts mogroside V further into mogroside IV E, mogroside III E, MG II E, mogroside I E1, and 11-oxo-mogroside I E1 [35]. To a great extent, Zhou et al. [48] applied enzymatic digestion of mogrosides with 62 different enzymes, including cellulase, dextranases, galactosidase, and pustulanases, to redistribute mogrol glycoside contents. The authors carefully screened the optimal conditions, including enzyme concentration, reaction temperature, and pH, for each individual enzyme to achieve the desired redistribution patterns of the mogrosides. Typically, a β-glucosidase, isolated from *Rhizobium etli*, exhibited high specificity in the conversion of mogroside V to mogroside IV

(59%) and siamenoside I (8%) under the condition of pH 5, at 37 °C. In contrast, a *β*-galactosidase from *Aspergillus oryzae* can convert mogrosides V into siamenoside I (44%) and mogroside IV (7%) with the same condition as mention above. However, if the same amounts of mogroside V were subject to 0.5 M H_2SO_4 at 78 °C for 6 h, only 5% and 2.9% of siamenoside I and mogroside III E were produced [51]. These results obviously demonstrate that enzymatic conversion, compared to chemical conversion, of mogrosides can lead to higher digestion efficiency and production of a specific mogroside. Additionally, enzymatic conversion also provides the protection of the environment by reducing the use of acid and solvents.

Enzymatic conversion has been used to produce various mogrosides, but several drawbacks have limited the application. Especially, β-glucosidase is widely distributed in living organisms, including microbes and plants, it has different degrees of specificity toward the substrates, mogrosides. Thus, searching for an appropriate enzyme with desired activities would be the key factor to perform the mogrosides conversion. In addition, the cost of commercial enzymes is relatively expensive for processing large volumes of mogrosides. However, the advanced development of enzyme immobilization techniques may offer an alternative solution to lower the cost of enzymes.

6.3 Bioconversion

Bioconversion is a biological process that can convert or transform organic materials into a new form. It has been applied to food system, particularly, food fermentation, for a long history. Cheese is one of the best-known microbial bioconversion examples. The flavors, aroma, and textures of mature cheese are contributed by microbial conversion [2]. Recently, bioconversion has been applied to manufacture pharmaceutical products or bioactive ingredients. For example, ginsenoside Rb2 can be bioconverted into compound K (20-O-β-D-glucopyranosyl-20(S)-protopanaxadiol), a major deglycosylated metabolite form of ginsenosides by *Aspergillus niger* [5, 45]. The coculture of probiotic bacteria and yeast, *Saccharomyces boulardii*, can greatly enrich the levels of isoflavones in soymilk fermentation [27].

The bioconversion of mogrosides primarily involves deglucosylation at the peripheral (1→6)-β -D-glucosyl linkages at the C-3 and C-24 positions of mogrol [49]. Glycoside hydrolases are prevalent enzymes in microorganisms. Hence, the properties of glucoside hydrolase could be potentially used to hydrolyze glycosidic bonds in mogrosides. Indeed, studies have demonstrated that rat and human intestinal bacteria are able to convert mogroside V into mogroside III E, mogroside II A, and mogrol [42, 50]. Our recent study further demonstrated that *Saccharomyces cerevisiae* is able to convert mogroside V into various mogrosides, including mogrolside IV, siamenoside I, and mogroside III E, during fermentation. Moreover, mogroside III E as the end product can be produced if long fermentation is applied. However, *Saccharomyces cerevisiae* is not able to convert mogroside V into mogroside II A or mogrol. This suggests that different enzymes secreted during whole cell incubation are responsible for mogroside conversion, and the underlying mechanism for mogrosides bioconversion needs to be further investigated.

6.4 Mogrosides Bioconversion in Yeast

Saccharomyces cerevisiae is a well-defined model organism. It exhibits several unique features, including rapid cell growth, easy for genetic manipulation. Most importantly, the reference genome sequence was completed and the *Saccharomyces genome* database (SGD; http://www.yeastgenome.org) was established with detailed gene and functional annotation. Remarkably, the availability of yeast deletion sets, containing more than 4,600 strains with single deletions of nonessential genes, enables us to identify specific genes required for biological functions and bioconversion of various compounds. In addition, individual gene open reading frame (ORF) is tagged with a TAP epitope and is expressed from its natural chromosomal location. These TAP-tagged strains can further facilitate protein functional analyses and protein-protein interaction studies [11]. By screening mutants from the deletion sets of yeast, we were able to identify Exgl as the enzyme responsible for the initiation of mogroside V conversion in yeast [49]. Unexpectedly, we also demonstrated that cells lacking Kre6, involving in glucan synthesis and cell wall integrity in yeast, could promote the release of Exgl into culture medium and facilitate the production of mogroside III E [39, 49]. Our results further demonstrate that yeast knockout mutants are a valuable tool for mogroside modification and for studying the specificity of enzyme function in mogroside conversion.

7 Conclusions and Future Perspectives

In recent years, biotechnology-derived production of flavors, fragrances, bioactive food ingredients, and drugs has expanded rapidly [10]. In particular, yeast and bacteria are considered with high potential to be good cell factories to produce valuable compounds. For example, through synthetic biology and genetic engineering techniques, scientists have engineered additional 21 genes in yeast to convert sugar into the makings of opioids for medical uses [9]. Similarly, several attempts to produce compound K and other ginsenosides have been successfully achieved in yeast [39, 43]. To provide further understanding of the metabolic pathways of mogroside production in S. grosvenorii, in depth cDNA library sequencing generated quantitative gene expression profiling of mogroside production pathways at different stages of S. grosvenorii fruit after flowering. Many of the key candidate genes encoding enzymes responsible for the mogroside biosynthesis were identified [36]. This study reveals the useful information for the biosynthesis of mogrosides. Indeed, two key enzymes for mogroside biosynthesis, cucurbitadienol synthase and CYP87D18, were co-expressed in yeast and generated three new mogroside derivatives [6, 46]. Despite the yields of the mogrosides are still limited, but this still offers an attractive tool to generate mogrosides.

With the increasing incidence of obesity and diabetes worldwide, the use of natural sweeteners is expanding globally. LHK extracts and mogrosides are both available in the market in many countries. The complexities of the mogroside structures hinder the purification or synthesis of mogrosides and result in difficulties for further producing specific mogrosides. Biotransformation of the LHK extracts and mogrosides V can provide a reliable method to greatly improve the sweet taste and enhance the flavor by converting mogroside V into other sweet mogrosides. *De nono* synthesis of mogrosides would also contribute to a stable production and avoid the insufficient supply of specific compounds due to changes of cultivation condition and environmental factors in the future.

References

1. Akihisa T, Hayakawa Y, Tokuda H, Banno N, Shimizu N, Suzuki T, Kimura Y (2007) Cucurbitane glycosides from the fruits of Siraitia gros venorii and their inhibitory effects on Epstein-Barr virus activation. J Nat Prod 70(5):783–788
2. Button JE, Dutton RJ (2012) Cheese microbes. Curr Biol 22(15):R587–R589
3. Chen Q, Yi X, Yu L, Uang R, Yang J (2005) Study on the variation of mogroside V and flavone glycoside in Luo Han Guo fresh fits in different growth periods. Guangxi Bot 03:274–277
4. Chen XB, Zhuang JJ, Liu JH, Lei M, Ma L, Chen J, Shen X, Hu LH (2011) Potential AMPK activators of cucurbitane triterpenoids from *Siraitia grosvenorii* Swingle. Bioorg Med Chem 19 (19):5776–5781
5. Chi H, Kim DH, Ji GE (2005) Transformation of ginsenosides Rb2 and Rc from *Panax ginseng* by food microorganisms. Biol Pharm Bull 28(11):2102–2105
6. Dai L, Liu C, Zhu Y, Zhang J, Men Y, Zeng Y, Sun Y (2015) Functional characterization of cucurbitadienol synthase and triterpene glycosyltransferase involved in biosynthesis of mogrosides from *Siraitia grosvenorii*. Plant Cell Physiol 56(6):1172–1182
7. Di R, Huang MT, Ho CT (2011) Anti-inflammatory activities of mogrosides from *Momordica grosvenori* in murine macrophages and a murine ear edema model. J Agric Food Chem 59 (13):7474–7481
8. FDA, U. S. (2010) GRAS notice 000301: Luo Han Fruit
9. Galanie S, Thodey K, Trenchard IJ, Filsinger Interrante M, Smolke CD (2015) Complete biosynthesis of opioids in yeast. Science 349(6252):1095–1100
10. Gallage NJ, Moller BL (2015) Vanillin-bioconversion and bioengineering of the most popular plant flavor and its de novo biosynthesis in the vanilla orchid. Mol Plant 8(1):40–57
11. Ghaemmaghami S, Huh WK, Bower K, Howson RW, Belle A, Dephoure N, O'Shea EK, Weissman JS (2003) Global analysis of protein expression in yeast. Nature 425(6959):737–741
12. Girisuta B, Danon B, Manurung R, Janssen LP, Heeres HJ (2008) Experimental and kinetic modelling studies on the acid-catalysed hydrolysis of the water hyacinth plant to levulinic acid. Bioresour Technol 99(17):8367–8375
13. Jia Z, Yang X (2014) Novel Sweetener ISO-Mogroside V. Application number: 20140170286, United States. Assignee: Givaudan SA (Vernier, CH)
14. Konoshima T, Takasaki M (2002) Cancer-chemopreventive effects of natural sweeteners and related compounds. Pure Appl Chem 74(7):1309–1316
15. Lee CH (1975) Intense sweeteners from Lo Han Kuo (*Momordica grosvenorii*). Experientia 31:533
16. Li DP, Ikeda T, Matsuoka N, Nohara T, Zhang HR, Sakamoto T, Nonaka GI (2006) Cucurbitane glycosides from unripe fruits of Lo Han Kuo (*Siraitia grosvenori*). Chem Pharm Bull (Tokyo) 54(10):1425–1428

17. Li DP, Ikeda T, Nohara T, Liu JL, Wen YX, Sakamoto T, Nonaka G (2007) Cucurbitane glycosides from unripe fruits of *Siraitia grosvenori*. Chem Pharm Bull 55:1082–1086
18. Makapugay HC, Nanayakkara NPD, Soejarto DD, Kinghorn AD (1985) High-performance liquid chromatographic analysis of the major sweet principle to Lo Han Kuo fruits. J Agric Food Chem 33(3):348–350
19. Matsumoto K, Kadai R, Ohtani K, Tanaka O (1990) Minor cucurbitane-glycosides from fruits of *Siraitia grosvenori* (Cucurbitaceae). Chem Pharm Bull 38:2030–2032
20. Murata Y, Ogawa T, Suzuki YA, Yoshikawa S, Inui H, Sugiura M, Nakano Y (2010) Digestion and absorption of *Siraitia grosvenori* triterpenoids in the rat. Biosci Biotechnol Biochem 74 (3):673–676
21. Murata Y, Ogawa T, Suzuki YA, Yoshikawa S, Inui H, Sugiura M, Nakano Y (2010) Digestion and absorption of *Siraitia grosvenori* triterpenoids in the rat. Biosci Biotech Bioch 74 (3):673–676
22. Murata Y, Yoshikawa S, Suzuki YA, Sugiura M, Inui H, Nakano Y (2006) Sweetness characteristics of the triterpene glycosides in *Siraitia grosvenori*. J Jpn Soc Food Sci Technol-Nippon Shokuhin Kagaku Kogaku Kaishi 53(10):527–533
23. Nohara T, Li D, Ikeda T, Huang Y, Liu J, Sakamoto T, Nonaka GI (2007) Seasonal variation of mogrosides in Lo Han Kuo (*Siraitia grosvenori*) fruits. J Nat Med 61(3):307–312
24. Puri M, Sharma D, Barrow CJ (2012) Enzyme-assisted extraction of bioactives from plants. Trends Biotechnol 30(1):37–44
25. Qi X, Chen W, Liu L, Yao P, Xie B (2006) Effect of a *Siraitia grosvenori* extract containing mogrosides on the cellular immune system of type 1 diabetes mellitus mice. Mol Nutr Food Res 50(8):732–738
26. Qi XY, Chen WJ, Zhang LQ, Xie BJ (2008) Mogrosides extract from *Siraitia grosvenori* scavenges free radicals in vitro and lowers oxidative stress, serum glucose, and lipid levels in alloxan-induced diabetic mice. Nutr Res 28(4):278–284
27. Rekha CR, Vijayalakshmi G (2010) Bioconversion of isoflavone glycosides to aglycones, mineral bioavailability and vitamin B complex in fermented soymilk by probiotic bacteria and yeast. J Appl Microbiol 109(4):1198–1208
28. Shi DF, Zheng MZ, Wang YM, Liu CM, Chen S (2014) Protective effects and mechanisms of mogroside V on LPS-induced acute lung injury in mice. Pharm Biol 52(6):729–734
29. Song FF, Qi XY, Chen WJ, Jia WB, Yao P, Nussler AK, Sun XF, Liu LG (2007) Effect of *Momordica grosvenori* on oxidative stress pathways in renal mitochondria of normal and alloxan-induced diabetic mice. Eur J Nutr 46(2):61–69
30. Suzuki YA, Tomoda M, Murata Y, Inui H, Sugiura M, Nakano Y (2007) Antidiabetic effect of long-term supplementation with *Siraitia grosvenori* on the spontaneously diabetic Goto-Kakizaki rat. Br J Nutr 97(4):770–775
31. Swingle WT (1941) *Momordica grosvenori* Sp. Nov., the Source of the Chinese Lo Han Kuo
32. Takasaki M, Konoshima T, Murata Y, Sugiura M, Nishino H, Tokuda H, Matsumoto K, Kasai R, Yamasaki K (2003) Anticarcinogenic activity of natural sweeteners, cucurbitane glycosides, from *Momordica grosvenori*. Cancer Lett 198(1):37–42
33. Takemoto T, Arihara S, Nakajima T, Okuhira M (1983) Studies on the constituents of fructus Momordicae. I. On the sweet principle. Yakugaku Zasshi-J Pharm Soc Jpn 103(11):1151–1154
34. Takemoto T, Arihara S, Nakajima T, Okuhira M (1983) Studies on the constituents of fructus Momordicae. II. Structure of sapogenin. Yakugaku Zasshi-J Pharm Soc Jpn 103(11):1155–1166
35. Takemoto T, Arihara S, Nakajima T, Okuhira M (1983) Studies on the constituents of fructus Momordicae. III. Structure of mogrosides. Yakugaku Zasshi-J Pharm Soc Jpn 103 (11):1167–1173
36. Tang Q, Ma X, Mo C, Wilson IW, Song C, Zhao H, Yang Y, Fu W, Qiu D (2011) An efficient approach to finding *Siraitia grosvenorii* triterpene biosynthetic genes by RNA-seq and digital gene expression analysis. BMC Genomics 12(1):343–355
37. Tang SQ, Bin XY, Peng YT, Zhou JY, Wang L, Zhong Y (2007) Assessment of genetic diversity in cultivars and wild accessions of Luohanguo (*Siraitia grosvenorii* [Swingle] A. M. Lu et Z. Y.

Zhang), a species with edible and medicinal sweet fruits endemic to southern China, using RAPD and AFLP markers. Genet Resour Crop Evol 54(5):1053–1061
38. Ukiya M, Akihisa O, Tokuda H, Toriumi M, Mukainaka T, Banno N, Kimura Y, Hasegawa J, Nishino H (2002) Inhibitory effects of cucurbitane glycosides and other triterpenoids from the fruit of Momordica grosvenori on Epstein-Barr virus early antigen induced by tumor promoter 12-O-tetradecanoylphorbol-13-acetate. J Agric Food Chem 50(23):6710–6715
39. Wang P, Wei Y, Fan Y, Liu Q, Wei W, Yang C, Zhang L, Zhao G, Yue J, Yan X, Zhou Z (2015) Production of bioactive ginsenosides Rh2 and Rg3 by metabolically engineered yeasts. Metab Eng 29:97–105
40. Xia Y, Rivero-Huguet ME, Hughes BH, Marshall WD (2008) Isolation of the sweet components from *Siraitia grosvenorii.* Food Chem 107(3):1022–1028
41. Xiong MJ, Tang Q, Ma XJ (2011) Biosynthesis of triterpene glycoside in Lo Han Kuo. Acad J Guangdong Coll Pharm 05:543–548
42. Xu F, Li DP, Huang ZC, Lu FL, Wang L, Huang YL, Wang RF, Liu GX, Shang MY, Cai SQ (2015) Exploring in vitro, in vivo metabolism of mogroside V and distribution of its metabolites in rats by HPLC-ESI-IT-TOF-MSn. J Pharm Biomed Anal 115:418–430
43. Yan X, Fan Y, Wei W, Wang P, Liu Q, Wei Y, Zhang L, Zhao G, Yue J, Zhou Z (2014) Production of bioactive ginsenoside compound K in metabolically engineered yeast. Cell Res 24(6):770–773
44. Yan XW, Zhang JY (2007) NMR structural elucidation of mogrol and its glycosides. Chin J Magn Reson 03:249–260
45. Yang XD, Yang YY, Ouyang DS, Yang GP (2015) A review of biotransformation and pharmacology of ginsenoside compound K. Fitoterapia 100:208–220
46. Zhang J, Dai L, Yang J, Liu C, Men Y, Zeng Y, Cai Y, Zhu Y, Sun Y (2016) Oxidation of cucurbitadienol catalyzed by CYP87D18 in the biosynthesis of mogrosides from *Siraitia grosvenorii.* Plant Cell Physiol 57(5):1000–1007
47. Zhang JY, Yang XW (2003) Assignments of ^{1}H and ^{13}C NMR signals of mogroside IVa. J Chin Pharm Sci 04:196–200
48. Zhou YPSP, Hoffman Estates, IL, 60192, US), ARMENTROUT, Richard W. (5450 Prairie Stone Parkway, Hoffman Estates, IL, 60192, US), WOODYER, Ryan D (5450 Prairie Stone Parkway, Hoffman Estates, IL, 60192, US), BRIDGES, John R (5450 Prairie Stone Parkway, Hoffman Estates, IL, 60192, US), SCHUNK, Timothy C (5450 Prairie Stone Parkway, Hoffman Estates, IL, 60192, US), FLETCHER, Joshua N (5450 Prairie Stone Parkway, Hoffman Estates, IL, 60192, US) (2014) Redistribution of mogrol glycoside content. TATE & LYLE INGREDIENTS AMERICAS, LLC (5450 Prairie Stone Parkway, Hoffman Estates, IL, 60192, US)
49. Chiu C-H, Wang R, Lee C-C, Lo Y-C, Lu T-J (2013) Biotransformation of mogrosides from *Siraitia grosvenorii* Swingle by *Saccharomyces cerevisiae*. J Agric Food Chem 61:7127–7134
50. Yang XW, Zhang J-Y, Xu W (2008) Biotransformation of mogroside III by human intestinal bacteria. J Peking Univ Health Sci 39(6):657–662
51. Zhou Y, Armentrout RW, Woodyer DR, Bridges JR, Schunk TC, Fletcher JN (2014) Redistribution of mogrol glycoside content. Application number: WO/2014/150127. Assignee: Tate & Lyle Ingredients Americas, LLC

Mass Production of the Taste-Modifying Protein Miraculin in Transgenic Plants

8

Hiroshi Ezura and Kyoko Hiwasa-Tanase

Abstract

Miraculin is a glycoprotein that is found in red berries, which are known as a miracle fruit (*Richadella dulcifica*; synonym *Synsepalum dulcificum*) and are produced by a tropical shrub native to West Africa. Miraculin itself is not sweet, but it can convert a sour taste into a sweet taste. Due to its unique properties and potential use as an alternative sweetener, the mass production of miraculin is of interest. However, the plant has low fruit productivity, and there are limited natural sources of miraculin protein. Therefore, heterologous miraculin production based on genetic engineering techniques has been attempted using plants such as tomato, lettuce, and strawberry. The recombinant miraculin protein has been successfully expressed in transgenic tomatoes and lettuce in a genetically stable manner. In addition, a plant factory, which is a closed cultivation system and may be suitable for producing transgenic plants expressing recombinant miraculin, has been developed. Finally, a simple method for purifying miraculin from transgenic tomato fruits was established. In this chapter, we introduce the mass production of recombinant miraculin protein in transgenic tomatoes and lettuce.

Keywords

Miraculin • Taste-modifying protein • Tomato • Lettuce • Plant factory • Transgenic plant • Purification

Contents

1 Introduction 168
2 What Is Miraculin? 168
3 Expression of Recombinant Miraculin Protein in Transgenic Tomatoes 171

H. Ezura (✉) • K. Hiwasa-Tanase
Faculty of Life and Environmental Sciences, Graduate School of Life, University of Tsukuba, Tsukuba, Japan
e-mail: ezura@gene.tsukuba.ac.jp; tanase.kyoko.gf@u.tsukuba.ac.jp

J.-M. Mérillon, K.G. Ramawat (eds.), *Sweeteners*, Reference Series in Phytochemistry,
https://doi.org/10.1007/978-3-319-27027-2_17

4 Expression of Recombinant Miraculin Protein in Transgenic Lettuce 175
5 Production of Transgenic Tomato Fruits at a Plant Factory 176
6 Purification of Recombinant Miraculin from Transgenic Tomato Fruits 177
7 Conclusions 178
References 180

1 Introduction

Sweet-tasting proteins that have activity as sweeteners and taste modifiers have been discovered mostly in tropical plants [1]. These proteins include thaumatin, monellin, mabinlin, pentadin, brazzein, neoculin (curculin), and miraculin. Their original functions remain unknown. However, because of their sweet-tasting properties, they have attracted research attention. These properties are classified into three categories: the protein itself is sweet (thaumatin, monellin, mabinlin, pentadin, and brazzein), the protein induces sweetness (miraculin), or both (neoculin).

In many countries, wealthy lifestyles and food satiety have increased the risk of adult disease from excessive sugar intake. Artificial sweeteners such as saccharin, aspartame, sucralose, and acesulfame-K are used worldwide as low-calorie sweeteners to prevent chronic lifestyle diseases and to help control sugar consumption in diabetics; however, some of these compounds may have carcinogenic side effects [1]. Sweet-tasting proteins may be able to replace these artificial sweeteners because of their properties as natural, safe, and low-calorie sweeteners. Due to their high industrial potential, the mass production of these sweet-tasting proteins has been attempted by using heterologous biological production systems such as bacteria, yeast, fungi, and transgenic plants [2]. In particular, the production of sweet proteins, including thaumatin and monellin, by microorganisms has been attempted. Attempts have also been made to produce thaumatin using transgenic plants such as potatoes [3], tomatoes [4], strawberries [5], pears [6], and cucumbers [7], whereas tomatoes and lettuce have been used to produce monellin [8]. However, the recombinant protein accumulation levels in those transgenic plants are quite low. Lamphear et al. [9] produced recombinant brazzein in transgenic corn at a concentration of 4.3 mg/80 g, and a field trial was conducted. However, none of these efforts, including the attempts with brazzein, have yet resulted in the commercialization of these proteins as sweeteners.

The mass production of miraculin, which is a taste-modifying protein, has also been attempted using heterologous organisms. Recently, progress has been made using transgenic plants to mass-produce miraculin. In this chapter, we provide a general overview of the published studies and assess the future commercial viability of miraculin.

2 What Is Miraculin?

Miraculin is a glycoprotein that is found in red berries, which are known as miracle fruit (*Richadella dulcifica*; synonym *Synsepalum dulcificum*) and are produced by a tropical shrub native to West Africa (Fig. 1). Miraculin itself is not sweet, but it has a

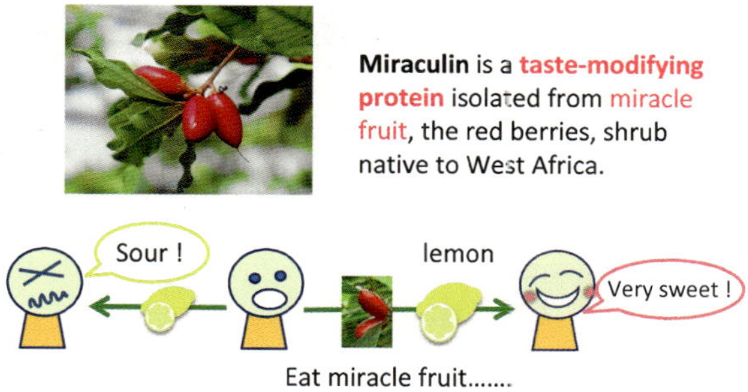

Fig. 1 What is miraculin? Miraculin by itself is not sweet, but it is able to turn a sour taste into a sweet taste

taste-modifying activity and is capable of converting a sour taste into a sweet taste. After chewing the miracle fruit, lemons taste as sweet as oranges. The name "miracle fruit" was derived from this unique and attractive property, and the isolated active substance was named miraculin [10]. Miraculin accumulates only in the miracle fruit, and its production begins 6 weeks after pollination, or at the turning stage, when the fruit color changes from green to orange; the accumulation peaks when the fruit is at the full-red stage.

Miraculin is able to elicit sweetness from various acids, such as HCl, oxalic acid, lactic acid, formic acid, acetic acid, and citric acid; the sweetening effect is dependent on the sourness and pH of the acid [11]. The sweetening effect of a miraculin solution reaches its maximum level after being held in the mouth for approximately 3 min. A concentration greater than 4×10^{-7} M is required for the maximum effect, and the sweetness corresponds to that of a 0.4 M sucrose solution. The taste-modifying effect of miraculin can be sustained for more than 1 h, although it depends on the concentration of the miraculin solution.

Miraculin is a glycoprotein that consists of 191 amino acid residues and two sugar chains that are linked to Asn-42 and Asn-186. The molecular weight of the single polypeptide chain calculated based on its amino acid sequence and carbohydrate content (13.9%) is 24.6 kDa [12, 13]. The nucleotide sequence encodes 220 amino acid residues, including a 29 amino acid signal sequence [14]. This indicates that mature miraculin protein is processed by posttranslational modification at its N terminus. The pure state of native miraculin protein forms a tetramer, whereas the crude non-reduced and denatured state forms a dimer [15]. Both the tetramer and native dimer forms of miraculin exhibit taste-modifying activity [15], although the monomer form has no activity [16]. Thus, miraculin dimerization, as mediated by a covalent linkage at Cys-138, is crucial for its taste-modifying activity at acidic pH.

Kurihara and Beidler [11] suggested that the miraculin protein has two binding sites: one binds the taste-receptor membrane and the other reacts with sweetness receptors. The conformation of the protein changes under acidic conditions to induce sweetness. This speculation comes from results that show that the thresholds for salty, bitter, sweet, and sour tastes remain unaffected after holding miraculin in the mouth. Furthermore, holding miraculin on the tongue does not turn sourness into sweetness when gymnemic acid, which depresses sweetness, is held in the mouth after applying the miraculin solution. Thus, miraculin binds to membrane receptors close to sweetness receptor sites and activates sweet receptors under acidic conditions. Some studies have revealed that miraculin does not possess taste-modifying properties as a monomer [16, 17]. Additionally, mutagenesis and simulation analyses have shown that two histidine residues play an important role in its taste-modifying activity [16, 18]. Moreover, the three-dimensional structure of the homodimer predicts a wide-open conformation at acidic pH and a closed form at neutral pH [19]; it has been suggested that the histidine residues are involved in this conformational change [18]. However, the details of the mechanisms of where and how miraculin binds on the tongue and how it modifies sourness to sweetness still require elucidation. Against this background, more recently, Koizumi et al. [20] quantitatively evaluated the acid-induced sweetness of miraculin using a cell-based assay system at the molecular level. They determined that miraculin activated the hT1R2–hT1R3 receptor, which is the human sweet taste receptor, when the pH decreased from 6.5 to 4.8 and suggested that miraculin binds to the hT1R2 receptor at the N-terminal region. In the near future, the taste-modifying mechanism from sourness to sweetness may also be clarified by the cell-based assay system for miraculin activity developed in that work.

Due to its unique properties and potential as a sweetener, the mass production of miraculin has been greatly anticipated. However, because it is a tropical plant, the miracle fruit plant cannot survive at less than 7 °C; additionally, it requires 3–4 years before it can bear fruit, and it has a low proportion of fruit setting after pollination. Furthermore, the taste-modifying activity is lost within 2–3 h after harvesting at room temperature [21]. Thus, there are limited natural sources of miraculin protein. Therefore, heterologous miraculin production based on genetic engineering techniques has been attempted using foreign hosts such as *Escherichia coli*, yeast, *Aspergillus oryzae*, and plants.

To utilize its unique properties and expand its market potential, the heterologous production of miraculin has been attempted using various hosts. Recombinant miraculin produced in *E. coli* was found to have no taste-modifying activity, even though the protein was detected by SDS-PAGE and western blot analysis [22]. The production of the recombinant miraculin was also attempted in *E. coli* by Matsuyama et al. [17], who produced an active miraculin dimer. However, its activity was only one-sixth that of native miraculin. This result showed that the glycosylation of miraculin is crucial for protein folding and/or stability. An active form of recombinant miraculin was also produced using *A. oryzae,* which has glycosylation capacity, as a host strain [16]. In the *Saccharomyces cerevisiae* system, recombinant miraculin was produced after optimizing codon usage and the signal sequence, although the activity was only detected when the sugar chains were removed [23]. The activity of recombinant miraculin in both *A. oryzae* and *S. cerevisiae* was evaluated at one-fifth the concentration of native miraculin. These results suggest that not only is glycosylation important but that the types of sugar chains are also crucial for the high and stable activity of miraculin. In the beginning of recombinant miraculin production in transgenic plants, the expression cassette usually incorporated a cauliflower mosaic virus 35S (35S) promoter and a nopaline synthase (NOS) terminator. Sun et al. [24] expressed recombinant miraculin in transgenic lettuce (*Lactuca sativa*). They

produced miraculin at a concentration of 43.5 ug/g fresh weight (FW), and the resulting recombinant miraculin had a taste-modifying activity that was equal in strength to that of native miraculin. However, transgene silencing occurred in the progeny line of the transgenic lettuce [24]. In transgenic strawberries (*Fragaria x ananassa*), which propagate vegetatively, recombinant miraculin is accumulated at an extremely low level of 2.0 μg/g FW, although the accumulation level was stable in the vegetative progeny [25]. In contrast, transgenic tomatoes (*Solanum lycopersicum*) have produced recombinant miraculin at a level of 90.7 μg/g FW in fruit tissue, with activity equal to that of native miraculin [26]. Additionally, the accumulation and gene expression levels were genetically stable from the T1 to T5 generations [27]. These reports suggest that plants are a good platform for producing a biologically active form of miraculin. It may be that the plant-mediated glycosylations are similar to those of miraculin produced from miracle fruit. Furthermore, transgenic tomatoes were found to be the most suitable hosts for recombinant miraculin production from these plants. Tomatoes also have an advantage in the established *Agrobacterium*-mediated transformation protocol using cotyledon explants from seedlings, and the transformation efficiency exceeds 40% of the explants [28].

3 Expression of Recombinant Miraculin Protein in Transgenic Tomatoes

Recombinant miraculin protein was expressed in transgenic tomatoes [26]. After the successful expression of recombinant miraculin in transgenic tomatoes, various approaches have been attempted to increase recombinant miraculin production in transgenic tomatoes. Recombinant miraculin accumulates to high levels in the overripe fruit during fruit development when the miraculin gene is driven by the 35S promoter [29]. Among fruit tissues, the miraculin levels in the exocarp, or epidermis, were found to be extremely high, at 928 μg/g FW compared to below 110 μg/g FW in other tissues. Why did recombinant miraculin accumulate to such a high amount in the exocarp? The deduced amino acids of miraculin include an N-terminal signal sequence of 29 amino acids [14]. Miraculin protein is transported to and accumulates in the intercellular layer space of both miracle fruit and transgenic tomatoes [30]. The cell size of the exocarp is considerably smaller than that of other fruit tissues, such as the mesocarp, septa, placenta, and jelly. The amount of intercellular layer space is affected by the number of cells per fresh weight. In fact, the amount of miraculin per dry weight, which is less affected by cell size, hardly differs between the exocarp and other tissues [29]. Therefore, the authors speculate that the high miraculin accumulation in the exocarp is caused primarily by cell size, which influences the amount of intercellular layer space [29].

The promoter is an essential element for inducing target gene expression, determining the target tissue and achieving the desired expression time [31–34]. There are multiple types of promoters that are suitable for a variety of purposes, including constitutive, inducible, and tissue-specific expression. A constitutive 35S promoter from cauliflower mosaic virus is frequently used to mass-produce recombinant proteins because it usually leads to higher expression than tissue-specific promoters in various plant tissues and organs [31, 35]. However, when the recombinant protein suppresses the growth or metabolism of the host plant in some tissues, the use of the constitutive promoter loses its advantage. Additionally, there have been some cases showing that gene silencing via co-suppression is caused by the utilization of a constitutive promoter. The compatibility of the promoter with the host plant is a crucial factor for the strong expression of a target gene [31, 36]. In fact, the miraculin gene, when driven by the 35S promoter, was silenced in transgenic lettuce [24]. Moreover, although gene silencing does not occur in transgenic strawberries using the 35S promoter, the production of miraculin is quite low [25]. Unlike these transgenic plants, the 35S promoter works in transgenic tomatoes and increases miraculin production [26, 27].

The tissue-specific E8 promoter has been used to control the expression of the miraculin gene in transgenic tomatoes [37]. This promoter, which was first identified in cherry tomatoes [38], activates transcription at the onset of tomato fruit ripening [39, 40]. Miraculin gene expression in red tomato fruit was found to be significantly lower under the control of the E8 promoter than under the 35S promoter, and the miraculin content was also lower [37]. This result indicates that the 35S promoter has higher transcriptional activity in tomatoes than the E8 promoter does. The 30-untranslated region and terminator sequence are also crucial factors for regulating the expression level of target genes. These sequences influence the gene expression level and the efficiency of translation by controlling transcriptional termination and posttranscriptional processes [41]. In fact, the terminator has a considerable effect on the gene expression level [42, 43]. The NOS terminator from the Ti (tumor-inducing) plasmid of *Agrobacterium* is universally applied in various expression vectors when genes are transformed into plants. When the miraculin gene was driven by the 35S promoter and terminated by the NOS terminator in transgenic tomatoes, the miraculin content was 1–1.5% of the total soluble protein [26, 44, 45]. In contrast, the miraculin content in miracle fruit is approximately 10% of the total soluble protein [12]. Because miraculin protein is produced more efficiently in miracle fruit, the native miraculin terminator from miracle fruit was cloned and used to produce miraculin in transgenic tomatoes [44]. As a result, the accumulation levels of recombinant miraculin in transgenic lines using the miraculin terminator were 1.5 times higher per fresh weight than they were when using the NOS terminator. Nagaya et al. [42] assessed the capacity of termination to support increased transgene expression using several terminators derived from Arabidopsis genes and showed that the heat shock protein 18.2 (HSP) terminator was the most effective at supporting high expression levels. In fact, the application of the HSP terminator increased the production of the B subunit of Shiga toxin 2e in transgenic lettuce at the T0 generation to levels approximately 40 times higher than those obtained with the NOS terminator [46]. The HSP terminator has also been used for miraculin production in transgenic tomatoes [47]. The miraculin accumulation level in the homozygous T1 generation reached 1726 μg/g FW, which is ten times higher than that obtained with the NOS terminator, equivalent to 17.1% of the total soluble protein. In most cases using nuclear transformation, the accumulation level of the recombinant protein does not exceed 1–2% of the total soluble protein [48–50]. Conversely, recombinant proteins derived from chloroplast transformation normally accumulate to 5–25% of the total soluble protein [51, 52]. It is difficult to use the chloroplast transgenic system for recombinant miraculin production because the protein requires posttranslational modifications such as glycosylation. However, the accumulation scale of recombinant miraculin in transgenic tomatoes using the 35S promoter/HSP terminator expression cassette matches the level obtained with chloroplast transformation.

Codon modifications are used to increase the translational efficiency of heterologous genes in their hosts. Preferred codon usage varies significantly among different organisms and even between plant species [53–59]. Therefore, this approach seeks to increase the accumulation of recombinant proteins by modifying their original codons to codons that are more suitable to the host without changing the amino acid sequence [60]. Codon optimization of the miraculin gene for tomatoes enhanced its translational efficiency more than twofold compared to that of the native miraculin gene in transgenic tomatoes [44]. The resulting accumulation of the codon-modified miraculin gene product exceeded that of native miraculin, and this effect was more pronounced when the miraculin terminator was used rather than the NOS terminator. These results indicate that it is important for productivity to both optimize codons and select a suitable terminator.

To produce recombinant proteins using plants, it is important to pay attention to not only the species but also the cultivar when selecting the genetic background. Kim et al. [61] created F1 hybrids from the transgenic tomato that accumulates miraculin (cv. "Moneymaker") by crossing with various tomato cultivars to elucidate the impact of genetic background on miraculin accumulation. The accumulation pattern in fruit tissues was unchanged; the protein is accumulated to high levels in the exocarp, followed by the mesocarp, just as in the original transgenic line. However, the accumulation of miraculin in fruits was different in the different genetic backgrounds. The ratio of fruit tissue to fruit weight is variable in tomato cultivars. As an example, the proportion of exocarp was 8.16% in "Micro-Tom," which was 5% higher than in the other varieties [29]. Therefore, miraculin accumulation could be affected by the ratio of each fruit tissue in the different genetic backgrounds because the accumulation levels differ depending on fruit tissues. The miraculin accumulation levels also differed in the exocarp and mesocarp among the hybrid lines [61]. Cell size and cell numbers per fresh weight are different even in the same exocarp, depending on the hybrid line, similar to the difference in fruit size. Such a different fruit structure is influenced by the final accumulation levels because cell numbers are roughly proportional to the intercellular layer space to which miraculin protein is transferred and stored. In addition, differences in harvesting time were found to affect miraculin accumulation in fruits in which the miraculin gene is driven by the 35S promoter; the miraculin levels in tomato fruits are higher at later harvesting times [37, 62]. Presumably, there are differences in the fruit maturation times of these cultivars. In the case of miraculin in transgenic tomatoes driven by the 35S promoter, this factor is important for increasing the miraculin amounts. Moreover, crossbreeding transgenic tomatoes with other cultivars expands the capacity for modulating the production level of recombinant miraculin. One of the best examples of this strategy is the cross between the transgenic tomato that accumulates miraculin (cv. "Moneymaker") and "Micro-Tom," which is a miniature dwarf tomato. The miraculin concentrations in the crossed lines were 2.5 times higher than in the transgenic line as a parent [45].

Recombinant protein content can be increased by increasing the gene dosage given to the host plant. However, increasing gene dosage by delivering multiple gene copies does not always yield a good result because multi-copy transformants often trigger low levels of transgene expression or silencing [63, 64]. In contrast, using homozygous transgenes can increase production compared to using heterozygous transgenes. In fact, the expression levels in the exocarp and mesocarp in homozygous lines given a single copy of the miraculin gene were 4.7 and 1.5 times higher than those in heterozygous transgenic tomatoes, respectively [61]. The recombinant miraculin levels in the exocarp and mesocarp were two times higher in homozygous lines than in heterozygous lines. These results indicate that homozygosity of a single-copy gene promotes the production of recombinant miraculin better than does heterozygosity. A single-copy transgenic plant tends to be more stable in its transgenic expression than a plant with multiple transgene copies [65]. Thus, it is crucial for the stable production of recombinant miraculin to select a transgenic plant in which the gene was introduced as a single copy and to maintain its homozygous state.

The intracellular localization of recombinant heterologous proteins is an important determinant of their accumulation levels. The subcellular site in which a recombinant protein accumulates can have a profound effect on its proper folding, assembly, and posttranscriptional processing [36]. When a heterologous protein is not correctly folded or synthesized in a plant, the synthesized proteins are easily attacked by proteases [31]. Thus, suitable subcellular targeting contributes to protein stability and, eventually, to the yield of the recombinant protein [30, 66]. In the same manner, recombinant miraculin is secreted to the intercellular layer in transgenic tomato fruits and leaves by virtue of the signal sequence of the native gene [30]. Furthermore, transgenic tomatoes in which the miraculin gene is driven by the 35S promoter and terminated by the HSP terminator were found to accumulate recombinant miraculin at up to 1,725 μg/g FW, equivalent to 17% of the total soluble protein [47]. This accumulation level exceeds the 400 μg/g FW of miraculin protein that is produced in miracle fruit. Thus, the intercellular layer space appears to be one of the best compartments for the accumulation of recombinant miraculin in transgenic tomato fruit, although it leaves the possibility that other subcellular spaces are also suitable for recombinant miraculin accumulation. The taste-modifying activity of miraculin is relatively stable under acidic conditions [12]. The fact that the apoplastic pH of tomatoes declines from an average of 6.7–4.4 during fruit ripening might also be advantageous [67].

4 Expression of Recombinant Miraculin Protein in Transgenic Lettuce

Lettuce is a popular and easy-to-grow leafy vegetable that is cultivated worldwide. Lettuce is a typical crop that is commercially cultivated in plant factories, which are indoor cultivation systems with a controlled light period, light intensity, temperature, and CO_2 concentration for the mass production of target plants [68]. In a plant factory, it takes approximately 3 weeks from seed sowing to harvest lettuce under optimal growth conditions; therefore, lettuce can be harvested more than 20 times per year. Thus, if we can stably express a target protein of interest, lettuce will be an alternative platform for the mass production of recombinant proteins.

Recombinant miraculin protein was expressed first in transgenic lettuce [24]. In that study, miraculin was produced at a concentration of 43.5 μg/g FW, and the resulting recombinant miraculin had taste-modifying activity that was equal in strength to that of native miraculin. However, transgenic silencing occurred in the progeny line of the transgenic lettuce. Many attempts have been made to produce transgenic lettuce, and useful traits have been introduced into the crop [24, 69–72]. A high degree of transgene silencing in lettuce is one of the major barriers to commercializing transgenic lettuce.

For this reason, it is important to develop a strategy for stably expressing transgenes in lettuce. The promoter is a major factor influencing the level and stability of transgene expression. Curtis et al. [69] compared several promoter-GUS gene fusions in transgenic lettuce plants and found that the plastocyanin gene (petE) promoter exhibited higher expression than did the mannopine synthase gene (MAS) [73], hybrid 35S/Mas promoter 50 region (Mac) [74], or cauliflower mosaic virus (CaMV) 35S promoters in first seed generation (T1) plants. The choice of both the promoter and transgene constructs is important for long-term stable expression of transgenes in lettuce [71]. Unstable gene expression is also often related to the integration of multiple copies of the transgene in the plant genome [75], position effects [76], and the extent of DNA methylation in the transgene loci [77].

Ubiquitin is a small, highly conserved protein that consists of 76 amino acid residues and is found in all eukaryotes. The ubiquitins are encoded by gene families that contain two types of structures: polyubiquitin genes and ubiquitin extension protein genes [78, 79]. Both types of genes are translated as polyprotein precursors and then proteolytically processed to ubiquitin monomers [80]. Polyubiquitin genes are constitutively expressed in all types of plant tissues, with higher levels found in young tissues [81, 82]. Various promoters from ubiquitin genes have been tested for their potential use in driving the expression of foreign genes in plant transformation systems. Ubiquitin promoters have been successfully used to transfer selected genes in many plants, including monocots and dicots (e.g., *Arabidopsis*, sunflower, potato) [83–86].

To achieve stable miraculin expression in lettuce, we compared the level of miraculin accumulation between the CaMV 35S promoter/nos terminator cassette and the native lettuce ubiquitin promoter/terminator cassette in transgenic lettuce

[87]. Transgenic lettuce plants using the 35S promoter/nos terminator cassette showed almost complete silencing of miraculin gene expression in the T2 generation, whereas those using the native ubiquitin promoter/terminator cassette showed stable miraculin expression, even in the progenies. The study indicates that the use of the endogenous lettuce ubiquitin promoter to drive the miraculin gene in a transformation system could overcome the transgene silencing problem. We found that all single-copy transgenic lines using the ubiquitin promoter expressed the miraculin gene in the T0 generation at a high level compared to using the 35S promoter. The expression of miraculin was also clear and stable in the T1 and T2 generations, as revealed by real-time PCR, western blotting, and ELISA. The transgenic lines showed stable expression and inheritance of the miraculin gene for up to three generations. The effectiveness of the maize ubiquitin promoter was reported by Chen et al. [88, 89]. Chen et al. [88] reported that the 35S promoter-derived gene was silenced in the T1 generation of transgenic maize and that the maize ubiquitin promoter-derived gene was expressed in the T1 generation of transgenic maize. In this study, we have demonstrated the genetically stable expression of miraculin in transgenic lettuce, which suggests that lettuce would be an alternative platform for mass production of recombinant miraculin protein.

5 Production of Transgenic Tomato Fruits at a Plant Factory

To improve recombinant miraculin production and stabilize its quality, it is essential to develop a cultivation system for transgenic tomatoes. Tomatoes are usually cultivated in fields or greenhouses for commercial use. However, these environments depend a great deal on the season and weather conditions, such as temperature, the amount of solar radiation, and the photoperiod; these influence the fruit quality and yield. In contrast, a closed cultivation system (also called a plant factory) makes it possible to stabilize fruit yield and quality in a constantly controlled environment [90]. In fact, the transgenic tomato that expresses the miraculin gene produces recombinant miraculin more stably in a closed tomato cultivation system than in a netted greenhouse [68]. Additionally, the yield has been estimated at approximately 45 ton FW of fruits and 4 kg of recombinant miraculin per 1,000 m^2 of cultivated area in a year. Closed cultivation systems also have other advantages: the spread of transgenic plants and pollen to the external environment is easy to prevent and, at the same time, the plants are protected from disease and pests from the external environment. However, the limited space available for cultivation and the high operating costs become bottlenecks for its practical application. It is important to breed a plant that is suitable for cultivation in a plant factory [91]. To breed tomatoes suitable for a closed cultivation system with the aim of effective cultivation space utilization, a transgenic tomato line based on a normal cultivar was crossed with a dwarf tomato cultivar, "Micro-Tom," that has a short life cycle and determinate-type growth [45]. In F2 plants, hybrids were selected by small plant size, miraculin accumulation, and determinate type; they were then propagated to the F6 or F7 generation by self-pollination. In the cross no. 2 line, it was not necessary to remove axial buds and leaves when cultivated in a two-layer closed cultivation system, in contrast to the original transgenic line. This growth habit helps reduce the labor costs. Additionally, this plant line has the potential to be grown in a three-layer cultivation system in the same space as a two-layer system because the plant height is significantly lower than in the other lines. The fruit yield of the cross no. 2 line in the two-layer cultivation system was equivalent to 45.9 ton FW per 1,000 m^2 of cultivation area in a year. The miraculin content in the tomato fruit significantly increased to 343 μg/g FW of pericarp (including exocarp and mesocarp), which is approximately 2.5 times higher than that of the original transgenic line, due to the effect of genetic background modification.

The high operating costs of closed cultivation systems mainly accrue from the electric bills arising from the artificial lighting and air conditioning. To reduce electricity costs, it is important to understand the effects of light intensity on both tomato fruit yield and the production of recombinant miraculin. The cross no. 2 line acquired some characteristics from the "Micro-Tom" dwarf tomato, such as its growth habit and its determinate type [45]. The ability to grow under lower intensity light was also gained by the cross no. 2 line. Normal-sized tomatoes cannot be cultivated under lower light conditions because low light makes processes such as flower-bud formation, flowering, photosynthesis, and healthy plant growth difficult. However, the cross no. 2 line grew normally even at a photosynthetic photon flux (PPF) of 100 μmol/m^2/s, as does "Micro-Tom," which can be cultivated under 80 μmol/m^2/s [62]. Kato et al. [62] evaluated the relationship between fruit yield and miraculin concentration under the different light conditions of PPF100, 200, 300, and 400 μmol/m^2/s and discussed the electricity costs of miraculin production. The miraculin concentrations were higher at low light intensity, but the fruit yield was higher at strong light intensity. Consequently, the miraculin production per unit area was highest at PPF300, but the miraculin production per unit of energy was best at PPF100. Thus, these results indicate that it is necessary to select a suitable light condition on the basis of market demand and the sales price of recombinant miraculin.

6 Purification of Recombinant Miraculin from Transgenic Tomato Fruits

The advantage of using an edible plant as a host lies in the fact that several processed and unprocessed types of recombinant proteins can be used, unlike in microbial and animal systems [36]. Thus, this practice can minimize purification costs.

Miraculin has a taste-modifying function that changes a sour taste into a sweet taste. This property of miraculin is helpful for preventing lifestyle diseases: it can be used as a low-calorie sweetener or as an additive for foods targeting diabetics because it removes the need for sugar. It might also find novel uses as a component of nutritionally enhanced snacks and drinks with the possibility of widespread consumer acceptance and new marketing opportunities. Additionally, recombinant miraculin could be used for safety assessment analysis and to identify the taste-modifying mechanism of miraculin. Therefore, the industrial uses of recombinant miraculin from transgenic tomatoes encompass the consumption of unprocessed raw fruit and processed fruit and its use as a food additive and as a reagent for research and development. To address this wide variety of uses, it is important to develop methods of purifying recombinant miraculin from transgenic tomatoes. Techniques for purifying miraculin from miracle fruit have been developed since 1968 [10, 92, 93]. However, these methods cannot remove the polyphenolic coloring agent and lead to a reduction in the taste-modifying activity due to the use of a carbonate extraction buffer (pH 10.5). In 1988, Theerasilp and Kurihara [12] presented a method in which a NaCl solution was used for crude extraction from the miracle fruit, followed by purification via CM-Sepharose ion-exchange chromatography and ConA-Sepharose 4B affinity chromatography. Armah et al. [94] also developed a method using a NaCl extraction buffer and hydrophobic interaction chromatography (Butyl-S-Sepharose 6 FF, Phenyl Sepharose 6 FF), ion-exchange chromatography (SP-Sepharose FF), and gel filtration chromatography (Sephacryl S-100); the authors received a US patent in 1999 (patent number 5886155). These methods permit the isolation of miraculin from polyphenolic pigments. Recently, a single-step purification system using nickel-immobilized metal-affinity chromatography (IMAC) was developed [95]. This method utilizes four histidine residues that are exposed on the surface of dimeric miraculin, as with a histidine tag. It is much simpler than previous methods, and the reduction in column steps makes the process less time-consuming, thus reducing the purification loss and costs. In addition, the method can purify the active form of miraculin free from the denatured form because the IMAC column binds only miraculin that has maintained the three-dimensional structure of its homodimer (which elicits its sweetness-inducing activity) and does not bind monomeric miraculin, which is the inactive form. However, recombinant miraculin purification is more difficult from tomato fruit than it is from miracle fruit because tomato fruit contains a wide variety of additional proteins. Therefore, this purification requires one additional column, an ion-exchange column (CM-Sepharose), for the production of highly purified miraculin [96]. Recombinant miraculin has also been purified using the method of Theerasilp and Kurihara [12] by the addition of molecular weight column chromatography (Sephacryl S-200 HR), but the method is time-consuming and has a low recovery rate due to the number of purification steps [26]. Although the purification method using IMAC is efficient compared to the other methods, in the future, it will require further development to be used on an industrial scale.

7 Conclusions

Transgenic plants are presumably the most suitable hosts for recombinant miraculin protein production with respect to posttranslational modifications such as glycosylation. A scheme for the mass production of recombinant miraculin protein using transgenic plants is summarized in Fig. 2. The selection of host plants; optimization

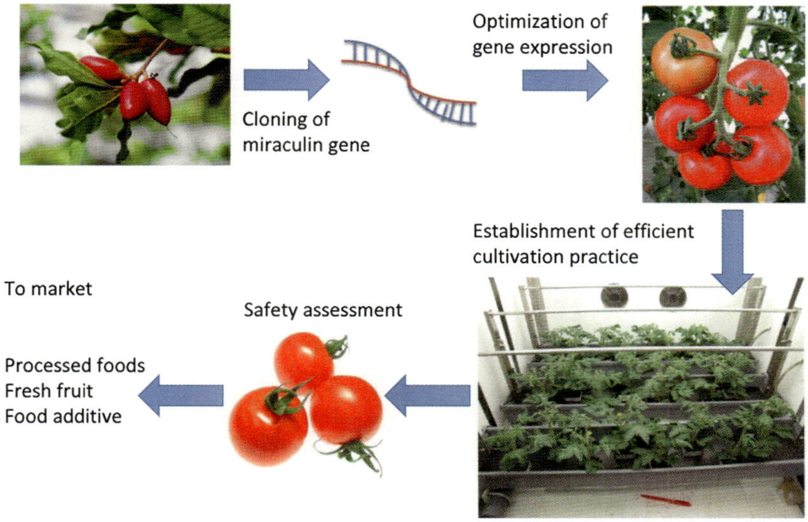

Fig. 2 A scheme for the effective and efficient mass production of recombinant miraculin using transgenic plant, a case of tomato

of expression vectors, especially promoters and terminators for each plant; and establishment of standard cultivation protocol for the transgenic plants are crucial for effective and efficient production of recombinant miraculin. In particular, the transgenic tomato has many advantages, including genetic stability, good accumulation levels, and good fruit-bearing capacity, and it is readily available to procure as a raw food. In addition, a newly produced transgenic lettuce was found to be able to stably accumulate recombinant miraculin in progeny lines using the ubiquitin promoter and terminator. Compared to the transgenic tomato, transgenic lettuce could also have some advantages such as scalability, a short life cycle, and an established cultivation system at a plant factory. Thus, it might be profitable as a processed food and purified powder. Future studies will assess the safety of recombinant miraculin: it is essential to investigate aspects such as its toxicity, allergenicity, digestibility, thermal stability, insertion position in the host genome, and processing status. Thermal stability of the recombinant miraculin protein is also crucial information for optimization and industrial use. Furthermore, scaling up the cultivation process and scaling up and improving the purification process are important for reducing the final cost. The commercialization of recombinant miraculin as a product will become a reality through these sustained development efforts in the near future.

Due to its unique properties and potential, recombinant miraculin proteins produced in transgenic plants such as tomatoes and lettuce will be an alternative sweetener that can help prevent chronic lifestyle diseases and control sugar consumption in diabetics.

References

1. Kant R (2005) Sweet proteins-potential replacement for artificial low calorie sweeteners. Nutr J 4:5
2. Masuda T, Kitabatake N (2006) Developments in biotechnological production of sweet proteins. J Biosci Bioeng 102:375–389
3. Witty M (1990) Preprothaumatin II is processed to biological activity in *Solanum tuberosum*. Biotechnol Lett 12:131–136
4. Bartoszewski G, Niedziela A, Szwacka M, Niemirowicz-Szczytt K (2003) Modification of tomato taste in transgenic plants carrying a thaumatin gene from *Thaumatococcus daniellii* Benth. Plant Breed 122:347–351
5. Schestibratov KA, Dolgov SV (2005) Transgenic strawberry plants expressing a thaumatin II gene from demonstrate enhanced resistance to *Botrytis cinerea*. Sci Hortic 106:177–189
6. Lebedev VG, Taran SA, Shmatchenko VV, Dolgov SV (2002) Pear transformation with the gene for super-sweet protein thaumatin II. Acta Hortic 596:199–202
7. Szwacka M, Kryzymowska M, Osuch A, Kowalczyk ME, Malepszy S (2002) Variable properties of transgenic cucumber plants containing the thaumatin II gene from *Thaumatococcus daniellii*. Acta Physiol Plant 24:173–185
8. Peñarrubia L, Kim R, Giovannoni J, Kim SH, Fischer RL (1992) Production of the sweet protein monellin in transgenic plants. Nat Biotechnol 10:561–564
9. Lamphear BJ, Barker DK, Brooks CA, Delaney DE, Lane JR, Beifuss K, Love R, Thompson K, Mayor J, Clough R, Harkey R, Poage M, Drees C, Horn ME, Streatfield SJ, Nikolov Z, Woodard SL, Hood EE, Jilka JM, Howard JA (2005) Expression of the sweet protein brazzein in maize for production of a new commercial sweetener. Plant Biotechnol J 3:103–114
10. Kurihara K, Beidler LM (1968) Taste-modifying protein from miracle fruit. Science 161:1241–1243
11. Kurihara K, Beidler LM (1969) Mechanism of the action of taste modifying protein. Nature 222:1176–1178
12. Theerasilp S, Kurihara Y (1988) Complete purification and characterization of the taste-modifying protein, miraculin, from miracle fruit. J Biol Chem 263:11536–11539
13. Theerasilp S, Hitotsuya H, Nakajo S, Nakaya K, Nakamura Y, Kurihara Y (1989) Complete amino acid sequence and structure characterization of the taste-modifying protein, miraculin. J Biol Chem 264:6655–6659
14. Masuda Y, Nirasawa S, Nakaya K, Kurihara Y (1995) Cloning and sequencing of a cDNA encoding a taste-modifying protein, miraculin. Gene 161:175–177
15. Igeta H, Tamura Y, Nakaya K, Nakamura Y, Kurihara Y (1991) Determination of disulfide array and subunit structure of taste modifying protein, miraculin. Biochim Biophys Acta 1079:303–307
16. Ito K, Asakura T, Morita Y, Nakajima K, Koizumi A, Shimizu-Ibuka A, Masuda K, Ishiguro M, Terada T, Maruyama J, Kitamoto K, Misaka T, Abe K (2007) Microbial production of sensory-active miraculin. Biochem Biophys Res Commun 360:407–411
17. Matsuyama T, Satoh M, Nakata R, Aoyama T, Inoue H (2009) Functional expression of miraculin, a taste-modifying protein in *Escherichia coli*. J Biochem 145:445–450
18. Paladino A, Colonna G, Facchiano AM, Constantini S (2010) Functional hypothesis on miraculin' sweetness by a molecular dynamics approach. Biochem Biophys Res Commun 396:726–730
19. Paladino A, Costantini S, Colonna G, Facchiano AM (2008) Molecular modeling of miraculin: structural analysis and functional hypothesis. Biochem Biophys Res Commun 367:26–32
20. Koizumi A, Tsuchiya A, Nakajima K, Ito K, Terada T, Shimizu-Ibuka A, Briand L, Asakura T, Misaka T, Abe K (2011) Human sweet taste receptor mediates acid-induced sweetness of miraculin. Proc Natl Acad Sci U S A 108:16819–16824
21. Kurihara Y, Nirasawa S (1997) Structures and activities of sweetness inducing substances (miraculin, curculin, strogin) and the heat stable sweet protein, mabinlin. FFI J Jpn 174:67–74

22. Kurihara Y (1994) Thaumatin. In: Witty M, Higginbotham JD (eds) Sweet proteins in general. CRC Press, Boca Raton, pp 1–18
23. Ito K, Sugawara T, Koizumi A, Nakajima K, Shimizu-Ibuka A, Shiroishi M, Asada H, Yurugi-Kobayashi T, Shimamura T, Asakura T, Masuda K, Ishiguro M, Misaka T, Iwata S, Kobayashi T, Abe K (2010) Bulky high-mannose-type N-glycan blocks the taste-modifying activity of miraculin. Biochim Biophys Acta 1800:986–992
24. Sun HJ, Cui ML, Ma B, Ezura H (2006) Functional expression of the taste-modifying protein, miraculin, in transgenic lettuce. FEBS Lett 580:620–626
25. Sugaya T, Yano M, Sun HJ, Hirai T, Ezura H (2008) Transgenic strawberry expressing a taste-modifying protein, miraculin. Plant Biotechnol 25:329–333
26. Sun HJ, Kataoka H, Yano M, Ezura H (2007) Genetically stable expression of functional miraculin, a new type of alternative sweetener, in transgenic tomato plants. Plant Biotechnol J 5:768–777
27. Yano M, Hirai T, Kato K, Hiwasa-Tanase K, Fukuda N, Ezura H (2010) Tomato is a suitable material for producing recombinant miraculin protein in genetically stable manner. Plant Sci 178:469–473
28. Sun HJ, Uchii S, Watanabe S, Ezura H (2006) A highly efficient transformation protocol for Micro-Tom, a model cultivar for tomato functional genomics. Plant Cell Physiol 47:426–431
29. Kim YW, Kato K, Hirai T, Hiwasa-Tanase K, Ezura H (2010) Spatial and developmental profiling of miraculin accumulation in transgenic tomato fruits expressing the miraculin gene constitutively. J Agric Food Chem 58:282–286
30. Hirai T, Sato M, Toyooka K, Sun HJ, Yano M, Ezura H (2010) Miraculin, a taste-modifying protein is secreted into intercellular spaces in plant cells. J Plant Physiol 167:209–215
31. Desai PN, Shrivastava N, Padh H (2010) Production of heterologous proteins in plants: strategies for optimal expression. Biotechnol Adv 28:427–435
32. Fluhr R, Kuhlemeier C, Nagy F, Chua NH (1986) Organ-specific and light-induced expression of plant genes. Science 232:1106–1112
33. Kaulen H, Schell J, Kreuzaler F (1986) Light-induced expression of the chimeric chalcone synthase-NPTII gene in tobacco cells. EMBO J 5:1–8
34. Stougaard J, Sandal NN, Gron A, Kuhle A, Marcker KA (1987) 50 Analysis of the soybean leghaemoglobin lbc3 gene: regulatory elements required for promoter activity and organ specificity. EMBO J 6:3565–3569
35. Williamson JD, Hirsch-Wyncott ME, Larkins BA, Gelvin SB (1989) Differential accumulation of a transcript driven by the CaMV 35S promoter in transgenic tobacco. Plant Physiol 90:1570–1576
36. Twyman RM, Stoger E, Schillberg S, Christou P, Fischer R (2003) Molecular farming in plants: host systems and expression technology. Trends Biotechnol 21:570–578
37. Hirai T, Kim YW, Kato K, Hiwasa-Tanase K, Ezura H (2011) Uniform accumulation of recombinant miraculin protein in transgenic tomato fruit using a fruit-ripening-specific E8 promoter. Transgenic Res 20:1285–1292
38. Deikman J, Fischer RL (1988) Interaction of a DNA binding factor with the 5**0**-flanking region of an ethylene-responsive fruit ripening gene from tomato. EMBO J 7:3315–3320
39. Lincoln JE, Fischer RL (1988) Diverse mechanisms for the regulation of ethylene-inducible gene expression. Mol Gen Genet 212:71–75
40. Lincoln JE, Fischer RL (1988) Regulation of gene expression by ethylene in wild-type and rin tomato (*Lycopersicon esculentum*) fruit. Plant Physiol 88:370–374
41. Lin HH, Huang LF, Su HC, Jeng ST (2009) Effect of the multiple polyadenylation signal AAUAAA on mRNA 3**0**-end formation and gene expression. Planta 230:699–712
42. Nagaya S, Kawamura K, Shinmyo A, Kato K (2010) The HSP terminator of *Arabidopsis thaliana* increases gene expression in plant cells. Plant Cell Physiol 51:328–532
43. Ingelbrecht ILW, Herman LMF, Dekeyser RA, Van Montagu MC, Depicker AG (1989) Different 30 end regions strongly influence the level of gene expression in plant cells. Plant Cell 1:671–680

44. Hiwasa-Tanase K, Nyarubona M, Hirai T, Kato K, Ichikawa T, Ezura H (2011) High-level accumulation of recombinant miraculin protein in transgenic tomatoes expressing a synthetic miraculin gene with optimized codon usage terminated by the native miraculin terminator. Plant Cell Rep 30:113–124
45. Kato K, Yoshida R, Kikuzaki A, Hirai T, Kuroda H, Hiwasa-Tanase K, Takane K, Ezura H, Mizoguchi T (2010) Molecular breeding of tomato lines for mass production of miraculin in a plant factory. J Agric Food Chem 58:9505–9510
46. Matsui T, Takita E, Sato T, Aizawa M, Ki M, Kadoyama Y, Hirano K, Kinjo S, Asao H, Kawamoto K, Kariya H, Makino S, Hamabata T, Sawada K, Kato K (2011) Production of double repeated B subunit of Shiga toxin 2e at high levels in transgenic lettuce plants as vaccine material for porcine edema disease. Transgenic Res 20:735–748
47. Hirai T, Kurokawa N, Duhita N, Hiwasa-Tanase K, Kato K, Kato K, Ezura H (2011) The HSP terminator of *Arabidopsis thaliana* induces extremely high-level accumulation of miraculin protein in transgenic tomato. J Agric Food Chem 59:9942–9949
48. Leelavathi S, Reddy VS (2003) Cloroplast expression of His-tagged GUS-fusions: a general strategy to overproduce and purify foreign proteins using transplastomic plants as bioreactors. Mol Breed 11:49–58
49. Meyers B, Zaltsman A, Lacroix B, Kozlovsky SV, Krichevsky A (2010) Nuclear and plastid genetic engineering of plants: comparison of opportunities and challenges. Biotechnol Adv 28:747–756
50. Staub JM, Garcia B, Graves J, Hajdukiewicz PT, Hunter P, Nehra N, Paradkar V, Schlittler M, Carroll JA, Spatola L, Ward D, Ye G, Russell DA (2000) High-yield production of a human therapeutic protein in tobacco chloroplasts. Nat Biotechnol 18:333–338
51. Daniell H (2006) Production of biopharmaceuticals and vaccines in plants via the chloroplast genome. Biotechnol J 1:1071–1079
52. Maliga P (2003) Progress towards commercialization of plastid transformation technology. Trends Biotechnol 121:20–28
53. Kawabe A, Miyashita NT (2003) Patterns of codon usage bias in three dicot and four monocot plant species. Genes Genet Syst 78:343–352
54. Liu Q, Xue Q (2005) Comparative studies on codon usage pattern of chloroplasts and their host nuclear genes in four plant species. J Genet 84:55–62
55. Murray EE, Lotzer J, Eberle M (1989) Codon usage in plant genes. Nucleic Acids Res 17:477–498
56. Peng RH, Yao QH, Xiong AS, Cheng ZM, Li Y (2006) Codonmodifications and an endoplasmic reticulum-targeting sequence additively enhance expression of an *Aspergillus* phytase gene in transgenic canola. Plant Cell Res 25:124–132
57. Perlak FJ, Fuchs RL, Dean DA, McPherson SL, Fischhoff DA (1991) Modification of the coding sequence enhances plant expression of insect control protein genes. Proc Natl Acad Sci U S A 88:3324–3328
58. Rouwendal GJ, Mendes O, Wolbert EJH, Boer AD (1997) Enhanced expression in tobacco of the gene encoding green fluorescent protein by modification of its codon usage. Plant Mol Biol 33:989–999
59. Xue GP, Patel M, Johnson JS, Smyth DJ, Vickers CE (2003) Selectable marker-free transgenic barley producing a high level of cellulose (1,4-**b**-glucanase) in developing grains. Plant Cell Rep 21:1088–1094
60. Gustafsson C, Govindarajan S, Minshull J (2004) Codon bias and heterologous protein expression. Trends Biotechnol 22:346–353
61. Kim YW, Hirai T, Kato K, Hiwasa-Tanase K, Ezura H (2010) Gene dosage and genetic background affect miraculin accumulation in transgenic tomato fruits. Plant Biotechnol 27:333–338
62. Kato K, Maruyama S, Hirai T, Hiwasa-Tanase K, Mizoguchi T, Goto E, Ezura H (2011) A trial of production of the plant-derived high-value recombinant protein in a plant factory:

photosynthetic photon fluxes affect the accumulation of recombinant miraculin in transgenic tomato fruits. Plant Signal Behav 8:1172–1179

63. Hobbs SLA, Kpodar P, DeLong CMO (1990) The effect of T-DNA copy number, position and methylation on reporter gene expression in tobacco transformants. Plant Mol Biol 15:851–864
64. Jorgensen RA, Cluste PD, English J, Que Q, Napoli C (1996) Chalcone synthase cosuppression phenotypes in petunia flowers: comparison of sense vs. antisense constructs and single copy vs. complex T-DNA sequences. Plant Mol Biol 31:957–973
65. Meyer P (1996) Repeat-induced gene silencing: common mechanisms in plants and fungi. Biol Chem Hoppe Seyler 377:87–95
66. Doran PM (2006) Foreign protein degradation and instability in plants and plant tissue cultures. Trends Biotechnol 24:426–432
67. Almeida DPF, Huber DJ (1999) Apoplastic pH and inorganic ion levels in tomato fruit: a potential means for regulation of cell wall metabolism during ripening. Physiol Plant 105:506–512
68. Hirai T, Fukukawa G, Kakuta H, Fukuda N, Ezura H (2010) Production of recombinant miraculin using transgenic tomato in a closed-cultivation system. J Agric Food Chem 58:6096–6101
69. Torres AC, Cantliffe DJ, Laughner B, Bienie M, Nagata R, Ashraf M, Ferl RJ (1993) Stable transformation of lettuce cultivar South Bay from cotyledon explants. Plant Cell Tissue Org Cult 34:279–285
70. Curtis IS, Power JB, McCabe M, de Laat A, Davey MR (1994) Promoter-GUS fusions in lettuce. In: Abstracts 4th Int Congr Plant Mol Biol (Abstract no. 1682). The International Society for Plant Molecular Biology, Amsterdam
71. McCabe MS, Schepers F, van der Arend A, Mohapatra U, de Laat AMM, Power JB, Davey MR (1999) Increased stable inheritance of herbicide resistance in transgenic lettuce carrying a petE promoter-bar gene compared with a CaMV 35S-bar gene. Theor Appl Genet 99:587–592
72. Park BJ, Liu Z, Kanno A, Kameya T (2005) Increased tolerance to salt- and water-deficit stress in transgenic lettuce (*Lactuca sativa* L.) by constitutive expression of LEA. Plant Growth Regul 45:165–171
73. Teeri TH, Lehvaslaiho H, Frank M, Uotila J, Heino P, Palva ET, Montagu VM, Herrera-Estrella L (1989) Gene fusions to lac Z reveal new expression patterns of chimeric genes in transgenic plants. EMBO J 8:343–350
74. Comai L, Moran P, Maslyar D (1990) Novel and useful properties of a chimeric plant promoter combining CaMV 35S and MAS elements. Plant Mol Biol 15:373–381
75. Muller E, Lorz H, Lutticke S (1996) Variability of transgene expression in clonal cell lines of wheat. Plant Sci 114:71–82
76. Weiler KS, Wakimoto BT (1995) Heterochromatin and gene expression in Drosophila. Annu Rev Genet 29:577–605
77. Srivastava V, Vasil V, Vasil IK (1996) Molecular characterization of the fate of transgenes in transformed wheat (*Triticum aestivum* L.) Theor Appl Genet 92:1031–1037
78. Monia BP, Ecker DJ, Crooke ST (1990) New perspectives on the structure and function of ubiquitin. Biotechnology 8:209–215
79. Ozkaynak E, Finley D, Solomon MJ, Varhavsky A (1987) The yeast ubiquitin genes: a family of natural gene fusions. EMBO J 6:1429–1439
80. Callis J, Vierstra RD (1989) Ubiquitin and ubiquitin genes in higher plants. Oxford Surv Plant Mol Biol 6:1–30
81. Burke TJ, Callis J, Viestra RD (1988) Characterization of a polyubiquitin gene from *Arabidopsis thaliana*. Mol Gen Genet 213:435–443
82. Cornejo MJ, Luth D, Blankenship KM, Anderson OD, Blechl AE (1993) Activity of a maize ubiquitin promoter in transgenic rice. Plant Mol Biol 23:567–581
83. Callis J, Raasch JA, Vierstra RD (1990) Ubiquitin extension proteins of *Arabidopsis thaliana*. J Biol Chem 265:12486–12493

84. Binet MN, Weil JH, Tessier LH (1991) Structure and expression of sunflower ubiquitin genes. Plant Mol Biol 17:395–407
85. Garbarino JE, Rockhold DR, Belknap WR (1992) Expression of stress responsive ubiquitin genes in potato tubers. Plant Mol Biol 20:235–244
86. Wang J, Jiang J, Oard JH (2000) Structure, expression and promoter activity of two poly ubiquitin genes from rice (*Oryza sativa* L.) Plant Sci 156:201–211
87. Hirai T, Shohael AM, Kim YW, Yano M, Ezura H (2011) Ubiquitin promoter-terminator cassette promotes genetically stable expression of the taste-modifying protein miraculin in transgenic lettuce. Plant Cell Rep 30:2255–2265
88. Chen WP, Gu X, Liang GH, Muthukrishnan S, Chen PD, Liu DJ, Gill BS (1998) Introduction and constitutive expression of a rice chitinase gene in bread wheat using biolistic bombardment and the bar gene as a selectable marker. Theor Appl Genet 97:1296–1306
89. Chen WP, Chen PD, Liu DJ, Kynast R, Friebe B, Velazhahan R, Muthukrishnan S, Gill BS (1999) Development of wheat scab symptoms is delayed in transgenic wheat plants that constitutively express a rice thaumatin-like protein gene. Theor Appl Genet 99:755–760
90. Morimoto T, Torii T, Hashimoto Y (1995) Optimal control of physiological processes of plants in a green plant factory. Control Eng Pract 3:505–511
91. Hiwasa-Tanase K, Ezura H (2016) Molecular breeding to create optimized crops: from genetic manipulation to potential applications in plant factory. Front Plant Sci 7:539
92. Brouwer JN, Van Der Wei H, Francke A, Hemming G (1968) Miraculin, the sweetness inducing protein miracle fruit. Nature 220:373–374
93. Giroux EL, Henkin RI (1974) Purification and some properties of miraculin, a glycoprotein from *Synsepalum dulcificum* which provokes sweetness and blocks sourness. J Agric Food Chem 22:594–601
94. Armah GE, Achel DG, Acquaah RA, Belew M (1999) Purification of miraculin glycoprotein using tandem hydrophobic interaction chromatography. United States Patent #5886155
95. Duhita N, Hiwasa-Tanase K, Yoshida S, Ezura H (2009) Single-step purification of native miraculin using immobilized metal affinity chromatography. J Agric Food Chem 57:5148–5151
96. Duhita N, Hiwasa-Tanase K, Yoshida S, Ezura H (2011) A simple method for purifying undenatured miraculin from transgenic tomato fruits. Plant Biotechnol 28:281–286

Transgenic Plants as Producers of Supersweet Protein Thaumatin II

9

A. P. Firsov, A. S. Pushin, and S. V. Dolgov

Abstract

Thaumatin II is a supersweet protein derived from the West-African plant *Thaumatococcus daniellii* Benth. It is a perspective low-calorie sugar substitute for food and pharmaceutical industries. Because of the limitations of its natural sources, obtaining recombinant thaumatin using plant-based expression systems is a promising field of research. This review summarizes many years of research focusing on the physicochemical properties of thaumatins I and II, their roles in plants as pathogenesis-related proteins, and the specific characteristics of their taste perception. A special attention is paid to the detailed description of the studies on obtaining transgenic plants that have been transformed with thaumatin II gene in order to improve their agronomic and consumer properties as well as to obtain recombinant thaumatin for industrial use. Further directions of the research focusing on such areas as obtaining transgenic plants to produce recombinant thaumatin and developing the technologies for its isolation and purification are discussed.

Keywords

Thaumatin • Sweet-tasting proteins • Sugar substitutes • Plant transformation • Transgenic plants • Plant expression systems • Recombinant thaumatin

Contents

1 Introduction 186
2 Thaumatins: Properties 188
2.1 Amino Acid and Nucleotide Sequences 188
2.2 Physicochemical Properties 189

A.P. Firsov (✉) • A.S. Pushin • S.V. Dolgov
Branch of the Shemyakin-Ovchinnikov Institute of Bioorganic Chemistry of the RAS, Russian Federation, Pushchino, Moscow Region, Russian Federation
e-mail: aleksey_firsov@mail.ru; aspushin@rambler.ru; dolgov@bibch.ru

J.-M. Mérillon, K.G. Ramawat (eds.), *Sweeteners*, Reference Series in Phytochemistry,
https://doi.org/10.1007/978-3-319-27027-2_11

2.3 The Sweet Taste 191
2.4 Biological Functions 192
2.5 Thaumatin in Transgenic Crops 193
3 Expression of the Recombinant Thaumatin in the Plant Systems 197
4 Expression of the Recombinant Thaumatin in Transgenic Tomatoes 198
5 Conclusion 204
References 205

1 Introduction

Thaumatin is a sweet protein produced from the fruit of West-African plant katemfe (*Thaumatococcus daniellii* Benth). Katemfe fruit has always been used by aboriginals as a sweetener, in particular in wine and baby food production. Thaumatin is present in the aril, surrounding katemfe seeds.

Katemfe plants grow in rainfall forests of Western Africa, from Sierra Leone to Zaire, and also in Angola. Katemfe prefers rich well-drained soils in the areas where annual precipitation is about 1500–2000 mm. Its fruit is a three-edged shape with one to three hard black seeds; mature fruit has bright red coloring. Katemfe generates fruit only in the areas of its natural habitat. Being grown in other regions or in the greenhouses, it does not produce fruit despite plentiful blossom. To some extent, it points out to the necessity of endemic pollinators and limits katemfe introduction to the other regions even though they may be appropriate in terms of soil and climate conditions. An average weight of katemfe fruit is about 16 g with variations from 6 to 30 g, the aril weight is 4–6% of the weight of the fruit [1, 2]. The amount of thaumatins in arils of mature fruit is 30–55 mg/g of fresh weight (about 50% of total soluble protein), varying within a wide range depending on the degree of fruit maturity and origin [3].

Small-scale farming is widespread in the areas of natural vegetation of katemfe. The scope of production is not very clear; it depends on a year and weather conditions of the current season. According to Higginbotham [4], the yield of katemfe fruit in Ghana is typically 2–8 tons per hectare increasing up to 30 t in favorable years if agrotechniques is properly applied. Production of katemfe arils in Cote d'Ivoire in 1984–1991 is estimated at 7,5–26 t annually [5]; Ghana produces approximately the same amount [2]. Practically all obtained raw material is sent to Europe for further processing.

The protein nature of the sweet substance, present in arils, was first discovered by Van Der Wel and Loeve [6]. This study revealed that the sweet taste is caused by a mixture of two proteins with similar properties called thaumatin I and thaumatin II. These proteins are 3000 times sweeter than sucrose based on weight equivalent and 100,000 times sweeter if molecular equivalent is taken for comparison. The cDNA nucleotide sequence encoding thaumatin II was cloned by Edens and coauthors in 1982 [7]; it made possible to determine amino acid sequence of this protein. Amino acid sequence of thaumatin I was determined using the method of direct sequencing [8]; nucleotide sequence of cDNA was first verified by Ide and coauthors [9].

Unique properties of thaumatins (sweet taste, unusual resistance to heat, and stability in environments with low pH) aroused much interest to these proteins. Historically most of the research was about thaumatin II. One of the reasons for that is its applicability in food processing as a low-calorie sugar substitute and a taste modifier. Now physical and chemical properties of thaumatin II are well-studied and its safety is approved [10]. It is allowed to use it as a sugar substitute in the European Union, Israel, and Japan (food addictive E957 [11]) and as a taste modifier in the USA [12].

It should be noted that thaumatins belong to PR5 protein family; and many members of this family are strong allergens [13, 14]. But to date, there is no evidence that either natural or recombinant thaumatins possess allergenic properties.

Based on the great scientific and practical interest, considerable efforts were directed to the expression of thaumatin II in heterologous systems. Attempts to express thaumatin II in bacteriological systems were not successful in general, because recombinant thaumatin was lacking the sweet taste. Attempts to get recombinant thaumatin II in expression systems based on yeast and filamentous fungus were more successful. The most successful were experiments with yeast species *Saccharomyces cerevisiae* and filamentous fungus *Aspergillus oryzae*, with the maximum accumulation of thaumatin being 140 and 150 mg/L of the culture medium correspondingly. However, the properties of the obtained thaumatin are not described in detail. In particular, it is not very clear if the obtained thaumatin II retained its properties as a sweetener and taste modifier and to what extent [15]. Despite of the availability of the producers, microbiological production of thaumatin II is not described in the literature. The challenge of thaumatin II expression in the microbiological systems is likely to be caused by the necessity to keep its correct conformation (in particular formation of eight disulfide bonds), which is not always effectively done. Notably, thaumatin belongs to the family of 5 pathogenesis-related proteins (PR5), which is characterized by pronounced antifungal activity [16–19]. Moreover activity of thaumatin against some types of phytopathogenic bacteria [20] has been detected. These factors may probably influence the effectiveness of microbiologic production of recombinant thaumatin.

The challenge of keeping correct conformation of thaumatin II and consequently its properties is likely to be solved by means of plant-based expression systems. To date a substantial amount of transgenic plants including agricultural plants, transformed by thaumatin II gene, have been developed [20–31]. Most of these studies are focused on improving agrotechnical properties of the plants – on improving the quality of the fruit (making them sweeter) and on increasing resistance of the plants to the phytopathogenic fungi. Transgenic plants were characterized by sweet taste of fruit and vegetative organs that indicated expression and correct processing of thaumatin. Moreover the study revealed a partial increase in the plant resistance to some phytopathogenic fungi, e.g., *Alternaria brassicae* and *Botrytis cinerea*. At the same time, almost there are no studies focusing on the exploring the potential of using transgenic plants as producers of recombinant thaumatin II and its further use in food and pharmaceutical industries.

Most of the known research lacks quantitative data on accumulation of thaumatin II in transgenic plants, which is essential for evaluation of their potential as producers of this protein. Methods of extraction and purification of recombinant thaumatin II, obtained in plain expression systems, are not developed, and the detailed analysis based on the high-resolution methods has not been performed; in particular there are no data on the taste characteristics of recombinant thaumatin. Thus development of expression platforms based on transgenic plans to obtain thaumatin II is a separate scientific task that needs to be tackled.

2 Thaumatins: Properties

2.1 Amino Acid and Nucleotide Sequences

Sweet katemfe proteins that were referred to as thaumatin I and thaumatin II (according to the order of their elution) were first extracted using the methods of ion-exchange chromatography by Van Der Wel and Loeve in 1972 [6].

Amino acid sequence of thaumatin I was determined using Edman sequencing protocol [8]. This study focused on the mature form of thaumatin I which was derived directly from katemfe fruit. cDNA nucleotide sequence of the mature form of thaumatin I (without N- and C-terminal signal peptides) was determined by Ide and coauthors only in 2007 [9]. In the case of thaumatin II, the cDNA was cloned first [7], allowing for the calculation of its protein amino acid sequence.

To validate the amino acid sequences of thaumatins I and II, Lee et al. [32] performed their resequencing (by Edman protocol). The resulting sequences derived by Lee and coauthors (referred to as thaumatin A and thaumatin B, correspondingly) were slightly different from those published by Iyengar et al. [8] and Edens et al. [7]. In 2007 an additional amino acid sequence of thaumatin I was determined by translation of the cloned cDNA sequence of thaumatin I [9]. Amino acid sequences of thaumatins I and II that were obtained by different researchers are shown in Fig. 1.

Amino acid sequence of thaumatin II [7] differs from that of thaumatin I [8] by five amino acid changes (K46N, R63S, R67K, Q76R, and D113N). On the other hand, amino acid sequence of thaumatin II published by Lee et al. [32] differs from that of thaumatin I by only two changes (K46N and D113N); according to these researchers, Ser, Lys, and Arg are located at positions 63, 67, and 76, correspondingly in both thaumatins. Amino acid sequences of thaumatin I that were derived by Iyengar et al. [8] and Lee et al. [32] are identical, whereas the sequence presented by Kaneko et al. [33] differs by a change of Asn to Asp at position 113, as in the case of thaumatin II. The reason for such a discrepancy in the amino acid sequences presented in the literature is not clear: it could be uncertainties in the sequencing methods or using different forms of katemfe fruit, which might vary in amino acid sequences of thaumatins. In any case, the differences in the reported amino acid sequences of thaumatin I and thaumatin II are minor.

The nucleotide sequence of preprothaumatin II cDNA was determined by Edens et al. [7]. In the case of thaumatin I, the cDNA sequence of the mature form was identified Ide et al. [9]. Moreover GenBank contains the nucleotide sequence of the mature form of thaumatin I determined by Kaneko et al. [33]. The comparison of these

```
                     Reference      1                                                                         75
Thaumatin II Lee et al., 1988   (1) ----------------------ATFEIVNRCSYTVWAAASKGDAALDAGGRQLNSGESWTINVEPGTKGGKIWAR
Thaumatin II Edens et al., 1982 (1) MAATTCFFFLFPFLLLLTLSRAATFEIVNRCSYTVWAAASKGDAALDAGGRQLNSGESWTINVEPGTKGGKIWAR
Thaumatin I Iyengar et al., 1979(1) ----------------------ATFEIVNRCSYTVWAAASKGDAALDAGGRQLNSGESWTINVEPGTNGGKIWAR
Thaumatin I Lee et al., 1988    (1) ---------------------MATFEIVNRCSYTVWAAASKGDAALDAGGRQLNSGESWTINVEPGTNGGKIWAR
Thaumatin I Kaneko et al., 2001 (1) ----------------------ATFEIVNRCSYTVWAAASKGDAALDAGGRQLNSGESWTINVEPGTNGGKIWAR

                                     76                                                                        150
Thaumatin II_Lee et_al., 1988    (54) TDCYFDDSGSGICKTGDCGGLLRCKRFGRPPTTLAEFSLNQYGKDYIDISNIKGFNVPMDFSPTTRGCRGVRCAA
Thaumatin II Edens et al., 1982  (76) TDCYFDDSGRGICRTGDCGGLLQCKRFGRPPTTLAEFSLNQYGKDYIDISNIKGFNVPMDFSPTTRGCRGVRCAA
Thaumatin I Iyengar et al., 1979(54) TDCYFDDSGSGICKTGDCGGLLRCKRFGRPPTTLAEFSLNQYGKDYIDISNIKGFNVPMNFSPTTRGCRGVRCAA
Thaumatin I Lee et al., 1988     (55) TDCYFDDSGSGICKTGDCGGLLRCKRFGRPPTTLAEFSLNQYGKDYIDISNIKGFNVPMNFSPTTRGCRGVRCAA
Thaumatin I Kaneko et al., 2001  (54) TDCYFDDSGSGICKTGDCGGLLRCKRFGRPPTTLAEFSLNQYGKDYIDISNIKGFNVPMDFSPTTRGCRGVRCAA

                                        151                                                                        225
Thaumatin II_Lee et_al., 1988     (129) DIVGQCPAKLKAPGGGCNDACTVFQTSEYCCTTGKCGPTEYSRFFKRLCPDAFSYVLDKPTTVTCPGSSNYRVTF
Thaumatin II Edens et al., 1982   (151) DIVGQCPAKLKAPGGGCNDACTVFQTSEYCCTTGKCGPTEYSRFFKRLCPDAFSYVLDKPTTVTCPGSSNYRVTF
Thaumatin I Iyengar et al., 1979  (129) DIVGQCPAKLKAPGGGCNDACTVFQTSEYCCTTGKCGPTEYSRFFKRLCPDAFSYVLDKPTTVTCPGSSNYRVTF
Thaumatin I Lee et al., 1988      (130) DIVGQCPAKLKAPGGGCNDACTVFQTSEYCCTTGKCGPTEYSRFFKRLCPDAFSYVLDKPTTVTCPGSSNYRVTF
Thaumatin I Kaneko et al., 2001   (129) DIVGQCPAKLKAPGGGCNDACTVFQTSEYCCTTGKCGPTEYSRFFKRLCPDAFSYVLDKPTTVTCPGSSNYRVTF

                                        226
Thaumatin II_Lee et_al., 1988     (204) CPTA------
Thaumatin II Edens et al., 1982   (226) CPTALELEDE
Thaumatin I Iyengar et al., 1979  (204) CPTA------
Thaumatin I Lee et al., 1988      (205) CPTA------
Thaumatin I Kaneko et al., 2001   (204) CPTA------
```

Fig. 1 Amino acid sequences of thaumatins I and II based on the data from different authors. The N-terminal signal peptide of thaumatin II is shown in *bold*, the C-terminal propeptide is shown in *bold italic*. The differences in the amino acid sequences are shown in *color*

sequences shows high percentage of identity between them (Fig. 2). cDNA nucleotide sequences corresponding to thaumatins I and II differ by only five nucleotide changes.

Not much is known about the thaumatin gene structure and organization in *T. daniellii*. Ledeboer et al. [34] have isolated five different clones homologous to thaumatin II from the *T. daniellii* DNA. It was shown that all isolated genes belong to the same gene family and contain two small (<100 base pairs) introns localized at different positions of the thaumatin genes, which is typical for the genes encoding seed storage proteins. There are no data available on the other structural elements and gene organization of thaumatins in the *T. daniellii* genome.

2.2 Physicochemical Properties

Thaumatin II is expressed in *T. daniellii* plants in the form of preprothaumatin which is 235 amino acid long. Preprothaumatin II amino acid sequence includes N-terminal signal peptide (aa 1–22) and C-terminal propeptide LELEDE (aa 230–235) (Fig. 1). The mature form of thaumatin II consists of a single 207 amino acid long chain (aa 23–229) with molecular weight of 22286.44 Da. It does not include any unusual amino acids or carbohydrates and has eight disulfide bonds (at positions 31↔226, 78↔88, 93↔99, 143↔215, 148↔199, 156↔167, 171↔180, and 181↔186) [35]. During the maturation process, the N-terminal signal peptide and C-terminal propeptide are cleaved off producing a mature form of thaumatin II. The function of the C-terminal propeptide LELEDE has not been elucidated yet; it is missing from the mature forms of thaumatins I and II.

The 3D structure of thaumatins I and II has been studied in great detail and resolution. The thaumatin molecule consists of three domains that are rich in disulfide bonds: core domain (domain I, aa 1–53, 85–127, and 178–207), large

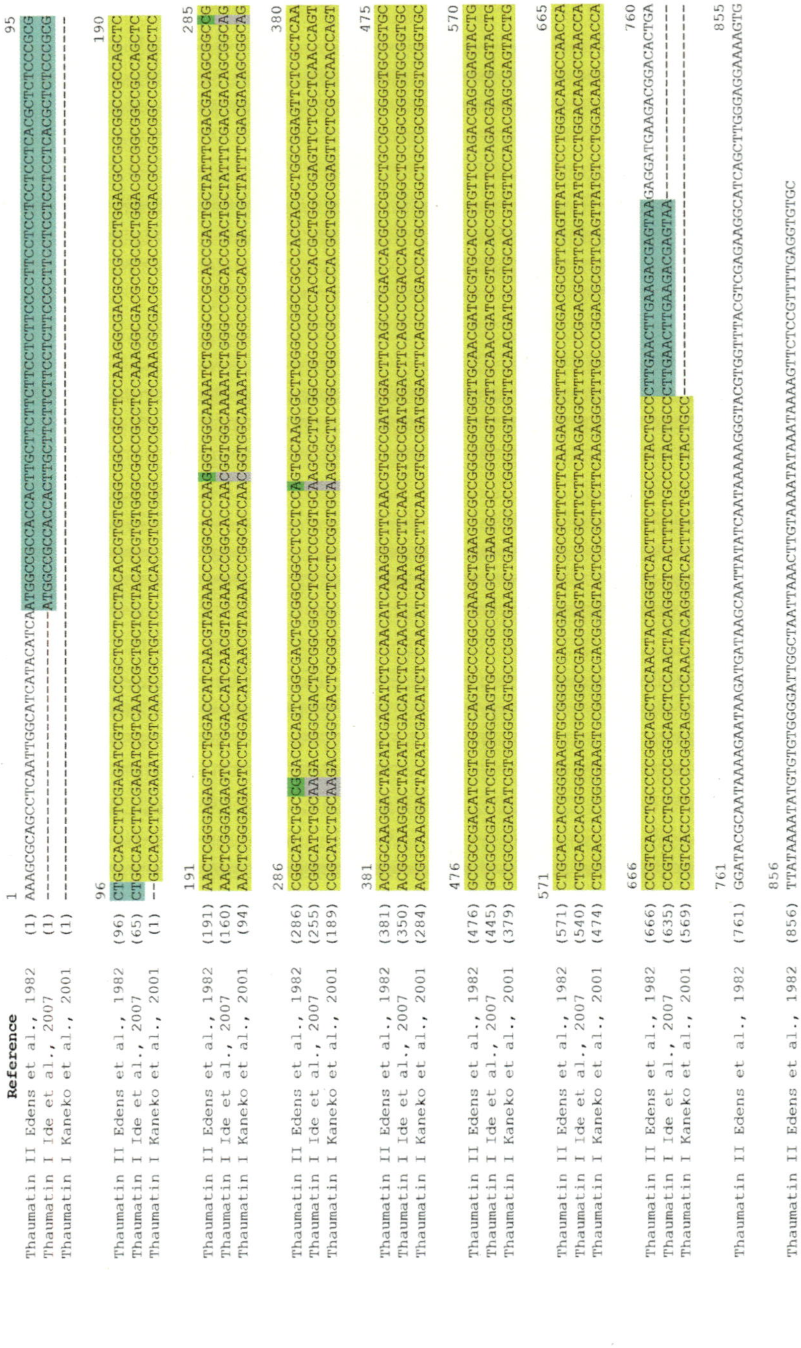

Fig. 2 Nucleotide sequences of thaumatins I and II based on the data from different authors. The differences in the nucleotide sequences are shown in *color*

domain (domain II, aa 128–177), and small domain (domain III, aa 54–84) [36–40]. However, these studies have not detected any direct link between the thaumatin conformation and its sweet taste.

Thaumatin II possesses an interesting property – a high heat resistance in acidic conditions. It has been demonstrated that thaumatin II was able to completely retain its sweet taste for 30 min at 80 °C in a buffer with pH2.0. If heated for 2 h, the thaumatin sweetness was decreased two times compared to the unheated protein, with the minimal concentration of the protein perceived to be sweet being 100 nM. Meanwhile, thaumatin II was losing its sweet taste within 30 min if pH was increased from 2.0 to 7.0 [41, 42]. The reason of such striking thermoresistance in acidic conditions is not exactly clear; however, it could be caused by the presence of multiple disulfide bonds within the protein. It is believed that the lysine residues at positions 137, 163, and 187 as well as free sulfhydryl residues that are being formed via the β-elimination of disulfide bonds by heating at pH > 7.0 play a special role in the process of thaumatin thermoinactivation.

There are a few reports of thaumatin possessing proteinase inhibitor properties [43] and that it could be resistant to proteinase degradation [44]; however, these properties have not been studied in detail yet.

2.3 The Sweet Taste

Thaumatin was the first protein found to have a sweet taste. The thaumatin sweetness differs from the saccharose sweetness. Firstly, the perception of the sweet taste caused by thaumatin is 10–15 s delayed compared to that caused by saccharose. Secondly, the perception of thaumatin sweetness is developed gradually until it reaches maximum, with the evident phase of perception growth. Thirdly, the sweet taste retains for a long time, in many cases for several minutes. Lastly, thaumatin possesses a long and characteristic aftertaste, with a tinge of liquorice [45]. Interestingly, the perception of thaumatin sweetness is much stronger if the mouth has been rinsed with water before thaumatin tasting [6]. Another interesting feature of the thaumatin is that the sweet taste can be perceived only by the higher primates (including humans), but not by the New World monkeys or rodents [46, 47].

Such unusual features of thaumatin taste might complicate its wide applications as a commercially available food sweetener. To facilitate commercialization of the product, certain modifications were performed to make its taste more sugar-like. One of these modifications involved site-directed mutagenesis of the thaumatin I gene to obtain a library of variants with different amino acid changes. Some of these variations had a decreased duration of aftertaste which was important for commercial use of thaumatin as a sweetener. However, in general, the results of such experiments were not significant [45, 48].

The threshold concentration of thaumatin II to cause perception of sweetness is 48 nM, meaning that thaumatin is about 100,000 times sweeter than saccharose (by molar ratio) [6]. Based on the weight equivalent, thaumatin is 1600–3000 times sweeter than saccharose [6, 45, 48]. The exact determination of this parameter is complicated because relative sweetness of thaumatin is quickly decreasing when the concentration is increased [45]. For example, thaumatin II is 8600 times sweeter compared to 2% saccharose solution, 6000 times sweeter compared to 6% saccharose solution, and 2000 times sweeter compared to 8% saccharose solution [49].

The exact molecular mechanism for thaumatin's sweetness has not been completely elucidated yet. It has been shown that the critical factor is the retaining of the intact molecular structure – even a partial reduction of disulfide bonds in thaumatin protein results in the loss of its sweet taste [6, 50]. In the meantime, refolding of recombinant thaumatin that was expressed in the inclusion bodies of *E. coli* resulted in a complete rescue of its sweetness [51].

It has been determined that the sweet taste perception is enabled through the interaction with the sweet receptor T1R2–T1R3 which is also responsible for the perception of the sweetness of the low molecular weight compounds – natural sugars and artificial sweeteners [52–56]. However, due to the significant difference in the molecular weight of the sweet proteins and the low molecular weight sweeteners, it has been proposed that the mechanisms of T1R2–T1R3 function might differ by the two types of ligands [53, 57].

The possible mechanism of the interaction between the sweet proteins such as thaumatin, monellin, or brazzein with the T1R2–T1R3 sweet receptor is presented as a "wedge" model. According to this model, the sweet protein binds to the secondary binding site present on the receptor surface thus activating it without the involvement of the low molecular weight ligand [57–59]. This model is supported by the importance of the distribution of the electrical potential on the surface of the sweet protein monellin: changing this parameter resulted in a decreased sweetness of the protein [60]. Multiple studies report that the mutations affecting thaumatin's surface charge resulted in a decrease or even a complete loss of its sweet taste [33, 61]. Lys at position 67 and Arg at position 82 are especially important [61, 62]. The other Lys and Arg residues (K49, R76, K78, K79, K97, K106, K137, K163, K187) also play an important role in the protein-receptor interaction [33, 61]. The "wedge" model is also supported by the fact that the mutations in the cysteine-rich domain of the human T1R3 resulted in a decreased perception of thaumatin I sweetness without affecting the perception of the sucralose (low molecular weight sweetener) sweetness [63].

The importance of the electrostatic potentials to the interaction between thaumatin and the receptor of sweet taste has been experimentally validated recently [64]. In this experiment, the authors obtained several variants of thaumatin I based on the "wedge" model (point mutations D21N, E42Q, D55N, D59A, D60A, and E89Q). One of these variants (D21N) appeared to be significantly sweeter compared to the native protein – the threshold concentration for the reliable sweet taste perception was determined to be 31 ± 4 nM. Thus, the "wedge" model gained an additional experimental support.

2.4 Biological Functions

To date the biological functions of thaumatin in *T. daniellii* plants have not been explored in detail. However, an educated guess can be made that the main function of thaumatin is to protect the seeds and fruit from the fungal pathogens, and, possibly, from abiotic stress.

Thaumatin belongs to a large family of "thaumatin-like" proteins (demonstrating a significant homology with the thaumatin amino acid sequence) which are universal in plants [65, 66]. Thaumatin-like proteins (TLP) are inducible proteins that are being expressed in response to different biological and abiological stress, including attacks of pathogens and pests, drought, cold temperatures, and mechanical damage. Based on that, thaumatin-like proteins are categorized as a family 5 pathogenesis-related proteins (PR5) [17].

PR5 proteins are a part of plant's defense mechanism against pathogen attacks, most often caused by fungus [67, 68]. The exact mechanism of the antifungal function of PR5 proteins is not clear yet. The majority of TLPs demonstrate a broad spectrum of the antifungal activities. It has been proposed that the antifungal activities of PR5 proteins, including thaumatin, are caused by their ability to disrupt the synthesis of the cell walls in the fungus causing their damage. It has also been proposed that the first step of the antifungal activity is the binding of TLP to the components of the fungal cell wall. This is indirectly supported by the fact that the changes in the fungal cell wall composition can result in decrease or increase in the TLP antifungal activity [69].

Some TLPs are able to bind to a water-insoluble β-1,3-glucans which are typical components of the fungal cell walls. Also, some TLPs have been shown to have a β-1,3-glucanase activity [70–72]. However, β-1,3-glucanase activity does not always correspond to their antifungal activity [17], suggesting the existence of the other mechanisms for the induction of their antifungal activity.

Xylanase inhibition activity which has been found in some of the TLPs might be one of such mechanisms [73]. Xylanases play an important role in the plant infections caused by some pathogenic fungus, for example, *Botrytis cinerea*. Inhibition of xylanase activity results in the attenuation of the infection process and thus to a higher resistance to fungal pathogens [73].

The exact mechanism of the antifungal activity of thaumatin II (the ability to bind to β-1,3-glucans, the presence of β-1,3-glucanase or xylanase activities, binding sites on the fungal cell wall, etc.) has not been studied in detail yet. However, a number of experiments have been performed to demonstrate the antifungal activity of thaumatin (discussed below).

2.5 Thaumatin in Transgenic Crops

The sweet taste and the antifungal activity of thaumatin made it an attractive target for genetic engineering of the agricultural plants in order to improve their agronomic properties. To date, a number of transgenic agricultural plants transformed with the thaumatin gene have been obtained (Table 1).

During the first phase of the research, the main attention was paid to the improvement of the taste of the fruit and the eatable parts of the plants by increasing their sweetness. In 1990 Witty was the first to show that thaumatin II expressed in potatoes resulted in a sweet taste, suggesting the correct synthesis and processing of thaumatin in the heterologous plant system. Transgenic cucumbers [23], tomatoes

Table 1 Transgenic plants transformed with thaumatin gene

Transgenic plant	Thaumatin expression level	Novel trait	Reference	Notes
Potatoes *Solanum tuberosum* L. cv. Iwa	~2.0 μg/g of the raw root mass	Sweet taste of vegetative tissues	[21, 22]	Preprothaumatin II promoter 35S CaMV
Cucumbers *Cucumis sativus* L. cv. Borszczagowski, line Bc	Up to 1 % of the total soluble protein	Sweet taste of fruits	[23]	Preprothaumatin II promoter 35S CaMV
Tomato *Lycopersicon esculentum* Mill. cv. 'Beta', line *nor*	Not determined	Sweet taste of fruits	[25]	Preprothaumatin II promoter 35S CaMV
Peer *Pyrus communis* L. cv. Burakovka	Not determined	Sweet taste of vegetative tissues	[24, 27]	Preprothaumatin II promoter 35S CaMV
Apple *Malus domestica* Borkh. cv. Melba	Not determined	Sweet taste of vegetative tissues	[26]	Preprothaumatin II promoter 35S CaMV
Tomato *Lycopersicon esculentum* Mill. cv. Yalf	0.007–0.17 % of the total soluble protein	Sweet taste of fruits and vegetative tissues	[28]	Preprothaumatin II promoter 35S CaMV
Garden strawberry *Fragaria* x *ananassa* Duch. cv. Firework	Not determined	Sweet taste of fruits Increased resistance to *Botrytis cinerea*	[18]	Preprothaumatin II promoter 35S CaMV
Garden strawberry *Fragaria* x *ananassa* Duch. cv. Selekta	Not determined	Sweet taste of fruits Increased resistance to *Botrytis cinerea*	[29]	Preprothaumatin II promoter 35S CaMV
Hyacinth *Hyacinthus orientalis* L. cvs. Edisson and Chine Pink	0.06–0.28 % of the total soluble protein	Increased resistance of bulbs to *Fusarium culmorum* and *Botrytis cinerea*	[19]	Preprothaumatin II promoter 35S CaMV
Tobacco *Nicotiana tabaccum* L. var. Xanthii	Not determined	Increased resistance to *Pythium aphanidermatum* and *Rhizoctonia solani*; increased resistance to salty	[30]	Preprothaumatin II promoter 35S CaMV

(*continued*)

Table 1 (continued)

Transgenic plant	Thaumatin expression level	Novel trait	Reference	Notes
		conditions and drought		
Carrot *Daucus sativus* L. cv. Nantskaya 4	Not determined	Increased resistance to *Fusarium avenaceum*	[31]	Preprothaumatin II promoter 35S CaMV
Carrot *Daucus sativus* L. cvs. Nantska, Korotel, Red Giant, and Perfektziya Celery *Apium graveolens* L. cvs. Paskal, Zephyr, Yablochnyi	Not determined	Antibacterial activities of the plant extracts, lack of antifungal activity, increased resistance to salty conditions and drought	[20]	Thaumatin II fused with the signal transport peptide into plastids promoter 35S CaMV
Barley *Hordeum vulgare* L.	More than 3 g/kg of grain	Obtaining recombinant thaumatin I	[77]	Preprothaumatin I under the control of barley D-hordein promoter or mature thaumatin I under the control of promoter and N-terminal signal peptide from barley D-hordein
Tomato *Lycopersicon esculentum* Mill. cv. Yalf	61–154 mg/kg of the raw mass of mature fruits	Obtaining recombinant thaumatin II	[83]	Preprothaumatin II promoter 35S CaMV
Tobacco *Nicotiana tabaccum* L. Hairy root culture	1.4–2.6 mg/L of the culture medium	Obtaining recombinant thaumatin I	[78]	Thaumatin I without the signal sequence or fused with the N-terminal signal peptide of calreticulin promoter 35S CaMV

[25, 28], and strawberries [18] transformed with thaumatin II gene also possessed a distinct sweet taste of the fruit and leaves. The sweet taste of root crop was also very distinct in transgenic carrots [31].

In these experiments, the plants were transformed with the preprothaumatin II gene which included its own N- and C-terminal peptides under the control of 35S promoter from the cauliflower mosaic virus. The sweet taste of the resulting transgenic plants supported the correct processing of preprothaumatin in a broad range of heterologous plant systems.

A number of perennial fruit plants were transformed with the thaumatin II gene, including apple and pear trees. Their leaves were also characterized by the sweet taste [24, 26]. The resulting transgenic trees were further cultivated in the field conditions and started to fructify in 2009–2010. The pears from the transgenic trees were characterized by a strong sweet taste, typical for thaumatin (based on the organoleptic analysis) [27].

The apples had a less pronounced sweetness; however, they were noticeably sweeter than the apples from the control untransformed trees (data not published).

The fruit and vegetative organs of cucumber plants that had been transformed with the thaumatin II gene have been studied in detail by Szwacka and coauthors [74]. In their study, fresh transgenic cucumbers were evaluated with higher scores than control ones. In the meantime, the taste of transgenic cucumbers in the form of salads and marinades received about the same score as the control ones. The same authors indicated that transgenic cucumbers were more flavorous compared to the control ones due to the increased amounts of (E,Z)-2,6 nonadienal which is the main flavor ingredient that is present in cucumbers [74, 75].

To date, no detailed analysis of the taste of other transgenic plants (apples, pears, tomatoes, strawberries, and carrots) has been performed. The researchers working with these plants mostly focused on the confirmation of the sweetness and the evaluation of the sweetness levels compared to control plants. The increased sweetness of the fruit as well as vegetative parts of the plants was confirmed in multiple studies. However, despite the successful expression, genetic engineering of the plants to express thaumatin in order to improve the taste has not gained a widespread use yet. One of the reasons might be the insufficient sugar-likeliness taste of thaumatin and the presence of the long aftertaste, especially in case of thaumatin high expression.

An important property of thaumatin is its ability to enhance the taste of a broad variety of products, including plant-derived ones [45, 76]. The ability of thaumatin to enhance the taste and odor of cucumbers, even at low levels of expression, has been confirmed by Zawirska-Wojtasiak et al. [75] and Szwacka et al. [74]. Our group indicates the taste improvement of the transgenic Melba apples and Selecta strawberries when the accumulation of thaumatin was low and the increased sweetness was not detected (data not published). Thus, taste improvement might be reached not only due to the increased sweetness per se but also due to the enhancement of the natural taste of the plant.

Another research focus is the resistance of agricultural plants to the fungal pathogens. An increased resistance of the transgenic plants transformed with thaumatin II was confirmed in a number of experiments. For example, transgenic strawberry plants demonstrated an increased resistance to *Botrytis cinerea* [18, 29]. Bulbs of the transgenic hyacinth had an increased resistance to *B. cinerea* and *Fusarium culmorum*: the development of symptoms was 7–8 days delayed in the transgenic plants compared to the control ones [19]. A significant increase in the resistance of the transgenic tobacco plants to *Rhizoctonia solani* (fungus causing rhizoctoniosis) and

Pythium aphanidermatum (causal agent of root rot) was reported by Rajam and coauthors [30].

Transgenic carrot expressing thaumatin II demonstrated a significant increase in the resistance to phytopathogenic fungus *F. avenaceum*. Interestingly, some transgenic lines were completely resistant to this pathogen [31].

In several studies, an increased resistance of the transgenic plants expressing thaumatin to abiotic stresses– drought and increased salt concentration in the soil – was demonstrated [20, 30]. The exact mechanism of that is not clear yet; however, it is known that different abiotic stresses are among the inducers of TLP expression. It has been reported that the plant-derived PR5 proteins might recognize different signal molecules and participate in the signaling interactions as a ligand or a receptor [66]. It is possible that the increased resistance of transgenic plants to abiotic stress is mediated through the participation of thaumatin in various signaling pathways that are important for the plant adaptation.

It is important to note that thaumatin is a well-studied protein which has been permitted to be used in the food industry. Its application in increasing the plant resistance to the fungal pathogens should not alarm consumers in terms of the safety of the transgenic plants.

3 Expression of the Recombinant Thaumatin in the Plant Systems

As it was noted above, the possibility of using thaumatin in the food industry and the limited natural sources of this protein have drawn a significant attention to the development of methods to obtain recombinant thaumatin in heterologous expression systems. We are sure that the most efficient approach for obtaining recombinant thaumatin would be using the expression systems based on the transgenic plants. Nevertheless, there are only a few studies focusing on the obtaining transgenic plants producing recombinant thaumatin.

Stahl and coauthors [77] obtained transgenic barley that has been transformed with the nucleotide sequence encoding thaumatin I. The expressed sequence was under the control of barley D-hordein promoter which enabled the accumulation of the recombinant thaumatin in the grain. Thaumatin was expressed in two different forms: as the native preprothaumatin I or as a mature thaumatin I fused with the N-terminal signaling peptide from the barley D-hordein. These forms did not differ in terms of levels of thaumatin accumulation, and it has been shown that the N- and C-terminal peptides of preprothaumatin I were cleaved normally in the barley plants. The levels of thaumatin accumulation in the best transgenic lines exceeded 3 g/kg of the grain. The authors also noted the relative simplicity of thaumatin extraction and purification from the barley grain.

Pham and coauthors [78] used tobacco hairy root cultures to produce recombinant thaumatin I. They used two variants of thaumatin for transformation – without the signal sequence or fused with the N-terminal signal sequence of calreticulin. In both cases, the genetic constructs were driven by 35S promoter from the cauliflower mosaic virus.

When thaumatin was fused with calreticulin signal peptide, the recombinant thaumatin was observed to be secreted into the culture medium, with the maximal level of thaumatin accumulation being 0.21 mg/L. When thaumatin did not contain the N-terminal signal sequence, the target protein has not been detected in the culture medium. The accumulation of thaumatin reached maximum on the 30th day of the hairy roots cultivation followed by its decreasing. Enrichment the medium with PVP (up to 1.5%) or NaCl (100 mM) resulted in the increased accumulation of the recombinant thaumatin in the medium up to 1.4 mg/L and 2.6 mg/L, correspondingly. Thus, the discussed study demonstrated the possibility of developing a closed system for recombinant thaumatin expression based on the rhizosecretion. However, further research is required to increase the levels of the target protein accumulation and to inhibit proteinase activity in the culture medium.

4 Expression of the Recombinant Thaumatin in Transgenic Tomatoes

Our group used tomatoes to express recombinant thaumatin II. To begin with, we determined the possible correlation between the intracellular localization of thaumatin and the level of its accumulation. The following versions of thaumatin II gene were used: th35 (the full-sized preprothaumatin II), mt (the mature form of thaumatin), mtc (thaumatin lacking the N-terminal sequence), ubnt (thaumatin lacking the C-terminal sequence), and ter (thaumatin with the C-terminal sequence LELEDE being exchanged for the endoplasmic reticulum (ER)-localizing signal HDEL) (Fig. 3). Nucleotide sequences encoding different versions of thaumatin II were cloned into the pBI121 vector [79] instead the GUS gene or into pGD vector [80], both under the control of 35S CaMV promoter. The resulting vectors (Fig. 3) were used for agrobacterial transformation of the tobacco plants and enabled accumulation of thaumatin in different cellular compartments – in the cytoplasm, in the ER, or in the apoplastic space – depending on the structure of the expressed sequence.

Western blot was performed to analyze the expression of thaumatin II in transgenic plants. Thaumatin has not been detected in the tobacco plants transformed with the pGD-mt vector (mature thaumatin), suggesting that there was no accumulation of the recombinant thaumatin in the cytoplasm. Also, thaumatin has not been detected in the transgenic plants transformed with the pGD-mtc vector (with the intact C-terminal peptide LELEDE). Most likely, the LELEDE peptide itself was not capable of facilitating transport of thaumatin to the cellular compartments where it could be accumulated at a detectable level.

In the plants transformed with pBI-th35, pGD-ter or pGD-ubnt vectors, the recombinant thaumatin was expressed in the majority of the analyzed lines (Fig. 4). Thaumatin was detected about 22 kDa and migrated along the thaumatin from *T. daniellii* (22.4 kDa). This indicated the cleavage of the N-terminal signal peptide and thus the correct processing of the thaumatin precursor in tobacco plants.

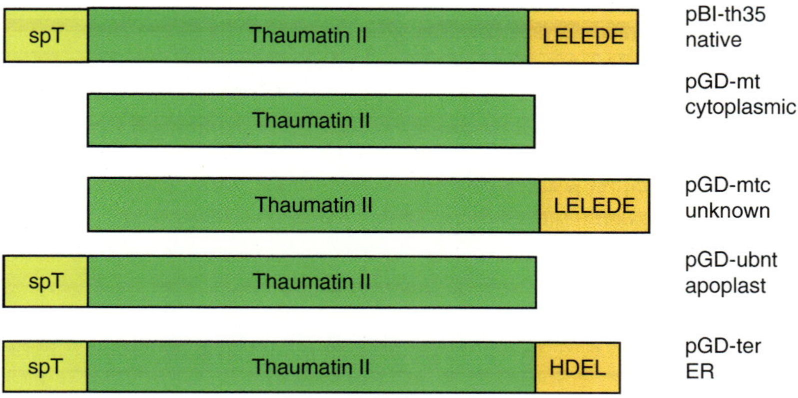

Fig. 3 The schematics of the thaumatin II genes used in the development of the expression system based on the tomato plants. *spT* N-terminal signal peptide of thaumatin II, *LELEDE* C-terminal propeptide of thaumatin II, *HDEL* the ER localization signal. The expected localization of the recombinant thaumatin is shown to the right of each construct

Fig. 4 Western blot analysis of thaumatin II expression in transgenic tobacco plants. (**a**) The plants transformed with the pBI-th35 vector. (**b**) The plants transformed with the pGD-ter vector. (**c**) The plants transformed with the pGD-ubnt vector. The numbers correspond to the different transgenic lines. *M* molecular size marker, *th* thaumatin (25 ng, Sigma, USA), *K* non-transgenic plant. The *arrow* indicates the band corresponding to thaumatin

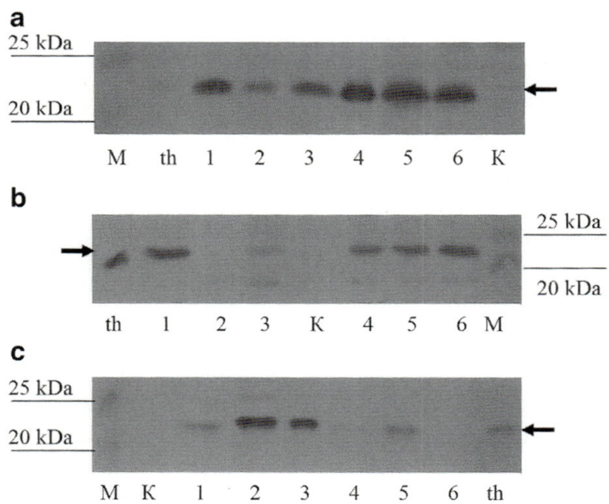

The leaves from the transgenic plants were characterized by obvious sweet taste accompanied by the thaumatin-specific aftertaste which also supports the correct thaumatin processing in the transgenic plants.

The results of the quantitative analysis of the thaumatin expression are presented in Fig. 5. On average, accumulation of thaumatin in the lines transformed with the pBI-th35, pGD-ter, and pGD-ubnt vectors was detected to be 0.61%, 0.64%, and 0.28% of the TSP, correspondingly. The differences between pBI-th35 and pGD-ter were not statistical significant. In the meantime, the removal of the C-terminal

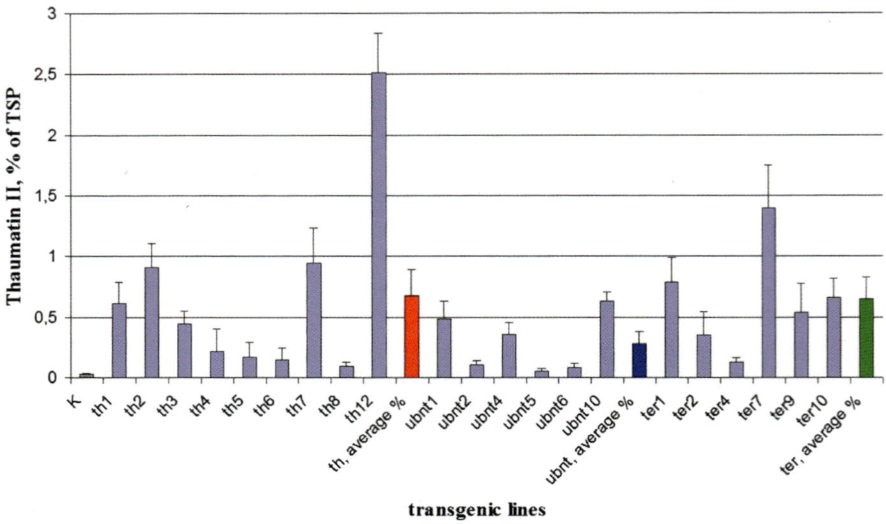

Fig. 5 Accumulation of the recombinant thaumatin II in transgenic tobacco plants transformed with pBI-th35, pGD-ubnt, and pGD-ter vectors. The average content of thaumatin II in plants transformed with the vectors pBI-th35, pGD-ubnt, and pGD-ter is also shown (in *red*, *blue*, and *green*, respectively). Error bars indicate ± SD. The numbers correspond to different lines of transgenic plants. *K* control non-transgenic plant

peptide LELEDE with the intact N-terminal sequence (in the case of pGD-ubnt) resulted in almost two times less accumulation of thaumatin. Notably, among of the plants transformed with pBI-th35 and pGD-ter were revealed lines (line th12, vector pBI-th35; line ter7, vector pGD-ter) with high accumulation of recombinant thaumatin (2.5% and 1.4% of TSP, correspondingly). It might be caused by the differences in the number of transgene copies or in the positions within the genome into which the foreign DNA has integrated.

The exact thaumatin compartmentalization in the *T. daniellii* cells is not known. Edens and coauthors [7] have shown that thaumatin is synthesized in a form of precursor which contains C- and N-terminal peptides. Notably, these terminal peptides are absent in the mature thaumatin extracted from katemfe. Preprothaumatin N-terminal sequence (amino acids 1–22) represents a signal sequence for protein translocation into the ER and thus plays an important role in the intracellular transport of thaumatin.

During preprothaumatin processing, C-terminal peptide LELEDE (amino acids 230–235), the function of which has not been identified yet, is also removed. Based on the evidence that (a) the mature form of thaumatin fails to accumulate in the cytoplasm or apoplast and (b) the removal of LELEDE peptide with the N-terminal signal peptide remaining intact results in the accumulation of thaumatin in the apoplast [81], we hypothesize that the LELEDE sequence represents a signal for the intracellular transport. Notably, the LELEDE sequence does not demonstrate any homology with the known ER retention signals or the signals required for the vacuole sorting. Thus, determination of the exact function of the C-terminal peptide LELEDE requires further research.

As mentioned earlier, thaumatin II was successfully used to obtain transgenic plants with the increased resistance to the phytopathogenic fungus. In those experiments, a gene that encodes preprothaumatin sequence was used; however, the compartmentalization of the mature thaumatin in the plant cells has not been determined. We hypothesized that the accumulation of the mature thaumatin in the intercellular space might enable a better contact with the fungal hyphae at the early stages of infection and thus result in an increase in its protective effect. However, as it was shown in experiments with tobacco plants, transformation with the pGD-ubnt vector resulted in a decrease in the accumulation of the recombinant thaumatin compared to the pBI-th35 and pGD-ter vectors. In order to enhance thaumatin expression, we attempted to optimize the target gene structure by exchanging the native thaumatin N-terminal peptide for the corresponding peptide derived from the radish defensin rs-afr2 [82]. In addition, we included the first intronic sequence of the rs-afp2 gene to the resulting chimeric thaumatin II sequence (Fig. 6a). The resulting construct referred to as pGD-lt was used to transform the Yalf tomato plants.

Western blot analysis showed that thaumatin was expressed in four out of six transgenic tomato lines that have been analyzed (Fig. 6b). As in the case of tobacco, thaumatin in the transgenic plants was found to be 22.4 kDa, supporting the cleavage of the N-terminal radish defensin signaling peptide as well as the correct splicing of the chimeric mRNA. The leaves and the fruit of the transgenic tomatoes were characterized by the sweet taste with the thaumatin-specific aftertaste which also supported the normal thaumatin processing in the transgenic plants.

However, the tomato plants transformed with the pGD-lt vector were significantly different from those of non-transformed plants. All transgenic lines were characterized by either a complete absence of fruit or small amounts of fruit (2–3 tomatoes per plant). The average weight of those fruit was 20–30 g, whereas the weight of the untransformed fruit varied from 60 to 80 g. Moreover, the fruit from the plants transformed with the pGD-lt vector almost lacked seeds. Due to these abnormalities, the transgenic tomato plant lines that have been transformed with the pGD-lt vector were not used in further experiments.

Based on the obtained data, we used pBI-th35 vector to produce a tomato plant expressing recombinant thaumatin. As a result, 17 transgenic lines of Yalf tomatoes were obtained [83]. Western blot analysis was used to confirm the expression of the recombinant thaumatin in both leaves and fruit (Fig. 6b).

As in the case of tobacco, the recombinant thaumatin migrated along with the mature thaumatin (22.4 kDa). Tomato fruit developed a sweet taste which varied from mild to very strong, as well as the characteristic aftertaste.

The qualitative evaluation of the accumulation of thaumatin in the ripe tomato fruit was performed using ELISA. The lines characterized by the most sweetness were selected for the analysis. The maximal thaumatin II expression was observed in line 91 and was determined to be 46.4 ± 10.5 μg/mg of the TSP (4.6%), which corresponded approximately to 100 mg of thaumatin/kg of fruit [83].

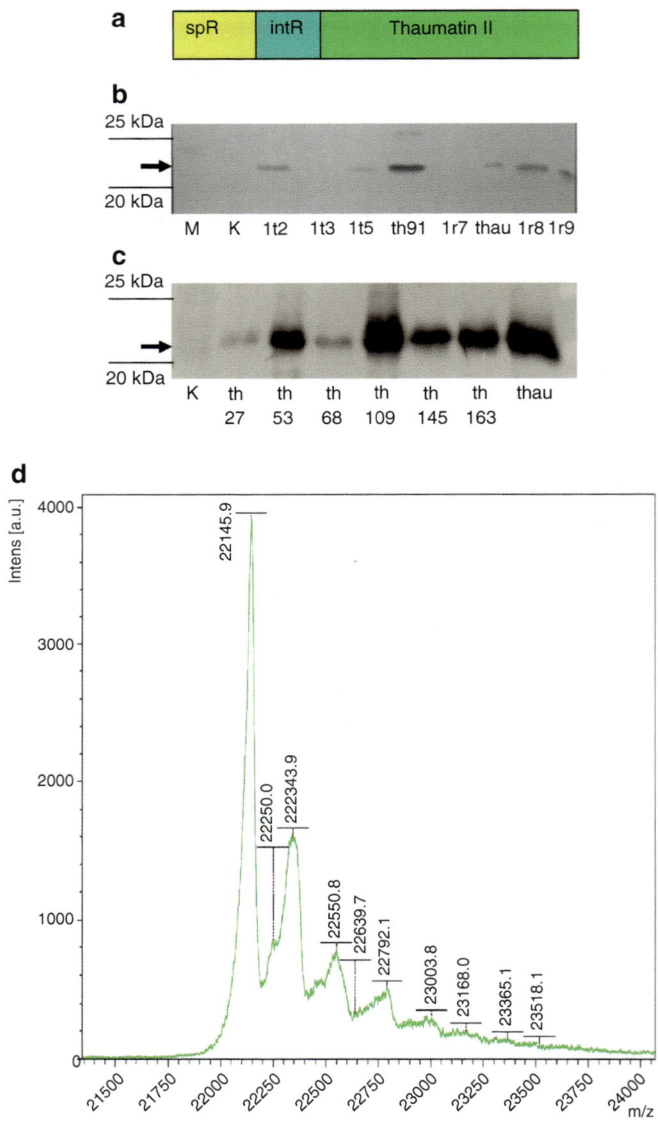

Fig. 6 Western blot analysis of thaumatin II expression in transgenic tomato plants. (**a**) The structure of the chimeric thaumatin II gene used in the pGD-lt vector. *spR* N-terminal signal peptide, *intR* the first intron of the radish defensin rs-afp-2. (**b**) Expression of thaumatin in the fruit of tomato plants transformed with the pGD-lt vector. lt2–lt9 are the transgenic lines transformed with pGD-lt vector; th91 is a transgenic line transformed with the pBI-th35 vector (line 91), *thau* thaumatin (20 ng, Sigma, USA). (**c**) Expression of thaumatin in the fruit of tomato plants transformed with the pBI-th35 vector. th27–th163 are the transgenic lines, *thau* thaumatin (100 ng, Sigma, USA). (**d**) MALDI-MS of recombinant thaumatin II purified from tomato fruit. *K* control non-transformed plant, *M* molecular size marker. The *arrow* indicated the band corresponding to thaumatin

The amount of thaumatin in different experiments varied from 60 to 150 mg/kg of line 91 fruit. This might be caused by the difference in the fruit ripeness or the conditions of growth. Such a variation is typical [84] and needs to be considered in the development of the corresponding technologies for obtaining transgenic plants.

Notably, all of the transgenic tomato lines that had been transformed with the pBI-th35 vector were not different from the non-transformed Yalf plants, based on their morphology and growth characteristics. They grew, developed, bloomed, and fruit normally. Also, the weight and the number of tomato fruit per tomato plant were not different from that of the control plants. Thus, the transgenic line of tomatoes #91 can be used to produce recombinant thaumatin II for the food and pharmaceutical industries.

For the widespread use of recombinant thaumatin as a sweetener, it is critical to confirm its structural and functional identity to the natural one. To address this concern, we extracted and purified recombinant thaumatin from the transgenic tomato line #91. Thaumatin was extracted using a simple low-salt buffer and purified with the ion-exchange chromatography on the SP-Sephacryl column. The yield of the recombinant thaumatin was 5.4 mg/100 g of the fruit fresh mass, with the purity being >90% [85].

Using MALDI-MS the molecular weight of the recombinant thaumatin was found to be 22,145 Da with the expected mass being 22,142. This result confirmed the correct cleavage of the N- and C-terminal signal peptides and the formation of all eight disulfide bonds in the molecule of thaumatin (Fig. 6d).

Organoleptic analysis of the resulting thaumatin confirmed its complete identity to the natural one. The taste perception threshold was determined to be 50 nM; the thaumatin solution was sweet, with a typical thaumatin aftertaste. The sweetness persisted for a long time – over 20 min for some of the testers [85].

The main limiting factors for the use of transgenic plants are the consumers' concerns regarding their safety. Particularly, consumers find the presence of the alien DNA of the viral or bacterial origin which is mostly represented as a sequence of genes corresponding to the selective antibiotics, promoters, and terminators of the target genes alarming. Such concerns might be addressed by obtaining of the cisgenic plants, where the genes for the selective antibiotics are absent and the regulatory elements are obtained from the corresponding sequences of the plant origin.

To obtain cisgenic plants, we used pMF1 (PRI, Wageningen) vector which includes the recombinase R and the bifunctional selectivity marker CodA-nptII. The Yalf tomatoes were transformed with the nucleotide sequence encoding preprothaumatin II under the control of the tomato-derived regulatory elements – fruit-specific promoter E8 and terminator of the gene rbcS. As a result we obtained 170 transgenic lines with the 35 lines being selected for further analysis. After recombinase activity was induced and the negative selection using 5-fluorocytosine was performed, 17 lines did not have any regenerants, while the rest 18 produced 121 sublines. All of them lost the kanamycin resistance; however, PCR and Southern blotting analyses confirmed the presence of the fragments corresponding to the gene nptII in 119 lines. Only two lines showed the correct removal of the unwanted DNA fragments [86].

Organoleptic analysis of the fruit derived from the cisgenic plants detected a sweet taste that was typical for thaumatin. The expression of the recombinant thaumatin has been confirmed with the Western blotting in both leaves and fruit. Based on the ELISA, the concentration of thaumatin reached 0.2% and 0.001% of the TSP in the fruit and leaves, correspondingly. The obtained cisgenic tomato lines would be used in further studies to improve the agronomic characteristics of tomato varieties.

Thus, to date, transgenic barley and tomato plants have been obtained, with the levels of thaumatin I and II expression sufficient for the development of the expression methods for producing this protein at the industrial scale. We are sure that further direction of research in this area should be the optimization of the technologies for extraction and purification of recombinant thaumatin and development of the ways to using it in the food and pharmaceutical industries.

5 Conclusion

As the result of more than 30 years of research, thaumatin has become one of the most well-studied plant proteins, especially in the context of its practical application. Using transgenic plants has proved the possibility to improve the taste of a crop and increase resistance of the plants to the phytopathogenic fungi. Physical and chemical properties of thaumatin crucial for its application in the industry are studied in detail; its safety for users is confirmed repeatedly.

The main challenge in thaumatin application as a sweetener in our opinion is the insufficiently sugar-like taste, in particular the presence of long and specific aftertaste. Unfortunately, efforts to improve thaumatin taste have not yet led to a desired result, and it is one of the research directions. It should be noted that in case of fruit and berry plants (especially strawberry and apple), thaumatin aftertaste is to a certain extent disguised by their own taste. This is especially the case for moderate levels of recombinant thaumatin accumulation. In our opinion, optimal level of expression of thaumatin in transgenic fruit can effectively compensate for its specific taste.

Moreover improving the taste of fruit and vegetables can be achieved not only through the increase of sweetness but also by intensification of their own taste. In this case, thaumatin is more likely to act as a flavor enhancer rather than a sweetener. However further research of the abovementioned feature of thaumatin in transgenic fruit is needed.

Nowadays thaumatin is approved in many countries as a low-calorie sugar substitute and a taste modifier. Potentially it can be widely used as a diet component for people suffering from diabetes, pathological forms of obesity, and various metabolic disorders. Taking into consideration limited natural sources of thaumatin, it presupposes the development of methods of obtaining recombinant thaumatin in industrial scale.

We believe that the most effective way is working out expressive systems based on transgenic plants. It enables not only production of large amounts of cheap raw material for further processing but also significantly simplifies the next stages of extraction and purification of recombinant thaumatin and makes the process cheaper.

In the case of using plant systems, there is no need for thorough purification of the obtained product from media components and various impurities which are essential in the case of microbiological production. In the case of using recombinant thaumatin in food production or as a component of special diets, the presence of minor contamination with the proteins from the host plant will hardly ever matter. The first steps in the development of plant-based expression systems for obtaining recombinant thaumatin are taken: the plant-producers, accumulating high levels of thaumatin have been obtained, identity of recombinant thaumatin to a natural one and its safety were proved.

The modern approaches in the field of genetic engineering of plants allow us to obtain cisgenic plants, i.e., plants transformed only by the target gene, without adding genes encoding resistance for selective antibiotics. It addresses one of the main concerns of consumers about genetically modified plants. There is no doubt that plant-based expression systems appropriate for large-scale production of recombinant thaumatin will emerge within the next few years.

References

1. Yeboa SO, Hilgerand TH, Kroschel J (2003) *Thaumatococcus danielli* (Benn.) Benth: a natural sweetener from the rain forest zone in west Africa with potential for income generation in small scale farming. In: Proceedings of international research on food security, National Resource Management and Rural Development. Georg-August-Universitat Gottingen. http://www.tropentag.de/2003/abstracts/full/305.pdf
2. Waliszewski WS, Oppong S, Hall JB, Sinclair FL (2005) Implications of local knowledge of the ecology of a wild super sweetener for its domestication and commercialization in West and Central Africa. Econ Bot 59:231–243
3. Mackenzie A, Pridham JB, Saunders NA (1985) Changes in the sweet proteins (Thaumatins) in *Thaumatococcus danielli* fruits during development. Phytochemistry 24:2503–2506
4. Higginbotham JD (1979) Protein sweeteners. In: Hough CAM, Parker KJ, Vlitos AJ (eds) Developments in sweeteners. I Applied Science Publications, London
5. Bonnéhin L (1997) Economic value and role of NTFP in the long term management of forests resources in the Cote d'Ivoire. In: Crafter SA, Awimbo J, Broekhoven AJ (eds) Non-timber forest products values, use and management issues in Africa including examples from Latin America. IUCN, The World Convention Union, Nairobi
6. van der Wel H, Loeve K (1972) Isolation and characterization of thaumatin I and II, the sweet-tasting proteins from *Thaumatococcus daniellii* Benth. Eur J Biochem 31:221–225
7. Edens L, Heslinga L, Klok R, Ledeboer AM, Maat J, Toonen MY, Visser C, Verrips CT (1982) Cloning of cDNA encoding the sweet-tasting plant protein thaumatin and its expression in *Escherichia coli*. Gene 18:1–12
8. Iyengar RB, Smits P, van der Ouderaa F, van der Wel H, van Brouwershaven J, Ravestein P, Richters G, van Wassenaar PD (1979) The complete amino-acid sequence of the sweet protein thaumatin I. Eur J Biochem Y6:193–204
9. Ide N, Kaneko R, Wada R, Mehta A, Tamaki S, Tsuruta T, Fujita Y, Masuda T, Kitabatake N (2007) Cloning of the thaumatin I cDNA and characterization of recombinant thaumatin I secreted by *Pichia pastoris*. Biotechnol Prog 23:1023–1030
10. Aquilina G, Bories G, Chesson A, Cocconcelli PS, Knecht J, Dierick NA, Gralak MA, Gropp J, Halle I, Hogstrand C, Kroker R, Leng L, Puente SL, Haldorsen A-KL, Mantovani A, Martelli G, Mézes M, Renshaw D, Saarela M, Sejrsen K, Westendorf J (2011) Scientific Opinion on the

Safety and Efficacy of thaumatin for all animal species. EFSA J 9:2354 doi:10.2903/j.efsa.2011.2354. www.efsa.europa.eu/efsajournal
11. Mortensen A (2006) Sweeteners permitted in the European Union: safety aspects. Scand J Food Nutr 50:104–116
12. Newberne P, Smith RL, Doull J, Feron VJ, Goodman JI, Munro IC, Portoghese PS, Waddell WJ, Wagner BM, Weil CS, Adams TB, Hallagan JB (2000) GRAS flavoring substances. The 19th publication by the flavor and extract manufacturers' Association's Expert Panel on recent progress in the consideration of flavoring ingredients generally recognized as safe under the Food Additives Amendment. Food Technol 54:66–84
13. Breiteneder H (2004) Thaumatin-like proteins – a new family of pollen and fruit allergens. Allergy 59:479–481
14. Veličković TĆ, Gavrović-Jankulović M (2014) Food allergens. Biochemistry and molecular nutrition. Springer Science + Business Media, New York
15. Masuda T, Kitabatake N (2006) Developments in biotechnological production of sweet proteins. J Biosci Bioeng 102:375–389
16. Selitrennikoff CP (2001) Antifungal proteins. Appl Environ Microbiol 67:2883–2894
17. van Loon LC, Rep M, Pieterse CM (2006) Significance of inducible defense-related proteins in infected plants. Annu Rev Phytopathol 44:135–162
18. Schestibratov KA, Dolgov SV (2005) Transgenic strawberry plants expressing a thaumatin II gene demonstrate enhanced resistance to *Botrytis cinerea*. Sci Hortic 106:177–189
19. Popowich EA, Firsov AP, Mitiouchkina TY, Filipenya VL, Dolgov SV, Reshetnikov VN (2007) *Agrobacterium*-mediated transformation of *Hyacinthus orientalis* with thaumatin II gene to control fungal diseases. Plant Cell Tissue Organ Cult 90:237–244
20. Luchakivska YS, Komarnytskii IK, Kurchenko IM, Yurieva OM, Zhytkevich NV, Kuchuk MV (2015) Construction and analysis of the transgenic carrot and celery plants expressing the recombinant thaumatin II protein. Biopolym Cell 31:285–293
21. Witty M (1990) Preprothaumatin II is processed to biological activity in *Solanum tuberosum*. Biotechnol Lett 12:131–136
22. Witty M, Harvey WJ (1990) Sensory evaluation of transgenic *Solanum tuberosum* producing r-thaumatin II. N Z J Crop Hortic Sci 18:77–80
23. Szwacka M, Krzymowska M, Osuch A, Kowalczyk ME, Malepszy S (2002) Variable properties of transgenic cucumber plants containing the thaumatin II gene from *Thaumatococcus daniellii*. Acta Physiol Plant 24:173–185
24. Lebedev VG, Taran SA, Shmatchenko VV, Dolgov SV (2002) Pear transformation with the gene for supersweet protein thaumatin II. Acta Hortic 596:199–202
25. Bartoszewski G, Niedziela A, Szwacka M, Niemirowicz-Szczytt K (2003) Modification of tomato taste in transgenic plants carrying a thaumatin gene from *Thaumatococcus daniellii* Benth. Plant Breed 122:347–351
26. Dolgov SV, Schestibratov KA, Mikhailov RV (2004) Apple transformation with the gene of supersweet protein thaumatin II. Acta Hortic 663:507–510
27. Dolgov SV, Lebedev VG, Firsov AP (2011) Pear fruit taste modification by thaumatin II gene expression. Acta Hortic 909:67–73
28. Korneeva I, Firsov A, Lebedev V, Schestibratov K, Pushin A, Shulga O, Dolgov S (2005) Expression and subcellular localization of PR-5 protein with different signal sequences in transgenic tomato and tobacco plants. Revue de Cytologie et Biologie végétales – Le Botaniste 28:260–267
29. Schestibratov KA, Dolgov SV (2007) Genetic engineering of strawberry for taste improvement and enhanced disease resistance by introduction of thau II gene. In: Xu Z, Li J, Xue Y, Yang W (eds) Biotechnology and sustainable agriculture 2006 and beyond: proceedings of the 11th IAPTC&B congress. Springer, Beijing/Dordrecht
30. Rajam MV, Chandola N, Saiprasad Goud P, Singh D, Kashyap V, Choudhary ML, Sihachakr D (2007) Thaumatin gene confers resistance to fungal pathogens as well as tolerance to abiotic stresses in transgenic tobacco plants. Biol Plant 51:135–141

31. Sidorova T, Miroshnichenko D, Dolgov S, Tjukavin G (2013) Transgenic carrot expressing thaumatin II gene has enhanced resistance against *Fusarium avenaceum*. Acta Hortic 974:123–130
32. Lee J-H, Weickmann JL, Koduri RK, Ghosh-Dastidar P, Saito K, Blair LC, Date T, Lai JS, Hollenberg SM, Kendall RL (1988) Expression of synthetic thaumatin genes in yeast. Biochemistry 27:5101–5107
33. Kaneko R, Kitabatake N (2001) Structure-sweetness relationship in thaumatin: importance of lysine residues. Chem Senses 26:167–177
34. Ledeboer AM, Theo Verrips C, Dekker BMM (1984) Cloning of the natural gene for the sweet-tasting plant protein thaumatin. Gene 30:23–32
35. Van der Wel H, Iyengar RB, Brouwershaven JV, Wassenaar PDV, Bel WJ, Ouderaa FJGVD (1984) Assignment of the disulfide bonds in the sweet-tasting protein thaumatin I. Eur J Biochem 144:41–45
36. Teixeira SCM, Blakeley MP, Leal RMF, Gillespie SM, Mitchell EP, Forsyth VT (2010) Sweet neutron crystallography. Acta Cryst D66:1139–1143
37. Masuda T, Ohta K, Mikami B, Kitabatake N (2011) High-resolution structure of the recombinant sweet-tasting protein thaumatin I. Acta Crystallogr Sect F: Struct Biol Cryst Commun F67:652–658
38. Masuda T, Ohta K, Tani F, Mikami B, Kitabatake N (2011) Crystal structure of the sweet-tasting protein thaumatin II at 1.27Å. Biochem Biophys Res Commun 410:457–460
39. Masuda T, Ohta K, Mikami B, Kitabatake N, Tani F (2012) Atomic structure of the sweet-tasting protein thaumatin I at pH 8.0 reveals the large disulfide-rich region in domain II to be sensitive to a pH change. Biochem Biophys Res Commun 419:72–76
40. Masuda T, Mikami B, Tani F (2014) Atomic structure of recombinant thaumatin II reveals flexible conformations in two residues critical for sweetness and three consecutive glycine residues. Biochimie 106:33–38
41. Kaneko R, Kitabatake N (2001) Sweetness of sweet protein Thaumatin is more thermoresistant under acid conditions than under neutral or alkaline conditions. Biosci Biotechnol Biochem 65:409–413
42. Elemo BO, Adu OB (2005) Studies of some functional properties of thaumatin, a protein sweetener. Jormar 2:48–55
43. Fan SG, Wu GJ (2005) Plant proteinase inhibitors against phytophagous insects. Bot Bull Acad Sin 46:273–292
44. Roberts W, Selitrennikoff CP (1990) Zeamatin, an antifungal protein from maize with membrane-permeabilizing activity. J Gen Microbiol 136:1771–1778
45. Lindley MG (2012) Natural high-potency sweeteners. In: O'Donnell K, Kearsley MW (eds) Sweeteners and sugar alternatives in food technology, 2nd edn. Wiley, Chichester
46. Hellekant G, Danilova V (1996) Species differences toward sweeteners. Food Chem 56:323–328
47. Liu B, Ha M, Meng X-Y, Kaur T, Khaleduzzaman M, Zhang Z, Jiang P, Li X, Cui M (2011) Molecular mechanism of species-dependent sweet taste toward artificial sweeteners. J Neurosci 31:11070–11076
48. Faus I, Sisniega H (2003) Sweet-tasting proteins. In: Fahnestock SR, Steinbüchel A (eds) Biopolymers, polyamides and complex proteinaceous materials II, vol 8. Wiley-VCH, New York
49. DuBois GE, Walters DE, Schiffman SS, Warwick ZS, Booth BJ, Pecore SD, Gibes K, Carr BT, Brands LM (1991) Concentration-response relationships of sweeteners. In: Walters DE, Orthoefer FT, DuBois GE (eds) Sweeteners, discovery, molecular design and chemoreception, vol 450, ACS symposium series. ACS, Washington, DC
50. van der Wel H, Bel WJ (1980) Enzymatic properties of the sweet-tasting proteins thaumatin and monellin after partial reduction. Eur J Biochem 104:413–418
51. Daniell S, Mellits KH, Faus I, Connerton I (2000) Refolding the sweet-tasting protein thaumatin II from insoluble inclusion bodies synthesised in *Escherichia coli*. Food Chem 71:105–110

52. Damak S, Rong M, Yasumatsu K, Kokrashvili Z, Varadarajan V, Zou S, Jiang P, Ninomiya Y, Margolskee RF (2003) Detection of sweet and umami taste in the absence of taste receptor T1r3. Science 301:850–853
53. Wintjens R, Viet TMVN, Mbosso E, Huet J (2011) Hypothesis/review: the structural basis of sweetness perception of sweet-tasting plant proteins can be deduced from sequence analysis. Plant Sci 181:347–354
54. Chandrashekar J, Hoon MA, Ryba NJ, Zuker CS (2006) The receptors and cells for mammalian taste. Nature 444:288–294
55. Assadi-Porter FM, Tonelli M, Maillet E, Hallenga K, Benard O, Max M, Markley JL (2008) Direct NMR detection of the binding of functional ligands to the human sweet receptor, a heterodimeric family 3 GPCR. J Am Chem Soc 130:7212–7213
56. Li X (2009) T1R receptors mediate mammalian sweet and umami taste. Am J Clin Nutr 9:733S–737S
57. Picone D, Temussi PA (2012) Dissimilar sweet proteins from plants: oddities or normal components? Plant Sci 195:135–142
58. Temussi PA (2002) Why are sweet proteins sweet? Interaction of brazzein, monellin and thaumatin with the T1R2-T1R3 receptor. FEBS Lett 526:1–4
59. Tancredi T, Pastore A, Salvadori S, Esposito V, Temussi PA (2004) Interaction of sweet proteins with their receptor. A conformational study of peptides corresponding to loops of brazzein, monellin and thaumatin. Eur J Biochem 271:2231–2240
60. Esposito V, Gallucci R, Picone D, Saviano G, Tancredi T, Temussi PA (2006) The importance of electrostatic potential in the interaction of sweet proteins with the sweet taste receptor. J Mol Biol 360:448–456
61. Ohta K, Masuda T, Ide N, Kitabatake N (2008) Critical molecular regions for elicitation of the sweetness of the sweet-tasting protein, thaumatin I. FEBS J 275:3644–3652
62. Ohta K, Masuda T, Tani F, Kitabatake N (2011) Introduction of a negative charge at Arg82 in thaumatin abolished responses to human T1R2–T1R3 sweet receptors. Biochem Biophys Res Commun 413:41–45
63. Masuda T, Taguchi W, Sano A, Ohta K, Kitabatake N, Tani F (2013) Five amino acid residues in cysteine-rich domain of human T1R3 were involved in the response for sweet-tasting protein, thaumatin. Biochimie 95:1502–1505
64. Masuda T, Ohta K, Ojiro N, Murata K, Mikami B, Tani F, Temussi PA, Kitabatake N (2016) A hypersweet protein: removal of the specific negative charge at Asp21 enhances thaumatin sweetness. Sci Rep 6:20255. doi:10.1038/srep20255
65. Hulo N, Bairoch A, Bulliard V, Cerutti L, Cuche BA, de Castro E, Lachaize C, Langendijk-Genevaux PS, Sigrist CJ (2008) The 20 years of PROSITE. Nucleic Acids Res 36(database issue):D245–D249
66. Liu J-J, Sturrock R, Ekramoddoullah AKM (2010) The superfamily of thaumatin-like proteins: its origin, evolution, and expression towards biological function. Plant Cell Rep 29:419–436
67. Velazhahan R, Datta SK, Muthukrishnan S (1999) The PR-5 family: thaumatin-like proteins in plants. In: Datta SK, Muthukrishnan S (eds) Pathogenesis-related proteins in plants. CRC Press, Boca Raton
68. Campos MA, Silva MS, Magalhaes CP, Ribeiro SG, Sarto RPD, Vieira EA, Grossi de Sa MF (2008) Expression in *Escherichia coli*, purification, refolding and antifungal activity of an osmotin from *Solanum nigrum*. Microb Cell Factories 7:7
69. Veronese P, Ruiz MT, Coca MA, Hernandez-Lopez A, Lee H, Ibeas JI, Damsz B, Pardo JM, Hasegawa PM, Bressan RA, Narasimhan ML (2003) In defense against pathogens: both plant sentinels and foot soldiers need to know the enemy. Plant Physiol 131:1580–1590
70. Trudel J, Grenier J, Potvin C, Asselin A (1998) Several thaumatinlike proteins bind to 1, 3-glucans. Plant Physiol 118:1431–1438
71. Grenier J, Potvin C, Trudel J, Asselin A (1999) Some thaumatin-like proteins hydrolyse polymeric β-1, 3-glucans. Plant J 19:473–480

72. Grenier J, Potvin C, Asselin A (2000) Some fungi express β-1,3-glucanases similar to thaumatin-like proteins. Mycologia 92:841–848
73. Fierens E, Rombouts S, Gebruers K, Goesaert H, Brijs K, Beaugrand J, Volckaert G, Van Campenhout S, Proost P, Courtin CM, Delcour JA (2007) TLXI, a novel type of xylanase inhibitor from wheat (*Triticum aestivum*) belonging to the thaumatin family. Biochem J 403:583–591
74. Szwacka M, Burza W, Zawirska-Wojtasiak R, Goslinski M, Twardowska A, Gajc-Wolska J, Kosieradzka I, Kiełkiewicz M (2012) Genetically modified crops expressing 35S-thaumatin II transgene: sensory properties and food safety aspects. Compr Rev Food Sci Food Saf 11:174–186
75. Zawirska-Wojtasiak R, Gośliński M, Szwacka M, Gajc-Wolska J, Mildner-Szkudlarz S (2009) Aroma evaluation of transgenic, thaumatin II-producing cucumber fruits. J Food Sci 74: C204–C210
76. Higginbotham JD (1983) Recent developments in non-nutritive sweeteners. In: Grenby TH, Parker KJ, Lindley MG (eds) Developments in sweeteners II. Applied Science, London
77. Stahl R, Luhrs R, Dargatz H (2009) Thaumatin from transgenic barley. US Patent 20,090,031,458
78. Pham NB, Schäfer H, Wink M (2012) Production and secretion of recombinant thaumatin in tobacco hairy root cultures. Biotechnol J 7:537–545
79. Jefferson RA, Kavanagh TA, Bevan MW (1987) GUS fusions: beta-glucuronidase as a sensitive and versatile gene fusion marker in higher plants. EMBO J 6:3901–3907
80. Goodin MM, Dietzgen RG, Schichnes D, Ruzin S, Jackson AO (2002) pGD vectors: versatile tools for the expression of green and red fluorescent protein fusions in agroinfiltrated plant leaves. Plant J 31:375–383
81. Pushin AS, Ovchinnikova EV, Shul'ga OA, Firsov AP, Dolgov SV (2008) Accumulation of a supersweet protein thaumatin II in apoplast of tobacco transgenic plants. Biotechnol Russia 6:40–54
82. Parashina EV, Serdobinskii LA, Kalle EG, Lavrova NV, Avetisov VA, Lunin VG, Naroditskii BS (2000) Genetic engineering of oilseed rape and tomato plants expressing a radish defensin gene. Russ J Plant Physiol 47:417–423
83. Firsov AP, Pushin AS, Korneeva IV, Dolgov SV (2012) Transgenic tomato plants as supersweet protein thaumatin II producers. Appl Biochem Microbiol 48:746–751
84. Floss DM, Falkenburg D, Conrad U (2007) Production of vaccines and therapeutic antibodies for veterinary applications in transgenic plants: an overview. Transgenic Res 16:315–332
85. Firsov A, Shaloiko L, Kozlov O, Vinokurov L, Vainstein A, Dolgov S (2016) Purification and characterization of recombinant supersweet protein thaumatin II from tomato fruit. Protein Expr Purif 123:1–5. doi:10.1016/j.pep.2016.03.002
86. Timerbaev V, Okuneva A, Pushin A, Dolgov S (2014) Production of marker-free transgenic tomato plants using inducible site-specific recombinase and a bifunctional selectable gene, IAPB 2014, 13rd international association of plant biotechnology congress 2014, Melbourne

Transgenics with Monellin

10

Tanushri Kaul, Chinreddy Subramanyam Reddy, and Saurabh Pandey

Abstract

Sweet proteins are potential substitutes to artificial sweeteners as they are natural and low calorie sweeteners. Monellin a naturally sweet protein from *Dioscoreophyllum cumminsii* elicits sweet flavour as a carbohydrate has been characterized to reveal that the native structure is essential for its sweet taste. It is a heterodimeric protein of 94 amino acid residues with a molecular weight of 10.5 kDa that loses its activity during denaturation. Monellin is proposed to be a promising sweetener that tastes sweet at pH 2–9, whereas high pH or heating beyond 70 °C or more can denature this protein. Structural studies showed that linking the two subunits of this heterodimer monellin confers stability than the native counterpart. Moreover, monellin comprised of a single polypeptide exhibits identical conformation and sweetness in flavour as the double-chain monellin. Monellin has dual applications as a flavour enhancer and a high-intensity sweetener and has been recommended for use in some countries. As provision of naturally existing monellin is limited, it has led to indepth research into its synthesis via transgenic organisms. Efforts have been directed towards production of this recombinant protein in different expression systems i. e., bacterial, yeast and transgenic plants. Production of monellin in transgenic fruits and vegetables offers a viable approach to improve their flavour and quality. We report the stable and enhanced expression of synthetic sweet protein monellin in prokaryotic (*E. coli*) and eukaryotic (tomato) systems. The recombinant monellin protein was thermostable and retained strong sweetness over a temperature range (up to 70 °C) and extreme pH levels. Transgenic tomato plants revealed stable gene integration in T_1 generation as verified by PCR analysis. Transcript profiling of T_2 lines revealed enhanced monellin expression, which correlated positively to its

T. Kaul (✉) • C.S. Reddy • S. Pandey
Nutritional Improvement of Crops Group, Plant Biology, International Centre for Genetic Engineering and Biotechnology (ICGEB), New Delhi, India
e-mail: tanushrikaul@gmail.com; tanushri@icgeb.res.in; suumani@gmail.com; saurabhpandey28@gmail.com

J.-M. Mérillon, K.G. Ramawat (eds.), *Sweeteners*, Reference Series in Phytochemistry,
https://doi.org/10.1007/978-3-319-27027-2_20

protein expression profiles as highlighted by western blots. Heterologous expression of monellin driven by the fruit-specific promoter (E8) did not cause any phenotypic anomalies either in vegetative or fruit growth parameters in the transgenic tomato lines. RT-PCR was performed and highest levels of transcript expression were observed in T_2-5 and T_2-14 lines. ELISA studies were ratified by immuno-blot analysis via comparison of the signal for each variant to that of purified monellin. Transgenic lines T_2-5 and T_2-14 showed maximum protein levels of 60 lg/mg (±2.06) and 54 lg/mg (±2.58), respectively. Thus, it is beneficial to earmark suitable monellin expression systems for large-scale production of this potent food supplement.

Keywords
Monellin • Heterologous • Transgenic tomatoes • Microbial, Yeast • Transcript expression • ELISA • Immunoblot analysis

Contents

1 Introduction 212
2 Microbial Expression Systems 215
3 Yeast Expression Systems 216
4 Transgenic Plants as an Expression System 217
5 Conclusions and Future Perspectives 219
References 220

1 Introduction

Globally, millions of people are affected by obesity and related problems, including type II diabetes, due to the consumption of high caloric food, which primarily comprises different sugars and/or carbohydrates. Worldwide-used low-calorie artificial sweeteners, for instance, aspartame, saccharin, cyclamate, and acesulfame K, by patients affected with diseases related to the sugar consumption, e.g., obesity, hyperlipidemia, diabetes, confer typically undesirable effects such as psychological- and mental disorders, bladder or brain cancers, and heart failure. Sweet proteins have the capacity to substitute these artificial sweeteners by being natural, good, low calorie sweeteners, as we know that proteins do not trigger a demand for insulin in these patients, whereas sucrose does (Table 1). Two rare proteins, identified from African berries, revealed an inherent high specificity for the sweet receptors. These proteins, monellin and thaumatin, were 100,000-fold sweeter than sugar on a molar basis and several thousand times sweeter on a weight basis. They neither contain carbohydrates nor modified amino acids. Monellin a naturally sweet protein as comprehended sweet by human taste receptor [1] is derived from West African berries of *Dioscoreophyllum cumminsii* and may be particularly beneficial to individuals clinically challenged for sugar intakes. Adding monellin as a flavor in the food could solve the problem to some extent. Thus, monellin has received sizeable attention from the food and beverage industry [2]. Previously, protein chemists have

Table 1 Structural details of important sweet-proteins known to mankind

	Thaumatin	Monellin	Mabinlin	Pentadin	Brazzein	Curculin	Miraculin
Source	*Thaumatococcus denielli* Benth	*Dioscoreophyllum cumminsii* Diels	*Capparis masakai* Levl	*Pentadiplantra brazzeana* Baillon	*Pentdaplandra brazzeana* Baillon	*Curulingo latifolia*	*Richadella dulcifica*
PDB	3WXS	1MOL	2DS2	2BRZ	1BRZ	2DPF	3IIR
	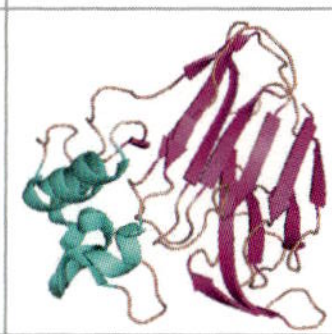	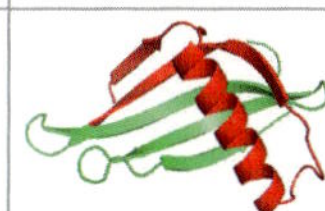		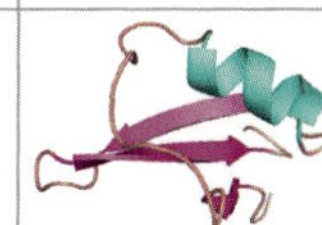			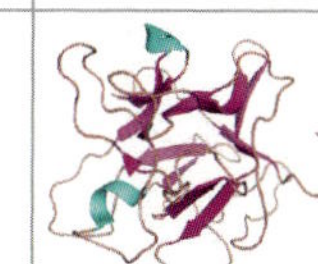
Sweetness (weight basis)	3000	3000	100	500	2000	550	–
Molecular wt. (KDa)	22.2	10.7	12.4	12	6.5	24.9	98.4
Residue length	207	45(A chain) 50(B chain)	33(A chain) 72 (B chain)	54	54	114	191
Active unit	Monomer	Dimer (A + B)	Dimer (A + B)	Monomer	Monomer	Dimer (A + A)	Tetramer (A + A + A + A)

characterized monellin [3, 4, 5, 32] and deciphered that its native structure is crucial for its sweet taste [6, 7]. However, production of this low-calorie natural sweetener is limited due to restricted habitats and due to its instability at high temperatures and acidic pH, causing loss of sweet flavor [8]. Linking the two subunits of this heterodimer can increase the stability of monellin [4, 5]. Structural studies have revealed that monellin that comprised of a single polypeptide exhibits identical conformation and sweetness in flavor as the double-chain monellin but stable than the native counterpart.

Monellin, a sweet protein from *Dioscoreophyllum cumminsii*, elicits sweet flavor as a carbohydrate was primarily reported in 1969. Subsequently, in 1972 it was isolated and functionally characterized as a protein by Monell, Chemical Senses Center in Philadelphia, USA. On the molar basis, it is 100 times sweeter than sugar [8]. It is a heterodimeric protein of 94 amino acid residues with a molecular weight of 10.5 kDa that loses its activity during denaturation [32]. Monellin is proposed to be a promising sweetener that tastes sweet at pH 2-9, whereas high pH or heating beyond 70 °C or more can denature it. Thereby, being unstable at high temperature or pH extremities limits its application in the food industry. Single chain monellin protein comprised of two polypeptides, i.e., chain A and B joined by a Gly-Phe dipeptide bond linkage. Site-directed mutagenesis has been extensively used to edit and improvise the functional characteristics of this protein [7, 9]. In vitro single chain monellin is as clearly revealed as a sweet wild type and more high temperature stable protein [10, 11]. Monellin and its single chain derivative (MNEI) comprise the group of sweetest proteins known to mankind (Table 1). Though less is known about this monellin protein, detailed studies have been carried out for the enhancement of sweetness and thermostability of the protein via gene editing and modification. In postcritical analysis of surface electrostatic potentials, Leone et al. [12] developed novel mutants of MNEI with improved sweetness. Further, they included the E23Q stabilizing mutation in the most encouraging variant, thereby obtaining a construct with enhanced properties that collated extreme sweetness to high, pH-independent, thermal stability. Resultant mutant possessed a sweetness threshold of only 0.28 mg/L (25 nM) and has been the strongest sweetener known. Moreover, the new proteins have been purified and structures of the most potent mutants were resolved by X-ray crystallography. Previously, some mutations as G1M, E2M, and E2N were reported for improvement of sweetness in this protein [13]. Another mutation in monellin protein like Y65R significantly improved sweetness and solubility in acidic medium [14, 15]. MNEI mutants showed enhanced thermostability, and their stable forms were produced through plastid transformation in tobacco. In order to confer thermostability, different MNEI-mutants after circular dichroism analysis were successfully transformed into chloroplasts of tobacco plant [4]. However, detailed investigation of these mutants for heat tolerance and thermal denaturation as well as sensory evaluation has not been examined. It has both A and B polypeptide chains, which are covalently, link with having 44 and 50 amino residues, respectively: Chain A (44 aa): REIKGYEYQLYVYASDKLFRADISED-YKTR GRKLLRFNGPVPPP and Chain B (50 aa): GEWEIIDIGPFTQNLGKFAV-DEENKIGQYG RLTFNKVIRPCMKKTIYEEN. It has been reported a secondary structural confirmation consisting of 5 sheets that form and a 17-residue right-handed alpha helix as clearly revealed in the resolved crystal structures of monellin [16]. These crystallographic structures were developed in vapor diffusion of 20% ethanol into buffer protein solution. The asymmetric carbon unit contains two whole molecules of monellin protein. Diffraction patterns of monellin protein crystal forms pull out to at least 2.5 Å, thus indicating that structural analysis via X-ray crystallography correlated with the atomic resolution [13, 33].

Monellin has dual applications both as a flavor enhancer and a high-intensity sweetener and has been approved for utilization in few countries. The supply of naturally occurring monellin is restricted leading to extensive research into its synthesis via transgenic organisms. The gene encoding monellin has been introduced into different microorganisms under transcriptional control of heterologous promoters. To date the yields have been low, nevertheless, the factors controlling high efficacy microbial production have been earmarked. Riveted research efforts shall render improved microbial yields with viable levels in future. The notable properties of monellin as a food additive may be capitalized by the food industry. Alternatively, the monellin gene may be genetically engineered into selected fruit and vegetable crops to enhance their flavor and sweetness.

2 Microbial Expression Systems

Monellin belongs to a family of intensely sweet proteins derived from tropical plants [2]. As extraction of monellin from *D. cumminsii* plant is a tedious task, efforts have been directed towards production of this recombinant protein in various expression systems (Table 2). Among the microbes (Table 3), one of the most effective secretion expression systems is found in *Bacillus subtilis* and has come up as a systematized expression host [21] due to numerous merits. The most desirable benefit is its capacity to secrete proteins right into the culture medium and assimilate them to a high level in a comparatively pure form [22]. In addition, the secreted foreign proteins remain elusively in biologically active states [23]. During the over-expression of heterologous genes in *Escherichia coli*, protein secretions may envelope the development of inactive inclusion bodies [24]. Additionally, as *B. subtilis* does not reside in humans as a pathogen, it poses to be biologically safe. Numerous transgenes have been efficiently expressed in *B. subtilis* [25, 26] and enhanced yields of recombinant monellin protein were obtained [3, 13, 27]. Single-chain monellin (SCM) protein has been synthesized by *E. coli* strain W3110 as host using expression vector with Trp promoter. The recombinant protein is as sweet compared with natural monellin; it is more heat tolerant and pH stable [13]. SCM gene (MNEI) was designed on the basis of biased codons of *E. coli* for maximized expression. The gene was cloned into the pET21 vector under the T7 promoter and expression, and purification of the recombinant monellin protein and its mutants was optimized. In the case of MNEI, the A chain was linked to the B chain by a Gly-Phe dipeptide [3]. High-level expression of monellin that accounted for 45% of the total soluble proteins and purified yields up to 43 mg protein/g dry cell wt [3] was achieved.

Table 2 Studies on ectopic expression of synthetic monellin protein in different expression systems

1.	[17]	Tobacco chloroplasts
2.	[18]	Strawberry fruits
3.	[3]	*Saccharomyces cerevisiae* (a-factor signal peptide)
4.	[31]	*Saccharomyces cerevisiae*
5.	[19]	*Candida utilis*
6.	[3]	*E. coli*
7.	[20]	Tomato fruits and *E. coli*
8.	[4]	Tobacco chloroplasts (MNEI variant)

Table 3 Protein expression levels of monellin in microbial expression systems

Host	Promoter (vector)	Gene	Expression level
E. coli (W3110)	Trp	SCM*	ND**
E.coli (BL21)	T7lac (pET21)	SCM	ND
E. coli (BL21)	T7lac (pET21)	MNEI	ND
E. coli (BL21)	T7lac (pET21)	MNEI	45% (total protein)
S. cerevisae	GAPDH	SCM	ND
S. cerevisae	GAP/DH2	SCM	54 g/l
S. cerevisae	GAP	SCM	5% (total protein)
C. utilis (AB110)	GAP	SCM	45% (total protein)
P. pastoris (GS115)	GAPDH	SCM	10 g/l

SCM* stands for single-chain monellin; ND** stands for not determined

Similarly, SCM monellin gene expressed in *B. subtilis* under the regulation of *B. subtilis* sacB promoter and signal peptide that effectively directed the secretion of recombinant monellin with a maximum yield of 0.29 g protein/l [27]. In both cases, the purity of the recombinant protein was verified by SDS-PAGE.

3 Yeast Expression Systems

Yeast expression systems have been extensively utilized for the synthesis of pharmaceutical and industrial proteins. Incidentally, recombinant protein expression and purification in non-Saccharomyces yeasts, for instance, methylotrophs *Pichia pastoris* and *Hansenula polymorpha* and lactose-producing *Kluyveromyces lactis*, has frequently surpassed that of *S. cerevisiae* (Table 3). Single-chain monellin was synthesized using a DNA synthesizer and cloned into the expression vector pGAP carrying the GAPDH promoter and terminator and was transformed into *S. cerevisiae* strains [6]. Monellin was expressed successfully in transformed yeast cells as glucose was depleted, and ethanol triggered the ADH2 gene [28]. Large-scale purification of single-chain monellin was achieved on expression in ABl 10, a recombinant yeast strain, by using process-scale equipment [28]. The expression level of monellin was estimated to be about 10% of the total protein by densitometry.

Moreover, its purification yield and the purity were estimated as 45% and 95%, respectively, as determined by HPLC gel filtration. Finally, 54 g of purified recombinant SCM that elicited a sweet-taste response was obtained. As monellin has been proved to be a highly basic protein [29], the basicity was exploited as an efficient purification method. Kondo et al. [19] expressed the SCM that possessed two polypeptides connected with a glycine residue in *Candida utilis* for improved thermal stability and sweetness of monellin. The SCM gene was inserted between the glyceraldehyde-3-phosphate decarboxylase gene promoter and terminator fragments cloned from *C. utilis*. This promoter-deficient marker gene allowed high-copy-number integrations of vectors into either URA3 or rDNA gene locus. SCM was produced at a high scale with an expression level was higher than 50% of the total soluble protein. The high-level expression and cost-effective purification method of recombinant monellin protein renders its industrial production economically viable. More recently, the production of SCM by *P. pastoris* was performed ([30], US patent 6780615). The SCM gene was synthesized and cloned into the pGAPZα vector that contained the *S. cerevisiae* α-factor secretion signal and GAPDH promoter from *P. pastoris* to generate pGWYS-1. The designed plasmid was transformed into *P. pastoris* GS115 strain and transformants were selected on the basis for their resistance to zeocin. The secreted recombinant monellin was observed as a band of 12 KDa. SDS-PAGE analysis revealed that approximately 10 g/l of purified SCM was obtained and it elicited a sweet taste. Chen et al. (2011) expressed the SCM gene in *S. cerevisiae* driven by *GAL1* promoter and alpha-factor signal peptide sequence of *S. cerevisiae*. The peptide effectively directed the secretion of active monellin from recombinant yeast cell with a maximum yield of 0.41 g/l of supernatant.

4 Transgenic Plants as an Expression System

Monellin protein reveals a glycemic index of zero, thereby utilized in food industry as a natural product and may be consumed by diabetics too [31]. However, as it is expensive to manufacture monellin from its native source therefore, in vitro synthesis and production of sweet protein "monellin" by transgenic approach and new gene editing technology holds great promise. Production of monellin in transgenic fruits and vegetables represents an alternative strategy to enhance their flavor and quality. Previous attempts to synthesize monellin via recombinant technology proved to be futile due to its low expression levels in plants [4]. Monellin was produced in transgenic lettuce and tomato plants [7]. A single-chain monellin gene, encoding both polypeptide chains linked by a hinge sequence, was driven by constitutive and fruit-ripening-specific promoters and transferred to lettuce and tomato. Expression of monellin gene in tomato and lettuce transgenics resulted in the high accumulation of monellin protein in leaf and fruit, respectively. The fruit-specific and whole plant expression was obtained by targeting monellin gene expression to tomato fruits. The latter was the CaMV promoter, which is active in diverse plant organs. The protein was extracted from the tomato pericarp or lettuce leaf and subjected to western blot analysis and ELISA. The recombinant monellin was detected in 50% red and red-ripe transgenic tomato fruits harboring the E8 promoter and monellin and in the transgenic lettuce leaf harboring the CaMV promoter and monellin gene. Transgenic tomato was exposed to ethylene to enhance monellin production in the tomato pericarp. In ethylene-treated transgenic tomato, monellin protein was estimated at 23.9 μg/g fresh weight and was notably higher in untreated transgenic fruit. Reddy et al. [20] reported the stable and enhanced expression of a synthetic sweet protein monellin in prokaryotic (*E. coli*) and eukaryotic (tomato) systems than reported, previously. The recombinant monellin protein was thermostable and retained strong sweetness over a temperature range (up to 70 °C) and extreme pH levels. Transgenic tomato plants revealed stable gene integration in T_1 generation as verified by PCR analysis. Transcript profiling of T_2 lines revealed enhanced monellin expression, which correlated positively to its protein expression profiles as highlighted by western blots [20]. Heterologous expression of monellin driven by the fruit-specific promoter (E8) did not lead to any phenotypic anomalies either in vegetative or fruit growth parameters in the transgenic tomato lines. RT-PCR was performed to analyze the mRNA expression levels of monellin in ripened fruits from T_2 lines. Similar amplification was not noted in WT tomatoes [20]. Abundance of monellin transcript was normalized against that of tubulin, an internal control. Maximal levels of transcript expression were monitored in two lines (T_2-5; T_2-14) at the exponential phase PCR cycle 24. Experiments were performed in triplicate with different biological replicates. ELISAs using polyclonal antibody against monellin were used to detect the presence of monellin protein in transgenic and WT lines. Independent experiments were performed in triplicate and each sample was measured in duplicate. The statistical errors were detected by standard deviations. Data from ELISA studies were confirmed by immuno-blot analysis by comparing the signal for each variant to that of purified monellin. Transgenic lines T_2-5 and T_2-14 showed maximum protein levels of 60 lg/mg (±2.06) and 54 lg/mg (±2.58), respectively. Other transgenic lines also showed expressions of monellin, i.e., T_2-4 (51 ± 1.54), T_2-7 (47 ± 1.78), T_2-8 (45 ± 1.96). Immuno-blotting of total fruit protein extract (after the major protein Rubisco removed) from wild type and transgenic plants expressing monellin was performed [20]. There was no detectable monellin protein in untransformed wild-type fruit, thereby confirming the absence of the monellin transgene. Estimated yield of monellin/mg protein was 60 lg that corresponded to 4.5% of the soluble proteins extracted from the transgenic tomato fruit expressing monellin. Evaluations for fruit taste, postripening (10 days) revealed that out of five transgenic lines, T_2-5 and T_2-14 exhibited distinct or noticeable sweetness compared with the WT. All five panelists described the sweetness of tomato pericarp of fruits from T_2-5 and T_2-14 transgenic lines as strongly distinguishable from the WT fruit [20]. A specific characteristic after-taste that lingered in the mouth after swallowing was noted for the fruits from linesT_2-5 and T_2-14. Transgenic tomatoes that expressed monellin heterologously were sweeter in flavor than WT (Figs. 1 and 2).

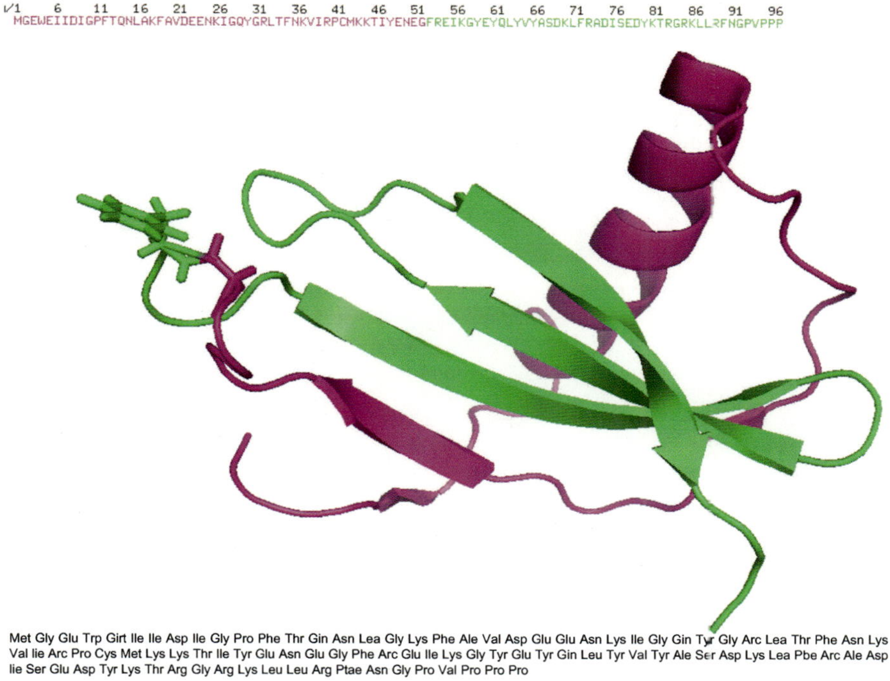

Fig. 1 Cartoon representation of monellin 3D structure (PDB-ID: 1M9G) showing its two polypeptide chains i.e., chain A (44 aa) and chain B (50 aa) in *magenta* and *green*, respectively. The Gly-Phe dipepetide linkage has been represented as stick forms (in *magenta* for Gly and *green* for Phe)

5 Conclusions and Future Perspectives

Consumption of high caloric food causes obesity and related problems mainly due to intake of refined sugars that eventually causes high risk of type II diabetes, heart disease, certain types of cancer, sleep apnea, and osteoarthritis. The main sources of these excess carbohydrates are sweetened beverages and other carbohydrate-rich foods. Though addition of artificial sugar-free sweeteners may substitute sweetness, they still have certain side effects. Consequently, use of monellin, a low-calorie, carbohydrate-free protein, as a natural sweetener, would be an ultimate option. However, monellin protein purification is a difficult task and limited to restricted habitats. Moreover, the two subunits of monellin, i.e., chain A (45 amino acids forming three anti-parallel strands) and chain B (50 amino acids forming two b strands separated by an alpha-helix), are bound by noncovalent interactions that lead to its instable nature.

To downplay these complications, reassembling of native monellin as recombinant single-chain protein confers enhanced stability at high temperatures and

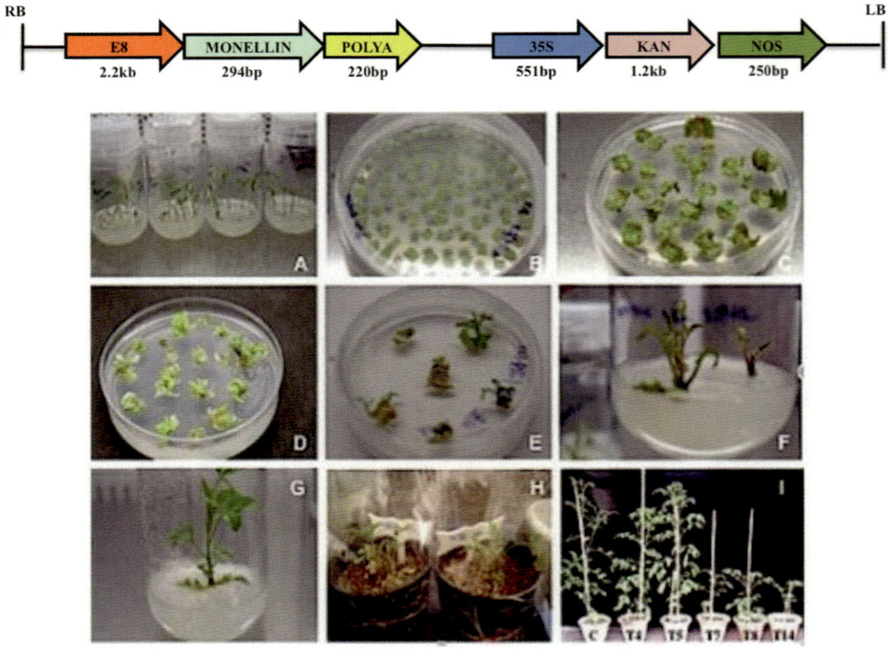

Fig. 2 (**a**) Schematic diagram of the T-DNA region of binary vector pMDC100 utilized for tomato transformation revealing synthetic monellin gene regulated by E8 fruit-specific promoter and the *kan* gene controlled by CaMV 35S promoter. (**b**) Overview of the agro-mediated protocol for tomato transformation exhibiting various stages i.e., seedlings grown aseptically (*A*), cotyledonary explants on pe-culture medium (*B*), regeneration of shoots from the explants (*C–E*), elongation of the regenerated shoots (*F* & *G*), young plantlets in hardening (*H*), and adult transgenic tomato lines expressing monellin (*I*)

extreme pH and facilitates retention of its sweet flavor. Thereby, it is beneficial to earmark suitable monellin expression systems for large-scale production of this potent food supplement.

References

1. Yarmolinsky DA, Zuker CS, Ryba NJ (2009) Common sense about taste: from mammals to insects. Cell 139(2):234–244
2. Dansby (1997) Monellin belongs to a family of intensely sweet proteins derived from tropical plants. Nat Biotechnol 15:419–420
3. Chen Z, Cai H, Lu F, Du L (2005) High-level expression of a synthetic gene encoding a sweet protein, monellin, in *Escherichia coli*. Biotechnol Lett 27(22):1745–1749
4. Lee SB, Kim Y, Lee J, Oh KJ, Byun MO, Jeong MJ, Bae SC (2012) Stable expression of the sweet protein monellin variant MNEI in tobacco chloroplasts. Plant Biotechnol Rep 6(4):285–295
5. Tancredi T, Iijima H, Saviano G, Amodeo P, Temussi PA (1992) Structural determination of the active site of a sweet protein A 1H NMR investigation of pMNEI. FEBS Lett 310(1):27–30

6. Kim SH, Kang CH, Kim R, Cho JM, Lee YB, Lee TK (1989) Redesigning a sweet protein: increased stability and renaturability. Protein Eng 2:571–575
7. Peñarrubia L, Kim R, Giovannoni J, Kim SH, Fischer RL (1992) Production of the sweet protein monellin in transgenic plants. Nat Biotechnol 10(5):561–564
8. Hung LW, Kohmura M, Ariyoshi Y, Kim SH (1999) Structural differences in D and L-monellin in the crystals of racemic mixture. J Mol Biol 285(1):311–321
9. Kohmura M, Nio N, Ariyoshi Y (1990) Complete amino acid sequence of the sweet protein monellin. Agric Biol Chem 54(9):2219–2224
10. Liu B, Ha M, Meng XY, Kaur T, Khaleduzzaman M, Zhang Z, Cui M (2011) Molecular mechanism of species-dependent sweet taste toward artificial sweeteners. J Neurosci 31(30):11070–11076
11. Liu B, Ha M, Meng XY, Khaleduzzaman M, Zhang Z, Li X, Cui M (2012) Functional characterization of the heterodimeric sweet taste receptor T1R2 and T1R3 from a New World monkey species (squirrel monkey) and its response to sweet-tasting proteins. Biochem Biophys Res Commun 427(2):431–437
12. Leone S, Pica A, Merlino A, Sannino F, Temussi PA, Picone D (2016) Sweeter and stronger: enhancing sweetness and stability of the single chain monellin MNEI through molecular design. Sci Rep 6:34045. doi:10.1038/srep34045
13. Sung YH, Shin J, Chang HJ, Cho JM, Lee W (2001) Solution structure, backbone dynamics, and stability of a double mutant single-chain monellin structural origin of sweetness. J Biol Chem 276(22):19624–19630
14. Esposito V, Gallucci R, Picone D, Saviano G, Tancredi T, Temussi PA (2006) The importance of electrostatic potential in the interaction of sweet proteins with the sweet taste receptor. J Mol Biol 360(2):448–456
15. Rega MF, Di Monaco R, Leone S, Donnarumma F, Spadaccini R, Cavella S, Picone D (2015) Design of sweet protein based sweeteners: hints from structure–function relationships. Food Chem 173:1179–1186
16. Ogata C, Hatada M, Tomlinson G, Shin WC, Kim SH (1987) Crystal structure of the intensely sweet protein monellin. Nature 328(6132):739–742
17. Roh KH, Shin KS, Lee YH, Seo SC, Park HG, Daniell H, Lee SB (2006) Accumulation of sweet protein monellin is regulated by thepsbA 5′ UTR in tobacco chloroplasts. J Plant Biol 49(1):34–43
18. Min SR, Ko SM, Lyu JI, Park JH et al (2015) Development of transgenic strawberry plants expressing monellin, a sweet protein. J Plant Biotechnol 42(3):180–185
19. Kondo K, Miura Y, Sone H, Kobayashi K, Iijima H (1997) High-level expression of a sweet protein, monellin, in the food yeast *Candida utilis*. Nat Biotechnol 15(5):453–457
20. Reddy CS, Vijayalakshmi M, Kaul T, Islam T, Reddy MK (2015) Improving flavour and quality of tomatoes by expression of synthetic gene encoding sweet protein monellin. Mol Biotechnol 57(5):448–453
21. Doi RH, Wong SL, Kawamura F (1998) Potential use of *Bacillus subtilis* for secretion and production of foreign proteins. Trends Biotechnol 4:232–235
22. Goeddle DV (1990) Expression in *B. subtilis*. Methods Enzymol 85:199–228
23. Nakayama A, Ando K, Kawamura K et al (1988) Efficient secretion of the authentic mature human growth hormone by *Bacillus subtilis*. J Biotechnol 8:123–134
24. Thomas JG, Baneyx F (1996) Protein misfolding and inclusion body formation in recombinant *Escherichia coli* cells overexpressing heat-shock proteins. J Biol Chem 271:11141–11147
25. Baillie L, Moir A, Manchee R (1998) The expression of the protective antigen of *Bacillus anthracis* in *B. subtilis*. J Appl Microbiol 84:741–746
26. Lee JJ, Hahm KS, Lee KY, Lee ST (1997) Characterization of an endoxylanase produced by an isolated strain of *Bacillus* sp. J Microbiol Biotechnol 7:114–120
27. Chen Z, Heng C, Li Z, Liang X, Xinchen S (2007) Expression and secretion of a single-chain sweet protein monellin in *Bacillus subtilis* by sacB promoter and signal peptide. Appl Microbiol Biotechnol 73(6):1377–1381
28. Kim IH, Lim KJ (1996) Large-scale purification of recombinant monellin from yeast. J Ferment Bioeng 82(2):180–182

29. Cagan RH (1973) Chemostimulatory protein: a new type of taste stimulus. Science 181 (4094):32–35
30. Duan L (2004) Production of recombinant monellin using methylotrophic yeast expression system. US 6780615 B1
31. Liu J, Yan DZ, Zhao SJ (2015) Expression of monellin in a food-grade delivery system in *Saccharomyces cerevisiae*. J Sci Food Agric 95(13):2646–2651
32. Aghera N, Udgaonkar JB (2011) Heterologous expression, purification and characterization of heterodimeric monellin. Protein Expr Purif 2:248–253
33. Wlodawer A, Hodgson KO (1975) Crystallization and crystal data of monellin. Proc Natl Acad Sci 72(1):398–399

Part III

Biological Effects

The Role of Dietary Sugars and Sweeteners in Metabolic Disorders and Diabetes

11

Motahar Heidari-Beni and Roya Kelishadi

Abstract

Sugar consumption has dramatically increased worldwide. A growing body of evidence suggests that sugars might have various adverse health effects. High intake of sugars may be related with an increased risk of several disorders including dental caries, obesity, cardiovascular disease, diabetes, gout, fatty liver disease, some cancers, components of the metabolic syndrome, and hyperactivity. Added sugar in processed foods are used to sweeten, to increase the flavour, to change the freezing or melting point or to protection of food spoilage. It is better to consume sugars in natural foods, since these foods provide useful micronutrients. Nowadays, there are questions as to whether excessive consumption of sugars, especially processed foods, might be correlated with metabolic syndrome or diabetes. However, insufficient study design, variety in evaluating dietary intake, contradictory findings and several definitions of sugars have inhibited definitive conclusions regarding these associations. However, limiting added sugars and monitoring carbohydrate consumption are serious strategy for keeping healthy weights and achieve glycemic control. This chapter describe different types of sweeteners in foods and beverages, as well as their effects on diabetes and metabolic disease. In addition, this chapter describes underlying mechanisms of sweeteners on health outcomes and how various types of sweeteners may threaten health.

Keywords

Sweeteners • Non-nutritive sweeteners • Sugar sweetened beverage • High-fructose corn syrup • Diabetes • Metabolic syndrome

M. Heidari-Beni (✉) • R. Kelishadi
Child Growth and Development Research Center, Research Institute for Primordial Prevention of Non Communicable Disease, Isfahan University of Medical Sciences, Isfahan, Iran
e-mail: heidari.motahar@gmail.com; motahar.heidari@nutr.mui.ac.ir; roya.kelishadi@gmail.com

J.-M. Mérillon, K.G. Ramawat (eds.), *Sweeteners*, Reference Series in Phytochemistry,
https://doi.org/10.1007/978-3-319-27027-2_31

Abbreviations

FAO	Food and agriculture organization
GI	Glycaemic index
GL	Glycemic load
GLP	Glucagon-like peptide-1
GLUT	Sodium-glucose transport proteins
HFCS	High fructose corn syrup
HPFS	Health professionals follow-up study
IMP	Inosine monophosphate
KHK	ketohexokinase
NHS	Nurses' health study
NNS	Non-nutritive sweeteners
NS	Natural sweeteners
SSB	Sugar-sweetened beverages
T1R	Taste receptors type 1
XO	Xanthine oxidase

Contents

1 Introduction 226
2 Concepts of Sugars and Sweeteners 227
3 Activation of Sweet Taste Receptors 228
4 Types of Sweeteners 229
4.1 Non-Nutritive Sweeteners (NNS) 229
4.2 Proposed Mechanisms Underlying NNS Metabolic Effects 230
4.3 Natural Sweeteners (NS) 231
5 Sugar Sweetened Beverage (SSB) 231
5.1 Sugar Sweetened Beverages and Blood Pressure 232
5.2 Sugar Sweetened Beverages and Diabetes 232
5.3 Sugar Sweetened Beverages and Metabolic Syndrome 233
6 High-Fructose Corn Syrup (HFCS) and its Relationship with Health 234
6.1 Fructose and Uric Acid and Metabolic Syndrome 236
7 Summary of Proposed Mechanisms Underlying SSB and Artificial Sweeteners Adverse Metabolic Effects 237
8 Conclusions and Recommendations 239
9 Cross-References 239
References 240

1 Introduction

Quality of life is markedly dependent on taste sensory system. We select foods primarily according to sentimentality of pleasure or displeasure experienced of their taste. Studies showed that sweet was an innately preferred sensation.

Carbohydrates are the main source of food energy. They are remarkable for their structure and well taste, which in sugars is sweet. Carbohydrates have two forms including complex molecules especially starches, dextrins and fibre or simple carbohydrates that generally called sugars. Fruit, some vegetables, milk, soft drinks and sweets are the main dietary sources of sugars. Structure of sugars is called monosaccharide. Glucose (also called dextrose), fructose, and galactose are monosaccharides. Disaccharide is formed by combination of two monosaccharides. For example, combination of glucose and fructose form disaccharide sucrose or table sugar. Two glucose molecules compose maltose and one molecule of glucose and one molecule of galactose make lactose (milk sugar). All sugars are carbohydrates and contain four calories per gram. The amount of sugars intake is so important in the daily diet. Sugars supply the rapid source of glucose for brain (cognitive functions) and muscles (physical activity) [1, 2].

Added sugar in processed foods are used to sweeten, increase the flavour, change the freezing or melting point or to protection of food spoilage. It is better to consume sugars in natural foods, since these foods provide other micronutrients. Nowadays, there are questions as to whether excessive consumption of sugars especially processed foods might be correlated with metabolic syndrome or diabetes [3, 4].

In a healthy diet, all types of carbohydrates must be consumed. According to recommendation of The Food and Agriculture Organization (FAO) and WHO, sugars or simple carbohydrates must be intake less than 10% of the total caloric value of the diet. Sugars must be part of a healthy diet with limited consumption of sugar-sweetened drinks. Also, for achievement greater benefits, it has been recommended that consumption of sugars or sugar- sweetened foods must be limited to less than three times/day not exceeding 6% of total energy intake [5].

2 Concepts of Sugars and Sweeteners

Sugar has been in human diets since ancient times. Earliest consumption has been reported from China and India and after that from Europe after the Crusades in the eleventh century. Natural dietary sugars (fructose, sucrose, and lactose) and added or extrinsic sugars (sugars and syrups that are added to foods and beverages during processing and preparation) are found in many foods. Over the past three decades, intake of sugars has dramatically increased in United State. The main reason of this increase is driven by high fructose corn syrup (HFCS) intake, that now consumption of HFCS is over 62 lb per person per year mainly in the form of sugar-sweetened beverages (SSB) [6, 7]. SSBs include soft drinks, fruit drinks, energy and vitamin water drinks. They are composed of naturally derived caloric sweeteners such as sucrose (50% glucose and 50% fructose), HFCS (most often 45% glucose and 55% fructose), or fruit juice concentrates [8].

According to evidence, sugars might have adverse effects on health condition. High intake of sugars may be related with an increased risk of dental caries, obesity, cardiovascular disease, diabetes, gout, fatty liver disease, some cancers, components of the metabolic syndrome and hyperactivity [9–12]. However, insufficient study design, variety in evaluating dietary intake, contradictory findings and several definitions of sugars have inhibited definitive conclusions regarding these associations [13].

There is metabolic interaction between different sugars, for example between fructose and glucose. When fructose ingests alone has different metabolic effect in

Table 1 Classification of sugars and sweeteners

Natural	Caloric	Sugars	Sucrose Glucose Galactose Fructose Lactose Maltose Trehalose
		Others	Honey Maple syrup, Palm sugar, Coconut sugar
Artificial	Caloric	Modified sugars	(High fructose) corn syrup Inverted sugar
		Sugar alcohols	Sorbitol Xylitol Mannitol Lactitol
	Non-caloric	Artificial sweeteners	Aspartame Sucralose Saccharin Neotame, Acesulfame K Cyclamic acid Alitame Advantame

comparison with consumption together with glucose. Fructose intolerance arise when it is consumed alone, however intolerance symptoms disappear when it consume with glucose. In addition, fructose intake alone leads to glycogen synthesis less than when fructose consumed with glucose [14].

There are different ways for classification sweeteners; using the glycaemic index (GI) or their energy contribution. Based on energy contribution, sweeteners are classified as "caloric" or "low calorie/non-caloric" or as "nutritive" or "non-nutritive" (Table 1) [3, 15].

3 Activation of Sweet Taste Receptors

Nutritive sweeteners and non-nutritive sweeteners (NNS) evoke sweet taste sensation in mammals. Sweet taste receptors type 1 (T1R) subunits 1 (T1R1) and 3 (T1R3) detect sweet sensation. These receptors are coupled to α-gustducin, a transducin-like heterotrimeric G-protein [16]. There is T1R in the lingual taste buds and also in the enteroendocrine cells of the small intestine that are involved in Glucagon-like peptide-1 (GLP-1) secretion [17]. Studies showed when intestinal T1R2 and T1R3 were activated by acesulfame-K, sucralose and saccharin in mice, upregulate the expression of the Na^+ dependent glucose transporter SGLT1. In turn, activation of this transporter enhances intracellular Ca^{2+} concentrations and consequently the translocation of the facilitative glucose trans-porter GLUT2 into the brush border membrane of the enterocyte [16–18]. In addition to, the promotion of intracellular Ca^{2+} levels activates a member of the transient receptor potential melastatin (TRPM), TRPM5, which is co-expressed with T1R in enterocytes and functions as a downstream component in sweet taste signal transduction pathway. TRPM5 activation then promotes Na^+ influx, whose role, although not understood yet, could hypothetically facilitate SGLT1-mediated glucose transport. As a whole, these data suggest that NNS leads to an increase in the intestinal absorptive capacity of glucose and consequently of its blood levels, eventually favoring an hyperglycemic state in NNS-exposed neonates. Interestingly enough, activation of T1R present in GLP-1-secreting cells can stimulate the secretion of incretins, which in turn increase insulin secretion, appetite and weight gain [19].

4 Types of Sweeteners

4.1 Non-Nutritive Sweeteners (NNS)

Consumption of NNS has increased among people of all ages. 28% of total population reported NNS intake that this amount is highly prevalent in children. Most of the NNS-containing foods intake among population is NNS-containing beverage. According to National Health and Nutrition Examination Survey, consumption of NNS-containing beverage increased from 6.1% to 12.5% among children and from 18.7% to 24.1% among adults [20]. Some foods contain NNS including beverage, ice cream, yogurt, chewing gum, chocolate, jams and chocolate. There has been much debate regarding the health advantages and disadvantages of artificial sweeteners [21].

NNS, also refer as very low-calorie sweeteners, artificial sweeteners, non-caloric sweeteners, and intense sweeteners. These sweeteners have a higher sweetness taste than caloric sweeteners such as sucrose, corn syrups and fruit juice. NNS are added in smaller quantities and provide no or few calories, no glycaemic effect and high sweetening power [22]. Among NNS, saccharin is prominent. It has great sweetening intensity. In addition to, Sucralose and aspartame are remarkable for their wide worldwide use, particularly in beverages. Methyl ester of phenylalanine and aspartic acid are components of aspartame. Its use is approved by the Food and Drug Administration (FDA) and the European Food Safety Authority (EFSA) [23, 24].

Currently, six sweeteners use in foods and beverages in US that are approved by the Food and Drug Administration (FDA). These sweeteners including aspartame, sucralose, saccharin, acesulfame potassium, neotame and advantame, with two plant-derived sweeteners (steviol glycosides, and Luo han guo extract) [25]. In most cases, these sweeteners provide little or no energy because they activate receptors of sweet taste at very low concentrations in comparison with sugar [26]. Although these sweeteners have various chemical structures, they can activate some of the multiple potential ligand binding sites of the heterodimeric T1R1 + T1R3 sweet-taste receptor in human subjects [27].

NNS provide little or no energy, so when NNS are used in place of caloric sweeteners in foods and beverages, the amount of calories intake will be reduced while maintaining high palatability. NNS provide sweet taste without calories or glycemic effects, so it is commonly believed that NNS are healthy substitutes for sugars [25]. However, scientific evidences do not actually support such a belief and there is a great deal of debate regarding the health consequences of NNS intake [26]. Studies suggested that NNSs are not physiologically inert, and may influence feeding and metabolism by different peripheral and central mechanisms [28].

Some studies showed positive relationship between NNS intake and weight gain, metabolic syndrome, and type II diabetes [29, 30], although other studies did not show any association [31, 32]. Nevertheless, several reviews and meta-analyses of epidemiological and experimental studies could not show a consensus result. Some studies reported potentially beneficial [33], harmful [34], or trivial [35] effects of NNSs. Taken as a whole, despite numerous studies showed an association between NNS intake and metabolic disorders in animal models [36, 37], there is no sufficient evidence in human subjects. However, findings from at least five various mammalian species (rats, mice, cows, pigs, human) reported that NNSs could be metabolically active [38, 39]. More research is needed to clarify the mechanisms of metabolic and potential effects of NNSs as commonly used food additives [25].

4.2 Proposed Mechanisms Underlying NNS Metabolic Effects

Physiological studies observed NNS decreased the release of the incretin hormone GLP-1. This hormone implicates in food intake regulation, blood sugar levels and protection of the cardiovascular system. When GLP-1 levels persistently decreased through intake of artificial sweeteners, risks of diabetes, cardiovascular disease and stroke would be increased in long term [40].

Modulation of the intestinal microbiota is another potential mechanism to illustrate the relationship between NNS intake and adverse metabolic outcomes. The intestinal microbiota of humans and rodents include only a few dominant and commensal bacterial phyla, exclusively Firmicutes, Bacteroidetes, Actinobacteria and Proteobacteria. These gut microbiota impact on important physiological functions of the host, especially immune system and dietary nutrients metabolism. A decrease in the proportion of Firmicutes has been found in type II diabetic patients and the proportion of Bacteroidetes has been decreased in obese individual in fecal samples [41, 42].

Animal studies showed that artificial sweeteners consist of saccharin, sucralose and aspartame modify gut microbiota and led to overeating, weight gain and impaired blood sugar regulation [43]. Findings of one animal study showed consumption of a mixture of sucralose with nutritive sweeteners decreased the number of commensal bacteria including Bifidobacterium (phylum Actinobacteria), Lactobacillus (phylum Firmicutes) and Bacteroides (phylum Bacteroidetes). These effects have been associated with the development of insulin resistance, hyperlipidemia, enhanced adiposity and inflammation [19].

In addition to, in a small human study, saccharin led to gut microbiota alteration and blood glucose dysregulation. Taken as a whole, gut microbiota have many roles in health and disease. If artificial sweeteners consume persistently, disrupt gut microbiota composition and lead to some adverse effects including weight gain and glucose intolerance [26].

4.3 Natural Sweeteners (NS)

Natural caloric sweeteners consist of sucrose, fructose, glucose and maltose. Fructose has been substitute with sucrose in diabetic patients. Recently, however, it has been reported that high fructose diets, particularly when the fructose is added to processed foods, can induce hyperinsulinaemia, hypertriglyceridaemia and insulin resistance, and this has led to limit fructose use in diabetic patients [3].

5 Sugar Sweetened Beverage (SSB)

SSB consumption has been enhancing worldwide. SSBs have become common beverages that consume around the globe. SSB consumption rise among Asians, however, it is lower than Western populations. Traditional diet of Asian populations that consist of rice or grains as staples and abundant vegetables substitute with the Western lifestyle include meat and sweet foods as desserts and beverages. The average energy intake from SSBs has increased from 22.3 to 20.0 kcal/day in adolescents and young adults in 1998 to 35.1 and 29.4 kcal/day in adolescents and young adults in 2009, respectively. In addition to, sugar consumption from soft drinks is the highest between all types of beverages [44].

Over 70% of adults consumed SSB include soft drinks or fruit drinks with added sugar with over 25% reporting daily intake in 2012 [45].

Recently, numerous large-scale cohort studies have reported that SSBs, that commonly include soft drinks, fruit drinks, and sports drinks, are related positively with development risk of metabolic dysfunctions including type II diabetes and cardiovascular disease [44]. In contrast, a beverage contains 100% fruit juice without any added sweeteners, is not recognized as SSB [46].

SSBs intake regularly has been directly related with variety of negative outcomes and adverse cardiometabolic effects, including weight gain and obesity, hypertension, diabetes, metabolic syndrome and stroke [47, 48]. According to recent meta-analysis, one serving per day increment in SSB intake was related to a 0.06 unit increase in body mass index in children and 0.22 kg weight gain in adults during 1 year [47]. Another meta-analysis found a clear association between SSB consumption and risk of metabolic syndrome and type II diabetes. These findings provide further support to restrict intake of SSBs and substitute healthy alternatives including water to decrease risk of obesity-related chronic disease. High intake of SSBs lead to 26% increase risk of diabetes and 20% increase risk of metabolic syndrome [46].

In addition to, frequent SSBs drinkers consume more total and saturated fat, carbohydrate, sodium, lower fiber, dairy products and as well as they have a sedentary lifestyle. All of these factors can be related to increase risk of metabolic disease [48].

The powerful and consistent relationship between SSB consumption, obesity and some diseases including diabetes and metabolic disorders lead to increasing emphasis on decreasing the availability and intake of sugars and SSBs among children and adults [49]. However, in individual that intake persistently high level of SSBs, reduction of sugary foods and beverages consumption is not simple.

The substitution of artificial sweetened beverages for SSBs may be considered. While, there is lack of clear and consistent supporting evidence that show artificial sweeteners will promote healthy outcomes [50].

According to findings of interventional studies, substitution of artificially sweetened versions for sugar sweetened versions of foods or beverages did not consistently show that artificial sweeteners promote weight loss in overweight individuals. These results suggest that less effectiveness of artificial sweeteners in comparison with sugar sweeteners. In other words, artificial sweeteners do not appear to make better outcomes than sugar sweeteners. Recently one meta-analysis showed that artificial sweeteners may be useful for short-term weight loss [51].

Taken as a whole, evidence suggests that sugar sweetened beverages consumption is problematic. However the data regarding whether artificial sweetener beverages are particularly beneficial as replacements is vague [26].

5.1 Sugar Sweetened Beverages and Blood Pressure

Recently, one systematic review reported SSBs consumption was correlated with higher blood pressure, so, increased incidence of hypertension [52]. In addition to, a recent meta-analysis has found that total fructose intake did not increase risk of hypertension [53]. A recent study investigated relationship between SSBs and the incidence of hypertension on data from Nurses' Health Study (NHS) I and II, and Health Professionals Follow-up Study (HPFS) that are three large American cohorts. Findings showed that SSBs were correlated with an increased risk of incident hypertension. Also, findings of cross-sectional study from the US National Health and Nutrition Examination Surveys (NHANES) 1999–2004 [54] and 2003–2006 [55], and the International Study of Macro/Micronutrients and Blood Pressure (INTERMAP) showed a positive correlation among SSB intake and directly measured blood pressure [56].

5.2 Sugar Sweetened Beverages and Diabetes

One meta-analysis study based on data from eight prospective cohort studies consists of 310,819 participants and 15,043 type II diabetic patients showed participants in the highest category of SSB consumption had a 26% greater risk of developing type II diabetes than participants in the lowest category of intake [46]. In the NHS II, a cohort of over 50,000 women, participants who intake more than one SSB per day had 83% greater risk of developing type II diabetes during 8 years compared to those who consumed less than one per month after adjusting for potential confounders [32].

Meta-analyses study showed a clear positive correlation between both sugar sweetened and artificially sweetened soft drinks and the increasing incidence of type II diabetes. The relationship was stronger and more consistent for sugar sweetened soft drinks than artificially sweetened soft drinks. According to findings, increase of 330 ml per day being correlated with approximately 20% increased risk of diabetes [57].

One 12 oz daily increase in SSB consumption was related with a 22% development in the risk of type II diabetes in European adults [44].

5.3 Sugar Sweetened Beverages and Metabolic Syndrome

Metabolic syndrome has developed in Western countries. Its components are obesity with other abnormalities such as alterations in glucose metabolism, hyperlipidemia and hypertension. Findings showed 20–25% of adult population in the world were diagnosed as metabolic syndrome. Obesity, unhealthy diet, and sedentary lifestyle lead to high prevalence of metabolic syndrome [58].

Study on 19,431 participants and 5,803 cases of metabolic syndrome (data from three prospective cohort studies) showed participants that consumed high amount of SSBs had 20% greater risk of developing metabolic syndrome than those in the lowest intake [46]. Findings of one cohort study on 6,000 adults showed those who consumed more than one soft drink per day had 39% greater risk of metabolic syndrome during 4 years [59].

Propensity of SSB to induce weight gain, rapidly absorption of large quantities of carbohydrates including sucrose or HFCS, rapidly enhance in blood glucose and insulin levels lead to metabolic consequences of SSB consumption. SSB often consume in the large volumes and contribute to a high dietary glycemic load (GL). According to findings, High GL diets stimulate appetite and lead to weight gain and are correlated with development of glucose intolerance and insulin resistance. In addition, high GL diets have been associated with lipid profiles abnormality and increase levels of inflammatory biomarkers including C-reactive protein that is a well-known marker for development of type II diabetes and cardiovascular disease risk [50].

Most epidemiological studies that investigate the association between SSBs and metabolic diseases have been carried out in Western populations and a few studies have considered Asian populations [44]. However study in Taiwan showed high SSB consumption associated with 1.9 and 2.7 times higher risk of metabolic syndrome in boys and girls, respectively [61]. One cohort study suggested that approximately one cup of soft drink per week was correlated with a 17% higher risk of metabolic syndrome in Korean populations [62].

6 High-Fructose Corn Syrup (HFCS) and its Relationship with Health

Fructose is generally called as fruit sugar because it presence in fruits. Fructose use in different forms including crystalline form in corn starch, liquid form in honey or liquid high-fructose corn syrup (HFCS) when combined with glucose. HFCS is used in the preparation of several beverages. Fructose is absorbed in the intestine and 60–70% of fructose is transported to the liver. Other 30–40% of fructose is transported to the kidney, adipose tissue and other organs. Fructose consumption has enhanced in countries adopting a Western diet during the past three decades. Although fructose has a lower glycemic load than glucose, it has numerous unpleasant effects on health status [48]. Diets containing large amounts of fructose from sucrose (50% fructose) or HFCS (42–55% fructose) has been associated with development of weight gain, visceral adiposity, dyslipidemia, insulin resistance, glucose intolerance, fatty liver and hypertension that are components of metabolic syndrome [58]. Recent findings have also reported that fructose intake may raise accumulation of visceral adiposity or ectopic fat deposition [46].

Different reviews have suggested that fructose or HFCS consumption were correlated with increased risk of obesity or metabolic syndrome [60, 63] However, others have not showed this conclusion [64, 65]. In addition to, all published meta-analyses studies have not showed a statistically significant association [66].

Clinical studies have supported that sucrose and especially fructose can stimulate weight gain and components of the metabolic syndrome. Findings showed serum triglyceride and insulin level increased in young men that consumed a diet with 200 g sucrose per day, whereas triglyceride levels did not enhanced when starch was consumed [9].

Fructose is rapidly absorbed from the diet and rapidly metabolized preferentially to lipid in the liver. Fructose can supply carbon atoms for the glycerol and the triglyceride acyl portions. Thus, fructose induces highly de novo lipogenesis, increase triglycerides levels and decrease HDL-C. High levels of fructose can serve as a relatively unregulated source of acetyl CoA. Unlike glucose, dietary fructose does not induce insulin or leptin secretion. These hormones involve in regulation of energy intake and body adiposity. In contrast, some findings have reported greater satiety and lower total energy intake after consumption of beverages contain fructose compared with glucose beverages [46].

Stimulated triglyceride synthesis by fructose lead to accumulate hepatic triglyceride, reduce hepatic insulin sensitivity, higher availability of substrate and thus increase formation of VLDL particles, increased apoB stability, and higher MTP (microsomal triglyceride transfer protein) (the critical factor in VLDL assembly) [22] (Fig. 1).

It has been suggested that fructose has adverse effects on blood pressure. Fructokinase phosphorylates fructose to fructose-1- phosphate. This results in a decrease in intracellular phosphate and ATP depletion, resulting in transient inhibition of protein synthesis. Adenosine monophosphate is produced and broken down by adenosine monophosphate deaminase, resulting in the generation of inosine

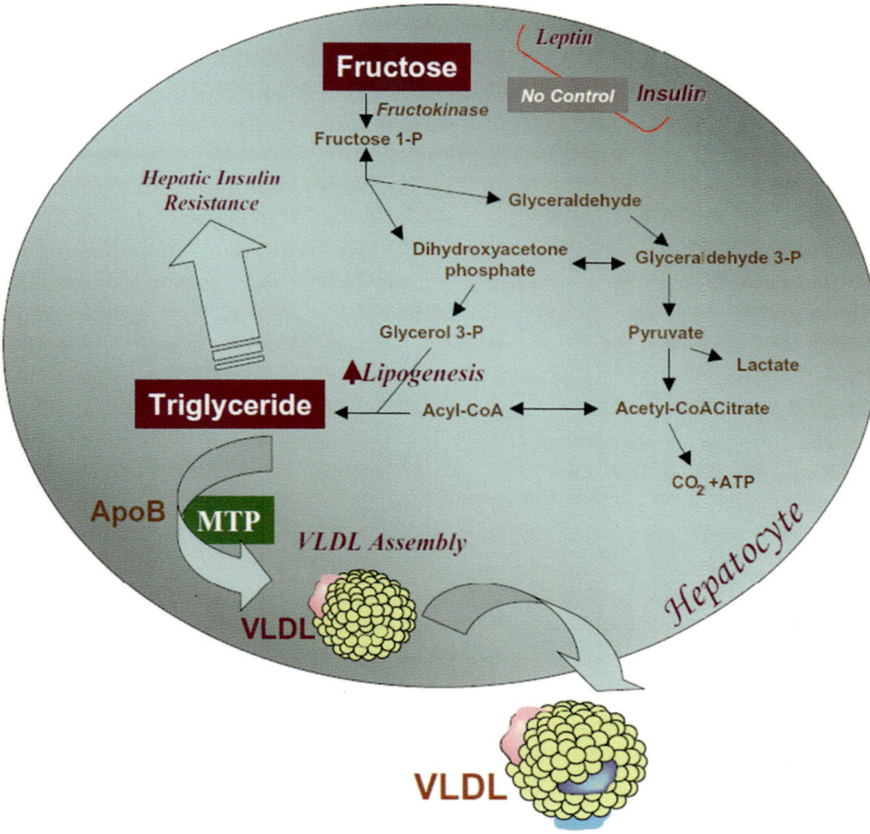

Fig. 1 Hepatic fructose metabolism: A highly lipogenic pathway [67]

monophosphate and eventually uric acid that it increases blood pressure. Thus, fructose consumption may stimulate hyperuricemia, and also enhance reabsorption of sodium and water; therefore the combination of sodium and fructose has a synergistic effect in e progress of hypertension [68].

In summary, according to findings, HFCS and sucrose are similar and one is not better or worse than the other. In addition to, small amounts of sugars containing fructose may be proper in diabetic patients because of the blunted glycemic response. However, they must be avoided in consumption of large amounts because of adverse effects on lipid profile levels; more researches are needed to identify thresholds for these effects [67]. Public notice about adverse effects of high fructose consumption on health status is so important. More efforts must be made to restrict the supplementation of foods with high fructose additives.

Chronic fructose consumption affects whole body. Multiple tissues such as liver, adipose, the gastrointestinal system and the central nervous system disturb following chronic fructose consumption. This disturbance lead to abnormalities influences

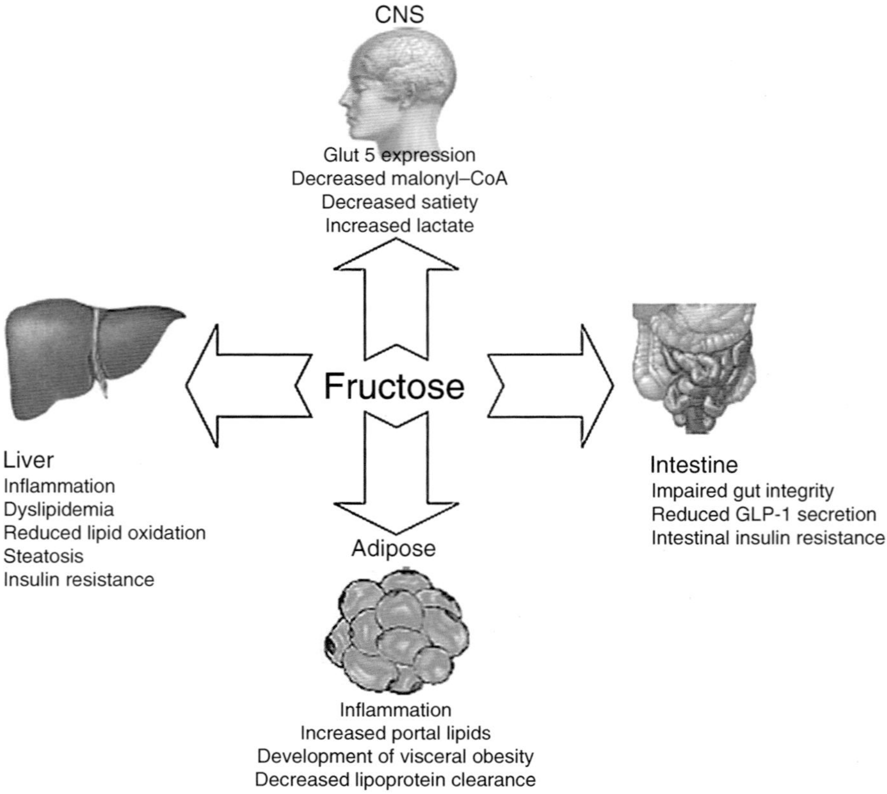

Fig. 2 Effect of fructose consumption on whole body [69]

including metabolic syndrome, dyslipidemia, insulin resistance and central adiposity. Impaired satiety, increased hepatic lipid deposition, inflammation and altered gastrointestinal integrity occur by fructose consumption (Fig. 2) [69].

6.1 Fructose and Uric Acid and Metabolic Syndrome

According to findings, fructose is the only sugar able to increase blood uric acid. fructokinase (also known as ketohexokinase [KHK]) phosphorylate fructose to fructose-1-phosphate in the hepatocyte, which ATP is used as a phosphate donor. It is lead to decrease intracellular phosphate (PO4) levels and stimulate the activity of AMP deaminase 2 (AMPD2). AMP is converted to inosine monophosphate (IMP) by AMPD2. 59 nucleotidase (59NT) metabolize IMP to inosine which is further degraded to xanthine and hypoxanthine by xanthine oxidase (XO), ultimately generating uric acid. Uric acid may associate with insulin resistance in the liver by stimulating mitochondrial oxidative stress and steatosis. In addition to, uric acid blocks the ability of insulin to induce vasodilation of blood vessels, which is serious for the transfer of glucose to the skeletal muscle. Uric acid also stimulates local inflammation in the adipose tissue with decrease in the production of adiponectin. Also, uric acid has directly effects on the islet cells that cause local oxidative stress and islet dysfunction [46, 70].

According to findings, it was identified that affluent diets in sucrose can rapidly persuade features of metabolic syndrome including hyperglycemia, insulin resistance, hyperlipidemia, hypertension, weight gain, and hyperuricemia in animals. More studies indicated that these metabolic changes were due to the fructose content. In addition, if rats are pair-fed equivalent amounts of fructose or glucose with same total energy intake, only the fructose-fed rats develop features of metabolic syndrome [71, 72].

Totally, fructose lead to hypertriglyceridemia, low HDL-C, weight gain, increase blood pressure, impaired glucose tolerance, endothelial dysfunction, oxidative stress, sympathetic nervous system activation, activation of the renin angiotensin system, systemic inflammation, fatty liver, increased intra-abdominal fat accumulation, leptin resistance, proteinuria, renal hypertrophy, glomerular hypertension, and renal microvascular disease (Fig. 3) [72].

7 Summary of Proposed Mechanisms Underlying SSB and Artificial Sweeteners Adverse Metabolic Effects

Numerous possible mechanisms have been considered to link soft drink consumption with metabolic syndrome risk.

- High amounts of added sugar in soft drinks rapidly absorbed in liquid form. SSBs are consumed in large quantities that have a moderate glycemic index (GI) and high the glycemic load (GL). High GL is related to impaired glycemic control, glucose intolerance, insulin resistance and increase levels of inflammatory biomarkers such as C-reactive protein [46].
- Studies reported that high GI and GL were positively related to components of metabolic syndrome in women, but not men in Asian populations. Japanese study showed that participants that consumed the highest quartile of dietary GL had a 52% higher risk of type II diabetes than those in the lowest quartile among women, but not men. Also, study of Koreans showed that GI and GL were positively correlated to metabolic syndrome in women but not men. In addition to, a Chinese cohort study reported that participants in the highest dietary GI and GL quintiles had a 21% and 34% higher risk of type II diabetes, respectively. According to these findings, most of the association between dietary sugar and metabolic syndrome has been reported in women. The exact mechanisms are not understood. However some studies indicated that estrogens and androgens affect lipoprotein metabolism in opposite ways and finally cause various responses to high carbohydrate diets by gender. These different responses to high sugar consumption in Asian populations certify more investigation [44].

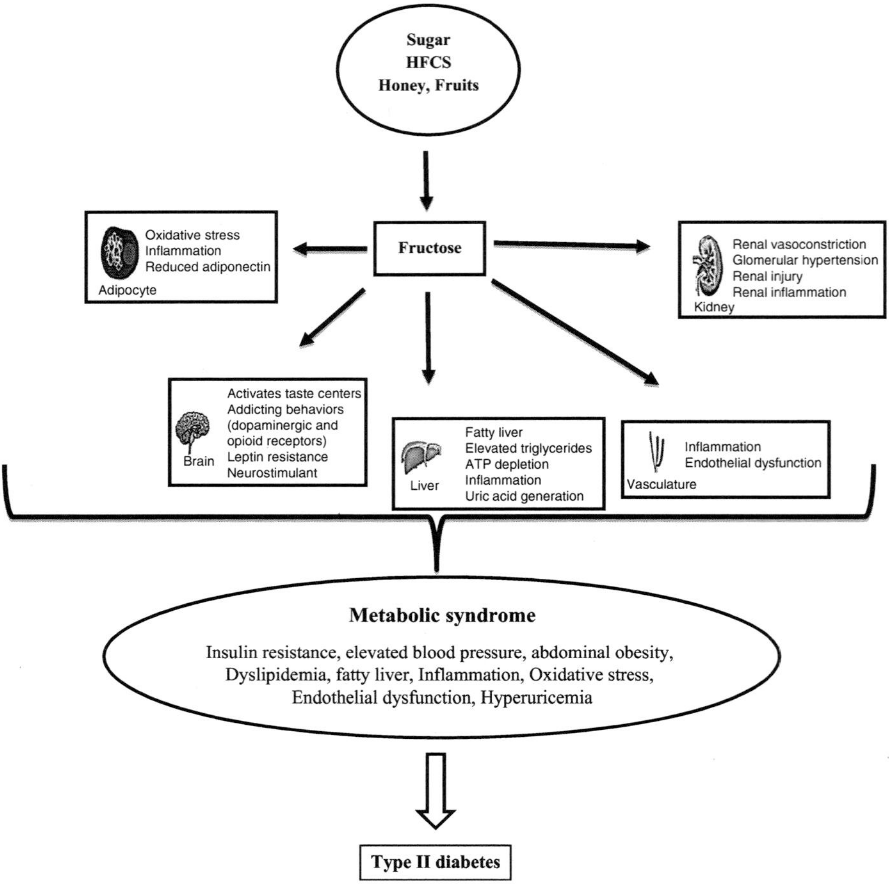

Fig. 3 Effect of sugar, HFCS, honey and fruits and finally fructose on various organ systems

- SSBs increase hepatic de novo lipogenesis and finally cause hypertension and raise accumulation of visceral adipose tissue and of ectopic fat [46].
- Fructose as a major component of sugar sweetened beverages increase triglyceride levels induces insulin resistant and lead to metabolic dysfunction [73]. A high flux of fructose to the liver disturbs normal metabolism of hepatic carbohydrate. Also, disturbs glucose metabolism and glucose uptake pathways and provides glycerol and acyl portions for increasing rate of *de novo* lipogenesis and triglyceride synthesis [67].
- Consumption of artificial sweeteners persistently can change gut microbiota that may cause metabolic abnormalities and glucose intolerance [43].
- SSBs change taste preferences and diet quality and may increase risk of disease indirectly. Persistently intake of highly sweetened beverages leads to accustomed to eating sugary foods [46].

8 Conclusions and Recommendations

There are inadequate evidence and data to determine conclusively whether NNS replace of caloric sweeteners in beverages and foods decrease added sugars or carbohydrate consumption or whether this substitution is beneficial for energy balance, body weight, cardiometabolic risk factors and metabolic disease. Limiting added sugars and monitoring carbohydrate consumption are serious strategy for keeping healthy weights and achieve glycemic control.

Some factors that may lead to the inconsistent findings about the effects of sweeteners on healthy outcome are [74]:

- Various conditions that the investigation take place
- Amount of the sweetener consumption (total and percentage of energy)
- Form and structure of the sweetener (monosaccharides, disaccharides or polysaccharides)
- The type of food in which it is present (liquid or solid)
- Presence of other saccharides and macronutrient composition in the basal diet
- Length of the study (short-term or long-term)
- Characteristics of the participants including sex, age, body size, physical activity level, energy balance status (weight gain, weight loss or weight stable)
- Health status including the presence of diabetes or metabolic syndrome
- Genetic differences among subjects.

Some findings suggested that NNS may be used to replace sources of nutritive sweetener and this replacement may reduce modestly energy intake and contribute to weight loos. Evidences showed when NNSs are consumed judiciously can promote useful effects on related metabolic parameters. However, these beneficial effects will not be fully obtained if there is a compensatory enhance in energy consumption from other sources [22].

We suggest that well-designed human trials investigating the potential role of NNS in healthy outcomes and cardiometabolic risk factors are scarce. Further researches including animal and human studies are needed to clarify the effects of different types of sweeteners on energy metabolism and chronic diseases' risk factors. However, findings of animal studies are difficult to extrapolate to humans. Many confounding variables must be considered to obtain more accurate results.

9 Cross-References

▶ Health Implications of Fructose Consumption in Humans
▶ Sugar Alcohols as Sugar Substitutes in Food Industry
▶ Sweeteners: Regulatory Aspects

References

1. Cummings J, Stephen A (2007) Carbohydrate terminology and classification. Eur J Clin Nutr 61(Suppl 1):S5–S18
2. Davies G, Williams S (2016) Carbohydrate-active enzymes: sequences, shapes, contortions and cells. Biochem Soc Trans 44(1):79–87
3. Gil-Campos M, San Jose Gonzalez M, Diaz Martin J (2015) Use of sugars and sweeteners in children's diets. Recommendations of the Nutrition Committee of the Spanish Paediatric Association. An Pediatr (Barc) 83(5):353e1–353e7
4. Welsh J, Cunningham S (2011) The role of added sugars in pediatric obesity. Pediatr Clin North Am 58(6):1455–1466
5. Lustig R, Schmidt L, Brindis C (2012) Public health: the toxic truth about sugar. Nature 482(7383):27–29
6. Duffey K, Popkin B (2008) High-fructose corn syrup: is this what's for dinner? Am J Clin Nutr 88(6):1722S–1732S
7. Karalius V, Shoham D (2013) Dietary sugar and artificial sweetener intake and chronic kidney disease: a review. Adv Chronic Kidney Dis 20(2):157–164
8. Malik V, Hu F (2012) Sweeteners and risk of obesity and type 2 diabetes: the role of sugar-sweetened beverages. Curr Diab Rep 12:195 [Epub ahead of print]
9. Johnson R, Segal M, Sautin Y, Nakagawa T, Feig D, Kang D, Gersch MS, Benner S, Sánchez-Lozada LG (2007) Potential role of sugar (fructose) in the epidemic of hypertension, obesity and the metabolic syndrome, diabetes, kidney disease, and cardiovascular disease. Am J Clin Nutr 86(4):899–906
10. van Baak M, Astrup A (2009) Consumption of sugars and body weight. Obes Rev 10(Suppl 1):9–23
11. Burt B, Pai S (2001) Sugar consumption and caries risk: a systematic review. J Dent Educ 65(10):1017–1023
12. Gross L, Li L, Ford E, Liu S (2004) Increased consumption of refined carbohydrates and the epidemic of type 2 diabetes in the United States: an ecologic assessment. Am J Clin Nutr 79(5):774–779
13. Te Morenga L, Mallard S, Mann J (2013) Dietary sugars and body weight: systematic review and meta-analyses of randomised controlled trials and cohort studies. BMJ 346:e7492
14. Coss-Bu J, Sunehag A, Haymond M (2009) Contribution of galactose and fructose to glucose homeostasis. Metabolism 58(8):1050–1058
15. American Dietetic Association (2004) Position of the American Dietetic Association: use of nutritive and nonnutritive sweeteners. J Am Diet Assoc 104(2):255–275
16. Mace O, Affleck J, Patel N, Kellett G (2007) Sweet taste receptors in rat small intestine stimulate glucose absorption through apical GLUT2. J Physiol 582(Pt 1):379–392
17. Margolskee R, Dyer J, Kokrashvili Z, Salmon K, Ilegems E, Daly K, Maillet EL, Ninomiya Y, Mosinger B, Shirazi-Beechey SP (2007) T1R3 and gustducin in gut sense sugars to regulate expression of Na + −glucose cotransporter 1. Proc Natl Acad Sci U S A 104(38):15075–15080
18. Zheng Y, Sarr M (2013) Effect of the artificial sweetener, acesulfame potassium, a sweet taste receptor agonist, on glucose uptake in small intestinal cell lines. J Gastrointest Surg 17(1):153–158
19. Araujo J, Martel F, Keating E (2014) Exposure to non-nutritive sweeteners during pregnancy and lactation: impact in programming of metabolic diseases in the progeny later in life. Reprod Toxicol 49:196–201
20. Sylvetsky A, Welsh J, Brown R, Vos M (2012) Low-calorie sweetener consumption is increasing in the United States. Am J Clin Nutr 96(3):640–646
21. Shankar P, Ahuja S, Sriram K (2013) Non-nutritive sweeteners: review and update. Nutrition 29(11–12):1293–1299
22. Gardner C, Wylie-Rosett J, Gidding S, Steffen L, Johnson R, Reader D, Lichtenstein AH (2012) Nonnutritive sweeteners: current use and health perspectives: a scientific statement from the American Heart Association and the American Diabetes Association. Diabetes Care 35(8):1798–1808

23. Garcia-Almeida J, Casado Fdez G, Garcia Aleman J (2013) A current and global review of sweeteners. Regulatory aspects. Nutr Hosp 28(Suppl 4):17–31
24. Ferrer I, Thurman E (2010) Analysis of sucralose and other sweeteners in water and beverage samples by liquid chromatography/time-of-flight mass spectrometry. J Chromatogr A 1217(25): 4127–4134
25. Pepino M (2015) Metabolic effects of non-nutritive sweeteners. Physiol Behav 152(Pt B): 450–455
26. Swithers S (2015) Artificial sweeteners are not the answer to childhood obesity. Appetite 93:85–90
27. Xu H, Staszewski L, Tang H, Adler E, Zoller M, Li X (2004) Different functional roles of T1R subunits in the heteromeric taste receptors. Proc Natl Acad Sci U S A 101(39):14258–14263
28. Burke M, Small D (2015) Physiological mechanisms by which non-nutritive sweeteners may impact body weight and metabolism. Physiol Behav 152(Pt B):381–388
29. Mattes R, Popkin B (2009) Nonnutritive sweetener consumption in humans: effects on appetite and food intake and their putative mechanisms. Am J Clin Nutr 89(1):1–14
30. Swithers S (2013) Artificial sweeteners produce the counterintuitive effect of inducing metabolic derangements. Trends Endocrinol Metab 24(9):431–441
31. Ludwig D, Peterson K, Gortmaker S (2001) Relation between consumption of sugar-sweetened drinks and childhood obesity: a prospective, observational analysis. Lancet 357(9255):505–508
32. Schulze M, Manson J, Ludwig D, Colditz G, Stampfer M, Willett W, Hu FB (2004) Sugar-sweetened beverages, weight gain, and incidence of type 2 diabetes in young and middle-aged women. JAMA 292(8):927–934
33. Fernstrom J (2015) Non-nutritive sweeteners and obesity. Annu Rev Food Sci Technol 6:119–136
34. Fowler S, Williams K, Resendez R, Hunt K, Hazuda H, Stern M (2008) Fueling the obesity epidemic? Artificially sweetened beverage use and long-term weight gain. Obesity 16(8): 1894–1900
35. Renwick A, Molinary S (2010) Sweet-taste receptors, low-energy sweeteners, glucose absorption and insulin release. Br J Nutr 104(10):1415–1420
36. Swithers S, Sample C, Davidson T (2013) Adverse effects of high-intensity sweeteners on energy intake and weight control in male and obesity-prone female rats. Behav Neurosci 127(2):262–274
37. Swithers S, Laboy A, Clark K, Cooper S, Davidson T (2012) Experience with the high-intensity sweetener saccharin impairs glucose homeostasis and GLP-1 release in rats. Behav Brain Res 233(1):1–14
38. Mitsutomi K, Masaki T, Shimasaki T, Gotoh K, Chiba S, Kakuma T, Shibata H (2014) Effects of a nonnutritive sweetener on body adiposity and energy metabolism in mice with diet-induced obesity. Metabolism 63(1):69–78
39. Moran A, Al-Rammahi M, Zhang C, Bravo D, Calsamiglia S, Shirazi-Beechey S (2014) Sweet taste receptor expression in ruminant intestine and its activation by artificial sweeteners to regulate glucose absorption. J Dairy Sci 97(8):4955–4972
40. Sivertsen J, Rosenmeier J, Holst J, Vilsboll T (2012) The effect of glucagon-like peptide 1 on cardiovascular risk. Nat Rev Cardiol 9(4):209–222
41. Abou-Donia M, El-Masry E, Abdel-Rahman A, McLendon R, Schiffman S (2008) Splenda alters gut microflora and increases intestinal p-glycoprotein and cytochrome p-450 in male rats. J Toxicol Environ Health A 71(21):1415–1429
42. Larsen N, Vogensen F, van den Berg F, Nielsen D, Andreasen A, Pedersen B, Al-Soud WA, Sørensen SJ, Hansen LH, Jakobsen M (2010) Gut microbiota in human adults with type 2 diabetes differs from non-diabetic adults. PLoS One 5(2):e9085
43. Suez J, Korem T, Zeevi D, Zilberman-Schapira G, Thaiss C, Maza O, Israeli D, Zmora N, Gilad S, Weinberger A, Kuperman Y, Harmelin A, Kolodkin-Gal I, Shapiro H, Halpern Z, Segal E, Elinav E (2014) Artificial sweeteners induce glucose intolerance by altering the gut microbiota. Nature 514(7521):181–186

44. Chung S, Ha K, Lee H, Kim C, Joung H, Paik H, Song Y (2015) Soft drink consumption is positively associated with metabolic syndrome risk factors only in Korean women: data from the 2007–2011 Korea National Health and Nutrition Examination Survey. Metabolism 64(11): 1477–1484
45. Kumar G, Pan L, Park S, Lee-Kwan S, Onufrak S, Blanck H (2014) Sugar-sweetened beverage consumption among adults – 18 states, 2012. Morb Mortal Wkly Rep 63(32):686–690
46. Malik V, Popkin B, Bray G, Despres J, Willett W, Hu F (2010) Sugar-sweetened beverages and risk of metabolic syndrome and type 2 diabetes: a meta-analysis. Diabetes Care 33(11): 2477–2483
47. Malik V, Pan A, Willett W, Hu F (2013) Sugar-sweetened beverages and weight gain in children and adults: a systematic review and meta-analysis. Am J Clin Nutr 98(4):1084–1102
48. Huang C, Huang J, Tian Y, Yang X, Gu D (2014) Sugar sweetened beverages consumption and risk of coronary heart disease: a meta-analysis of prospective studies. Atherosclerosis 234(1):11–16
49. Hu F (2013) Resolved: there is sufficient scientific evidence that decreasing sugar-sweetened beverage consumption will reduce the prevalence of obesity and obesity-related diseases. Obes Rev 14(8):606–619
50. Swithers S, Martin A, Davidson T (2010) High-intensity sweeteners and energy balance. Physiol Behav 100(1):55–62
51. Miller P, Perez V (2014) Low-calorie sweeteners and body weight and composition: a meta-analysis of randomized controlled trials and prospective cohort studies. Am J Clin Nutr 100(3): 765–777
52. Malik A, Akram Y, Shetty S, Malik S, Yanchou Njike V (2014) Impact of sugar-sweetened beverages on blood pressure. Am J Cardiol 113(9):1574–1580
53. Jayalath V, Sievenpiper J, de Souza R, Ha V, Mirrahimi A, Santaren I, Blanco Mejia S, Di Buono M, Jenkins AL, Leiter LA, Wolever TM, Beyene J, Kendall CW, Jenkins DJ (2014) Total fructose intake and risk of hypertension: a systematic review and meta-analysis of prospective cohorts. J Am Coll Nutr 33(4):328–339
54. Nguyen S, Choi H, Lustig R, Hsu C (2009) Sugar-sweetened beverages, serum uric acid, and blood pressure in adolescents. J Pediatr 154(6):807–813
55. Kim Y, Abris G, Sung M, Lee J (2012) Consumption of sugar-sweetened beverages and blood pressure in the United States: the national health and nutrition examination survey 2003–2006. Clin Nutr Res 1(1):85–93
56. Brown I, Stamler J, Van Horn L, Robertson C, Chan Q, Dyer A, Huang CC, Rodriguez BL, Zhao L, Daviglus ML, Ueshima H, Elliott P (2011) Sugar-sweetened beverage, sugar intake of individuals, and their blood pressure: international study of macro/micronutrients and blood pressure. Hypertension 57(4):695–701
57. Greenwood D, Threapleton D, Evans C, Cleghorn C, Nykjaer C, Woodhead C, Burley VJ (2014) Association between sugar-sweetened and artificially sweetened soft drinks and type 2 diabetes: systematic review and dose–response meta-analysis of prospective studies. Br J Nutr 112(5):725–734
58. Rabie E, Heeba G, Abouzied M, Khalifa M (2015) Comparative effects of Aliskiren and Telmisartan in high fructose diet-induced metabolic syndrome in rats. Eur J Pharmacol 760:145–153
59. Dhingra R, Sullivan L, Jacques P, Wang T, Fox C, Meigs J, D'Agostino RB, Gaziano JM, Vasan RS (2007) Soft drink consumption and risk of developing cardiometabolic risk factors and the metabolic syndrome in middle-aged adults in the community. Circulation 116(5):480–488
60. Hu F, Malik V (2010) Sugar-sweetened beverages and risk of obesity and type 2 diabetes: epidemiologic evidence. Physiol Behav 100(1):47–54
61. Chan T, Lin W, Huang H, Lee C, Wu P, Chiu Y, Huang CC, Tsai S, Lin CL, Lee CH (2014) Consumption of sugar-sweetened beverages is associated with components of the metabolic syndrome in adolescents. Nutrients 6(5):2088–2103

62. Baik I, Lee M, Jun N, Lee J, Shin C (2013) A healthy dietary pattern consisting of a variety of food choices is inversely associated with the development of metabolic syndrome. Nutr Res Pract 7(3):233–241
63. Stanhope K, Havel P (2010) Fructose consumption: recent results and their potential implications. Ann N Y Acad Sci 1190:15–24
64. Dolan L, Potter S, Burdock G (2010) Evidence-based review on the effect of normal dietary consumption of fructose on development of hyperlipidemia and obesity in healthy, normal weight individuals. Crit Rev Food Sci Nutr 50(1):53–84
65. Tappy L, Le K, Tran C, Paquot N (2010) Fructose and metabolic diseases: new findings, new questions. Nutrition 26(11–12):1044–1049
66. Khitan Z, Kim D (2013) Fructose: a key factor in the development of metabolic syndrome and hypertension. J Nutr Metab 2013:682673
67. Basciano H, Federico L, Adeli K (2005) Fructose, insulin resistance, and metabolic dyslipidemia. Nutr Metab (Lond) 2(1):5
68. Sayon-Orea C, Martinez-Gonzalez M, Gea A, Alonso A, Pimenta A, Bes-Rastrollo M (2015) Baseline consumption and changes in sugar-sweetened beverage consumption and the incidence of hypertension: the SUN project. Clin Nutr 34(6):1133–1140
69. Dekker M, Su Q, Baker C, Rutledge A, Adeli K (2010) Fructose: a highly lipogenic nutrient implicated in insulin resistance, hepatic steatosis, and the metabolic syndrome. Am J Physiol Endocrinol Metab 299(5):E685–E694
70. Johnson R, Nakagawa T, Sanchez-Lozada L, Shafiu M, Sundaram S, Le M, Ishimoto T, Sautin YY, Lanaspa MA (2013) Sugar, uric acid, and the etiology of diabetes and obesity. Diabetes 62(10):3307–3315
71. Reungjui S, Roncal CA, Mu W, Srinivas TR, Sirivongs D, Johnson RJ, Nakagawa T (2007) Thiazide diuretics exacerbate fructose-induced metabolic syndrome. J Am Soc Nephrol 18:2724–2731
72. Johnson RJ, Perez-Pozo SE, Sautin YY, Manitius J, Sanchez-Lozada LG, Feig DI et al (2009) Hypothesis: could excessive fructose intake and uric acid cause type 2 diabetes. Endocr Rev 30(1):96–116
73. Bray G (2012) Fructose and risk of cardiometabolic disease. Curr Atheroscler Rep 14(6): 570–578
74. Murphy S (2009) The state of the science on dietary sweeteners containing fructose: summary and issues to be resolved. J Nutr 139(6):1269S–1270S

The Pharmacological Activities of Glycyrrhizinic Acid ("Glycyrrhizin") and Glycyrrhetinic Acid

12

Cedric Stephan Graebin

Abstract

Glycyrrhizin or, more correctly, Glycyrrhizinic acid is a triterpenoid saponin obtained from the root and rhizome extracts of Licorice (*Glycyrrhiza glabra*), being commonly used as a sweetener, being reported as – at least – 30 times sweeter than sucrose. This natural product, along with its aglycone glycyrrhetinic acid, is known in the literature for its several pharmacological and biological activities. This chapter summarizes the activities reported in the literature for the saponin and its aglycone since 2010.

Keywords

Glycyrrhizinic acid • Glycyrrhizin • Glycyrrhetinic acid • *Glycyrrhiza glabra* • Liquorice

Contents

1 Introduction 246
2 Pharmacological Activities of Glycyrrizinic Acid ("Glycyrrhizin") and Related Compounds 248
 2.1 Anticancer 248
 2.2 Anti-inflammatory 249
 2.3 Hepatoprotective Effects 250
 2.4 Antiviral 251
 2.5 Other Activities 252
3 Pharmacological Activities of Glycyrrhetinic Acid (**2**) and Related Compounds 253
 3.1 Anticancer Activity 253
 3.2 Hepatoprotective Activity 254
 3.3 Antiparasitic Activity 254

C.S. Graebin (✉)
Chemistry Department, Federal Rural University of Rio de Janeiro, Seropédica, RJ, Brazil
e-mail: cedric@ufrrj.br

J.-M. Mérillon, K.G. Ramawat (eds.), *Sweeteners*, Reference Series in Phytochemistry,
https://doi.org/10.1007/978-3-319-27027-2_15

3.4 Antibacterial Activity 255
3.5 Anti-inflammatory Activity 255
3.6 Other Activities 256
4 Concluding Remarks 257
References 258

1 Introduction

Glycyrrhizinic acid (**1**, Fig. 1), also known as Glycyrrhizin [1, 2], is a triterpenoid saponin obtained from the root and rhizome extracts of Liquorice (or "licorice" in US English) (*Glycyrrhiza glabra*). Normally, the dried licorice root extract may contain around 4–25% of the saponin, along with other compounds such as polyphenols, saponins, triterpenes, etc. One of the triterpenes found in the extract is the triterpenic aglycone, glycyrrhetinic acid, or "enoxolone" (**2**). Further purification using several extraction techniques gives usually the monoammonium glycyrrhetinate salt as the purified product [2].

The World Health Organization (WHO) report on Glycyrrhizin recommends that the compound must be named "Glycyrrhizinic acid." According to the report, "Glycyrrhizin" is the name given, more correctly, to the Licorice extract and not the saponin itself, and "Glycyrrhizin" and "Glycyrrhizinic acid" should not be used as synonyms [1], although this is common practice in the literature. This chapter will follow this recommendation and use the second term as the saponin name.

Both the saponin (as a carboxylic acid) and its monoammonium salt are commonly employed as sweeteners, being reported as 30–50 [3] or 33–200 [2] times sweeter than sucrose. The saponin is also widely employed in several traditional Chinese, Tibetan, and Indian medicinal preparations [3]. "Stronger

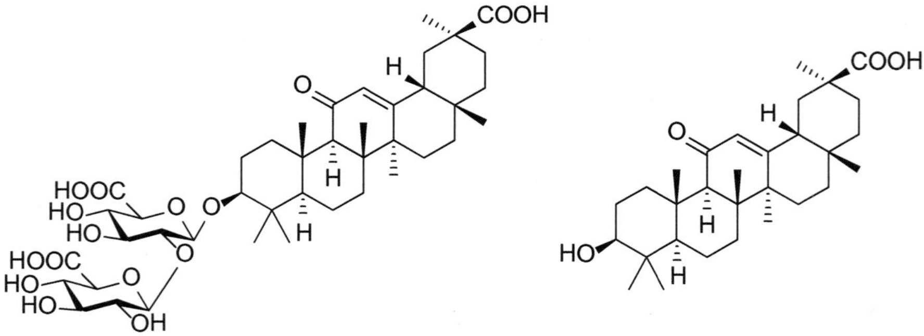

Glycyrrhizinic acid "Glycyrrhizin" **1**

Glycyrrhetinic acid or 18β-Glycyrrhetinic acid **2**

Fig. 1 Chemical structures of Glycyrrhizinic (**1**) and Glycyrrhetinic acid (**2**)

Neo-Minophagen C," a glycyrrhetinic acid-containing i.v. preparation, is employed in the treatment of chronic hepatic diseases, being marketed in Japan, China, Korea, Taiwan, Indonesia, India, and Mongolia [4].

The intent of this chapter is to report the wide range of pharmacological activities reported in the literature for the saponin (**1**) and its aglycone (**2**). The inclusion of the aglycone in this chapter is justified due to the fact that the saponin has no oral bioavailability, being absorbed as glycyrrhetinic acid (**2**) after hydrolysis of its carbohydrate moiety by intestine bacteria [5]. Since there are several reports published regarding those compounds until *ca.* 2010 [3, 6, 7], the time period covered by this chapter is from 2010 onwards.

Using SciFinder scientific literature search service [8], several articles mentioning "Glycyrrhizin" or "Glycyrrhetinic acid" in its titles and/or abstracts were found (Fig. 2). As mentioned earlier, using "Glycyrrhizinic acid" as search term gives a reduced number of articles as response. Due to the enormous number of publications in the selected period, it was decided to focus this chapter in the publications reporting pharmacological activities to the saponin, its aglycone, and some closely related compounds. Several Japanese, Korean, and (mostly) Chinese patents found in the bibliographic research mentioning **1** are related to its use as a sweetener/flavoring agent in herbal/drug preparations and, therefore, were left out of this review.

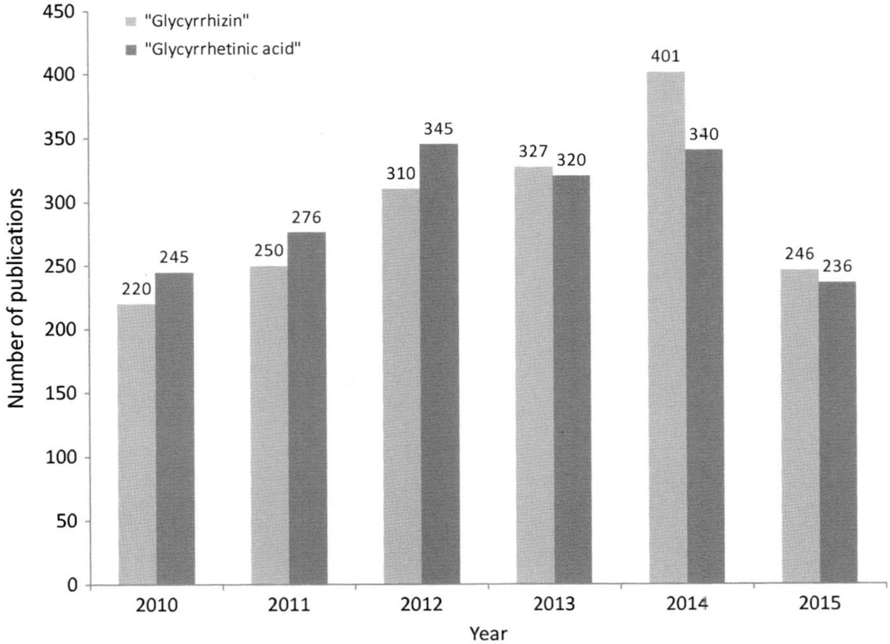

Fig. 2 Publications mentioning the terms "Glycyrrhizin" or "Glycyrrhetinic acid" in its titles and/or abstract in SciFinder database (year period: 2010–2015)

2 Pharmacological Activities of Glycyrrizinic Acid ("Glycyrrhizin") and Related Compounds

2.1 Anticancer

Several findings were reported in the last few years regarding the anticancer properties of Glycyrrhizinic acid. Wang et al. reported the in vitro activity of **1** against gastric cancer cell line BCG-823. The saponin was able to prevent cell proliferation, adhesion, and migration at a 40 μM dose and they observed also that the expression of β-catenin, Bcl-2, CyclinD1, and survivin was significantly decreased [9].

In order to evaluate the carbohydrate moiety importance in the anticancer properties of **1**, Yang et al. synthesized two monoglucuronides from **1** by means of a glucuronidase from *Aspergillus* sp. (Scheme 1). The authors reported that 18-α-monoglucuronide (**5**) obtained was more effective against HepG2, HeLa, and A549 cancer cell lines (IC_{50} values: 6.67, 7.43, and 15.76 μM, respectively) than 18β-monoglucuronide (**3**), **1,** and 18α-glycyrrhizinic acid (**4**). The monoglucuronide **3** was also the best Epidermal Growth Factor Receptor (EGFR) inhibitor, with a IC_{50} of 0.028 μM. Molecular docking simulations indicate that **3** also is able to bind with EGFR, suggesting that the anticancer properties of this glucuronide are related to EGFR inhibition [10].

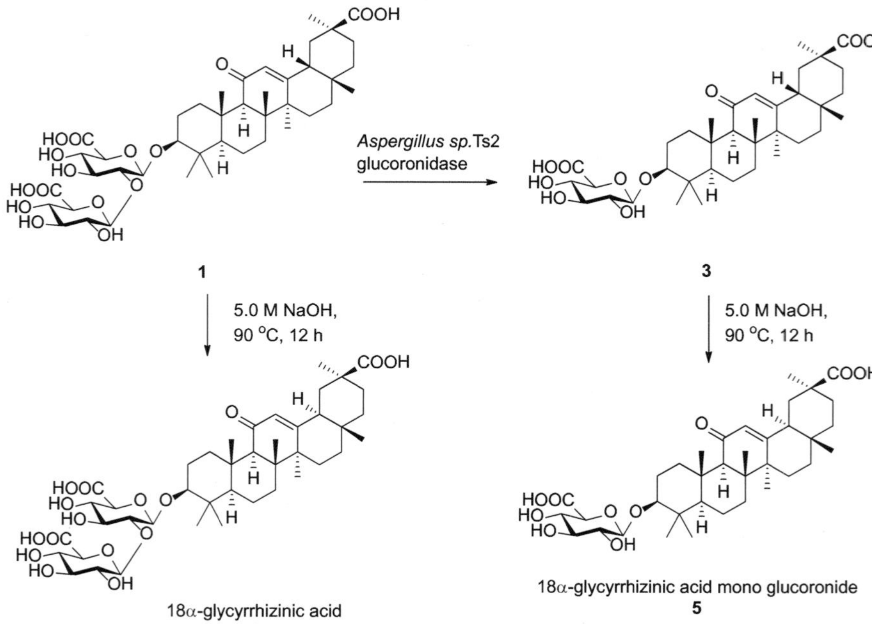

Scheme 1 Synthesis of 18α-epimers of Glycyrrhizinic acid

Huang et al. in order to clarify the mechanism of the anticancer activity of **1** against lung adenocarcinoma A549 cell lines, evaluated the expression of several enzymes in an in vitro model and found that the anticancer activity of **1** against this cancer cell line can be, according to the authors, related to the inhibition of Thromboxane Synthase by **1**. The same results were also found in an in vivo mouse model using a western blot assay [11].

The inhibition of High-Mobility Group Box 1 (HMGB1), a protein that acts outside the cellular milieu as a proinflammatory cytokine, can also be related to the anticancer properties of **1**. Smolarczyk et al. reported that, when employing in an in vivo tumor model, the use of **1** inhibited cell growth and proliferation in mice. The authors concluded that the cell growth inhibition was probably due to the inhibition of HMGB1 by **1** in the extracellular media [12].

2.2 Anti-inflammatory

Wang et al. determined that **1** was able to attenuate levels of Tumor Necrosis Factor-α (TNF-α), interleukin 1β (IL-1β), and activation of the HMGB1/NFκB (Nuclear factor-kB) signaling pathway in the hippocampus of neonatal rats after exposure to isoflurane (a general anesthetic drug). Moreover, treatment with **1** also prevented the deficits in spatial memory related to isoflurane [13].

Using a pancreatic trauma model in male Wistar rats, Xiang et al. reported that treatment with **1**, 15 min after inducing the trauma, led to lower serum levels of HMGB-1, TNF-α, and IL-6 when compared to the control group. The treatment with the saponin also increased the seven-day survival rate of the animal subjects in this experimental model.

Zhang et al. reported that **1** has a neuroprotective effect in the postischemic brain in mice through the HMGB1-TLR4-IL-17A pathway [14].

Glycyrrhizinic acid (**1**) and Licorice flavonoids, tested in vitro in separate groups, were found by Zhao et al. to modulate the secretion of several cytokines by macrophages (RAW 264.7 cells) induced by lipopolysaccharides (LPS) [15]. Fu et al. reported that **1** presented its anti-inflammatory activity in LPS-stimulated RAW 264.7 cells by disrupting lipid rafts and inhibiting Toll-like receptor-4 (TLR-4) translocation to those rafts [16].

Wang et al. reported that **1** was able to reduce xylene, carrageenan, and acetic acid-induced rat paw edema and the nociceptions induced by formalin and acetic acid, but it had no effects in the hot plate test. The authors also observed the downregulation of levels of TNF-α, IL-6, inducible Nitric Oxide Synthase (iNOS), and Cyclooxigenase-2 (COX-2) [17].

Investigating the effects of **1** in the subarachnoid hemorrhagic (SAH) rat model, Chang et al. reported that the saponin exerted anti-inflammatory effects reducing SAH-induced vasospasm, mostly through the attenuation of peroxisome proliferator-activating receptor gamma (PPAR-γ) [18]. The same research group reported that, in the same SAH model, treatment with **1** was also able to attenuate the ultrashort time expression of Toll-like receptors (TLR) 2 and 4 [19].

The saponin was also subject of some clinical trials regarding its anti-inflammatory activity. Xiao et al. conducted a randomized trial with 39 children presenting Henoch-Schonlein purpura (HSP) in order to verify the functions of regulatory T-cells (Treg) and Th17 cells in the peripheral blood of those patients. The group treated with **1** had significantly different Interleukin-17 (IL-17) serum levels after the treatment but no difference in Transforming Growth Factor-beta (TGF-β) and Interleukin-10 (IL-10) serum levels after the treatment. The authors concluded that **1** was able to reduce Th17 function, without noticeable effects in Treg function [20].

Glycyrrhizinic acid was also trialed in patients with chronic urticaria. Eighty four patients with the disease were randomized in two groups: one with **1** (50 mg oral tablets, three times per day) and the other with levocetirizine (an antiallergic drug). Both compounds were administered during a four-week period. The authors concluded that **1** was superior to levocetirizine and could be used to treat chronic urticaria [21]. It is worth noticing that **1** has no oral bioavailability, being hydrolyzed to form glycyrrhetinic acid (**2**) before being absorbed [1]. Therefore, the observed pharmacological activity is probably due to the known anti-inflammatory activity of **2** (see Sect. 3 of this chapter) rather than **1**.

2.3 Hepatoprotective Effects

Magnesium isoglycyrrhizinate, the magnesium salt of 18α-glycyrrhizinic acid (**4**), was found by Luo et al. to be an effective treatment against hepatitis E with severe jaundice during a clinical trial with 78 patients. The patients received an intravenous injection of 150 mg of the magnesium once a day during 6 weeks [22].

Zhang et al. conducted a clinical trial with 84 patients with digestive tract cancer in order to verify the hepatoprotective activity of **1** during standard cancer chemotherapy. The authors observed that the group undergoing chemotherapy and treated with the saponin (160 mg i.v. once a day) presented significantly lower liver transaminase levels and increased levels of neutrophile, granulocytes, and white blood cells when compared with the control group (standard chemotherapy only) [23].

Also while investigating the hepatoprotective activity of **1**, Hsiang et al. investigated the gene expression of HepG2 cells treated in vitro independently with **1**, silymarin, and ursodeoxycholic acid, natural products also known by their hepatoprotective properties. The authors concluded that the compounds inhibited NF-κB activities in a dose-dependent manner and that those compounds regulated the expression of genes related to oxidative stress and apoptosis in those cells [24].

Meng et al. conducted a meta-analysis in 12 randomized clinical trials regarding the hepatoprotective activity of **1** in patients with alcoholic liver disease. Their analysis concluded that in those trials the levels of alanine and aspartate aminotransferases, as well as γ-glutaryl transpeptidase were lower than the control groups after treatment with the saponin [25].

During an investigation with patients with chronic hepatitis B, Yin et al. found that a three-week treatment with **1** did not reduce the expressions of inflammatory cytokine IL-17 and antinuclear antibodies (ANA) in the treated patients when compared to the control group [26].

Treating chronic hepatitis B with either Magnesium isoglicyrrhizinate or **1** was compared in a clinical trial reported by Cai et al. The trial involved 64 patients which were divided in two groups (one group being treated intravenously with the Magnesium salt and the other with the saponin, also i.v., once a day for 4 weeks). In both cases liver function was improved in the patients, but the authors found no statistical difference between both treatments [27].

Ding et al. investigated the effect of 18α-glycyrrhizinic acid (**4**) in rats with carbon tetrachloride (CCl_4)-induced liver fibrosis. It was found that treatment with **4** increased the activity of enzymes glutathione peroxidase and superoxide dismutase, which reduced lipid peroxidation and the levels of malonodialdehyde and hydroxynonenal in the liver, protecting it from damage caused by the aldehydes [28].

Wang et al. found that the treatment of HepG2 cells with **1** caused the increase in both CYP3A mRNA and protein levels. The CYP3A gene encodes monooxygenases responsible for drug metabolism and steroid biosynthesis [29]. The authors also found that the induction of CYP3A was mediated through the activation of Pregnane X receptor (PXR), resulting in the induction of CYP3A11 expression and CYP7A1 inhibition [30].

The hepatoprotective effect of **1** was also studied using HL-77C2 (normal human liver cell line) through acetaminophen-induced damage. According to the report by Chen et al., **1**, liquirtin, and isoliquirtigenin (two polyphenols also found in the Liquorice root extract) were effective as hepatoprotecting agents against acetaminophen-induced damage [31].

Gwak et al., investigating the role of HMGB-1 in hepatocyte apoptosis, found that **1**, known as a HMGB-1 inhibitor, was able to prevent HMGB-1-induced apoptosis, cytochrome C release, and p38 activation in Huh-BAT (human hepatocellular carcinoma transfected with bile acid transporter) cell line [32].

2.4 Antiviral

Glycyrrhizinic acid (**1**) is known in the literature as an antiviral agent against clinically relevant viruses, such as HIV and the SARS-Coronavirus [3, 6]. Duan et al. reported that **1** is active against porcine reproductory and respiratory syndrome virus (PRRSV), relevant in veterinary medicine and in the swine industry. The saponin was able to inhibit mainly the penetration stage of the virus cycle [33].

Investigating the known anti-HIV activity of the saponin, Li et al. found that **1**, among 27 compounds isolated from Liquorice extract, has the high binding constant for R15K, the conserved core sequence in the V3 loop region of gp120. This region, according to the authors, is where the binding of glycophospholipids occurs, leading to the virus entry in the host cell, which can indicate that the saponin is having anti-HIV activity by inhibiting the virus entry stage [34].

Using a Hepatitis C virus (HCV)-infected Huh7 cell model, Matsumoto et al. reported that **1** inhibits the release of HCV particles. Suppression of viral entry and reproduction stages were not observed [35]. Ashfaq et al., also investigating the anti-HCV activity of the saponin, reported that treatment of HCV-infected hepatic cells led to the suppression of HCV 3a core gene both at mRNA and protein levels. Coadministration of **1** with interferon alpha-3a caused a synergistic effect [36].

2.5 Other Activities

Xiao and Zhou reported that **1** reduced the Human neutrophil elastase-induced Mucin 5 AC (MUC5AC) overproduction in human bronchial epithelial cell culture (16HBE) [37].

The interaction of **1** with human hemoglobin (Hb) was studied by Sil et al. The authors reported that the interaction between the saponin and the protein reduces Hb-mediated oxidative damage without affecting the oxygen binding properties of Hb [38].

The antithrombin activity of **1** was already known since 1997 [39]. Glycyrrhizinic acid was found to be a weak in vitro and in vivo thrombin inhibitor but, unlike other antithrombin agents, the saponin acts as an allosteric inhibitor [40]. Glycyrrhetinic acid (**2**), however, has no thrombin inhibiting activity, indicating that the carbohydrate moiety of the saponin is important to the allosteric site interaction. With these results in mind, Paula et al. synthesized some derivatives aiming to simplify the carbohydrate scaffold. From all the synthesized compounds, the hemiphtalate **6** (Scheme 2) presented better in vitro antithrombin activity against thrombin and increased the thrombin time, indicating that the carboxylic acid moiety present in the saponin (and in the phthalate) is important to the recognition at the allosteric site [41].

Investigating the free-radical scavenging abilities of **1**, Imai et al. reported that **1** and glycyrrhetinic acid (**2**) were able to scavenge 1,1-diphenyl-2-picrylhydrazyl (DPPH) radicals but not hydroxyl and superoxide anion radicals [42].

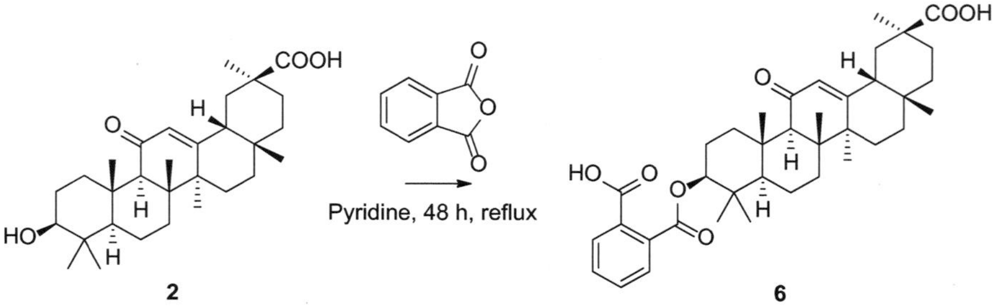

Scheme 2 Synthesis of glycyrrhetinic acid 3-*O*-hemiphtalate

3 Pharmacological Activities of Glycyrrhetinic Acid (2) and Related Compounds

As mentioned before, **2** is the compound absorbed in the human blood (due to the hydrolysis of **1** by bacterial metabolism) after oral administration of the saponin [1]. It was decided, therefore, to report the new pharmacological properties described for the aglycone in the literature below.

3.1 Anticancer Activity

There are several reports regarding the design, synthesis, and pharmacological evaluation of glycyrrhetinic acid derivatives towards compounds with improved anticancer properties. Since this chapter is focused mainly in the pharmacological properties of **1** and **2** itself (along with some closely related derivatives), only reports regarding the anticancer activity of **2** itself will be detailed below. Readers interested in more in-depth articles regarding chemical modifications in the aglycone structure are recommended to the review article by Czuk [43] as well as other review articles previously mentioned in the chapter [3, 6].

Kim et al. reported that **2** presented in vitro cytotoxic activity against A549 (human lung adenocarcinoma) cell line and also was able to reduce in vivo tumors induced by the same cell culture injected in Balb/c nu/nu mice [44].

Employing r/m HM-SFCE-1 cells (highly metastatic ras/myc-transformed serum-free mouse embryo), Yamaguchi et al. found that the cytotoxic effect of **2** in this model was due to the downregulation of glutathione, disrupting the redox balance in those cells [45].

Xu et al. using the human myeloma cell line (U266) reported that **2** induced in vitro apoptosis in this cell line by downregulating the survivin gene expression and arresting the cells in G_0/G_1 phase [46].

Aiming to study the potential use of **2** as a chemoprotective agent, Kowsalya et al. induced buccal pouch carcinogenesis in hamsters using 7,12-dimethylbenz(a) anthracene (DMBA) as a tumor initiator. Oral administration of **2** (45 mg/kg) along with DMBA prevented tumor formation in this animal model [47].

Charma et al. reported that the apoptotic effect of **2** in human breast cancer cell line MCF-7 was due to the activation of caspase-9 and modulation of the Protein Kinase B/Forkhead box O 3a transcription factor (Akt/FOXO3a) pathway [48].

Working with hepatic stellate cells (HSC), cells with immunosuppressive capabilities and related to the development of hepatocellular carcinomas, Kuang et al. found that **2** is able to reduce in vitro HSC-mediated immunosuppression by reducing T-cell apoptosis and regulatory T (Treg) cell expression, which in turn lead to an increase in T-cell activity against hepatocellular carcinoma cells [49].

Xie et al. reported that **2** inhibited the epidermal growth factor (EGF)-induced proliferation of HaCaT cells (human keratinocyte cell line), likely by suppressing the extracellular signal-regulated kinase (ERK1/2) signaling pathway [50].

An in vitro study with prostate cancer cell line LNCaP conducted by Li et al. indicates that the activity of **2** in those cells is probably due to the suppression of 17β-hydroxysteroid dehydrogenase type III (17β-HSD3) mRNA expression via activation of eukaryotic initiation factor 2α (eIF2α) [51].

Wang et al. investigated the cytotoxic effect of **2** in pituitary adenoma cancer cell lines MMQ and GH3. The authors reported that the triterpene induced apoptosis in those cells by activating mitochondria-mediated reactive oxygen species (ROS)/ mitogen-activated protein kinase (MAPK) pathways [52].

Huang et al. reported that **2** is able to suppress proliferation of non-small cell lung cancer (NSCLC) lines cells A549 and NCI-H460 through inhibition of thromboxane synthase and activation of ERK/CREB signaling [53].

Investigating the effect of **2** in the metastatic and cell-invading ability of several cancer cell lines, Jayasooriya et al. found that **2** is capable of downregulating matrix metalloproteinase-9 (MMP-9) and vascular endothelial growth factor (VEGF) via inhibition of phosphatidyl-inositol 3 kinase (PI3K)/Akt-dependent NF-κB activity [54].

3.2 Hepatoprotective Activity

As reported with Glycyrrhizinic acid (**1**), **2** also presents hepatoprotective activity. Mahmoud et al. observed that **2** was able to counteract cyclophosphamide-induced hepatotoxicity in rats via activation of nuclear factor-erythroid 2 (NF-E2) related factor 2 (Nrf2) and PPAR-γ [55].

The factor Nrf2 is also involved in the chemoprotective mechanism of **2** against carbon tetrachloride-induced liver fibrosis in mice. Chen et al. reported that the triterpene is able to upregulate Nrf2 and increase the activity of antioxidant enzymes in the liver [56].

Glycyrrhetinic acid is also a chemoprotective agent against 2-acetylamino-fluorene-induced hepatotoxicity in Wistar rats via attenuation of oxidative stress, inflammation, and hyperproliferation, according to Hasan et al [57].

3.3 Antiparasitic Activity

Investigating the antileishmanial activity of **2** in experimental visceral leishmaniasis models, Gupta et al. reported that this activity occurs through nitric oxide (NO) upregulation, proinflammatory cytokine expression and NF-κB activation through p38 kinase [58]. The same research group later reported that this activity also depends on phosphatase-dependent modulation of cellular MAP kinases [59].

Kalani et al. in an effort to find new affordable antimalarial agents, reported that **2** has in vitro anti-*P. falciparum* activity, with $IC_{50} = 1.69$ μg/mL, and in vivo activity during an 8-day course treatment [60]. The same research group found that **2** is also an antifilarial agent, being active in vitro against microfilariae but not against adult worms. Some **2**-related amides synthesized by the group were able to affect both microfilariae and adult worms [61].

3.4 Antibacterial Activity

Long et al. reported that **2** is active against Methicilin-resistant *Staphylococcus aureus* (MRSA), a major source of complicated infections in hospital environments. According to the authors, **2** presented in vitro bactericidal activity at 0.223 μM and, in sublethal concentrations, it is able to reduce the expression of genes related to the *S. aureus* virulence, such as saeR and hla [62].

Investigating the effect of 18*α*-glycyrrhetinic acid (**7**, Fig. 3) on the periodontal pathogen *Porphyromonas gingivalis*, Kim et al. reported that **7** reduces the bacterial LPS-induced permeability by suppressing repressing NF-κB-dependent endothelial IL-8 production [63].

3.5 Anti-inflammatory Activity

Glycyrrhetinic acid (**2**) is also known for its anti-inflammatory activity, acting as a 11*β*-hydroxysteroid dehydrogenase inhibitor. This enzyme is responsible for the conversion of cortisol (**8**, Fig. 4) to cortisone and, as a result, **2** is an indirect glucocorticoid anti-inflammatory agent by extending cortisol half-life [1, 6]. Carbenexolone (**9**), the hemisuccinate ester of **2**, is used as an anti-inflammatory drug in United Kingdom [3, 6]. The chemical structures of **2, 8**, and **9** are shown in Fig. 4.

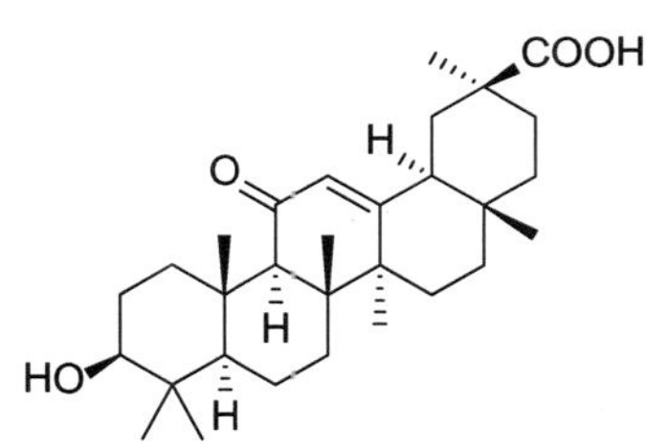

18α-glycyrrhetinic acid
7

Fig. 3 Chemical structure of 18*α*-glycyrrhetinic acid (**7**)

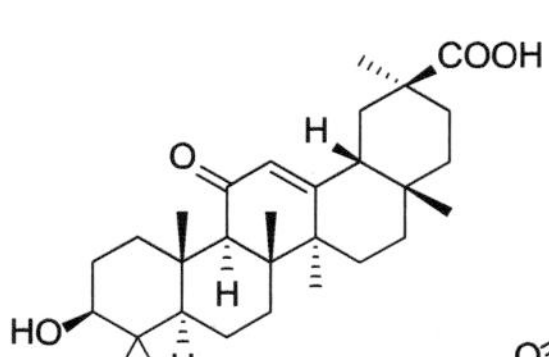

Glycyrrhetinic acid
2

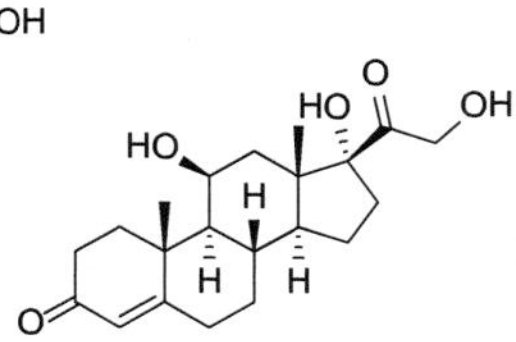

Cortisol
8

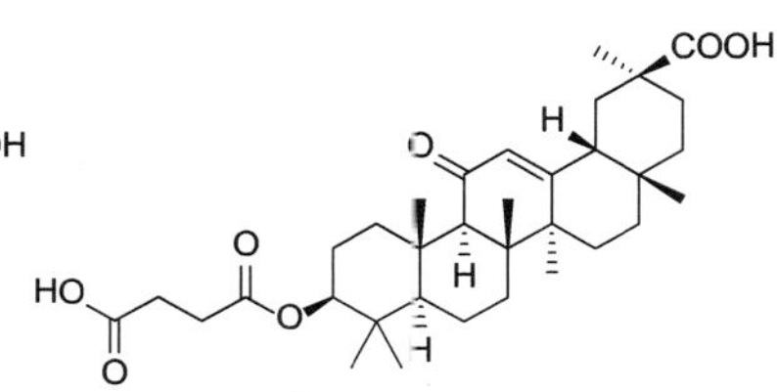

Carbenexolone
9

Fig. 4 Structures of Glycyrrhetinic acid (**2**), Cortisol (**8**), and Carbenexolone (**9**)

Since there are several reports regarding the design, synthesis, and pharmacological evaluation of glycyrrhetinic acid (**2**) derivatives towards new anti-inflammatory agents with improved properties regarding the natural product, it was decided, in order to not extend this review size, to focus in the new reports of the anti-inflammatory activity of **2** itself.

Puchner et al. tested the efficacy of **2** in a model of rheumatoid arthritis (hTNFtg mice) and found that the triterpene was not effective in this model when compared with the positive control (TNF inhibitors) [64].

Investigating the anti-inflammatory activity of **2** in HepG2 cells, Chen et al. reported that it suppresses TNF-induced inflammation in this cell culture by diminishing NF-κB activation [65].

Wang et al. reported that using **2** in RAW264.7 macrophages induced with LPS did not alter cell viability and suppressed the activity of NF-κB and PI3K, inhibiting the expression of several proinflammatory genes and attenuating the production of NO, prostaglandin E_2 (PGE_2), and reactive oxygen species (ROS) [66].

Studying the proinflammatory cytokine effect of **1** and **2**, Kao et al. reported that both compounds are capable of suppressing cytokine production but by different mechanisms: while **1** work via PI3K/Akt/GSK3β pathway, **2** acts by dissociating the glucocorticoid receptor (GR)-HSP90 complex [67].

3.6 Other Activities

Zhang et al. reported that **2** is a competitive Glyoxalase I inhibitor, with $K_i = 0.29$ μM. The authors also studied the crystal structure of the complex Glyoxalase I-glycyrrhetinic acid, indicating that the carboxylic acid moiety of the triterpene is important for the inhibitor–enzyme interaction [68].

Inspired in previous reports regarding the chemoprotective effect of **1** and **2**, Wu et al. reported that **2** can act as a chemoprotective agent against cisplatin-induced nephrotoxicity in BALB/c mice. This effect is, according to the authors, related to upregulation of Nrf2 and downregulation of Nf-κB [69].

Kong et al. reported that **2** can act as a skin protector against UV-induced photoaging in a mouse model. The authors also found that this protective effect is probably due to the anti-inflammatory and antioxidative properties of **2** [70].

Moon et al. investigated the effects of **2** in in vitro models of adipogenesis and found that the triterpene alters fat mass by affecting adipogenesis in maturing preadipocytes and lipolysis in matured adipocytes [71]. Also investigating the antiobesity properties of Liquorice, Park et al. reported that **2** suppresses the activation of cannabinoid type 1 receptor (CB1R) induced by the endogenous agonist anandamide [72].

While researching the effects of **2** in the hemostasis, Jiang et al. found that glycyrrhetinic acid is a Factor Xa (FXa) inhibitor, with in vitro $IC_{50} = 32.6 \pm 1.24\ \mu M$. The authors also tested the in vivo effect in rats using two protocols (tail bleeding and venous stasis models) and reported that **2** can increase prothrombin time and reduce thrombus weight (at a 50 mg/kg dose) when compared to the control group [73].

Hardy et al. reported that **2** was able to inhibit rotavirus replication in in vitro assays, while **1** was ineffective in this assay. The authors also reported that the antiviral activity of **2** occurs after the entry stage of the viral cycle [74].

4 Concluding Remarks

A great range of pharmacological and biological activities have been reported in the literature in the last 5 years, ranging from the exploration of known activities (such as the signaling pathways responsible for the anticancer and anti-inflammatory activity of **1** and **2**), scope broadening (the investigation of the antibacterial and antiviral properties against new, unreported species of virus and bacteria) and also new, unreported activities (such as the antiparasitic activities, glyoxalase I, factor Xa, etc.). Several clinical trials regarding the anti-inflammatory and hepatoprotective activity of **1** were also discussed.

The wide range of reported activities of **1** and **2** can lead to a very important question: how safe it is the use of **1** as a sweetener? Glycyrrhizinic acid is currently considered "Generally recognized as safe" by the US Food and Drug Administration, and there is a guideline for maximum permitted levels of the saponin in several preparations [75]. Also, the European Commission report on Glycyrrhizin states that the saponin is safe for consumption with the warning that the maximum daily dose of **1** should not be higher than 100 mg because of the glucocorticoid effects of glycyrrhetinic acid (2), given that the saponin is absorbed by the intestine as the aglycone, after hydrolysis of its carbohydrate moiety [2].

Regarding the reported properties for both products, most of the published articles explored in this chapter describe the in vitro or experimental in vivo activities, many of them in μM levels, which are considered too high to be useful as a drug [76] unless a very high dose is administered – way over the quantities permitted by the regulatory boards in both European Union [2] and USA [74]. Those activities should be regarded as potential ones, being useful as prototypes for the design and synthesis of new, improved compounds structurally related to **1** and/or **2**.

Glycyrrhizinic acid can be, described, eventually, as more than a sweetener. Along with glycyrrhetinic acid, it can be considered as scaffold molecule for the design and development of new bioactive compounds.

Acknowledgments This work is partly supported by grants from the Brazilian Government Science and Research Council (in Portuguese: *Conselho Nacional de Pesquisa* or CNPq) and the Rio de Janeiro State Research Foundation (*Fundação de Apoio à Pesquisa do Estado do Rio de Janeiro* or FAPERJ).

References

1. World Health Organization (2006) Safety evaluation of certain food additives. WHO Press, Geneva
2. European Comission Scientific Comiteee on Food (2003) Opinion of the scientific committee on food on glycyrrhizinic acid and its ammonium salt. European Commission. http://ec.europa.eu/food/safety/docs/fs-improv-flavourings-out186_en.pdf. Accessed 29 Jan 2016
3. Graebin CS, Verli H, Guimarães JA (2010) Glycyrrhizin and glycyrrhetic acid: scaffolds to promising new pharmacologically active compounds. J Braz Chem Soc 21:1595–1615
4. Ijeda K, Arase Y, Kobayashi M, Saitoh S, Someya T, Hosaka T, Sekazi H, Akuta N, Suzuki S, Suzuki F, Kumada H (2006) A long-term glycyrrhizin injection therapy reduces hepatocellular carcinogenesis rate in patients with interferon-resistant active chronic hepatitis C: a cohort study of 1249 patients. Dig Dis Sci 51:603–609
5. Glavac KN, Kreft S (2012) Excretion profile of glycyrrhizin metabolite in human urine. Food Chem 131:305–308
6. Baltina LA (2003) Chemical modification of glycyrrhizic acid as a route to new bioactive compounds for medicine. Curr Med Chem 10:155–171
7. Ming LJ, Yin ACY (2013) Therapeutic effects of glycyrrhizic acid. Nat Prod Comm 8:415–418
8. American Chemical Society (2016) SciFinder. http://scifinder.cas.org. Accessed 29 Jan 2016.
9. Wang H, Chu Z, Gao Y, Huang M (2015) Effect of glycyrrhizin on gastric cancer BGC-823 cell proliferation. Shijie Huaren Xiaohua Zazhi 23:2868–2873
10. Yang YA, Tang WJ, Zhang X, Yuan JW, Liu XH, Zhu HL (2014) Synthesis, molecular docking and biological evaluation of glycyrrhizin analogs as anticancer agents targeting EGFR. Molecules 19:6368–6381
11. Huang RY, Chu YL, Jiang ZB, Chen XM, Zhang X, Zeng X (2014) Glycyrrhizin suppresses lung adenocarcinoma cell growth through inhibition of thromboxane synthase. Cell Phys Biochem 33:375–388
12. Smolarczyk R, Cichon T, Matuszczak S, Mitrus I, Lesiak M, Kobusinska M, Kamysz W, Jarosz M, Sieron A, Szala S (2012) The role of glycyrrhizin, an inhibitor of HMGB1 protein, in anticancer therapy. Arch Immun Therap Exper 60:391–399
13. Wang W, Chen X, Zhang J, Zhao Y, Li S, Tan L, Gao J, Fang X, Luo A (2015) Glycyrrhizin attenuates isoflurane-induced cognitive deficits in neonatal rats via its anti-inflammatory activity. Neuroscience 316:328–336
14. Zhang J, Wu Y, Weng Z, Zhou T, Feng T, Lin Y (2014) Glycyrrhizin protects brain against ischemia-reperfusion injury in mice through HMGB1-TLR4-IL-17A signaling pathway. Brain Res 1582:176–186
15. Liu Z, Zhong J, Gao E, Yang H (2014) Effects of glycyrrhizin acid and licorice flavonoids on LPS-induced cytokines expression in macrophage. Zhongguo Zhongyao Zazhi 39:3841–3845
16. Fu Y, Zhou E, Wei Z, Song X, Liu Z, Wang T, Wang W, Zhang N, Liu G, Yang Z (2014) Glycyrrhizin inhibits lipopolysaccharide-induced inflammatory response by reducing TLR4 recruitment into lipid rafts in RAW264.7 cells. Biochimica Biophys Acta Gen Subjects 1840:1755–1764
17. Wang HL, Li YX, Niu YT, Zheng J, Wu J, Shi GJ, Ma L, Niu Y, Sun T, Yu JQ (2015) Observing anti-inflammatory and anti-nociceptive activities of glycyrrhizin through regulating COX-2 and pro-inflammatory cytokines expressions in mice. Inflammation 38:2269–2278
18. Chang CZ, Wu SC, Kwan AL (2015) Glycyrrhizin attenuates proinflammatory cytokines through a peroxisome proliferator-activated receptor-γ-dependent mechanism and experimental vasospasm in a rat model. J Vasc Res 52:12–21
19. Chang CZ, Wu SC, Kwan AL (2014) Glycyrrhizin attenuates toll like receptor-2, -4 and experimental vasospasm in a rat model. J Immun Res 740549:1–9
20. Xiao J, Liu J, Zhang Q (2014) Influence of compound glycyrrhizin on the functions of Treg and TH17 cells in children with henochschonlein purpura. Qingdao Daxue Yixueyuan Xuebao 50:315–317
21. Zhou J, Ma N, Qie Y, Shen J (2014) Efficiency of compound glycyrrhizin in treatment of 42 cases with chronic urticaria. Zhongguo Yaoye 23:117

22. Luo X, Li X, Yang L, Wang X (2014) Efficacy of magnesium isoglycyrrhizinate in treatment of hepatitis E with severe jaundice. Linchuang Gandanbing Zazhi 30:537–539
23. Zhang XX, Wu J, Xu Y, Zou X, Hu SY (2014) Clinical observation of compound glycyrrhizin for the prevention of liver and hematological system toxicity after chemotherapy for digestive tract cancer. Zhongguo Yaofang 25:4364–4366
24. Hsiang CY, Lin LJ, Kao ST, Lo HY, Chou ST, Ho TY (2015) Glycyrrhizin, silymarin, and ursodeoxycholic acid regulate a common hepatoprotective pathway in HepG2 cells. Phytomedicine 22:768–777
25. Meng W, Yu LZ, Wang L (2013) Effects of compound glycyrrhizin on liver function in patients with alcoholic liver disease: a meta-analysis. Zhongguo Yaofang 24:1116–1118
26. Yin J, Feng H, Chen S, Yang L (2013) Effects of compound glycyrrhizin preparation on IL-17 and ANA in patients with chronic hepatitis B. Jiangsu Yiyao 39:1280–1281
27. Cai L, Wu Y, Zhang J (2012) Comparison of the efficacy of two glycyrrhizin in treatment of chronic hepatitis B. Yaowu Liuxingbingxue Zazhi 21:160–161
28. Ding TP, Qu Y, Lu LG, Shi GL, Wang RF (2012) 18α-Glycyrrhizin ameliorates oxidative stress in rats with CCl_4-induced liver fibrosis. Shijie Huaren Xiaohua Zazhi 20:1819–1823
29. Smith G, Stubbins MJ, Harries LW, Wolf CR (1999) Molecular genetics of the human cytochrome P450 monooxygenase superfamily. Xenobiotica 28:1129–1165
30. Wang YG, Zhou JM, Ma ZC, Li H, Liang QD, Tan HL, Xiao CR, Zhang BL, Gao Y (2012) Pregnane X receptor mediated-transcription regulation of CYP3A by glycyrrhizin: a possible mechanism for its hepatoprotective property against lithocholic acid-induced injury. Chemico-Biol Interact 200:11–20
31. Chen Y, Wan X, Sun J, Wang W, Zhang S (2012) Hepato-protective activity of glycyrrhizin, liquiritin and isoliquiritigenin on HL-7702 cells injury induced by acetaminophen. Zhongguo Shiyan Fangjixue Zazhi 18:245–248
32. Gwak GY, Moon TG, Lee DH, Yoo BC (2012) Glycyrrhizin attenuates HMGB1-induced hepatocyte apoptosis by inhibiting the p38-dependent mitochondrial pathway. World J Gastr 18:679–684
33. Duan E, Wang D, Fang L, Ma J, Luo J, Chen H, Li K, Xiao S (2015) Suppression of porcine reproductive and respiratory syndrome virus proliferation by glycyrrhizin. Antivir Res 120:122–125
34. Li Z, Zhao Y, Lin W, Ye M, Ling X (2015) Rapid screening and identification of active ingredients in licorice extract interacting with V3 loop region of HIV-1 gp120 using ACE and CE-MS. J Pharm Biom Anal 111:28–35
35. Matsumoto Y, Matsuura T, Aoyagi H, Matsuda M, Hmwe SS, Date T, Watanabe N, Watashi K, Suzuki R, Ichinose S, Wake K, Suzuki T, Miyamura T, Wakita T, Aizaki H (2013) Antiviral activity of glycyrrhizin against hepatitis C virus in vitro. PLoS One 8:e68992
36. Ashfaq UA, Nawaz Z, Riazuddin S (2011) Glycyrrhizin as antiviral agent against Hepatitis C virus. J Trans Med 9:112
37. Xiao Q (2014) Zhou X (2014) Glycyrrhizin inhibits human neutrophil elastase-induced mucin 5 AC overproduction in human bronchial epithelial cells. Zhongnan Daxue Xuebao, Yixueban 39:252–257
38. Sil R, Sen S, Chakraborti AS (2015) Interaction of glycyrrhizin with human haemoglobin. Curr Sci 108:364–371
39. Francischetti IMB, Monteiro RQ, Guimarães JA (1997) Identification of glycyrrhizin as a thrombin inhibitor. Biochem Biophys Res Comm 9:259–263
40. Mendes-Silva W, Assafim M, Ruta B, Monteiro RQ, Guimarães JA, Zingali RB (2003) Antithrombotic effect of glycyrrhizin, a plant-derived thrombin inhibitor. Thromb Res 112:93–98
41. Paula FT, Frauches PQ, Pedebos C, Berger M, Gnoatto SC, Gossmann G, Verli H, Guimarães JA, Graebin CS (2013) Improving the thrombin inhibitory activity of glycyrrhizin, a triterpenic saponin, through a molecular simplification of the carbohydrate moiety. Chem Biol Drug Design 82:756–760
42. Imai K, Takagi Y, Iwazaki A, Nakanishi K (2014) Radical scavenging ability of glycyrrhizin. Free Rad Antioxid 3:40–42

43. Csuk R (2014) Recent developments in the synthesis of antitumor-active glycyrrhetinic acid derivatives. Mini-Rev Org Chem 11:253–261
44. Kim H, Kim SY, Lee JH, Han Y (2011) Antitumor effect of 18β-glycyrrhetinic acid against human tumor xenografts caused by A549 cancer cell. Yakhak Hoechi 55:39–44
45. Yamaguchi H, Yu T, Kidachi Y, Akitaya T, Yoshida K, Kamiie K, Noshita T, Umetsu H, Ryoyama K (2011) Selective toxicity of glycyrrhetinic acid against tumorigenic r/m HM-SFME-1 cells is potentially attributed to downregulation of glutathione. Biochimie 93:1172–1178
46. Xu S, Zhou L, Liu Z, Chen B, Li Y (2011) Glycyrrhetinic acid induces apoptosis and alters survivin gene expression in human myeloma cell line U266. Zhongguo Shiyan Xueyexue Zazhi 19:652–655
47. Kowsalya R, Vishwanathan P, Manoharan S (2011) Chemopreventive potential of 18β-glycyrrhetinic acid: an active constituent of liquorice, in 7,12-dimethylbenz(a)anthracene induced hamster buccal pouch carcinogenesis. Pakistan J Biol Sci 14:619–626
48. Sharma G, Kar S, Palit S, Das PK (2012) 18β-glycyrrhetinic acid (concur) induces apoptosis through modulation of Akt/FOXO3a/Bim pathway in human breast cancer MCF-7 cells. J Cell Phys 227:1923–1931
49. Kuang P, Zhao W, Su W, Zhang Z, Zhang L, Liu J, Ren G, Yin Z, Wang X (2013) 18β-glycyrrhetinic acid inhibits hepatocellular carcinoma development by reversing hepatic stellate cell-mediated immunosuppression in mice. Int J Cancer 132:1831–1841
50. Xie F, Cao Y, Liu GR, Yang XH, Dai Q, Chen W (2013) Mechanism underlying the inhibitory effect of glycyrrhetinic acid on epidermal growth factor-induced proliferation of HaCaT cells. Zhonghua Pifuke Zazhi 46:278–281
51. Li ZL, Sun D, Yang YC, Xu XL, Kong CZ (2012) 18β-glycyrrhetinic acid inhibits human 17β-hydroxysteroid dehydrogenase type III and induces cell apoptosis via activation of eIF2α in prostate cancer. Zhongguo Yike Daxue Xuebao 41:882–885
52. Wang D, Wong HK, Feng YB, Zhang ZJ (2014) 18beta-Glycyrrhetinic acid induces apoptosis in pituitary adenoma cells via ROS/MAPKs-mediated pathway. J Neuro-Oncology 116:221–230
53. Huang RY, Chu YL, Huang QC, Chen XM, Jiang ZB, Zhang X, Zeng X (2014) 18β-Glycyrrhetinic acid suppresses cell proliferation through inhibiting thromboxane synthase in non-small cell lung cancer. PLoS One 9:e93690
54. Jayasooriya RGPT, Dilshara MG, Park SR, Choi YH, Hyun JW, Chang WY, Kim GY (2014) 18β-Glycyrrhetinic acid suppresses TNF-α induced matrix metalloproteinase-9 and vascular endothelial growth factor by suppressing the Akt-dependent NF-κB pathway. Toxicol In Vitro 28:751–758
55. Mahmoud AM, Al Dera HS (2015) 18β-glycyrrhetinic acid exerts protective effects against cyclophosphamide-induced hepatotoxicity: potential role of PPARγ and Nrf2 upregulation. Genes Nutr 10:1–13
56. Chen S, Zou L, Li L, Wu T (2013) The protective effect of glycyrrhetinic acid on carbon tetrachloride-induced chronic liver fibrosis in mice via upregulation of Nrf2. PLoS One 8:e53662
57. Hasan SK, Khan R, Ali N, Khan AQ, Rehman MU, Tahir M, Lateef A, Nafees S, Mehdi SJ, Rashid S, Shalid A, Sultana S (2015) 18-β glycyrrhetinic acid alleviates 2-acetylaminofluorene-induced hepatotoxicity in Wistar rats: role in hyperproliferation, inflammation and oxidative stress. Human Exp Toxicol 34:628–641
58. Gupta P, Das PK, Ukil A (2015) Antileishmanial effect of 18β-glycyrrhetinic acid is mediated by toll-like receptor-dependent canonical and noncanonical p38 activation. Antimicrob Ag Chemother 59:2531–2539
59. Ukil A, Kar S, Srivastav S, Ghosh K, Das PK (2011) Curative effect of 18β-glycyrrhetinic acid in experimental visceral leishmaniasis depends on phosphatase-dependent modulation of cellular MAP kinases. PLoS One 6:e29062
60. Kalani K, Agarwal J, Alam S, Khan F, Pal A, Srivastava SK (2013) In silico and in vivo anti-malarial studies of 18β glycyrrhetinic acid from *Glycyrrhiza glabra*. PLoS One 8:e74761

61. Kalani K, Kushwaha V, Verma R, Murthy PK, Srivastava SK (2013) Glycyrrhetinic acid and its analogs: a new class of antifilarial agents. Bioorg Med Chem Lett 23:2566–2570
62. Long DR, Mead J, Hendricks JM, Hardy ME, Voyich JM (2013) 18β-glycyrrhetinic acid inhibits methicillin-resistant *Staphylococcus aureus* survival and attenuates virulence gene expression. Antimicrob Ag Chemother 57:241–247
63. Kim SR, Jeon HJ, Park HJ, Kim MK, Choi WS, Jang HO, Bae SK, Jeong CH, Bae MK (2013) Glycyrrhetinic acid inhibits *Porphyromonas gingivalis* lipopolysaccharide-induced vascular permeability via the suppression of interleukin-8. Inflammation Res 62:145–154
64. Puchner A, Hayer S, Niederreiter B, Hladik A, Blueml S, Bonelli M, Scheinecker C, Smolen J, Redlich K (2012) Effects of 18*β*-glycyrrhetinic acid in hTNFtg mice – a model of rheumatoid arthritis. Wiener Klinische Wochenschrift 124:170–176
65. Chen HJ, Kang SP, Lee IJ, Lin YL (2014) Glycyrrhetinic acid suppressed NF-κB activation in TNF-α-induced hepatocytes. J Agric Food Chem 62:618–625
66. Wang CY, Kao TC, Lo WH, Yen GC (2011) Glycyrrhizic acid and 18β-glycyrrhetinic acid modulate lipopolysaccharide-induced inflammatory response by suppression of NF-κB through PI3K p110δ and p110γ inhibitions. J Agric Food Chem 59:7726–7733
67. Kao TC, Shyu MH, Yen GC (2010) Glycyrrhizic acid and 18*β*-glycyrrhetinic acid inhibit inflammation via PI3K/Akt/GSK3β signaling and glucocorticoid receptor activation. J Agric Food Chem 58:8623–8629
68. Zhang H, Huang Q, Zhai J, Zhao Y, Zhang L, Chen Y, Zhang R, Li Q, Hu X (2015) Structural basis for 18-*β*-glycyrrhetinic acid as a novel non-GSH analog glyoxalase I inhibitor. Acta Pharmacol Sinica 36:1145–1150
69. Wu CH, Chen AZ, Yen GC (2015) Protective effects of glycyrrhizic acid and 18*β*-glycyrrhetinic acid against cisplatin-induced nephrotoxicity in BALB/c mice. J Agric Food Chem 63:1200–1209
70. Kong SZ, Chen HM, Yu XT, Zhang X, Feng XX, Kang XH, Li WJ, Huang N, Luo H, Su ZR (2015) The protective effect of 18*β*-glycyrrhetinic acid against UV irradiation induced photo-aging in mice. Exp Geront 61:147–155
71. Moon MH, Jeong JK, Lee YJ, Seol JW, Ahn DC, Kim IS, Park SY (2012) 18*β*-glycyrrhetinic acid inhibits adipogenic differentiation and stimulates lipolysis. Biochem Biophys Res Comm 420:805–810
72. Park M, Lee JH, Choi JK, Hong YD, Bae IH, Lim KM, Park YH, Ha H (2014) 18β-glycyrrhetinic acid attenuates anandamide-induced adiposity and high-fat diet induced obesity. Mol Nutr Food Res 58:1436–1446
73. Jiang L, Wang Q, Shen S, Xiao T, Li Y (2014) Discovery of glycyrrhetinic acid as an orally active, direct inhibitor of blood coagulation factor Xa. Thromb Res 133:501–506
74. Hardy ME, Hendricks JM, Paulson JM, Faunce NR (2012) 18*β*-glycyrrhetinic acid inhibits rotavirus replication in culture. Virology J 9:96
75. Food and Drug Administration (2015) CFR – code of federal regulations title 21. US Food and Drug Administration. https://www.accessdata.fda.gov/scripts/cdrh/cfdocs/cfcfr/CFRSearch.cfm?fr=184.1408. Accessed 29 Jan 2016
76. Dougall IG, Unitt J (2015) Evaluation of the biological activity of compounds: techniques and mechanism of action studies. In: Wermuth CG, Aldous D, Raboisson P, Rognan D (eds) The practice of medicinal chemistry, 4th edn. Elsevier, London

Beneficial Effects of *Stevia rebaudiana* Bertoni and Steviol-Related Compounds on Health

13

Alan Talevi

Abstract

Stevia rebaudiana (Bertoni) extracts and/or leaves have gained almost worldwide approval as a low-calorie sweetener. This fact has stimulated the research on potential additional functionalities of stevia as food additives and possible beneficial effects on health. The chapter analyzes the abundant evidence related to the potential effects of stevia in diabetes, cancer, and hypertension, among others. It also discusses, on the basis of the biopharmaceutical and pharmacokinetic knowledge on stevia and its glycosides, possible causes of the majorly disappointing results obtained at clinical studies when examining their effects in humans. Finally, the use of steviol and related compounds as starting points for drug discovery campaigns is overviewed.

Keywords

Stevia rebaudiana • Sweeteners • Stevioside • Rebaudioside • Steviol • Steviol glycosides • Isosteviol • Pharmacology • Pharmacokinetics

Contents

1 Introduction 264
2 Absorption and Pharmacokinetic Studies 265
3 Potential Effects of Stevia in Diabetes and Related Disorders 267
4 Antimicrobial Activity 273
5 Anticancer Effects 274
6 Effects on Blood Pressure 275
7 Other Effects 277

A. Talevi (✉)
Medicinal Chemistry, Department of Biological Sciences, Faculty of Exact Sciences, National University of la Plata (UNLP) - Argentinean National Council of Scientific and Technical Research (CONICET), La Plata, Buenos Aires, Argentina
e-mail: atalevi@biol.unlp.edu.ar

J.-M. Mérillon, K.G. Ramawat (eds.), *Sweeteners*, Reference Series in Phytochemistry,
https://doi.org/10.1007/978-3-319-27027-2_24

8 Some General Considerations to Interpret and Appraise Pharmacological Studies on *Stevia rebaudiana* and Its Constituents 278
9 Conclusions 279
10 Cross-References 280
References 280

1 Introduction

Stevia rebaudiana Bertoni, a member of the Asteraceae family, is a plant native to South America whose leaves have a long traditional use to sweeten food and beverages in South America. Back in 1971, Japan approved a commercial stevia-based sweetener as an alternative to artificial sweeteners. Afterward, the use of stevia leaves and/or their purified extracts expanded to China (1984), Brazil (1986), and Korea (2000). In 2004, the Joint Food and Agriculture Organization of the United Nations (FAO) and the World Health Organization (WHO) Expert Committee on Food Additives established a temporary acceptable daily intake of 2 mg/kg body weight for steviol glycosides expressed as steviol [1]. Later, in 2008, the same Committee set the safety of steviol glycosides as food additives at daily intakes of 4 mg/kg, expressed as steviol [2], which facilitated the approval of stevia leaves and/or extracts as additives and/or dietary supplements by the health authorities of many other countries, including some major markets in the United States (2008), Australia, New Zealand and most Latin American countries (2008), the European Union (2011), and Canada (2012). Consequently, research on possible additional functionalities and therapeutic uses of stevia has intensified in the last few years. Most research focuses on the potential antimicrobial and antioxidant effects of stevia constituents (which could expand their functionalities as food additives; antioxidant effects are extensively reviewed in a separate chapter of this volume) or study of the effects of stevia or stevia constituents on blood pressure and diabetes; interestingly, the combined antidiabetic, antihypertensive, and antioxidant effects could have a beneficial impact in metabolic syndrome [3]. A minority of papers report potential anticancer and central nervous system effects that deserve consideration and will be also reviewed in this chapter.

It is worth mentioning that the regulatory status of stevia and its extracts varies from county to country, with some regulatory authorities allowing the use of stevia leaves and extracts, and others authorizing the use of specific highly purified extracts only. The variability of the legal status of stevia products from country to country is not trivial, since it impacts on the scope of reports dealing with possible pharmacological and toxicological effects of the plant or its extracts. Accordingly, we have decided to specify, throughout the text, the nature and quantity of the plant organ and/or plant extract which has been tested in the reviewed articles.

Pharmacokinetic studies on stevia constituents are essential to appraise the extent of pharmacological studies (e.g., by providing clues on interspecies variability in absorption and metabolism that could compromise the predictability of animal models regarding stevia effects). Thus, we will include an overview on pharmacokinetic knowledge of stevia constituents before addressing studies on its potential health benefits, in the belief that absorption and disposition data could help interpreting the extent of the findings emerging from preclinical pharmacological studies and clinical trials.

2 Absorption and Pharmacokinetic Studies

Owing to their relatively high molecular weight and polarity, steviol glycosides are poorly absorbed as such after oral administration [4, 5]. Note that steviol glycosides do not comply with Lipinski rules nor do they obey other similar rules related to passive oral drug absorption (e.g., Veber's) [6, 7], a central point of future discussions throughout this chapter. In contrast, the aglycon steviol is more effectively absorbed, its apparent permeability coefficient being around 300–400 times higher than those of the glycosides [4]. That said, are steviol glycosides hydrolyzed to steviol within the intestinal tract? Stevioside is unchanged after incubation with human salivary α-amylase, porcine pancreatic α-amylase, human saliva, porcine pepsin, human gastric secretion, porcine pancreatin, and intestinal brush border membrane enzymes of mice, rats, and hamsters, suggesting stability against gastrointestinal digestive enzymes [8]. Similar results have been obtained for rebaudiosides A and D [9]. Steviol glycosides can be nevertheless hydrolyzed to steviol (which is then readily absorbed) by intestinal microbiota from a diversity of species, including humans [4, 8, 10–12]. While Hutapea et al. reported that stevioside was partially biotransformed to steviol-16,17-epoxide by mice and human microbiota, such finding was not confirmed in subsequent studies and has been later attributed to lack of specificity in the analytical method employed by Hutapea or a non-sufficiently anaerobic experimental procedure [13]. In 2008, Roberts and Renwick performed a comparative pharmacokinetic and biotransformation study in rats, by administering labeled 4.2 mg/kg stevioside, 5 mg/kg rebaudioside A, and 1.6 mg/kg steviol to rats, p. o., using both intact and bile duct-cannulated animals and discriminating between males and females [14]. Peak plasma concentrations were observed at much shorter times (and, in fact, at a very short time after administration) for steviol than for steviol glycosides, probably because glycosides should undergo hydrolysis to the aglycon before their steviol content can be absorbed. While plasma profiles were similar for the three treatments for males and females, steviol plasma concentration was higher in females compared to males in the case of steviol glycosides, suggesting sex differences in the rate of absorption or biotransformation. Secondary plasma peaks were observed when administering labeled steviol, suggesting some degree of enterohepatic circulation, as previously observed by Nakayama et al. for stevioside [15]. Of upmost importance for the subsequent discussions, the predominant radioactive component in all plasma samples was identified as steviol. Lower amounts of steviol glucuronide(s) and even lower quantities of unidentified metabolites of slightly higher polarity than steviol (possibly, phase I metabolites) were observed after single-dose administration of each of the tested compounds. The mean recoveries of radioactivity in urine, feces, gastrointestinal tract, and carcass were measured for intact animals. 97–98% of the dose was recovered in feces for rebaudioside A and stevioside, while 90% of the dose was recovered in feces for steviol. In cannulated rats, most of the radioactivity was recovered in the bile rather than in the feces. About 70–80% of the administered steviol glycosides were recovered in bile; though there was limited excretion of radioactivity in the first 3 h after dosing, most of the biliary excretion occurred within the first 12 h. In contrast, the biliary excretion after steviol administration was much more rapid (consistently with the much faster absorption), with 50–70% of the dose eliminated within the first 3 h and the total recovery in bile being 97–98% of the administered dose of steviol. For both rebaudioside A and stevioside, most of the remaining radioactivity was recovered in the feces (21–30%). The extent of absorption was estimated by summing the mean total amount of radioactivity in the bile, urine, liver, and remaining carcass. Approximately 70.7% and 82.0% of the dose of rebaudioside A was absorbed by males and females, in that order, whereas 77.9% and 80.5% of the stevioside dose was absorbed by males and females, respectively, and 97.2% and 99.4% of the steviol was absorbed by males and females, respectively. The peak plasma concentration and area under the curve for stevioside were slightly higher than those for rebaudioside A, indicating slightly greater formation of steviol from stevioside than from rebaudioside A, which can be anticipated from the fact that rebaudioside A contains an additional glucose unit in relation to stevioside; such additional glycoside unit has to be cleaved from the aglycon prior to absorption. The predominant radioactive components in the bile were found to be steviol glucuronides, while the predominant component in the feces was, again, the aglycon. These findings are indicative of rapid first-pass phase II metabolism (glucuronidation); in rats, the glucuronides are mainly eliminated in the bile and deconjugated in the gastrointestinal tract prior to excretion in the feces, with some of the released steviol probably undergoing reabsorption.

Though the metabolism of steviol glycosides seems similar in humans (glucuronidation being the main biotransformation reaction), important differences exist regarding excretion routes and steviol exposure after oral administration. In a study on ten healthy male and female subjects, Geuns et al. administered doses of 250 mg of stevioside three times a day for 3 days [16]. Steviol glucuronide was isolated as the only metabolite detected in the urine on the third day, whereas no free steviol was detected. Later, the same group performed an extensive metabolism study of oral stevioside in nine human healthy subjects (five females, four males) [17]. Neither stevioside nor steviol could be detected in blood samples at any time of the experimental period, in contrast with previously discussed findings in rats. Stevioside and steviol were also absent in urine samples, where high levels of steviol glucuronide were however detected. Only free steviol was observed in feces. Unlike previous findings in rats, no differences related to sex of the volunteers were observed. No other steviol metabolites could be detected in blood or urine, suggesting that the phase I hepatic metabolism of steviol is low or nonexistent at the studied doses or occurs at a very slow rate compared to glucuronidation, which is in good agreement with previous in vitro studies with human microsomes [18]. Similarly, administration of single doses of 5 mg/kg rebaudioside A and 4.2 mg/kg stevioside to nine healthy males in a randomized double-blind crossover study did not result in detectable levels of free steviol in plasma in seven out of eight subjects [19]. Only small amounts of steviol were detected in urine, while larger amounts were found in feces. Steviol glucuronide was primarily excreted in the urine, accounting for 59% and 62% of the rebaudioside A and stevioside doses, in that order. The previous findings indicate that stevioside and rebaudioside A follow similar metabolic and excretion pathways in humans, while significant differences have been established compared to rats when administering comparable doses to both species, including exposure to free steviol, detection of phase I steviol metabolites, sex differences, and excretion route. It is possible that the absence of free steviol and its phase I metabolites in humans could be explained by non-saturation of the glucuronidation systems in the human liver at the concentrations resulting from the administered doses. Such increased steviol biotransformation capacity could explain the absence of the free aglycon in blood after first-pass metabolism. On the other hand, the differences in the excretion route could arise from dissimilarities in molecular weight thresholds for urine and biliary excretion in both species. These results could lead to significant interspecies variability in pharmacological studies involving oral administration of stevia extracts, since exposure to different metabolites of steviol glycosides is observed in different species at the doses used in the studies, a relevant point since, as it will be seen throughout the chapter, the glycosides and the aglycon do not always display identical or similar intrinsic activity and display considerable differences regarding their biodistribution. A schematic representation of the differences between the disposition of steviol glycosides in rats and humans is presented in Fig. 1.

Steviol (but not stevioside) is transported by uptake transporters OAT1 and OAT3 [20], while transport of steviol glucuronide was mainly mediated by OAT3 and to a lesser extent by OATP1B1, OATP1B3, or OATP2B [21]. Consequently, steviol and its glucuronide may be involved in drug interactions involving other substrates for these transporters, which could impact on steviol renal secretion. In contrast, steviol glucuronide (at 10 μM concentrations) is not transported by efflux transporters BCRP, MDR2, P-gp, or MATE1, as emerges from studies with MDCKII transfected cells. Remarkably, though as we will discuss later stevia extracts have antimicrobial effects, it has been shown that rebaudioside A exerts little pressure on the growth and composition of microbes, suggesting it is safe for gut microbes (mouse) [22]. Note that this fact is relevant not only from a safety perspective but also from the pharmacokinetics point of view, since as previously discussed gut microbiota is involved in steviol absorption and disposition.

3 Potential Effects of Stevia in Diabetes and Related Disorders

Steviol glycosides do not induce a glycemic response when ingested, making them attractive as natural sweeteners to diabetic patients on carbohydrate-controlled diets; what is more, a number of in vitro and in vivo studies show that they have antihyperglycemic, insulinotropic, insulin-mimetic, and glucogonostatic effects

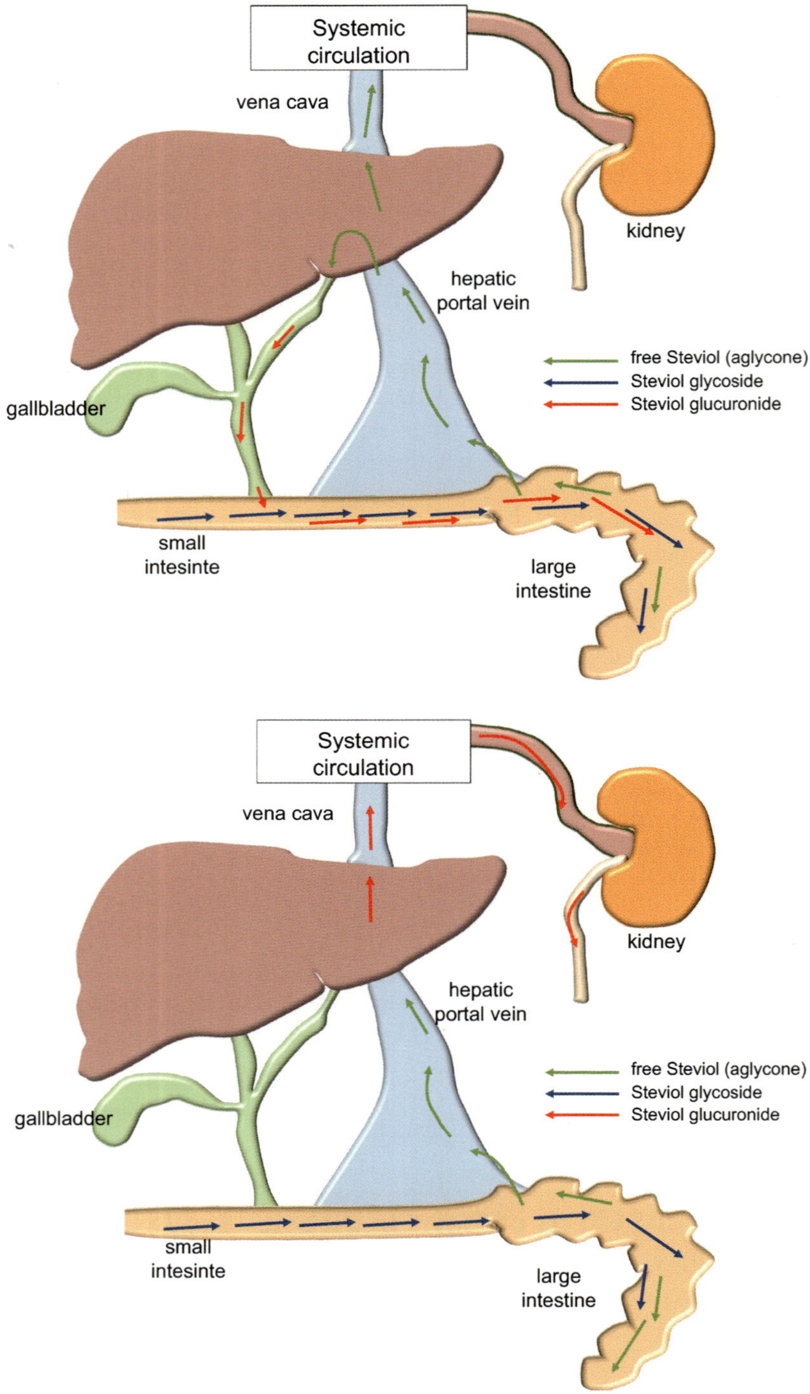

Fig. 1 (continued)

[23], though such results have so far failed to translate into randomized controlled clinical trials.

Jeppesen et al. examined the effect of stevioside and steviol (1 nM–10 mM) on the insulin release from isolated mouse pancreatic islets and β-cell line INS-1 [24]. Stevioside potentiated the insulin release evoked by 16.7 mM glucose with an apparent maximal effect at 1 mM concentrations; steviol displayed insulinotropic effects as well, but a maximal effect at 1 μM concentrations. The stimulatory effect on insulin release was observed at concentrations of at least 0.1 nM, but only at glucose levels above 8.3 mM, while no effect was observed at normal or subnormal glucose levels. The authors suggested that this feature could imply an antihyperglycemic effect without hypoglycemic activity. Very similar results were found for rebaudioside A, which stimulated insulin secretion in mice isolated islets starting from very low concentrations (10^{-14} M) achieving a maximum effect around 0.1 nM, but only upon exposure to high glucose levels [25]. In studies in diabetic type-2 Goto-Kakizaki rats [26, 27], these authors tested the acute and long-term antihyperglycemic effects of stevioside. In the first study, they tested the effect of intravenous 200 mg/kg stevioside in an intravenous glucose tolerance test. Stevioside reduced the glucose response and increased the insulin response. A decrease in glucagon was observed 15 min after the injection of stevioside, and control levels were not recovered until 120 min after treatment. In comparison, no effect on glucose response or glucagon levels was observed when the experiment was repeated in Wistar rats, though a significant increase in insulin response was still observed. When feeding Goto-Kakizaki rats with 25 mg/kg stevioside for 6 weeks, the drug displayed, once again, an antihyperglycemic effect, enhanced the first-phase insulin response, and suppressed the glucagon levels, besides a pronounced suppression of both the systolic and the diastolic blood pressure. Bolus injections of stevioside did not induce hypoglycemia, and 6-week administration of the glycoside did not alter fasting blood glucose, insulin, and glucagon, in consistence with the previously described in vitro results. The observed insulinotropic effects were linked to the induction of genes involved in glycolysis and the improvement of nutrient-sensing mechanisms. Ferreira and colleagues compared the effects of oral stevia leaves (20 mg/kg/day during 15 days) and stevioside (5.5 mg/kg/day during 15 days) on 15-h fasted Wistar rats [28]. Glycemia and gluconeogenesis were examined in perfused liver and isolated hepatocytes. Stevia leaves, but not stevioside, decreased

Fig. 1 Differences between the elimination pathway of steviol glycosides in rats (*up*) and humans (*down*). Steviol glycosides (*blue arrow*) are poorly absorbed, but are hydrolyzed to free steviol (*green arrow*) in the gut. Part of the free steviol is absorbed and undergoes first-pass metabolism, mainly glucuronidation. In the study by Roberts and Renwick, part of the free steviol survives first-pass metabolism and reaches systemic circulation (rats), while in the studies by Geuns et al. and Wheeler et al. no or almost no free steviol is detected in blood (humans). In rats, steviol glucuronide undergoes biliary excretion and is deconjugated by microbiota; in contrast, in humans, the glucuronide undergoes urinary excretion. In the study by Roberts and Renwick, the administered doses of steviol glycosides seemingly saturated rats' phase II metabolism, while the administered doses in humans did not reach the saturation point

glycemia and glucose production. These results seem to be in good agreement with the previously described from Jeppesen et al., where long-term exposure to stevioside did not exert effects on fasting blood glucose, insulin, or glucagon in diabetic rats, and acute exposure did not alter glucose response or glucagon levels in Wistar rats. Apparently, the effects of stevioside are more evident in rodent models of diabetes than in nondiabetic animals and after a glucose load than in fasted animals. Simultaneously, it is worth highlighting that other constituents of stevia leaves seem to exert different actions (and/or to display different potency) than purified stevioside.

In 2004, Lailerd et al. studied the in vivo and in vitro effects of stevioside on glucose tolerance and skeletal muscle glucose transport in insulin-sensitive lean Zucker rats and insulin-resistant obese Zucker rats (the latter constitutes a model of obesity-associated insulin resistance and marked hyperinsulinemia, but no hyperglycemia, which resembles a prediabetic state) [29]. For the in vivo studies, the rats were administered either 200 or 500 mg/kg. Whereas doses of 200 mg/kg did not modify either glucose or insulin response, doses of 500 mg/kg significantly reduced the plasma insulin response in lean rats 15 and 30 min after glucose load and the plasma glucose response in obese rats 30 min after the load. Both the total and incremental areas under the curve for glucose were unaltered in the lean group, while the total and incremental areas under the curve for insulin were reduced after stevioside treatment. While the total urea under the curve for glucose was slightly decreased in the obese rats, the incremental areas for glucose and insulin and the glucose-insulin index using incremental areas were reduced, suggesting that acute stevioside treatment causes an increase in insulin sensitivity. Basal uptake of labeled deoxyglucose in rat epitrochlearis dissected from lean rats was not modified by low doses (0.01–0.1 mM) of stevioside, while higher doses (0.5–5.0 mM) decreased basal uptake. In contrast, 0.01 mM stevioside enhanced the rate of insulin-stimulated deoxyglucose uptake. In the soleus, the basal uptake was not affected by any dose but the rate of insulin-stimulated uptake was increased by low doses; at 0.5 mmol/L or higher, this positive effect of SVS was no longer detected. The influence of stevia constituents on glucose uptake was also studied in other cells and species. Rizzo and coworkers examined the effect of different commercial stevia extracts with a total steviol glycoside content of at least 95% on the cellular glucose transport using cultured cells (SH-SY5Y neuroblastoma and HL-60 myeloid leukemia human cells) [30]. The authors tested 0.5–5 mg/mL of extract, which corresponded to about 0.5–5 mM in the case of the purest extract tested (rebaudioside A 97–98%). They also tested stevioside and rebaudioside A standards. Both the extracts (1 mg/mL) and the pure glycosides (1 mM) increased glucose transport in both cell types in a similar manner than insulin 100 nM. Cotreatment with insulin and stevia extracts caused a rise of glucose transport significantly higher than the increase due to insulin alone. By examining the phosphorylation status of PI3K and Akt following stevia extract treatment and the effect of preincubation with methylglyoxal (an inhibitor of the of the insulin receptor/PI3K/Akt pathway) on glucose transport, the authors proposed that steviol glycosides acted by inducing glucose transporter translocation from cytosolic vesicles to the plasma membrane through the PI3K/Akt pathway. A very recent study from Bhasker and coworkers examined the influence of stevioside, steviol, and a crude preparation of stevia leaf extract (extraction technique unspecified) on the mRNA and protein expression level of glucose transporter (GLUT4) in rat L6 myoblast and mouse 3 T3-L1 fibroblast cells [31]. Steviol 1 μM and stevioside 100 μM displayed the most significant enhancement on GLUT4 gene transcripts, in both cell lines, at both mRNA and protein levels. The crude extract (1 μL) also showed a positive effect on the expression levels. Interestingly, the effects of stevioside were positively correlated with its concentration, while an inverse correlation was found for steviol.

Shivanna et al. tested the effects of stevia leaf powder, its polyphenol extract, and its fiber extract on streptozotocin-induced diabetic rats (a type-1 diabetes model) [32]. For this purpose, they incorporated 4% of stevia leaf powder to the diet or administered the equivalent amount of extracts through force-feeding. Whereas the treatment with streptozotocin led to polydypsia, polyphagia, and polyuresis, pre-feeding the rats with stevia leaves and polyphenol extracts prevented these changes significantly; both treatments also reduced the streptozotocin-induced elevated blood glucose levels, enhanced the serum insulin level, and increased cellular insulin sensitivity. No protective effects were observed after treatment with stevia fiber extract. Previously, Chang et al. had examined the effects of repeated oral administration of stevioside to streptozotocin-induced diabetic rats (0.2 mg/kg three times a day during 10 days) and the acute effect of stevioside (5 mg/kg) on the glucose-insulin index of fructose-rich chow-fed rats (a model of type-2 diabetes) [33]. They also studied the effect of stevioside on the development of insulin resistance by repeatedly administering the glycoside during the 4-week period of fructose-rich feeding. Acute stevioside improved sensitivity to insulin in the insulin-resistant, fructose-rich fed rats; repeated administration of stevioside delayed the time associated to the loss of response to tolbutamide and increased the response to exogenous insulin in the type-1 diabetes model.

More recent studies suggest that the effect of stevia on diabetes may go way beyond the direct impact on insulin secretion, glucose disposition, and insulin sensitivity; research using obese rodent models indicates that stevia extracts could even be used as etiologic/preventive treatment to type-2 diabetes and other obesity-related disorders such as cardiovascular diseases and fatty liver disease. Obesity is linked to oxidative stress and mild inflammation in a number of tissues (liver, adipose tissue, skeletal muscle, and pancreatic islets); adipose tissue is the main source of pro-inflammatory factors, which may cause systemic insulin resistance [34, 35]. For instance, Geeraert et al. studied the influence of 12-week treatment with oral 10 mg/kg stevioside on combined leptin and LDL-receptor deficiency double-knockout mice (obese insulin-resistant mice that exhibit most of the metabolic syndrome components) [35]. Such treatment was associated with improved insulin signaling, antioxidant defense in the vascular wall and the adipose tissue, and inhibition of atherosclerotic plaque development. Wang and collaborators examined the impact of 1-month treatment with orally administered 10 mg/kg/day stevioside on high-fat diet mice [34]. It was found that stevioside ameliorates insulin resistance, downregulates inflammatory cytokines mRNA levels and macrophage infiltration, and suppresses the NF-kB signaling pathway in adipose tissue. Recently, the effects of 12-week administration of stevioside (10 mg/kg/day, oral administration), rebaudioside A (12 mg/kg/day, oral administration), and steviol (5 mg/kg/day, oral administration) were evaluated in a mouse model of obesity and insulin resistance (ob/ob and LDRL-double-knockout mice) [36]. It was shown that the three compounds decreased the level of hepatic steatosis without weight loss. They mostly affected metabolic pathways associated with lipotoxicity, improving glucose metabolism in the liver, fat catabolism, lipid storage and transport, and reducing oxidative stress. Nuclear receptors PPARα, PPARγ, and PPARδ were identified as central regulators of the effects of steviol and its glycosides; notably, the three treatments seem to be linked to different pathophysiological changes.

Regarding clinical evidence related to possible effects of stevia on diabetes, back in 1986, Curi and coworkers observed a positive effect on glucose tolerance in 16 healthy adults [37]. In an acute crossover study involving 12 type-2 diabetic patients, it was shown that supplementing a test meal with 1 g stevioside slightly reduced the postprandial glucose levels and increased the insulinogenic index [38]. Similar results were later obtained in a study involving 19 healthy lean and 12 obese individuals. Receiving a preload sweetened with stevia (the quantity of the stevia extract is not disclosed in the original article) before standardized meals significantly reduced postprandial glucose levels and postprandial insulin levels compared to sucrose-sweetened preloads; no effect on the insulinogenic index was however observed in this study [39]. In contrast, no effect was registered in the studies on healthy subjects by Geuns et al. (250 mg of stevioside three times a day, for 3 days) [17] or Chan et al. (250 mg os stevioside three times a day, for 1 year) [40]. So far, the overviewed results are not necessarily in conflict with previously discussed in vitro and preclinical data, since stevia extracts seem to display a more pronounced effect on diabetic animals and some articles report that the observed effects tend to disappear at large doses. Furthermore, the effects of stevia in in vitro and animal models seem to appear when blood glucose levels are elevated (antihyperglycemic effect), and no hypoglycemic effect is observed.

However, the firmest evidence on the potential impact of long-term consumption of stevia glycosides on diabetes comes from two disappointing randomized controlled clinical trials on diabetic subjects that failed to demonstrate pharmacological effects of the glycosides. Barriocanal and coworkers conducted a study on three groups of subjects [41]: 30 with type-2 diabetes, 16 with type-1 diabetes, and 30 without diabetes. During the study, diabetic patients continued their treatment with insulin (type 1) or oral hypoglycemic drugs (type 2), but they received steviol glycoside capsules (250 mg) or matching placebo during 3 months. No statistically significant change was observed in the posttreatment mean glucose and glycated hemoglobin levels in the steviol glycoside group. For its part, Maki et al. conducted a randomized trial comparing the effects of 16-week consumption of 1000 mg rebaudioside A with placebo in volunteers with type-2 diabetes receiving one to three hypoglycemic agents [42]. Again, no differences between glycosylated hemoglobin were found between treatment and placebo.

Why the in vitro and preclinical data does not seem to translate into clinical trials? While the key for answering this question could lie in a diversity of factors, from inadequate dosage to pharmacodynamics and/or pharmacokinetics variability between species (including interspecies differences in gene expression regulation upon exposure to stevia constituents), the author would like to draw the reader's attention to the fact that, at the preclinical level, most evidence on the potential benefits of stevia in diabetes come either from in vitro studies or from studies in which steviol or its glycosides are orally administered to rodents. It has already been established that glycosides are poorly absorbed through the gastrointestinal tract. Therefore, even if they had antihyperglycemic properties per se, they could hardly manifest it after oral administration (which is the administration route that has been chosen in clinical trials conducted so far). However, in the case of rodents, steviol is the dominant chemical species in circulation after oral administration of steviol glycosides. In humans, most of the absorbed steviol appears as glucuronide, while no (or negligible free aglycon) absorbed steviol is observed at the studies discussed in the *Absorption and Pharmacokinetics Studies* section. If steviol and/or its phase I biotransformation products were the active entities responsible for the observed effects on glycemia observed at preclinical models, such effects would probably not manifest in clinical trials unless phase II metabolism capacity was overridden (e.g., at high doses). Note that, with some exceptions, phase I metabolites are more likely to present biological activity than phase II biotransformation products (i.e., glucuronides), which in general tend to be devoid of pharmacological activity [43]. Identification of the identity of the chemical species with intrinsic activity on glycemia and quantification of the exposure to steviol and its phase I metabolites in clinical trials are thus important aspects to explain apparent differences in stevia pharmacology in humans and animal models.

Finally, it is worth noting that steviol and the steviol-16a,17-epoxide have been used as starting points to obtain analogs with plasma glucose-lowering activity in diabetic rats through microbial metabolism [44], suggesting the therapeutic potential of steviol functionalization biotransformation products, as will be also discussed in other sections of the chapter.

4 Antimicrobial Activity

Jayaraman and coworkers have examined the antibacterial, anti-yeast, and antifungal activity of different stevia leaf extracts (ethyl acetate, acetone, chloroform, and water) against a number of microorganisms including *Escherichia coli*, *Bacillus subtilis*, *Staphylococcus aureus*, *Salmonella typhi*, *Vibrio cholera*, *Candida albicans*, *Cryptococcus neoformans*, *Trichophyton mentagrophytes*, and *Epidermophyton* [45]. Each extract was tested at a final concentration of 50 mg/mL. All the extracts showed some level of activity against the tested species with the exception of the water extract, which failed to elicit antimicrobial effects. The precedent results are consistent with the report from Debnath, who tested methanol, chloroform, and aqueous stevia leaf crude extracts against several bacteria and fungi: *E. coli*, *B. subtilis*, *S. aureus*,

Staphylococcus mutans, *Sclerotonia minor*, *Curvularia lunata*, *Alternaria alternata*, *Aspergillus niger*, *Microsporum gypseum*, and *Rhizopus* [46]. Methanol and chloroform extracts were the most effective against the four bacteria species; only two fungi were affected by the methanol extract and one by the chloroform extract. Aqueous extract was ineffective in all cases. Singh et al. assayed a series of methanolic extracts of root, stem, leaves, and flower (100–500 mg/ml) against *E. coli* and *B. subtilis* [47], finding inhibitory activity on both species. Puri et al. examined the antimicrobial activity of organic extracts from stevia and stevioside against a wide spectrum of food-borne pathogens, including *B. subtilis*, *Bacillus cereus*, *Staphylococcus xylosus*, *Alcaligenes denitrificans*, *Klebsiella pneumoniae*, *Pseudomona aeruginosa*, *Salmonella typhimurium*, and *Serratia marcescens* [48]. The extracts were tested at two concentrations, 500 and 1000 μg/mL, and displayed a dose-dependent inhibition across all the microorganisms tested. Ethanol and acetone extracts displayed the broader spectra of inhibitory activity. The purified stevioside displayed dose-dependent activity against all bacteria but *S. marcescens*, at concentrations between 31 and 124 μM. On the basis of the study outcome, the authors hypothesized that stevioside could be useful to enhance the durability and quality of foods, potentially expanding its functionalities as food additive. Though the purified stevioside seems to be only modestly potent as antibacterial, the previous results suggest that it should be interesting to isolate other constituents of stevia extracts to test their antimicrobial effect. What is more, stevia constituents could be subjected to structural optimizations in order to enhance their inhibitory effects. In this line, Sharipova et al. examined the effects of stevioside and steviolbioside plus hybrid compounds synthesized from steviolbioside and isoniazid or nicotinic acid hydrazide *on Mycobacterium tuberculosis*, finding that these compounds exhibit moderate activity against this microorganism (MIC below 10 μg/mL) [49]. Many other reports have been published describing the preparation of steviolbioside, isosteviol, and steviol derivatives with relatively high inhibitory effect on *Mycobacterium tuberculosis* [50–52]. Other isosteviol analogs have also shown activity against hepatitis B virus [53].

5 Anticancer Effects

A number of studies have pointed at the potential therapeutic and preventive effects of stevia and steviol glycosides against cancer. The worldwide approval of steviol glycosides as food additives and the benign medicinal properties assigned to steviol and its derivatives have recently encouraged many organic and medicinal chemists to explore the synthesis of stevioside and steviol analogs [54], an interest that as we have seen applies to virtually every branch of medicine in which stevia has demonstrated potential, but with emphasis in the field of anticancer and antitumor agents development [55–60].

The first report we have knowledge about related to potential antitumor activity of stevia extracts dates from 1997, when, while studying the carcinogenicity of stevioside in F344 rats, Toyoda et al. observed that, unexpectedly, the incidence of adenomas of the mammary gland in the stevioside-treated female rats was significantly lower than that of the controls [61]. In 2002, Yasukawa et al. discovered the strong inhibitory effect of four steviol glycosides (stevioside, rebaudioside A and C, and dulcoside A) against 12-*O*-tetradecanoylphorbol-13-acetate (TPA)-induced inflammation (ear edema) in mice [62]. The ID_{50} was in the range 0.06–0.36 μM/ear. The topically applied mixture of steviol glycosides (0.1 or 1 mg per application, twice a week 30 min before TPA) also displayed an inhibitory effect on skin tumor promotion in a two-stage carcinogenesis test model by application of 7,12-dimethylbenz[a]anthracene and TPA. At week 30, the percentage of tumor-bearing animals with TPA was 80%, while the proportion with 0.1 and 1 mg steviol glycosides was only 20% and 13%, in that order. The potential cancer chemopreventive effects of stevioside, steviol, and isosteviol were later identified by Takasaki et al. using an in vitro screening test for inhibition of Epstein–Barr virus early antigen activation promoted by TPA [63]. The three compounds also displayed protective effects in two two-stage carcinogenesis models on mouse skin, in all cases with comparable potency. Boonkaewwan and coworkers proved that high concentrations of stevioside (2–5 mM) and steviol (0.2–0.8 mM) decreased cell viability in T84, Caco-2, and HT29 colon carcinoma cell lines [64]. More recently, it was found that stevioside induces antiproliferation effects on MCF-7 breast cancer cell line [65]. In 5 and 10 μM concentrations, stevioside induces apoptosis, reduces mitochondrial transmembrane potential, and enhances the production of reactive oxygen species.

6 Effects on Blood Pressure

The first evidence on the effects of stevia on blood pressure was reported by Humboldt and Beech, who observed that stevioside and stevia aqueous extracts induced a marked decrease in blood pressure and heart rate as well as diuresis, in rats [66, 67]. These results were later confirmed by Melis and Sainati's findings and demonstrated that infusion of stevioside at 8 and 16 mg/kg/h rate produced diuresis, kaliuresis, and natriuresis in rats, while the 16 mg/kg/h rate also displayed significant hypotensive effects [68]. Since infusion of CaCl in rats pretreated with stevioside induced a pronounced reduction on the vasodilating response to stevioside, the authors hypothesized that the glycoside could be acting in a similar way to verapamil (calcium antagonist). It was later shown that infusion of stevioside at 16/mg/kg/h reduced mean arterial pressure not only in normal but also in renal hypertensive rats; the effect of stevioside on renal plasma flow and glomerular filtration rate was more pronounced in hypertensive rats [69]. Similar results were obtained by the same author with stevia aqueous extracts. In one study, treated animals were given orally an aqueous extract of *Stevia rebaudiana* (corresponding to 66.7 g of dried leaves/100 ml final solution) twice a day to normal rats for 20, 40, and 60 days [70]. No effects in mean arterial pressure or renal function were observed for the 20-day group. However, a significant decrease in the first parameter was observed in the 40-day group, with still no differences regarding filtration rate or renal plasma flux, while an enhanced diuresis was observed. Sixty-day treatment also reduced mean arterial pressure and increased diuresis and natriuresis. Later, it was reported that a crude extract of stevia (2.67 g of dry leaves for 30 days) resulted in a significant decrease in mean arterial pressure in both normotensive and renal hypertensive rats. Glomerular filtration rate was constant in the normal rats and increased significantly in the hypertensive rats after stevia treatment, whereas both groups presented an increase in renal plasma and urinary flows following oral stevia administration and natriuresis [71]. The hypotensive and antihypertensive effects of stevioside (50–200 mg/kg, intraperitoneally) were later confirmed in normotensive, renal hypertensive, and deoxycorticosterone acetate salt-sensitive hypertensive rats, with a more pronounced effect on hypertensive animals [72]. Isosteviol (25 mg/kg, intraperitoneally) also proved to decrease mean arterial pressure in spontaneously hypertensive rats [73]. Isolated aortic strips and vascular smooth muscle cells were employed to evaluate the mechanism of hypotension. Isosteviol was able to relax vasopressin-induced vasoconstriction in a dose-dependent manner, but failed to modify the vasopressin-induced vasoconstriction in Ca^{2+}-free medium. Reduction in the vasopressin-induced vasoconstriction was also obtained in the absence of endothelium, suggesting the treatment effect was not related to nitric oxide or the endothelium system. Isosteviol produced no effect on phenylephrine-induced vasoconstriction. Altogether, these results indicated that isosteviol vasorelaxatory activity occurred via an inhibition of Ca^{2+} influx. Very similar conclusions were later obtained for stevioside [74]. In the presence of potassium chloride, the vasodilator effect of isosteviol on arterial strips disappeared, and specific inhibitors for the ATP-sensitive potassium channel and the small-conductance calcium-activated potassium channel suppressed the vasodilator effect of isosteviol in isolated aortic rings [75]. Other studies also suggest a possible role of prostaglandins in the effects of stevioside on blood pressure [76].

With reference to clinical trials, Chan et al. performed a multicenter, randomized, placebo-controlled study in 106 Chinese hypertensive individuals [40]. The treatment arm received capsules of 250 mg of stevioside, three times a day, during 1 year. Stevioside was found to be a safe and effective compound in the treatment of hypertension with a smaller blood pressure-lowering effect than other antihypertensive drugs. The authors suggested its use as a supplementary antihypertensive therapy. In contrast, the same dose of stevioside, administered for 3 days, seems to lack blood pressure-lowering effects in healthy subjects [17]. In another multicenter, randomized, placebo-controlled study, 82 mildly hypertensive Chinese people in the treatment arm received capsules containing 500 mg stevioside 3 times a day for 2 years and showed significant decreases in mean systolic and diastolic pressure compared with baseline and with placebo [77]. Using a randomized, placebo-controlled design, Ferri et al. studied the effect of oral stevioside on a group of previously untreated mild hypertensive patients [78]. The active arm of the study received capsules containing standardized stevioside during 24 weeks, following the next scheme: phase 1 (3.75 mg/kg/day during 7 weeks), phase 2 (7.5 mg/kg/day during 11 weeks), and phase 3 (15.0 mg/kg/day during 6 weeks). Despite systolic and diastolic blood pressure significantly decreased during the treatment with stevioside, a similar effect was observed in the placebo group, concluding that stevioside up to 15.0 mg/kg/day did not show an antihypertensive effect.

250 mg/kg steviol glycosides (92% purity) administered t.i.d. during 3 months failed to display blood pressure-lowering effects on healthy subjects and normotensive and hypertensive diabetic patients [41]. Similarly, multicenter, randomized, placebo-controlled trials assessing the effect of 1000 mg/day of rebaudioside A on type-2 diabetic patients (16 weeks) or healthy volunteers (4 weeks) did not revealed effects on blood pressure [42]. Neither acute stevioside (1000 mg) nor acute rebaudioside A (500, 750, 1000 mg) modified blood pressure in type-2 diabetic subjects [38].

In 2015, Onakpoya and Heneghan published a meta-analysis on double-blind, randomized controlled trials related to possible antihypertensive or hypotensive effects of steviol glycosides [79]. They conclude that the evidence of the analyzed trials suggested that stevioside may generate reductions in blood pressure, but the small sizes of the effects and the substantial heterogeneity of the interventions (discrepancies in the active composition of the stevia intervention, variation in daily dosages of steviol glycosides, and differences in lifestyle adjustments) limited the robustness of the conclusions. The effects seem to be accentuated in long-term studies. They also underlined that there is still no established dose at which these changes occur and it should not be recommended as a substitute for current management of hypertension.

As in the case of other potential therapeutic uses of stevia described in the chapter, analogs of steviol with potent vasorelaxant effects have been obtained through chemical modification [80]. In this very interesting study, Wonganan et al. used the fungus *Cunninghamella echinulata*, to obtain active metabolites of isosteviol and two related compounds. The reported series of analogs display vasorelaxant effects in nM concentrations on the relaxation of phenylephrine-induced contraction in rat aorta.

7 Other Effects

In 2010, Sharma et al. reported that pretreatment with oral stevioside (250 mg/kg) reversed scopolamine-induced learning and memory impairment in rats (assessed through Morris water maze) and attenuated scopolamine-induced rise in brain acetylcholinesterase activity and brain oxidative stress [81]. Later, after reporting the anticonvulsant effects of a series of artificial sweeteners [82], Talevi and coworkers explored the anticonvulsant effects of stevia and confirmed such effect for an aqueous infusion of stevia and isolated stevioside (from 30 mg/kg, intraperitoneally) and rebaudioside A (100 mg/kg, intraperitoneally) [83]. Interestingly, different computer models predicted that the chemical species with intrinsic activity were not the glycosides, but steviol and its phase I metabolites, suggesting steviol glycosides would be acting as a natural prodrug. In line with this hypothesis, while investigating the effects of stevioside on mu opioid receptors, Yang et al. observed that stevioside (1.5 mg/kg) did not affect the feeding behavior of rats when administered intraperitoneally, but, in contrast, the food intake was markedly increased in rats receiving the direct injection of stevioside (0.1 μM, intracerebroventricular) into the brain, and this action was also inhibited by naloxonazine pretreatment [84]. This led the authors to conclude that stevioside has intrinsic activity on mu opioid receptors but it has poor brain bioavailability, which seems consistent with the prediction by Talevi and colleagues suggesting that stevioside or rebaudioside A are not directly responsible for the detected anticonvulsant activity.

Matera et al. reported weak sedative effects of a stevia aqueous leaf extract (mice, intraperitoneally) and antispasmodic effects of aqueous leaf extract and stevioside on rat intestine [85]. The aqueous extract reduced the number of crossed lines (spontaneous locomotion) at 30 min at doses of 40, 100, 450, and 975 mg lyoph/kg and the number of rearings at 450 and 975 mg lyoph/kg. At 1 to 3 mg lyoph/mL, the extract of *Stevia rebaudiana* inhibited the acetylcholine dose–response curve in a noncompetitive way, with an IC50 of 1.22 ± 0.16 mg lyoph/mL. Regarding the pure glycoside, the results showed that it is a noncompetitive inhibitor of acetylcholine and Ca^{2+} influx, with a IC50 on the acetylcholine dose–response curve similar to 1 mM.

8 Some General Considerations to Interpret and Appraise Pharmacological Studies on *Stevia rebaudiana* and Its Constituents

Owing to the increasing public and commercial interest about potential health benefits of stevia, accumulating scientific evidence regarding its possible therapeutic benefits should be read with the utmost care to avoid false or unfounded expectations. A relevant question is whether the interest lies in additional benefits of stevia as food additive (in which case only the studies that use oral doses, equal or below the acceptable daily intake, will be relevant) or as potential source of therapeutic agents (in which case higher doses and alternative routes of administration could be considered). Under a medication regimen, the level of exposure to therapeutic agents rarely exceeds low micromolar concentrations; thus, studies reporting possible therapeutic effects at much higher concentrations (e.g., millimolar concentrations) are not much relevant, simply because it is highly unlikely to achieve such levels in vivo in a therapeutic setting. Note that, in the study by Geuns et al. [17], 750 mg/day were administered during 3 days and the maximum plasma levels observed corresponded approximately to 26 μM concentrations (expressed as steviol equivalents), and all of it as glucuronide.

Identification of the active chemical species is of the highest importance to assess the value of a pharmacological report and to explain interspecies variability. A large body of evidence demonstrates that orally administered steviol glycosides and other stevia extracts are bioactive in certain species. It is also clear from pharmacokinetic studies, however, that due to poor absorption, steviol glycosides are not bioavailable as such, but, depending on the species, as steviol, steviol glucuronides, and other metabolites. So, even if steviol glycosides do have intrinsic activity, it is unlikely that they are directly responsible for in vivo responses to stevia extracts after oral administration. Many reviewed reports show that the intrinsic potency of steviol or its phase I metabolites can be comparable or greater than the potency of its glycosides (for both pharmacokinetic and pharmacodynamics reasons that lie outside the scope of the chapter, glycosides are less likely to present high potency per se). While in rodents steviol appears in plasma in measurable quantities after oral administration of the acceptable daily intake, the same does not apply to humans, probably due to a higher capacity of the phase II biotransformation route (glucuronidation). Consequently, if steviol or any of its phase I derivatives are the chemical entities responsible for a biological activity in rodents, one should not expect to find a correlation in humans unless higher doses or other routes of administration that at least avoid first-pass metabolism are employed. It is known that phase I metabolites can usually be active and sometimes even more active than the parent drug (a process known as bioactivation) [86], and recently this fact has been deliberately exploited to obtain active biotransformation products of steviol and related compounds.

Finally, note that most of the reported potential benefits of stevia and its extracts are in relation with lifelong or chronic conditions (e.g., diabetes, hypertension, epilepsy). Chronic conditions, in general, require long-term treatments. Studies reporting pharmacological effects following acute exposure (acute models) are useful for primary screening purposes (to reveal an ignored bioactivity and provide initial insight into possible action mechanisms), but long-term studies are much more informative and correlate better with the clinical results, a perspective shared by Onakpoya and Heneghan in their recent meta-analysis study on blood pressure-lowering effects of steviol glycosides. In the age of systems pharmacology, it is well known that, for complex health conditions, the evolution of the cell gene signature upon long-term exposure to a drug may be equally (or even more) informative and therapeutically relevant than the initial biological response.

9 Conclusions

An enormous body of preclinical evidence demonstrates the great pharmacological potential of *Stevia rebaudiana* Bertoni and its crude and purified extracts. With the exception of the blood pressure-lowering effects and the effects on fasting plasma glucose, however, the pharmacological effects evidenced at preclinical stage have so far failed to replicate in randomized clinical trials, thus limiting the conclusions we can draw regarding the therapeutic or preventive effects of stevia. Disregarding some fundamental knowledge on steviol glycosides' biopharmaceutical properties and pharmacokinetics may partially explain such disappointing results. First, orally administered steviol glycosides are not absorbed as such, but as the aglycon. Thus, any intrinsic activity of the glycosides will be lost due to lack of absorption and pre-systemic biotransformation. Second, at the most common doses used in clinical trials, most of the aglycon undergoes rapid glucuronidation. Exposure to steviol and its phase I metabolites is very limited or inexistent after oral administration. Thus, if steviol or its phase I metabolites do have intrinsic activity (as they have proved to have in preclinical models), it will go unnoticed owing to their very low bioavailability. Exploitation of steviol beneficial effects could then require extensive revision of the dosage forms, including doses to be administered, route of administration, and use of advanced pharmaceutical carriers which protect from excessive metabolism. As many pharmaceutical chemists have already proven, using the very interesting steviol scaffold as starting point to prepare active analogs with better pharmacokinetic profile could constitute an alternative solution.

10 Cross-References

▶ Cultivation of *Stevia rebaudiana* Bertoni and Associated Challenges
▶ Nonnutritive Sweeteners and Their Role in the Gastrointestinal Tract
▶ *Stevia rebaudiana*'s Antioxidant Properties
▶ Steviol Glycosides: Natural Noncaloric Sweeteners
▶ The Role of Dietary Sugars and Sweeteners in Metabolic Disorders and Diabetes

Acknowledgments The author is a researcher from the University of La Plata and the Argentinean National Council of Scientific and Technical Research (CONICET). The author declares no conflict of interest.

References

1. Joint FAO/WHO Expert Committee on Food Additives (2005) Evaluation of certain food additives. WHO technical report series 928
2. Joint FAO/WHO Expert Committee on Food Additives (2009) Evaluation of certain food additives. WHO technical report series 60
3. Geuns JMC (2009) Stevioside: a safe sweetener and possible new drug for the treatment of the metabolic syndrome. In: Weerasinghe DK, DuBois GE (eds) Sweetness and sweeteners. Biology, chemistry and psychophysics. American Chemical Society, Washington, DC
4. Geuns JMC, Augustijns P, Mols R et al (2003) Metabolism of stevioside in pigs and intestinal absorption characteristics of stevioside, rebaudioside A and steviol. Food Chem Toxicol 41:1599–1607
5. Koyama E, Sakai N, Ohori Y et al (2003) Absorption and metabolism of glycosidic sweeteners of Stevia mixture and their aglycone, steviol in rats and humans. Food Chem Toxicol 41:875–883
6. Lipinski CA (2004) Lead- and drug-like compounds: the rule-of-five revolution. Drug Discov Today Technol 1:337–341
7. Veber DF, Johnson SR, Cheng HY et al (2002) Molecular properties that influence the oral bioavailability of drug candidates. J Med Chem 45:2615–2623
8. Hutapea AM, Toskukao C, Buddhasukh D et al (1997) Digestion of stevioside, a natural sweetener, by various digestive enzymes. J Clin Biochem Nutr 23:177–186
9. Nikiforov AI, Rihner MO, Eapen AK et al (2013) Metabolism and toxicity studies supporting the safety of rebaudioside D. In J Toxicol 34:261–273
10. Wingard JP, Brown F, Enderlin J et al (1980) Intestinal degradation and absorption of the glycosidic sweeteners stevioside and rebaudioside A. Experientia 36:519–520
11. Koyama E, Kitazawa K, Ohori Y et al (2003) In vitro metabolism of the glycosidic sweeteners, stevia mixture and enzymatically modified stevia in human intestinal microflora. Food Chem Toxicol 41:359–374

12. Gardana C, Simonetti P, Canzi E et al (2003) Metabolism of stevioside and rebaudioside A from *Stevia rebaudiana* extracts by human microflora. J Agric Food Chem 51:6618–6622
13. Renwick AG, Tarka SM (2008) Microbial hydrolysis of steviol glycosides. Food Chem Toxicol 46(Suppl 7):S70–S74
14. Roberts A, Renwick AG (2008) Comparative toxicokinetics and metabolism of rebaudioside A, stevioside, and steviol in rats. Food Chem Toxicol 46(Suppl 7):S31–S39
15. Nakayama K, Kasahara D, Yamamoto F (1986) Absorption distribution metabolism and excretion of stevioside in rats. Shokugin Eiseigaku Zasshi 27:1–8
16. Geuns JMC, Buyse J, Vankeirsbilck A et al (2006) Identification of steviol glucuronide in human urine. J Agric Food Chem 54:2794–2798
17. Geuns JMC, Buyse J, Vankeirsbilck A et al (2007) Metabolism of stevioside by healthy subjects. Exp Biol Med (Maywood) 232:164–173
18. Koyama E, Sakai N, Ohori Y et al (2003) Absorption and metabolism of the glycosidic sweeteners, Stevia related compounds in human and rat. Food Chem Toxicol 41:875–883
19. Wheeler A, Boileau AC, Winkler PC et al (2008) Pharmacokinetics of rebaudioside A and stevioside after single oral doses in healthy men. Food Chem Toxicol 46(Suppl 7): S54–S60
20. Srimaroeng C, Chatsudthipong V, Aslamkhan AG et al (2005) Transport of the natural sweetener stevioside and its aglycone steviol by human organic anion transporter (hOAT1; SLC22A6) and hOAT3 (SLC22A8). J Pharmacol Exp Ther 313:621–628
21. Wang M, Qi H, Li J et al (2015) Transmembrane transport of steviol glucuronide and its potential interaction with selected drugs and natural compounds. Food Chem Toxicol 86:217–224
22. Li S, Chen T, Dong S et al (2014) The effect of rebaudioside A on microbial diversity in mouse intestine. Food Sci Technol Res 20:459–467
23. Bastaki S (2015) Pharmacotherapy of nonnutritive sweeteners in diabetes mellitus. Int J Diabetes Metab 23:11–22
24. Jeppesen PB, Gregersen S, Poulsen C et al (2000) Stevioside acts directly on pancreatic beta cells to secrete insulin: actions independent of cyclic adenosine monophosphate and adenosine triphosphate-sensitive K^+-channel activity. Metabolism 49:208–214
25. Abudula R, Jeppesen PB, Rolfsen SE et al (2004) Rebaudioside A potently stimulates insulin secretion from isolated mouse islets: studies on the dose-, glucose-, and calcium-dependency. Metabolism 53:1378–1381
26. Jeppensen PB, Gregersen S, Alstrup KK et al (2002) Stevioside induces antihyperglycaemic, insulinotropic and glucagonostatic effects in vivo: studies in the diabetic Goto-Kakizaki (GK) rats. Phytomedicine 9:9–14
27. Jeppensen PB, Gregersen S, Rolfsen SE (2003) Antihyperglycemic and blood pressure-reducing effects of stevioside in the diabetic Goto-Kakizaki rat. Metabolism 52:372–378
28. Ferreira EB, de Assis Rocha Neves F, da Costa MA et al (2006) Comparative effects of *Stevia rebaudiana* leaves and stevioside on glycaemia and hepatic gluconeogenesis. Planta Med 72:691–696
29. Lailerd N, Saengsirisuwan V, Sloniger JA et al (2004) Effects of stevioside on glucose transport activity in insulin-sensitive and insulin-resistant rat skeletal muscle. Metabolism 53:101–107
30. Rizzo B, Zambonin L, Angeloni C et al (2013) Steviol glycosides modulate glucose transport in different cell types. Oxid Med Cell Longev 2013:348169
31. Bhasker S, Madhay H, Chinnamma M (2015) Molecular evidence of insulinomimetic property exhibited by steviol and stevioside in diabetes induced L6 and 3T3L1 cells. Phytomedicine 22:1037–1044
32. Shivanna N, Naika M, Khanum F et al (2013) Antioxidant, anti-diabetic and renal protective properties of *Stevia rebaudiana*. J Diabetes Complications 27:103–113
33. Chang JC, Wu MC, Liu IM et al (2005) Increase of insulin sensitivity by stevioside in fructose rich chow-fed rats. Horm Metab Res 37:610–616
34. Wang Z, Xue L, Guo C et al (2012) Stevioside ameliorates high-fat diet-induced insulin resistance and adipose tissue inflammation by downregulating the NF-kB pathway. Biochem Biophys Res Commun 416:1280–1285

35. Geeraert B, Crombé F, Hulsmans M et al (2010) Stevioside inhibits atherosclerosis by improving insulin signaling and antioxidant defense in obese insulin-resistant mice. Int J Obes (Lond) 34:569–577
36. Holvoet P, Rull A, García-Heredia A et al (2015) Stevia-derive compounds attenuate the toxic effects of ectopic lipid accumulation in the liver of obese mice a transcriptomic and metabolomic study. Food Chem Toxicol 77:22–33
37. Curi R, Alvarez M, Bazorre RB et al (1986) Effect of *Stevia rebaudiana* on glucose tolerance in normal adult humans. Braz J Med Biol Res 19:771–774
38. Gregersen S, Jeppensen PB, Holst JJ et al (2004) Antihyperglycemic effects of stevioside in type 2 diabetic subjects. Metabolism 53:73–76
39. Anton SD, Martin CK, Han H et al (2010) Effects of stevia, aspartame, and sucrose on food intake, satiety, and postprandial glucose and insulin levels. Appetite 55:37–43
40. Chan P, Tomlinson B, Chen YJ et al (2000) A double-blind placebo-controlled study of the effectiveness and tolerability of oral stevioside in human hypertension. Br J Clin Pharmacol 50:215–220
41. Barriocanal LA, Palacios M, Benitez G et al (2008) Apparent lack of pharmacological effect of steviol glycosides used as sweeteners in humans. A pilot study of repeated exposures in some normotensive and hypotensive individuals and in Type 1 and Type 2 diabetics. Regul Toxicol Pharmacol 51:37–41
42. Maki KC, Curry LL, Reeves MS et al (2008) Chronic consumption of rebaudioside A, a steviol glycoside, in men and women with type 2 diabetes mellitus. Food Chem Toxicol 46(Suppl 7): S47–S53
43. Talevi A (2016) The importance of bioactivation in computer-guided drug repositioning. Why the parent drug is not always enough. Curr Top Med Chem 16(19):2078–2087
44. Yang LM, Hsu FL, Chang SF et al (2007) Microbial metabolism of steviol and steviol-16alpha,17-epoxide. Phytochemistry 68:562–570
45. Jayaraman S, Manoharan MS, Illanchezian S (2008) In-vitro antimicrobial and antitumor activity of *Stevia rebaudiana* (Asteraceae) leaf extracts. Trop J Pharm Res 7:1143–1149
46. Debnath M (2008) Clonal propagation and antimicrobial activity of an endemic medicinal plant *Stevia rebaudiana*. J Med Plant Res 2:45–51
47. Singh S, Garg V, Yadav D et al (2012) In vitro antioxidative and antibacterial activities of various parts of *Stevia rebaudiana* (Bertoni). Int J Pharm Pharm Sci 4:468–473
48. Puri M, Sharma D (2011) Antibacterial activity of stevioside towards food-borne pathogenic bacteria. Eng Life Sci 11:326–329
49. Sharipova RR, Strobykina IY, Mordovskoi GG et al (2011) Antituberculosis activity of glycosides from *Stevia rebaudiana* and hybrid compounds of steviolbioside and pyridinecarboxylic acid hydrazides. Chem Nat Compd 46:902–905
50. Kataev VE, Strobykina IY, Andreeva OV et al (2011) Synthesis and antituberculosis activity of derivatives of *Stevia rebaudiana* glycoside steviolbioside and diterpenoid isosteviol containing hydrazone, hydrazide, and pyridinoyl moieties. Russ J Bioorg Chem 37:483
51. Khaybullin RN, Strobykina IY, Dobrynin AB (2012) Synthesis and antituberculosis activity of novel unfolded and macrocyclic derivatives of ent-kaurane steviol. Bioorg Med Chem Lett 22:6909–6913
52. Sharipova RR, Lodochnikova OA, Strobykina IY et al (2013) Synthesis, structures, and properties of 15-oxoisosteviol thiosemicarbazone and oxime. Russ Chem Bull 62:175–182
53. Huang TJ, Chou BH, Lin CW et al (2014) Synthesis and antiviral effects of isosteviol-derived analogues against the hepatitis B virus. Phytochemistry 99:107–114
54. Moons N, Borggraeve W, Dehaen W (2012) Stevioside and steviol as starting materials in organic synthesis. Curr Org Chem 16:1986–1995
55. Lin LH, Lee LW, Sheu SY et al (2004) Study on the stevioside analogues of steviolbioside, steviol, and isosteviol 19-alkyl amide dimers: synthesis and cytotoxic and antibacterial activity. Chem Pharm Bull (Tokyo) 53:1117–1122

56. Wu Y, Dai GF, Yang JH et al (2009) Stereoselective synthesis of 15- and 16-substituted isosteviol derivatives and their cytotoxic activities. Bioorg Med Chem Lett 19:1818–1821
57. Zhu SL, Wu Y, Liu CJ et al (2013) Synthesis and in vitro cytotoxic activity evaluation of novel heterocycle bridged carbothioamide type isosteviol derivatives as antitumor agents. Bioorg Med Chem Lett 23:1343–2346
58. Ukiya M, Sawada S, Kikuchi T et al (2013) Cytotoxic and apoptosis-inducing activities of steviol and isosteviol derivatives against cancer cell lines. Chem Biodivers 10:177–188
59. Malki A, Laha R, Bergmeier SC (2014) Synthesis and cytotoxic activity of MOM-ether analogs of isosteviol. Bioorg Med Chem Lett 24:1184–1187
60. Strobykina IY, Belenok MG, Semenova MN et al (2015) Triphenylphosphonium cations of the diterpenoid isosteviol: synthesis and antimitotic activity in a sea urchin embryo model. J Nat Product 78:1300–1308
61. Toyoda K, Matsui H, Shoda T et al (1997) Assessment of the carcinogenicity of stevioside in F344 rats. Food Chem Toxicol 35:597–603
62. Yasukawa K, Kitanaka S, Seo S (2002) Inhibitory effect of stevioside on tumor promotion by 12-*O*-tetradecanoylphrbol-13-acetate in two-stage carcinogenesis in mouse skin. Bio Pharm Bull 25:1488–1490
63. Takasaki M, Konoshima T, Kozuka M et al (2009) Cancer preventive agents. Part 8: chemopreventive effects of stevioside and related compounds. Bioorg Med Chem 17:600–605
64. Boonkaewwan C, Ao M, Toskulkao C et al (2008) Specific immunomodulatory and secretory activities of stevioside and steviol in intestinal cells. J Agric Food Chem 56:3777–3784
65. Paul S, Sengupta S, Bandyopadhyay TK (2012) Stevioside induced ROS-mediated apoptosis through mitochondrial pathway in human breast cancer cell line MCF-7. Nutr Cancer 64: 1087–1094
66. Beech EMA, Humboldt G (1981) Cardio-circulatory effects of total water extract in normal persons and of stevioside in rats. Cienciu e Cultura 32:208–210
67. Humboldt G, Beech EMA (1977) Efeito do edulcorante natural, (stevioside) e sintetico (sacarina) sobre o ritmo cardisco em ratos. Arq Bras Cardiol 30:275–277
68. Melis MS, Sainati AR (1991) Effect of calcium and verapamil on renal function of rats during treatment with stevioside. J Ethnopharmacol 33:257–262
69. Melis MS (1992) Stevioside effect on renal function of normal and hypertensive rats. J Ethnopharmacol 36:213–217
70. Melis MS (1995) Chronic administration of aqueous extract of *Stevia rebaudiana* in rats: renal effects. J Ethnopharmacol 47:129–134
71. Melis MS (1996) A crude extract of *Stevia rebaudiana* increases the renal plasma flow of normal and hypertensive rats. Braz J Med Biol Res 28:669–675
72. Hsu YH, Liu JC, Kao PF et al (2002) Antihypertensive effect of stevioside in different strains of hypertensive rats. Zhonghua Yi Xue Za Zhi (Taipei) 65:1–6
73. Liu JC, Kao PF, Ming HH et al (2000) Antihypertensive effect of stevioside derivative isosteviol in spontaneously hypertensive rats. Acta Cardiol Sinica 17:133–140
74. Liu JC, Kao PF, Chan P et al (2003) Mechanism of the antihypertensive effect of stevioside in anesthetized dogs. Pharmacology 67:14–20
75. Wong KL, Chan P, Yang HY et al (2004) Isosteviol acts on potassium channels to relax isolated aortic strips of Wistar rat. Life Sci 75:2379–2387
76. Melis MS, Sainati AR (1991) Participation of prostaglandins in the effect of stevioside on rat renal function and arterial pressure. Braz J Med Biol Res 24:1269–1276
77. Hsieh MH, Chan P, Sue YM et al (2003) Efficacy and tolerability of oral stevioside in patients with mild essential hypertension: a two-year, randomized, placebo-controlled study. Clin Ther 25:2797–2808
78. Ferri LA, Alves-Do-Prado W, Yamada SS et al (2006) Investigation of the antihypertensive effect of oral crude stevioside in patients with mild essential hypertension. Phytother Res 20:732–736

79. Onakpoya IJ, Heneghan CJ (2015) Effect of the natural sweetener, steviol glycoside, on cardiovascular risk factors: a systematic review and meta-analysis of randomised clinical trials. Eur J Prev Cardiol 22:1575–1587
80. Wonganan O, Tocharus C, Puedsing C et al (2013) Potent vasorelaxant analogs from chemical modification and biotransformation of isosteviol. Eur J Med Chem 62:771–776
81. Sharma D, Puri M, Tiwari AK et al (2010) Antiamnesic effect of stevioside in scopolamine-treated rats. Indian J Pharmacol 42:164–167
82. Talevi A, Enrique AV, Bruno-Blanch LE (2012) Anticonvulsant activity of artificial sweeteners: a structural link between sweet-taste receptor T1R3 and brain glutamate receptors. Bioorg Med Chem Lett 22:4072–4074
83. Di Ianni ME, Del Valle ME, Enrique AV et al (2015) Computer-aided identification of anticonvulsant effect of natural nonnutritive sweeteners stevioside and rebaudioside A. Assay Drug Dev Technol 13:313–318
84. Yang PS, Lee JJ, Taso CW et al (2009) Stimulatory effect of stevioside on peripheral mu opioid receptors in animals. Neurosci Lett 454:72–75
85. Matera S, Piersante MV, Ragone MI, Consolini AE (2011) Sedative and antispasmodic effects of *Stevia rebaudiana* and noncompetitive inhibition of intestinal contractility by stevioside. Pharmacol online 1:1–8
86. Fura A (2006) Role of pharmacologically active metabolites in drug discovery and development. Drug Discov Today 11:133–142

Health Implications of Fructose Consumption in Humans

14

Luc Tappy

Abstract

Fructose is a component of natural (sucrose) or industrial (high-fructose corn syrup) sugars. It has a caloric content similar to glucose and a higher sweetening power. All cells of the human body can use glucose as an energy substrate, and most can also use fatty acids. In contrast, most cells cannot use directly fructose, and this substrate needs first to be converted into lactate, glucose, or fatty acids in the gut, liver, and kidneys. Cells of these three organ express a set of fructolytic enzymes: fructokinase, aldolase B, and triokinase which convert fructose into two triose phosphate. Since there is no negative feedback on the activity of these fructolytic enzymes, the ingested fructose is almost completely and immediately metabolized in the gut and the liver. When a large amount of fructose is ingested, splanchnic organs are faced with an overproduction of triose phosphate, which they first release into the systemic circulation as lactate and glucose. When these pathways become saturated, fructose is converted into fatty acids through de novo lipogenesis, and newly synthetized fatty acids are either secreted into the blood as very low density lipoproteins – triglycerides – or temporarily stored as intra-hepatic triglycerides. The metabolic fate of ingested fructose is dependent on whole body energy output. In resting conditions, glucose and lactate oxidation are limited by the low whole body energy expenditure, and excess fructose intake leads to a moderate increase in splanchnic glucose output to increased fasting and postprandial blood triglyceride concentration and to increased intrahepatic fat concentrations. During exercise, whole body energy output is high; fructose and glucose and lactate synthetized from glucose are essentially oxidized in skeletal muscle.

L. Tappy (✉)
Physiology Department, Faculty of Biology and Medicine, University of Lausanne UNIL, Lausanne, Switzerland
e-mail: luc.tappy@unil.ch

J.-M. Mérillon, K.G. Ramawat (eds.), *Sweeteners*, Reference Series in Phytochemistry,
https://doi.org/10.1007/978-3-319-27027-2_29

Keywords
Sweet taste receptors • Fructolysis • Gluconeogenesis • De novo lipogenesis • Intrahepatic fat • Insulin resistance • Nonalcoholic fatty liver disease • Dyslipidemia • Gut malabsorption • Cardiometabolic risk factors

Contents

1 Introduction ... 286
2 Specific Fructose Metabolism and Possible Consequences on Risk Factors for Non-Communicable Diseases ... 287
2.1 Dietary Carbohydrates and Sugars ... 287
2.2 Effects of Sugars on Sweet Taste Receptors and Their Potential Physiological Roles ... 289
2.3 Caloric Content of Dietary Carbohydrates ... 290
2.4 Dietary Intake of Carbohydrate and Sugars in Modern Diets ... 290
2.5 How Does Fructose's Metabolism Differ from That of Glucose? ... 291
2.6 Monosaccharides Can Be Transported into Cells Through Several Families of Hexoses Transporter ... 292
2.7 Splanchnic Fructose Processing ... 293
2.8 Metabolic Fate of Ingested Fructose ... 294
2.9 Consequences Linked to Splanchnic Fructose Metabolism in Subjects Fed with a High-Fructose Diet ... 296
2.10 Effects of Exercise on Fructose Metabolism ... 300
2.11 Proposed Integrated Model of the Effects of Fructose on Cardiometabolic Risk Factors ... 301
2.12 Gender-Related Differences: An Under-Investigated Area of Fructose Metabolism ... 304
2.13 Fructose Malabsorption and Gut-Related Symptoms: A Potentially Underestimated Adverse Effect of Dietary Fructose ... 304
3 Conclusions ... 305
References ... 305

1 Introduction

Fructose is a natural sugar, which is found in its "free" monomeric hexose form in fruits and honey. Our main dietary source for fructose is refined cane or beet sugar, which is constituted of one molecule of glucose linked to one molecule of fructose through a beta 1–2 bound. There is presently much confusion regarding this sugar. On one hand, fructose is naturally present in fruits, the consumption of which is highly encouraged by most national and international nutrition associations [1, 2]. By extension, and because of its name, it is considered as "fruit sugar" and hence as healthy. Agave contains particularly high-fructose content, and agave syrups, which are almost 100% fructose, have found their way to the shelves of healthy food stores. On the other hand, there is increasing concern that refined sugars, and more specifically sugar-sweetened beverages, may play a causal role in the current epidemic of obesity and of associated metabolic diseases such as diabetes, atherosclerosis, and nonalcoholic fatty liver disease. It is further suspected that this is due to pathogenic, "toxic" effects of the fructose component of refined sugar [3]. Even more than sugar, high-fructose corn syrup (HFCS) which is commonly used as a sugar substitute in the confection of sugar-sweetened beverages has been pointed as a prime culprit and as an unhealthy component of foods by many scientists and laypeople [4]. How can the same sugar be "healthy" when present in fresh fruits and become "unhealthy" in refined sugar and frankly toxic when industrially produced from corn starch?

2 Specific Fructose Metabolism and Possible Consequences on Risk Factors for Non-Communicable Diseases

2.1 Dietary Carbohydrates and Sugars

Glucose is the prime energy substrate for almost all cells of the human organism. It can be synthesized from our endogenous protein or amino acids and to a lesser extent from the glycerol component of triglycerides present in our body fat stores. The main source for blood glucose is dietary carbohydrate, however.

Dietary carbohydrates include a variety of chemical compounds constituted of monomers or polymers of a basic unit, the ose, or monosaccharide (Fig. 1).

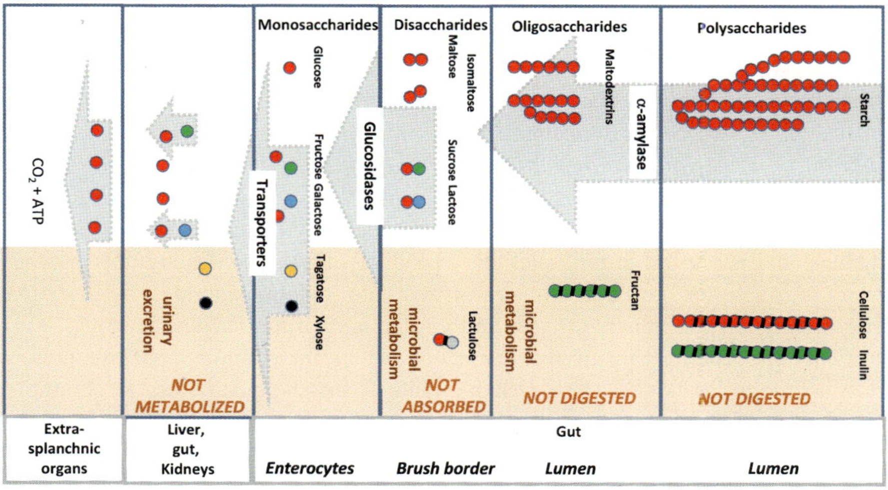

Fig. 1 Digestion and absorption of various dietary carbohydrates in the intestinal lumen during and at the luminal surface of enterocytes. Compounds on the *left* part of the diagram are digestible and absorbable carbohydrates: starch, maltodextrins, and sugars. Compounds on the bottom (light brown) part of the diagram are undigestible carbohydrates and sugars, which are not absorbed as such by the enterocytes (although their fermentation products generated by colonic bacteria may to some extent be absorbed): dietary fibers and undigestible sugars such as lactulose. On this side of the diagram are also represented sugars, which can be absorbed, but are not metabolized by human cells: examples are xylose and tagatose

2.1.1 Monosaccharides

Monosaccharides are polyhydroxy aldehydes or polyhydroxy ketones with usually three to nine carbon atoms. The most common biological monosaccharides are hexoses, made up of six carbon atoms. Glucose, fructose, and galactose are all hexoses, sharing the same formula $C_6H_{12}O_6$; glucose and galactose are aldoses, while fructose is a ketose. They are present in various proportions in fruits and vegetables [5]. Monosaccharides are also present in some processed dietary products, such as fructose-glucose syrup prepared from the hydrolysis and enzymatic conversion of cereal or tuberous starch [6, 7]. These products are denominated "isoglucose" in Europe. In North America, isoglucose is mainly prepared from corn starch and is widely known under the name of high-fructose corn syrup (HFCS).

Glucose, fructose, and galactose, as monosaccharides, can be directly absorbed by enterocytes without requiring the intervention of any digestive enzyme [8]. This is also the case for other rare dietary monosaccharides such as tagatose [9] or xylose [10]. Once absorbed into the systemic circulation, these two compounds however cannot be metabolized by human cells and undergo renal excretion.

2.1.2 Sugar Alcohols

The compounds included in this class have a chemical structure close to monosaccharides but lack the aldose or ketone function. They are naturally present in small amounts in fruits, and some of them can be synthesized from glucose within the human body [11]. They share with mono- and disaccharide the ability to activate sweet receptors and, for practical reasons, are included in "dietary carbohydrates."

2.1.3 Disaccharides

Monosaccharides can be polymerized into larger units, called polysaccharides in plants. Dimers are called disaccharides. The most common dietary disaccharides are sucrose (glucose-fructose) and lactose (glucose-galactose). Sucrose is present in large quantities in sugar beets and sugar cane, from which it is refined to yield table sugar. Lactose is almost exclusively produced by mammary glands of mammalians and is present in milk and dairy products. Disaccharides are not absorbed intact through the gut but are hydrolyzed into monosaccharides by digestive enzymes synthesized in enterocytes and located at their luminal surface [12].

Other disaccharides are also present in our diet, as, for example, trehalose, present in mushrooms or honey. Other can be synthesized chemically, as, for example, lactulose, which is made up of one molecule of galactose and one molecule of fructose. Lactulose cannot be hydrolyzed by brush border enzymes and hence is not absorbed by the small intestine. As an osmotically active substance, its presence in the small intestinal lumen attracts water and produces osmotic diarrhea. Lactulose then moves to the large intestinal where it undergoes bacterial fermentation [13].

2.1.4 Polysaccharides

Polysaccharides are polymers made up of 3 to more than 10^4 monosaccharide units. They are present in large amounts in vegetables, fruits, cereals, and pulses, which synthesize them from glucose and sucrose produced by photosynthesis.

Polysaccharides represent energy stores for the plant but are also major constituents of vegetal cells walls. From a nutritional perspective, plant polysaccharides are classified as digestible and nondigestible carbohydrate.

Digestible polysaccharides are almost exclusively represented by starch, made up of polymers of glucose linked through alpha 1–4 and alpha 1–6 bounds. It can be digested by salivary and pancreatic amylases, which specifically hydrolyze every second alpha 1–4 bounds, thus yielding maltose, maltotrioses, and short polymers of glucose called maltodextrins (made up of 3–9 units of glucose) [14]. Amylases leave alpha 1–6 bounds intact and hence do not digest isomaltose (made up of two molecules of glucose linked through an alpha 1–6 bound) or ramified maltodextrins. Maltodextrins are present in very small amounts in natural foods but can be produced industrially and are included in products such as infant formulas and sport drinks. Maltodextrins, maltose, isomaltose, and triomaltose are further digested by digestive enzymes present at the luminal pole of enterocytes, thus eventually releasing glucose for absorption.

Nondigestible polysaccharides are polymers of monosaccharides, which resist digestion by amylase. Plants synthesize of variety of such compounds, as, for instance, cellulose, made up of chains of glucose linked through beta 1–4 bounds, or inulin, and made up of chains of fructose. Nondigestible polysaccharides are called "dietary fibers" and transit through the proximal gastrointestinal tract essentially unchanged, until they reach the large intestine where they undergo bacterial fermentation.

2.2 Effects of Sugars on Sweet Taste Receptors and Their Potential Physiological Roles

From a sensorial perspective, sugars are characterized by the sweet taste they elicit when ingested. Sweet taste is produced by activation of one single sweet taste receptor, T1R2/T1R3, which is expressed in gustatory cells located in the oral mucosa. Many monosaccharides, disaccharides, and oligosaccharides when present in solute form in the oral cavity activate sweet receptors and hence may be referred to as "sugars" [15]. The intensity of the sweet sensation produced varies between sugars. Sweetness depends not only on sugars concentrations but also of physico-chemical properties of the fluid in which they are dissolved and on the presence of other dietary components [16, 17]. Generally, however, fructose is characterized by a slightly higher sweetness than sucrose, while sweetness relative to sucrose is lower with glucose, polyols, maltose lactose, and maltodextrins. Sweet receptors can also be activated by various natural or artificial substances not belonging to the oses' groups and the metabolism of which does not release free energy in humans. These are referred to as noncaloric sweeteners.

Large polysaccharides do not activate sweet taste receptors by themselves. However, mastication of starchy foods can be associated with a mild sweet sensation due to partial digestion of starch by salivary amylase, releasing maltodextrins, maltotrioses, and maltose in the oral cavity.

Sweet taste receptors are present in many animal species. The sweet sensation elicited by ingestion of sugars is associated with an activation of the dopaminergic reward pathways located in the brain. Sweet taste is generally resented as pleasant, and hence the presence of sweet receptors may have the function of signaling the presence of energy-rich, readily digestible substrate and reinforcing consumption of sugary foods [18]. This response is however likely to be modulated by individual experience, since subjects with hereditary fructose, who experience severe adverse effect when exposed to even small amounts of fructose, spontaneously learn to avoid sweet products (and interestingly have a very low incidence of dental caries!) [19, 20].

Interestingly sweet taste receptors are not exclusively expressed in oral taste buds but are also present at the surface of intestinal cells, and even on non-digestive cells, such as pancreatic endocrine cells, endothelial cells, or some brain cells. Their function remains presently hypothetical. Stimulation of sweet taste receptors on intestinal endocrine cells has been shown to enhance glucoincretin release and hence potentiate glucose-induced insulin secretion. Brain sweet taste receptors have been proposed to play a role in food intake control [15]. The physiological ligands which activate sweet receptors located outside the oral cavity have however not been clearly identified. In the gut, one may speculate that maltose and glucose issued from starch digestion may have a larger impact than dietary sugars. In non-digestive cells, glucose would most likely be the most prominent sugar able to activate sweet receptor. How this contributes to overall glucose and energy homeostasis remains an open question.

2.3 Caloric Content of Dietary Carbohydrates

Oxidation of glucose to CO_2 and water in cells releases 3.73 kcal/g. The same is true for fructose and galactose. Pure anhydrous glucose therefore contains 3.73 kcal/g. One gram of disaccharides (sucrose, lactose) will release 1× g monosaccharides and hence contain 3.9 kcal/g. One gram starch will release 1× g monosaccharide, corresponding to a starch energy content of 4.1 kcal/g [21] .

Dietary fibers have so far been considered as containing no "metabolizable" energy. There is however increasing evidence that their fermentation by gut microbiota releases substantial amounts of short-chain volatile fatty acids (acetate, lactate, propionate, butyrate), which are absorbed and used as energy substrate [22]. The total amount of energy actually made available from dietary fibers remains presently disputed.

2.4 Dietary Intake of Carbohydrate and Sugars in Modern Diets

Macronutrients of human diets have long been essentially dependent on social organization and local availability of foods. Accordingly, there are tremendous differences between traditional diets in various part of the world, ranging from very high-fat diets in Inuit communities to very high-carbohydrate, high-fiber diets in native American and African populations living inland or to high-unsaturated fat intake in the Mediterranean area [23–25]. It has also shown continuous changes throughout history [26]. One can postulate that early hunter-gatherers' diet relied largely on animal protein and fat and on a mixture of complex and simple sugars found in wild vegetables, fruits, and berries. In the Neolithic period, the development of agriculture had for consequence that human diets became increasingly based on cereals and hence on complex carbohydrate. Cereal products remained a major building block of human diet in Europe, with little intake of sugars from fruits vegetables and honey until sugar became largely available in the colonial times. Since then, consumption of sugars has increased tremendously. Sugar consumption in Europe and North America and the Middle East is currently estimated to represent up to 15–20% of average total energy intakes [27–29]. This represents a more than 50-fold increase within three centuries. Major sources of dietary sugar intake are sugar-sweetened beverages, processed cereal and dairy products, fruit and vegetables, and sugary snacks, ice creams and candies.

2.5 How Does Fructose's Metabolism Differ from That of Glucose?

Glucose was the key energy substrate in early forms of aerobic life and remains a prime energy substrate for many cell types and in many specific conditions. Brain cells can use (almost) only glucose and glucose-derived compounds such as pyruvate or lactate for its energy production. However, glucose cannot be stored in any large amounts due to the osmotic effects of free glucose in solution, and to the large volume, it occupies within the cells where it is stores as glycogen. As a consequence to these limitations, fat emerged as the main energy stores in many animals due to the fact that it has no osmotic effect and that it has a high-energy density allowing to store large amount of energy without excessive increases in body volume and weight. To get an idea, storing about 1000 kcal as glycogen requires that 250 g glycogen + about 1 L of water be stored for a total 1.25 kg increase in body weight. In contrast, the same amount of energy can be stored as about 120 g fat. Evolution of energy storage as fat was of course associated with the development of fat-oxidation pathways in addition to glucose oxidation pathways in nearly all cells. In humans, only a small number of "special" cells, such as neurons, glial cells, and erythrocytes, use almost exclusively glucose, while all other cells can indifferently use glucose or fatty acids as an energy source.

Glucose and fatty acids can therefore be coined as "primary energy substrates" for humans. In contrast, all other energy-containing substrates produced from food digestion, such as amino acids, fructose, galactose, or alcohol, cannot be directly used as energy substrate by human cells. This may be explained by the fact that highly specialized cells which constitute our organs cannot afford to express a full set of all metabolic enzyme in addition to the many specific proteins they need to carry out their specialized functions. As a consequence, non-primary substrates, or "subsidiary substrates," have first to be processed in specialized metabolic organs where they are converted into primary substrates. Such processing of subsidiary substrates is indeed one of the major metabolic function of the liver, which converts many amino acids into glucose and which can in special circumstances convert the carbon skeleton of carbohydrate or amino acids into fat through de novo lipogenesis [30].

After ingestion of sucrose, its glucose component is absorbed in the hepatic portal blood and reaches first the liver. A small part of it is taken up by the liver, but the rest reaches the systemic circulation. This leads to a rise in arterial blood glucose concentration, which subsequently triggers the secretion of insulin. Hyperinsulinemia in turn shuts down cellular fat oxidation and promotes glucose transport, oxidation and storage in many cell types, including the muscle and adipose tissue.

In contrast, the fructose portion of sucrose is absorbed as free fructose in the hepatic portal blood but is almost completely extracted and metabolized in splanchnic tissues. This occurs first to some extent in the gut and then in the liver. Only a small amount of fructose escapes gut and liver uptake and reaches the systemic circulation. The small amount of fructose present in arterial blood is then thought to be metabolized by either kidney proximal tubule cells or by gut and liver cells at second pass. The metabolism of fructose in these specialized metabolic cells is made possible by the presence in them of specific fructolytic enzymes: fructokinase, aldolase B, and triokinase [31]. These enzymes, which are not expressed in other cell types, allow the rapid degradation of fructose into triose phosphate, i.e., dihydroxyacetone phosphate and glyceraldehyde 3-phosphate, which are normal metabolic products of glucose. Unlike glycolysis, the conversion of fructose into triose phosphate is not dependent on insulin, and hence is not regulated by insulinemia, as is glucose metabolism. In addition, fructokinase and aldolase B have a high affinity for fructose a high maximal velocity and are not inhibited by cellular fructose metabolites or ATP or citrate [31]. As a consequence, the major portion of fructose ingested is immediately taken up in fructose-metabolizing cells, where it is immediately converted into triose phosphate irrespective of the cells energy need.

2.6 Monosaccharides Can Be Transported into Cells Through Several Families of Hexoses Transporter

All cells express membrane glucose transporters of the GLUT family. These transporters operate a facilitated diffusion of glucose across membranes according to glucose's gradient concentration. Major isoforms of this family of receptors are GLUT1 and GLUT3, expressed in brain cells, fibroblasts, and many other cell types; GLUT4, which are present in skeletal muscle fibers and adipose tissue and operate an insulin-sensitive glucose transport; or GLUT2, which is present in glucose-sensing cells such as pancreatic endocrine beta-cells, liver cells, and neurons located in discrete brain areas. Although GLUTs show substrate specificity, GLUT2 can transport either glucose or fructose [32].

Sodium-*gl*ucose *t*ransporters are expressed in a limited number of specialized cells and operate hexose transport mechanisms irrespective of intracellular-extracellular glucose concentration gradient [33]. Pumping hexoses sometime against their concentration gradient is made possible because these transporters operate a secondary active transport, i.e., hexose/sodium transport is driven by low intracellular sodium concentration maintained inside cells by means of the continuous active extrusion of sodium from the cells by membrane Na-K ATPase pumps. One isoform of this family, SGLT2, is expressed at the apical surface of small bowel enterocytes and cotransport glucose or galactose and sodium from the intestinal lumen into the enterocytes. Another isoform, SGLT1, is expressed in tubular kidney cells and operate the reabsorption of filtered sodium from primary urine into proximal renal tubular cells. Glucose then diffuses into the blood stream through GLUT2 transporters located at the capillary pole of the cells.

SGLT2 does not transport fructose, and fructose absorption from the gut relies on mechanisms distinct from those used for glucose. Unexpectedly, one transporter of this family, GLUT5, ultimately turned out to be the specific transporter involved in fructose absorption from the intestinal lumen into the enterocytes [34]. In contrast, the efflux of fructose from the enterocyte, or the uptake of fructose from the portal vein into hepatocytes, is operated by the same transporter GLUT2 than glucose [35].

Intestinal fructose absorption therefore requires that fructose concentration inside the enterocyte remains relatively low. This may be one of the reasons why enterocyte express fructolytic enzymes and immediately degrade part of the absorbed fructose into triose phosphate (Fig. 2).

Surprisingly, GLUT5 transporters are also expressed in some cells located outside the splanchnic area, such as subcutaneous adipose tissue or some neurons of brain nuclei [34]. Their function remains unknown, since these cells receive their blood supply from systemic arteries and hence from arterial blood in which fructose concentration remains very low compared to fructose. Furthermore, these cells have so far not been shown to express fructolytic enzymes, but classical hexokinases, which have an affinity for fructose several fold slower than for glucose. The functional roles of these GLUT5 expressed in these areas, as well as the metabolic and physiological effects of dietary fructose on the cells, which express them, remain presently controversial.

2.7 Splanchnic Fructose Processing

Splanchnic organs metabolizing fructose express either GLUT2 (liver, kidney) or GLUT5 (small bowel, kidney) membrane fructose transporters [34, 36]. Fructose transported into these cells is metabolized by fructokinase and aldolase B into fructose-1-phosphate and triose phosphates. Due to the low Km of fructokinase for fructose and its high Vmax, intracellular fructose, and the absence of negative feedback mechanisms on fructolysis, intracellular fructose is then completely and rapidly converted into triose phosphate. Triose phosphate are then either oxidized in the tricarboxylic acid cycle to produce energy locally or channeled into lactate and glucose production or into de novo lipogenesis to synthesize fatty acids. Lactate is essentially secreted into the blood stream, while glucose is either secreted or, in the liver, stored as glycogen. Newly synthesized fatty acids are either stored temporarily

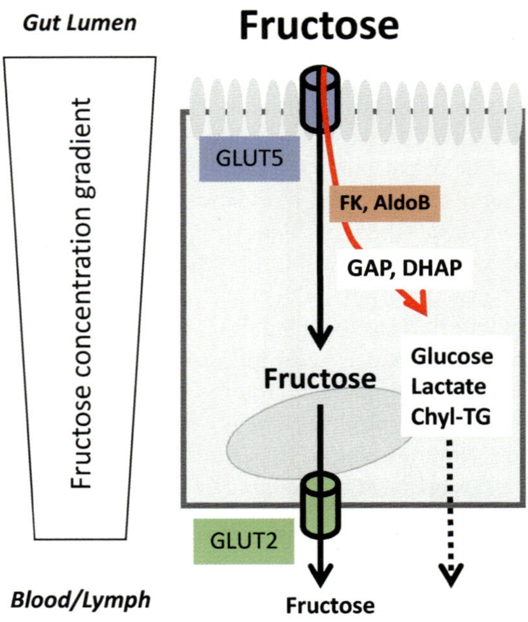

Fig. 2 Absorption of fructose in small intestinal enterocytes. Fructose present in the gut lumen near is transported into enterocytes by facilitated diffusion through GLUT5 located in the enterocytes'apical membrane. Intracellular fructose is then transported into the blood by facilitated diffusion through GLUT2 located in the basolateral portion of enterocytes. Gut fructose absorption therefore requires the presence of a gut lumen to blood fructose gradient (luminal > intraenterocyte >blood fructose concentrations); the presence of fructolytic enzymes within he enterocyte, by converting part of intracellular fructose into lactate, glucose, or fat, contributes building up this fructose gradient

as intrahepatic triglycerides or secreted as VLDL triglycerides. In order to carryout these metabolic processes, small bowel enterocytes, hepatocytes, and kidney tubule cells also express, in addition to fructolytic enzymes, a complex set of proteins coding for metabolic enzymes which are not expressed in most other cells of the organism (Fig. 3). These include glucose-6-phosphatase, which is required for free glucose release into the blood stream, de novo lipogenic enzymes such as fatty acid synthase, and fatty acid elongases, apolipoprotein B48 or B100, and enzymes operating the assembly and secretion of lipoproteins [30].

2.8 Metabolic Fate of Ingested Fructose

The relative contributions of splanchnic lactate and glucose production, hepatic and extrahepatic glycogen synthesis, intrahepatic lipid storage, and secretion of triglyceride-rich VLDLs in response to ^{13}C-labelled fructose ingestion have been reported in several human and animal studies (reviewed in [37]). It is generally accepted that the fate of a pure fructose load fructose is partitioned into about 15% hepatic

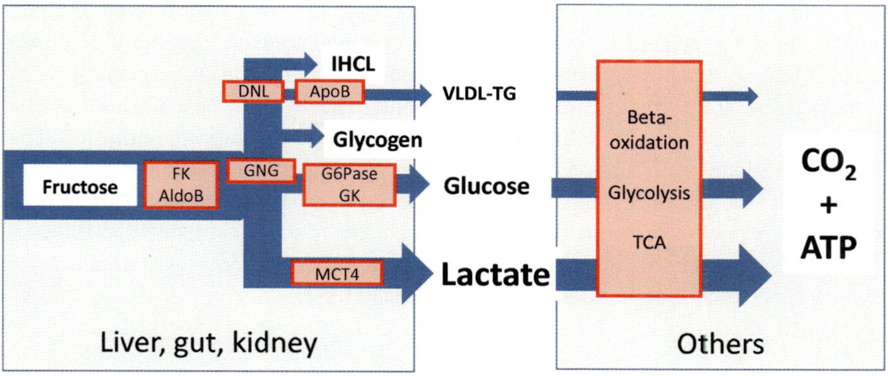

Fig. 3 Enzymatic machinery of splanchnic fructose-metabolizing cells. Small intestinal enterocytes, hepatocytes, and renal proximal tubule cells all express sets of enzyme which are not synthesized in most other cell types, i.e.: (1) fructolytic enzymes, (2) gluconeogenic enzymes and glucose-6-phosphatase, and (3) lipogenic enzymes, and enzymes involved in the secretion of apoB-containing lipoproteins

glycogen, 50% hepatic glucose production, and 25% lactic acid production in healthy resting subjects and that the remaining 10% correspond to splanchnic fructose oxidation and lipid synthesis in fasted, resting normal subjects. In addition, 50% of fructose carbons are recovered as breath $^{13}CO_2$ over the 4–6 h following its ingestion. This represents the sum of fructose oxidation in splanchnic organs and of oxidation of fructose-derived blood glucose and lactate in extrasplanchnic tissues. This also suggests that 5% of fructose carbons are temporarily stored within the body, presumably mainly as hepatic and muscle glycogen but possibly also as hepatic and adipose fat. After ingestion of a mixed meal containing isocaloric amounts of glucose and fructose, recovery of labelled fructose carbons in breath $^{13}CO_2$ and endogenous glucose production remain the major pathways for fructose disposal, while de novo lipogenesis does not appear to be markedly enhanced compared to fructose alone [38].

Of importance, the relative contributions of these various fructose disposal pathways are markedly affected by physical exercise. In subjects ingesting ^{13}C-labelled fructose during strenuous aerobic exercise, nearly 100% labelled fructose carbons were recovered as breath $^{13}CO_2$, indicating that ingested fructose was completely oxidized to produce energy. Interestingly, fructose energy was indirectly made available to skeletal muscle as both blood lactate and blood glucose.

The mechanisms regulating the partition of fructose carbons into blood glucose/hepatic glycogen, blood lactate, and intrahepatic fat/VLDL triglyceride remain unknown. Noteworthy, fructose conversion into glucose or lipids is associated with, respectively, about 10 and 25% the energy content of fructose being lost, respectively. It is therefore tempting to speculate that lactate is the preferential pathway for fructose disposal, that glucose will be produced when lactate production is close to maximal, and that de novo synthesis from fructose will become activated when the two former pathways are saturated. This hypothesis would imply that splanchnic lactate production is quantitatively limited through yet unknown mechanisms, possibly related to the kinetics and maximal transport capacity of monocarboxylate-4 transporter, which are the lactate carriers presumably involved in lactate efflux from fructose-metabolizing cells. The hypothesis also implies that glucose production is quantitatively limited, at least under resting conditions. That splanchnic glucose production hardly increases after ingestion or intravenous administration of fructose in resting subjects has indeed been demonstrated. Although the mechanisms remain incompletely understood, there is some evidence that this may be related to the relative insulin to glucagon ratio in portal blood regulating glucose exit of the cells at the glucose-6-phosphatase step [39].

2.9 Consequences Linked to Splanchnic Fructose Metabolism in Subjects Fed with a High-Fructose Diet

Many short-term studies have addressed the metabolic effects of a high-fructose diet in healthy and in obese and overweight subjects. A complete description of these effects can be found in several well-documented reviews [40–42] and will only be summarized here.

2.9.1 Effects on Blood Triglycerides

Almost all studies performed in healthy normal weight or overweight subjects and in patients with type 2 diabetes concord to show that a high-fructose diet increases blood triglyceride concentration. Dietary fructose mainly affects triglyceride circulating with the VLDL lipoprotein fraction. It increases both fasting and postprandial triglyceride concentration, most likely through distinct mechanisms. Fasting hypertriglyceridemia may be related to increased hepatic de novo lipogenesis and enhanced secretion of VLDL triglyceride from the liver [43]. Of interest, this effect is observed within a few days of fructose exposure and remains thereafter unchanged [44]. Postprandial hypertriglyceridemia may result, not only from hepatic but also from intestinal de novo lipogenesis, and hence from secretion of both VLDL and chylomicrons triglycerides [38, 45]. In addition, fructose may impair the extrahepatic clearance of triglyceride-rich lipoproteins. This may be in part related to the fact that fructose does not increase insulin secretion and hence that a blunted postprandial hyperinsulinemia would ineffectively activate adipose-tissue lipoprotein lipase [46]. Alterations of lipoprotein structure and function have also been proposed to play a role [47].

Fructose-induced hypertriglyceridemia has been extensively documented, and several meta-analyses indicate that it occurs with dietary fructose intake higher than 50–60 g/day for postprandial triglycerides and higher than 100 g/day for fasting triglycerides [48, 49]. Furthermore, these effects are primarily observed in subjects fed both a high amount of fructose and an excess energy relative to needs [50]. It is however related to fructose per se since short-term overfeeding with fructose increases fasting blood triglyceride within a few days while overfeeding excess energy, as fat has no such effects [51]. Furthermore, replacing starch with a high

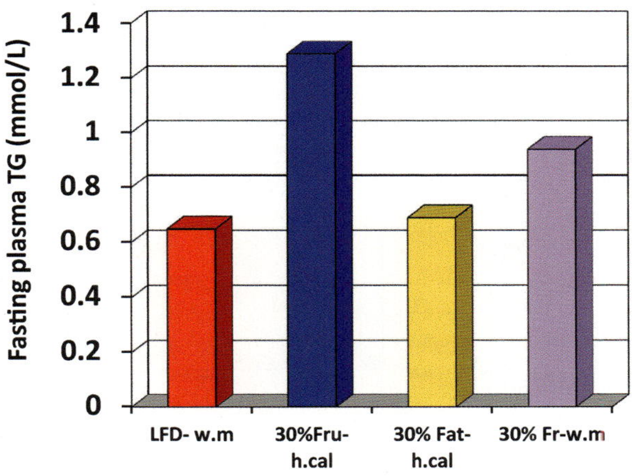

Fig. 4 Effects of dietary fructose and/total energy intake on fasting blood triglyceride. Fasting blood triglyceride concentration with a low-fructose diet providing weight-maintenance energy levels (LFD-w.m), a 30% fructose diet "hypercaloric" (30%Fru-h.cal) providing 30% excess energy provided as fructose, a 30% fat diet "hypercaloric" (30%Fat-h.cal) providing 30% excess energy provided as fat, or a 30% fructose diet providing weight-maintenance energy levels (30%Fr-w.m). Data are from Refs. [52–54]. Duration of high-fructose diet consumption ranged from 4 to 7 days; total energy as fat and fructose

amount of fructose in the diet without increasing total energy intake slightly but significantly increased fasting blood triglyceride [52] (Fig. 4).

2.9.2 Effects on Intrahepatic Fat Stores

An increase in dietary fructose intake has been shown to increase intrahepatic fat content in both normal weight and overweight/obese subjects. The main mechanisms identified so far for fructose-induced hepatic fat accumulation are stimulation of hepatic de novo lipogenesis together with an inhibition of intrahepatic fat oxidation [55]. The effect is observed with high daily fructose doses (higher than 100 g/day); however even with such high doses [56] (Fig. 5), intrahepatic fat increases about threefold in normal weight people, but absolute intrahepatic fat concentrations remain markedly lower than what is observed in most patients with the metabolic syndrome.

One obvious limitation of all high-fructose diet trials is a short duration of exposure to fructose, ranging from a few days to 12 weeks. Some authors express their concern that increases of intrahepatic fat concentrations (and of other cardiometabolic risk markers, such as hypertriglyceridemia and insulin resistance) may develop slowly over months or even years. This concern is well founded since high-fructose diets are often also hypercaloric and hence can be expected to be associated with a progressive gain of weight. There is therefore legitimate concern that dietary fructose, when associated with chronically increased energy intake relative to requirements, will cause the progressive development of excess body fat mass and by the same token of metabolic complications associated with excess body fat, such as hepatic steatosis and insulin

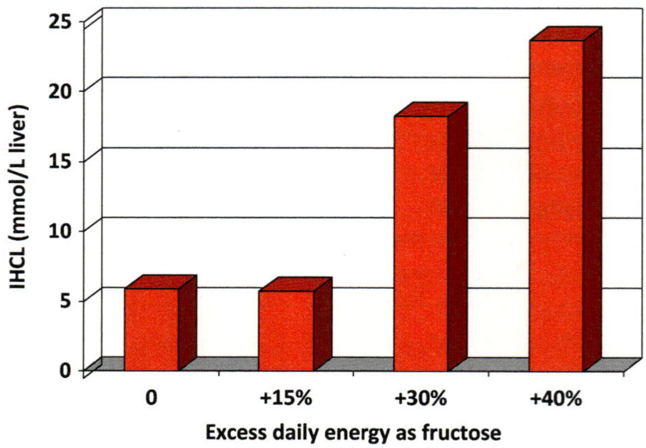

Fig. 5 Effects of fructose overfeeding on intrahepatic fat concentration. Intrahepatocellular lipid concentrations (expressed in millimole lipid per liter of liver), measured by ^{1}H-MRS spectroscopy, in normal weight subjects consuming a low-fructose, weight-maintenance diet (0) or 15%, 30%, and 40% excess daily energy as fructose for 6–7 days. Data redrawn from Ref. [56]

resistance. This concern is even strengthened by some studies suggesting that consuming excess calories as fructose may favor the extension of visceral rather than subcutaneous fat stores and hence favor the development of metabolic complications associated with visceral obesity. This effect of fructose to promote visceral fat deposition was only reported in two small sized studies [57, 58]. In one of these studies, it was observed only in males [57].

For obvious ethical reasons, there is no intervention study comparing the very long-term effects of a high-fructose hypercaloric diet and high-fat or high-glucose hypercaloric diets nor of a weight-maintenance, high-fructose diet. Furthermore, there is no study documenting that the effects of a high-fructose diet is actually associated with a progressive increase in cardiometabolic risk markers with time. At the opposite, in one 4-week study, fructose overfeeding increased blood triglyceride concentration during the first week; during the next 3 weeks, blood triglyceride remained elevated, but did not further increase [44]. Similarly, there was no progressive increase in intrahepatic fat over time [44]. Furthermore, postulating that alterations of metabolic parameters such as blood triglyceride, insulin sensitivity, or intrahepatic fat require several months to years of exposure to fructose is at odds with the general concept of metabolic adaptations to changes in nutritional composition. Thus, it has been well documented that protein oxidation will adapt within a few days to changes in dietary protein intake and carbohydrate:fat oxidation ratio to changes in dietary carbohydrate content [59, 60] but then will remain steady after having reached new equilibrium values. One would therefore expect the same type of response to a high-fructose diet.

There remain some controversies regarding the respective roles of fructose and excess energy intake in increasing intrahepatic fat. Several studies showed that overfeeding healthy nonobese subjects with glucose, fructose, or fat resulted in the same increase [61, 62], indicating that excess energy intake, more than fructose per se, may be involved. The effects of fructose ingested together with a weight-maintenance diet remain controversial. One short-term study observed that it increased intrahepatic fat [63], while another one failed to observe any significant increase [61].

Beside its effect on intrahepatic fat storage, there is also concern that fructose may exert additional deleterious effects on the liver. Fructose has indeed been shown to produce an oxidative stress, which may promote subacute liver inflammation. Whether this may promote the progression of nonalcoholic fatty liver disease from benign steatosis to more aggressive steatohepatitis remains to be evaluated [64].

2.9.3 Effects Intravenous Fructose on Insulin Sensitivity

It has been recognized long ago that the intravenous fructose infusion fails to increase hepatic glucose production in fasting subjects [65] but acutely increases hepatic glucose production when administered together with intravenous glucose and insulin [66, 67]. These apparently contradictory observations can be readily explained when looking at splanchnic fructose metabolism in an integrated perspective:

- In fasting conditions, increased intracellular triose phosphate concentration activates gluconeogenesis and increases glucose-6-phosphate concentrations. Glucose export from the cell is however regulated by the relative fluxes through glucose-6-phosphatase and glucokinase. Any increase of glucose-6-phosphate flux through glucose-6-phosphatase results in an increase in free glucose concentration, which in turn increases glucose flux through glucokinase. As a consequence, net glucose efflux remains constant, and the release of glucose synthesized from fructose carbons into the systemic circulation is compensated by a net decrease of glycogen breakdown [68].
- During glucose and insulin infusion, hyperinsulinemia simultaneously inhibits of glycogenolysis and gluconeogenesis in splanchnic organs, resulting in an inhibition of splanchnic glucose output.
- During fructose administration, however, hyperinsulinemia is not sufficient to prevent the stimulation of gluconeogenesis induced by the unregulated, massive flux of triose phosphate, and net splanchnic glucose output increases [66, 69].

2.9.4 Effects High-Fructose Diet on Hepatic Insulin Sensitivity

The same mechanism, which regulates glucose production during IV fructose, appears to also effective with oral fructose. As a consequence, consumption of a high-fructose diet does not increase fasting plasma glucose and insulin concentration, suggesting minimal alterations of fasting glucose homeostasis. In contrast, it blunts the inhibition of endogenous, splanchnic glucose production by insulin [43, 70]. Such impaired inhibition of hepatic glucose production by insulin clearly bears some similitude with the hepatic insulin resistance observed in type 2 diabetes mellitus. However insulin-resistant subjects with type 2 diabetes mellitus in addition display increased fasting hyperglycemia and hyperinsulinemia and increased fasting endogenous glucose production.

2.9.5 Effects High-Fructose Diet on Extra-Hepatic Insulin Sensitivity

In contrast to rapid changes in hepatic insulin sensitivity, almost all studies which directly assessed extrahepatic effects of insulin (whole body glucose uptake in response to high physiological hyperinsulinemia) failed to show any significant reduction after a few days or weeks on a high-fructose diet (reviewed in [71]). That fructose per se does not impair insulin-stimulated glucose transport in extrasplanchnic organs is completely expected, since blood fructose concentration is very low even after ingestion of very large fructose loads. Furthermore, extrasplanchnic organs contain no fructokinase, and their hexokinases have much lower affinity for fructose than glucose. As a consequence, they are presently considered not to take up and metabolize fructose, which, by the same token, is unlikely to exert direct effects on them.

2.9.6 Other Effects of Fructose

Besides its effect on glucose and lipid homeostasis, fructose has been suspected to be associated with many other metabolic and hemodynamic alterations such as hyperuricemia, endothelial dysfunction, high blood pressure, and many others, which will not be reviewed here [72]. Fructose has also been suspected to have lower satiating effects than other carbohydrate due to the fact that it does not increase insulin secretion and that it does not stimulate release of gut peptides PYY and GLP-1, which normally inhibit food intake [73]. Whether these mechanisms actually lead to substantial overfeeding when fructose is present in everyday diet remain controversial, other studies have documented that a liquid fructose preload significantly reduces subsequent food intake in both lean and overweight subjects [74, 75]. There is also growing evidence that fructose may impact gut and metabolic health by altering gut microbiota. Detailed critical description of these effects can be found in a recent review [76].

2.10 Effects of Exercise on Fructose Metabolism

The energy expenditure increases in proportion with energy intensity, or muscle power output, and is met by increased oxidation of both glucose and fatty acids. There is however ample evidence that glucose is the prime energy substrate for high-intensity exercise. Studies relying on administration of ^{13}C-labelled substrate have however shown that pure fructose can be used as an alternate substrate during exercise [77], although with suboptimal effects on exercise performance. More recently, it has been observed that maximal carbohydrate oxidation was higher in subjects ingesting mixtures of glucose and fructose than pure fructose during a high-intensity exercise [78]. This effect has been attributed to two main factors; first, gut glucose absorption may be saturated at the high rate of glucose administration used during exercise, with the consequence that any further increase in glucose rate fails to efficiently increase glucose absorption into the blood [78]. Second, muscle glucose uptake may be near maximal during exercise, but fructose converted into blood lactate may provide additional, non-glucose carbohydrate for oxidation in working muscles [79].

The pathways used for disposing of fructose in splanchnic organs are markedly influenced by exercise: fructose ingested immediately before an exercise was almost completely metabolized to CO_2 during exercise, while a substantial portion of fructose administered in resting subjects and in subjects immediately after an exercise was stored, presumably as liver and muscle glycogen. In both conditions, there was some fructose that was converted into fat in the liver, but the amount of fructose undergoing this pathway was likely minor compared to glucose and lactate production [80].

Of special interest, consumption of a high-fructose diet increased hepatic de novo lipogenesis and blood triglyceride concentrations when study subjects were maintained under strictly sedentary conditions, but these effects were completely offset when subjects simultaneously performed daily physical exercise [52, 81, 82]. Furthermore, exercise reverted these effects even when the excess energy expended with exercise was balanced by increased food intake (including increased fructose intake compared to sedentary conditions) [52]. This strongly suggests that the effects on fructose cardiometabolic risk markers is dependent on both dietary fructose and energy intake on one hand and on whole body total energy output on the other hand.

2.11 Proposed Integrated Model of the Effects of Fructose on Cardiometabolic Risk Factors

The increase in blood triglyceride concentrations, intrahepatic fat concentrations, and splanchnic glucose production induced by high dietary fructose intakes is related to the intra-splanchnic conversion of fructose into glucose and fatty acids in the gut, liver, and kidney. They result in an imbalance between the rates of triose phosphate synthesis and oxidation in these organs. Of importance, fructose conversion into lactate, glucose, and fatty acid has consequences on whole body energetics. Synthesis of glucose from triose phosphate is associated with an energy loss corresponding to about 5% fructose energy content; similarly, synthesis of fatty acids is associated with about 25% energy loss. For both processes, the energy consumed for biosynthetic processes is finally lost as heat and accounts for a higher thermic effect of fructose compared to other carbohydrates [83]. It implies that fructose energy efficiency is somewhat lower with fructose than other carbohydrate. It also implies that energy efficiency decreases when de novo lipogenesis from fructose is stimulated, due to the very high-energy cost of this process.

In contrast, conversion of fructose into lactate does not increase overall energy expenditure. It nonetheless alters energy distribution within the body. In splanchnic tissues, fructose degradation to pyruvate, like glycolysis, results in a net gain of two molecules of ATP which are produced directly in the cytosol, i.e., independently of oxygen consumption. It also produces one molecule of cytosolic NADH, which may be oxidized in the respiratory chain to produce ATP; this NADH is however reduced to NAD during conversion of pyruvate to lactate. In extrasplanchnic tissues, lactate conversion into pyruvate produces conversion of one molecule of NAD to NADH; thereafter, pyruvate is converted into acetyl-CoA and oxidized in the tricarboxylic acid cycle, where it produce ATP at no additional energy cost. The overall pathways are therefore associated with the same net ATP gain than glucose oxidation; however

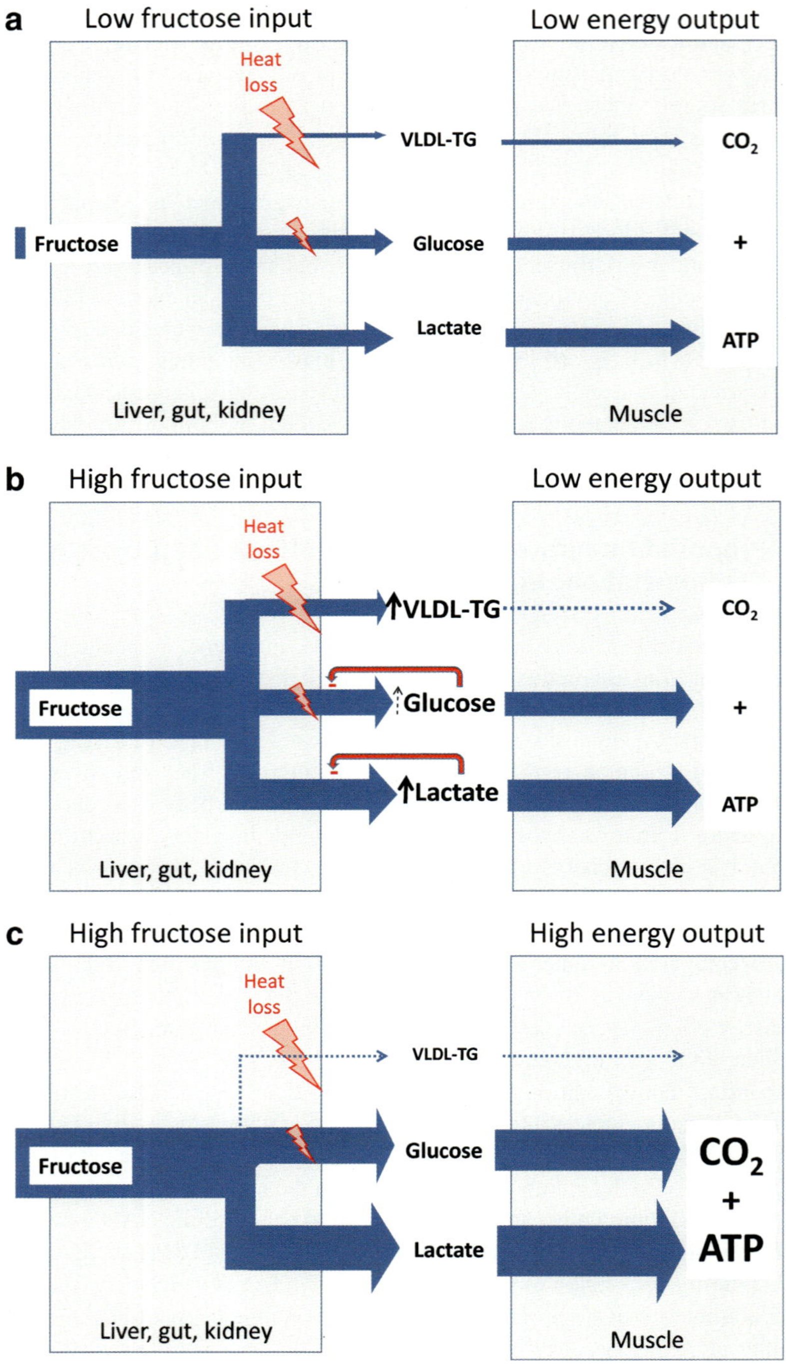

Fig. 6 (continued)

the ATP synthesized at the substrate level are made available to splanchnic organs, and those synthetized from "glycolytic" NADH and by pyruvate oxidation to extrasplanchnic tissues.

One may therefore propose the following model to account for the effects of dietary fructose on lipid and glucose homeostasis (Fig. 6):

- Fructose uptake and conversion to triose phosphate are directly proportional to dietary fructose absorption from the gut. The higher the fructose intake, the higher the rate and amount of triose phosphate generated in fructose-metabolizing splanchnic organs.
- Since the amount of triose phosphate oxidized within fructose-metabolizing organs is limited by their energy needs, excess triose phosphate is already generated by relatively small doses of fructose. One can postulate that the less energy-requiring pathways will be favored. According to this scheme, lactate production is expected to be initially stimulated; the maximal rate of lactate synthesis from fructose and its limiting factors remain unknown. One may however postulate that lactate efflux through membrane lactate transporters MCTs may become limiting when blood lactate concentration increases (Fig. 6a). Since glucose synthesis from fructose is associated with a significant, yet relatively modest energy cost, one can further postulate that it is already activated with small doses of fructose and increases progressively when lactate production reaches near maximal values. Glucose output from splanchnic organs is however closely regulated by both insulin and blood glucose itself hence is quantitatively limited in resting conditions (Fig. 6b). De novo lipogenesis is highly energy inefficient and therefore becomes mainly activated when the former two pathways become saturated (Fig. 6b). It can be increased already with administration of a single large dose of fructose but will really become quantitatively important when fructose intake remains high over several days and lipogenic enzymes expression increases.

Fig. 6 Temptative model for regulation of fructose disposal in splanchnic organs according to fructose intake and energy output. Fructose is preferentially converted into lactate, since this is the most energy-efficient pathway for fructose carbons' disposal. At low-fructose intake (**a**), the amount of lactate and glucose synthesized from fructose is small and can be metabolized by extrasplanchnic organs and tissues even when total energy output is low. When fructose intake is high, but total energy output is low (**b**), it is hypothesized that the amount of fructose carbons converted into lactate and glucose becomes saturated when blood lactate and glucose increase and inhibit the diffusion of glucose and lactate into the blood. Since blood lactate and glucose concentration depend on extrasplanchnic (mainly muscle) extraction and metabolism, low extrasplanchnic glucose and lactate oxidation is the major limiting factor under such conditions. The excess of triose phosphate generated from fructose is massively directed toward lipid synthesis in spite of the high-energy cost (and hence large energy loss to heat) of the metabolic pathway. When fructose intake and total energy output are both high, as during exercise (**c**), the rate of lactate and glucose oxidation in muscle increases, thus removing the brake on lactate and glucose output from fructose-metabolizing splanchnic cells. Fructose carbons are preferentially directed into lactic acid and glucose synthesis at a lower energy cost, and stimulation of de novo lipogenesis is prevented

According to this model, for any given fructose intake, the fructose converted into lipid depends essentially on the amount of fructose which can be disposed of as lactate and glucose. If these processes are inhibited when blood lactate and glucose increase as postulated above, their maximal rate will be mainly determined by the clearance of lactate and glucose from the circulation. During physical activity, muscle energy output is high, and muscle lactate and glucose uptake are enhanced. This decreases blood lactate and glucose concentration, thus decreasing the brake on glucose and lactate synthesis in splanchnic organs. In such conditions, the bulk of fructose can be oxidized of as lactate and glucose in skeletal muscle (Fig. 6c).

This model rests on many assumptions, however, and one has to keep in mind that several major questions remain unanswered presently. This is essentially due to methodological limitations inherent to tracer methods and the fact that we have limited access to splanchnic organs or splanchnic vessels in human trials. Future progress in noninvasive methods allowing to assess organ-specific metabolism will most likely bring important novel information in the future and may by the same token require revision if this hypothetical model.

2.12 Gender-Related Differences: An Under-Investigated Area of Fructose Metabolism

There is evidence that the effects of fructose differ according to gender. Thus, fructose clearly increases blood triglyceride and impairs hepatic insulin sensitivity in males but much less in females [84, 85]. Animal experiments indicate that these gender-related differences may be directly or indirectly mediated by estrogens [86]. The physiologically higher body fat mass in females than males may also have an important impact on the clearance of newly formed lipids from the blood and on their long-term consequences. More generally, the role of adipose tissue as an active player in the disposition of a fructose load remains largely unexplored.

2.13 Fructose Malabsorption and Gut-Related Symptoms: A Potentially Underestimated Adverse Effect of Dietary Fructose

Beside the abovementioned effects of fructose and their possible relationship with increased cardiometabolic risk, there is increasing concern that a high-fructose diet may be associated with more trivial yet unpleasant adverse effects on the gastroenteric system. Due to the slow and sometimes incomplete absorption of fructose by GLUT2, some subjects experience moderate to severe gastrointestinal symptoms after ingestion of fructose-containing foods [87, 88]. Since the presence of glucose in the intestine facilitates GLUT5-mediated intestinal fructose absorption, the symptoms may be mainly present or particularly severe after eating foods with high fructose:glucose ratio, such as agave syrup, some honeys, or apples. Dietary sugar restriction is effective in reverting the symptoms.

3 Conclusions

Fructose has always been present in the human diet, at least in most part of the world. The fact that most vertebrates synthesize sweet taste receptors and fructolytic enzymes clearly indicate that it constituted a valuable source of energy, at least at some time points during evolution.

Fructose is always ingested together with glucose in natural foods, and these two hexoses have a synergic effect on energy storage in the liver. Glucose enhances fructose intestinal absorption, while fructose enhances hepatic glucose uptake and glycogen synthesis. Fructose is also by far the best lipogenic substrate in liver cells. This makes fructose-containing sugars prime anabolic substrate. Before industrialization, fructose-containing foods, such as fruits and berries, were mainly available at the end of summer and fall in many areas of the world and were therefore likely to be prime sources for the rapid constitution of energy stores before winter. The unique anabolic properties of fructose were certainly advantageous in this context. These same properties are still advantageous to sustain high-energy output during high-intensity exercise and to allow for a rapid replenishment of body energy stores at the end of exercise. However, these anabolic effects may well become deleterious when high amounts of fructose are continuously consumed throughout the year by individuals with low physical activity.

Acknowledgments The author's research in this field has been supported by grant from the Swiss National Science Foundation 26074044 and by grants from the Swiss Federal Office for Sport Science. I warmly thank Dr. Kevin Seyssel for having reviewed and commented the manuscript.

References

1. Liu S, Manson JE, Lee IM, Cole SR, Hennekens CH, Willett WC, Buring JE (2000) Fruit and vegetable intake and risk of cardiovascular disease: the Women's Health Study. Am J Clin Nutr 72:922–928
2. Freedman DA, Choi SK, Hurley T, Anadu E, Hebert JR (2013) A farmers' market at a federally qualified health center improves fruit and vegetable intake among low-income diabetics. Prev Med 56:288–292
3. Lustig RH, Schmidt LA, Brindis CD (2012) Public health: the toxic truth about sugar. Nature 482:27–29
4. Bray G, Nielsen S, Popkin B (2004) Consumption of high-fructose corn syrup in beverages may play a role in the epidemic of obesity. Am J Clin Nutr 79:737–743
5. Macdonald I (1986) Dietary carbohydrate and energy balance. Prog Biochem Pharmacol 21:181–191
6. Johnson R, Moorthy SN, Padmaja G (2010) Production of high fructose syrup from cassava and sweet potato flours and their blends with cereal flours. Food Sci Technol Int 16:251–258
7. Hanover LM, White JS (1993) Manufacturing, composition, and applications of fructose. Am J Clin Nutr 58:724S–732S
8. Rosensweig NS (1972) Dietary sugars and intestinal enzymes. J Am Diet Assoc 60:483–486
9. Lu Y, Levin GV, Donner TW (2008) Tagatose, a new antidiabetic and obesity control drug. Diabetes Obes Metab 10:109–134

10. Cook GC (1979) The D-xylose absorption test in different ethnic groups. Trop Geogr Med 31:93–97
11. Wang YM, van Eys J (1981) Nutritional significance of fructose and sugar alcohols. Annu Rev Nutr 1:437–475
12. Robayo-Torres CC, Quezada-Calvillo R, Nichols BL (2006) Disaccharide digestion: clinical and molecular aspects. Clin Gastroenterol Hepatol 4:276–287
13. Clausen MR, Mortensen PB (1997) Lactulose, disaccharides and colonic flora. Clinical consequences. Drugs 53:930–942
14. Gray GM (1992) Starch digestion and absorption in nonruminants. J Nutr 122:172–177
15. Laffitte A, Neiers F, Briand L (2014) Functional roles of the sweet taste receptor in oral and extraoral tissues. Curr Opin Clin Nutr Metab Care 17:379–385
16. Drewnowski A (1999) Sweetness, appetite, and energy intake: physiological aspects. World Rev Nutr Diet 85:64–76
17. Beauchamp GK (1999) Factors affecting sweetness. World Rev Nutr Diet 85:10–17
18. Treesukosol Y, Smith KR, Spector AC (2011) The functional role of the T1R family of receptors in sweet taste and feeding. Physiol Behav 105:14–26
19. Newbrun E, Hoover C, Mettraux G, Graf H (1980) Comparison of dietary habits and dental health of subjects with hereditary fructose intolerance and control subjects. J Am Dent Assoc 101:619–626
20. Burmeister LA, Valdivia T, Nuttall FQ (1991) Adult hereditary fructose intolerance. Arch Intern Med 151:773–776
21. Flatt JP (1978) The biochemistry of energy expenditure. In: Bray GA (ed) Recent advances in obesity research. Newman Publishing, London, pp 211–228
22. Raninen KJ, Lappi JE, Mukkala ML, Tuomainen TP, Mykkanen HM, Poutanen KS, Raatikainen OJ (2016) Fiber content of diet affects exhaled breath volatiles in fasting and postprandial state in a pilot crossover study. Nutr Res 36:612–619
23. Trichopoulou A, Lagiou P (1997) Healthy traditional Mediterranean diet: an expression of culture, history, and lifestyle. Nutr Rev 55:383–389
24. Sheehy T, Kolahdooz F, Roache C, Sharma S (2015) Traditional food consumption is associated with better diet quality and adequacy among Inuit adults in Nunavut, Canada. Int J Food Sci Nutr 66:445–451
25. Boyce VL, Swinburn BA (1993) The traditional Pima Indian diet. Composition and adaptation for use in a dietary intervention study. Diabetes Care 16:369–371
26. Cordain L, Eaton SB, Miller JB, Mann N, Hill K (2002) The paradoxical nature of hunter-gatherer diets: meat-based, yet non-atherogenic. Eur J Clin Nutr 56(Suppl 1):S42–S52
27. Sluik D, van Lee L, Engelen AI, Feskens EJ (2016) Total, free, and added sugar consumption and adherence to guidelines: the Dutch National Food Consumption Survey 2007–2010. Nutrients 8:70
28. Marriott BP, Olsho L, Hadden L, Connor P (2010) Intake of added sugars and selected nutrients in the United States, National Health and Nutrition Examination Survey (NHANES) 2003–2006. Crit Rev Food Sci Nutr 50:228–258
29. Lei L, Rangan A, Flood VM, Louie JC (2016) Dietary intake and food sources of added sugar in the Australian population. Br J Nutr 115:868–877
30. Campos VC, Tappy L (2016) Physiological handling of dietary fructose-containing sugars: implications for health. Int J Obes (Lond) 40(Suppl 1):S6–S11
31. Mayes PA (1993) Intermediary metabolism of fructose. Am J Clin Nutr 58:754S–765S
32. Mueckler M, Thorens B (2013) The SLC2 (GLUT) family of membrane transporters. Mol Aspects Med 34:121–138
33. Kinne RK, Castaneda F (2011) SGLT inhibitors as new therapeutic tools in the treatment of diabetes. Handb Exp Pharmacol:105–126
34. Douard V, Ferraris RP (2008) Regulation of the fructose transporter GLUT5 in health and disease. Am J Physiol Endocrinol Metab 295:E227–E237
35. Barone S, Fussell SL, Singh AK, Lucas F, Xu J, Kim C, Wu X, Yu Y, Amlal H, Seidler U et al (2009) Slc2a5 (Glut5) is essential for the absorption of fructose in the intestine and generation of fructose-induced hypertension. J Biol Chem 284:5056–5066

36. Sugawara-Yokoo M, Suzuki T, Matsuzaki T, Naruse T, Takata K (1999) Presence of fructose transporter GLUT5 in the S3 proximal tubules in the rat kidney. Kidney Int 56:1022–1028
37. Sun SZ, Empie MW (2012) Fructose metabolism in humans – what isotopic tracer studies tell us. Nutr Metab 9:89
38. Theytaz F, de Giorgi S, Hodson L, Stefanoni N, Rey V, Schneiter P, Giusti V, Tappy L (2014) Metabolic fate of fructose ingested with and without glucose in a mixed meal. Nutrients 6:2632–2649
39. Paquot N, Schneiter P, Jequier E, Gaillard R, Lefebvre PJ, Scheen A, Tappy L (1996) Effects of ingested fructose and infused glucagon on endogenous glucose production in obese NIDDM patients, obese non-diabetic subjects, and healthy subjects. Diabetologia 39:580–586
40. Bizeau ME, Pagliassotti MJ (2005) Hepatic adaptations to sucrose and fructose. Metabolism 54:1189–1201
41. Tappy L, Le KA (2010) Metabolic effects of fructose and the worldwide increase in obesity. Physiol Rev 90:23–46
42. Samuel VT (2011) Fructose induced lipogenesis: from sugar to fat to insulin resistance. Trends Endocrinol Metab: TEM 22:60–65
43. Faeh D, Minehira K, Schwarz J, Periasami R, Seongus P, Tappy L (2005) Effect of fructose overfeeding and fish oil administration on hepatic de novo lipogenesis and insulin sensitivity in healthy males. Diabetes 54:1907–1913
44. Le KA, Faeh D, Stettler R, Ith M, Kreis R, Vermathen P, Boesch C, Ravussin E, Tappy L (2006) A 4-wk high-fructose diet alters lipid metabolism without affecting insulin sensitivity or ectopic lipids in healthy humans. Am J Clin Nutr 84:1374–1379
45. Haidari M, Leung N, Mahbub F, Uffelman KD, Kohen-Avramoglu R, Lewis GF, Adeli K (2002) Fasting and postprandial overproduction of intestinally derived lipoproteins in an animal model of insulin resistance. Evidence that chronic fructose feeding in the hamster is accompanied by enhanced intestinal de novo lipogenesis and ApoB48-containing lipoprotein overproduction. J Biol Chem 277:31646–31655
46. Teff KL, Grudziak J, Townsend RR, Dunn TN, Grant RW, Adams SH, Keim NL, Cummings BP, Stanhope KL, Havel PJ (2009) Endocrine and metabolic effects of consuming fructose- and glucose-sweetened beverages with meals in obese men and women: influence of insulin resistance on plasma triglyceride responses. J Clin Endocrinol Metab 94 1562–1569
47. Jeppesen J, Chen YI, Zhou MY, Schaaf P, Coulston A, Reaven GM (1995) Postprandial triglyceride and retinyl ester responses to oral fat: effects of fructose. Am J Clin Nutr 61:787–791
48. Livesey G, Taylor R (2008) Fructose consumption and consequences for glycation, plasma triacylglycerol, and body weight: meta-analyses and meta-regression models of intervention studies. Am J Clin Nutr 88:1419–1437
49. Sievenpiper JL, Carleton AJ, Chatha S, Jiang HY, de Souza RJ, Beyene J, Kendall CW, Jenkins DJ (2009) Heterogeneous effects of fructose on blood lipids in individuals with type 2 diabetes: systematic review and meta-analysis of experimental trials in humans. Diabetes Care 32:1930–1937
50. David Wang D, Sievenpiper JL, de Souza RJ, Cozma AI, Chiavaroli L, Ha V, Mirrahimi A, Carleton AJ, Di Buono M, Jenkins AL et al (2014) Effect of fructose on postprandial triglycerides: a systematic review and meta-analysis of controlled feeding trials. Atherosclerosis 232:125–133
51. Sobrecases H, Le KA, Bortolotti M, Schneiter P, Ith M, Kreis R, Boesch C, Tappy L (2010) Effects of short-term overfeeding with fructose, fat and fructose plus fat on plasma and hepatic lipids in healthy men. Diabetes Metab 36:244–246
52. Egli L, Lecoultre V, Theytaz F, Campos V, Hodson L, Schneiter P, Mittendorfer B, Patterson BW, Fielding BA, Gerber PA et al (2013) Exercise prevents fructose-induced hypertriglyceridemia in healthy young subjects. Diabetes 62:2259–2265
53. Le KA, Ith M, Kreis R, Faeh D, Bortolotti M, Tran C, Boesch C, Tappy L (2009) Fructose overconsumption causes dyslipidemia and ectopic lipid deposition in healthy subjects with and without a family history of type 2 diabetes. Am J Clin Nutr 89:1760–1765

54. Bortolotti M, Kreis R, Debard C, Cariou B, Faeh D, Chetiveaux M, Ith M, Vermathen P, Stefanoni N, Le KA et al (2009) High protein intake reduces intrahepatocellular lipid deposition in humans. Am J Clin Nutr 90:1002–1010
55. Roglans N, Vila L, Farre M, Alegret M, Sanchez RM, Vazquez-Carrera M, Laguna JC (2007) Impairment of hepatic Stat-3 activation and reduction of PPARalpha activity in fructose-fed rats. Hepatology 45:778–788
56. Lecoultre V, Egli L, Carrel G, Theytaz F, Kreis R, Schneiter P, Boss A, Zwygart K, Le KA, Bortolotti M et al (2013) Effects of fructose and glucose overfeeding on hepatic insulin sensitivity and intrahepatic lipids in healthy humans. Obesity (Silver Spring) 21:782–785
57. Stanhope KL, Schwarz JM, Keim NL, Griffen SC, Bremer AA, Graham JL, Hatcher B, Cox CL, Dyachenko A, Zhang W et al (2009) Consuming fructose-sweetened, not glucose-sweetened, beverages increases visceral adiposity and lipids and decreases insulin sensitivity in overweight/obese humans. J Clin Invest 119:1322–1334
58. Maersk M, Belza A, Stodkilde-Jorgensen H, Ringgaard S, Chabanova E, Thomsen H, Pedersen SB, Astrup A, Richelsen B (2012) Sucrose-sweetened beverages increase fat storage in the liver, muscle, and visceral fat depot: a 6-mo randomized intervention study. Am J Clin Nutr 95:283–289
59. Elwyn DH, Bursztein S (1993) Carbohydrate metabolism and requirements for nutritional support: part I. Nutrition 9:50–66
60. Bursztein S, Elwyn DH, Askanazi J, Kinney JM (1989) Energy metabolism, indirect calorimetry, and nutrition. Williams & Wilkins, Baltimore, 266 pp
61. Johnston RD, Stephenson MC, Crossland H, Cordon SM, Palcidi E, Cox EF, Taylor MA, Aithal GP, Macdonald IA (2013) No difference between high-fructose and high-glucose diets on liver triacylglycerol or biochemistry in healthy overweight men. Gastroenterology 145(1016–1025): e1012
62. Ngo Sock ET, Le KA, Ith M, Kreis R, Boesch C, Tappy L (2010) Effects of a short-term overfeeding with fructose or glucose in healthy young males. Br J Nutr 103:939–943
63. Schwarz JM, Noworolski SM, Wen MJ, Dyachenko A, Prior JL, Weinberg ME, Herraiz LA, Tai VW, Bergeron N, Bersot TP et al (2015) Effect of a high-fructose weight-maintaining diet on lipogenesis and liver fat. J Clin Endocrinol Metab 100:2434–2442
64. Tappy L, Le KA (2012) Does fructose consumption contribute to non-alcoholic fatty liver disease? Clin Res Hepatol Gastroenterol 36:554–560
65. Tounian P, Schneiter P, Henry S, Jéquier E, Tappy L (1994) Effects of infused fructose on endogenous glucose production, gluconeogenesis and glycogen metabolism in healthy humans. Am J Physiol 267:E710–E717
66. Schwarz JM, Acheson KJ, Tappy L, Piolino V, Muller MJ, Felber JP, Jequier E (1992) Thermogenesis and fructose metabolism in humans. Am J Physiol 262:E591–E598
67. Dirlewanger M, di Vetta V, Guenat E, Battilana P, Seematter G, Schneiter P, Jequier E, Tappy L (2000) Effects of short-term carbohydrate or fat overfeeding on energy expenditure and plasma leptin concentrations in healthy female subjects. Int J Obes Relat Metab Dis: J Int Assoc Study Obes 24:1413–1418
68. Mevorach M, Giacca A, Aharon Y, Hawkins M, Shamoon H, Rossetti L (1998) Regulation of endogenous glucose production by glucose per se is impaired in type 2 diabetes mellitus. J Clin Invest 102:744–753
69. Delarue J, Normand S, Pachiaudi C, Beylot M, Lamisse F, Riou JP (1993) The contribution of naturally labelled 13C fructose to glucose appearance in humans. Diabetologia 36:338–345
70. Aeberli I, Hochuli M, Gerber PA, Sze L, Murer SB, Tappy L, Spinas GA, Berneis K (2013) Moderate amounts of fructose consumption impair insulin sensitivity in healthy young men: a randomized controlled trial. Diabetes Care 36:150–156
71. Tappy L, Le KA (2015) Health effects of fructose and fructose-containing caloric sweeteners: where do we stand 10 years after the initial whistle blowings? Curr Diab Rep 15:627

72. Johnson RJ, Nakagawa T, Sanchez-Lozada LG, Shafiu M, Sundaram S, Le M, Ishimoto T, Sautin YY, Lanaspa MA (2013) Sugar, uric acid, and the etiology of diabetes and obesity. Diabetes 62:3307–3315
73. Teff KL, Elliott SS, Tschop M, Kieffer TJ, Rader D, Heiman M, Townsend RR, Keim NL, D'Alessio D, Havel PJ (2004) Dietary fructose reduces circulating insulin and leptin, attenuates postprandial suppression of ghrelin, and increases triglycerides in women. J Clin Endocrinol Metab 89:2963–2972
74. Soenen S, Westerterp-Plantenga MS (2007) No differences in satiety or energy intake after high-fructose corn syrup, sucrose, or milk preloads. Am J Clin Nutr 86:1586–1594
75. Rodin J, Reed D, Jamner L (1988) Metabolic effects of fructose and glucose: implications for food intake. Am J Clin Nutr 47:683–689
76. Payne AN, Chassard C, Lacroix C (2012) Gut microbial adaptation to dietary consumption of fructose, artificial sweeteners and sugar alcohols: implications for host-microbe interactions contributing to obesity. Obes Rev 13:799–809
77. Jandrain BJ, Pallikarakis N, Normand S, Pirnay F, Lacroix M, Mosora F, Pachiaudi C, Gautier JF, Scheen AJ, Riou JP et al (1993) Fructose utilization during exercise in men: rapid conversion of ingested fructose to circulating glucose. J Appl Physiol 74:2146–2154
78. Jeukendrup AE (2010) Carbohydrate and exercise performance: the role of multiple transportable carbohydrates. Curr Opin Clin Nutr Metab Care 13:452–457
79. Lecoultre V, Benoit R, Carrel G, Schutz Y, Millet GP, Tappy L, Schneiter P (2010) Fructose and glucose co-ingestion during prolonged exercise increases lactate and glucose fluxes and oxidation compared with an equimolar intake of glucose. Am J Clin Nutr 92:1071–1079
80. Egli L, Lecoultre V, Cros J, Rosset R, Marques AS, Schneiter P, Hodson L, Gabert L, Laville M, Tappy L (2016) Exercise performed immediately after fructose ingestion enhances fructose oxidation and suppresses fructose storage. Am J Clin Nutr 103:348–355
81. Bidwell AJ, Fairchild TJ, Redmond J, Wang L, Keslacy S, Kanaley JA (2014) Physical activity offsets the negative effects of a high-fructose diet. Med Sci Sports Exerc 46:2091–2098
82. Koutsari C, Karpe F, Humphreys SM, Frayn KN, Hardman AE (2001) Exercise prevents the accumulation of triglyceride-rich lipoproteins and their remnants seen when changing to a high-carbohydrate diet. Arterioscler Thromb Vasc Biol 21:1520–1525
83. Tappy L, Egli L, Lecoultre V, Schneider P (2013) Effects of fructose-containing caloric sweeteners on resting energy expenditure and energy efficiency: a review of human trials. Nutr Metab 10:54
84. Tran C, Jacot-Descombes D, Lecoultre V, Fielding BA, Carrel G, Le KA, Schneiter P, Bortolotti M, Frayn KN, Tappy L (2010) Sex differences in lipid and glucose kinetics after ingestion of an acute oral fructose load. Br J Nutr 104:1139–1147
85. Couchepin C, Le KA, Bortolotti M, da Encarnacao JA, Oboni JB, Tran C, Schneiter P, Tappy L (2008) Markedly blunted metabolic effects of fructose in healthy young female subjects compared with male subjects. Diabetes Care 31:1254–1256
86. Busserolles J, Mazur A, Gueux E, Rock E, Rayssiguier Y (2002) Metabolic syndrome in the rat: females are protected against the pro-oxidant effect of a high sucrose diet. Exp Biol Med (Maywood) 227:837–842
87. Wilder-Smith CH, Li X, Ho SS, Leong SM, Wong RK, Koay ES, Ferraris RP (2014) Fructose transporters GLUT5 and GLUT2 expression in adult patients with fructose intolerance. U Eur Gastroenterol J 2:14–21
88. Fedewa A, Rao SS (2014) Dietary fructose intolerance, fructan intolerance and FODMAPs. Curr Gastroenterol Rep 16:370

Potential Carcinogenic Risks of Aspartame 15

Morando Soffritti, Angela Guaragna, and Marco Manservigi

Abstract

Aspartame is an intense artificial sweetener with a sweet test approximately 200 times that of sucrose and used as additive in more than 6,000 products.

Aspartame was invented by GD-Searle in 1965 and submitted for pre-marketing safety evaluation in early 1980s. The studies conducted by GD-Searle to evaluate the potential carcinogenic risks of aspartame did not show any effect. Because of the great commercial diffusion of aspartame, in 1997 the Ramazzini Institute started a large experiments project on rodents to test the carcinogenic effects of aspartame in our experimental model with more sensitive characteristics, namely large number of rats and mice, observation until natural death. Overall the project included the study of 2,270 rats and 852 mice starting the treatment from prenatal life or in mature age and lasting all life.

These studies have shown that aspartame is a carcinogenic agent inducing a significant dose-related increased incidence of several types of malignant tumors and, among them, haematological neoplasias. Later this effect was confirmed by an epidemiological study conducted by a group of the Harvard University.

Keywords

Aspartame • Food additive • Carcinogenic bioassay • Rat • Mice • Artificial sweetener • Carcinogenic effects • EFSA • FDA • GD-SEARLE

M. Soffritti (✉)
Clinical Center of Cancer Prevention, Ramazzini Institute, Bologna, Italy
e-mail: soffrittim@ramazzini.it

A. Guaragna
Clinical Center of Cancer Prevention, Bologna, Italy

M. Manservigi
Cesare Maltoni Cancer Research Center, Bologna, Italy

J.-M. Mérillon, K.G. Ramawat (eds.), *Sweeteners*, Reference Series in Phytochemistry,
https://doi.org/10.1007/978-3-319-27027-2_5

Abbreviations

ADI	Admitted daily dose
APM	Aspartame
CI	Confidence interval
DKP	5-benzyl-3, 6-dioxo-2 piperazine acetic acid
EFSA	European Food Safety Authority
FDA	Food Drug Administration
GR	Glutathione reductase
GSH	Glutathione
IARC	International Agency for Research on Cancer
MPL	Maximum permitted level
RR	Relative risk
US-NCI	United States – National Cancer Institute
US-NTP	United States – National Toxicology Program
WHO	World Health Organization

Contents

1 Introduction 312
2 Technical Characteristics, Production, Uses, and Stability of APM in Food and Beverages 313
3 Absorption, Distribution, and Metabolism of APM 315
4 Acute, Subacute Toxicity, and Genotoxicity 316
5 The Historical Available Data on the Chronic Toxicity and Carcinogenicity of Aspartame 317
6 The Carcinogenicity Bioassays on APM Conducted by Ramazzini Institute on Rats and Mice 323
6.1 Carcinogenicity Bioassay on APM Administered in Feed to Sprague-Dawley Rats from 8 Weeks of age until Natural Death (Exp. BT6008) 323
6.2 Carcinogenicity Bioassay on APM Administered in Feed to Sprague-Dawley Rats from Prenatal Life until Natural Death (Exp. BT6009) 328
6.3 Carcinogenicity Bioassay on APM Administered in Feed to Swiss Mice from Prenatal Life until Natural Death (Exp. BT6010) 333
7 Epidemiological Studies on Cancer Risks and Exposure to Aspartame 338
8 Conclusions 342
References 345

1 Introduction

The use of artificial sweeteners as substitutes for sucrose began during World Wars I and II when the use of saccharin became prevalent due to its low cost and the wartime shortage of table sugar. Since the 1970s, the growing obesity problem in industrialized countries, due in part to fast food and soft drink consumption, has led to an increased demand for reduced-calorie foodstuffs. Given the lucrative market for these so-called "diet" or "light" products, additional intense artificial sweeteners have emerged, including aspartame, cyclamate, acesulfame-K, sucralose, and neotame. These substances are used in various food/beverages/drugs labeled as diet/sugar-free, including candy, chewing gum, ice cream, beverages, yogurt, and baked goods. They are also present in almost 500 drugs (including pediatric drugs) as cough drops and syrups, vitamins, antibiotics, etc. [1, 2].

With the expansion of the artificial sweetener market, concern has arisen among consumers regardings the safety of these sweeteners and their possible long-term health effects, in particular the potential carcinogenic risks.

Given the fact that wide consumption of artificial sweeteners emerged in the 1980s and 1990s, epidemiological studies are by definition limited, the exposure to such compounds being widely diffused among the population.

Moreover, most long-term carcinogenicity bioassays on artificial sweeteners performed on rodents in the past have not been adequately designed to assess the carcinogenic risk. The sensitivity of these studies in detecting risk has been greatly limited by the following factors: (1) the number of animals per sex per group was usually 50 or less; (2) experiments were usually truncated at 104 weeks (or earlier) from the start of the experiment, thus not allowing the tested compound to express its carcinogenic potential; (3) conduct of the experiments was often inadequate with incomplete or nonsystematic histopathological analysis for all organs and tissues.

The purpose of this review is to summarize the existing literature regarding general data on the technical characteristics and the biological and toxicological effects of aspartame (APM). Unpublished studies dealing with these issues were performed by GD-Searle in the 1970s. These studies were provided for EFSA in 2011 after a public call was launched. The results of the GD-Searle studies are derived from the documentation posted online by EFSA and made available to the scientific community and interested stakeholders [3]. The references of the GD-Searle studies are here identified with the letter E plus the code number and the year (i.e., E18, 1972). In this review, particular emphasis will be deserved on what we know now about the carcinogenic potential of APM. Indeed, this is a major public health concern, if we consider that children and women in childbearing age are among the major consumers, and this prolonged experience may represent a potential risk not only for children but also in adult age.

2 Technical Characteristics, Production, Uses, and Stability of APM in Food and Beverages

Aspartame (APM) is a dipeptide of L-phenylalanine methyl ester and L-aspartic acid bearing an amino group at the α-position from the carbon of the peptide bond (Fig. 1). APM was first synthesized in 1965 by the US GD-Searle and first sold under the name "Nutrasweet." APM has a molecular formula of $C_{14}H_{18}N_2O_5$ with a molecular weight of 294.31 g/mol and the CAS Registry Number is 22839-47-0. APM is a white odorless, crystalline powder with a sweet taste approximately 200 times that of sucrose. At room temperature and at pH7, the solubility in water is 10 g/l and it is insoluble in oil [4]. The major degradation products of APM are 5-benzyl-3,6-dioxo-2-piperazine acetic acid (DKP), methanol, L-phenylalanine, and L-aspartic acid. Aspartame is used in >6,000 products, and the maximum permitted levels

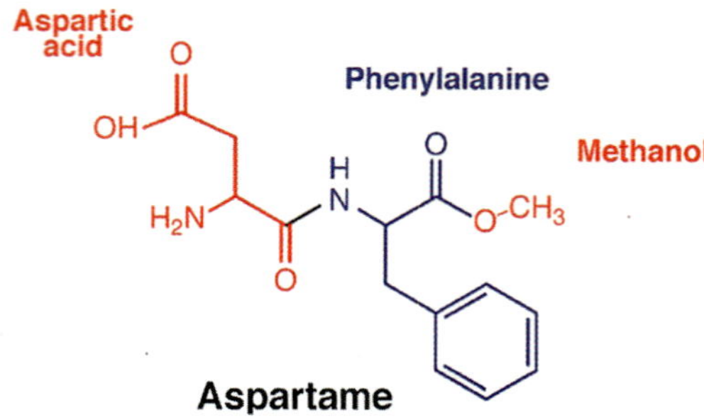

Fig. 1 Molecular structure of aspartame and its three components

Table 1 Estimated exposure (mg/kg b.w./day) to Aspartame (*APM*) based on its use as a food additive using the maximum permitted levels (*MPLs*), in five population groups[a]

Estimated exposure (mg/kg b. w./day)	Toddlers (12–35 months)	Children (3–9 years)	Adolescents (10–17 years)	Adults (18–64 years)	The elderly (>65 years)
> Mean level	3.2–16.3	2.3–12.8	0.8–4.0	0.8–8.6	0.5–4.4
> High level	11.8–36.9	7.1–32.9	2.3–13.3	2.5–27.5	1.5–23.5

[a]From EFSA, 2013

(MPL) of APM in food and beverage products range from 25 to 6,000 mg/kg, except for tabletop sweeteners for which there is *quantum satis* authorization [5]. The highest MPL of APM is for chewing gum (MPL 5,500 mg/kg) and breath refreshing microsweets (6,000 mg/kg); coca, chocolate, and dietary food products have a range of 1,000–2,000 mg/L (Commission Regulation-EU-No 1129/2011). Moreover, it has been reported that in the USA, the annual use of APM in food products was estimated at 5,000–5,500 tons and that most (~ 85%) was used for diet soda [6]. Overall, hundreds of millions of people in the world consume APM on a daily basis. An estimated exposure to APM based on its use as a food additive in the population stratified in five groups by age is reported in Table 1 [5].

The US Food and Drug Administration (FDA) approved the use of APM in a limited number of dry foods in 1981 [7], in soft drinks in 1983 [8], and in all foods in 1996 [9]. In 1994, the use of APM was approved by Europe [10]. Currently, the daily admitted dose (ADI) in the USA and in Europe is 50 mg/kg b.w. and 40 mg/kg b.w., respectively. In 2013, EFSA carried out a reevaluation of APM concluding that APM is safe for human consumption [5].

The stability of APM in solid or solution state is affected mainly by temperature, moisture, pH, and storage time. The major degradation products of APM are L-phenylalanine, aspartic acid, methanol, and 5-benzyl-3,6-dioxo-2-piperazine acetic acid (DKP).

Solid-state stability studies showed that at temperatures higher than 80 °C, APM may release methanol to form DKP [11]. Under dry conditions, conversion of APM into DKP is slow (5% per 100 h at 105 °C), while at higher temperatures, the rate of conversion to DKP increases (around 50% per 80 h at 120 °C and 100% per 30 h at 150 °C) [12].

The degradation products of APM in solution have been tested at different pH levels. The main degradation product found at pH 2–6 was L-phenylalanine methyl ester; at pH 7–10, the chief degradation product was DKP and at pH 12, the L-aspartyl-phenylalanine [13].

The content of APM and its degradation products have been tested in several products. The degradation of APM was high in fruit cream (40%), milk chocolate (26%), and fruit yogurt; 12–21% decomposition of APM in cola drink and fruit cream was in the form of DKP [14]. In soft drinks, the degradation of APM into DKP at a temperature of 25 °C for 2 months has been found four times higher than at 4–5 °C [15].

3 Absorption, Distribution, and Metabolism of APM

The metabolism of APM has been investigated in mouse, rat, rabbit, dog, and monkey, using different radiolabeled forms of ^{14}C-APM located either on the phenylalanine or aspartic acid moiety or on the methyl group.

In mice, [14 C(U)-phe]-APM was administered by gavage at the dose of 20 mg/kg b.w. The radioactivity found in plasma, urine, feces, and expired air clearly demonstrated that APM was hydrolyzed in the gut before the absorption of its phenylalanine moiety occurred (E 18, 1972)[1]. Studies conducted on rats, primates, and man receiving APM by oral ingestion have shown that APM is hydrolyzed in the lumen of the gastrointestinal tractor within the mucosal cells lining the GI tract, its half-life in the GI tract being in the order of a few minutes [16]. This short time was mediated in the GI tract by enzymes of the intestine, including esterases and peptidases [17, 18]. The metabolites are methanol, aspartic acid, and phenylalanine. APM releases a maximum of 10% methanol by weight [19]. The metabolic process is very efficient and the amount of APM that enters the blood stream is undetectable [17, 20]. Methanol is oxidized in the liver to formaldehyde and then to formic acid. The enzymes involved depend on the species: in rats the metabolism of methanol to formaldehyde is mediated through a catalase-peroxidase system; in primates and humans, an alcohol dehydrogenase is responsible [19].

In humans, APM is hydrolyzed in the intestine as in animals. Qualitatively and quantitatively, the metabolites are also the same. Doses of APM higher than 100–200 mg/kg b.w. peak blood levels were around 13–26 mg methanol/l blood and persisted for several hours in the blood. After 24 h, the methanol was under the limit of detection [21].

Stegink [22] found that, following a single oral administration of APM at various dose levels ranging from 34 to 200 mg/kg b.w., a significant dose-related increase in plasma phenylalanine was detected between 30 min and 2 h following dining; concerning plasma aspartate, no significant dose-related increase was observed after dosing. The author concluded that a rapid metabolization of aspartic acid occurred at all doses. In another study conducted to evaluate the effects of APM on its metabolism after repeated ingestions, no accumulation of plasma phenylalanine or aspartate or methanol concentrations was observed [23]. Other studies were

[1]Code of the reference as reported on the documentation posted on EFSA website [3].

conducted to test the effects of oral administration of APM (at doses ranging from 600 to 8,100 mg per person per day for 27 weeks) on plasma levels of phenylalanine and aspartic acid, in obese individuals, diabetics, and children (E 23, 1972; E 24, 1972; E 60, 1973; E 61, 1972; E 64, 1972). No significant differences were observed in plasma amino acid concentrations between treated subjects and controls.

In conclusion, pharmacokinetic studies on APM showed that unchanged APM was not detectable in plasma. After ingestion, APM is hydrolyzed in the intestinal tract to methanol, phenylalanine and aspartic acid, and metabolites which are then absorbed and metabolized.

4 Acute, Subacute Toxicity, and Genotoxicity

The acute effects of APM were tested in mice, rats, and rabbits (E 46, 1973). Male mice were treated by gavage at 1,000 or 5,000 mg/kg b.w. or by i.p. injection at 200 or 450 mg/kg b.w.; rats were treated by gavage at 2,033 or 5,000 mg/kg b.w. or by i.p. injection at 2,033 mg/kg b.w.; rabbits were treated by gavage in the dose range of 2,000–5,000 mg/kg b.w. The animals were kept under clinical observation for 7 days, and no remarkable changes of the behavior and motility were registered. No mortality was observed. The LD_{50} dose level was estimated by the authors higher than the ones treated in the various species.

Subacute tests were conducted on mice, rats, and beagle dogs. Mice of 8 weeks of age were treated with APM in feed at the dose levels of 0, 3,000, 5,000, and 13,000 mg/kg b.w. for 4 weeks. No significant difference in body weight and no clinical changes or mortality was observed (E 2, 1972). Groups of five male and five female 7 weeks old CD rats were treated with APM at the dose of 0, 2,000, 4,000, and 10,000 ppm in feed for 8 weeks. Decreased feed consumption at the highest dose was observed in females, without consequence for body weight and without mortality (E 3, 1972). Groups of ten male and ten female CD rats were treated with APM in feed at the dose of 0, 5, and 125 mg/kg b.w. for 8 weeks. No changes in hematological and clinical chemistry tests in urine analysis were observed, as well as no clinical alterations or mortality. Among males of the highest dose level, an increased liver to body weight ratio compared to controls was observed. Bile duct hyperplasia and pericholangitis were present in all treated and untreated rats (E 20, 1969). Three groups of male Wister rats were daily exposed for 6 months to 500 or 1,000 mg/kg b.w. of APM dissolved in water and administered by gavage. Rats that received 1,000 mg/kg b.w. showed a significant serum increase in activities of alanine aminotransferase, aspartate aminotransferase, alkaline phosphatase, and gammaglutamyltransferase. The concentration of reduced glutathione (GSH) and the activities of glutathione peroxidase and glutathione reductase (GR) were significantly decreased in the liver of rats exposed to 1,000 mg/kg b.w./daily. The levels of GSH were significantly decreased in both treated groups and controls. Concerning the brain, it was reported that the concentration of GSH and GR activity were significantly reduced in rats exposed to 1,000 mg/kg b.w./day of APM. GSH resulted significantly decreased also at the level of exposure to 500 mg/kg b.w./day. On the basis of these data, the authors concluded that long-term consumption of APM may induce liver toxicity and a dose-related correlation between toxic effects of APM and alteration in the glutathione-dependent system of the brain [24, 25].

The genotoxicity of APM was evaluated in the in vitro and in vivo studies.

Various strains of *Salmonella* Typhimurium (TA 1535; TA 1537; TA 1538; TA 98 and TA 100) were tested in the absence and presence of a rat liver metabolic activation system at dose levels from 10 to 5,000 mg/plate (E 97, 1978; E 101, 1978). Aspartame was not mutagenic in that tests. Other studies conducted on *Salmonella* Typhimurium resulted negative for genotoxicity [26–28]. A study to evaluate DNA damaging activity in the in vitro primary rat hepatocyte/DNA repair assay resulted negative [29].

In an in vivo bone marrow micronucleus test conducted on male Fischer 344 rats a rally exposed to 0, 500, 1,000, or 2,000 mg/kg b.w. for 3 days, no increase in the number of micronucleated erythrocytes was observed [26]. In other peripheral blood micronucleus tests, conducted in male and female transgenic mice (Tg.AC hemizygous, p53 haploinsufficient, or Cdkn2a deficient), after 9 months of exposure to APM at doses from 3.1 mg to 50 mg/kg diet, no clastogenic activity was observed in male and female Tg.AC hemizygous and Cdkn2a-deficient mice and in male p53 haploinsufficient mice. In female p53 haploinsufficient mice, the results of the test were considered positive [26].

Four male mice were treated with a single dose of 2,000 mg/kg b.w. APM and analyzed the DNA migration in the cells of the stomach, colon, liver, kidney, bladder, lung, brain, and bone marrow. No significant increase in the DNA migration was observed [1].

To assess the genotoxicity of APM, five groups of mice were treated with APM administered by gavage at the dose levels of 0, 250, 455, 500, and 1,000 mg/kg b.w. The treatment induced (1) micronuclei in bone marrow erythrocytes, (2) micronuclei in peripheral blood, and (3) chromosome aberrations in bone marrow erythrocytes [30].

5 The Historical Available Data on the Chronic Toxicity and Carcinogenicity of Aspartame

Food additives and any other ingredient used for human consumption must be submitted before marketing to several laboratory tests for regulatory safety evaluation. Guidelines to performing these tests, including chronic toxicity and carcinogenicity bioassays, were published in the mid-1960s by the US National Cancer Institute. The main recommendations concerned (1) the use of two species of animals, namely, rats and mice; (2) using at least 50 males and 50 females for each dose level, without restrictions on increasing the number in order to improve the statistical power of the studies; (3) duration of the treatment and observation for at least 104 weeks; and (4) complete necropsy of all animals in the experiment, a complete histopathological evaluation of control and high-dose-treated animals and a limited evaluation of the animals of other groups. Updating of the guidelines has been performed by the Environmental Protection Agency [31], by the FDA [32] and the Organization for Economic Cooperation and Development [33]. Meanwhile, bioassays have been conducted by academic and independent institutions not for regulatory purposes but to evaluate if, following a different protocol design and conduct, the sensitivity and specificity of the studies might be improved [34–36].

In the early 1970s, GD-Searle submitted three 2-year carcinogenicity bioassays, codified as E-33/34, E-70 on rats, and E-75 on mice, to the FDA with the intent to gain market approval for APM. These studies are summarized in Table 2.

In the first experiment (E-33/34, 1973), four groups of 40 male and 40 female Charles River Sprague-Dawley rats were treated at the doses of 1, 2, 4, 6, and 8% of APM in the feed starting from mature age and lasting to 104 weeks. One group of 60 males and 60 females served as control. At the highest dose level, both males and females ate significantly less as well as the body weight was significantly decreased. The survival in males was higher at the highest dose than among controls (52.5% vs. 38.4%), the reverse with females treated at the highest dose versus controls (25% vs. 46.7%). Incidence of malignant tumors in males was higher in controls than in the animals treated at the highest dose (16.7% vs. 7.5%). No differences in females were observed.

In a second experiment (E-70, 1974), two groups of 40 male and 40 female Charles River Sprague-Dawley rats were exposed to APM in the feed at the dose levels of 2 or 4% from prenatal life and lasting to 104 weeks. Both males and females exposed at the higher doses ate significantly less than controls. The body weight was significantly lower in females exposed at the highest dose compared to controls. No difference was observed in survival among the groups of both genders. No difference was observed in survival among the groups. No significant difference in the incidences of malignant tumors was observed in treated males and females compared to controls.

A third experiment was performed on four groups of male and female Swiss CD1 mice treated with APM in feed at the dose of 0, 1, 2, and 4% for 104 weeks starting at 4 weeks of age. Compared to controls, it showed (1) a significant decreased feed consumption among males of all treated groups, (2) no substantial difference in mean body weight among the groups, and (3) no significant difference in the incidence of malignant tumors observed between treated and control rats.

During the period 1981–2005, other chronic toxicity/carcinogenicity bioassays were conducted. These studies are summarized on Table 2.

The first experiment [37] was conducted on five groups of 86 male and 86 female Wistar rats treated with APM in feed at the daily dose levels of 0, 1,000, 2,000, 4,000, and 4,000 mg/kg b.w. (the last as a mixture of APM plus DKP, 3:1) for 104 weeks starting from 6 weeks of age and then sacrificed. Two interim sacrifices of 10 M and 10 F and 16 M and 16 F were performed after 26 and 52 weeks of treatment, respectively. A dose-dependent decrease in feed consumption was observed in males treated at 2,000 and 4,000 mg/kg b.w. daily dose levels of APM and in all treated females as well as concerns mean body weights. No significant differences of brain tumor incidence were observed among the groups. In 2006, a reevaluation of all neoplastic and nonneoplastic lesions of the Ishii study was performed. No differences were observed in the neoplastic incidences among males and females treated with APM versus controls [38]. In 2005, the US NTP

Table 2 Long-term carcinogenicity bioassays on ASPARTAME administered orally with feed to male (M) and female (F) mice or rats: summary of experimental design and results of major studies

	Animals				Treatment		Histopath. evaluation		Biophase results of the studies			
Authors	Species/strain	Group	No. M/F	Age at start	Dose (mg/kg/bw) in feed day	Duration (wks/LS[a])	(Complete/Limited)[b]	Endpoints	Mean feed consumption differences (%) M/F	Mean body weights differences (%) M/F	Survival (%) at the end of the study M/F	Major carcinogenic results and statistical significance (pvalue) when available
GD Searle & Co., 1973 (E-33/34)	Rat Charles River (CD) Sprague-Dawley	I	60/60	NA[c]	0	104	Complete	Tumor incidence	–	–	38.4/46.7	*Animals bearing malignant tumors (%)*:
		II	40/40		1,000		Limited		−1.4/+1.0	0/+4.5	45.0/57.5	M: 16.7 (I, control); 17.5 (II); 7.5 (III); 12.5 (IV);7.5 (V)
		III	40/40		2,000		Limited		+2.6/+3.0	1.8/−3.2	52.5/50.0	F: 23.3 (I, control); 20.0 (II); 35.0 (III); 22.5 (IV); 25.0 (V)
		IV	40/40		4,000		Complete		−3.1/−8.2*	− 0.1/−3.5	57.5/35.0	*Animals bearing mammary cancers (%)*:
		V	40/40		6,000–8,000		Complete		−5.7*/−14.6* *(pvalueNA)	−12.3*/−15.1* *(pvalueNA)	52.5/25.0* *($p<0.05$)	F: 11.7 (I, control); 5.0 (II); 15.0 (III); 17.5 (IV); 17.5 (V)
												Final sacrifice probably at 110 weeks of age No statistically significant differences between controls and tested animals based on Life-table Technique and T test

(continued)

Table 2 (continued)

	Animals				Treatment		Histopath. evaluation		Biophase results of the studies			
Authors	Species/strain	Group	No. M/F	Age at start	Dose (mg/kg/bw) in feed day	Duration (wks/LS[a])	(Complete/Limited)[b]	Endpoints	Mean feed consumption differences (%) M/F	Mean body weights differences (%) M/F	Survival (%) at the end of the study M/F	Major carcinogenic results and statistical significance (pvalue) when available
GD Searle & Co., 1974 (E-70)	Rat Charles River (CD) Sprague-Dawley	I	60/60	Prenatal	0	104	Complete	Tumor incidence	–	–	41.7/48.4	*Animals bearing malignant tumors (%)*:
		II	40/40		2,000		Complete		+1.3/−1.7*	+ 2.2/−0.6	50.0/45.0	M:18.3 (I, control); 10.0 (II); 20.0 (III)
		III	40/40		4,000		Complete		−8.7*/−3.5* *(pvalueNA)	−2.4/−5.5* *(pvalueNA)	57.5/52.5	F: 18.3 (I, control); 20.0 (II); 27.5 (III) Final sacrifice probably at 106 weeks of age
GD Searle & Co., 1974 (E-75)	Mice CD1 Swiss	I	72/72	4 weeks	0	104	Complete	Tumor incidence	–	–	32.5/41.7	*Animals bearing malignant tumors (%)*:
		II	36/36		1,000		Limited		−3.5*/−2.6	−0.5*/+0.8	27.8/38.9	M: 16.9 (I, control); 9.4 (II); 8.6 (III); 22.2 (IV)
		III	37/35		2,000		Limited		−4.7*/−8.3	+ 0.7*/+ 0.5	25.8/41.7	F: 16.7 (I, control); 12.9(II); 12.9(III); 16.1(IV)
		IV	37/35		4,000		Complete		−7.5*/−9.0*	−0.2*/+ 1.4 *(pvalue NA)	25.0/41.7	Total number of mice available for histopathological evaluation: M: 65 (I, control); 32 (II); 35 (III); 27 (IV) F: 66 (I, control); 31 (II); 31 (III); 31 (IV) Final sacrifice at 108 weeks of age

Ishii [37]; Iwata [38]	Rat, Wistar	I	86/86	6 weeks	0	104	1. 10M+10F × group killed at 26 weeks; 2. 16M+16F × group killed at 52 weeks; 3. 60M+60F × group followed until 104 weeks	Brain tumors (6 section each brain)	Dose dependent decrease at 2 and 4 g/kg b.w. and at 4 g/kg b.w. APM+DKP in males and in all treated females (numerical data NA)	A dose-dependent decrease at 2 and 4 g/kg b.w. APM and at 4 g/kg b.w. APM*DKP in males and in all treated females (Numerical data NA)	43.3/81.7	No brain tumors were observed in animals killed at 26 and 52 weeks – Brain tumors in rats observed until 104 weeks M: 0 (I, control); 1.7 (II); 0 (III); 1.7 (IV); 0 (V) F: 1.7 (I, control); 0 (II); 3.3 (III); 0 (IV); 1.7 (V)
		II	86/86		1,000						26.7/66.7	
		III	86/86		2,000						46.7/85.0	
		IV	86/86		4,000						41.7/71.7	
		V	86/86		4,000						48.3/68.3	Final sacrifice at 110 weeks of age
					ASP+DKP[d] (3:1)							
National Toxicology Program [26]	Mice Tg-AC hemizygous[e]	I	15/15	6 weeks	0	40	Complete	Development of papillomas/carcinomas of skin and/or forestomach	Generally similar among control and treated groups	No differences among the groups	–	There were no neoplasms or non neoplastic lesions related to exposure to APM
		II	15/15		520		Limited				–	
		III	15/15		1,040		Limited				–	Final sacrifice at 46 weeks of age
		IV	15/15		2,110		Limited				–	
		V	15/15		4,190		Limited				–	
		VI	15/15		7,920		Complete				–	

(*continued*)

Table 2 (continued)

	Animals				Treatment		Histopath. evaluation		Biophase results of the studies			
Authors	Species/strain	Group	No. M/F	Age at start	Dose (mg/kg/bw) in feed day	Duration (wks/LS[a])	(Complete/Limited)[b]	Endpoints	Mean feed consumption differences (%) M/F	Mean body weights differences (%) M/F	Survival (%) at the end of the study M/F	Major carcinogenic results and statistical significance (pvalue) when available
National Toxicology Program [26]	Mice p53 haploinsufficient[f]	I	15/15	7 weeks	0	40	Complete	Development of lymphomas or sarcomas	Generally similar among control and treated groups	No differences among the groups	–	There were no neoplasms or non neoplastic lesions related to exposure to APM
		II	15/15		520		Limited				–	
		III	15/15		1,040		Limited				–	Final sacrifice at 46 weeks of age
		IV	15/15		2,110		Limited				–	
		V	15/15		4,190		Limited				–	
		VI	15/15		7,920		Complete				–	
National Toxicology Program [26]	Mice Cdkn2a deficient[g]	I	15/15	7–9 weeks	0	40	Complete	Development of brain tumors, lymphomas and fibrosarcomas	Similar among control and treated groups	No differences among the groups	–	There were no neoplasms or non neoplastic lesions relate to exposure to APM
		II	15/15		520		Limited				–	
		III	15/15		1,040		Limited				–	Final sacrifice at 46 weeks of age
		IV	15/15		2,110		Limited				–	
		V	15/15		4,190		Limited				–	
		VI	15/15		7,920		Complete				–	

[a] *LS* Life Span
[b] Histopathological evaluation: *C* complete; *L* limited
[c] *NA* not available
[d] *DKP* diketopiperazine
[e] Model susceptible for development of high incidence of skin papillomas in response to topical application of TPA (12-O-tetradecanoyl-phorbol-13-acetate)
[f] Model susceptible for development of high incidence of lymphomas or sarcomas
[g] Model susceptible for development of high incidence of brain tumors, lymphomas and fibrosarcomas

published the results of three studies conducted on genetically modified mice models [26]. According to the NTP report, there were no neoplasms or nonneoplastic lesions related to APM exposure (see Table 2).

Overall, these long-term studies did not show that exposure to APM in feed could produce carcinogenic effects in rats and mice.

6 The Carcinogenicity Bioassays on APM Conducted by Ramazzini Institute on Rats and Mice

In the early 1997, the Ramazzini Institute started a large project of life-span carcinogenicity bioassays on rats and mice to test the carcinogenic effects of exposure to APM administered with feed. The reasons for performing these studies were to test the potential of APM for carcinogenicity using an experimental model with more sensitive characteristics, namely, a large number of animals per sex per group, exposure, and observation until their natural death. This model used at the Cesare Maltoni Cancer Research Center of the Ramazzini Institute in the past 40 years allowed us to show for the first time that several agents widely used in the working place and diffused in the general environment are carcinogenic, such as vinyl chloride, benzene, formaldehyde, xylene, trichlorethylene, and others [34, 35].

The Ramazzini Institute project on APM included three life-span bioassays on APM already published, the first on rats exposed from 8 weeks of age until death [39, 40], the second on rats exposed from prenatal life until death [41], and the third on mice exposed prenatally until death [42]. The studies were conducted according to the procedures usually followed for the studies performed in Good Laboratory Practices, which means periodical measuring of feed consumption and body weight, daily clinical control of health and behavior, full necropsy of each animal after death, full histopathological evaluation of all organs and tissues of all animals in each group, and statistical elaboration of the results.

6.1 Carcinogenicity Bioassay on APM Administered in Feed to Sprague-Dawley Rats from 8 Weeks of age until Natural Death (Exp. BT6008)

Six groups of 200–300 male and female Sprague-Dawley rats were exposed to APM administered in feed at the concentration of 100,000, 50,000, 10,000, 2,000, 400, and 80 ppm from 8 weeks of age until natural death; 300 males and females served as controls [40]. Overall, the study encompassed 1,800 rats and the biophase proceeded smoothly and well. A slight decrease in feed consumption was observed at the highest dose in both males and females without affecting the body weight. No difference in survival was observed among males; a slight decrease in females was observed in controls versus treated rats from 104 weeks of age. The results of the carcinogenic effects are reported in Tables 3, 4, and 5.

Table 3 Long-term carcinogenicity bioassay on aspartame, administered with the feed supplied ad libitum, from 8 weeks of age until natural death, to male (M) and female (F) Sprague-Dawley rats (Experiment BT6008)

Results: benign and malignant tumors												
Group No.	Concentration (ppm)	Animals		Benign tumors				Malignant tumors				
				Tumor-bearing animals[a,b]		Tumors[c]		Tumor-bearing animals[a,b]		Tumors[c]		
		Sex	No.	No.	%	No.	Per 100 animals	No.	%	No.	Per 100 animals	
I	100,000	M	100	66	66.0	92	92.0	43	43.0	55	55.0	
		F	100	88	88.0	245	245.0	51	51.0	64	64.0	
		M+F	200	154	77.0	337	168.5	94	47.0	119	59.5	
II	50,000	M	100	71	71.0	120	120.0	38	38.0	45	45.0	
		F	100	87	87.0	231	231.0	58	58.0[##]	84	84.0	
		M+F	200	158	79.0	351	175.5	96	48.0	129	64.5	
III	10,000	M	100	77	77.0[**##]	127	127.0	34	34.0	42	42.0	
		F	100	85	85.0	221	221.0	40	40.0	62	62.0	
		M+F	200	162	81.0	348	174.0	74	37.0	104	52.0	
IV	2,000	M	150	87	58.0	131	87.3	60	40.0	69	46.0	
		F	150	121	80.7	265	176.7	67	44.7	86	57.3	
		M+F	300	208	69.3	396	132.0	127	42.3	155	51.7	
V	400	M	150	99	66.0	148	98.7	48	32.0	52	34.7	
		F	150	120	80.0	278	185.3	70	46.7	95	63.3	
		M+F	300	219	73.0	426	142.0	118	39.3	147	49.0	

VI	80	M	150	82	54.7	120	80.0	44	29.3	49	32.7
		F	150	112	74.7	256	170.7	64	42.7	85	56.7
		M+F	300	194	64.7	376	125.3	108	36.0	134	44.7
VII	0 (control)	M	150	93	62.0	124	82.7	53	35.3*	59	39.3
		F	150	123	82.0*	310	206.7	55	36.7**	69	46.0
		M+F	300	216	72.0	434	144.7	108	36.0	128	42.7

[a]Near the dosed group incidence are the p-values corresponding to pairwise comparisons between the controls and that dosed group
[b]Near the control incidence are the p-values associated with the trend test
[c]Multiple tumors of different type and site, or of different type in the same site, or of the same type in bilateral organs, in the skin and in the subcutaneous tissues (i.e., bones, skeletal muscles, etc.), were plotted as single/independent tumors. Multiple tumors of the same type in the same tissue and organ, a part those above mentioned, were plotted only once
*Statistically significant ($p \leq 0.05$) using Cochran-Armitage test
**Statistically significant ($p \leq 0.01$) using Cochran-Armitage test
##Statistically significant ($p \leq 0.01$) using poly-k test (k=3)

Table 4 Long-term carcinogenicity bioassay on ASPARTAME, administered with the feed supplied ad libitum, from 8 weeks of age until natural death, to male (M) and female (F) Sprague-Dawley rats (Experiment BT6008)

Results: lesions of oncological interest of the TCE[a] of the kidney pelvis and ureter

Group No.	Concentration (ppm)	Animals		No. of organs examined	Animals bearing tumors of the pelvis and ureter[b,c]			
					PAA[a]		CA[a,d]	
		Sex	No.		No.	%	No.	%
I	100,000	M	100	100	0	–	1	1.0
		F	100	100	3	3.0	4	**4.0**[#]
		M+F	200	200	3	1.5	5	2.5
II	50,000	M	100	100	0	–	1	1.0
		F	100	99	1	1.0	3	**3.0**
		M+F	200	199	1	0.5	4	2.0
III	10,000	M	100	100	0	–	1	1.0
		F	100	100	1	1.0	3(4)	**3.0**
		M+F	200	200	1	0.5	4	2.0
IV	2,000	M	150	150	0	–	1	0.7
		F	150	150	1	0.7	3(4)	**2.0**
		M+F	300	300	1	0.3	4	1.3
V	400	M	150	149	1	0.7	0	–
		F	150	150	1	0.7	3	**2.0**
		M+F	300	299	2	0.7	3	1.0
VI	80	M	150	149	0	–	0	–
		F	150	150	1	0.7	1	**0.7**
		M+F	300	299	1	0.3	1	0.3
VII	0 (control)	M	150	150	0	–	0	–
		F	150	150	0	–	0	–[*#]
		M+F	300	300	0	–	0	–

[a] *TCE* transitional cell epithelium, *PAA* papilloma with atypia, *CA* carcinoma
[b] Near the dosed group incidence are the p-values corresponding to pairwise comparisons between the controls and that dosed group
[c] Near the control incidence are the p-values associated with the trend test
[d] Between brackets the number of tumors (one animal can bear bilateral tumors)
[*] Statistically significant ($p \leq 0.05$) using Chocran-Armitage test
[#] Statistically significant ($p \leq 0.05$) using poly-k test ($k = 3$)

The data showed (1) a significant dose-related increased incidence of total malignant tumors in males and females (Table 3); (2) a significant dose-related increased incidence of transitional cell carcinomas of the renal pelvis in females, particularly at the highest dose (Table 4). Some of the lesions appeared highly invasive, extending from the pelvis surface right into the renal parenchyma, showing anaplastic cellular morphology, a lobular growth pattern originating from a broad base of papillary growth with some areas of mineralization (Fig. 2a, b);

Table 5 Long-term carcinogenicity bioassay on aspartame, administered with the feed supplied ad libitum, from 8 weeks of age until natural death, to male (M) and female (F) Sprague-Dawley rats (Experiment BT6008)

Results: malignant schwannomas of the peripheral nervous system and hematological neoplasias (HN)[a]

Group No.	Concentration (ppm)	Animals		Tumor-bearing animals							
				Peripheral nerves						Animals bearing HN[a, b]	
				Cranial		Others		Total[b]			
		Sex	No.	No.	%	No.	%	No.	%	No.	%
I	100,000	M	100	3	3.0	1	1.0	4	4.0	29	29.0
		F	100	1	1.0	1	1.0	2	2.0	25	25.0##
		M+F	200	4	2.0	2	1.0	6	3.0	54	27.0
II	50,000	M	100	3	3.0	0	–	3	3.0	20	20.0
		F	100	1	1.0	0	–	1	1.0	25	25.0##
		M+F	200	4	2.0	0	–	4	2.0	45	22.5
III	10,000	M	100	2	2.0	0	–	2	2.0	15	15.0
		F	100	1	1.0	0	–	1	1.0	19	19.0#
		M+F	200	3	1.5	0	–	3	1.5	34	17.0
IV	2,000	M	150	2	1.3	0	–	2	1.3	33	22.0
		F	150	1	0.7	2	1.3	3	2.0	28	18.7#
		M+F	300	3	1.0	2	0.7	5	1.7	61	20.3
V	400	M	150	1	0.7	2	1.3	3	2.0	25	16.7
		F	150	0	–	0	–	0	–	30	20.0##
		M+F	300	1	0.3	2	0.7	3	1.0	55	18.3
VI	80	M	150	1	0.7	0	–	1	0.7	23	15.3
		F	150	1	0.7	1	0.7	2	1.3	22	14.7
		M+F	300	2	0.7	1	0.3	3	1.0	45	15.0
VII	0 (control)	M	150	1	0.7	0	–	1	0.7*#	31	20.7*#
		F	150	0	–	0	–	0	–	13	8.7**#
		M+F	300	1	0.3	0	–	1	0.3	44	14.7

[a] *HN* hematological neoplasias
[b] Near the control incidence are the p-values associated with the trend test
* Statistically significant ($p \leq 0.05$) using Cochran-Armitage test
** Statistically significant ($p \leq 0.01$) using Cochran-Armitage test
\# Statistically significant ($p \leq 0.05$) using poly-k test (k = 3)
\#\# Statistically significant ($p \leq 0.01$) using poly-k test (k = 3)

(3) a significant dose-related increased incidence of malignant schwannomas of the peripheral nervous system in males (Table 5). The tumors originated mainly from Schwann cells of cranial nerves and, as shown in Fig. 3a–e, were composed of cystic cavities containing proteinaceous fluid and blood cells (a feature called Antoni type B pattern); the neoplastic cells were arranged in fascicles with dense cellularity and

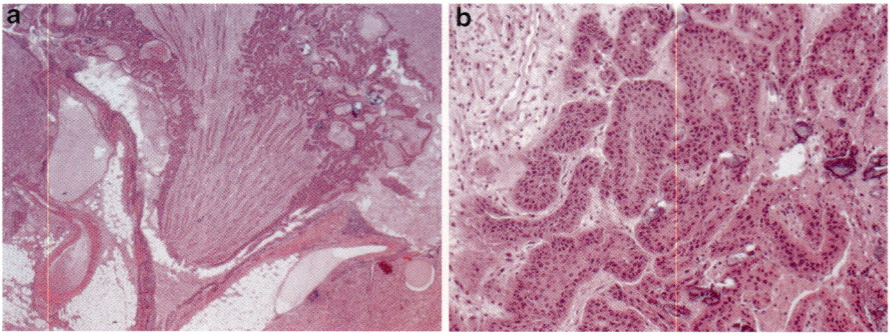

Fig. 2 Transitional cell carcinoma of the renal pelvis. (**a**) Highly invasive carcinoma extending from the pelvis surface to the renal parenchyma (HE 25X). (**b**) Higher magnification (HE 200X)

nuclear palisading (called Antoni type A); moreover, the neoplastic cells proved positive for S-100 proteins as evidenced by the dark brown staining and by Verocay bodies (Fig. 3e); (4) a significant dose-related increased incidence of lymphomas/leukemias in females, particularly in the range of APM exposure from 100,000 to 400 ppm (Table 5). The characteristics of the hematopoietic neoplasms were defined by morphology as lymphocytic lymphoma, lymphoimmunoblastic lymphoma, histiocytic sarcoma, and myeloid and monocytic leukemia. In some cases, immunohistochemical stainings were used to confirm the kind of neoplastic lesion.

The hematopoietic neoplasms involved the thymus and mediastinal lymph nodes (Fig. 4a, b), the spleen (Fig. 5a, b), the lung (Fig. 6a–f), the liver with invasion of blood vessels as in the case of myeloid leukemias (Fig. 7a, b), multiple organs as in the case of histiocytic sarcoma (Fig. 8a–c), or multiple organs with invasion of the blood vessels as in the case of monocytic leukemias (Fig. 9a, b).

This experiment showed for the first time that APM administered in feed to rats induces a significant dose-related increased incidence of several malignant tumors, even at a dose of 400 ppm (simulating an APM daily assumption of 20 mg/kg b.w. in rat).

6.2 Carcinogenicity Bioassay on APM Administered in Feed to Sprague-Dawley Rats from Prenatal Life until Natural Death (Exp. BT6009)

Two groups of 140 male and female Sprague-Dawley rats were exposed to APM administered in the feed at the concentration of 2,000 and 400 ppm from prenatal life (11th day of gestation) until natural death; one group of 190 males and females served as controls [41]. Overall, the study encompassed 470 rats. No

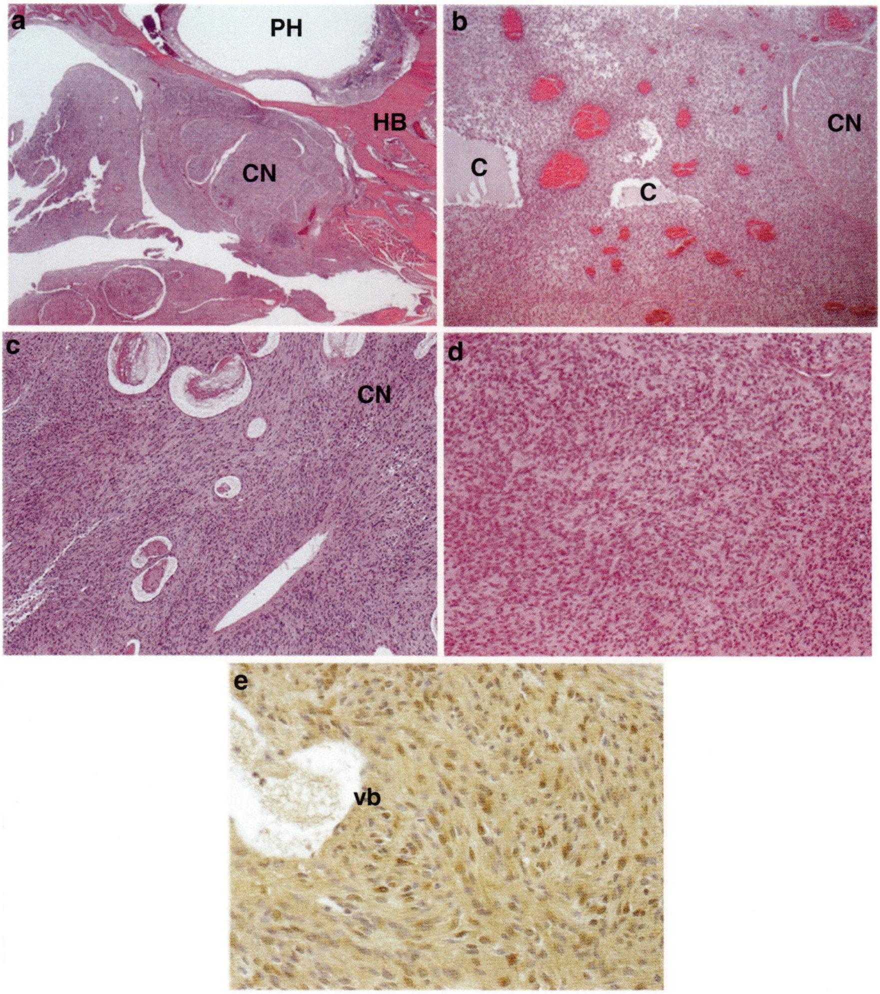

Fig. 3 Malignant schwannoma of cranial nerves. (**a**) The tumor originates from Schwann cells of the cranial nerves (*CN*) observed in this case along the sides of the pharynx (*PH*) in the head section (*HB*, head bone). (**b**) The tumor is composed of cystic cavities (*C*) (this feature is known as the Antoni type B pattern). (**c**) The neoplastic cells are arranged in fascicles with dense cellularity. (**d**) Nuclear palisading may be observed (this pattern is known as Antoni type A). (**e**) Positivity for S-100 protein evidenced by the dark brown staining of the neoplastic cells; Verocay bodies (*Vb*) are also present (**a**: HE 25X; **b**: HE 100X; **c**: HE 100X; **d**: HE 100X; **e**: 200X)

substantial difference in feed consumption was observed in males and females of treated rats compared to controls or again in mean body weights and survival [41].The results of the carcinogenic effects of APM are reported in Tables 6, 7, and 8.

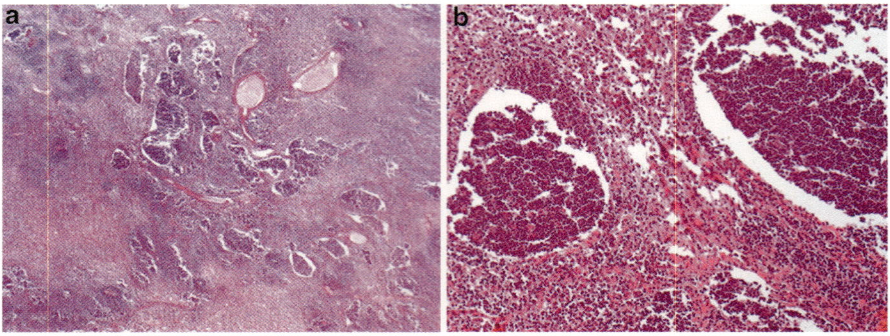

Fig. 4 Lymphocytic lymphoma in the thymus. (**a**) The architecture of the thymus is not maintained (HE 25X). (**b**) The cells are small to medium size resembling normal circulating lymphocytes that are arranged in high-density cell packages of probably monoclonal origin (HE 200X)

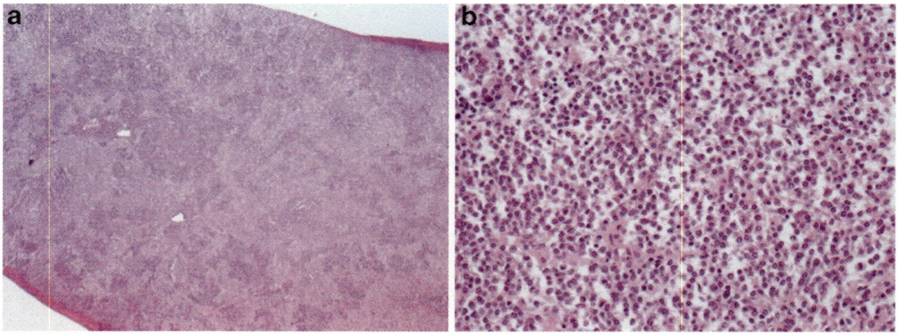

Fig. 5 Lymphoblastic lymphoma in the spleen. (**a**) The architecture of the spleen is not maintained (HE 25X). (**b**) The cells are non-cohesive, medium-sized to large lymphoblasts. The nuclear to cytoplasmic ratio is high and the cytoplasm is moderate in amount and often basophilic (HE 200X)

The data showed (1) a significant increased incidence (dose-related) of malignant tumors in males treated at the highest dose (Table 6), (2) a significant increased incidence (dose-related) of mammary adenocarcinomas (Fig.10a, b) in females treated at the highest dose (Table 7), and (3) a significant increased incidence (dose-related) of hematological neoplasias both in males and females treated at the highest dose (Table 8).

This study confirmed the carcinogenic effects of APM in rats and moreover clearly demonstrated an increased carcinogenic effect when life-span exposure to APM started from prenatal life.

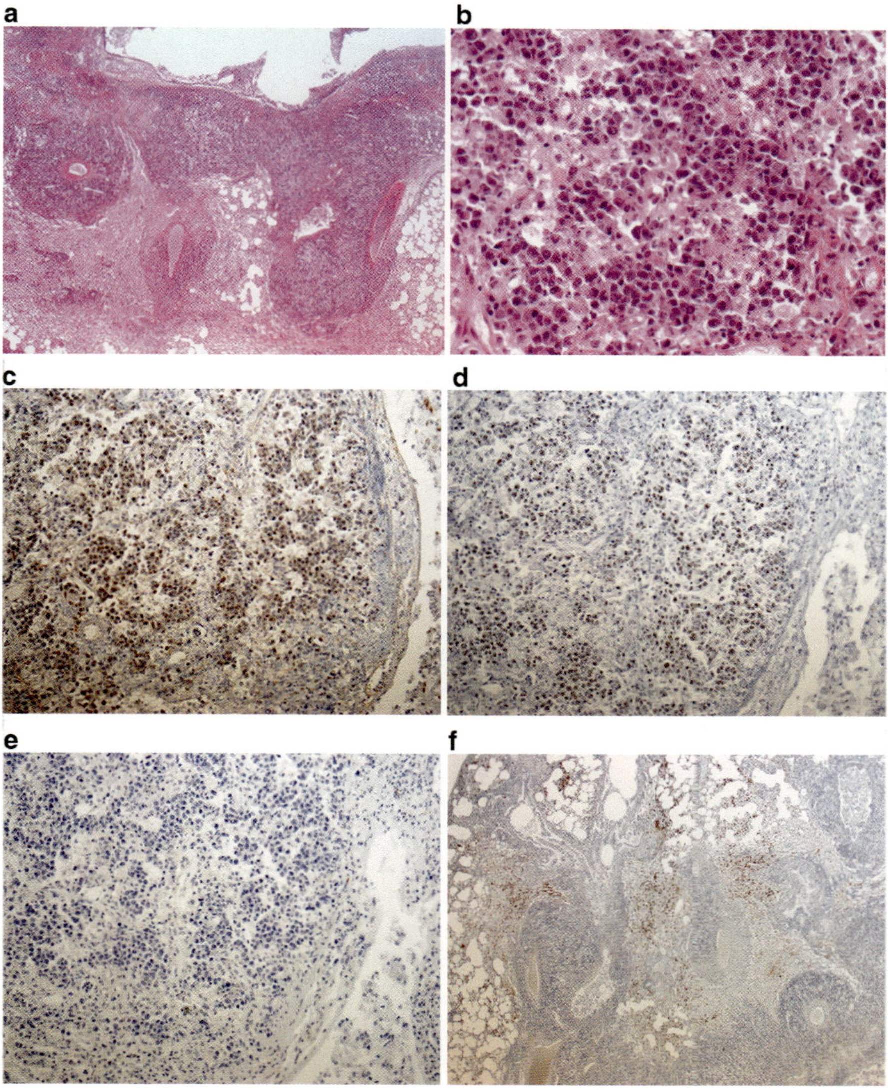

Fig. 6 Lymphoimmunoblastic lymphoma in the lung. (**a**) The pattern of organ involvement is aggressive with diffuse infiltration along the vascular tree in the lung (HE 25X). (**b**) The cells are large, non-cohesive, and monotypic; sometimes plasma cells may be present (HE 400X). (**c**) Pax5-positive staining in the zone occupied by the tumor shows the origin to be B-lymphocytes (200X). (**d**) Ki67-positive staining indicates a high index of cellular proliferation in the same area (200X). (**e**) CD68-negative staining in the zone occupied by the tumor (25X). (**f**) CD68-positive staining in the areas surrounding the lymphoma indicates that the macrophages are external to the tumor, and the lymphoma is not confused or mixed with inflammatory reaction (200X)

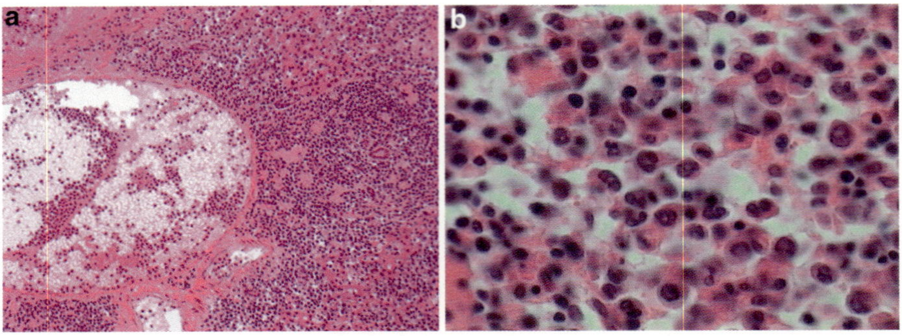

Fig. 7 Granulocytic or myeloid leukemia in the liver. (**a**) Basophilic cells are invading the blood vessels (HE 400X). (**b**) The cells are large and resemble developing or immature granulocytes. The cells may show some ring or lobed forms and others may be blastic in appearance (HE 1000X oil)

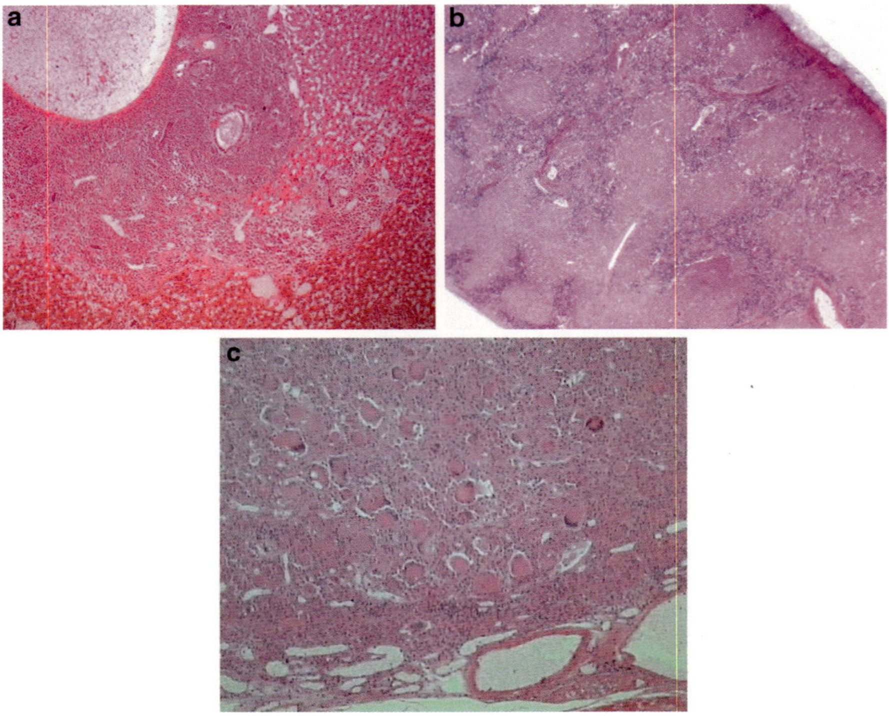

Fig. 8 Histiocytic sarcoma. The neoplasm involves the liver (**a**), spleen (**b**), and thymus (**c**). (**a**) In the liver, the tumor is characterized by a uniform monomorphic population of rounded cells with foamy eosinophilic cytoplasm (HE 100X). (**b**) In the spleen, the neoplastic cells invade the white pulp giving a pink color to the organ (HE 25X). (**c**) In the thymus, many multinucleated giant cells are scattered throughout the tumor (HE 200X)

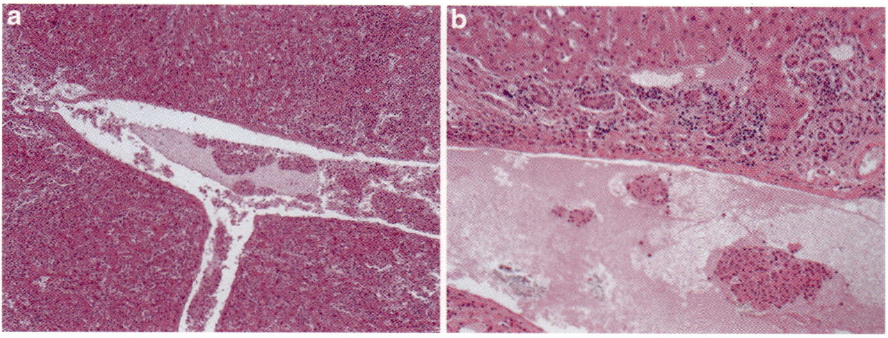

Fig. 9 Monocytic leukemia in the liver. The neoplastic cells appear immature and invade the blood vessels (HE 25X). (**b**) Higher magnification (HE 400X)

6.3 Carcinogenicity Bioassay on APM Administered in Feed to Swiss Mice from Prenatal Life until Natural Death (Exp. BT6010)

Four groups of 128–225 male and female Swiss mice were treated with APM in feed at the concentrations of 32,000, 16,000, 8,000, 2,000, or 0 ppm from prenatal life until natural death [42]. The biophase proceeded smoothly and without any problem. During the study, no substantial differences among the male and female treated groups compared to controls were observed in feed consumption and mean body weights. Compared to the control group, slight decreased survival was observed in treated males and females from 104 weeks of age until the end of the study [42].

The carcinogenic effects of APM in treated mice compared to controls are reported in Tables 9, 10, and 11. The data show (1) no significant difference among the treated groups versus control group in the incidences of animals bearing benign and malignant tumors (Table 9); (2) a significant increased incidence (dose-related) of alveolar carcinomas of the lung in males treated with APM at 32,000 and 16,000 ppm and a significant dose-related difference in the incidence of alveolar/bronchial adenoma in treated groups versus control group (Table 10). Microscopically, alveolar/bronchial adenomas presented different patterns of growth with a sharp demarcation from the adjacent parenchyma. The most common feature was papillary type with deeply basophilic cells (Fig. 11a, b). Alveolar/bronchial carcinomas presented an irregular nodular growth that tended to occupy the entire lobe invading the surrounding parenchyma. The most frequent feature was the papillary growth pattern (Fig. 12a, b); (3) a significant increased (dose-related) incidence of hepatocarcinoma of the liver in males treated at 32,000 and 16,000 ppm of APM compared to controls, as well as a dose-related (not

Table 6 Long-term carcinogenicity bioassay on ASPARTAME, administered with the feed, supplied ad libitum, from prenatal life until natural death, to male (M) and female (F) Sprague-Dawley rats (Experiment BT6009)

Results: benign and malignant tumors												
Group No.	Concentration (ppm)	Animals		Benign tumors				Malignant tumors				
				Tumor-bearing animals		Tumors		Tumor-bearing animals		Tumors		
		Sex	No.	No.	%	No.	Per 100 animals	No.	%	No.	Per 100 animals	
I	2,000	M	70	44	62.9	60	85.7	28	40.0**	31	44.3	
		F	70	58	82.9°	146	208.6	37	52.9	60	85.7	
		M+F	140	102	72.9	206	147.1	65	46.4	91	65.0	
II	400	M	70	43	61.4	65	92.9	18	25.7	19	27.1	
		F	70	54	77.1	166	237.1	31	44.3	44	62.9	
		M+F	140	97	69.3	231	165.0	49	35.0	63	45.0	
III	0 (control)	M	95	66	69.5	98	103.2	23	24.2***	26	27.4	
		F	95	75	78.9****	190	200.0	42	44.2	48	50.5	
		M+F	190	141	74.2	288	151.6	65	34.2	74	38.9	

°Statistically significant ($P \leq 0.05$) using logistic regression with a time covariate
**Statistically significant ($P \leq 0.01$) using Cox Regression Model
***Near the control incidence are the p-values ($P \leq 0.01$) associated with the Cox Regression Model for the analysis of the trend
****Near the control incidence are the p-values ($P \leq 0.05$) associated with the logistic regression with a time covariate for the analysis of the trend

Table 7 Long-term carcinogenicity bioassay on ASPARTAME, administered with the feed, supplied ad libitum, from prenatal life until natural death, to male (M) and female (F) Sprague-Dawley rats (Experiment BT6009)

Results: mammary benign and malignant tumors											
Group No.	Concentration (ppm)	Animals		Mammary benign tumors				Carcinomas			
				Tumor-bearing animals		Tumors		Tumor-bearing animals		Tumors	
		Sex	No.	No.	%	No.	Per 100 animals	No.	%	No.	Per 100 animals
I	2,000	M	70	0	–	0	–	2	2.9	2	2.9
		F	70	34	48.6	54	77.1	11	15.7*	15	21.4
		M+F	140	34	24.3	54	38.6	13	9.3	17	12.1
II	400	M	70	3	4.3	3	4.3	0	–	0	–
		F	70	38	54.3	52	74.3	5	7.1	6	8.6
		M+F	140	41	29.3	55	39.3	5	3.6	6	4.3
III	0 (control)	M	95	2	2.1	3	3.2	0	–	0	–
		F	95	45	47.4	65	68.4	5	5.3**	6	6.3
		M+F	190	47	24.7	68	35.8	5	2.6	6	3.2

*Statistically significant ($P \leq 0.05$) using Cox Regression Model
**Near the control incidence are the p-values ($P \leq 0.05$) associated with the Cox Regression Model for the analysis of the trend

Table 8 Long-term carcinogenicity bioassay on ASPARTAME, administered with the feed, supplied ad libitum, from prenatal life until natural death, to male (M) and female (F) Sprague-Dawley rats (Experiment BT6009)

Results: haematological neoplasias					
Group No.	Concentration (ppm)	Animals		Animals with Haematological neoplasias	
		Sex	No.	No.	%
I	2,000	M	70	12	17.1*
		F	70	22	31.4**
		M+F	140	34	24.3
II	400	M	70	11	15.7
		F	70	12	17.1
		M+F	140	23	16.4
III	0 (control)	M	95	9	9.5
		F	95	12	12.6***
		M+F	190	21	11.1

*Statistically significant ($P \leq 0.05$) using Cox Regression Model
**Statistically significant ($P \leq 0.01$) using Cox Regression Model
***Near the control incidence are the p-values ($P \leq 0.01$) associated with the Cox Regression Model for analysis of the trend

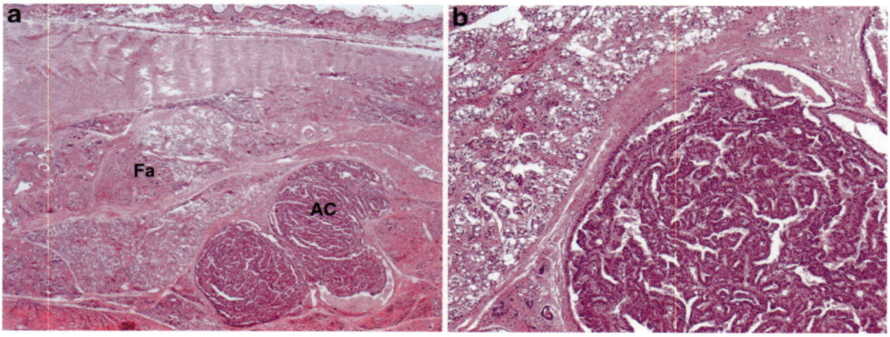

Fig. 10 Mammary gland adenocarcinoma. (**a**) Adenocarcinoma (*AC*) in fibroadenoma (*Fa*) diagnosed in a mammary lump from a female rat (HE 25X). (**b**) Note the papillary and ductular growth with anaplastic glandular features (HE 100X)

significant) increased incidence of hemangiosarcomas (Table 11). The liver tumors were usually grossly described as nodules, varying in size and color. Microscopically, they were distinguished as adenomas and hepatocarcinomas. Hepatocellular adenoma appeared as a very-well-demarcated nodular lesion causing distinct compression of the adjacent parenchyma. It appeared solid and moderately trabecular and the cells occurred in irregular plates two to three layers thick (Fig. 13a, b). The hepatocellular carcinoma was not well demarcated from the surrounding tissue and was characterized by an abnormal growth pattern with cellular atypia. The trabecular pattern was the most frequent type (Fig. 14a, b).

Table 9 Long-term carcinogenicity bioassay on ASPARTAME, administered with the feed, supplied ad libitum, from prenatl life until natural death to male (M) and female (F) Swiss mice (Experiment BT6010)

Results: benign and malignant tumors											
Group No.	Concentration (ppm)	Animals		Benign tumors				Malignant tumors			
				Tumor-bearing animals		Tumors		Tumor-bearing animals		Tumors	
		Sex	No.	No.	%	No.	Per 100 animals	No.	%	No.	Per 100 animals
I	32,000	M	83	22	26.5	29	34.9	57	68.7	87	104.8
		F	62	32	51.6	64	103.2	40	64.5	55	88.7
		M+F	145	54	37.2	93	64.1	97	66.9	142	97.9
II	16,000	M	64	26	40.6	33	51.6	39	60.9	63	98.4
		F	64	30	46.9	51	79.7	44	68.8	61	95.3
		M+F	128	56	43.8	84	65.6	83	64.8	124	96.9
III	8,000	M	62	20	32.3	26	41.9	45	72.6	66	106.5
		F	73	30	41.1	49	67.1	47	64.4	61	83.6
		M+F	135	50	37.0	75	55.6	92	68.1	127	94.1
IV	2,000	M	103	35	34.0	49	47.6	58	56.3	76	73.8
		F	122	53	43.4	81	66.4	90	73.8	117	95.9
		M+F	225	88	39.1	130	57.8	148	65.8	193	85.8
V	0 (control)	M	117	42	35.9	53	45.3	66	56.4	94	80.3
		F	102	39	38.2	56	54.9	69	67.6	92	90.2
		M+F	219	81	37.0	109	49.8	135	61.6	186	84.9

Table 10 Long-term carcinogenicity bioassay on ASPARTAME, administered with the feed, supplied ad libitum, from prenatal life until natural death to male (M) and female (F) Swiss mice (Experiment BT6010)

Results: tumors of the lung

Group No.	Concentration (ppm)	Animals		Tumor-bearing animals[a]					
				Alveolar/ bronchiolar Adenomas		Alveolar/ bronchiolar Adenocarcinomas		Total	
		Sex	No.	No.	%	No.	%	No.	%
I	32,000	M	83	6	7.2	11	13.3**	17	20.5
		F	62	3	4.8	2	3.2	5	8.1
		M+F	145	9	6.2	13	9.0	22	15.2
II	16,000	M	64	7(1)	10.9	8(3)	12.5*	15(4)	23.4
		F	64	2	3.1	7	10.9	9	14.1
		M+F	128	9	7.0	15	11.7	24	18.8
III	8,000	M	62	7(1)	11.3	7	11.3	14(1)	22.6
		F	73	3	4.1	6	8.2	9	12.3
		M+F	135	10	7.4	13	9.6	23	17.0
IV	2,000	M	103	9(2)	8.7	6	5.8	15(2)	14.6
		F	122	9	7.4	10	8.2	19	15.6
		M+F	225	18	8.0	16	7.1	34	15.1
V	0 (control)	M	117	8(1)	6.8[b]	7	6.0*	15(1)	12.8***
		F	102	4	3.9	7	6.9	11	10.8
		M+F	219	12	5.5	14	6.4	26	11.9

[a]Between brackets are reported the number of animals bearing multiple tumors
[b]Near the control incidence are the p-values associated with the trend test
*Statistically significant ($P \leq 0.05$), using Cox proportional hazard model
**Statistically significant ($P \leq 0.01$), using Cox proportional hazard model
***Statistically significant ($P \leq 0.05$) using logistic analysis

7 Epidemiological Studies on Cancer Risks and Exposure to Aspartame

Not until the early 2000s, there were two epidemiological case-control studies conducted to evaluate the carcinogenic risks among people who consumed products containing APM. The first study [43] showed an increased (nonsignificant) risk of brain tumors correlating with consumption of drinks containing APM. A second study [44] showed an increased trend of medulloblastoma in children born by mothers who frequently consumed diet soda during the preconception period and during pregnancy.

After the publication of the results of the Ramazzini Institute bioassays, some groups were motivated to perform epidemiological studies to evaluate the potential carcinogenic risks among consumers of products containing APM, particularly in

Table 11 Long-term carcinogenicity bioassay on ASPARTAME, administered with the feed, supplied ad libitum, from prenatal life until natural death to male (M) and female (F) Swiss mice (Experiment BT6010)

Results: preneoplastic and neoplastic lesions of the liver													
Group No.	Dose (ppm)	Animals		Bearing animals									
		Sex	No.	Foci		Hepatocellular adenoma		Hepatocellular carcinoma		Hemangioma		Hemangiosarcoma	
				No.	%	No.	%	No.	%	No.	%	No.	%
I	32,000	M	83	2	2.4	2	2.4	15	18.1**	0	–	8	9.6
		F	62	0	–	0	–	0	–	0	–	1	1.6
		M+F	145	2	1.4	2	1.4	15	10.3	0	–	9	6.2
II	16,000	M	64	2	3.1	6	9.4	10	15.6*	1	1.6	5	7.8
		F	64	1	1.6	0	–	2	3.1	2	3.1	3	4.7
		M+F	128	3	2.3	6	4.7	12	9.4	3	2.3	8	6.3
III	8,000	M	62	2	3.2	4	6.5	9	14.5	1	1.6	5	8.1
		F	73	1	1.4	2	2.7	0	–	0	–	0	–
		M+F	135	3	2.2	6	4.4	9	6.7	1	0.7	5	3.7
IV	2,000	M	103	2	1.9	10	9.7	12	11.7	2	1.9	5	4.9
		F	122	1	0.8	6	4.9	2	1.6	3	2.5	3	2.5
		M+F	225	3	1.3	16	7.1	14	6.2	5	2.2	8	3.6
V	0 (control)	M	117	5	4.3	9	7.7	6	5.1***	2	1.7	5	4.3
		F	102	2	2.0	1	1.0	0	–	0	–	4	3.9
		M+F	219	7	3.2	10	4.6	6	2.7	2	0.9	9	4.1

*Statistically significant ($P \leq 0.05$), using Cox proportional hazard model
**Statistically significant ($P \leq 0.01$), using Cox proportional hazard model
***Near the control incidence is the p-values ($P \leq 0.01$) associated with trend test

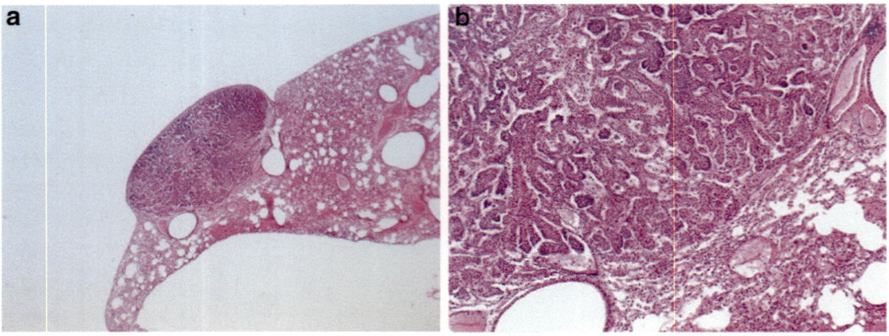

Fig. 11 Bronchiolo-alveolar adenoma in mice. (**a**) Papillary structure (HE 25X). (**b**) Higher magnification (HE 200X)

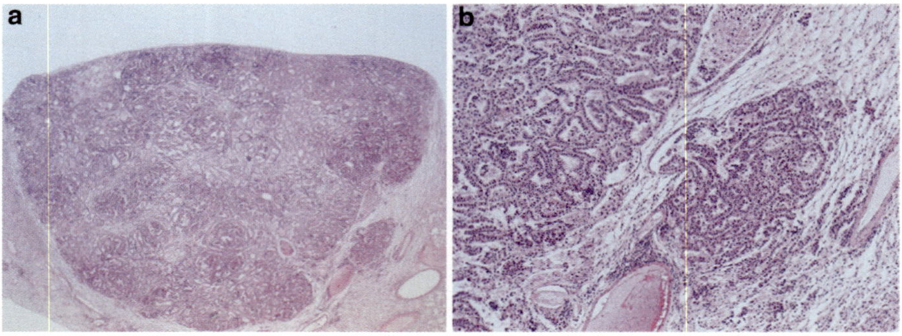

Fig. 12 Bronchiolo-alveolar carcinoma in mice. (**a**) Carcinoma with irregular growth invading large parts of the lobe (HE 25X). (**b**) The most frequent feature is the papillary growth pattern (HE 200X)

diet beverages. Of these, two prospective studies, one conducted at the US National Cancer Institute (NCI) and the second at the University of Harvard, dealt with hematopoietic cancers.

The NCI study was conducted on 473,984 males and females aged 50–71 who were surveyed in 1995 and followed up until 2000 for signs of gliomas (315 cases) and hematopoietic tumors (1,885 cases). The authors reported that for a daily intake of APM $>$ 900 mg/day, no significant increased risk of hematopoietic neoplasms (RR 0.98, 95% CI 0.76–1.27) or gliomas (RR 0.73, 95% CI 0.46–1.19) was observed [45]. However, it must be noted that the limited duration of exposure, the limited follow-up, and the low exposure levels greatly reduce the power to detect an effect.

The Harvard Study [6] conducted a prospective study on diet soda containing APM in relation to cancers in two cohorts: (1) the Nurses' Health Study, which started in 1976, and includes 121,701 female registered nurses; and (2) the Health Professional Follow-up Study which began in 1986 and includes 51,529 male

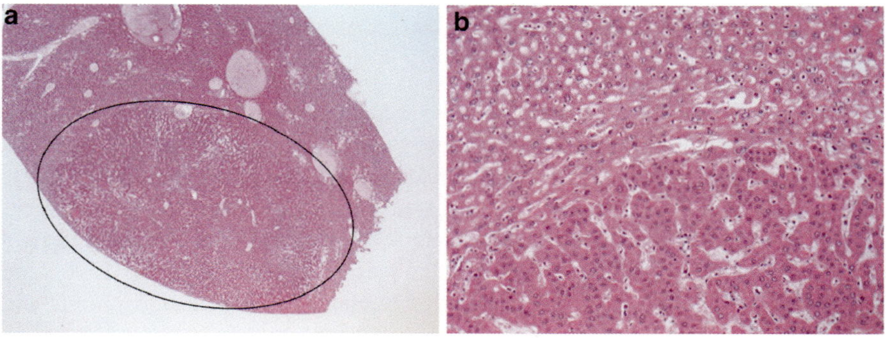

Fig. 13 Hepatocellular adenoma. (**a**) Hepatocellular adenoma (HE 25X). (**b**) Higher magnification (HE 200X)

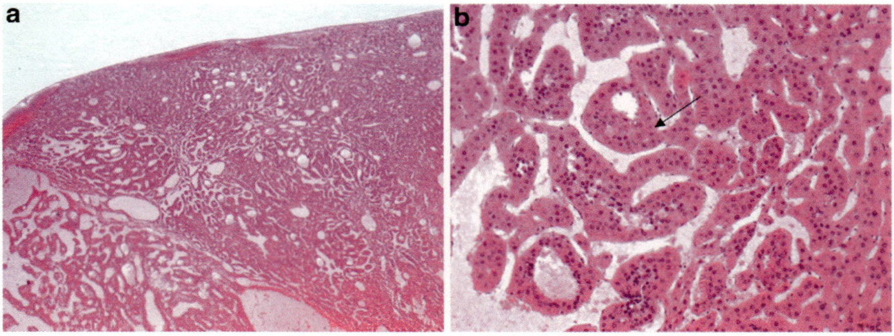

Fig. 14 Hepatocellular carcinoma. (**a**) Hepatocellular carcinoma (HE 25X). (**b**) Higher magnification (HE 200X)

health professionals. The study design covered diet soda consumption from 1984 for women and for both genders from 1986. The diet was reassessed every 4 years until 2006. The final study population included 77,218 women and 47,810 men. The authors concluded that in men a statistically significant increase was observed in the risk of non-Hodgkin lymphoma in individuals who consumed >1 serving of diet soda/day. Moreover, in men, the risk of multiple myeloma increased linearly with increased consumption, and a statistically significant increase was observed in individuals who consumed >1 diet soda/day. The authors concluded that their data provide some support for the evidence that APM induces hematopoietic neoplasias in animals chronically exposed to APM. However, the authors claimed that because this was the first demonstration of the carcinogenic effects of APM in humans, and because in the experimental studies [40, 41] the hematopoietic neoplasias were shown in females and not in males, the results need confirmation by other cohort studies in order to rule out chance as a possible explanation.

8 Conclusions

Aspartame is an artificial sweetener discovered by GD-Searle in 1965. The tests to evaluate its safety before commercialization were performed by the producer in the 1970s. Queries regarding the conduct of these studies were raised in the 1980s and are still present today. Because of the large-scale use of APM in more than 6,000 products and because pregnant women and children are the main consumers, it became urgent we perform new studies to evaluate the potential carcinogenic effects of APM using a more sensitive experimental model. For this reason, at the end of the 1990s, the Ramazzini Institute started a large project of carcinogenicity bioassays on rats and mice using the experimental methods employed from more than 40 years [34, 35].

The results of these studies showed that APM may induce a significant dose-related increased incidence of multiple malignant tumors in two species, rats and mice, and in both genders. The consistency and robustness of the evidence as to the carcinogenic potential of APM are supported by the following considerations due to the distinctive characteristics of the method followed by the Ramazzini Institute in conducting the carcinogenicity bioassays: (1) life-span observation of the animals enabled us to detect the tumor trend in the last part of the life-span which is the time that brings out the difference between treated and untreated experimental animals. As shown by the cumulative prevalence of carcinomas and atypical lesions of the pelvis and ureter, represented in Fig. 15, if the experiment had been truncated at 110 weeks of age, probably the dose-related significance of the increased incidences of these lesions would not have been so clear. The same pattern holds true for what concerns hematological neoplasias in females. The

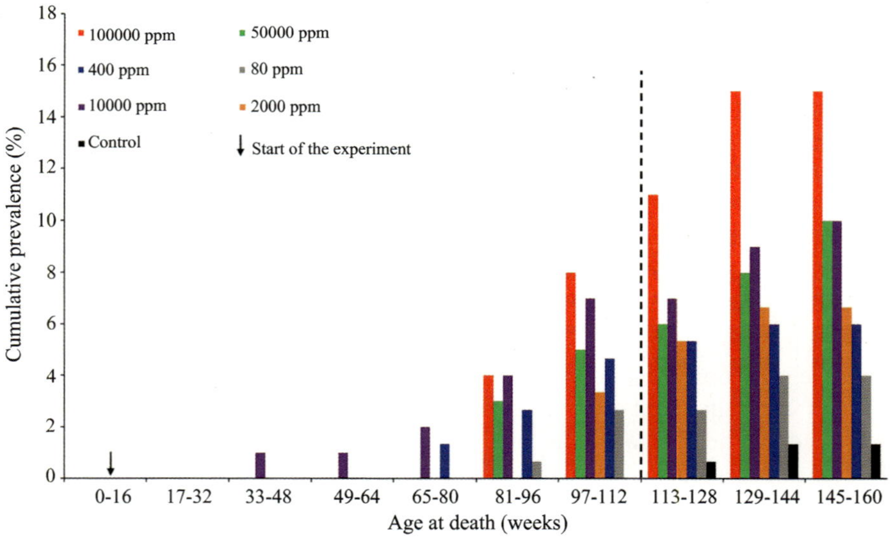

Fig. 15 Cumulative prevalence of female animals with carcinoma and atypical lesions of the pelvis and ureter, by age at death

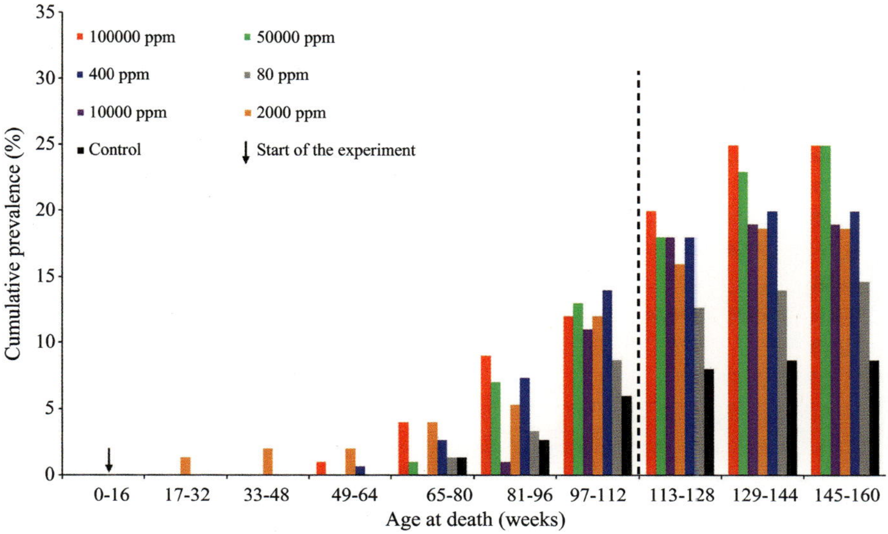

Fig. 16 Cumulative prevalence of female animals with hemolymphoreticular neoplasias, by age at death

cumulative prevalence of females bearing hematological neoplasias shows the difference between treated and untreated animals after 112 weeks of age (Fig. 16); (2) exposure starting from prenatal life increases the carcinogenic effects of APM. Indeed as shown in Fig. 17, the cumulative prevalence of hematological neoplasias in females exposed from prenatal life is higher than the cumulative prevalence in females exposed from mature age; (3) if the animals are allowed to survive until natural death, it is possible to observe the diffusion of hematological neoplasias into multiple organs and tissues. In the study where exposure to 2,000 ppm APM started from prenatal life, out of 22 females bearing hematological neoplasias, 14 involved the lung and other organs, while eight involved other organs though not the lung. Concerning lymphoimmunoblastic lymphoma in lung, the histopathological diagnosis of this type of lesions was confirmed by immunohistochemical stainings (Fig. 6c-f); (4) moreover, the prospective epidemiological study performed by the Harvard group supported the findings of the Ramazzini Institute in showing a positive association between APM exposure and hematological neoplasias in humans.

However, it cannot be forgotten that the evidence of the carcinogenic effects of APM documented by the experimental studies of the Ramazzini Institute and by the epidemiological results from the group at Harvard generated intense and coordinated criticism by spokespersons for the chemical industry in Europe, Japan, and the USA, as well as from EFSA and FDA, which led to a prompt reply by the Ramazzini Institute in a commentary published in 2014 [46].

Finally, if we consider that in our experimental conditions, the carcinogenic effects of APM are evident in female rats even at 400 ppm in feed, equivalent at a

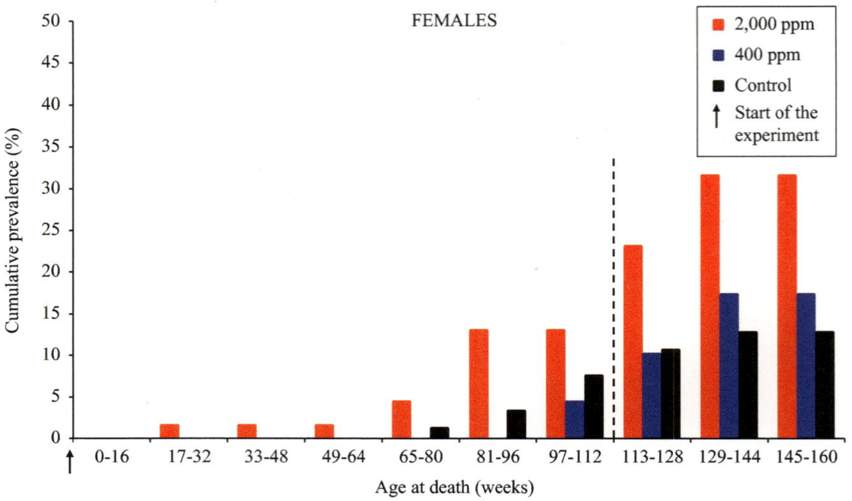

Life span feed carcinogenicity s tu d y of A**SPARTAME (BT 6009)**: cumulative prevalence of animals with **hemolymphoreticular neoplasias**, by age at death

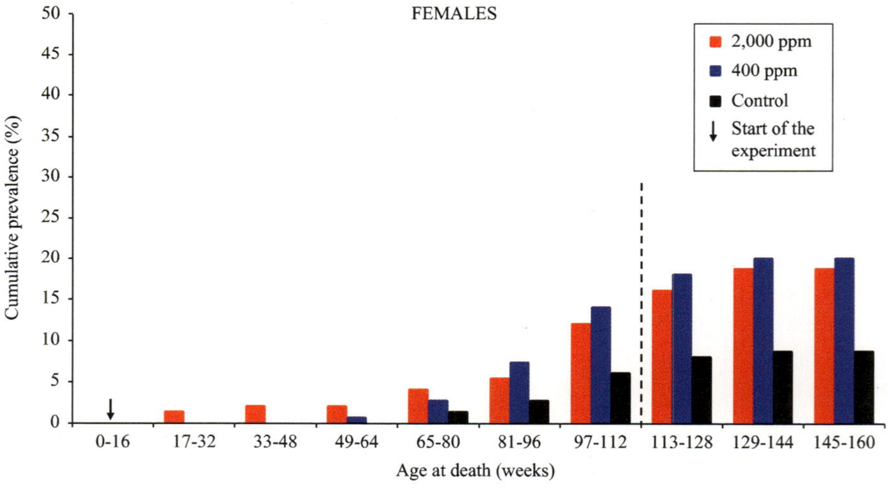

Life span feed carcinogenicity study of **ASPARTAME (BT 6008)**: cumulative prevalence of animals with **hemolymphoreticular neoplasias**, by age at death

Fig. 17 Comparison of cumulative prevalence of female animals with hemolymphoreticular neoplasias, by age at death, in experiments starting at 8 weeks of age or during fetal life

daily dose of 20 mg/kg b.w., much less than the current ADI for humans in Europe (40 mg/kg b.w.), we must conclude that it is urgent that agencies, such as the WHO, EFSA, FDA, international scientific institutions like IARC, and national public health institutions, take a stand and re-examine their evaluation on APM. Moreover, we recommend pregnant women and children to abstain from consuming products containing APM.

Acknowledgements A special thanks to Dr. Eva Tibaldi and Dr. Erica Tommasini for their extraordinary technical support during the preparation of the paper.

References

1. Sasaki YF, Kawaguchi S, Kamaya A, Ohshita M, Kabasawa K, Iwama K, Taniguchi K, Tsuda S (2002) The comet assay with 8 mouse organs: results with 39 currently used food additives. Mutat Res 519:103–119
2. Weihrauch MR, Diehl V (2004) Artificial sweeteners-do they bear a carcinogenic risk? Ann Oncol 15:1460–1465
3. EFSA (European Food Safety Authority) (2011) Call for scientific data on Aspartame (E 951). Available online http://www.efsa.europa.eu//en/dataclosed/call/110601.htm
4. Prankerd RJ (2002) Aspartame. Anal Profiles Drug Subst Excip 29:7–55
5. EFSA (European Food Safety Authority) (2013) Scientific opinion on the re-evaluation of aspartame (E 951) as a food additive. EFSA J 11(12):3496, 263 pp. doi:10.2903/j.efsa.2013.3496
6. Schernhammer ES, Bertrand KA, Birmann BM, Sampson L, Willett WW, Feskanich D (2012) Consumption of artificial sweetener-and sugar-containing soda and risk of lymphoma and leukemia in man and woman. Am J Clin Nutr 96:1419–1428
7. Food and Drug Administration (1981) Aspartame: commissioner's final decision. Fed Regist 46:38285–38308
8. Food and Drug Administration (1983) Food additives permitted for direct addition to food for human consumption: aspartame. Fed Regist 48:31376–31382
9. Food and Drug Administration (1996) Food additives permitted for direct addition to food for human consumption: aspartame. Fed Regist 61:33654–33656
10. European Parliament and Council (1994) Directive 94/35/EC of 30 June 1994 on sweeteners for use in foodstuffs. Off J L 237:3–12
11. Leung SS, Grant DJ (1997) Solid state stability studies of model dipeptides: aspartame and aspartylphenylalanine. J Pharm Sci 86:64–71
12. Homler (1984) Aspartame: implications for the food scientist. In: Stegink LD, Filer LJ Jr (eds) Aspartame: physiology and biochemistry, Marcel Dekker Inc. New York
13. Pattanaargson S, Chuapradit C, Srisukphoraruk S (2001) Aspartame degradation in solutions at various pH conditions. J Food Sci 66:808–809
14. Prodolliet J, Bruelhart M (1993) Determination of aspartame and its major decomposition products in foods. J AOAC Int 76:275–282
15. Saito K, Horie M, Hoshino Y, Nose N, Nakazawa H, Fujita M (1989) Determination of diketopiperazine in soft drinks by high-performance liquid chromatography. J Liq Chromatogr 12:571–582
16. Prankerd RJ, Stone HW, Sloan KB, Perrin JH (1992) Degradation of aspartame in acidic aqueous media and its stabilization by complexation with cyclodextrins or modified cyclodextrins. Int J Pharm 88:189–199
17. Opperman J (1984) Aspartame metabolism in animals. In: Stegink LD, Filer LJ Jr (eds) Aspartame. Physiology and biochemistry. Marcel Dekker Inc, New York
18. Hooper NM, Hesp RJ, Tieku S (1994) Metabolism of aspartame by human and pig intestinal microvillar peptidases. Biochem J 298:635–639
19. Magnuson BA, Burdock GA, Doull J, Kroes RM, Marsh GM, Pariza MW, Spencer PS, Waddele WJ, Walker R, Williams GM (2007) Aspartame: a safety evaluation based on current use levels, regulations, and toxicological and epidemiological studies. Crit Rev Toxicol 37:629–727
20. Burgert SL, Anderson DW, Stegink LD, Takeuchi H, Schedl HP (1991) Metabolism of aspartame and its L-phenylalanine methyl ester decomposition products by the porcine gut. Metab Clin Exp 40:612–618

21. Horton VL, Higuchi MA, Rickert DE (1992) Physiologically based pharmacokinetic model methanol for rats, monkeys and humans. Toxicol Appl Pharmacol 117:26–36
22. Stagink LD (1984) Aspartame metabolism in humans: acute dosing studies. In: Stagink LD, Filer LJ Jr (eds) Aspartame: physiology and biochemistry. Marcel Dekker Inc, New York
23. Leon AS, Unninghake DB, Bell C, Rassin DK, Tephly TR (1989) Safety of long-term large doses of aspartame. Arch Intern Med 149:2318–2124
24. Abhilash M, Sauganth Paul MV, Varghese MV, Nair RH (2011) Effects of long-term intake of aspartame on antioxidant defense status in liver. Food Chem Toxicol 49:1203–1207
25. Abhilash M, Sauganth Paul MV, Varghese MV, Nair RH (2013) Long-term consumption of aspartame and brain antioxidant defense status. Drug Chem Toxicol 39:135–140
26. NTP (US National Toxicology Program) (2005) NTP report on the toxicology studies of aspartame (CAS No. 22839-47-0) in genetically modified (FVB Tg.AC hemizygous) and B6.129-Cdkn2atm1Rdp (N2) deficient mice and carcinogenicity studies of aspartame in genetically modified [B6.129-Trp53tm1Brd (N5) haploinsufficient] mice (feed studies). National Toxicology Program genetically modified model report, pp 1–222
27. Rencuzogullarj E, Tuylu BA, Topaktas M, Ila HB, Kayraldiz A, Arslan M, Diler SB (2004) Genotoxicity of aspartame. Drug Chem Toxicol 27:257–268
28. Bandypadhyay A, Ghoshal S, Mukherjee A (2008) Genotoxicity testing of low-calorie sweeteners: aspartame, acesulfame-K and saccharin. Drug Chem Toxicol 31:447–457
29. Jeffrey AM, Williams GM (2000) Lack of DNA –damaging activity of five non-nutritive sweeteners in the rat hepatocyte/DNA repair assay. Food Chem Toxicol 38:335–338
30. Kamath S, Vijaynarayana K, Shetty DP, Shetty P (2010) Evaluation of genotoxic potential of aspartame. Pharmacol Online 1:753–769
31. EPA (1984) Good laboratory practices standards in environmental protection agency – fungicides, insecticides and rodenticides Act. 48 federal register 53946. Environmental Protection Agency
32. FDA (2000) Toxicological principles for the safety assessment of food ingredients: FDA Redbook 2000: section IV.B.1. General guidelines for designing and conducting toxicity studies. Nov 2003. Food and Drug Administration, Washington, DC, p 18
33. OECD (2009) Guideline for the testing of chemicals. Test guideline 451: carcinogenicity studies. Organization for Economic Cooperation and Development, Sep 2009. Paris, p 15
34. Soffritti M, Belpoggi F, Minardi F, Bua L, Maltoni C (1999) Mega-experiments to identify and assess diffuse carcinogenic risks. Ann N Y Acad Sci 895:34–55
35. Soffritti M, Belpoggi F, Minardi F, Maltoni C (2002) Ramazzini foundation cancer program: history and major projects, life-span carcinogenicity bioassay design, chemicals studied, and results. Ann N Y Acad Sci 982:26–45
36. Melnick RL, Thayer KA, Bucher JR (2008) Conflicting views on chemical carcinogenesis arising from the design and evaluation of rodent carcinogenicity studies. Environ Health Perspect 116:130–135
37. Ishii H (1981) Incidence of brain tumors in rats fed aspartame. Toxicol Lett 7:433–437
38. Iwata S (2006) Histopathological examination in a carcinogenicity study of SC-18862 in Wistar rats. Ajinomoto, unpublished report
39. Soffritti M, Belpoggi F, Degli Esposti D, Lambertini L (2005) Aspartame induces lymphomas and leukemias in rats. Eur J Oncol 10:107–116
40. Soffritti M, Belpoggi F, Degli Esposti D, Lambertini L, Tibaldi E, Rigano A (2006) First experimental demonstration of the multipotential carcinogenic effects of aspartame administered in the feed to Sprague-Dawley rats. Environ Health Perspect 114:379–385
41. Soffritti M, Belpoggi F, Tibaldi E, Degli Esposti D, Lauriola M (2007) Life-span exposure to low doses of aspartame beginning during pre-natal life increases cancer effects in rats. Environ Health Perspect 115:1293–1297
42. Soffritti M, Belpoggi F, Manservigi M, Tibaldi E, Lauriola M, Falcioni L, Bua L (2010) Aspartame administered in feed, beginning pre-natally through life-span, induces cancers of the liver and lung in male Swiss mice. Am J Ind Med 53:1197–1206

43. Hardell L, Mild KH, Pahlson A, Hallquist A (2001) Ionizing radiation, cellular telephones and the risk for brain tumours. Eur J Cancer Prev 10:523–529
44. Bunin GR, Kushi LH, Gallagher PR, Rorke-Adams LB, McBride ML, Cnaan A (2005) Maternal diet during pregnancy and its association with medulloblastoma in children: a children's oncology group study (United States). Cancer Causes Control 16:877–891
45. Lim U, Subar AF, Mouw T, Hartge P, Morton LM, Stolzenberg-Solomon R, Campbell D, Hollenbeck AR, Schatzkin A (2006) Consumption of aspartame-containing beverages and incidence of hematopoietic and brain malignancies. Cancer Epidemiol Biomark Prev 15:1654–1659
46. Soffritti M, Padovani M, Tibaldi E, Falcioni L, Manservisi F, Belpoggi F (2014) The carcinogenic effects of aspartame: the urgent need for regulatory re-evaluation. Am J Ind Med 57:383–397

Stevia Rebaudiana's Antioxidant Properties 16

Cecilia Bender

Abstract

Stevia rebaudiana Bertoni is a perennial herb native to South America and commonly known as "stevia." The leaves of stevia have been used for centuries in Paraguay and Brazil to sweeten food and beverage. Nowadays, the leaves of stevia and the steviol glycosides extracted from them are commonly used to sweeten beverages and foods. The steviol glycosides are a non-nutritive substitute of sugar that does not provide energy. Besides their sweetening properties, the antioxidant capacity of stevia and its sweet diterpenes are reported in the literature.

This chapter aims to review the antioxidant properties of *S. rebaudiana* reported in the literature, the future directions on this topic is discussed taking into consideration its implication on human health.

Keywords

Stevia • Stevioside • Reb A • Antioxidant activity in vitro • Antioxidant activity in vivo

Abbreviations

ABTS	2,2′-azino-bis(3-ethylbenzthiazoline-6-sulphonic acid)
CAA	Cellular antioxidant activity
CQA	Caffeoyl quinic acid
DPPH	1,1-diphenyl-2-picrylhydrazyl radical
EFSA	European food safety authority
FRAP	Ferric reducing/antioxidant power
GPx	Glutathione peroxidase

C. Bender (✉)
Institut Kurz GmbH, Köln, Germany

Istituto Kurz Italia SRL, Parma, Italy
e-mail: c.bender@kurz-italia.com

J.-M. Mérillon, K.G. Ramawat (eds.), *Sweeteners*, Reference Series in Phytochemistry,
https://doi.org/10.1007/978-3-319-27027-2_6

GR	Glutathione reductase
GSH	Reduced glutathione
JECFA	Joint FAO/WHO expert committee on food additives
ORAC	Oxygen radical absorbance capacity
Ox-LDL	Oxidized low density lipoprotein
Reb A	Rebaudioside A
ROS	Reactive oxygen species
SCF	Scientific committee for food
SOD	Superoxide dismutase

Contents

1 Introduction 350
2 About Stevia 351
3 International Legislation 352
4 Importance of *Stevia Rebaudiana* to Human Health 353
4.1 Nutritional Implications 353
4.2 Obesity 353
4.3 Diabetes and Blood Glucose Maintenance 353
4.4 Hypertension 354
4.5 Antimicrobial Properties 354
4.6 Anti-cariogenic Activity 355
4.7 Atherosclerosis 355
4.8 Cancer 355
5 Functional Components 355
5.1 Nutritional Composition 355
5.2 Stevioside and Rebaudioside A 356
5.3 Minor Steviol Glycosides 356
6 Antioxidants 357
6.1 Phenolic Composition 358
6.2 Flavonoids 360
6.3 Antioxidant Activity of *Stevia Rebaudiana* 360
6.4 Antioxidant Activity of Steviol Glycosides 363
6.5 Antioxidant Activity in Biological Systems 364
6.6 Antioxidant Activity In Vivo 367
7 Discussion 368
References 371

1 Introduction

Stevia rebaudiana (Stevia genus) is a composite herb, native to tropical South America, which is still in the wild in Paraguay. Nevertheless, it can grow fairly well and can be adapted to a variety of terrains and climates.

Stevia is known since the sixteenth century; the plant was known and often used by the Guaraní native population from Paraguay by the name of "Ka'á He'é" (literally sweet herb). This plant has excellent sweetening potential and medicinal properties, although it is yet not internationally concluded the controversy about the possible risks of its consumption. Stevia leaves are about 30 times sweeter than table sugar, depending upon the quality of the leaf, and are the source of a non-caloric sweetener. The natural stevia, unrefined, contains more than 100 chemical elements and volatile oils identified [1].

Stevia and its glycosides are currently used in several ways such as a mere infusion, in the liquid form or as soluble crystals, each of those having different properties or applications.

Stevioside ($C_{38}H_{60}O_{18}$), a steviol glycoside compound extracted from stevia plants, was identified as the main sweet principle of stevia, it has a relative sweetness estimated to be 300 times as sweet as table sugar. Notwithstanding, several steviol glycosides constituents from stevia are extracted to prepare formulations containing besides stevioside, rebaudioside A (Reb A) and other minor steviol glycosides constituents such as rebaudiosides B, C, D, E, F and M, steviolbioside, rubusoside and dulcoside.

The rapid development of the sweeteners industry and the research for less risky alternatives to the traditional synthetic sweeteners has incentivized stevia cultivation in various countries; its consumption has spread nearly globally as dried leaves as well in a wide variety of products such as sweeteners, additive for beverages and food, dietary supplements as well as for cosmetic formulations [2]. Steviol glycosides-preparations of high purity are between 200 and 350 times sweeter than sucrose at 5% sucrose equivalency [3]. In the international markets stevia derived sweeteners are commonly commercialized in the form of liquid extracts or as stevioside crystals. These compounds are becoming more interesting since they can be particularly beneficial for persons affected by heart disease, diabetes, overweight, and tooth decay among other health problems.

Beyond the interest demonstrated for its sweet diterpenes, several data reported in the literature point out the potential of stevia not only as the source of natural sweetness but also as an interesting source of other bioactive molecules (e.g., polyphenols) that are as well of commercial importance. Of particular interest is the antioxidant potential exerted by stevia's crude extracts and its derivatives, important for health purposes as well as natural preservatives for the food and cosmetic industries. Hence, the antioxidant properties of stevia and its steviol glycosides are an attractive argument for producers and consumers.

This chapter is focused on the antioxidant properties of the stevia plant and its steviol glycosides, a deepening on the data available in the literature is provided and discussed, thereby aims to a better understanding of stevia as a natural raw material and to its importance for the health food industry.

2 About Stevia

Stevia plant, scientifically named *S. rebaudiana* (Bertoni), belongs to the *Asteraceae* family and comprises about 230 species of perennial herbs and shrubs in the genus Stevia. *S. rebaudiana* is one of the two species that produce steviol glycosides [4]. The plant is well identified botanically, holds simple elliptic leaves, in early spring exposes small tubular white flowers, their fruits are achenes that have a papus which facilitates wind transport.

The dried leaves, which are about 30 times more sweet than sucrose, have been used for centuries by Guaraní indigenous from Paraguay as a traditional sweetener, added to yerba mate, medicinal teas and other food and beverages.

Most recently, stevia has become an important source of natural calorie-free sweeteners of which stevioside and reb A are the most abundant steviol glycosides [2, 5–9].

However, the production and marketing of stevia became difficult due to its safety, which conditioned the legislation in different countries. Another great challenge is inherent to the intrinsic characteristics of the plant, such its complex reproductive mechanism, poor seed germination and cross-fertilization, a very important characteristic when the commercial propagation is desired, since in the area of origin there are another 200 species of Stevia, which can fertilize the *S. rebaudiana*, varying unpredictably the characteristics of the offspring, regarding the content of the sweetener components [1] and other metabolites.

3 International Legislation

The international legislation on food products for human consumption is very strict, the sweeteners are part of these products, and hence their approval is conditioned by extensive studies regarding the viability of its consumption in short and long term and its performance at biological and chemical levels.

In the case of stevia, as well as for many other sweeteners, its regulation has involved great controversies over the safety of use, as food additive or as a dietary supplement, what resulted in total or partial ban on commercialization of stevia and/or its derivatives, such is the case of the United States (permitted as a dietary supplement). Despite, other international legislations not just allowed its commercialization, yet fostered their production, as in Paraguay, Argentina and Brazil. Nowadays, among the main producers of stevia plants and dried leaves are Paraguay, Argentina, Brazil, Israel, China, Thailand, and Japan [1].

Of note, in the European Union (EU) the use of stevia dried leaves and leaf crude extracts are currently not authorized as dietary supplement. However, the use of steviol glycosides as food sweetener (additive E960) was approved in 2011, and its use is regulated by the European Parliament and Council Regulation (EC) No 1333/2008 on food additives.

During the past three decades, several evaluations were made for stevioside by the Scientific Committee for Food (SCF), the Joint FAO/WHO Expert Committee on Food Additives (JECFA) and the European Food Safety Authority (EFSA). JECFA reviewed the safety of steviol glycosides and established an acceptable daily intake for steviol glycosides of 4 mg/kg body weight/day, expressed as steviol equivalents. In 2010, EFSA adopted a scientific opinion on the safety of steviol glycosides and concluded that under JECFA specifications these are not carcinogenic, genotoxic or associated with any reproductive/developmental toxicity [10]. More recently, in December 2015, EFSA reviewed the safety of steviol glycosides as food additives and concluded that provided the final product contain not less than 95% of steviol glycosides on a dry basis, the specific combination and ratio of steviol glycosides wouldn't be of safety concern. Moreover, being that there is no evidence of absorption for intact glycosides at realistic exposure levels, the acceptable daily intake was confirmed whether total steviol glycosides (stevioside, reb A, B, C, D, E, F and M; steviol biocide; rubusoside and/or dulcoside) comprise more than 95% of the whole mixture [3].

4 Importance of *Stevia Rebaudiana* to Human Health

4.1 Nutritional Implications

The medicinal and nutritional properties of stevia are being proclaimed since several years. A spectrum of biological effects has been attributed to stevia extracts and steviol glycosides derivatives, such as antioxidant, oxidative stress reducers, antimicrobial, anti-hypertensive and anti-hyperglycemic properties. In addition, stevia leaves and its sweet derivatives have been used as sugar substitutes to alleviate and treat a range health problems including obesity, diabetes, hypertension, infections, and dental caries.

A deepening about the use of stevia and its influence on human health is proposed in the following sections.

4.2 Obesity

The calorie contribution of stevia leaves was calculated as 2.7 kcal/g on dry weight basis [11], thus is considered a low-calorie natural sweetener. The pure steviol glycosides, however, do not contribute to calories intake.

The sweet aqueous extract from stevia contains various steviol glycosides that are 100–400 times sweeter than sucrose. Stevia does not substitute the sugar natural present in foods, but moderates the use of additional sugar to sweeten foods [12]. In this way, stevia and its steviol glycosides are recommended to reduce the caloric intake, as a sugar substitute, for overweight individuals with restricted diet or controlled calorie intake.

4.3 Diabetes and Blood Glucose Maintenance

Stevioside was introduced in England already in 1942 and adopted as a sugar substitute; in 1945, the Argentine Biological Institute had tested and achieved some preparations for diabetics, obtaining the first results in the treatment of diabetes, what motivated the scientific interest about the stevioside [1].

Presently, stevia is among the plants used for the treatment of diabetes, it helps in the prevention of diabetes type 2 firstly by moderating the added sugar intake.

In a feeding study in rats using a 10% of dried leaves of stevia it was reported a continued significant decrease in the liver glycogen after 2 weeks, and a significant reduction on blood glucose level after 4 weeks [13].

In another study in induced-diabetic mice daily treated with a methanolic extract of stevia leaves (300 mg/kg), it has been recorded a significant reduction on blood glucose level and an elevated hepatic glycogen content after 3 weeks treatment [14].

In a long-term study, a mixture of dried stevia plants has been used in the form of tea for some 45 years to reduce the blood sugar level in diabetic patients at a daily dose of up to 5 g, with a maintenance daily dose 1 g [15]. In a human study, Curi et al. evaluated the effect of aqueous extracts from stevia leaves on glucose tolerance [16]. It was found an increased glucose tolerance and decreased glucose levels in plasma of healthy volunteers, upon stevia treatment.

Stevioside counterbalances hypersecretion of alpha cells caused by fatty acids, it also up-regulates the expression of fatty acid metabolism genes [17]. Moreover, stevioside maintains blood glucose level by increasing glucose utilization, enhancing the secretion of insulin and down-regulating gluconeogenesis process [18, 19].

However, by regulating gluconeogenesis key enzymes, the leaves of stevia have shown a higher anti-hyperglycemic activity than stevioside [20].

4.4 Hypertension

In animal studies, it has been reported that stevioside can act as an anti-hypertensive substance. The intravenous administration of stevioside had an anti-hypertensive effect in hypertensive rats and anesthetized dogs [21, 22].

Additionally, several human studies promote stevioside as an interesting alternative for hypertensive subjects. In a 3 month clinical trial, stevioside shown a marked hypotensive effect [23]. In addition, two long-term clinical studies, conducted with hypertensive subjects who had consumed stevioside daily, shown evidence that stevioside has constantly reduced the systolic and diastolic blood pressures during the whole trial [23, 24].

4.5 Antimicrobial Properties

In ethnobotanical practices, stevia has been used to alleviate and treat wounds and skin sores. Indeed, stevia has the ability to inhibit the growth of certain microorganisms; these antimicrobial properties are mainly attributed to primary and secondary plant metabolites.

Several data have been reported in the literature providing evidence that stevia crude extracts and steviol glycosides have antimicrobial effects, which comprise bactericidal [25–27], antifungal [27–30] and antiviral [31] activities against several pathogen microorganisms.

4.6 Anti-cariogenic Activity

Sugar substitutes play a significant role in preventing dental caries. Based on data available from oral route exposure experiments, steviol glycosides are not toxic, have anti-microbial activity and are safe for consumption.

In a cariogenicity experiment conducted with a murine model, it was observed that stevioside and rebaudioside have the ability to reduce caries scores and *Streptococcus sobrinus* growth when compared to sucrose feed, it was concluded that steviol glycosides are not cariogenic [25]. These results are in agreement with other reports [32, 33].

4.7 Atherosclerosis

It has been reported that stevioside inhibits the atherosclerotic plaque in obese insulin-resistant mouse model by improving insulin signaling and antioxidant defense. Stevioside does not induce a glycemic response in this model [34].

4.8 Cancer

In vivo studies using murine models seem to indicate an inhibitory effect exerted by stevia leaf extracts and its derivatives on skin tumor development [35–37]. The presented studies suggest that stevia and related compounds could be an important source of chemoprotective agents. This notwithstanding, prudence is mandatory since these results must be confirmed with clinical trials.

5 Functional Components

The presence of many functional components has been confirmed in the extracts from *S. rebaudiana*; including diterpene glycosides, labdane-type glycosides, triterpenes and phytosterols. Stevia is also rich in phenolic compounds (flavonoids and phenolic acids) and their derivatives; their composition widely depends on the genotype, ecological conditions, as well as the source of plant material.

5.1 Nutritional Composition

The moisture of dried stevia leaves is ~ 8.4% and dry weight composition ~91.6% (Institut Kurz GmbH, unpublished results) is given as ~ 49.7% of total carbohydrates, up to 20% steviol glycosides, ~ 13.4% of proteins, ~ 3.7% of lipids and ~ 13.83% of crude fibers (summarized in [38]). The content of water-soluble substances has been indicated ~ 42% [39].

Unrefined stevia contains more than 100 chemical elements and volatile oils [1]. The dried leaves of the stevia are 20–35 times sweeter than sugar. Indeed, the leaves contain a complex mixture of natural sweet diterpene glycosides which do not provide calories. The content of steviol glycosides in leaves ranges from 8.4 to 14.3% in dry weight, being much higher in leaves than in stems (2.3% in dry stems weight) [40].

Additional not-sweet constituents have been identified in stevia leaves such as polyphenols, flavonoids, labdane-type diterpenes, sterebins, triterpenes, sterols, pigments, gums and inorganic matter [38, 39]. Several antioxidants compounds have been isolated from stevia, such as apigenin, quercetin, isoquercitrin, luteolin, chlorogenic acid, miocene, kaempferol and caffeic acid [41].

Between water soluble vitamins, the more represented has been folic acid (~ 520 μg/g), followed by ascorbic acid (~ 100 μg/g) and vitamin B2 (~ 40 μg/g) [42]. Excluding tryptophan, all essential amino acids have been identified in stevia plants [43], with a content of 7.7 g/100 g, while total nonessential amino acids content have been reported to 3.7 g/100 g of dried leaves [27].

Mineral content is mostly represented by potassium, calcium, magnesium, phosphorus sodium, iron and zinc [27, 44, 45].

5.2 Stevioside and Rebaudioside A

Most of the steviol glycosides consist of stevioside, a glycoside of the diterpene derivate steviol. Stevioside is an abundant component of the leaf, it has a molecular weight of 804.8, formula $C_{38}H_{60}O_{18;}$ and is described as a glycoside comprising three glucose molecules attached to an aglycone, the steviol moiety [43]. Stevioside is extracted and refined mainly from leaves; it has an intense sweetness being nearly 300 times sweeter than sucrose, thus making stevioside an excellent non-caloric substitute of sugar.

Stevioside is also the substrate for rebaudioside A (reb A) synthesis [46], the second most representative glycoside component of stevia leaves. Reb A has a sweetening potential still superior to stevioside (250–450 times) [47].

Table 1 provides an overview of the two main steviol glycosides contents variation on dried leaves (w/w %), as reported by several authors.

5.3 Minor Steviol Glycosides

Besides reb A and stevioside, the leaves of Stevia accumulate a combination of at least 30 steviol glycosides, differing in the substitution of R1, R2 and/or R3 of the entkaurene body [38]. Among these less abundant diterpene glycosides, the most representative are rubusoside (concentrations ranging from traces to 1.8%), Reb C (0.2–1.1%), Reb F (up to 1.4%), steviolbioside (up to 0.8%), dulcoside A (up to 0.2%), Reb E (up to 0.2%), Reb D (up to 0.1%) and Reb B (up to 0.06%) [40].

Table 1 Variation of stevioside (Stv) and rebaudioside A (reb A) levels, total content on steviol glycosides (SvG), and reb A to Stv ratio on different leaves samples from S. rebaudiana. nr: not reported

Stv %	Reb A %	SvG content %	Reb A/Stv	Reference
6.8–16.0	4.6–11.3	19.9–26.0	0.3–1.5	[48]
4.0–13.0	2.0–4.0	7.4–19.7	n.r	[39]
5.5–7.5	2.0–4.7	8.4–14.3	0.3–0.7	[40]
5.9–10.9	3.3–7.6	12.8–15.4	0.3–1.3	[49]
6.2 ± 0.6	2.0 ± 0.3	9.6 ± 1.1	0.3 ± 0.1	[50]
1.4–9.3	1.0–11.2	10.0–17.0	0.1–6.1	[51]
1.8–7.8	4.7–11.7	13.5–20.1	0.75–3.9	[52]

Of note, another minor steviol glycoside constituent of the stevia plant was recently identified as reb M (also known as reb X) [53], with concentration levels reaching up to 0.26% in leaves [54]. Reb M is even sweeter than table sugar (approximately 350 times) and has clean sweet taste that resembles sucrose more closely [3, 54].

The accumulation of steviol glycosides in the leaves is a dynamic process, widely influenced by genetic variability, the development stage, photoperiod, ontogenesis and of course by the cultivation conditions [38, 55]. However, it showed that improving glycosides biosynthesis is feasible, for example, its modulation through nitrogen fertilization significantly increased the reb A content, hence reb A to stevioside ratio [56].

6 Antioxidants

Free radicals and their reactive metabolites are essential intermediates for cell signaling, infection control, and apoptosis; however, free radicals have been implicated in many pathological processes (progression or development) comprising cancer, heart disease, Parkinson's and Alzheimer diseases or aging process [57–59].

Free radicals are mostly very reactive substances, that can pair their unpaired electron by oxidizing another compound [57]. Then, free radicals are oxidants.

Conversely, antioxidants are electron donors, able to counteract free radicals and their reactive metabolites and thus are considered to be health promoters. Antioxidants exert their effect via several basic mechanisms; these are capable of scavenge free radicals and ROS (reactive oxygen species) by directly transferring electron free radicals or by modulating the antioxidant enzyme system (i.e., activation of antioxidant enzymes or inhibition of cellular oxidases).

The antioxidants can be of different sources, including endogenous enzymatic systems (such as superoxide dismutase, catalase and glutathione peroxidase enzymes), phenol compounds, non-enzymatic proteinaceous (like ferritin and albumin) and low molecular weight molecules (phenolic compounds, glutathione, vitamin A, C and E, among others) [57, 60, 61].

Oxidative stress is an imbalance between the production of oxidants and their elimination by antioxidant defenses, leading to damage to key biomolecules (DNA, lipids, and proteins), cell dysfunction and ultimately cell death, which in turns influence the whole organism [57, 58].

In healthy conditions nearly 1% of the ROS produced evade the control of the endogenous antioxidative systems, thus contributing to the oxidative damage [59].

The contribution to the protection of cells and their constituent molecules against oxidation can be beneficial for the cells, in addition, antioxidant-rich formulations often exhibit health benefits such as anti-inflammatory, anti-diabetic and anti-aging. The exogenous antioxidants can be provided by dietary intake at adequate doses, common dietary antioxidants include supplementation with vitamins C, A, E, and polyphenols [61].

Several studies were conducted to investigate the potential of stevia as a source of antioxidant compounds with promising benefits to human health. The stevia plant contains antioxidant compounds with various biochemical functions, including, among others, essential antioxidants (e.g., vitamin C and amino acids) [27, 42], phenolic compounds [40, 62–64]; flavonoids [27, 45, 62, 65–67], and tannins [11].

6.1 Phenolic Composition

Polyphenols are very important plant metabolites because of their ability in trapping free radicals. As antioxidants, polyphenols may contribute to the antioxidative action and thus are believed to play a protective role in various degenerative diseases associated with oxidative stress.

Different authors evaluated the phenolic content of stevia. Detailed information on individual phenolic acid composition is summarized in Table 2.

By means of HPLC-UV, the TPC has been found varying from 17.49 to 195.30 mg chlorogenic acid/g dry leaves [40].

In a study conducted by Karaköse et al. (2011), the chlorogenic acids of methanolic extracts from stevia leaves were profiled by means of LC-MS and LC-TOF methods. A total of 24 chlorogenic acids were detected (370 μg/g of dry leaves). Among all caffeoylquinic acids (CQAs), 3,5-di-CQA was the most abundant (145.6 μg/g); while among the mono-CQAs, 4-CQA was found to be the most abundant (70.3 μg/g). Additionally, tri-CQA were found with three regio-isomers for the first time reported in nature [64].

In agreement with this findings, in a recent study it was found by means of UHPLC-UV technology that the main hydroxybenzoic acid ester in stevia leaves is 3,5-di-O-CQA (ranging from 10.04 to 97.99 mg/g), followed by 4-O-CQA (ranging from 4.81 to 41.56 mg/g) and 4,5-di-O-CQA (ranging from 1.51 to 40.40 mg/g) [40].

According to a large agricultural trial, counting 166 *S. rebaudiana* samples including 7 botanical varieties and many commercial samples, a unique data set of the phenolic metabolome of stevia have been recently obtained [62]. In this study, a quantitative analysis was carried out by LC-MS with methanolic extracts from leaves. Similar to their previous report, these authors have detected 29 chlorogenic

Table 2 Total phenolic content (TPC) and phenolic composition of S. rebaudiana

Main phenolic components	TPC	Separation/detection modes	Reference
29 Chlorogenic acids, mainly 4-CQA, 5-CQA, 3,5-di-CQA and 4,5-di-CQA	53.42 mg chlorogenic acid/g	LC-ESI/MS and LC-TOF/UV (methanol extracts)	[64], [62]
Pyrogallol, 4-methoxybenzoic acid, p-coumaric acid, 4-methylcatechol, sinapic and cinnamic acids	130.67 mg catechin/g	HPLC/DAD (water extracts)	[42]
Hydrobenzoic acid esters: 4-O-CQA, 3,5-di-O-CQA, 4,5-di-O-CQA	17.49–195.30 mg chlorogenic acid/g	UHPLC/UV. MS/MS peaks identification (water extract)	[40]
Flavonoids (0.83 mg quercetin/mg): quercetin-3-O-arabinoside, quercitrin, apigenin, apigenin-4-O-glucoside, luteolin, and kaempferol-3-O-rhamnoside	860 mg gallic acid/g	LC-MS and HNMR (ethyl acetate extract)	[65]
di-CQA, chlorogenic acid, quercetin 3-*O*-xyloside, apigenin-7-*O*-glucoside, 3,4-dimethoxycinnamic acid, luteolin-7-*O*-rutinoside, and caffeic acid	91 mg gallic acid/g	LC-MS (ethanol extracts)	[68]
Ferulic acid derivatives, rosmarinic acid derivatives, caffeic acid derivatives	13.35 mg gallic acid/g	HPLC/DAD (glycol-aqueous extract)	[69]

acid derivatives, conversely, the quantitative values obtained with fresh leaves were found to be 3–5 times higher than values previously reported. Three di-CQA isomers were identified and allocated as 3,5- di-CQAs, 3,4-di-CQAs, and 4,5-di-CQAs. Total di-CQA content was in average 2.435 g per 100 g of dried leaf material, being 4,5-di-CQA the most abundant among this group. Also, mono-CQA content was in average 2.91 g/100 g, and among this group was 5- CQA the most abundant. Tri-CQAs and caffeoyl shikimates were present as minor compounds.

The quantity of each mono- and di-CQA as well as the total mono- and di-CQAs have shown normal distribution among all the samples. The data reveal significant variations in average chlorogenic acid concentrations between different origins, varying from 3.09 to 1.64 g/100 g for total mono-CQAs, and from 2.89 to 1.14 g/100 g for di-CQAs [62].

Additionally, a total of 15 peaks corresponding to flavonoid glycosides were identified in stevia samples. Unexpectedly, none correlation was found between the concentrations of each individual flavonoid glycosides and chlorogenic acids, as well as between total contents of chlorogenic acids and flavonoids [62].

Authors noted that in terms of absolute quantities of chlorogenic acids and among all dietary materials, the leaves from stevia rank third, following green coffee beans and yerba mate leaves [62].

Gaweł-Bęben et al. [69] analyzed the content of polyphenols on aqueous, ethanolic and glycol-aqueous extracts from stevia leaves. The highest amount of phenols (15.50 mg/g) and flavonoids (3.85 mg/g) were found in glycol-aqueous extract, especially ferulic (5.50 mg/g) and rosmarinic (4.95 mg/g) acids derivatives.

By means of a mass spectrometry assay, conducted with ethanolic extracts from stevia powder, it has been identified the main phenolic compounds as di-CQA, chlorogenic acid, quercetin 3-O-xyloside, apigenin-7-O-glucoside, 3,4-dimethoxycinnamic acid, luteolin-7-O-rutinoside, and caffeic acid [68]. While, in another study, pyrogallol was identified as the main phenolic compound in leaf extract (9.51 mg/g dry base water extract), followed by 4-methoxybenzoic acid (0.33 mg/g) [42].

6.2 Flavonoids

Among phenolic compounds, flavonoids are one of the largest groups widely distributed in plants. Flavonoids exert a significant antioxidant capacity in vitro, they are effective in chelate metals, oxidize/reduce free radicals, and regenerate antioxidants.

Although the absorption from dietary-flavonoids is low and have also a short half-life in plasma [57], flavonoids can positively modulate antioxidant enzymes and inhibit oxidases in vivo.

All the flavonoids identified in *S. rebaudiana* belong to the flavonols and flavones subgroups. By means of the aluminum chloride colorimetric method, the flavonoid content in water extracts from leaves has been quantified in 15.64 mg quercetin [42].

In a further study, the isolated flavonoids (0.83 mg quercetin/mg) from ethyl acetate extracts from leaves were characterized by LC-MS and HNMR analyzes as quercetin-3-O-arabinoside, quercitrin, apigenin, apigenin-4-O-glucoside, luteolin, and kaempferol-3-O-rhamnoside [65].

Interestingly, in a recent study performed by our group with UHPLC equipment, the flavonoid constituents have been detected in trace amounts, through MS but not by UV, this low detection could be explained by the polar extraction performed in this study since flavonoids are more efficiently extracted using organic solvents [40].

With the largest report about the phenolic metabolome of *S. rebaudiana*, it was confirmed that methanolic extracts from stevia leaf samples contain a series of flavonoid glycosides; additional hydrolysis experiments followed by LC-MS analyses confirmed the presence of aglycones of kaempferol, quercetin, luteolin, and apigenin [62], thus in agreement with previous findings [65, 68]. No new additional phenolic compounds have been identified in any of the 166 samples analyzed by these researchers.

6.3 Antioxidant Activity of *Stevia Rebaudiana*

The methods employed to determine the antioxidant capability of stevia extracts are mainly based on chemical reactions. These chemical assays provide important basic information on the in vitro antiradical activity exerted by extracts.

Table 3 Methodologies used for measurement of the radical scavenging antioxidant activities of different S. rebaudiana extracts from leaves, callus and stems

Assay	Target	Reference
Hydroxyl radical scavenging test modified using TPA	Hydroxyl radical (OH)	[42, 70, 71]
Superoxide anion radical scavenging activity	Superoxide anion radical ($O_2^{\cdot -}$)	[42, 63, 72]
ORAC	Peroxyl radical (ROO·)	[40, 73, 74]
Deoxyribose assay	·OH	[63, 65, 72]
DPPH	DPPH radical	[42, 45, 63, 65, 66, 75, 76]
ABTS	ABTS radical	[65, 73, 77]
FRAP	Ferrous ion	[45]
Folin-Ciocalteu assay	Total phenolic	[42, 45, 63, 65, 66]
Peroxynitrite ion	Peroxynitrite ion (ONO_2^-)	[72]
Greiss test	Nitric oxide (NO·)	[63, 72]

Table 3 lists several free radical scavenging assays which have been reported in the literature for in vitro evaluation of the antioxidant activity of stevia.

Tadhani et al. presented the first report showing that leaf and callus extracts from *S. rebaudiana* are able to effectively scavenge free radicals in vitro. Leaves extracts exhibited a slightly higher total antioxidant activity than callus when assessed by ferric reducing/antioxidant power (FRAP) assay while the highest percent of inhibition of 1,1-diphenyl-2-picrylhydrazyl (DPPH) radical was recorded with methanolic extract of callus. In their work, authors also evaluated the total phenols and flavonoids content according to the Folin–Ciocalteu method, amounts were found to be nearly 1.4 times greater in callus extracts when compared with leaves extracts [45].

The same year, investigators from Thailand studied the antioxidant capacities of five medicinal plants including *S. rebaudiana*, by means of ABTS (2,2′-azino-bis (3-ethylbenzthiazoline-6-sulfonic acid)) assay, utilizing water extracts as well as different organic extractions [77]. These investigators reported that stevia has the highest antioxidant capacity (1.67–2.96 μmol of trolox/mg), ranging from 1.9 to 9.3 fold higher with regard of *Cassia alata* > *Curcuma longa* > *Andrographis paniculata* and > *Pueraria mirific* plants. In this study, it was also found that the acetone extraction from stevia gives the highest antioxidant capacity compared to other solvents (acetone ≈ methanol > ethanol > acetic acid > water).

Gantha et al. (2007) evaluated the total antioxidant capacity of an ethyl acetate extract derived from of stevia leaves by means of DPPH (IC_{50} = 9.26 μg/mL), ABTS (IC_{50} = 3.04 μg/mL) and deoxyribose assay (IC_{50} = 3.08 μg/mL), the total polyphenols resulted to be 0.86 mg gallic acid/mg leaves [65].

Shukla et al. (2009) studied the antioxidant activity of ethanolic extracts from stevia leaves using five complementary methods. It was found that the ethanolic extract is rich in polyphenols (61.5 mg gallic acid/g) and exerts, in a dose dependent manner, a higher antioxidant activity when compared to ascorbic acid and against hydroxyl radicals (IC_{50} = 81.08 μg/mL), superoxide anion (IC_{50} = 109.01 μg/mL), DPPH (IC_{50} = 93.46 μg/mL) and nitric oxide (IC_{50} = 132.05 μg/mL) [63].

Jahan et al. (2010) assessed the antioxidant capacity of stevia leaves using aqueous and ethanolic extracts and different in vitro methods. The total phenolic concentration in an 80% ethanol extract (65.21 mg of gallic acid equivalent/g) was 1.57 fold higher than the water extracts, the total flavonoid concentration in the ethanolic extract was 1.23 fold higher than in water. Also, the ethanol extract showed the higher reducing power and DPPH free radical scavenging values [66].

Kim et al. (2011) found that water extracts of stevia leaves have a relatively high dose-dependent antioxidant activity, by inhibiting the hydroxyl radical and superoxide anion activities in vitro. Their work provided evidence that the radical scavenging capacities, against free radicals, hydroxyl radicals and superoxide anion radicals, are higher in leaves extracts than those of the calli extracts. Additionally, it was shown that leaves water extracts have $\sim$ 10 and $\sim$ 3 fold higher content of flavonoids and phenolics, respectively, when compared with calli water extracts [42].

Three independent studies reported that stevia leaves and calli have a good DPPH radical scavenging activity [27, 42, 45]. However, the first two publications reported a higher inhibition of DPPH radical in calli while the third one in leaves. This divergence could possibly be explained by the plant material used, the extraction procedures as well as the growing conditions. In this regard, it is known that environmental conditions and genetic characteristics of plants certainly influence the yield of polyphenols and antioxidants compounds, among other secondary metabolites. Also, the growth conditions (e.g., light exposure), as well as the harvest period, can influence the contents of polyphenols and other antioxidants compounds found in stevia crop [78].

More recently, in a report investigating the antioxidant properties of four different medicinal plant extracts, it was found that *S. rebaudiana* extract has a very good ability to scavenge superoxide radical and a significant dose-dependent effect against hydroxyl radicals [72]. In addition, stevia extracts showed the ability to decompose nitric oxide.

Another comparative study evaluated the antioxidant properties of 28 fresh herbal teas based on total phenolic content, primary antioxidant properties of free radical scavenging activity and ferric reducing power, as well as the secondary antioxidant ability to chelate ferrous ion. The results showed that the *C. sinensis* teas outperform all herbal teas on antioxidant properties; however *S. rebaudiana* match those of *C. sinensis*. Moreover, stevia showed a stronger ferrous ion chelating ability than the *C. sinensis* teas. [79].

By means of ORAC (Oxygen Radical Absorbance Capacity) assay, results obtained by our group revealed that the crude aqueous extracts from stevia leaves exhibit a promising antioxidant activity, superior to the stem extracts and ranging from 958.8 to 1071.1 μmol trolox equivalents/g on dried basis (μmol TE/g) [40]. Comparison with a previous study shows that ascorbic acid, which is content in stevia, has an ORAC value comparable (1058 μmol TE/g) with that of the stevia leaves extracts [80].

The effects of addition of stevia crude extracts (up to 2.5% w/v) to high-pressure-processed fruits have been recently described. The addition of stevia significantly increased the total phenolic content, the inhibition of absorption of ABTS, and ORAC values [73]. Furthermore, it was showed that a 2.5% w/v addition of stevia reduces the microbial load of the product (inactivation of *Listeria monocytogenes* over 5 log cycles) as well as improves the inactivation of polyphenol oxidase and peroxidase, the main enzymes involved in the oxidation of phenolic compounds. Authors concluded that 2.5% w/v of stevia minimizes the oxidative reactions and maintains higher levels of antioxidant compounds which in turn succeed in reducing the microbial load in commercial fruit beverages.

6.4 Antioxidant Activity of Steviol Glycosides

Although if less studied, the radical scavenging activities of the steviol derivatives have been also reported. By means of ORAC, it was found that stevioside and reb A are active peroxyl radical scavengers [40]. Moreover, through a competitive reaction between terephthalic acid (TPA) and these steviol glycosides the hydroxyl radical scavenger activity was demonstrated [70, 71, 81]. Stevioside and reb A have a significant-higher antioxidant activity when compared with sugars (e.g., glucose, sucrose). Similar results were found with steviol glucuronide, a product of the stevioside metabolism which is found in peripheral blood after its consumption and subsequently is excreted in urine [70].

However, while purified extracts shown an antioxidant activity in vitro, by means of the ORAC assay this activity is lower (from 2 to 10%) when compared to the activity exhibited by crude extracts of stevia, being 23.8 μmol TE/g and 22.8 μmol TE/g of stevioside and reb A, respectively [40]. These data confirm previous findings from Hajihashemi and Geuns [70], which also have found a better ROS scavenging capacity for crude extracts.

All the studies mentioned above have dealt with the antioxidant and the free radical scavenging properties of *S. rebaudiana* extracts, by utilizing different sources, different approaches or calculation modes. In any case, researchers concluded that the crude extracts from *S. rebaudiana* exert a remarkable antioxidant activity. Due to the difference in the techniques used, the different solvents for extraction used (including water, ethanol, acetone, methanol and acetic acid) and the diversity of the standards used to measure the antioxidant capacity, it is not surprising that quantitative results are not comparable the one to the other.

Among the different chemical methods, DPPH has been the most frequently used for the analysis of stevia crude extracts. The assay protocol is quite fast and inexpensive, uses an end point measurement, with a shorter reaction time when compared to other tests. The DPPH assay provides basic information on the anti-DPPH activity of the extracts. However, DPPH is a stable organic nitrogen radical that rarely occurs in the human body. Also, interpretation of results is difficult, as natural pigments, in particular carotenoids, interfere with the spectroscopic measurement [70, 82].

The need to find a gold standard for the detection and quantification of the antioxidant activity of foods has been discussed for long in the scientific community. However, since the antioxidants employ several modes of action, a single assay cannot account for all the different mechanisms. As a consequence, the standardization of in vitro screening still remains essential to obtain reliable results during the first steps of screening for the antioxidant activity.

One standardized method for determining the antioxidant capacity of a substance, both in foods and physiological systems, is the ORAC assay [82–84]. The method concerns the inhibition of the peroxyl radical-induced oxidation which is initiated by thermal decomposition of azo-compounds [85]; it measures the capacity of an antioxidant to directly quench free radicals, has high specificity and responds to numerous antioxidants [86]. Currently, the ORAC test is a relevant measure used to assess the radical scavenging capacity in foods as well in clinical studies [82]. It uses a biological radical source (the peroxyl radical is the most abundant free radical in the human body), mimics the reactions between antioxidants and lipids and combines inhibition time and degree of inhibition into one quantity [82, 86].

6.5 Antioxidant Activity in Biological Systems

The in vitro evaluation remains crucial in the first steps of the screening for the antioxidant capacity; however, as discussed above, the antioxidant potential should not be concluded based on a single test model. Moreover, questions concerning whether the results obtained with in vitro chemical methods may support the efficiency in the human body are leading to the development of supplementary methods that may be more suitable for screening potential antioxidant materials. More relevant data can be provided by using biological systems, for example, with human cells and enzymes as the test model.

Clearly, clinical studies are the most reliable methods for determining the actual efficacy of antioxidants in the human body. However, they are expensive, time-consuming and not suitable for initial screening of food antioxidants. In comparison with in vivo assays, in vitro biological assays provide a cost-effective and fast turnaround preclinical data.

Few scientific works have provided data in relation to the antioxidant action of stevia by means of a relevant biological system. Table 4 summarizes data available from the literature.

In this respect, Kelmer et al. assessed whether steviol glycosides influence oxidative phosphorylation in isolated rat liver mitochondria [30]. In their work, it was reported an inhibition of the oxidative phosphorylation. Data obtained revealed that the activities of ATPase, NADH oxidase, succinate oxidase, as well as succinate and glutamate dehydrogenases were inhibited in this system upon treatment with stevioside, isosteviol, steviolbioside or steviol. Among the extracts studied, stevioside resulted in the less active. The authors observed that steviol and isosteviol (and to a lesser extent steviolbioside and stevioside) probably act as uncouplers.

Table 4 Methodologies used for measurement of the antioxidant activities of stevia extracts using different biological systems

Assay	Model	Reference
Cellular antioxidant activity	Human liver cells	[40, 74]
Antioxidant enzymes activities	Isolated rat liver mitochondria Human liver cells	[72] [40]
Reduced glutathione level	Isolated rat liver mitochondria	[72]
Oxidative phosphorylation	Rat liver mitochondria	[30]
DNA oxidative damage protection	Isolated DNA	[65, 75]
Lipid peroxidation (TBARS)	Rat liver homogenates	[65, 75]

More recently, Vaško et al. studied the modulation of antioxidant enzymes in isolated rat liver mitochondria treated with stevia leaf extracts [72]. Their work provided evidence that the treatment caused no changes in the activity of superoxide dismutase (SOD), one of the most important antioxidative enzymes, which catalyzes the dismutation of the superoxide anion into hydrogen peroxide and molecular oxygen. Additionally, no changes were recorded in glutathione peroxidase (GPx) and glutathione reductase (GR) activities in comparison to control, although the levels of reduced glutathione (GSH) were dose-dependent and several times higher than that in controls.

Similar results have been provided by Bender et al., measuring the activity of catalase (another key antioxidant enzyme that converts hydrogen peroxide to water) in a human liver cell line treated with stevia crude extracts [40]. This work showed that the modulation of catalase is not significantly different when compared to untreated cells ($p > 0.05$).

Overall, these two studies suggest that the antioxidant activity of stevia leaf extracts arises from its ability to directly reduce oxidizing free radicals, rather than a potential to modulate the endogenous enzymatic system.

A study conducted by Gantha et al. has provided information on the preventive effect of the DNA oxidative damage, exerted by flavonoid-rich extracts from stevia leaves [65]. This work showed that the crude extract prevents the strand scission of plasmid DNA, by hydroxyl radicals generated via the Fenton reaction. However, such preventive activity was not observed with stevioside. Authors concluded that stevia crude extracts are able to counteract the DNA oxidative damage in vitro. In addition, the ethyl acetate extract from stevia leaf inhibit the lipid peroxidation by means of TBARS (thiobarbituric acid reactive substances) assay in rat liver homogenates ($IC_{50} = 2.1 \pm 1.07$ mg/mL).

These results are in accordance with a recent study of Mediesse et al., in which it was evaluated the effect of the cell wall polysaccharide fractions from stevia leaves, on human DNA protection and lipid peroxidation inhibition [75]. By means of a DNA migration assay, these authors reported that the polysaccharide fractions can completely revert the DNA damage induced by the radical hydroxide generated in vitro. Moreover, these fractions also inhibit lipid peroxidation in rat liver homogenates.

Overall, these research groups have concluded that stevia leaves extracts and polysaccharide fractions are potent protective agents against oxidative stress.

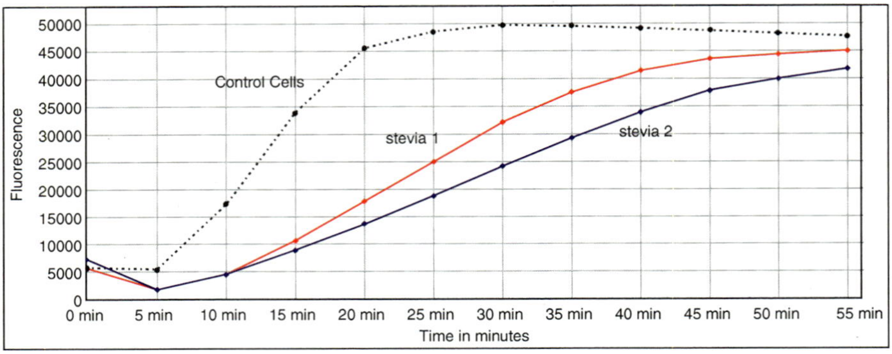

Fig. 1 Intracellular antioxidant capacity of stevia extracts evaluated by the CAA assay. The curves shown are representative from a single experiment (mean values, $n = 3$). The antioxidant activities of the crude extracts from stevia are dose dependent (stevia 2 > stevia 1), the higher the dosage the lower the fluorescence recorded over time and greater the CAA value (stevia 2 > stevia 1). Control: untreated cells

To further characterize the antioxidant activity of stevia extracts in a valid biological system, a recent study from our group has provided data on the intracellular antioxidant activity of stevia extracts (from leaves and stems) and the main pure steviol glycosides (stevioside and Reb A) [40]. In this study the cellular antioxidant activity (CAA) assay was used, a method similar to the ORAC test performed within cells. Some of the advantages of using this method are that it considers some aspects of cell uptake, metabolism, and distribution of the antioxidants within cells [87]. The method measures the capacity of the antioxidants in counteracting the oxidation of a fluorescent probe by peroxyl radicals produced inside human cells. The decrease in cellular fluorescence compared to control cells indicates the intracellular antioxidant capacity of the samples (Fig. 1).

It was found that stevia leaves extracts have a remarkable capacity in scavenging peroxyl radicals also at the intracellular level. The CAA values, calculated as μmol quercetin equivalents/g, have ranged from 61.3 to 14.5 (n = 5), being much lower in the stems sample (8 μmol quercetin equivalents/g). Under our experimental conditions, the association between ORAC and CAA assays has resulted positively correlated by means of regression analysis ($R^2 = 0.84$, $p < 0.01$). Interestingly, the CAA values of the *C. sinensis* green tea, a well-known antioxidant, are in line (38.6- 17.2 μmol quercetin equivalents/g, n = 5) with the CAA values from stevia leaves samples on weight basis; although the ORAC value from green tea resulted from 2.2 to 11.2 fold higher than those of stevia crude extracts.

Inversely, stevioside and rebA, which have exhibited a low ORAC value, do not elicit any intracellular antioxidant activity in the HepG2 human hepatocytes model. This result suggests that steviol glycosides do not have an antioxidant effect or in an insufficient extent to be detected. Nevertheless, by its high molecular weight, it is likely that stevioside and reb A can hardly be absorbed without further metabolization by liver cells, as also suggested by a study on transport experiments with Caco-2 intestinal monolayers [88].

The efficiency of dietary antioxidants depends mainly on their bioavailability, which at the same time is affected by digestion and complex metabolic reactions occurring in the intestine [89]. Besides, the food matrix also plays an important role in the antioxidant efficacy together with the synergistic and antagonistic interactions and effects that occur between food ingredients [89].

It was previously reported that steviol glycosides undergo hydrolysis to steviol [90, 91], the aglycone of stevioside, a metabolism that takes place in the intestine by human microflora before the absorption occurs [92]. Orally administered stevioside and reb A, are very little or are not absorbed by the human body [88, 93]. Usually, in the human body the steviol glycosides do not reach the hepatocytes. However, no information is available in the literature regarding metabolization of these molecules in human hepatocytes.

Overall, the results obtained with the HepG2 cellular model have shown that no antioxidant activity is detected in human hepatocytes treated with steviol glycosides. And because absorption is independent from metabolization, it is possible both, that hepatocytes are not able to metabolize steviol glycosides and/or none significant absorption occurs to display a detectable measurement. Consequently, neither stevioside nor rebA may exert any measurable antioxidant effect at the intracellular level. In particular, the results obtained indicate that the intracellular antioxidant activity measured by the CAA test is not caused by steviol glycosides, but rather by other stevia constituents from leaves and stem extracts. Also, stevia crude extracts are efficiently absorbed by human cells, acting as an effective antioxidant within the cells, minimizing oxidative damages by peroxyl radicals under physiological conditions.

6.6 Antioxidant Activity In Vivo

Due the numerous evidence supporting the antioxidant activity based on in vitro experiments, it has been suggested that stevia could help in counteracting oxidative stress in vivo. However, human studies on the antioxidant role of stevia supplementation have not been so far established. As a result, most of the studies with significant findings come from animal models.

Some evidence has been provided by Geeraert et al. [34], as a result of an animal study with stevioside. These authors found that purified stevioside modulates the expression of catalase and SOD mRNAs in hyperlipidemic and hyperglycemic obese insulin-resistant mice. The treatment with stevioside (10 mg/kg) has shown to decrease the lipid and oxidized low-density lipoprotein (ox-LDL) contents, possible by an increase in the antioxidant defense in the vascular wall, as supported by increased gene expression of SOD. Additionally, in visceral adipose tissue the stevioside treatment increased the expression of catalase.

Moreover, it was found that even if stevioside treatment has no effect on weight and triglycerides, its supplementation lowers glucose and insulin levels as well as improves glucose transport, insulin signaling, and antioxidant defense in both the adipose tissue and the vascular wall of obese insulin resistant mice.

Table 5 Summary of S. rebaudiana's antioxidant properties reported in the literature and assessed through biological assays and animal models

Extract	Assay	In vitro	In vivo
Stevia	DNA oxidative damage	Reverted	–
	Lipid peroxidation (TBARS)	Decreased	Decreased
	Peroxyl radical (CAA)	Counteracted	–
	GSH	Increased	Increased
	Catalase	Not modulated	Up-regulated
	SOD	Not modulated	Up-regulated
	GPx	Not modulated	Up-regulated
	GR	Not modulated	–
Stevioside	DNA oxidative damage	Not counteracted	–
	Peroxyl radical (CAA)	Not counteracted	–
	Catalase (mRNA)	–	Increased
	SOD (mRNA)	–	Increased
	Lipid peroxidation (ox-LDL)	–	Decreased
Reb A	Peroxyl radical (CAA)	Not counteracted	–

Additional data have been presented through a study in streptozotocin-induced diabetic rats [68]. In this work Shivanna et al. have described the effects of stevia feeding on lipid peroxidation by means of TBARS assay, conjugated dienes, and hydroperoxides measurements. It was found that the lipid peroxidation was reduced by 25% in rat livers, upon feeding supplementation with 4% stevia leaf powder. In addition, the induction of diabetes, by streptozotocin administration, resulted in SOD and catalase reduction by 50%, activity that was restored towards normal by pre-feed with stevia.

The findings were confirmed by Singh et al., in a study performed with alloxan-induced diabetic mice, daily treated with methanolic extracts from stevia leaves [14]. These investigators observed a decline of the antioxidant enzymes in diabetic animals; however, upon treatment, the hepatic SOD level was reversed towards normal. The treatment significantly increased GPx, while lipid peroxidation (TBARS concentration) and GSH were restored towards normal (Table 5).

7 Discussion

This chapter reviews the current findings on the antioxidant properties of the *S. rebaudiana*, its sweet diterpenes and their implication for human health. Many research studies on their antioxidant potential have been done in recent years. The data provided from the available literature indicate that stevia crude extracts have a remarkable antioxidant potential in vitro.

Data obtained with mere chemical methods, as well with more relevant biological models have reported that the crude extracts, as a whole, have a higher antioxidant capacity when compared with the purified steviol glycosides that are more frequently found in commerce. Comparing crude stevia extracts with purified steviol glycosides, it was found that the crude extracts are very potent ROS scavengers; in particular, in superoxide radical scavenging activity, DPPH scavenging activity, as well as peroxyl radical scavenging activity, even when the steviol glycosides were much less active [40, 70]. These results are not surprising if considering that stevia leaves contain additional molecules, not only steviol glycosides, that have a recognized antioxidant capabilities (e.g., vitamin C, polyphenols, flavonoids and carotenoids along with others) [38, 43].

Stevia crude extracts are rich in phenolic compounds, among them the chlorogenic acids are the most representative [40, 62, 64]. The phenolic compounds are mainly responsible for the antioxidant potential of plant extracts. In terms of chlorogenic acid concentrations, stevia leaves ranks third, following green coffee beans and yerba mate leaves [62].

In vitro, *S. rebaudiana* crude extracts exert, in a dose dependent manner, a high antioxidant activity against hydroxyl radicals, superoxide anion, DPPH, and nitric oxide, also higher than those of vitamin C [63]. While assessed by ORAC, stevia leaves have a comparable value with that of vitamin C [40, 80]. Also, *S. rebaudiana* has comparable antioxidant properties to *C. sinensis* teas assessed by total phenolic content, primary antioxidants activity against free radicals, and ferric reducing power. Moreover, stevia has shown a strongest ferrous ion chelating ability than that of *C. sinensis*. [79].

In fruit beverage products, the addition of stevia crude extracts at 2.5% w/v showed to minimize the oxidative reactions, retain higher levels of antioxidant compounds and improve the inactivation of polyphenol oxidase and peroxidase enzymes, which in turn succeed in reducing the microbial load of the final product [73].

In food, flavonoids can be free monomers or oligomers, which after ingestion can be metabolized to other derivatives, which in turns can be measured in plasma, reaching concentration around 1 μM (discussed in [57]). A polyphenol-rich dietary intake is correlated with a lesser occurrence of heart disease, neurodegenerative disorders, and cancer among other health disorders (reviewed in [57]). Thus, dietary polyphenols have been suggested to improve health condition through primary and secondary antioxidant mechanisms.

S. rebaudiana leaves may be a good source for food preservation, as a food additive, due to their antioxidant activities. Exhaustive work has been done in vitro, but still there is need for further research on their in vivo antioxidant effects in humans.

Little data were reported about the biological significance of these in vitro assays for the in vivo physiology. Data obtained with a liver mitochondria model revealed that steviol glycosides influence oxidative phosphorylation [30] by inhibiting the ATPase, NADH oxidase, succinate-oxidase, succinate and glutamate dehydrogenases; being stevioside the less active compound, among isosteviol, steviolbioside, and steviol extracts.

With two different biological systems, it was showed that the extracts from stevia leaves do not modulate the antioxidant enzymes activities [40, 72]. Overall, the in vitro antioxidant activity of the stevia leaves extracts seems to arise from its ability to directly reduce oxidizing free radicals, rather than a potential to modulate the endogenous enzymatic system.

Also, studies performed with two different biological systems demonstrated that stevia crude extracts are potent protective agents against oxidative stress, by counteracting the DNA oxidative damage as well as inhibiting the lipid peroxidation [65, 75]. However, stevioside extract does not exert a protective role against DNA oxidative damage [65, 75].

By means of the cellular antioxidant activity assay it was showed that stevia crude extracts are efficiently absorbed by human cells and can act as an effective antioxidant within the cells, minimizing oxidative damage by peroxyl radicals under physiological conditions [40]. Also, when assessed in the same cellular model, the intracellular antioxidant capability of stevia crude extracts was comparable to those of green tea. On the contrary, stevioside and reb A purified extracts do not elicit an intracellular antioxidant activity in the human cells [40]. The data provided by means of the CAA assay indicate that the intracellular antioxidant activity is not caused by steviol glycosides, but rather by other stevia constituents; in part, this can be supported by the lack of evidence of in vivo absorption for intact glycosides at realistic exposure levels. In vivo, the radical scavenging of steviol glycosides is probably due to its conversion product steviol glucuronide, that is found in the peripheral blood. Fewer data were so far produced with respect the antioxidant potential of stevia in vivo, all data available are based on animal models. On this regard two reports described that stevia feeding can reduce the lipid peroxidation and restore the levels of antioxidant enzymes, in diabetic mice models [14, 68]. Similar findings were obtained as a result of a study in hyperlipidemic and hyperglycemic obese insulin-resistant mice treated with stevioside extract [34].

There are several markers of oxidative damage that can be measured in vivo, determining at least one of these markers can contribute to defining the antioxidant role of stevia.

Among reliable markers of lipid peroxidation are F2-isoprostanes, ox-LDL and phosphatidylcholine hydroperoxide [94]; thus the oxidative damage to lipids can be measured in vivo by measuring changes in F2-isoprostanes in urine and plasma, as well as oxidised LDL particles and phosphatidylcholine hydroperoxide in blood. Detection vary from immunoassays with adequate specificity to gas chromatography and HPLC techniques with various detection modes.

In vitro studies by means of chemical methods are useful in the first step of screening for antioxidants, but by themselves are not enough. The methods proposed using biological relevant models, for example CAA or antioxidant enzymes modulation within a biological system, could be used as supportive evidence that more closely resembles the in vivo effect. However, further in vivo studies are recommended to suitably substantiate the role of stevia in the protection of body cells and key bio-molecules (e.g., DNA, proteins and lipids) from oxidative damage.

In summary, several in vitro results and few in vivo results are reported in the literature that provide evidence about the promising role of *S. rebaudiana* as a source of natural antioxidants for foods and dietary supplements. *S. rebaudiana* leaves or crude extracts have strong antioxidant activity and may be rich sources of antioxidants. Also, the incorporation of stevia in processed foods can decrease its oxidation rate, thus might be considered as healthy ingredient for hypocaloric functional foods instead of its purified steviol glycosides.

References

1. Marín W (2004) Sondeo de mercado de la Estevia. Instituto de Investigación de Recursos Biológicos Alexander von Humboldt. Bogotá, Colombia
2. Yadav SK, Guleria P (2012) Steviol glycosides from Stevia: biosynthesis pathway review and their application in foods and medicine. Crit Rev Food Sci Nutr 52(11):988–998
3. EFSA Panel on Food Additives and Nutrient Sources added to food (ANS) (2015) Scientific opinion on the safety of the proposed amendment of the specifications for steviol glycosides (E 960) as a food additive. EFSA J 13(12):4316 [29 pp]
4. Madan S et al (2010) *Stevia rebaudiana* (Bert) Bertoni: a review. Indian J Nat Prod Resour 1:267–286
5. Dacome AS et al (2005) Sweet diterpenic glycosides balance of a new cultivarof *Stevia rebaudiana* (Bert) Bertoni: isolation and quantitative distribution by chromatographic, spectroscopic and electrophoretic methods. Process Biochem 44:3587–3594
6. Goettemoeller J, Ching A (1999) Seed germination in *Stevia rebaudiana*. In: Perspectives on new crops and new users. J Janick, Alexandria
7. Sekaran T, Giridhar P, Ravishankar GA (2007) Production of steviosides in ex vitro and in vitro grown *Stevia rebaudiana* Bertoni. J Sci Food Agric 87:420–424
8. Singh S, Rao G (2005) Stevia: the herbal sugar of 21st Century. Sugar Tech 71:17–24
9. Heikal AH, Badawy OM, Hafez AM (2008) Genetic relationships among some Stevia (*Stevia Rebaudiana* Bertoni) accessions based on ISSR analysis. Res J Cell Mol Biol 2(1):1–5
10. EFSA Panel on Food Additives and Nutrient Sources (ANS) (2010) Scientific opinion on safety of steviol glycosides for the proposed uses as a food additive. EFSA J 8(4):1537 [85 pp.]
11. Savita SM et al (2004) *Stevia rebaudiana* A: functional component for food industry. J Hum Ecol 15:261–264
12. European Stevia Association (2007) *Stevia rebaudiana* bertoni plants and dried leaves as novel food. In: Summary of the Application and Specifications of the EUSTAS Quality Label. http://www.eustas.org/images/Documents/Steviol_glycosides_summary_application.pdf p 7
13. Alvarez M et al (1981) In XII Semináreo Brasiliero de *Stevia rebaudiana* Bertoni 25–26/6/81. Campinas, Brazil
14. Singh S, Garg V, Yadav D (2013) Antihyperglycemic and antioxidative ability of *Stevia rebaudiana* (Bertoni) leaves in diabetes induced mice. Int J Pharm Pharm Sci 5(2):297–302
15. European Commission SCoF (1999) Opinion on *Stevia rebaudiana* Bertoni plants and leaves. SCF/CS/NF/STEV/3 Final 17/6/1999
16. Curi R et al (1986) Effect of *Stevia rebaudiana* on glucose tolerance in normal adult humans. Braz J Med Biol Res 19(6):771–774
17. Hong J et al (2006) Stevioside counteracts the α-cell hypersecretion caused by long-term palmitate exposure. Am J Physiol Endocrinol Metab 290:E416–E422
18. Chen TH et al (2005) Mechanism of hypoglycemic effect of stevioside, a glycoside of *Stevia rebaudiana*. Planta Med 71:108–113
19. Jeppesen PB et al (2000) Stevioside acts directly on pancreatic β-cells to secrete insulin: Actions independent of cyclic adenosinemonophosphate and adenosine triphosphate-sensitiveK+ channel activity. Metabolism 49:208–214

20. Ferreira EB et al (2006) Comparative effects of *Stevia rebaudiana* leaves and stevioside on glycaemia and hepatic gluconeogenesis. Planta Med 72:691–696
21. Chan P et al (1998) The effect of stevioside on blood pressure and plasma catecholamines in spontaneously hypertensive rats. Life Sci 63:1679–1684
22. Liu JC et al (2003) Mechanism of the antihypertensive effect of stevioside in anesthetized dogs. Pharmacology 67:14–20
23. Chan P et al (2000) A double-blind placebo-controlled study of the effectiveness and tolerability of oral stevioside in human hypertension. Br J Clin Pharmacol 50(3):215–220
24. Hsieh MH, Chan P, Sue YM (2003) Efficacy and tolerability of oral stevioside in patients with mild essential hypertension: a two-year, randomized, placebo-controlled study. Clin Ther 25:2797–2808
25. Das S et al (1992) Evaluation of the cariogenic potential of the intense natural sweeteners stevioside and rebaudioside A. Caries Res 26:363–366
26. Tomita T et al (1997) Bactericidal activity of a fermented hot-water extracts from *Stevia rebaudiana* Bertoni and other food-borne pathogenic bacteria. Microbiol Immunol 41: 1005–1009
27. Abou-Arab EA, Abu-Salem FM (2010) Evaluation of bioactive compounds of *Stevia rebaudiana* leaves and callus. Afr J of Food Sci 4:627–634
28. Ghosh S, Subudhi E, Nayak S (2008) Antimicrobial assay of *Stevia rebaudiana* Bertoni leaf extracts against 10 pathogens. Inter J Integr Biol 2:27–31
29. Jayaraman S, Manoharan M, Illanchezian S (2008) In-vitro antimicrobial and antitumor activities of *Stevia rebaudiana* (Asteraceae) leaf extracts. Trop J of Pharm Res 7:1143–1149
30. Kelmer Bracht A, Alvarez M, Bracht A (1985) Effects of *Stevia rebaudiana* natural products on rat liver mitochondria. Biochem Pharmacol 34(6):873–882
31. Takahashi K et al (2001) Analysis of anti-rotavirus activity of extract from *Stevia rebaudiana*. Antiviral Res 49:15–24
32. Kinghorn AD et al (1998) Non-cariogenic intense natural sweeteners. Med Res Rev 18: 347–360
33. Matsukubo T, Takazoe I (2006) Sucrose substitutes and their role in caries prevention. Int Dent J 56:119–130
34. Geeraert B et al (2009) Stevioside inhibits atherosclerosis by improving insulin signaling and antioxidant defense in obese insulin-resistant mice. Int J of Obes 34:1–9
35. Konoshima T, Takasaki M (2002) Cancer-chemopreventive effects of natural sweeteners and related compounds. Pure Appl Chem 74:1309–1316
36. Takasaki M et al (2009) Cancer preventive agents. Part 8: chemopreventive effects of stevioside and related compounds. Bioorg Med Chem 17(2):600–605
37. Yasukawa K, Kitanaka S, Seo S (2002) Inhibitory effect of stevioside on tumor promotion by 12-O-tetradecanoylphorbol-13-acetate in two-stage carcinogenesis in mouse skin. Biol Pharm Bull 25:1488–1490
38. Wölwer-Rieck U (2012) The leaves of *Stevia rebaudiana* (Bertoni), their constituents and the analyses thereof: a review. J Agric Food Chem 60:886–895
39. Kinghorn AD (1992) Food Ingredient safety review. *Stevia rebaudiana* leaves. Unpublished report submitted to the European Commission, scientific committee on food (16.3.1992)
40. Bender C, Graziano S, Zimmermann BF (2015) Study of *Stevia rebaudiana* Bertoni antioxidant activities and cellular properties. Int J Food Sci Nutr 66(5):553–558
41. Ruiz-Ruiz JC, Moguel-Ordoñez YB, Segura-Campos MR (2015) Biological activity of *Stevia rebaudiana* Bertoni and their relationship to health. Crit Rev Food Sci Nutr. doi:10.1080/10408398.2015.1072083
42. Kim S et al (2011) The antioxidant activity and the bioactive compound content of *Stevia rebaudiana* water extracts. Food Sci Technol 44:1328–1332
43. Lemus-Mondaca R et al (2012) *Stevia rebaudiana* Bertoni, source of a high-potency natural sweetener: a comprehensive review on the biochemical, nutritional and functional aspects. Food Chem 132:1121–1132

44. Goyal SK, Samsher GRK (2010) Stevia (*Stevia rebaudiana*) a bio-sweetener: a review. Int J Food Sci Nutr 61:1–10
45. Tadhani MB, Patel VH, Subhash R (2007) In vitro antioxidant activities of *Stevia rebaudiana* leaves and callus. J Food Compos Anal 20:323–329
46. Shibata H et al (1995) Steviol and steviol glycoside. Glucosyl transferase activities in S. rebaudiana Bertoni. Purification and partial characterization. Arch Biochem Biophys 321:390–396
47. Barriocanal L et al (2008) Apparent lack of pharmacological effect of steviol glycosides used as sweeteners in humans, a pilot study of repeated exposures in some normatensive and hypotensive individuals and in type 1 and type 2 diabetics. Regul Toxicol Pharmacol 51:37–41
48. Turgut K et al (2015) *Stevia rebaudiana* Bertoni could be an alternative crop in the mediterranean region of Turkey. In: Stevia: growth in knowledge and taste. Proceedings of the 8th Stevia symposium. Ceunen A and Geuns JMC, Bonn
49. Grevsen K, Sørensen JN,et al (2015) A danish research project for the development of "Green Stevia" – a natural sweetener for organic food products. In: Stevia: growth in knowledge and taste. In: Proceedings of the 8th Stevia symposium 2015. Ceunen A and Geuns JMC, Bonn
50. Hajihashemi S, Geuns JMC, Ehsanpour AA (2015) Response of *Stevia rebaudiana* Bertoni to polyethylene glycol treatment under in vitro culture conditions. In: Stevia: growth in knowledge and taste. Proceedings of the 8th Stevia symposium. Ceunen A and Geuns JMC, Bonn.
51. Lankes C, Grosser P (2015) Evaluation of *Stevia rebaudiana* genotypes at a location in the Antelejo region in Portugal. In: Stevia: growth in knowledge and taste. Proceedings of the 8th Stevia symposium. Ceunen A and Geuns JMC, Bonn
52. Yücesan B, Mohammed A,et al (2015) Comparison of coneventional and biotechnological approaches for *Stevia rebaudiana* production with elevated rebaudioside A content. In: Stevia: growth in knowledge and taste. Proceedings of the 8th Stevia Symposium 2015. Ceunen A and Geuns JMC, Bonn
53. Ohta M, Sasa S (2010) Characterization of Novel Steviol Glycosides from Leaves of *Stevia rebaudiana* Morita. Phytochemistry 57:199–209
54. Prakash I, Chaturvedula VS, Markosyan A (2013) Isolation, characterization and sensory evaluation of a Hexa beta-D-glucopyranosyl diterpene from *Stevia rebaudiana*. Nat Prod Commun 8(11):1523–1526
55. Ceunen S, Geuns JM (2013) Steviol glycosides: chemical diversity, metabolism, and function. J Nat Prod 76(6):1201–1228
56. Tavarini S, Guidi L,et al (2015) Nitrogen affects growth, steviol glycosides, leaf gas exchanges and chlorophill fluorescence in *Stevia rebaudiana* Bert. In: Stevia: growth in knowledge and taste. Proceedings of the 8th Stevia Symposium. Ceunen A and Geuns JMC, Bonn
57. Ďuračková Z (2014) Free radicals and antioxidants for non-experts. In: Systems biology of free radicals and antioxidants. Laher I. Springer, Berlin/Heidelberg
58. Vranesic-Bender D (2010) The role of nutraceuticals in anti-aging medicine. Acta Clin Croat 49 (4):537–544
59. Fusco D et al (2007) Effects of antioxidant supplementation on the aging process. Clin Interv Aging 2(3):377–387
60. Huang D, Ou B, Prior RL (2005) The chemistry behind antioxidant capacity assays. J Agric Food Chem 53:1841–1856
61. Chong-Han K (2010) Dietary lipophilic antioxidants: implications and significance in the aging process. Crit Rev Food Sci Nutr 50(10):931–937
62. Karaköse H, Müller A, Kuhnert N (2015) Profiling and quantification of phenolics in *Stevia rebaudiana* leaves. J Agric Food Chem 63:9188–9198
63. Shukla S et al (2009) In vitro antioxidant activity and total phenolic content of ethanolic leaf extract of *Stevia rebaudiana* Bertoni. Food Chem Toxicol 47:2338–2343
64. Karaköse H, Jaiswal R, Kuhnert N (2011) Characterisation and quantification of hydrocinnamate derivatives in *Stevia rebaudiana* leaves by LC-MSn. J Agric Food Chem 59:10143–10150

65. Ghanta S et al (2007) Oxidative DNA damage preventive activity and antioxidant potential of *Stevia rebaudiana* (Bertoni) Bertoni, a natural sweetener. J Agric Food Chem 55:10962–10967
66. Jahan IA et al (2010) Antioxidant activity of *Stevia rebaudiana* Bertoni leaves from Bangladesh. Bangladesh Pharm J 13:67–75
67. Rajbhandari A, Roberts MF (1983) The Flavonoids of *Stevia rebaudiana.* J Nat Prod 46:194–195
68. Shivanna N et al (2013) Antioxidant, anti-diabetic and renal protective properties of *Stevia rebaudiana.* J Diabetes Complications 27:103–113
69. Gaweł-Bęben K et al (2015) *Stevia rebaudiana* Bert. Leaf extracts as a multifunctional source of natural antioxidants. Molecules 20:5468–5486
70. Hajihashemi S, Geuns JMC (2013) Free radical scavenging activity of steviol glycosides, steviol glucuronide, hydroxytyrosol, metformin, aspirin and leaf extract of *Stevia rebaudiana.* Free Radicals Antioxid 3(supplement):34–41
71. Geuns JMC, Struyf T (2010) Radical scavenging activity of steviol glycosides and steviol glucoronide. In: Proceedings of the 4th Stevia symposium. Heverlee, Belgium
72. Vaško L et al (2014) Comparison of some antioxidant properties of plant extracts from Origanum vulgare, Salvia officinalis. *Eleutherococcus senticosus* and *Stevia rebaudiana.* In Vitro Cell Dev Biol Anim 50:614–622
73. Barba FJ et al (2014) *Stevia rebaudiana* Bertoni as a natural antioxidant/antimicrobial for high pressure processed fruit extract: processing parameter optimization. Food Chem 148:261–267
74. Bender C (2015) Evaluation of *Stevia rebaudiana* Bertoni antioxidant properties at cellular level, cytotoxicity and proliferative effects. In: Stevia: growth in knowledge and taste. Proceedings of the 8th Stevia Symposium. Ceunen A and Geuns JMC, Bonn
75. Mediesse KF et al (2014) Antioxidant properties of cell wall polysaccharides of *Stevia rebaudiana* leaves. J Coast Life Med 2(12):962–969
76. Muanda F et al (2010) Study on chemical composition and biological activities of essential oil and extracts from *Stevia rebaudiana* Bertoni leaves. Food Sci Technol 44:1865
77. Phansawan B, Poungbangpho S (2007) Antioxidant capacities of *Pueraria mirifica*, *Stevia rebaudiana* Bertoni, *Curcuma longa* Linn., *Andrographis paniculata* (Burm.f.) Nees and *Cassia alata* Linn for the development of dietary supplement. Kasetsart J Nat Sci 41:548–554
78. Tavarini S, Angelini LG (2013) *Stevia rebaudiana* Bertoni as a source of bioactive compounds: the effect of harvest time, experimental site and crop age on steviol glycoside content and antioxidant properties. J Sci Food Agric 93:2121–2129
79. Chiang Chan EW, Wong SK (2015) Herbs and herbal teas with antioxidant properties comparable to or superior than those of camellia sinensis. Int J Pharmacognosy 2(1):33–37
80. Bender C et al (2014) Antioxidant potential of aqueous plant extracts assessed by the cellular antioxidant activity assay. Am J Biol Life Sci 2:72–79
81. Jüttner MM et al (2015) ROS – scavenging capabilities of steviol glycosides and derivatives thereof. In: Stevia: Growth in knowledge and taste. Proceedings of the 8th Stevia Symposium. Ceunen A and Geuns JMC. Euprint, Bonn
82. Prior R, Wu X, Schaich K (2005) Standardized methods for the determination of antioxidant capacity and phenolics in foods and dietary supplements. J Agric Food Chem 53:4290–4302
83. Cao G, Alessio HM, Cutler RG (1993) Oxygen-radical absorbance capacity assay for antioxidants. Free Radic Biol Med 14:303–311
84. Ou B, Hampsch-Woodill M, Prior RL (2001) Development and validation of an improved oxygen radical absorbance capacity assay using fluorescein as the fluorescent probe. J Agric Food Chem 49:4619–4926
85. Glazer AN (1990) Phycoerythrin flurorescence-based assay for reactive oxygen species. Methods Enzymol 186:161–168
86. Cao G, Prior RL (1998) Comparison of different analytical methods for assessing total antioxidant capacity of human serum. Clin Chem 44(6 (Pt 1)):1309–1315
87. Wolfe KL, Liu RH (2007) Cellular antioxidant activity (CAA) assay for assessing antioxidants, foods and dietary supplements. J Agric Food Chem 55:8896–8907

88. Geuns JMC et al (2003) Metabolism of stevioside in pigs and intestinal absorption characteristics of stevioside, rebaudioside A and steviol. Food Chem Toxicol 41:1599–1607
89. Brewe MS (2011) Natural antioxidants: sources, compounds, mechanisms of action, and potential applications. Compr Rev Food Sci Food Saf 10:221–247
90. Gardana C et al (2003) Metabolism of Stevioside and rebaudioside A frcm *Stevia rebaudiana* extracts by human microflora. J Agric Food Chem 51:6618–6622
91. Wheeler A et al (2008) Pharmacokinetics of rebaudioside A and stevicside after single oral doses in healthy men. Food Chem Toxicol 46:54–60
92. Simonetti P et al (2004) Bioavailability of Stevioside from *Stevia rebaudiana* in human volunteers: preliminary report. In: Proceedings of the first symposium on the safety of stevioside. KULeuven/Heverlee-Leuven, Belgium, Euprint Editions
93. Koyama E et al (2003) Absorption and metabolism of the glycosidic sweeteners, Stevia related compounds in human and rat. Food Chem Toxicol 41:875–883
94. EFSA NDA Panel (EFSA Panel on Dietetic Products NaA) (2011) Guidance on the scientific requirements for health claims related to antioxidants, oxidative damage and cardiovascular health. EFSA J 9(12):2474 [13 pp.]

Nonnutritive Sweeteners and Their Role in the Gastrointestinal Tract

17

Allison C. Sylvetsky, Jenna Abdelhadi, Najy Issa, and Kristina I. Rother

Abstract

Non-nutritive sweeteners (NNS) were once thought to be metabolically inert, but have recently been shown to exert physiologic activity in the gastrointestinal tract. Mechanisms underlying their activity include binding to sweet taste receptors in enteroendocrine L-cells and pancreatic beta cells, through influencing glucose transport, and through altering the gut microbiota. The majority of in vitro studies demonstrate that NNS elicit gut hormone secretion and stimulate insulin release; and, findings from rodent models largely support these data. However, whether NNS affect gut hormones, insulin responses, glucose absorption, or microbiota in humans is not clear. Further research investigating the extent to which NNS exert clinically relevant activity in the gastrointestinal tract is required to determine whether these commonly consumed replacements for added sugars are beneficial or detrimental to human health

---

Disclosures: The authors have no conflicts of interest.

A.C. Sylvetsky (✉)
Section on Pediatric Diabetes and Metabolism, National Institute of Diabetes, Digestive, and Kidney Diseases, Bethesda, MD, USA

Department of Exercise and Nutrition Sciences, The George Washington University, Washington, DC, USA
e-mail: asylvets@email.gwu.edu

J. Abdelhadi • N. Issa
Department of Exercise and Nutrition Sciences, The George Washington University, Washington, DC, USA
e-mail: jennahadi@gwmail.gwu.edu; najyissa@gwmail.gwu.edu

K.I. Rother
Section on Pediatric Diabetes and Metabolism, National Institute of Diabetes, Digestive, and Kidney Diseases, Bethesda, MD, USA
e-mail: kr58q@nih.gov; kristinar@intra.niddk.nih.gov

J.-M. Mérillon, K.G. Ramawat (eds.), *Sweeteners*, Reference Series in Phytochemistry,
https://doi.org/10.1007/978-3-319-27027-2_4

Keywords
Artificial sweeteners • Diet soda • Metabolism • Obesity • Diabetes

Abbreviations

FDA	United states food and drug administration
GRAS	Generally recognized as safe
NNS	Non-nutritive sweetener

Contents

1 Introduction 378
2 Are all NNS the Same? 379
2.1 Aspartame 379
2.2 Advantame 379
2.3 Acesulfame-Potassium 383
2.4 Neotame 383
2.5 Saccharin 383
2.6 Stevia and Rebaudioside A 383
2.7 Sucralose 383
3 How Might Different NNS Influence Metabolism, Weight, and Health? 384
4 How are Sweet Tasting Compounds Recognized in the Body? 385
5 Do NNS Affect Gut Hormone Secretion in Vivo? 386
6 Do NNS Affect Insulin Responses in Vivo? 387
7 Do NNS Affect Intestinal Glucose Absorption? 387
8 Do NNS Induce Xenobiotic Enzymes in the Gastrointestinal Tract? 387
9 Do NNS Influence the Gut Microbiota? 388
10 Conclusions 388
References 389

1 Introduction

Non-nutritive sweeteners (NNS) including acesulfame-potassium, aspartame, saccharin, and sucralose provide sweet taste with few or no calories, and therefore reduce the calorie content of foods and beverages. However, the role of NNS in metabolism and health is controversial [30, 37, 38, 51]. NNS were once believed to be metabolically inert, yet recent evidence suggests that NNS do in fact exert metabolic activity [3, 20].

Associations between NNS consumption, weight gain, and metabolic abnormalities are reported in epidemiologic studies [11, 12, 14, 33, 34]; and, a causal role of NNS in inducing weight gain, altering the gut microbiota, and impairing glycemia has been consistently observed in rodent models [1, 36, 48, 52, 53]. In contrast, human intervention studies demonstrate that NNS consumption may be beneficial for weight management [40, 43], with little human data available on changes in other metabolic outcomes (e.g., glycemia) following chronic NNS exposure.

Understanding the effects of NNS in the human gastrointestinal tract is integral to elucidate whether NNS are advantageous or detrimental to health [3]. These effects may include stimulation of gut hormone secretion [20], upregulation of intestinal glucose absorption [26], enhancement of insulin secretion [8, 32], alteration of the gut microbiome [1, 36, 48], and induction of key enzymes in first-pass drug metabolism [26].

The objective of this chapter is therefore to summarize data from in vitro, rodent, and human studies evaluating the activity of NNS in the gastrointestinal tract and to highlight the urgent need for additional research to elucidate their effects in humans.

2 Are all NNS the Same?

There are six NNS approved by the United States Food and Drug Administration (FDA) as food additives, and two additional NNS are permitted for use in foods and beverages due to their Generally Recognized as Safe (GRAS) designation [37, 58]. The chemical structures and trade names of these NNS are shown in Table 1.

While all NNS are potently sweet relative to sucrose, each is chemically distinct and is therefore used differently in food and beverage applications [41]. Importantly, each NNS is absorbed, distributed, metabolized, and excreted differently [61], suggesting that different NNS are likely to have different effects in the gastrointestinal tract. Chemical and pharmacokinetic properties of each NNS are summarized below.

2.1 Aspartame

Aspartame is a synthetic dipeptide methyl ester composed of aspartic acid and phenylalanine [27]. Following oral ingestion, aspartame is rapidly hydrolyzed into its constituent amino acids: aspartate and phenylalanine. This is achieved through complete hydrolysis in the gastrointestinal tract, after which, aspartame and phenylalanine enter the circulation. Phenylalanine enters the plasma free amino acid pool, while aspartate is metabolized in the enterocyte, reducing the amount of aspartate entering the circulation. Thus, aspartame is not detectable in the plasma or other bodily fluids. In addition, aspartame is different from other NNS in that it provides the same caloric intake as sucrose, yet because little is needed to provide sweetness due its high potency, the energy intake contributed by aspartame is negligible.

2.2 Advantame

Advantame, one of the newest NNS, approved by the FDA in 2014, is a white, crystalline powder that is stable at high temperatures and approximately 20,000 times more potent than sucrose [35, 60]. Approximately 15% of the advantame ingested is absorbed in the small intestine, with considerable variability between individuals. Advantame has several metabolites, all of which are rapidly excreted after ingestion, both in the urine and in the feces [60].

Table 1 Chemical formulas and structures of NNS available for use in the United States

Chemical Name	Chemical Formula	Trade Name	Structure
Aspartame	$(C_{14}H_{18}N_2O_5)$	Nutrasweet™ Equal™	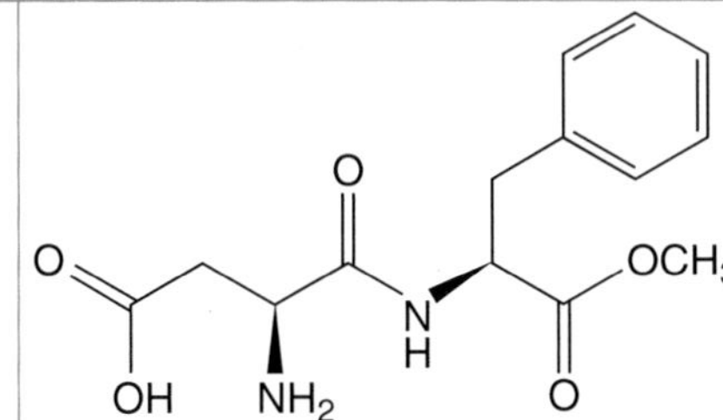
Acesulfame-Potassium	$(C_4H_4KNO_4S)$	Sunett™ SweetOne™	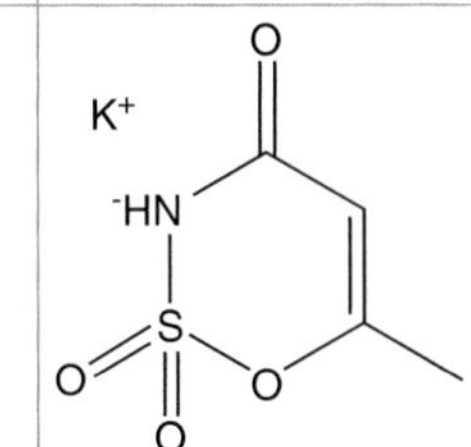
Advantame	$C_{24}H_{30}N_2O_7$	N/A	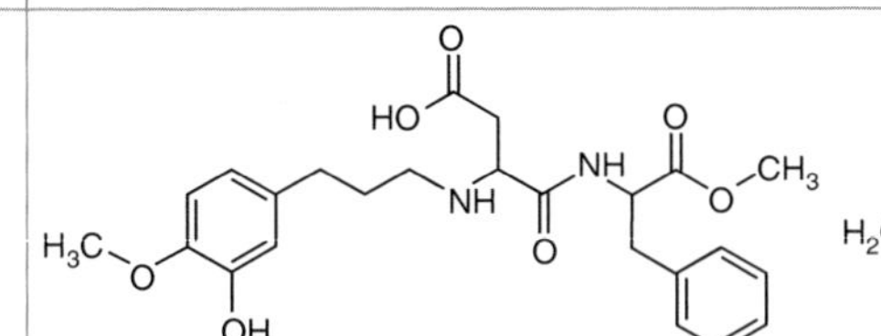

Saccharin	($C_7H_5NO_3S$)	Sweet N'Low™	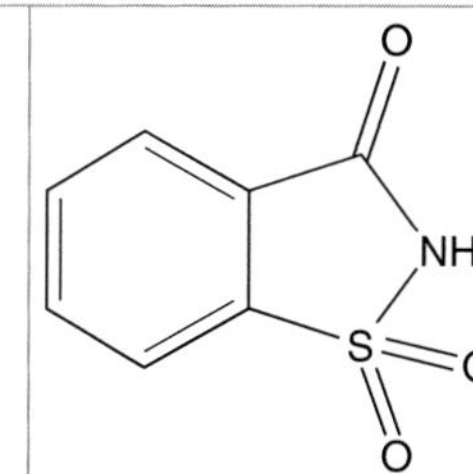
Sucralose	($C_{12}H_{19}Cl_3O_8$)	Splenda™	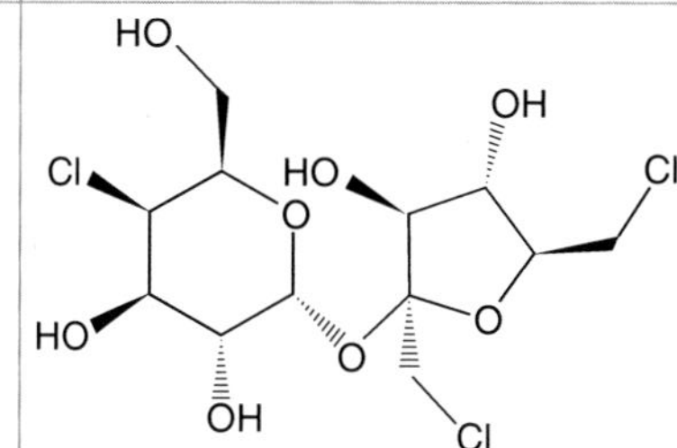

(continued)

Table 1 (continued)

Chemical Name	Chemical Formula	Trade Name	Structure
Stevioside (Stevia)	$(C_{38}H_{60}O_{18})$	Truvia™	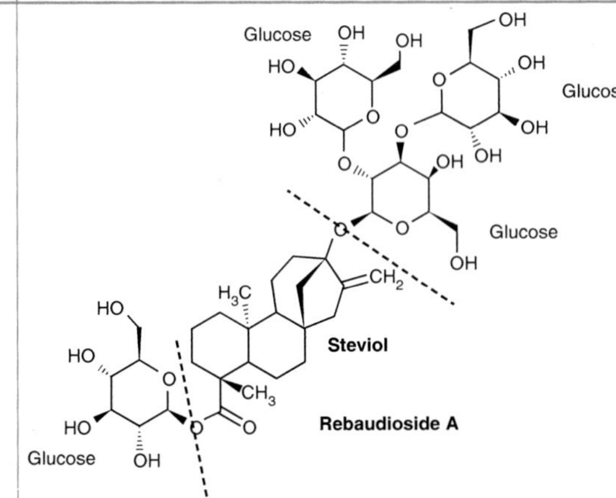

2.3 Acesulfame-Potassium

Acesulfame-potassium is a white, odorless crystalline powder that is stable in normal conditions and is approximately 200 times sweeter than sucrose [18]. After ingestion, acesulfame-potassium is rapidly absorbed in the digestive tract and is excreted unchanged, predominantly in the urine. No accumulation of acesulfame-potassium is reported in body tissues [18].

2.4 Neotame

Neotame is a dipeptide methyl ester, produced in a one-step chemical synthesis from aspartame. It is white powder with a sweetness potency of 8,000–13,000 times greater than sucrose [59]. After ingestion, approximately half is converted into de-esterified neotame and absorbed into circulation, prior to rapid excretion in both the urine and the feces [61]. There is no accumulation of neotame reported in body tissues.

2.5 Saccharin

Saccharin is a synthetically produced, bicyclic, sulfur-containing compound. It is a white, odorless powder that is approximately 300 times sweeter than sucrose. Saccharin is unchanged following absorption [19] and it is predominantly excreted in the urine, with no evidence of bioaccumulation in body tissues [31].

2.6 Stevia and Rebaudioside A

Stevioside (stevia) and rebaudioside A are both steviol glycosides and are extracted from the S. Rebaudiana Bertoni plant [63]. Steviol glycosides are not absorbed in the small intestine, but rather, are converted to steviol by bacteria in the lower gastrointestinal tract [6], which is then released into the circulation [7]. The absorbed glycoside metabolites are excreted primarily in the feces, with no bioaccumulation in body tissues [7].

2.7 Sucralose

Sucralose is a synthetic, trichlorinated disaccharide that is approximately 600 times sweeter than sucrose and does not contain calories [45]. It is a white crystalline solid that is stable in both its solid form and in aqueous solution [15]. There are two known hydrolysis products of sucralose [42], however there is no documented evidence of sucralose hydrolysis in vivo. The majority (~80%) of ingested sucralose is excreted unchanged in the feces, with the remainder excreted in the urine [16]. However, the extent to which sucralose is absorbed in the gastrointestinal tract has been challenged [45].

3 How Might Different NNS Influence Metabolism, Weight, and Health?

One explanation for epidemiologic associations between NNS consumption, body weight, and metabolic impairments is reverse causality, in that individuals who are already overweight or who are at risk for metabolic complications may use NNS as a strategy to reduce their calorie intake, lose weight, and/or delay the onset of metabolic disease. While reverse causality is a likely contributor to these associations, several other mechanisms (Fig. 1), may explain epidemiologic links between NNS, weight gain, and metabolic impairments; and, several of these proposed mechanisms are well-supported in rodent models [48, 51, 53, 64]. While sweet-taste mediated changes are likely generalizable across different NNS, gut microbiota and transporter mediated effects are likely to be sweetener specific.

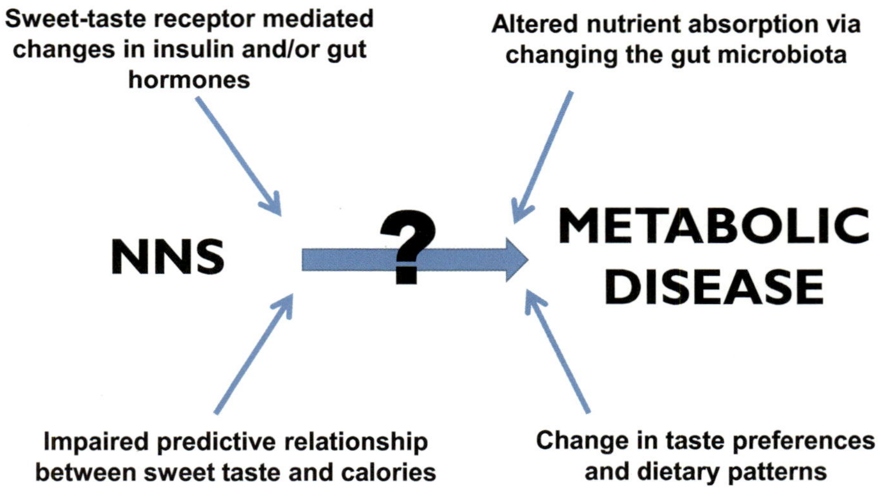

Fig. 1 In addition to reverse causality, several biologically plausible mechanisms have been proposed to explain epidemiologic findings connecting NNS consumption to various adverse metabolic outcomes. Several of these mechanisms, which have received scientific support in rodent models, are described in Fig. 1

4 How are Sweet Tasting Compounds Recognized in the Body?

Due to their sweetness, NNS bind to sweet-taste receptors, as do other sweet tasting compounds including caloric sugars, sweet proteins, and sugar alcohols. The sweet taste receptor is a heterodimeric G-protein coupled receptor, which consists of two sub-units: T1R2 and T1R3. These receptors are located both within the oral cavity, primarily on taste receptor cells within taste buds of the tongue and in other parts of the body including intestinal mucoca [10, 11, 13], pancreatic β-cells [32], preadipocytes [29], the biliary tract [57], the testes [22], and the lungs [9].

When NNS bind to sweet-taste receptors, whether in the oral cavity or extra-orally, α-gustducin, the alpha subunit, is activated. This results in increased phospholipase Cβ2, which then promotes increases in inositol trisphosphate and diacylglycerol, resulting in activation of the taste-transduction channel, TRPM5 (transient receptor potential cation channel subfamily M member 5). In the mouth, this leads to the release of neurotransmitters which convey the sensation of sweetness to the brain. However, in the intestinal enteroendocrine cells and pancreatic beta cells, in vitro studies demonstrate that activation of sweet taste receptors results in secretion of gut hormones and insulin release, respectively [20, 32].

Jang et al. [20] demonstrated that sucralose stimulated GLP-1 secretion from a human L-cell line (NCI-H716 cells) in a dose dependent fashion. This study also showed that GLP-1 release was suppressed when sucralose was administered with lactisole, a sweet taste receptor inhibitor. Margolskee et al. [28] published similar findings, showing sucralose-induced GLP-1 and GIP secretion from a murine enteroendocrine cell line (GLUTag cells). Gut hormone secretion was then suppressed when gurmarin, also a sweet taste receptor inhibitor, was administered along with sucralose. Given that gut hormone release was inhibited in the presence of sweet taste receptor inhibitors, these data support sweet-taste receptor mediated activity of NNS.

Although located outside of the intestinal epithelium, pancreatic β-cells may also be considered enteroendocrine cells and also express sweet-taste receptors [32]. In MIN6 cells (a mouse insulinoma cell line frequently used to study β-cell function), sucralose, saccharin, and acesulfame-K stimulated insulin secretion, and similar findings were reported in isolated rat islets when sucralose was mixed with low concentrations of glucose. Interestingly, this effect was also blocked by the sweet-taste receptor antagonist gurmarin, implying that the augmentation of insulin release was sweet-taste receptor mediated. Similar findings have also been reported by Corkey et al., where a high but physiologic concentration of saccharin was also shown to enhance insulin secretion [8]. Insulinotropic effects of stevia have also been reported in rodent and human cells lines [21].

While binding of NNS to sweet-taste receptors in intestinal enteroendocrine cells and pancreatic beta cells in vitro stimulates downstream actions such as GLP-1 and insulin release, whether this translates to hormone release in vivo is not clear.

5 Do NNS Affect Gut Hormone Secretion in Vivo?

The majority of the existing in vivo data have not confirmed the sweet-taste receptor mediated activity of NNS demonstrated in vitro. For example, Fujita et al. [13] reported that administration of acesulfame-K, stevia, d-tryptophan, and sucralose by gastric gavage did not stimulate GLP-1 or GIP release in rats. A similar lack of an effect was observed in humans, where intragastric administration of sucralose, aspartame, or acesulfame-K did not stimulate GLP-1, PYY, ghrelin, or GIP secretion [24, 47]. Augmentation of GLP-1 response was also not observed following intra-duodenal administration of sucralose, even when administered in combination with glucose [25].

These null results in both humans and rodents suggest that stimulation of gastrointestinal sweet taste receptors, while bypassing lingual exposure, may be insufficient to prompt hormone release. Meanwhile, several experiments in healthy humans have also demonstrated no effects of oral NNS administration (where lingual taste receptors were in fact exposed to NNS) on gut hormone release, when administered alone [2, 10]. Taken together, these data support the notion that NNS in isolation are not a sufficient stimulus to induce gut hormone secretion in vivo, when administered in isolation (without the presence of nutrients or calories).

Interestingly however, NNS have been shown to induce gut hormone secretion when administered in combination with caloric sugars [5]. We have reported a 34% increase in GLP-1 release following ingestion of a caffeine-free Diet Rite Cola™ compared to carbonated water (control), when administered 10 min prior to an oral glucose tolerance test [4, 5]. This finding was then replicated in adolescents with type 1 diabetes, in whom GLP-1 area under the curve was 43% higher in the diet soda condition, but not in those with type 2 diabetes [4]. Despite augmentation of GLP-1 release following sucralose and acesulfame-potassium containing diet soda [5], downstream effects of increased GLP-1 secretion were not observed in these studies. Enhancement of GLP-1 response following sucralose administered prior to a glucose load in healthy adults was also reported by Temizkan et al., yet no differences in GLP-1 were observed when the same experiment was conducted in patients with newly diagnosed type 2 diabetes [55].

Taken together, findings of acute studies testing NNS effects on gut hormone secretion are inconsistent; and, non-sweet taste receptor dependent mechanisms are also likely. Inconsistencies in study design emphasize the need to consider the composition of the beverages in which NNS are administered, the dose and volume provided, as well as the timing between ingestion of the NNS containing preload and the glucose load or mixed meal [30, 54]. Furthermore, the above described studies involve administering a single dose of NNS, where responses are then measured over several hours. Thus, the clinical significance of these findings requires further study following chronic NNS exposure.

6 Do NNS Affect Insulin Responses in Vivo?

Despite compelling evidence that NNS induce insulin secretion in vitro, this effect has also been inconsistently observed in vivo. While serum insulin increased in rodents given access to a solution of eryrthitol and aspartame (99% and 1%, respectively) for 4 weeks, most human studies have found no differences in insulin levels between groups among humans treated with sucralose or acesulfame-K and those treated with a placebo [2, 24, 62]. However, Pepino et al. [39] did observe higher glucose and insulin responses during an oral glucose tolerance test, when sucralose was administered 10 min prior to glucose load, compared to plain water. In contrast, intravenous injection of type 2 diabetic rats with stevioside during an intravenous glucose tolerance test (IVGTT) significantly lowered glucose levels and suppressed glucagon [21]. This supports hypothesized insulin mimetic effects of stevia and its potential as a therapeutic anti-hyperglycemic agent [7].

7 Do NNS Affect Intestinal Glucose Absorption?

A possible role of NNS in the regulation of intestinal glucose absorption has been proposed and is supported by upregulation of sodium-glucose linked transporter 1 (SGLT-1) in vitro by sucralose, acesulfame-K, and saccharin [28]. Mace et al. [26] also demonstrated that sucralose increased the rate of intestinal glucose absorption in rats in vivo, by stimulating activity of GLUT2 in the enterocytes. However, no differences in intestinal glucose absorption were observed in humans during intraduodenal infusion of sucralose versus saline, administered in combination with glucose [25]. Thus, there is currently insufficient evidence to support NNS stimulates increases in intestinal glucose absorption.

8 Do NNS Induce Xenobiotic Enzymes in the Gastrointestinal Tract?

Sucralose (administered as Splenda™) has been shown to increase intestinal expression of P-glycoprotein P-gp and cytochrome P450 enzymes (CYP3A) in rodents [1]. P-gp (also known as MDR1 or ABCB1) is a membrane efflux protein expressed throughout the body and appears to be localized to the apical surface of cells in the lower intestine [56], where sucralose (and other NNS excreted in the feces) accumulate before elimination. CYP3A isoforms are highly expressed in intestine and liver, and metabolize a wide variety of drugs. Increased expression of P-gp or CYP3A results in decreased bioavailability of their substrates [50, 56], often with clinically significant consequences [17, 23, 46]. It is therefore critical to evaluate whether NNS induce these xenobiotic enzymes in humans and to determine the clinical relevance of these effects.

9 Do NNS Influence the Gut Microbiota?

A large body of evidence now demonstrates that micro-organisms in the gastrointestinal tract, known as the gut microbiota, play a critical role in the development of obesity and metabolic disease [44]. Abou-Donia et al. [1] first suggested that NNS may alter the gut microbiome of rats, in demonstrating that sucralose exposure for 12 weeks reduced bacterial abundance. A subsequent study then demonstrated that 11 weeks of saccharin exposure altered gut microbiome composition in rodents, leading to the development of glucose intolerance [48, 49]. In the same study, a causal role of LCS-induced changes in the gut microbiota in the promotion of glucose intolerance was further demonstrated by microbiome transplantation. Microbiota from saccharin-exposed mice were implanted into germ-free mice. The recipients of saccharin-exposed microbiota developed glucose intolerance, while recipients of the control group did not [48, 49].

A third rodent study demonstrated that 8 weeks of aspartame consumption shifted gut bacterial composition, led to elevated fasting glucose, and impaired insulin-stimulated glucose disposal [36]. Compositional changes observed in this study also suggest that NNS may increase the abundance of microbial species associated with inflammation and insulin resistance. Aspartame-induced microbial alterations were also proposed to favor propionate production, which may play a key role in regulating metabolic processes including insulin sensitivity. Changes in microbial composition following NNS exposure are also suggested to promote increased energy extraction from food, potentially disturbing energy balance and increasing fat storage.

Little is known about NNS-induced alterations in the gut microbiota in humans [48]. In one small human study, saccharin exposure for 1 week was associated with changes in the microbiome and glucose metabolism. However, the lack of a control group makes these findings less persuasive and reiterates the need for further research in this area.

10 Conclusions

NNS are in fact physiologically active in the gastrointestinal tract, and may act through various mechanisms, which are likely not mutually exclusive [37]. The elucidation of the presence of extra-oral sweet taste receptors, the potential role of NNS in altering the gut microbiota, and the findings that NNS stimulate transporters involved in both glucose and drug absorption, have prompted the generation of highly significant translational research questions, which require rigorous investigation in humans. Given the in vitro, epidemiologic and rodent data suggesting that NNS may induce metabolic impairments, further study of the gastrointestinal activity of NNS in vivo is integral to determining whether NNS are helpful or harmful for human health.

Acknowledgements This work was supported in part by the intramural research program of the National Institute of Diabetes and Digestive and Kidney Diseases and in part by the Department of Exercise and Nutrition Sciences at the George Washington University.

References

1. Abou-Donia MB, El-Masry EM, Abdel-Rahman AA et al (2008) Splenda alters gut microflora and increases intestinal p-glycoprotein and cytochrome p-450 in male rats. J Toxicol Environ Health A 71:1415–1429
2. Brown AW, Bohan Brown MM, Onken KL et al (2011) Short-term consumption of sucralose, a nonnutritive sweetener, is similar to water with regard to select markers of hunger signaling and short-term glucose homeostasis in women. Nutr Res 31:882–888
3. Brown RJ, Rother KI (2012) Non-nutritive sweeteners and their role in the gastrointestinal tract. J Clin Endocrinol Metab 97:2597–2605
4. Brown RJ, Walter M, Rother KI (2012) Effects of diet soda on gut hormones in youths with diabetes. Diabetes Care 35:959–964
5. Brown RJ, Walter M, Rother KI (2009) Ingestion of diet soda before a glucose load augments glucagon-like peptide-1 secretion. Diabetes Care 32:2184–2186
6. Carakostas MC, Curry LL, Boileau AC et al (2008) Overview: the history, technical function and safety of rebaudioside A, a naturally occurring steviol glycoside, for use in food and beverages. Food Chem Toxicol 46(Suppl 7):S1–S10
7. Chatsudthipong V, Muanprasat C (2009) Stevioside and related compounds: therapeutic benefits beyond sweetness. Pharmacol Ther 121:41–54
8. Corkey BE (2012) Banting lecture 2011: hyperinsulinemia: cause or consequence? Diabetes 61:4–13
9. Finger TE, Kinnamon SC (2011) Taste isn't just for taste buds anymore. F1000 Biol Repord 3:20
10. Ford HE, Peters V, Martin NM et al (2011) Effects of oral ingestion of sucralose on gut hormone response and appetite in healthy normal-weight subjects. Eur J Clin Nutr 65:508–513
11. Fowler SP, Williams K, Hazuda HP (2015) Diet soda intake is associated with long-term increases in waist circumference in a biethnic cohort of older adults: the san antonio longitudinal study of aging. J Am Geriatr Soc 63:708–715
12. Fowler SP, Williams K, Resendez RG et al (2008) Fueling the obesity epidemic? Artificially sweetened beverage use and long-term weight gain. Obesity (Silver Spring) 16:1894–1900
13. Fujita Y, Wideman RD, Speck M et al (2009) Incretin release from gut is acutely enhanced by sugar but not by sweeteners in vivo. Am J Physiol Endocrinol Metab 296:E473–E479
14. Gardener H, Rundek T, Markert M et al (2012) Diet soft drink consumption is associated with an increased risk of vascular events in the Northern Manhattan Study. J Gen Intern Med 27:1120–1126
15. Grice HC, Goldsmith LA (2000) Sucralose – an overview of the toxicity data. Food Chem Toxicol 38(Suppl 2):S1–S6
16. Grotz VL, Munro IC (2009) An overview of the safety of sucralose. Regul Toxicol Pharmacol 55:1–5
17. Gurley BJ, Swain A, Williams DK et al (2008) Gauging the clinical significance of P-glycoprotein-mediated herb-drug interactions: comparative effects of St. John's wort, Echinacea, clarithromycin, and rifampin on digoxin pharmacokinetics. Mol Nutr Food Res 52:772–779
18. International Programme on Chemical Safety Acesulfame Potassium. Available at http://www.inchem.org/
19. International Programme on Chemical Safety (Inchem) Saccharin. Available at http://www.inchem.org/

20. Jang HJ, Kokrashvili Z, Theodorakis MJ et al (2007) Gut-expressed gustducin and taste receptors regulate secretion of glucagon-like peptide-1. Proc Natl Acad Sci U S A 104:15069–15074
21. Jeppesen PB, Gregersen S, Alstrup KK et al (2002) Stevioside induces antihyperglycaemic, insulinotropic and glucagonostatic effects in vivo: studies in the diabetic Goto-Kakizaki (GK) rats. Phytomedicine 9:9–14
22. Li F (2013) Taste perception: from the tongue to the testis. Mol Hum Reprod 19:349–360
23. Lown KS, Bailey DG, Fontana RJ et al (1997) Grapefruit juice increases felodipine oral availability in humans by decreasing intestinal CYP3A protein expression. J Clin Invest 99:2545–2553
24. Ma J, Bellon M, Wishart JM et al (2009) Effect of the artificial sweetener, sucralose, on gastric emptying and incretin hormone release in healthy subjects. Am J Physiol Gastrointest Liver Physiol 296:G735–G739
25. Ma J, Chang J, Checklin HL et al (2010) Effect of the artificial sweetener, sucralose, on small intestinal glucose absorption in healthy human subjects. Br J Nutr 104:803–806
26. Mace OJ, Affleck J, Patel N et al (2007) Sweet taste receptors in rat small intestine stimulate glucose absorption through apical GLUT2. J Physiol 582:379–392
27. Magnuson BA, Burdock GA, Doull J et al (2007) Aspartame: a safety evaluation based on current use levels, regulations, and toxicological and epidemiological studies. Crit Rev Toxicol 37:629–727
28. Margolskee RF, Dyer J, Kokrashvili Z et al (2007) T1R3 and gustducin in gut sense sugars to regulate expression of Na^+-glucose cotransporter 1. Proc Natl Acad Sci U S A 104: 15075–15080
29. Masubuchi Y, Nakagawa Y, Ma J et al (2013) A novel regulatory function of sweet taste-sensing receptor in adipogenic differentiation of 3T3-L1 cells. PLoS One 8:e54500
30. Mattes RD, Popkin BM (2009) Nonnutritive sweetener consumption in humans: effects on appetite and food intake and their putative mechanisms. Am J Clin Nutr 89:1–14
31. Miller SA, Frattali VP (1989) Saccharin. Diabetes Care 12:74–80
32. Nakagawa Y, Nagasawa M, Yamada S et al (2009) Sweet taste receptor expressed in pancreatic beta-cells activates the calcium and cyclic AMP signaling systems and stimulates insulin secretion. PLoS One 4:e5106
33. Nettleton JA, Lutsey PL, Wang Y et al (2009) Diet soda intake and risk of incident metabolic syndrome and type 2 diabetes in the Multi-Ethnic Study of Atherosclerosis (MESA). Diabetes Care 32:688–694
34. O'connor L, Imamura F, Lentjes M et al (2015) Prospective associations and population impact of sweet beverage intake and type 2 diabetes, and effects of substitutions with alternative beverages. Diabetologia 58:1474–1483
35. Otabe A, Fujieda T, Masuyama T et al (2011) Advantame – an overview of the toxicity data. Food Chem Toxicol 49(Suppl 1):S2–S7
36. Palmnas MS, Cowan TE, Bomhof MR et al (2014) Low-dose aspartame consumption differentially affects gut microbiota-host metabolic interactions in the diet-induced obese rat. PLoS One 9:e109841
37. Pepino MY (2015) Metabolic effects of non-nutritive sweeteners. Physiol Behav 152:450–455
38. Pepino MY, Bourne C (2011) Non-nutritive sweeteners, energy balance, and glucose homeostasis. Curr Opin Clin Nutr Metab Care 14:391–395
39. Pepino MY, Tiemann CD, Patterson BW et al (2013) Sucralose affects glycemic and hormonal responses to an oral glucose load. Diabetes Care 36:2530–2535
40. Peters JC, Wyatt HR, Foster GD et al (2014) The effects of water and non-nutritive sweetened beverages on weight loss during a 12-week weight loss treatment program. Obesity (Silver Spring) 22:1415–1421
41. Renwick AG (2006) The intake of intense sweeteners – an update review. Food Addit Contam 23:327–338

42. Roberts A, Renwick AG, Sims J et al (2000) Sucralose metabolism and pharmacokinetics in man. Food Chem Toxicol 38(Suppl 2):S31–S41
43. Rogers PJ, Hogenkamp PS, De Graaf C et al (2015) Does low-energy sweetener consumption affect energy intake and body weight? A systematic review, including meta-analyses, of the evidence from human and animal studies. Int J Obes (Lond) 40:381
44. Sanz Y, Santacruz A, Gauffin P (2010) Gut microbiota in obesity and metabolic disorders. Proc Nutr Soc 69:434–441
45. Schiffman SS, Rother KI (2013) Sucralose, a synthetic organochlorine sweetener: overview of biological issues. J Toxicol Environ Health B Crit Rev 16:399–451
46. Schmiedlin-Ren P, Edwards DJ, Fitzsimmons ME et al (1997) Mechanisms of enhanced oral availability of CYP3A4 substrates by grapefruit constituents. Decreased enterocyte CYP3A4 concentration and mechanism-based inactivation by furanocoumarins. Drug Metab Dispos 25:1228–1233
47. Steinert RE, Frey F, Topfer A et al (2011) Effects of carbohydrate sugars and artificial sweeteners on appetite and the secretion of gastrointestinal satiety peptides. Br J Nutr 105: 1320–1328
48. Suez J, Korem T, Zeevi D et al (2014) Artificial sweeteners induce glucose intolerance by altering the gut microbiota. Nature 514:181–186
49. Suez J, Korem T, Zilberman-Schapira G et al (2015) Non-caloric artificial sweeteners and the microbiome: findings and challenges. Gut Microbes 6:149–155
50. Suzuki H, Sugiyama Y (2000) Role of metabolic enzymes and efflux transporters in the absorption of drugs from the small intestine. Eur J Pharm Sci 12:3–12
51. Swithers SE (2013) Artificial sweeteners produce the counterintuitive effect of inducing metabolic derangements. Trends Endocrinol Metab 24:431–441
52. Swithers SE, Davidson TL (2008) A role for sweet taste: calorie predictive relations in energy regulation by rats. Behav Neurosci 122:161–173
53. Swithers SE, Laboy AF, Clark K et al (2012) Experience with the high-intensity sweetener saccharin impairs glucose homeostasis and GLP-1 release in rats. Behav Brain Res 233:1–14
54. Sylvetsky A, Rother KI, Brown R (2011) Artificial sweetener use among children: epidemiology, recommendations, metabolic outcomes, and future directions. Pediatr Clin North Am 58:1467–1480, xi
55. Temizkan S, Deyneli O, Yasar M et al (2015) Sucralose enhances GLP-1 release and lowers blood glucose in the presence of carbohydrate in healthy subjects but not in patients with type 2 diabetes. Eur J Clin Nutr 69:162–166
56. Thiebaut F, Tsuruo T, Hamada H et al (1987) Cellular localization of the multidrug-resistance gene product P-glycoprotein in normal human tissues. Proc Natl Acad Sci U S A 84: 7735–7738
57. Toyono T, Seta Y, Kataoka S et al (2007) CCAAT/Enhancer-binding protein beta regulates expression of human T1R3 taste receptor gene in the bile duct carcinoma cell line, HuCCT1. Biochim Biophys Acta 1769:641–648
58. US Food and Drug Administration (2004) Guidance for industry: frequently asked questions about GRAS. In: Center for Food Safety and Applied Nutrition (ed) United States Department of Health and Human Services. Silver Spring, Maryland
59. United States Food and Drug Administration (2015) Additional information about high-intensity sweeteners permitted for use in food in the United States. Silver Spring, Maryland
60. United States Food and Drug Administration (2014) Food additives permitted for direct addition to food for human consumption; Advantame. In: Department of Health and Human Services (ed) United States Department of Health and Human Services. Silver Spring, Maryland
61. Whitehouse CR, Boullata J, Mccauley LA (2008) The potential toxicity of artificial sweeteners. AAOHN J 56:251–259, quiz 260–251

62. Wu T, Zhao BR, Bound MJ et al (2012) Effects of different sweet preloads on incretin hormone secretion, gastric emptying, and postprandial glycemia in healthy humans. Am J Clin Nutr 95:78–83
63. Yadav SK, Guleria P (2012) Steviol glycosides from Stevia: biosynthesis pathway review and their application in foods and medicine. Crit Rev Food Sci Nutr 52:988–998
64. Zheng Y, Sarr MG (2013) Effect of the artificial sweetener, acesulfame potassium, a sweet taste receptor agonist, on glucose uptake in small intestinal cell lines. J Gastrointest Surg 17:153–158; discussion p 158

Sweet-Taste Receptor Signaling Network and Low-Calorie Sweeteners

18

Menizibeya O. Welcome, Nikos E. Mastorakis, and Vladimir A. Pereverzev

Abstract

Over the past decades, low-calorie sweeteners (LCSs) have emerged as the main source of sweetening food products and beverages. Evidences suggest mixed effects of LCSs intake. In addition, there is a paucity of data on long-term health outcomes of LCSs intake as well as their usefulness in certain groups of the population (e.g., pregnant women, children, and the elderly). Moreover, available unbiased studies conducted to date are fairly scanty, thereby restraining far-reaching conclusions. LCSs are biomolecules with cognate cellular receptors ubiquitously localized in many organs and tissues of the body. The T1R2-T1R3 receptor heterodimer is one of the widely known receptors of LCSs. Upon activation by LCSs, T1R2-T1R3 signals downstream that result in several cellular responses that characterize the biochemical and physiological effects of LCSs. In the gut, for instance, signaling of LCSs may be mediated via humoral and neural mechanisms (involving the gut-brain axis, gut-adipose tissue axis, gut-adipose tissue-brain triangle, gut-pancreas-adipose tissue or entero-adipo-insular triangle, gut-brain-adipose tissue triangle) influence obesity, overweight, appetite, satiety,

M.O. Welcome (✉)
Department of Physiology, Faculty of Basic Medical Sciences, Madonna University, Elele, Rivers State, Nigeria

World Scientific and Engineering Society and Academy, Athens, Greece
e-mail: menimed1@yahoo.com

N.E. Mastorakis
World Scientific and Engineering Society and Academy, Athens, Greece

Technical University of Sofia, Sofia, Bulgaria
e-mail: mastor@tu-sofia.bg

V.A. Pereverzev
Department of Normal Physiology, Belarusian State Medical University, Minsk, Belarus
e-mail: Pereverzev2010@mail.ru

J.-M. Mérillon, K.G. Ramawat (eds.), *Sweeteners*, Reference Series in Phytochemistry,
https://doi.org/10.1007/978-3-319-27027-2_25

cognition, and memory. Several interconnected signaling pathways are important in modulating the physiological outcomes of LCS intake. In this chapter, we discuss the signaling mechanisms of LCSs, and their relationship to metabolism, obesity, weight gain, satiety, appetite, cognition, and memory.

Keywords
Low calorie sweeteners • Non-calorie sweeteners • Non-nutritious sweeteners • Artificial sweeteners • Sugar alcohols • Carbohydrate derived sweeteners • Sweet proteins • Sweet taste receptors • Sweet taste receptor signaling • Sweet taste receptor heterodimer • T1R2-T1R3 • Obesity • Overweight • Appetite • Satiety • Diabetes • Cognition • Memory • Enteroendocrine cell • Enterocyte

Abbreviations

AC	Adenylate cyclase
ACTH	Adrenocorticotropic Hormone
AgRP	Agouti-related protein
AMPK	AMP-activated protein kinase
AP	Area postrema
ATP	Adenosine triphosphate
CALHM1	Calcium homeostasis modulator 1
CAMK	Calmodulin-dependent protein kinase
cAMP	Cyclic adenosine monophosphate
CART	Cocaine and amphetamine regulated transcript
CREB	cAMP-related element binding protein
CRH	Corticotropin releasing hormone
DMN	Dorsomotor nucleus
GABA	Gamma amino butyric acid
GLP-1	Glucagon-like peptide-1
GLUT	GLUcose Transporter
IP3	1,4,5-inositol trisphosphate
Low-calorie sweeteners	LCSs
MC4R	Melanocortin receptor type-4
mTOR	Mammalian target of rapamycin
mTORC1	Mammalian target of rapamycin 1
NPY	Neuropeptide Y
NTS	Nucleus tractus solitarius
Ob-Rb	Adiposity (obesity) receptor type b
PDE	Phosphodiesterase
PI3K	Phosphoinositide-3-kinase
PKB	Protein kinase B
PLCβ	Phospholipase Cβ
POMC	Proopiomelanocortin
PVN	Paraventricular nucleus
PYY	Peptide YY
SGLT	Sodium-dependent glucose cotransporter

T1R2-T1R3	Sweet-taste receptor heterodimer
TRH	Thyrotropin releasing hormone
TRPM5	Transient receptor potential cation channel, subfamily M, member 5
α-MSH	Alpha-melanin-stimulating hormone

Contents

1 Introduction 395
2 General View on Sweet-Taste Receptor Signaling Network 397
3 Cells and Receptors for LCSs 398
3.1 Gastrointestinal Signaling of LCSs 398
4 Functional Implication of Activation of Taste Receptors by LCSs 401
4.1 Obesity, Weight Gain, and the Taste Receptor Signaling Network of LCSs 401
4.2 Cognition, Memory, and the Taste Receptor Signaling Network of LCSs 403
4.3 Metabolic Functions and the Taste Receptor Signaling Network of LCSs 404
5 Conclusion 405
6 Future Directions 405
References 405

1 Introduction

The sweetness of sugar (a natural product) was discovered over two centuries ago, and some decades later, precisely in the 1870s, saccharin, the first artificial sweetener was synthesized. The intake of these sweeteners significantly increased over a few decades. It also served as a boost to the food industry. Concomitantly, there was a recorded surge in the prevalence of obesity especially in the Western nations where these sweet products are widely used. Around the mid-twentieth century, researchers thought that the major factor responsible for the association between sugar intake and overweight and obesity was related to the amount of calories. Thus, both the industry and researchers sought a way in solving this problem. Substitutes for sugar and saccharin with even hundreds of times sweeter than sugar were discovered [1, 2]. Some of the newly discovered sweeteners (artificial and natural) were termed low-calorie sweeteners (LCSs). The use of LCSs in food and food products have increased over the past four decades. The list of LCSs is also increasing due to the economic advantages and industrial application. Low-calorie sweeteners include aspartame, erythritol, isomalt, xylitol, saccharin, sorbitol, sucralose, maltitol, acesulfame potassium, neotame, and rebaudioside A. These are carbohydrate-derived low-calorie sweeteners. Sweet proteins are excellent substitutes for carbohydrate-derived low-calorie sweeteners and they include brazzein, curculin, monelin, miraculin, mabinlin, pentadin, and thaumatin [2, 3]. Sweet proteins are even sweeter and tastier than carbohydrate-derived or sugar alcohol sweeteners. Thaumatin, for instance, provides only about 4 kcal/g and when compared with the same weight of sucrose and sugar alcohols, it is 100-fold and 2000-fold sweeter [4]. These sweeteners have found extensive application in beverages and food products, serving as alternatives to sugar. LCSs as the name indicates, when consumed provide only a very small amount of energy or negligible energy content compared with glucose or sugar. Some sweeteners are believed to have a zero glycemic index. For this reason, LCSs is used to control energy intake, diet quality, manage weight and obesity, and recommended for diabetics for the purpose of sweetening of food [5–8].

The substantial increase in the use of LCSs in food and beverages is not comparative with the wide prevalence of overweight and obesity even in those nations where LCSs are increasingly been used [3, 5], suggesting that overweight and obesity have multiple causative factors. Besides, investigations of the impact of LCSs on these indices are largely inconclusive or biased. Most LCSs have been approved by Food and Drug Administration, further boosting the market [5]. While the benefits of use of LCSs have been widely researched, there is a paucity of research data evaluating both the disadvantages and advantages of using LCSs in different groups of individuals and population. In most instances, the studies that evaluate the impact of LCSs on the body are biased as these sweeteners are consumed usually in combination with other food items, or used in a relatively short duration, so evaluating the benefits of LCSs in an individual over a relatively long period of time is often than not, flawed. Only a very few studies tend to substantially reduce bias to the very possible minimum. In this regard, meta-analysis and systemic reviews on high-quality publications may provide additional information on the benefits of LCSs to the body.

In a review of the literature, Anderson et al. [9] reported that there was no evidence that the use of LCSs caused higher body weights or served as a source of weight management in adults. The authors noted that confounding factors in most studies, such as short duration of experimental studies, sample size limitation, multiple determinants (environmental and behavioral) of obesity, weight gain, etc., were possibly necessary for mixed results of the different studies conducted to date.

In a recent meta-analysis of randomized controlled trials (RCTs) and prospective cohort (PC) studies, Miller and Perez [5] examined the relation between LCSs and body weight. The author's results showed that LCSs significantly reduced body weight, body mass index (BMI), fat mass, and waist circumference in RCTs. Whereas in PC studies, LCSs did not have any impact on body weight or fat mass but was significantly associated with slightly higher BMI.

Besides the effect of LCSs on weight and obesity, there is also concern on the effect of LCSs on the occurrence of cancers, especially, when used in excess or for a very long duration. For instance, high doses of saccharin induce bladder cancer in laboratory animals. Epidemiological, experimental, and clinical data are not adequate to establish any evidence on the likely carcinogenic risks of new generation sweeteners (LCSs) [1]. Moreover, obesity and overweight are risks for other comorbidities including cancer, diabetes, cardiovascular disease, and hypertension [5]. Surprisingly, LCSs are now widely used by children and pregnant women, though there is no enough epidemiological evidences suggesting their harm on children, embryo, and fetus, the negative outcome of using LCSs by these group of individuals cannot be ruled out completely [10–13].

The consensus statement on benefits of low-calorie sweeteners produced at the International Sweeteners Association Conference held on 1–2 April 2014 in Brussels, Belgium, concluded on five major aspects of the impact of LCSs [14], which were largely positive benefits. Unfortunately, there are still concerns about prolonged intake of LCSs and their use by certain group of people in the population.

Despite the negligible or zero energy content of LCSs, their effects on body weight of adults remains unclear. Accumulating scientific data over the past decades have opened our understanding on the mechanisms of obesity, weight gain, satiety, and numerous other measures and outcomes for which the impact of LCS are currently been evaluated. Thus, the cellular and molecular mechanisms of signaling of LCS may provide useful information to their likely physiological impact. It should be noted, however, that there are multiple mechanisms and numerous confounders responsible for the pathogenesis of obesity, weight gain, etc. The literature database on this is increasing on a yearly basis. In essence, it follows that the mechanisms of signaling of LCS is important, at least, in understanding the physiological outcomes of their application in food or beverages.

Preference for sweet tasting substances is innate and universal, and probably represents a biological response to safe sources of energy such as mother's milk and fruit. There is some evidence that sweetness preference declines with age. Children and newborns have higher level of preference for sweet taste compared with adults. Threshold for taste also has cultural peculiarities and may be heritable [1, 14–18]. These differences may be due to the genetic polymorphisms of cellular receptors to which tasty substances bind. The expression of these receptors in different cells of the body may, to a large degree, determine the level of preference for sweet taste and thus have some impact on many cellular processes and signaling.

In humans, sweet substances such as LCSs activate sweet-taste receptor types "T1R2-T1R3." Functional T1R2-T1R3 receptor is a dimer of T1R2 and T1R3. This heterodimer recognizes both natural and synthetic sweet substances [2]. Importantly, the impact of LCSs intake is due to their signaling, which we shall discuss here – sweet-taste receptor signaling network [5].

2 General View on Sweet-Taste Receptor Signaling Network

The type-2 taste receptor gene family encodes for several taste receptors and are ubiquitously located in different cells and tissues of the body [19]. Taste receptors are expressed in gastrointestinal tract, pancreas, respiratory track, some blood cells, and brain. These different localizations of taste receptors are associated with specific roles. For instance, in the brain (hypothalamus) they are believed to be involved in sensing extracellular glucose level and modulate neuronal signaling and excitability. In addition, they may be important in cognition and memory. In the respiratory tract, sweet-taste receptors play significant role in the maintenance of the mucosal and ciliary functioning, mainly by ensuring adequate and supportive role for the clearance of glucose via GLUT-1 and -10. (Because sweeteners previously listed also activate receptors of glucose, GLUTs, while sweet-taste receptors also sense glucose in the surrounding environment of the cell, the receptors are sometimes, generally, called glucosensors). In blood cells such as macrophages, they modulate the degree of immune response. In the pancreas, sweet-taste receptors modulate the sensing of glucose and secretion of hormones and other paracrine factors that may be involved in the pathogenesis of diabetes. In the gastrointestinal tract, these receptors play a supportive role in glucose absorption and metabolism. In these and other organs and tissues, sweet-taste receptors are involved in the initiation and progression of some pathological processes including diabetes, inflammation, and asthma.

Figure 1 shows signaling mechanisms of sweet-taste receptors. The signaling is initiated by sweet substances (e.g., LCSs) activating the cognate membrane receptors and may also modulate the functioning of non-sweet taste cell via paracrine mechanisms. The signaling events lead to the modulation of numerous cellular activities, [20–22], including the release of peptides and several biomolecules [23]. Increasing evidences suggest that sweet-taste signaling mechanisms in the gastrointestinal tract are similar to those in the pancreas, respiratory tract, and other parts of the body which expresses taste receptors [24]. It should be noted that there may be yet unidentified taste receptors in taste cells [25]

3 Cells and Receptors for LCSs

Cells involved in sensing of LCSs include enteroendocrine cells, neuro- or endocrine cells, neurons, and some blood cells. These cells are known to express various types of taste receptors including the T1R2-T1R3. Receptor types other than T1R2-T1R3, such as SGLT, GLUT family [26, 27] may be involved in LCSs signaling. This hypothesis is based on the role of these receptor types in the signaling of glucose (a sweet-taste molecule) and numerous other molecules. Further research on this direction may yield important results. It is possible that there are currently identified membrane transporters for LCSs.

3.1 Gastrointestinal Signaling of LCSs

Gastrointestinal signaling of LCSs is the identification of luminal LCSs by gut epithelial cells, activation of membrane receptors that initiate the further activation of membrane enzymes involved in downstream signaling via production of second messengers resulting to the change in cell potential or production or exocytosis of molecules (e.g., mediators) that take part in a variety of functioning. Figure 2 summarizes the mechanisms of LCS signaling in the gastrointestinal tract.

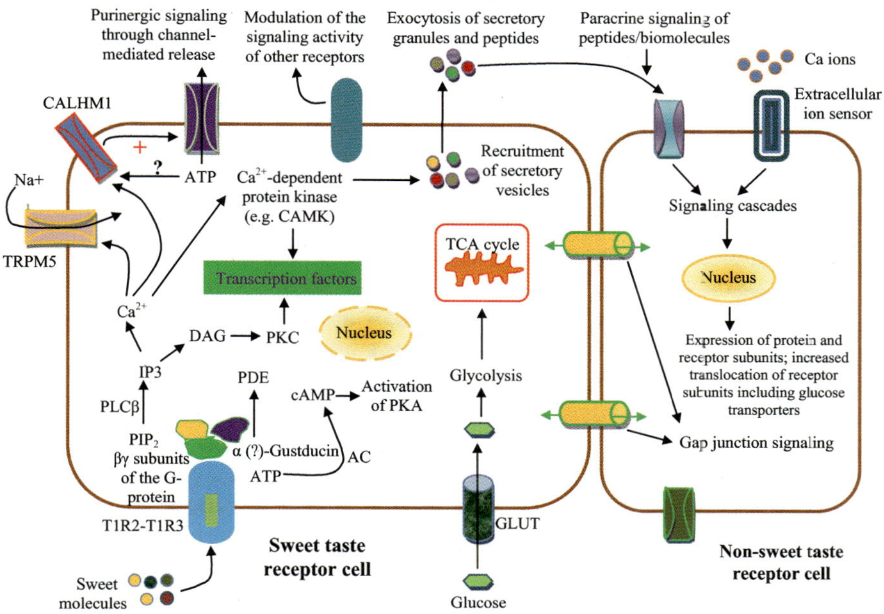

Fig. 1 A general model of sweet-taste signaling network. Sweet-taste receptors are G-protein-coupled receptors that possess multiple binding sites and mode of interaction for sweet molecules. Sweet molecules activate the G-protein by downstream signaling leading to the dissociation of the α-subunit gustducin from the βγ subunits. Dissociated βγ subunits of the G-protein activate phospholipase Cβ (PLCβ), leading to the formation of 1,4,5-inositol trisphosphate (IP3). IP3 is responsible for the release of Ca^{2+} from intracellular stores through its binding to IP3-receptor in these stores. There may be calcium-induced calcium release. Increase in intracellular Ca^{2+} activates calcium-dependent kinase, selective cation channel, TRPM5 (transient receptor potential cation channel, subfamily M), as well as other receptors. TRPM5 may also play a role in capacitance-mediated calcium entry into taste cells. Modulation of purinergic signaling by taste receptors also plays a key role in taste sensation. The voltage-gated ion channel, calcium homeostasis modulator 1 (CALHM1) is indispensable for taste-stimuli-evoked ATP release from sweet (and bitter, umami) taste cells. Dissociated α subunit referred to Gα-gustducin activates a phosphodiesterase (PDE) thereby decreasing intracellular cAMP levels. Gα-gustducin is also reported to activate adenylate cyclase (AC) to increase cAMP level. α-Gustducin is necessary to maintain low-level of cAMP level, which is necessary to maintain the adequate signaling of Ca^{2+} by disinhibition of cyclic nucleotide-inhibited channels to elevate intracellular Ca^{2+}. Changes in cAMP levels also affect other ion channels, including K^+ channels. The events resulting in activation/modulation of ion channels lead to membrane depolarization and formation of action potentials. Potential-dependent release of mediators (ATP, serotonin, etc.) and peptides and calcium-dependent release of peptides/biomolecules are some of the results of sweet-taste receptor signaling. A hallmark of sweet-taste receptor signaling is the activation of transcription factors and gene expression, which might be dependent on intracellular calcium waves and activity-dependent activation of calcium-dependent kinases, including the calmodulin-dependent protein kinase (CAMK)

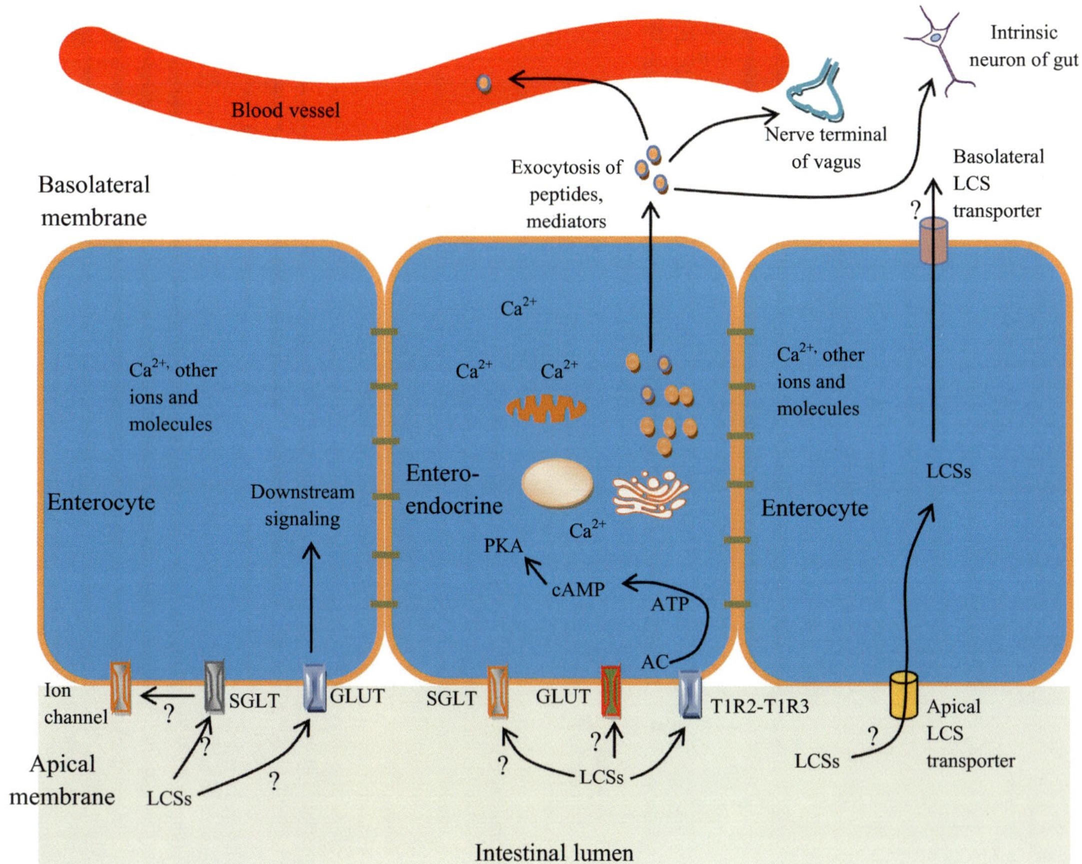

Fig. 2 (continued)

4 Functional Implication of Activation of Taste Receptors by LCSs

4.1 Obesity, Weight Gain, and the Taste Receptor Signaling Network of LCSs

The association between LCSs intake and obesity or weight gain may be an indirect one. The mechanism may be, in part, related to regulation of appetite and satiety by substances released in the gut in response to TIR2-TIR3 signaling. The signaling may be related to different axes and triangles including gut-adipose tissue, gut-adipose tissue-brain axis, gut-pancreas-adipose tissue, gut-brain, and gut-brain-adipose.

The initiation of signaling under the activation of mediator synthesis by LCSs is the first stage in the reaction mechanism. LCS signaling in the gut releases a host of molecules which possess anoxerigenic (appetite suppression) and orexigenic (appetite promotion) properties.

Anorexigenic hormones and molecules (also called satiety signals) include leptin, amylin, adiponectin, insulin, glucagon-like peptide 1 (GLP-1), peptide YY (PYY), pancreatic polypeptide, cholecystokinin, oxyntomodulin, cocaine, and amphetamine-regulated transcript (CART), and proopiomelanocortin (POMC) [30–35]. Orexigenic hormones or molecules include ghrelin, cannabinoids, agouti-related peptide (AgRP), and neuropeptide Y (NPY) [30, 34]. The expression of taste receptors, anoxerigenic, and orexigenic receptors may vary in different cells and also have individual and cultural variations. Thus, the response from their signaling may as well vary in different groups of cells, tissues, and organs of the body.

Research evidences indicate that PYY and many other orexins may be responsible for pathophysiology of obesity through multiple mechanisms that involve regulation of energy homeostasis via axes and triangles previously outlined [36]. The mechanism of action of PYY on central nervous system involves modulation of neurons activity. The molecule leptin suppresses appetite; whereas ghrelin is orexigenic as it stimulates appetite [37]. Ghrelin is released in the stomach during hunger to signal specific group of neurons located in various brain regions. The hormone ghrelin stimulates the production of other orexigenic peptides of arcuate nucleus such as NPY and AgRP and modulates anorexigenic proopiomelanocortin neurons.

Fig. 2 Mechanism of LCs signaling in the gastrointestinal tract. Enteroendocrine cell may be K- or L- cell. In the gastrointestinal tract, LCSs activate their cognate receptors, TIR2-TIR3, usually located on the plasma membrane of enteroendocrine cell, resulting to downstream signaling that subsequently lead to exocytosis of secretory vesicles containing mediators and peptides. The mediators and peptides localize their receptors on nerve terminal and the impulses are transmitted to the central nervous system. The mediators may be transported into the blood stream from where they reach different tissues and organs of the body including the brain, adipose tissues, pancreas, etc. It is currently not clear whether or not gastrointestinal epithelial cells (enterocyte, enteroendocrine cell) also contain other receptor or transporters other than the cognate sweet-taste receptors of LCSs. Owing to the involvement of GLUTs and SGLTs in glucose sensing [28, 29], it is possible to hypothesize that these receptors may contribute to LCSs signaling in the gut and probably, elsewhere in the body

The anorexigenic mediator, leptin, selectively suppresses sweet taste via the adipocyte receptor, Ob-Rb, in sweet-taste cells; whereas the orexigenic mediators, endocannabinoids, selectively enhance sweet taste via cannabinoid receptor type-1 in sweet-taste cells [38]. (Leptin is also produced in the salivary glands, stomach, adipose tissue, etc.). These mediators act centrally in the hypothalamus and limbic system to modulate energy balance and homeostasis. It has been suggested that increase in insulin secretion may lead to suppression of sweet taste, while glucagon will enhance sweet-taste perception in sweet-taste cells.

NPY is an orexigenic molecule expressed in arcuate neurons that are activated by food deprivation and inhibited by feeding in a nutrient-dependent manner. PYY and leptin also prevent fasting-induced activation of the arcuate neurons. The anorexigenic hormone amylin secreted mainly by the pancreas and stomach acts via the area postrema to control energy intake [32]. The mechanism for anorexigenic and orexigenic regulation may be related to AMP-activated protein kinase (AMPK) activity and mammalian target of rapamycin 1 (mTORC1). For instance, anorexigenic effect has been related to the activation mTORC1 in a phosphoinositide-3-kinase (PI3K)- and protein kinase B (PKB)-dependent manner, while orexigenic effect – increased AMPK activity [30, 37]. Interestingly, increased AMPK activity blocks mTORC1 activity.

The secretion of PYY increases postprandially (immediately following food intake) and activates vagus afferent nerve fibers or via the circulatory system (blood vessels) activate NPY neurons of the arcuate nucleus, which also contains receptors for orexigenic molecules [39]. However, PYY may activate anorexigenic circuits by reducing the synaptic inhibition mediated by Y_2 receptor [40]. In the hypothalamus, PYY acts as an agonist in the neuron of the arcuate nucleus by activating Y2 receptor, a presynaptic inhibitory receptor that blocks the orexigenic effects of the NPY/AgRP neurons. One of the effects of the inhibition of hypothalamic neurons (arcuate POMC neurons) by PYY is the release of the inhibitory effect of the GABA on these groups of neurons, which results to the removal of the inhibitory effect on POMC neurons. So, POMC neurons now release alpha-MSH, a powerful inhibitor of appetite. Therefore, appetite is inhibited.

The arcuate nucleus and the paraventricular nucleus of the hypothalamus as well as the brainstem play a significant role in integration of humoral and metabolic signals to maintain energy homeostasis [39]. Major neurons in the arcuate nucleus controlling food intake are POMC and CART (anorexigenic) referred to as POMC/CART neurons control appetite, and AgRP, NPY (orexigenic) referred to as NPY/AgRP/GABA-neurons control satiety signals [41]. The anorexigenic and orexigenic neurons project to the paraventricular hypothalamus and to the lateral and dorsomedial hypothalamus, which contains high proportion of leptin and POMC neurons. The population of NPY/AgRP/GABA neurons of arcuate nucleus, which project to the PVN and send inhibitory GABA collaterals to arcuate nucleus of POMC neurons. GABA blocks the anorexic effect of α-MSH, whereas a GABA antagonist increases anorexigenic effects.

These neurons form synapses with alpha-MSH neurons located in the arcuate nucleus which inhibits food intake. The alpha-MSH neurons produce POMC, a precursor of melanocortins which include melanin-stimulating hormone, adrenocorticotropin (ACTH), and β-endorphin. Melanin-stimulating hormone is a strong orexigenic molecule. The receptors for this hormone are widely distributed in the hypothalamus, brainstem. Paraventricular nucleus of the hypothalamus has high expression of melanocortin receptors. Moreover, this hypothalamic region also produces anorexigenic peptides including thyrotropin-releasing hormone (TRH), corticotropin releasing hormone (CRH), and oxytocin. The lateral hypothalamus and perifornical area produce the orexigenic substances orexin-A and melanin concentrating hormone [33].

In the brainstem, areas involved in appetite and satiety regulation include dorsal vagal complex (dorso motor nucleus – DMN), area postrema (AP), nucleus tractus solitarius (NTS), parabrachial, hypoglossal, trigeminal, lateral reticular, cochlear nuclei, locus coeruleus, and inferior olive. These areas contain leptin receptors to which the hormone leptin localizes to suppress feeding. The main areas believed to control feeding are the NTS, DMN, AP, which have high expression of melanocortin receptor type-4, MC4R. Neurons secreting POMC from the arcuate nucleus extend to the NTS. These neurons can be activated by CCK or other mediator. Neurons of the dorsal vagal complex also extend to the NTS. Moreover, dorsal motor complex of the vagus receive fibers from the POMC neurons of the arcuate nucleus. It is important to note that neurons of the dorsal motor complex reach the NTS. Activation of the sensory receptors extending to the vagus in the gastrointestinal tract allows the transmission of satiety signals to the vagus motor complex and then to the NTS. The afferent neurons of the NTS project to the arcuate nucleus of the hypothalamus. This is where satiety signals are integrated with appetite signals. It should be noted, however, that axons of the neurons of the hypothalamus also reach the nucleus accumbens, suggesting emotional influence of food intake [33].

Is the physiological state of satiety related to signaling of taste receptors? Ren et al. experimentally observed variations in the expression of taste receptors in the brain. The expression of the taste receptors, T1R1 and T1R2, increases with fasting. However, the level of T1R1 and T1R2 expression returns to normal when sweet molecules are administered. This suggests that signaling of orexigenic and anorexigenic molecules released from the gut may indirectly modulate the expression of sweet-taste receptors via changes in the level of appetite or satiety, which necessarily affect body mass via weight loss or weight gain.

4.2 Cognition, Memory, and the Taste Receptor Signaling Network of LCSs

The impact of LCSs on cognition is due to signaling via gut-brain axis. Substances (mediators) released from the gut upon stimulation of the membrane receptors of epithelial cells by LCSs can diffuse into the blood stream from where they transported to many organs and tissues including the brain. These molecules easily cross the blood-brain barrier to influence neural information processing. Through this pathway, LCSs may modulate cognition and memory [42, 43].

The recent review [43] suggests that T1R2-T1R3 functioning as a hypothalamic glucosensor serves to modulate cognitive functions in the neuro-astroglial system. (The hypothalamus serves as a site of regulation of feeding, central and peripheral metabolism, hormones secretion, and functions. This brain region is also involved in satiety and appetite). Thus, specific LCS that may modulate neuro-astroglial cooperativity may affect cognition and memory. The mechanisms of this modulatory effect of LCSs on neuro-astroglial system or cooperativity may largely involve signaling through metabolic coupling; activity-dependent signaling; cross-signaling, initiated by downstream effectors; and receptor cooperativity and associativity. In addition, these processes may occur or are mediated via paracrine signaling and/or homeostasis of intracellular ions – which are also responsible for modulating neuro-astroglial system or cooperativity [42, 43].

Besides T1R2-T1R3 receptor heterodimer, taste related genes are known to be highly expressed in the cortex and hippocampus. These brain regions are largely involved in cognition and memory [44].

4.3 Metabolic Functions and the Taste Receptor Signaling Network of LCSs

The T1R2-T1R3 receptor heterodimer contributes to maintaining the activity of the cell by controlling glucose absorption and transport. This is possibly achieved by the associativity and/or cooperativity between T1R2-T1R3 and cognate glucose receptors such as GLUT2 and the SGLT members. The mechanism for this process is not known, but it may be suggested that the signaling mechanisms might involve downstream effectors that subsequently modulate both the transport activity and the expression of glucose transporters. In the gastrointestinal tract, for instance, T1R2-T1R3 receptor has been shown to influence the activity and expression of GLUT2 and SGLT1, possibly through auto-paracrine signaling. The cooperative signaling of T1R2-T1R3, GLUT2, and SGLT1 may also involve modulation of mediator (e.g., ATP) synthesis and release [43].

The functional link between the gut and brain (hypothalamus) provides a means of regulating not only behavior but also peripheral and central metabolic functions [45]. Downstream effectors of sweet-taste receptor signaling may involve energy sensors such as glucokinase, GLUT2, AMPK (AMP-activated protein kinase), CREB (cAMP related element binding protein), mTOR (Mammalian Target of Rapamycin), sirtuins, and PASK (also called PASKIN – a kinase protein). The activation or inhibition of these membrane and intracellular sensors of energy balance may consequently affects overall cellular metabolism. This is a functional relationship (or cross-talk) between metabolic sensors, which are useful in maintaining metabolic homeostasis. PASK, for instance, is required for the normal functioning of AMPK and mTOR/S6 K1. In this regard, previous data have consistently shown that AMPK, CREB, and mTOR are involved in both glucose metabolism as well as higher brain functions (cognition and memory). Interestingly, these metabolic sensors may cross-signal with T1R2-T1R3 in taste cells or interact with mediators released from the gut [42, 43, 45–48]. Hurtado-Carneiro et al. [46] observed that GLP-1 can attenuate the activity of AMPK and mTOR induced by fluctuations in glucose levels in hypothalamic areas involved in feeding behavior. Thus, there are myriads of signaling pathways that could be initiated by LCSs resulting to a range of physiological outcomes.

5 Conclusion

LCSs are increasingly been used in the population for sweetening food products and beverages. However, adequate data on the impact of these sweeteners on long-term health outcomes and as well as their usefulness in certain groups of the population (e.g., pregnant women, newborns, and children) are lacking. In addition, available unbiased studies conducted to date are fairly scanty. The cognate receptors of LCSs, T1R2-T1R3 heterodimer, via humoral and neural mechanisms influence obesity, overweight, appetite, satiety, cognition, and memory. Several interconnected signaling pathways are important in modulating these outcomes.

6 Future Directions

Several aspects on the impact of LCS intake are yet to be investigated. Therefore, further research in unraveling the signaling mechanisms of LCSs, receptor polymorphisms, and possible cellular transporters is necessary to understand the association among receptors of LCSs, features of their downstream signaling, and relationship to metabolism, obesity, weight gain, satiety, appetite, diabetes, cognition, and memory in the short, medium, and long-term duration of use in different compositions.

References

1. Weihrauch MR, Diehl V (2004) Artificial sweeteners – do they bear a carcinogenic risk? Ann Oncol 15:1460–1465
2. Kant R (2005) Sweet proteins – potential replacement for artificial low calorie sweeteners. Nutr J 4:5
3. Roberts A (2016) The safety and regulatory process for low calorie sweeteners in the United States. Physiol Behav. doi:10.1016/j.physbeh.2016.02.039
4. Grembecka M (2015) Natural sweeteners in a human diet. Rocz Panstw Zakl Hig 66(3): 195–202
5. Miller PE, Perez V (2014) Low-calorie sweeteners and body weight and composition: a meta-analysis of randomized controlled trials and prospective cohort studies. Am J Clin Nutr 100:765–777
6. Gardner C (2014) Non-nutritive sweeteners: evidence for benefit vs. risk. Curr Opin Lipidol 25(1):80–84

7. Fitch C, Keim KS, Academy of Nutrition and Dietetics (2012) Position of the academy of nutrition and dietetics: use of nutritive and nonnutritive sweeteners. J Acad Nutr Diet 112(5):739–758
8. Piernas C, Mendez MA, Ng SW, Gordon-Larsen P, Popkin BM (2014) Low-calorie- and calorie-sweetened beverages: diet quality, food intake, and purchase patterns of US household consumers. Am J Clin Nutr 99(3):567–577
9. Sylvetsky AC, Rother KI (2016) Trends in the consumption of low-calorie sweeteners. Physiol Behav. doi:10.1016/j.physbeh.2016.03.030
10. Foreyt J, Kleinman R, Brown RJ, Lindstrom R (2012) the use of low-calorie sweeteners by children: implications for weight management. J Nutr 142:1155S–1162S
11. Raben A, Richelsen B (2012) Artificial sweeteners: a place in the field of functional foods? Focus on obesity and related metabolic disorders. Curr Opin Clin Nutr Metab Care 15(6): 597–604
12. Shankar P, Ahuja S, Sriram K (2013) Non-nutritive sweeteners: review and update. Nutrition 29(11–12):1293–1299
13. Anderson GH, Foreyt J, Sigman-Grant M, Allison DB (2012) The use of low-calorie sweeteners by adults: impact on weight management. J Nutr. 142(6):1163S–1169S
14. Gibson S, Drewnowski A, Hill J, Raben AB, Tuorila H, Widström E (2014) Consensus statement on benefits of low-calorie sweeteners. Nutr Bull 39:386–389
15. Mennella JA, Pepino MY, Reed DR (2005) Genetic and environmental determinants of bitter perception and sweet preferences. Pediatrics 115(2):e216–e222
16. Tepper BJ, White EA, Koelliker Y, Lanzara C, d'Adamo P, Gasparini P (2009) Genetic variation in taste sensitivity to 6-*n*-propylthiouracil and its relationship to taste perception and food selection. Ann N Y Acad Sci 1170:126–139
17. Ventura AK, Mennella JA (2011) Innate and learned preferences for sweet taste during childhood. Curr Opin Clin Nutr Metab Care 14(4):379–384
18. Liem DG, de Graaf C (2004) Sweet and sour preferences in young children and adults: role of repeated exposure. Physiol Behav 83(3):421–429
19. El-Sohemy A, Stewart L, Khataan N, Fontaine-Bisson B, Kwong P, Ozsungur S, Cornelis MC (2007) Nutrigenomics of taste – impact on food preferences and food production. Forum Nutr 60:176–182
20. Margolskee RF (2002) Molecular mechanisms of bitter and sweet taste transduction. J Biol Chem 277:1–4
21. McLaughlin SK, McKinnon PJ, Robichon A, Spickofsky N, Margolskee RF (1993) Gustducin and transducin: a tale of two G proteins. Ciba Found Symp 179:186–200
22. Lindemann B (1996) Chemoreception: tasting the sweet and the bitter. Curr Biol 6(10): 1234–1237
23. Margolskee RF (1993) The molecular biology of taste transduction. Bioessays 15(10):645–650
24. Depoortere I (2014) Taste receptors of the gut: emerging roles in health and disease. Gut 63(1):179–190
25. Ohkuri T, Yasumatsu K, Horio N, Jyotaki M, Margolskee RF, Ninomiya Y (2009) Multiple sweet receptors and transduction pathways revealed in knockout mice by temperature dependence and gurmarin sensitivity. Am J Physiol – Regul, Integr Comp Physiol 296(4): R960–R971
26. Wright EM, Turk E (2004) The sodium/glucose cotransport family SLC5. Pflugers Arch 447(5):510–518
27. Harada N, Inagaki N (2012) Role of sodium-glucose transporters in glucose uptake of the intestine and kidney. J Diab Invest 3(4):352–353
28. Takata K, Hirano H, Kasahara M (1997) Transport of glucose across the blood-tissue barriers. Int Rev Cytol 172:1–53
29. Janssen S, Depoortere I (2013) Nutrient sensing in the gut: new roads to therapeutics? Trends in Endocrin Metab 24(2):92–100

30. Folmes CD, Lopaschuk GD (2007) Role of malonyl-CoA in heart disease and the hypothalamic control of obesity. Cardiovasc Res 73(2):278–287
31. Riediger T, Zuend D, Becskei C, Lutz TA (2004) The anorectic hormone amylin contributes to feeding-related changes of neuronal activity in key structures of the gut-brain axis. Am J Physiol Regul Integr Comp Physiol 286(1):R114–R122
32. Riediger T (2012) The receptive function of hypothalamic and brainstem centres to hormonal and nutrient signals affecting energy balance. Proc Nutr Soc 71(4):463–477
33. Valassi E, Scacchi M, Cavagnini F (2008) Neuroendocrine control of food intake. Nutr Metab Cardiovas Dis 18:158–168
34. Cui H, Cai F, Belsham DD (2005) Anorexigenic hormones leptin, insulin, and alpha-melanocyte-stimulating hormone directly induce neurotensin (NT) gene expression in novel NT-expressing cell models. J Neurosci 25(41):9497–9506
35. Wellman PJ (2000) Norepinephrine and the control of food intake. Nutrition 16(10):837–842
36. Hanusch-Enserer U, Roden M (2005) News in gut-brain communication: a role of peptide YY (PYY) in human obesity and following bariatric surgery? Eur J Clin Invest 35(7):425–430
37. Watterson KR, Bestow D, Gallagher J, Hamilton DL, Ashford FB, Meakin PJ, Ashford ML (2013) Anorexigenic and orexigenic hormone modulation of mammalian target of rapamycin complex 1 activity and the regulation of hypothalamic agouti-related protein mRNA expression. Neurosignals 21(1–2):28–41
38. Yoshida R, Niki M, Jyotaki M, Sanematsu K, Shigemura N, Ninomiya Y (2013) Modulation of sweet responses of taste receptor cells. Sem in Cell Dev Biol 24(3):226–231
39. Gil-Campos M, Aguilera CM, Cañete R, Gil A (2006) Ghrelin: a hormone regulating food intake and energy homeostasis. Br J Nutr 96(2):201–226
40. Ghamari-Langroudi M, Colmers WF, Cone RD (2005) PYY3–36 inhibits the action potential firing activity of POMC neurons of arcuate nucleus through postsynaptic Y2 receptors. Cell Metab 2(3):191–199
41. Druce M, Bloom SR (2006) The regulation of appetite. Arch Dis Child. 91(2):183–187
42. Welcome MO, Pereverzev VA (2013) A mini-review of the mechanisms of glucose memory enhancement. Int J Med Pharm Sci 4(1):17–30
43. Welcome MO, Mastorakis NE, Pereverzev VA (2015) Sweet taste receptor signaling network: possible implication for cognitive functioning. Neurol Res Int (Hindawi Publishing Corporation), Article ID 606479, 13 pages. doi:10.1155/2015/606479. https://www.hindawi.com/journals/nri/2015/606479/
44. Ren X, Zhou L, Terwilliger R, Newton SS, de Araujo IE (2009) Sweet taste signaling functions as a hypothalamic glucose sensor. Front Integr Neurosci 3:12
45. de Araujo IE, Ren X, Ferreira JG (2010) Metabolic sensing in brain dopamine systems. Res Prob Cell Different 52:69–86
46. Hurtado-Carneiro V, Sanz C, Roncero I, Vazquez P, Blazquez E, Alvarez E (2012) Glucagon-like peptide 1 (GLP-1) can reverse AMP-activated protein kinase (AMPK) and S6 kinase (P70S6K) activities induced by fluctuations in glucose levels in hypothalamic areas involved in feeding behavior. Mol Neurobiol 45(2):348–361
47. Xu J, Ji J, Yan XH (2012) Cross-talk between AMPK and mTOR in regulating energy balance. Crit Rev Food Sci Nutr 52(5):373–381
48. Ehninger D, de Vries PJ, Silva AJ (2009) From mTOR to cognition: molecular and cellular mechanisms of cognitive impairments in tuberous sclerosis. J Intellect Disabil Res 53(10): 838–851

Xylitol: One Name, Numerous Benefits

19

Chika Ifeanyi Chukwuma and Md. Shahidul Islam

Abstract

Xylitol among other sugar alcohols have been extensively studied, showing numerous beneficial effects and potential clinical uses other than being used as a sweetener. The present chapter focuses on the beneficial effects of xylitol and its potential clinical relevance. It also elaborated the several beneficial effects of xylitol that requires more investigation, especially at clinical levels to ascertain its clinical and therapeutic applications. Information from different sources, mainly from "PubMed" journals, were reviewed, focusing on the beneficial effects and potential therapeutic values of this unique sugar alcohol in oral health care, glycemic control, lipid metabolism, weight management, bone metabolism, skin care, ear and upper respiratory tract infection and oxidative stress. There were consistent reports showing beneficial effects and potential clinical application of xylitol in the above-mentioned areas. There is a global acceptance of the use of xylitol in dental caries prevention and glycemic control, but more investigations are required to ascertain the clinical application of xylitol in the other areas. It has been approved as a safe sweetener by the Food and Drug Administration (FDA), USA, thus xylitol may be used not only as a sweetener but also as a therapeutic additive in the abovementioned health-related issues.

Keywords

Xylitol • Health benefits • Dental caries • Glycemic index • Lipid metabolism • Obesity • Bone • Acute otitis media • Atopic dermatitis • Oxidative stress

C.I. Chukwuma (✉) • M.S. Islam (✉)
Department of Biochemistry, School of Life Sciences, University of KwaZulu-Natal, Durban, South Africa
e-mail: chykochi@yahoo.com; islamd@ukzn.ac.za

J.-M. Mérillon, K.G. Ramawat (eds.), *Sweeteners*, Reference Series in Phytochemistry,
https://doi.org/10.1007/978-3-319-27027-2_33

Abbreviations

AEP	Acquired enamel pellicle
AOM	Acute otitis media
ChREBP	Carbohydrate response element binding protein
EPS	Extracellular polysaccharides
FFA	Free fatty acid
GI	Glycemic index
II	Insulinemic index
NEFA	Non-esterified fatty acid
NFBG	Non-fasting blood glucose
PPARα	Peroxisome proliferator-activated receptor alpha
PPARγ	Peroxisome proliferator-activated receptor gamma
SREBP-1c	Sterol response element binding protein 1c
T2D	Type 2 diabetes

Contents

1 Introduction 410
2 Metabolism of Xylitol 411
3 Health Benefits of Xylitol 411
3.1 Xylitol in Oral Health Care 414
3.2 Xylitol in Glycemic Control 418
3.3 Xylitol in Lipid Metabolism 420
3.4 Xylitol in Body Weight Management 422
3.5 Xylitol in Oxidative Stress 423
3.6 Xylitol in Bone Metabolism 424
3.7 Xylitol in Ear and Upper Respiratory Tract Infections 425
3.8 Xylitol in Skin Care 427
3.9 Xylitol as a Sweetener 428
3.10 Other Beneficial Effects of Xylitol 428
4 Safety and Toxicity of Xylitol 429
5 Conclusions 429
References 429

1 Introduction

Xylitol is a five-carbon sugar alcohol (Fig. 1) first discovered in 1891 by Emil Fischer, a German chemist [1]. Usually less than 1% xylitol is known to occur naturally in many fruits and vegetables such as yellow plums, strawberries, raspberries, spinach, lettuce, onion, carrots, etc. [1–3]. However, progress in research has revealed the feasibility of producing xylitol from xylose – a polymer

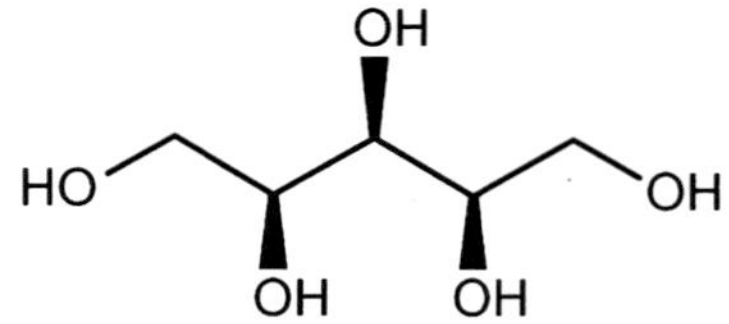

Fig. 1 Chemical structure of xylitol

Table 1 Glycemic and insulinemic indices, caloric value, and relative sweetness of sucrose and some commonly used sugar alcohols

Sweeteners	GI (glucose = 100)^{i}	*n*GI	II (glucose = 100)^{i}	*n*II	Caloric value (Kcal g^{-1})^{ii}	Relative sweetness (sucrose = 1)^{iii}
Sucrose	69 ± 13	10	48 ± 14	8	4	1
Erythritol	0 ± 6	2	1	1	0	0.7
Xylitol	13 ± 4	6	11 ± 5**	4	2.4	0.95
Mannitol	0	1	0	1	2.4	0.5
Isomalt	5 ± 4	2	7 ± 2	2	2.4	0.4
Lactitol	5 ± 3	2	4 ± 4	2	2.4	0.4
Maltitol	36 ± 9	6	29 ± 4	4	2.4	0.9
Sorbitol	10 ± 3	7	9 ± 5**	4	2.4	0.5

i: Modified from Livesey 2003 [5] (average of relative glucose and insulinemic responses of different studies on normal subjects after intake of 20 g to 100 g glucose) with some values (**) excluded due to high standard deviation; *ii*: Values reported by Flambeau et al. [113]; *iii*: Values reported by Sadler and Stowell [4]; GI: mean glycemic index; II: mean insulinemic index; "*n*GI" and "*n*II" mean number of studies considered for GI and II, respectively.

of xylan – by catalytic or enzymatic hydrogenation (reduction) and has become a major source of xylitol [1]. Xylitol appears as a white crystalline solid. It has about similar sweetness as sucrose (Table 1) but gives a taste characterized by an after-cooling effect [1, 4].

2 Metabolism of Xylitol

Xylitol is partially absorbed in the small intestine by passive diffusion. Only about 25–50% of ingested xylitol is absorbed in the small intestine, while about 50–75% is passed to the large intestine [1, 5]. Thus, xylitol has a lower glycemic index (13 vs. 69), insulinemic response, and caloric value (2.4 vs. 4.0 kcal g^{-1}) compared to sucrose [5] (Table 1).

Digested and absorbed xylitol is mostly metabolized in the liver to become part of normal carbohydrate metabolism via the glucuronic acid-pentose phosphate shunt of the pentose phosphate pathway. In the liver, xylitol is dehydrogenated by a nonspecific cytoplasmic NAD-dependent polyol dehydrogenase and phosphorylated via a specific xylulose kinase to produce xylulose-5-phosphate, an intermediate of the non-oxidative branch of the pentose phosphate pathway [6]. Unabsorbed xylitol reaching the large intestine is fermented by colonic bacteria to short-chain fatty acids, which can be absorbed and further metabolized for energy production [1].

3 Health Benefits of Xylitol

Xylitol is associated with several health benefits, which cannot be overemphasized. These include dental protection, weight control, glycemic control, and antimicrobial effects among others, which have been summarized in Tables 2, 3, 4, and 5. In this

Table 2 Beneficial effects of xylitol and potential clinical relevance in oral health care

Beneficial effects	Comment/remarks	References[abcd]
Dental plaque formation and oral pH	Decreases the oral synthesis of insoluble polysaccharide by *S. mutans*	[12][a]
	Reduces polysaccharide-mediated *S. mutans* adherence and oral biofilm formation	[15–17][a], [18][c]
	Reduces the growth of dental plaque	[28, 32][c]
	Reduces dental plaque acidity and enhance plaque or saliva alkalinity	[21][a], [22][c], [20][d]
Oral and dental plaque bacteria	Inhibits key metabolic processes like glycolysis in *S. mutans*	[21][a]
	Inhibits growth and reduces count of *S. mutans*	[24][a], [26][c]
	Reduces mother-to-child *S. mutans* transmission	[29][c]
Tooth demineralization and remineralization	Enhances stability and prolonged solubility of saliva calcium phosphate and enamel calcium absorption	[32][c], [33, 34][d]
	Reduces tooth demineralization but promotes tooth remineralization	[36, 37][a], [38][b], [41][c]
Dental caries	Reduces caries incidence	[41, 42][c]

"abc" superscript letters mean the information was based on the original research "a"; in vitro, "b"; in vivo and "c"; clinical/human studies, while "d" means information was based on the literature review of a research or review article

Table 3 Potential clinical relevance of xylitol in glycemic control

Beneficial effects of xylitol	Comment/remarks	References[abcd]
Blood glucose	Reduces blood glucose and improves glucose tolerance in normoglycemic and diabetic conditions	[43, 45, 46][b]
	Exerts minimal effect on postprandial rise in blood glucose in both normoglycemic and diabetic states	[47–49][c], [5][d]
Carbohydrate digestion and intestinal glucose absorption	Inhibits carbohydrate-hydrolyzing enzymes activity and intestinal glucose absorption	[51, 53][a], [52][ab]
Insulin secretion	Exhibit low insulin response to control hyperinsulinemia in obesity or prediabetes	[54][c], [5][d]
	Improves insulin secretion in diabetic conditions	[55][a], [45, 46][b]
Insulin resistance and pancreatic beta-cell regeneration	Improves insulin-mediated muscle glucose uptake and NEFA-induced insulin resistance	[52][a], [59][b]
	Improves pancreatic beta-cell damage in diabetic state	[46][b]

"abc" superscript letters mean the information was based on an original research "a"; in vitro, "b"; in vivo and "c"; clinical/human studies, while "d" means information was based on the literature review of a research or review article

Table 4 Potential clinical relevance of xylitol in lipid metabolism, weight gain, and obesity management

Beneficial effects of xylitol	Comment/remarks	References[abcd]
Blood lipids and NEFA	Reduces serum LDL cholesterol without increasing serum triglycerides in normoglycemic and type 2 diabetic states	[43, 45, 46][b], [62][c]
	Reduces NEFA levels and NEFA-related pathophysiology in T2D and obesity	[63][a], [59][b], [64, 65][c]
	Reduces dietary fat-induced plasma cholesterol and triglycerides increase	[2][b]
Lipid metabolism and lipid-metabolizing genes	Suppresses dietary fat-induced hepatic lipogenic processes while upregulating lipolytic processes and fatty oxidation	[2][b]
Appetite control	Delays gastric emptying and increases satiety	[52][b], [48, 73][c]
Body fat weight gain and obesity	Reduces dietary fat-induced body weight gain and visceral fat accummulation	[2, 43][b]

"abc" superscript letters mean the information was based on the original research "a"; in vitro, "b"; in vivo and "c"; clinical/human studies, while "d" means information was base on the literature review of a research or review article

Table 5 Potential clinical relevance of xylitol in oxidative stress and bone metabolism

Beneficial effects of xylitol	Comment/remarks	References[abcd]
Antioxidant	Exhibits in vitro antioxidant activities	[51, 79][a]
	Reduces nonenzymatic glycosylation of acid-soluble collagen in diabetic rats	[80][b]
	Maintains an active glutathione-dependent antioxidant system	[81, 82][a], [33][d]
Bone strength, volume, density, and mineral composition	Improves deteriorating bone physiochemical properties	[83–86][b]
Bone remineralization and recalcification	Promoting the absorption and bone retention of calcium	[33][d]
	Bone recalcification of calcium deficiency-induced metabolic changes	[88][b]
Bone resorption	Retards detrimental bone resorption	[89, 90][b]
Bone infection	Improving the therapeutic efficacy of PMMA-based antibiotic (daptomycin) therapy for treatment of postsurgical osteomyelitis	[91][b]

"abc" superscript letters mean the information was based on the original research "a"; in vitro, "b"; in vivo and "c"; clinical/human studies, while "d" means information was base on the literature review of a research or review article

chapter the beneficial effects and potential clinical relevance of this unique sugar alcohol are thoroughly reviewed.

3.1 Xylitol in Oral Health Care

Other than being a sugar replacer, xylitol consumption is closely associated with dental and oral health care, which has been proven in many recent and previous studies that are discussed below. These studies revealed that xylitol does not promote the various processes that lead to dental plaque formation, caries development, and tooth decay (Table 2).

3.1.1 Microbial Oral Biofilm and Dental Plaque Formation, and pH

Dental biofilms produce acids from carbohydrates via bacterial fermentation process that facilitates dental caries [7]. Sucrose and many other common sugars are considered to be the most cariogenic dietary agents because they can serve as a precursor for extracellular and intracellular polysaccharide (EPS and IPS) synthesis in dental plaque by plaque bacteria [8, 9]. Insoluble EPS promotes adherence of mouth bacteria to the tooth surface [8, 9] and contributes to dental biofilm structural integrity [10]. Additionally, EPS increases biofilm porosity, allowing sugar to penetrate into the biofilm [9], which would result in acidic plaque pH due to bacterial catabolic activities.

On the other hand, xylitol cannot be fermented by plaque bacteria [11]. Additionally, xylitol does not serve as a precursor for the synthesis of detrimental plaque EPS. Previous in vitro studies have demonstrated the lower synthesis of insoluble polysaccharides by *Streptococcus mutans* in the presence of xylitol [12] (Table 2). Accordingly, the plaque of regular xylitol users seems to contain smaller amount of insoluble polysaccharide than the plaque of nonusers [13] (Table 2). This is because polysaccharide exposed to xylitol becomes more water soluble, and this can affect the adherence of bacteria to the tooth surface and bacterial biofilm or plaque growth [14]. Previous in vitro studies demonstrated that the presence of xylitol decreased polysaccharide-mediated streptococci cell adherence to smooth glass [15] (Table 2), inhibited the formation of a multispecies (*S. mutans*, *S. sobrinus*, *Lactobacillus rhamnosus*, *Actinomycesviscosus*, *Porphyromonas gingivalis*, *and Fusobacterium nucleatum*) biofilms [16] (Table 2) and reduced the biomass of mouth bacterial biofilm [17] (Table 2). In human subjects, 12-month consumption of xylitol chewing gum weakened the adherence of *S. mutans* colonies to dental surface biofilm [18] (Table 2). These anti-biofilm properties of xylitol are of therapeutic importance in preventing dental plaque formation, growth, and acidity.

The pH of dental plaque is a major factor in the development of dental caries. Acid production resulting from carbohydrate metabolism by plaque bacteria and the subsequent decrease in environmental pH are responsible for the demineralization of tooth surfaces [19]. Xylitol enhances plaque alkalinity by increasing saliva nitrogen and bicarbonate [20] or by reducing acid production of plaque bacteria, as demonstrated in both in vitro [21] and in vivo [22] studies (Table 2), which is partly due to its antagonistic effects on acidogenic plaque microbes.

3.1.2 Oral and Plaque Pathogenic Bacteria

The bacteria flora on the surface of healthy enamel contains mainly nonmutant *Streptococci* and *Actinomyces*, but frequent sugar fermentation and acid production will trigger a shift in the balance of the resident plaque microflora toward the emergence of more cariogenic bacteria [7]. Although additional microorganisms may be involved, *S. mutans* has been identified as a primary etiological agent in the pathogenesis of dental caries [7].

On the other hand, xylitol hinders the growth and metabolism of several species of bacteria, but among the mouth bacterial flora, *S. mutans* has been reported to be the most affected [13]. Studies have reported that accumulation of xylitol 5-phosphate, a xylitol metabolite, could inhibit glycolysis and growth of *S. mutans* [21] (Table 2). Xylitol-sensitive *S. mutans* strains are mostly inhibited by xylitol. Xylitol-resistant *S. mutans* strain, which emerges after long-term xylitol consumption, is not inhibited by xylitol, but is less cariogenic and less virulent than the xylitol-sensitive strains [23]. Several in vitro studies have demonstrated the - concentration-dependent growth inhibition of *S. mutans* by xylitol in appropriate growth media [24, 25]. In fact as low as 0.1% xylitol caused considerable amount of growth inhibition, while 1% xylitol caused mean growth inhibition percentages ranging from 61% to 76% at log phase [24] (Table 2). Additionally, the clinical relevance of xylitol in the inhibition of dental *S. mutans* growth has been demonstrated in school children (age 10–12 years) who showed significant reductions ($p < 0.025$) in *S. mutans* count after having daily chewing gum containing 100% or 55% xylitol as sweetener for 90 days [26] (Table 2), which is consistent with results of several other clinical trials [27, 28], including interference of maternal-to-child transmission of *S. mutans* [29] (Table 2). These results suggest the efficacy of xylitol in preventing or ameliorating *S. mutans*-induced tooth demineralization or dental caries development.

3.1.3 Tooth Demineralization and Remineralization

Demineralization and remineralization play important role in the hardness and strength of tooth enamel. Demineralization weakens the tooth enamel, which may promote dental caries development and tooth decays; hence, in healthy condition remineralization occurs to protect the tooth. Important chemical prerequisites of tooth remineralization include saturated concentrations of salivary calcium and phosphate ions and high salivary pH [20], which is available under normal condition. This condition will promote precipitation of calcium phosphate, which will facilitate remineralization process in the presence of normal, physiologic, salivary concentration range of fluoride ion, required organic and inorganic matrix, and natural salivary glycoproteins (mucins) and peptides [20]. Natural salivary glycoproteins (mucins) and peptides serve as a precursor for residual protein structures that forms part of the acquired pellicle film on enamel surface (AEP) [20, 30]. The AEP plays an important role in the maintenance of oral health by protecting the tooth from oral detrimental environment and also regulating processes such as demineralization and remineralization [30].

When pH falls in acidic condition, calcium ions are being released to the plaque, thus demineralizing the enamel. Previous study has shown that food acidulants like sugars reduce phosphorous and fluoride ion concentrations and chelate calcium ion (demineralize enamel) in a manner that is directly proportional to their acidity [31], thus enhancing the mineral loss of enamel.

Contrarily, xylitol has shown some important properties that can enhance tooth remineralization or reduce demineralization. Xylitol not only maintains a high pH value in saliva and plaque fluid, but also a supersaturated calcium level in saliva. Studies have suggested that xylitol consumption is associated with an increase of plaque calcium levels [32] (Table 2), which may be linked to the chemical nature of xylitol. Due to the polyhydroxy structure and hydrophilic nature (Fig. 2a), xylitol molecules are readily soluble in saliva [1]. It also contains a tridentate ligand [(H–C–OH)$_3$], which can react with polyvalent ions like calcium ion (Fig. 2b); hence it can compete with water molecules for the primary hydration layer of calcium to form a "xylitol-calcium" complex (Fig. 2c) [33]. This complexation interaction contributes to the stabilization of salivary calcium phosphate systems [33] (Table 2). This enhances the prolonged solubility of calcium phosphate system in saliva, even at supersaturated concentrations [34] (Table 2), which is a prerequisite of tooth remineralization. Through complexation interaction, xylitol also acts as a carrier of calcium to the teeth for remineralization, which enhances calcium absorption due to enamel permeability of xylitol [33] (Table 2). Consequently, xylitol can promote remineralization of deeper layers of demineralized enamel by enhancing calcium ion movement and accessibility. Additionally, Western blot analysis of protein deposition on an artificial hydroxyapatite disc [35] suggests that xylitol may enhance AEP formation.

There are several evidences supporting the remineralization properties of xylitol. The Japanese authors [36] (Table 2) presented evidences on the recuperative process

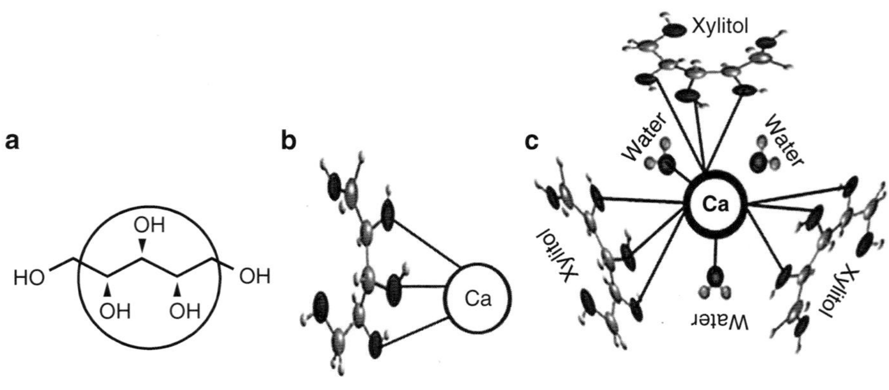

Fig. 2 Structure of xylitol showing the tridentate ligand {(H–C–OH)$_3$} (**a**) complexation reaction between xylitol and calcium ion in aqueous solution to form a xylitol-calcium complex (**b**) competition of xylitol with water molecule for primary hydration layer of calcium (**c**)

of xylitol-associated remineralization using contact microradiography, a multipurpose image processor, and high-resolution electron microscopy techniques. Chunmuang et al. [37] showed that xylitol can reduce acid-induced enamel demineralization in vitro (Table 2). In rats, dentinal molar fissure caries produced by initial exposure to dietary sucrose were significantly reversed (remineralized) by subsequent exposure to 3 % or 6 % xylitol supplementation [38] (Table 2). Overall, it has been found that the remineralization effects of xylitol have been associated with reduction in dental caries incidence in humans.

3.1.4 Dental Caries

The study on the effects of xylitol on dental caries prevention was pioneered in Turku, Finland, between late 1960s and early 1970s [20]. Results from these studies reported 50% less dental plaque formation in the xylitol-consuming subjects compared to sucrose-, D-glucose-, or fructose-consuming subjects. Following these results, a 2-year clinical trial and 1-year chewing gum study on caries prevention potential of xylitol (also known as the Turku sugar studies) was conducted and reported in "Turku sugar studies I-XXI" reports [39]. These results supported the association of xylitol consumption with impressive caries reduction in humans. By the mid- and late 1970s, a novel concept of the importance of xylitol in the dental health was revealed, which was subsequently verified and has been expanding since the last several decades.

According to a review article published in 2010, about 19 clinical trials and 300 short-term biological oral studies have been conducted which claim the usefulness of xylitol in dental and oral health care [20]. These trials have presented results that have partly contributed to the endorsement of xylitol by the public, worldwide. The summary of human caries studies on xylitol that in part have constituted to the justifications for public endorsements of xylitol have been previously reported [20]. In fact, regarding caries prevention, it was stated that "The most impressive results have been achieved when a xylitol program has been added to existing prevention efforts" [40].

Other notable recent studies have also demonstrated the efficacy of xylitol usage in the prevention of caries incidences. In a recently conducted 26–28-month trial at Finnish Public Health Centre, it was reported that 45% solution of xylitol swabbed (daily) onto all available deciduous teeth of children at the age of approximately 6–8 months (equivalent to 13.5 mg per each deciduous tooth) resulted in a significantly reduced ($p < 0.001$) incidence of enamel and dentine caries compared to the control subjects [41] (Table 2). Additionally, a study in Italy confirmed the efficacy of a 6-month daily consumption of xylitol chewing gum (36.6% xylitol) in long-term control of caries formation in children with high risk of dental carries [42] (Table 2).

From the above discussion, it can be said that dental protection efficacy of xylitol is bi-facet (Fig. 3). Xylitol could act as a non-cariogenic or an anticariogenic agent. As a non-cariogenic agent, xylitol does not contribute to the caries disease process such as plaque development and microbial activities [21, 24]; thus it serves as a preventive measure against caries development. As an anticariogenic agent, it reverses the caries disease process (remineralization of caries-like lesions) [38].

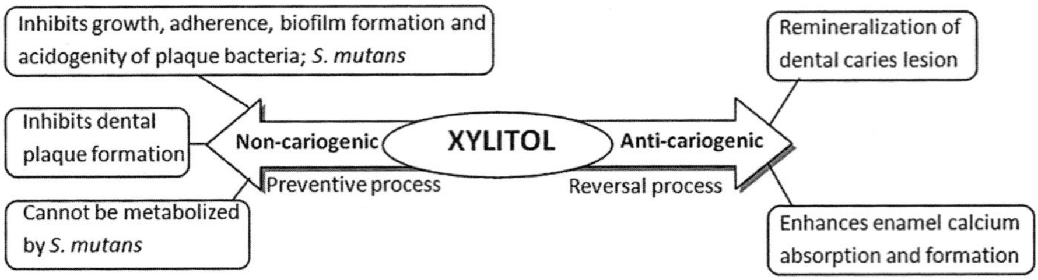

Fig. 3 Bi-facet therapeutic efficacy of xylitol on dental caries. Non-cariogenic: Does not contribute to caries disease process. Anticariogenic: reverses caries disease process

3.2 Xylitol in Glycemic Control

Although xylitol is derived from a simple carbohydrate precursor, it merely contributes to glycemic or caloric effects, which could partly be due to its partial digestion and slow absorption [5] (Table 3). In a number of previous and recent studies, the importance of xylitol in glycemic control has been extensively studied, showing evidence of the potential use of xylitol in the management of diabetes and its associated complications (Table 3).

3.2.1 Blood Glucose

Xylitol is a known source of energy in parenteral nutrition and as a sweetener in the diabetic diet since it contributes very less to the integrated glycemic response [33]. Several studies have demonstrated the glycemic control potentials of xylitol in both normal [43, 44] and diabetic conditions [44–46]. Among these, the studies conducted by Islam and coauthors have shown significant impact of xylitol on blood glucose levels in both normal [43] and diabetic rats [45, 46]. According to one of their studies, normal rats which received 10% xylitol solution ad libitum as a replacement of drinking water for 3 weeks showed significantly lower ($p < 0 .05$) NFBG levels as early as the first 2 weeks of intervention period as well as significantly better glucose tolerance ($p < 0.05$) compared to the rats that received 10% sucrose solution [43] (Table 3). In a subsequent investigation, Islam and Indrajit [45] reported that supplementation of 10% xylitol solution ad libitum to type 2 diabetic rats for 5 weeks significantly reduced ($p < 0.05$) NFBG levels and improved glucose tolerance ability compared to the type 2 diabetic control animals [45] (Table 3). In a more recent study, Rahman and Islam (2014) reported that a 4-week consumption of 10% xylitol solution ad libitum caused appreciable improvements in NFBG levels, glucose tolerance, pancreatic histopathology, and most diabetes-related parameters in a T2D rat model compared to lesser concentrations (2.5% and 5%) and the diabetic control group [46] (Table 3). From this study, it has been concluded that 10 % xylitol solution can be more effective compared to lower dosages although no higher dosages were included in their study. Although these studies were mainly conducted on animals, some clinical studies have also been reported.

In an old clinical study, the infusion of xylitol at a rate of 6 mg kg^{-1} min^{-1} for 120 min followed by (90 mins after the insulin infusion) epinephrine 6 microgram/

min and propranolol 0.08 mg/min to healthy human subjects caused no change in blood glucose levels, contrary to glucose infusion [47] (Table 3). However, carbohydrate oxidation was higher with xylitol infusion than glucose, which suggests that oxidation of xylitol does not cause significant rise in glycemia during suppression of endogenous insulin secretion. Consistent with the abovementioned study, a recently conducted study in lean and obese normoglycemic subjects showed that 50 g xylitol in 300 mL of water supplied via nasogastric tube sparing increased plasma glucose compared to similar administration of 75 g glucose in 300 mL of water [48] (Table 3). Additionally, investigation in insulin-dependent diabetics showed that sucrose diet induced a greater postprandial rise in blood glucose level than xylitol diet despite of counter-regulation by a glucose-controlled insulin infusion system [49] (Table 3).

3.2.2 Carbohydrate Digestion and Intestinal Glucose Absorption

Another facet of the glycemic control potential of xylitol is strongly associated with its effect on other dietary carbohydrate metabolism. It is a known fact that alpha-glucosidase inhibitors control postprandial blood glucose rise by inhibiting the action of carbohydrate-hydrolyzing enzymes, alpha-glucosidase, thus limiting intestinal absorption of glucose from dietary carbohydrates [50]. Recent studies have suggested that xylitol may also have similar effect on dietary carbohydrate metabolism as well as postprandial glucose increase. A recent in vitro study reported that among five sugar alcohols (xylitol, erythritol, maltitol, isomalt, mannitol, sorbitol, and inositol) tested, xylitol exhibited the highest alpha-glucosidase inhibitory activity [51] (Table 3). Additionally, Chukwuma and Islam (2015) recently demonstrated the concentration-dependent (2.5–40%) inhibitory effect of xylitol on α-amylase and α-glucosidase activities in vitro and glucose absorption in isolated rat jejunum [52] (Table 3). In the same study, they showed the consistent inhibition of small intestinal glucose absorption in normal and diabetic rats 1 h after the co-ingestion of 1 g/kg bw xylitol and 2 g/kg bw glucose. In line with the above-mentioned data, Frejnagel et al. [53] also reported that xylitol infusion (2.5–20 g L^{-1}) at a flow rate of 1 mL min^{-1} in situ, through 70 cm of the small intestine of anesthetized normal rats, significantly reduced ($p < 0.05$) intestinal glucose absorption compared to similar infusion of 2 g L^{-1} glucose (Table 3). These data suggest the potency of xylitol to reduce postprandial glucose increase via inhibiting carbohydrate digestion and intestinal glucose absorption.

3.2.3 Insulin Secretion

The achievement of lower postprandial glycemia without hyperinsulinemia is an ideal glycemic control. It is a well-established fact that postprandial insulin response (secretion) after consumption of xylitol (enterally and parenterally) is lower or slower compared to the equimolar amount of sucrose, glucose, or most glucose-yielding carbohydrates either in normal or in diabetic condition [5] (Table 3). Accordingly, xylitol has a lower insulinemic index of 11, which is due to its lower glycemic index (Table 1). This characteristic of xylitol can be very important in the control of hyperinsulinemic condition in obese and prediabetic individuals [54] (Table 3).

Furthermore, improvement of insulin secretion is of great importance in the later state of type 2 diabetes due to impaired glucose tolerance caused by pancreatic beta-cell inefficiency. Although there is no reported clinical evidence on xylitol-induced improvement of insulin secretion in diabetic patients, the insulinotropic effects of xylitol has been reported in a previous in vitro study [55] and recently in normal [43] and diabetic animals [45, 46] and normoglycemic human subjects [56] (Table 3). In one of the abovementioned studies, it was noted that serum insulin concentration was initially reduced due to the induction of diabetes in rats but was normalized by a 5-week daily consumption of xylitol [46] (Table 3). However, the insulinotropic effect of xylitol still requires more investigations, especially at clinical level.

3.2.4 Insulin Resistance and Pancreatic Beta-Cell Regeneration

Insulin resistance is a metabolic disorder that is associated with common diseases such as diabetes, obesity, and coronary heart disease [57], which among other insulin-associated metabolic defects can diminish insulin-mediated glucose uptake in peripheral tissues [52]. It is widely accepted that plasma non-esterified fatty acid (NEFA) or free fatty acid (FFA) can partly play a major role in the pathogenesis of insulin resistance in obesity and non-insulin-dependent diabetes mellitus [57, 58]. However, data from recent studies suggest that xylitol may be useful in the amelioration of NEFA-induced insulin resistance [59] and in the enhancement of muscle glucose uptake [52] (Table 3).

Pancreatic beta-cell dysfunction, which may be influenced by the increased insulin requirements imposed by insulin resistance, and toxicities from hyperglycemia and elevated FFA play a major role in the progression of type 2 diabetes (T2D) [60]. Currently, there is no adequate evidence available to confirm the pancreatic beta-cell-regenerating ability of xylitol in diabetic condition. However, the improved insulin secretion in diabetic rat after a short-term xylitol feeding [45, 46] may be directly or indirectly associated with the improvement of pancreatic beta-cell histology and function. Supporting evidences presented in the studies reported by Rahman and Islam [46] revealed that the histological deterioration of the pancreas due to the induction of T2D was recovered after a 5-week feeding of 10% xylitol solution [46] (Table 3). Islet size was bigger and number of beta cells was higher in the diabetic rats that received 10% xylitol compared to the diabetic control rats. Data from these studies suggest the potency of xylitol on improving peripheral insulin sensitivity and pancreatic beta-cell regeneration. However, further studies are warranted to understand the effects of xylitol on pancreatic beta-cell regeneration.

3.3 Xylitol in Lipid Metabolism

Regulation of synthesis, breakdown, or storage of lipids affects the level of circulating lipid, and xylitol has been shown to have a positive impact on lipid metabolism, regulation, and circulating levels in subjects with some metabolic disorders or diseases (Table 4).

3.3.1 Blood Lipids and Nonesterified Fatty Acids

Major functional blood lipids include triglyceride and lipoprotein cholesterol, which have physiological functions but could be detrimental at abnormally high levels. Abnormally high levels of blood LDL cholesterol and triglycerides and reduced HDL cholesterol are very common metabolic states in T2D, obesity, and metabolic syndrome, which are the major contributing factors of cardiovascular complications like hypertension, cardiac failure, and stroke [57, 58, 61]. A number of previous studies reported that these disease conditions may be ameliorated by regular consumption of dietary xylitol.

According to studies conducted by Islam and coauthors, 3–5 weeks consumption of 10% xylitol solution significantly reduced ($p < 0.05$) serum levels of total and LDL cholesterol in normal [43] and T2D model of rats [45, 46] (Table 4), when it did not significantly affect serum HDL cholesterol levels in both normal and diabetic and serum triglyceride levels only in diabetic rats. Amo et al. [2] reported that normal rats fed with high fat diet containing 2.0 g xylitol 100 $kcal^{-1}$ energy for 8 weeks showed significantly lower ($p < 0.05$) plasma total cholesterol and triglyceride compared to normal rats fed with high-fat diet without xylitol (control) (Table 4). In human subjects, daily consumption of 40–100 g xylitol for 18 days appreciably decreased plasma cholesterol in healthy human subjects, but did not increase plasma triglyceride levels [62] (Table 4). Data from these studies suggest that xylitol may possess therapeutic potential in the prevention of lipid-induced cardiovascular complications of diabetes, obesity, and related metabolic disorders.

Furthermore, elevated plasma NEFA or FFA in obesity, T2D, and metabolic syndrome can mediate adverse metabolic effects like insulin resistance and pancreatic beta-cell deterioration [57–60]. However, xylitol has been shown to inhibit fat mobilization in vitro [63] (Table 4) and also cause significant reduction in plasma NEFA levels in high-fat diet-fed rats [59] (Table 4), without causing any change in blood glucose level. In human subjects, a 90-min 10% xylitol infusion significantly reduced plasma NEFA concentration for up to 180 mins, which was about 30% of the initial level in normal and diabetic subjects [64] (Table 4). Additionally, intravenous injection of xylitol (100 mg kg^{-1} bw) appreciably decreased plasma FFA concentration compared to baseline (before xylitol administration) in the diabetic patient and in the cerebral stroke patients with hyperlipidemia (serum triglycerides and cholesterol over 100 and 200 mg dL^{-1}, respectively) [65] (Table 4). These data suggest that xylitol may have therapeutic value in the amelioration of FFA-induced metabolic disorders in disease states like diabetes, obesity, and cardiovascular diseases.

3.3.2 Lipid Metabolism and Lipid-Metabolizing Genes

Excess dietary fat can adversely affect physiological metabolism of lipids via increasing lipogenesis and fat deposition, which causes obesity, hyperglycemia, and dyslipidemia and consequently hyperinsulinemia, insulin resistance, and T2D [58, 66, 67].

Liver cells and adipocytes are major sites of lipogenesis in the body. De novo lipogenesis in the liver and adipocytes is driven by two transcriptional regulators –

sterol response element binding protein 1c (SREBP-1c) and carbohydrate response element binding protein (ChREBP). Both regulate the expression of key lipogenic genes, such as fatty acid synthase, acetyl coenzyme A carboxylase, and ATP-citrate lyase genes [68]. Insulin stimulates SREBP-1c expression in the liver, while glucose and other carbohydrates regulate ChREBP activity.

In a previous in vitro study, it was reported that carbohydrate (glucose)-associated activation of ChREBP and lipogenic genes was mediated by xylulose-5-phosphate, a xylitol metabolite [69], which suggests that xylitol may also affect lipid-metabolizing transcription factors in vivo. In another study, Amo et al. [2] observed that xylitol significantly increased ($p < 0.05$) mRNA levels of peroxisome proliferator-activated receptor gamma (PPARγ), adiponectin, and lipolytic enzymes such as hormone-sensitive lipase and adipose triglyceride lipase in adipose tissue of rat fed with high fat diet for 8 weeks [2] (Table 4). These results suggest an increased miniaturization of adipocytes and lipolysis in the adipose tissue, thus significantly reducing ($p < 0.05$) the visceral fat mass in xylitol-fed rats. Additionally, while mRNA levels of glucose-/carbohydrate-dependent lipogenic transcription factor, ChREBP and transcription factors, PPARα, and PPARγ coactivator 1α were significantly higher ($p < 0.05$), the insulin-dependent lipogenic transcription factor, SREBP-1c, was significantly lower ($p < 0.05$) in the liver of xylitol-fed rats compared to the control group. Although expression of both lipogenic transcriptional regulators (ChREBP and SREBP-1c) are elevated in the liver during obesity-related insulin resistance, SREBP-1c appears to be a dominant regulator of lipogenesis in the liver but not in the adipose tissue, because SREBP-1c knockout reduces hepatic but not adipose lipogenic enzyme expression [70]. Accordingly, the significantly lower ($p < 0.05$) SREBP-1c expression levels and higher mRNA levels of lipolytic and fatty acid oxidation transcription factors (PPARα and PPARγ coactivator 1α) in the liver of xylitol-fed animals can be considered as a potential beneficial effect of xylitol that can contribute to the improvement of insulin resistance in obesity and T2D, although additional studies are needed to ascertain this effect of xylitol in humans. Data of the abovementioned studies strongly support the potency of dietary xylitol in the management of obesity, metabolic syndrome, and related disorders.

3.4 Xylitol in Body Weight Management

Maintaining a healthy body weight is important for overall health and can help in preventing and controlling many detrimental chronic metabolic disorders. The chronic positive energy balance and weight gain resulting from unhealthy dietary habits and sedentary life style is a major contributor to overweight or obesity, which can increase the risk of developing some serious health problems, including cardiovascular diseases, T2D, hypertension, and metabolic syndrome [71]. Accordingly, dietary changes involving the use of less or noncaloric food additives such as sugar substitutes have been reported to be useful in managing weight gain and obesity [72]. Data from several studies have suggested that dietary xylitol could be a potent sweetener to control food and calorie intake [45, 73] as well as weight gain or obesity [2].

3.4.1 Food Intake and Appetite Control

It has been well documented that low GI food causes a stronger satiety sensation and more reduction in food intake than high GI food [74], and delayed gastric emptying may facilitate this effect [75]. Xylitol, a known low GI sweetener (Table 1), can slow gastric emptying, which could produce a strong satiety effect and reduce calorie or food intake subsequently. Shafer et al. [73] observed significantly longer ($p < 0.01$) gastric emptying time and significantly lower ($p < 0.01$) food or calorie intake due to pre-consumption of 25 g of xylitol compared to the control (Table 4). Recently, Chukwuma and Islam observed that a bolus dose (1 g kg^{-1} bw) of xylitol significantly reduced gastric emptying rate in both normal and T2D model of rats [52] (Table 4). In lean and obese normoglycemic human subjects, it was recently reported that 50 g xylitol significantly increased the levels of glucagon-like peptide-1, a gut hormone that modulates glucose homeostasis via delaying gastric emptying and promoting satiety, which consequently delayed gastric emptying [48] (Table 3). The abovementioned studies suggest that xylitol may be potentially useful in dietary and calorie control.

3.4.2 Body Fat, Weight Gain, and Obesity

High glycemic index food could be counterproductive to control weight gain, because they promote postprandial carbohydrate oxidation at the expense of fat oxidation, which can lead to fat accumulation, a major contributor to detrimental weight gain and obesity. On the other hand, the diets that produce low glycemic response may encourage weight control because they promote satiety and use of fat as energy source [76]. The lower caloric value and GI of xylitol compared to sucrose, glucose, and some other sugar alcohols (Table 1) is consistent with the objective of controlling weight gain and obesity. In a recent study, Islam reported that a 3-week ad libitum feeding of 10% xylitol in normal rats caused significantly lower ($p < 0.05$) body weight gain compared to 10% sucrose [43] (Table 4), which is consistent with the anti-obesogenic effects of xylitol demonstrated in normal rats fed with high-fat diets [2] (Table 4). The results of these studies suggest the potency of dietary xylitol in weight gain and obesity control.

3.5 Xylitol in Oxidative Stress

Imbalance between prooxidants or free radicals and antioxidants is the major cause of oxidative stress in physiological system. These reactive oxygen species can cause oxidative damage at cellular and tissue levels, which can worsen the disease condition like diabetes, microvascular disease, and so on [77]. Hence, dietary antioxidants have been employed to ameliorate or prevent oxidative stress-related disease conditions [78].

The antioxidative activities of polyols and their potentials as quenchers of reactive oxygen species have been recently demonstrated in vitro [51] (Table 5). In another previous study, xylitol exhibited the highest in vitro antioxidant activity compared to some other commercially available sweeteners [79] (Table 5). Additionally, in vivo study also suggested that 10% xylitol reduced elevated nonenzymatic glycosylation of acid-soluble collagen in diabetic rats [80] (Table 5). Some other old studies demonstrated that xylitol metabolism produces antioxidative capacity and keeps the glutathione antioxidant system active via generating NADPH [33, 81] (Table 5) and decreasing oxidized/reduced glutathione ratio [82] (Table 5). This can help to quench free radicals and thereby reducing oxidative damage in the liver as well as in the muscle and blood cells. Thus, xylitol may be further investigated as a potential antioxidative agent to confirm its antioxidative effects in various disease states.

3.6 Xylitol in Bone Metabolism

A defect in bone metabolism can result to abnormal bone bio- and physiochemical properties, which can cause most degenerative bone diseases like osteomalacia, osteoarthritis, osteoporosis, etc. These disease conditions are characterized by poor bone cortical and trabecular volume and mineral density, deteriorated bone microarchitecture, inadequate bone formation, defective bone mineralization, abnormal bone resorption, deficient bone calcium and phosphorus levels, and abnormal levels of bone metabolism marker like bone alkaline phosphatase and tartrate-resistant acid phosphatase activities, 1,25-dihydroxycholecalciferol, parathyroid hormone, hydroxyproline, and osteocalcin. Interestingly, dietary xylitol has been shown to either prevent or ameliorate most of these abnormalities associated with bone metabolism.

3.6.1 Bone Strength, Volume, Density, and Mineral Composition

The mineral composition and strength, density, and volume of bones are important markers to evaluate a healthy bone. The effect of xylitol on these parameters has been extensively studied in normal, diabetic, ovariectomized, and aging rats. Mattila et al. [83] demonstrated that 10 or 20 % dietary xylitol supplementation significantly increased shear stress of the tibia and femoral neck and shaft as well as trabecular bone volume of normal rats (Table 5). Sato et al. [84] also confirmed that similar concentrations of xylitol supplied to normal rats for 40 days were able to significantly increase bone mineral density and calcium level compared to control (Table 5). The consistent results of other studies further suggest the potency of xylitol to improve bone biochemical properties in osteoporotic conditions due to aging [85] and diabetes [86] (Table 5).

3.6.2 Bone Remineralization and Recalcification

Calcium forms an integral part of the bone, and its deficiency is usually accompanied by some metabolic changes as well as poor formation and structural weakness of the bone. Major metabolic markers during calcium deficiency include reduced serum calcium level; reduced bone calcium, magnesium, and phosphorous; elevated levels of bone hydroxyproline and serum alkaline phosphatase, 1,25-dihydroxycholecalciferol, parathyroid hormone, and osteocalcin; and increased activity of bone and serum tartrate-resistant acid phosphatase [87]. These calcium deficiency-induced metabolic changes have been previously reported in rats fed with calcium-deficient diet, but were significantly improved when calcium was supplemented with dietary xylitol [88] (Table 5). In fact, co-supplementation with calcium ($CaCO_3$) or the supplementation of xylitol alone improved calcium deficiency-related metabolic changes including bone recalcification than the supplementation of $CaCO_3$ alone. This effect may be due to the interaction of xylitol with certain dietary multivalent metal cations like calcium ion (Fig. 2b and c), thereby promoting the absorption, retention, and bioavailability of calcium [33] (Table 5), which suggests the advantage of the co-supplementation of xylitol.

3.6.3 Bone Resorption

Bone formation and resorption are two important bone metabolic processes that help in the modeling and shaping of bones. However, excessive bone resorption can lead to osteoporotic changes like low bone mass, structural deterioration of bone tissue, and increased bone fragility. The effect of xylitol on bone resorption has been studied in normal [89] (Table 5) and calcium-deficient rats [90] (Table 5) by measuring the amount of "3H" (radioactive marker) excreted in urine or preserved in bone of [3H]-tetracycline-pre-labeled rat models fed with xylitol. Data of these studies showed that xylitol supplementation significantly decreased urinary excretion of "3H" and significantly increased the amount of "3H" in the bone of the different rat models. This suggests that dietary xylitol may retard detrimental bone resorption and improve the imbalance of bone metabolism caused by increased bone resorption.

3.6.4 Bone Infection

Other bone-related beneficial properties of xylitol include its usefulness in enhancing the therapeutic efficacy of polymethyl methacrylate-based antibiotic in the treatment of chronic bone diseases like osteomyelitis. In rabbit model, incorporation of xylitol into polymethyl methacrylate bone cement enhanced the elution of daptomycin antibiotics, when used in the treatment of osteomyelitis infection following surgical debridement, thus enhancing the therapeutic efficacy of daptomycin [91] (Table 5).

In summary, data of the studies discussed above confirm the therapeutic potency of xylitol in the prevention and treatment of bone metabolic diseases.

3.7 Xylitol in Ear and Upper Respiratory Tract Infections

Upper respiratory tract infection is caused by a bacterium, *Streptococcus pneumonia*, which is a major contributing factor in the development of acute otitis media (AOM), a common middle ear infection among children due to the nasopharyngeal carriage of this bacterium [92]. Xylitol exhibits antimicrobial effects on pathogens of ear and upper respiratory infections like AOM and rhinosinusitis and, consequently, shows potential clinical significance in preventing and reducing the prevalence of AOM and rhinosinusitis [93–95].

3.7.1 Ear and Upper Respiratory Tract Infections

Streptococcus pneumoniae, *Haemophilus influenza*, and *Moraxella catarrhalis* are the most common bacteria isolated from the middle ear in AOM. Previous in vitro studies demonstrated that 1% or 5% xylitol appreciably reduced the growth and adherence of *S. pneumoniae* and *H. influenza* in cell culture media [96, 97] (Table 6). Other in vitro studies showed that 0.5% and/or 5% xylitol significantly lowered the capsular gene expression and also damaged the cell wall and polysaccharide capsule of viable pneumococci [98, 99] (Table 6). Furthermore, additional evidences suggest that xylitol, an osmolyte with low transepithelial permeability, lowers the salt concentration of airway epithelium in vitro and consequently enhances the innate antimicrobial defense at the airway surface of normal subjects [95] (Table 6). These antibacterial effects of xylitol on ear and upper respiratory tract pathogens have clinical relevance in the prevention of ear and upper respiratory tract infection like AOM and rhinosinusitis.

3.7.2 Middle Ear Infection (Acute Otitis Media) and Rhinosinusitis

Premier clinical studies that demonstrated the efficacy of xylitol in the prevention of middle ear infection were those reported by Uhari et al. [93, 94]. According to their chewing gum trial reported in 1996, children who received xylitol chewing gum (equivalent to 8.4 g xylitol) daily for 2 months showed lesser occurrence of AOM (12% vs. 20.8%) and received significantly fewer ($p = 0.04$) antimicrobial

Table 6 Potential clinical relevance of xylitol in ear and upper respiratory tract infection and skin care

Beneficial effects of xylitol	Comment/remarks	References[abcd]
Ear and upper respiratory infections	Inhibits pneumococcal growth, adherence, and biofilm formation	[96, 97][a]
	Causes damage to pneumococcal cell wall and polysaccharide capsule	[98, 99][a]
	Enhances the innate antimicrobial defense at the airway surface	[95][ac]
	Reduces the occurrence of acute otitis media in children	[93, 94][c]
	Improves sinusitis symptoms	[101][b], [102][c]
Atopic dermatitis	Inhibit colonization of *S. aureus*, on the lesioned skin of atopic dermatitis	[104][a], [105][c]
Collagen	Improves collagen synthesis and stability but reduces collagen damage	[106, 107][a], [80][b]

"abc" superscript letters mean the information was based on the original research "a"; in vitro, "b"; in vivo and "c"; clinical/human studies, while "d" means information was based on the literature review of a research or review article

prescription compared to children that received sucrose chewing gum for similar duration [93] (Table 6). In 1998, they reported that a 3-month daily dose of xylitol in chewing gum (containing 8.4 g xylitol) or syrup (containing 10 g xylitol) caused 40% and 30% decrease, respectively, in the occurrence of AOM in children [94] (Table 6). Although a more convenient vehicle and dose still needs to be clinically investigated, advances in research have suggested the prophylactic daily use of 10 g xylitol (5 × 2 g) via a chewing gum or syrup vehicle to be generally accepted, due to its safety and fewer side effects [100].

Xylitol has also shown potential clinical therapeutic effects against rhinosinusitis. In a previous in vivo study, xylitol administered to the maxillary sinus, simultaneously with *Pseudomonas aeruginosa* bacteria, reduced sinusitis in experimental rabbits [101] (Table 6). In a recent short-term clinical pilot study, 10-day sinonasal irrigation using xylitol solution was well tolerated by human subjects and caused significantly more improvement of chronic rhinosinusitis symptoms than saline sinonasal irrigation [102] (Table 6).

3.8 Xylitol in Skin Care

The several benefits and potential clinical relevance of xylitol have prompted researchers to explore other areas like skin care and possible dermatological applications.

3.8.1 Atopic Dermatitis

Healthy skin is normally colonized by beneficial microorganism like *Staphylococcus epidermidis*, a resident microflora. On the other hand, transient microflora of the skin include microorganisms like *Staphylococcus aureus*, which temporarily harbor on skin surface due to external contact. *S. aureus* has been found to be the major pathogen of atopic dermatitis, constituting about 90% of bacterial microflora on lesioned skin of atopic dermatitis; hence, the clinical severity of atopic dermatitis is a function of the colonization rate and density of *S. aureus* on skin lesions [103]. Among other factors, biofilm formation and increased adherence of *S. aureus* on skin lesion are major contributing factors to lesioned skin colonization and antibiotic resistance of *S. aureus* [103]. *S. aureus* forms a unique biofilm composition of fibrin fiber and glycocalyx, a glycoprotein-polysaccharide that strengthens its adhesion on lesioned skin [103].

Xylitol exhibited strong antimicrobial activity against *S. aureus* by inhibiting glycocalyx production and colonization of *S. aureus* on the horny cells of atopic dermatitis lesions without affecting *Staphylococcus epidermidis* – a major constituent of skin microflora on healthy human skin that provides protection against the growth of pathogenic bacteria [104, 105] (Table 6). The results of the above-mentioned studies suggest the therapeutic potentials of xylitol against atopic dermatitis.

3.8.2 Collagen

Collagen is the most abundant protein in the body, which is important for normal biological functions. Studies have shown that dietary xylitol may have a positive impact on collagen synthesis, glycosylation, and stability against degradation. Increased skin collagen glycosylation in rats due to the induction of diabetes was significantly reversed ($p < 0.01$) by a 3- or 20-month supplementation of 10% xylitol [80] (Table 6). Additionally, in vitro treatment with xylitol improved collagen stability [106] (Table 6) and offered protection against guanidine hydrochloride-induced collagen denaturation [107] (Table 6). The results of the abovementioned studies suggest the potency of xylitol in the preservation of the functional and structural integrity of physiological collagen, even in disease states like diabetes.

3.9 Xylitol as a Sweetener

Given the innate human appetite for sweetness, it has become a major determining factor in the choice of a suitable sugar-free product or sugar substitute. Xylitol is sweeter than all other polyols with about similar level of sweetness with sucrose (Table 1). Xylitol can also be used in combination with other polyols and high-intensity sweeteners to produce a unique sweetness compared to sucrose. It is used to improve the sweetness quality of intense nonnutritive sweeteners, especially when the sweetness of the nonnutritive sweetener is lost due to interaction with some flavors [1]. Xylitol readily dissolves in water and provides a stronger cooling effect upon consumption than other polyols, which contributes to its better taste compared to other polyol and sugar [1]. Accordingly, it has become a preferred sweetener for mint-flavored products, where a refreshing cooling sensation is required upon consumption. Xylitol is also heat and pH stable so it can be used in baking or cooking and over a wide pH range.

Furthermore, xylitol occurs naturally in several fruits and vegetables and can also be naturally sourced [1, 3]. Considering the innate human desire or preference for natural products over synthetic products, xylitol can be a better sugar substitute than synthetic sweeteners. Moreover, there have been several controversies regarding the use of synthetic sweeteners because of their short- and long-term side effects, so xylitol can be a safer alternative as a sweetener [45].

3.10 Other Beneficial Effects of Xylitol

Other than the above-elaborated beneficial effects, xylitol has also demonstrated other promising effects, which may be further investigated for the following potential clinical uses: treatment of xerostomia (dry mouth) [108], as an oral antifungal agent for oral candidiasis [109], as an adjunct in gastric ulcer therapy [110], protection of blood cell membranes to reduce hemolysis [111], treatment of pseudomembranous colitis (antibiotic-associated diarrhea) [112], and improvement of nonpathogenic microflora in the colon (prebiotic effect) [1].

4 Safety and Toxicity of Xylitol

Xylitol has been reported to be a safer sugar alcohol even at a higher dose compared to many other sugar alcohols, e.g., sorbitol and lactitol, although tolerance ability is varied from person to person. Gastrointestinal symptoms like diarrhea, laxation, watery stool, and bloating are commonly experienced when large amounts of polyols are consumed [113]. Storey et al. [114] reported that a bolus dose of 35 g or 50 g of xylitol in water caused significant gastrointestinal symptoms in young adults. After a review of five randomized studies, the Academy of Nutrition and Dietetics reported that up to 10 or 15 g/day of polyols including xylitol can be tolerated [115], although this may vary among different individuals. Nevertheless, there are evidences that the gastrointestinal tolerance of xylitol may increase with regular consumption, which may be due to its lower laxative threshold compared to most sugar alcohols [116].

Xylitol has a wide range of safety and very low order of toxicity regardless of the route of administration. Results from several in vivo and in vitro toxicity studies have shown that xylitol is not associated with embryotoxicity, teratogenicity, and reproductive toxicity, as well as mutagenicity and clastogenicity [117]. In dogs, however, oral ingestion of xylitol is believed to induce excessive insulin secretion that can cause hypoglycemia and hepatotoxicity [118] and even death. Xylitol is an FDA-approved food additive, safe for use in foods for special dietary uses [115]. It has been used in numerous pharmaceutical, food, and oral health-care products worldwide for many decades.

5 Conclusions

There are clear evidences of the beneficial effects and potential clinical relevance of xylitol in oral health care, glycemic control, lipid metabolism, weight management, bone metabolism, skin care, ear and upper respiratory tract infection, and oxidative stress. Some areas however still require more investigation especially at clinical levels to ascertain the clinical application of xylitol. Being a natural sweetener, the safety of xylitol used at moderate and recommended doses cannot be overemphasized. This unique sugar alcohol may be used not only as a sweetener but also as a therapeutic additive for several disorders that is void of some side effects associated with most synthetic drugs.

References

1. Zacharis C (2012) Xylitol. In: O'Donnell K, Kearsley MW (eds) Sweeteners and sugar alternatives in food technology, 2nd edn. Wiley & Sons, UK
2. Amo K, Arai H, Uebanso T et al (2011) Effects of xylitol on metabolic parameters and visceral fat accumulation. J Clin Biochem Nutr 49:1–7
3. Peterson ME (2013) Xylitol. Top Companion Anim Med 28:18–20

4. Sadler M, Stowell JD (2012) Calorie control and weight management. In: O'Donnell K, Kearsley MW (eds) Sweeteners and sugar alternatives in food technology, 2nd edn. Wiley & Sons, UK
5. Livesey G (2003) Health potential of polyols as sugar replacers, with emphasis on low glycaemic properties. Nutr Res Rev 16:163–191
6. Baessler KH (1978) Absorption, metabolism and tolerance of polyol sugar substitutes. Pharmacol Ther Dent 3:85–93
7. Takahashi N, Nyvad B (2011) The role of bacteria in the caries process: ecological perspectives. J Dent Res 90:294–303
8. Bowen WH (2002) Do we need to be concerned about dental caries in the coming millennium? Crit Rev Oral Biol Med 13:126–131
9. Paes Leme AF, Koo H, Bellato CM, Bedi G, Cury JA (2006) The role of sucrose in cariogenic dental biofilm formation – new insight. J Dent Res 85:878–887
10. Banas JA, Vickerman MM (2003) Glucan-binding proteins of the oral streptococci. Crit Rev Oral Biol Med 14:89–99
11. Takahashi N, Washio J (2011) Metabolomic effects of xylitol and fluoride on plaque biofilm in vivo. J Dent Res 90:1463–1468
12. Söderling E, Alaräisänen L, Scheinin A, Mäkinen KK (1987) Effect of xylitol and sorbitol on polysaccharide production by and adhesive properties of *Streptococcus mutans*. Caries Res 21:109–116
13. Söderling EM (2009) Xylitol, mutans streptococci, and dental plaque. Adv Dent Res 21:74–78
14. Wood JM (1997) An investigation of the effect of xylitol on the production of polysaccharide from sucrose by *Streptococcus mutans*. Rep Spec Proj 506:1–9
15. Söderling EM, Hietala-Lenkkeri A (2010) Xylitol and erythritol decrease adherence of polysaccharide-producing oral streptococci. Curr Microbiol 60:25–29
16. Badet C, Furiga A, Thébaud N (2008) Effect of xylitol on an in vitro model of oral biofilm. Oral Health Prev Dent 6:337–341
17. Alves FR, Neves MA, Silva MG, Rôças IN, Siqueira JF Jr (2103) Antibiofilm and antibacterial activities of farnesol and xylitol as potential endodontic irrigants. Braz Dent J 24:224–229
18. Lee YE, Choi YH, Jeong SH, Kim HS, Lee SH, Song KB (2009) Morphological changes in *Streptococcus mutans* after chewing gum containing xylitol for twelve months. Curr Microbiol 58:332–337
19. Marsh PD, Nyvad B (2008) The oral microflora and biofilms on teeth. In: Fejerskov O, Kidd EAM (eds) Dental caries: the disease and its clinical management, 2nd edn. Blackwell Munksgaard, UK
20. Mäkinen KK (2010) Sugar alcohols, caries incidence, and remineralization of caries lesions: a literature review. Int J Dent. doi:10.1155/2010/981072
21. Miyasawa-Hori H, Aizawa S, Takahashi N (2006) Difference in the xylitol sensitivity of acid production among *Streptococcus mutans* strains and the biochemical mechanism. Oral Microbiol Immunol 21:201–205
22. Splieth CH, Alkilzy M, Schmitt J, Berndt C, Welk A (2009) Effect of xylitol and sorbitol on plaque acidogenesis. Quintessence Int 40:279–285
23. Assev S, Stig S, Scheie AA (2002) Cariogenic traits in xylitol-resistant and xylitol-sensitive *mutans streptococci*. Oral Microbiol Immunol 17:95–99
24. Söderling EM, Ekman TC, Taipale TJ (2008) Growth inhibition of *Streptococcus mutans* with low xylitol concentrations. Curr Microbiol 56:382–385
25. Radmerikhi S, Formantes B, Fajardo KR, Azul E (2013) Antimicrobial effect of different xylitol concentrations on *Streptococcus mutans* and *Lactobacillus acidophilus* count. J Res Dent 1:95–98
26. Thaweboon S, Thaweboon B, Soo-Ampon S (2004) The effect of xylitol chewing gum on *mutans streptococci* in saliva and dental plaque. Southeast Asian J Trop Med Public Health 35:1024–1027

27. Loesche WJ, Grossman NS, Earnest R, Corpron R (1984) The effect of chewing xylitol gum on the plaque and saliva levels of *Streptococcus mutans*. J Am Dent Assoc 108:587–592
28. Söderling E, Trahan L, Tammiala-Salonen T, Häkkinen L (1997) Effects of xylitol, xylitol-sorbitol, and placebo chewing gums on the plaque of habitual xylitol consumers. Eur J Oral Sci 105:170–177
29. Söderling E, Isokangas P, Pienihäkkinen K, Tenovuo J, Alanen P (2001) Influence of maternal xylitol consumption on mother-child transmission of *mutans streptococci*: 6-year follow-up. Caries Res 35:173–177
30. Zimmerman JN, Custodio W, Hatibovic-Kofman S, Lee YH, Xiao Y, Siqueira WL (2013) Proteome and peptidome of human acquired enamel pellicle on deciduous teeth. Int J Mol Sci 14:920–934
31. Kleber CJ, Putt MS, Muhler JC (1978) Enamel dissolution by various food acidulants in a sorbitol candy. J Dent Res 57:447–451
32. Steinberg LM, Odusola F, Mandel ID (1992) Remineralizing potential, antiplaque and antigingivitis effects of xylitol and sorbitol sweetened chewing gum. Clin Prev Dent 14:31–34
33. Mäkinen KK (2000) Can the pentitol-hexitol theory explain the clinical observations made with xylitol. Med Hypotheses 54:603–613
34. Mäkinen KK (1985) New biochemical aspects of sweeteners. Int Dent J 35:23–35
35. Vacca Smith AM, Bowen WH (2000) In situ studies of pellicle formation on hydroxyapatite discs. Arch Oral Biol 45:277–291
36. Miake Y, Saeki Y, Takahashi M, Yanagisawa T (2003) Remineralization effects of xylitol on demineralized enamel. J Electron Microsc (Tokyo) 52:471–476
37. Chunmuang S, Jitpukdeebodintra S, Chuenarrom C, Benjakul P (2007) Effect of xylitol and fluoride on enamel erosion *in vitro*. J Oral Sci 49:293–297
38. Leach SA, Green RM (1980) Effect of xylitol-supplemented diets on the progression and regression of fissure caries in the albino rat. Caries Res 14:16–23
39. Scheinin A, Mäkinen KK (1975) Turku sugar studies I-XXI. Acta Odontol Scand 33:1–351
40. Mäkinen KK (1991) Prevention of dental caries by xylitol. Environ Manag Health 2:6–11
41. Mäkinen KK, Järvinen KL, Anttila CH, Luntamo LM, Vahlberg T (2013) Topical xylitol administration by parents for the promotion of oral health in infants: a caries prevention experiment at a Finnish Public Health Centre. Int Dent J 63:210–224
42. Campus G, Cagetti MG, Sale S, Petruzzi M, Solinas G, Strohmenger L, Lingström P (2013) Six months of high-dose xylitol in high-risk caries subjects – a 2-year randomised, clinical trial. Clin Oral Investig 17:785–791
43. Islam S (2011) Effects of xylitol as a sugar substitute on diabetes-related parameters in non-diabetic rats. J Med Food 14:505–511
44. Mushtaq Z, Imran M, Salim-ur Rehman, Zahoor T, Ahmad RS, Arshad MU (2014) Biochemical perspectives of xylitol extracted from indigenous agricultural by-product mung bean (*vignaradiata*) hulls in a rat model. J Sci Food Agric 94:969–974
45. Islam S, Indrajit M (2012) Effects of xylitol on blood glucose, glucose tolerance, serum insulin and lipid profile in a type 2 diabetes model of rats. Ann Nutr Metab 61:57–64
46. Rahman MA, Islam MS (2014) Xylitol improves pancreatic islets morphology to ameliorate type 2 diabetes in rats: a dose response study. J Food Sci 79:1436–1442
47. de Kalbermatten N, Ravussin E, Maeder E, Geser C, Jéquier E, Felber JP (1980) Comparison of glucose, fructose, sorbitol, and xylitol utilization in humans during insulin suppression. Metabolism 29:62–67
48. Woelnerhanssen BK, Cajacob L, Keller N et al (2016) Gut hormone secretion, gastric emptying and glycemic responses to erythritol and xylitol in lean and obese subjects. Am J Physiol Endocrinol Metab 310:E1053-E1061
49. Hassinger W, Sauer G, Cordes U, Krause U, Beyer J, Baessler KH (1981) The effects of equal caloric amounts of xylitol, sucrose and starch on insulin requirements and blood glucose levels in insulin-dependent diabetics. Diabetologia 21:37–40

50. Patel DK, Kumar R, Laloo D, Hemalatha S (2012) Natural medicines from plant source used for therapy of diabetes mellitus: an overview of its pharmacological aspects. Asian Pac J Trop Dis 2:239–250
51. Kang Y, Jo S, Yoo J, Cho J, Kim E, Apostolidis E, Kwon K (2014) Anti-hyperglycemic effect of selected sugar alcohols. FASEB J 24:829–832
52. Chukwuma CI, Islam MS (2015) Effects of xylitol on carbohydrate digesting enzymes activity, intestinal glucose absorption and muscle glucose uptake: a multi-mode study. Food Funct 6:955–962
53. Frejnagel SS, Gomez-Villalva E, Zduńczyk Z (2003) Intestinal absorption of xylitol and effect of its concentration on glucose and water absorption in the small intestine of rat. Pol J Food Nutr 12(53):32–34
54. Solomon TP, Haus JM, Kelly KR et al (2010) A low–glycemic index diet combined with exercise reduces insulin resistance, postprandial hyperinsulinemia, and glucose dependent insulinotropic polypeptide responses in obese, prediabetic humans. Am J Clin Nutr 92:1359–1368
55. Khalid Q, Rahman MA (1984) Studies on xylitol-induced insulin secretion in vitro. Biomed Pharmacother 38:175–178
56. Seino Y, Taminato T, Inoue Y, Goto Y, Ikeda M (1976) Xylitol: stimulation of insulin and inhibition of glucagon responses to arginine in man. J Clin Endocrinol Metab 42:736–743
57. Bessesen DH (2001) The role of carbohydrates in insulin resistance. J Nutr 131:2782–2786
58. Karpe F, Dickmann JR, Frayn KN (2011) Fatty acids, obesity, and insulin resistance: time for a reevaluation. Diabetes 60:2441–2449
59. Kishore P, Kehlenbrink S, Hu M et al (2012) Xylitol prevents NEFA-induced insulin resistance in rats. Diabetologia 55:1808–1812
60. Marchetti P, Lupi R, Del Guerra S, Bugliani M, Marselli L, Boggi U (2010) The beta-cell in human type 2 diabetes. Adv Exp Med Biol 654:501–514
61. Krauss RM (2004) Lipids and lipoproteins in patients with type 2 diabetes. Diabetes Care 27:1496–1504
62. Förster H, Quadbeck R, Gottstein U (1982) Metabolic tolerance to high doses of oral xylitol in human volunteers not previously adapted to xylitol. Int J Vitam Nutr Res Suppl 22:67–88
63. Opitz K (1966) On the action of sugars, sugar alcohols and related substances on the fat mobilization. Naunyn Schmiedebergs Arch Pharmakol Exp Pathol 255:192–199
64. Yamagata S, Goto Y, Ohneda A et al (1965) Clinical effects of xylitol on carbohydrate and lipid metabolism in diabetes. Lancet 286:918–921
65. Goto Y, Nagano S, Nakamura H (1969) Comment: effect of xylitol on individual free fatty acids in the plasma. In: Horecker BL, Lang K, Takagi Y (eds) International symposium on metabolism, physiology, and clinical use of pentoses and pentitols, 1st edn. Springer, Heidelberg
66. Bray GA, Lovejoy JC, Smith SR (2002) The influence of different fats and fatty acids on obesity, insulin resistance and inflammation. J Nutr 132:2488–2491
67. Strable MS, Ntambi JM (2010) Genetic control of de novo lipogenesis: role in diet-induced obesity. Crit Rev Biochem Mol Biol 45:199–214
68. Postic C, Girard J (2008) Contribution of de novo fatty acid synthesis to hepatic steatosis and insulin resistance: lessons from genetically engineered mice. J Clin Invest 118:829–838
69. Kabashima T, Kawaguchi T, Wadzinski BE, Uyeda K (2003) Xylulose 5-phosphate mediates glucose-induced lipogenesis by xylulose 5-phosphate-activated protein phosphatase in rat liver. Proc Natl Acad Sci U S A 100:5107–5112
70. Shimano H, Shimomura I, Hammer RE, Herz J, Goldstein JL, Brown MS, Horton JD (1997) Elevated levels of SREBP-2 and cholesterol synthesis in livers of mice homozygous for a targeted disruption of the SREBP-1 gene. J Clin Invest 100:2115–2124
71. Artham SM, Lavie CJ, Milani RV, Ventura HO (2009) Obesity and hypertension, heart failure, and coronary heart disease – risk factor, paradox, and recommendations for weight loss. Ochsner J 9:124–132

72. Brown RJ, De Banate MA, Rother KI (2010) Artificial Sweeteners: a systematic review of metabolic effects in youth. Int J Pediatr Obes 5:305–312
73. Shafer RB, Levine AS, Marlette JM, Morley JE (1987) Effects of xylitol on gastric emptying and food intake. Am J Clin Nutr 45:744–747
74. Jiménez-Cruz A, Gutiérrez-González AN, Bacardi-Gascon M (2005) Low glycemic index lunch on satiety in overweight and obese people with type 2 diabetes. Nutr Hosp 20:348–350
75. Bergmann JF, Chassany O, Petit A, Triki R, Caulin C, Segrestaa JM (1992) Correlation between echographic gastric emptying and appetite: influence of psyllium. Gut 33: 1042–1043
76. Brand-Miller JC, Holt SHA, Pawlak DB, McMillan J (2002) Glycemic index and obesity. Am J Clin Nutr 76:281–285
77. Fowler MJ (2008) Microvascular and macrovascular complications of diabetes. Clin Diabetes 26:77–82
78. Parke DV (1999) Nutritional antioxidants and disease prevention: mechanisms of action. In: Basu TK, Temple NJ, Garg ML (eds) Antioxidants in human health and disease, 1st edn. CABI Publishing, New York
79. Al-Shahrani MM, Zaman GS, Amanullah A (2013) Measurement of antioxidant activity in selected food products and nutraceuticals. J Nutr Food Sci. doi:10.4172/2155-9600.1000205
80. Knuuttila ML, Kuoksa TH, Svanberg MJ, Mattila PT, Karjalainen KM, Kolehmainen E (2000) Effects of dietary xylitol on collagen content and glycosylation in healthy and diabetic rats. Life Sci 67:283–290
81. Pösö AR (1979) Shuttles for the translocation of NADH in isolated liver cells from fed rats during oxidation of xylitol. Acta Chem Scand B B33:93–99
82. Asakura T, Adachi K, Yoshikawa H (1970) Reduction of oxidized glutathione by xylitol. J Biochem 67:731–733
83. Mattila P, Knuuttila M, Kovanen V, Svanberg M (1999) Improved bone biomechanical properties in rats after oral xylitol administration. Calcif Tissue Int 64:340–344
84. Sato H, Ide Y, Nasu M, Numabe Y (2011) The effects of oral xylitol administration on bone density in rat femur. Odontology 99:28–33
85. Mattila PT, Svanberg MJ, Jämsä T, Knuuttila ML (2002) Improved bone biomechanical properties in xylitol-fed aged rats. Metabolism 51:92–96
86. Mattila PT, Knuuttila ML, Svanberg MJ (1998) Dietary xylitol supplementation prevents osteoporotic changes in streptozotocin-diabetic rats. Metabolism 47:578–583
87. Akesson K, Lau KH, Johnston P, Imperio E, Baylink DJ (1998) Effects of short-term calcium depletion and repletion on biochemical markers of bone turnover in young adult women. J Clin Endocrinol Metab 83:1921–1927
88. Hämäläinen MM, Knuuttila M, Svanberg M, Koskinen T (1990) Comparison of the effect of gluconate, lactose, and xylitol on bone recalcification in calcium-deficient rats. Bone 11:429–438
89. Mattila PT, Svanberg MJ, Mäkinen KK, Knuuttila ML (1996) Dietary xylitol, sorbitol and D-mannitol but not erythritol retard bone resorption in rats. J Nutr 126:1865–1870
90. Svanberg M, Knuuttila M (1994) Dietary xylitol retards bone resorption in rats. Miner Electrolyte Metab 20:153–157
91. Beenken KE, Bradney L, Bellamy W (2012) Use of xylitol to enhance the therapeutic efficacy of polymethyl methacrylate-based antibiotic therapy in treatment of chronic osteomyelitis. Antimicrob Agents Chemother 56:5839–5844
92. Faden H, Waz MJ, Bernstein JM, Brodsky L, Stanievich J, Ogra PL (1991) Nasopharyngeal flora in the first three years of life in normal and otitis-prone children. Ann Otol Rhinol Laryngol 100:612–615
93. Uhari M, Kontiokari T, Koskela M, Niemelä M (1996) Xylitol chewing gum in prevention of acute otitis media: double blind randomised trial. BMJ 313:1180–1184
94. Uhari M, Kontiokari T, Niemelä M (1998) A novel use of xylitol sugar in preventing acute otitis media. Pediatrics 102:879–884

95. Zabner J, Seiler MP, Launspach JL (2000) The osmolyte xylitol reduces the salt concentration of airway surface liquid and may enhance bacterial killing. Proc Natl Acad Sci U S A 97:11614–11619
96. Kontiokari T, Uhari M, Koskela M (1995) Effect of xylitol on growth of nasopharyngeal bacteria *in vitro*. Antimicrob Agents Chemother 39:1820–1823
97. Kontiokari T, Uhari M, Koskela M (1998) Antiadhesive effects of xylitol on otopathogenic bacteria. J Antimicrob Chemother 41:563–565
98. Tapiainen T, Sormunen R, Kaijalainen T, Kontiokari T, Ikäheimo I, Uhari M (2004) Ultrastructure of *Streptococcus pneumoniae* after exposure to xylitol. J Antimicrob Chemother 54:225–228
99. Kurola P, Tapiainen T, Sevander J, Kaijalainen T, Leinonen M, Uhari M, Saukkoriipi A (2011) Effect of xylitol and other carbon sources on *Streptococcus pneumoniae* biofilm formation and gene expression *in vitro*. APMIS 119:135–142
100. Danhauer JL, Johnson CE, Corbin NE, Bruccheri KG (2010) Xylitol as a prophylaxis for acute otitis media: systematic review. Int J Audiol 49:754–761
101. Brown CL, Graham SC, Cable BB, Ozer EA, Taft PJ, Zabner J (2004) Xylitol enhances bacterial killing in the rabbit maxillary sinus. Laryngoscope 114:2021–2024
102. Weissman JD, Fernandez F, Hwang PH (2011) Xylitol nasal irrigation in the management of chronic rhinosinusitis: a pilot study. Laryngoscope 121:2468–2472
103. Niebuhr M (2011) Infections and bacterial colonization including treatment. In: Werfel T, Spergel JM (eds) Atopic dermatitis in childhood and adolescence, vol 15. Pediatr Adolesc Med, Karger, Basel
104. Masako K, Hideyuki I, Shigeyuki O, Zenro I (2005) A novel method to control the balance of skin microflora. Part 1. Attack on biofilm of *Staphylococcus aureus* without antibiotics. J Dermatol Sci 38:197–205
105. Masako K, Yusuke K, Hideyuki I, Atsuko M, Yoshiki M, Kayoko M, Makoto K (2005) A novel method to control the balance of skin microflora Part 2. A study to assess the effect of a cream containing farnesol and xylitol on atopic dry skin. J Dermatol Sci 38:207–213
106. Usha R, Raman SS, Subramanian V, Ramasami T (2006) Role of polyols (erythritol, xylitol and sorbitol) on the structural stabilization of collagen. Chem Phys Lett 430:391–396
107. Usha R, Ramasami T (2008) Stability of collagen with polyols against guanidine denaturation. Colloids Surf B Biointerfaces 61:39–42
108. Olsson H, Spak CJ, Axéll T (1991) The effect of a chewing gum on salivary secretion, oral mucosal friction and the feeling of dry mouth in xerostomic patients. Acta Odontol Scand 49:273–279
109. Abu-Elteen H, Khaled H (2005) The influence of dietary carbohydrates on in vitro adherence of four Candida species to human buccal epithelial cells (BEC). Microb Ecol Health Dis 17:156–162
110. Assouline G, Danon A (1981) Hyperosmotic xylitol, prostaglandins and gastric mucosal barrier. Prostaglandins Med 7:63–70
111. Ukab WA, Sato J, Wang YM, van Eys J (1981) Xylitol mediated amelioration of acetylphenylhydrazine-induced hemolysis in rabbits. Metabolism 30:1053–1059
112. Naaber P, Lehto E, Salminen S, Mikelsaar M (1996) Inhibition of adhesion of Clostridium difficile to caco-2 cells. FEMS Immunol Med Microbiol 14:205–209
113. Flambeau M, Respondek F, Wagner A (2012) Maltitol syrups. In: O'Donnell K, Kearsley MW (eds) Sweeteners and sugar alternatives in food technology, 2nd edn. Wiley & Sons, UK
114. Storey D, Lee A, Bornet F, Brouns F (2007) Gastrointestinal tolerance of erythritol and xylitol ingested in a liquid. Eur J Clin Nutr 61:349–354
115. Fitch C, Keim KS (2012) Position of the academy of nutrition and dietetics: use of nutritive and non nutritive sweeteners. J Acad Nutr Diet 112:739–758

116. Wang YM, van Eys J (1981) Nutritional significance of fructose and sugar alcohols. Annu Rev Nutr 1:437–475
117. World Health Organization (1977) Summary of toxicology data of certain food additives. Twenty-first report of the joint FAO/WHO Expert Committee on Food Additives, Technical Report Series 617, Geneva, Switzerland.
118. Murphy LA, Coleman AE (2012) Xylitol toxicosis in dogs. Vet Clin North Am Small Anim Pract 42:307–312

Part IV

Methods of Analysis

Analytical Strategies to Determine Artificial Sweeteners by Liquid Chromatography-Mass Spectrometry

20

Rosa A. Lorenzo Ferreira, María Asunción Lage Yusty, and Antonia María Carro Díaz

Abstract

The interest of consumers and government organizations on improving the quality of food and its impact on human health is growing in recent years. Consuming low-calorie foodstuffs are increasingly demanded, and the presence of artificial sweeteners plays an important role not only in food but also in waste reaching the environment. Analysis of these compounds is important in assessing food safety and quality. This chapter reviews the analytical approaches for the extraction and reliable identification and quantification of most commonly used artificial sweeteners in food, pharmaceutical, and environmental related matrices. The advantages and disadvantages of determination techniques used are described, with special emphasis on liquid chromatography coupled to mass spectrometry with different approaches, including applications of high resolution mass spectrometry. The possibility of using new materials for efficient extraction, miniaturized extraction techniques, and the potential for quantification of LC-MS/MS techniques are highlighted as future prospects, based on achievements.

Keywords

Sweeteners • Liquid chromatography-mass spectrometry • Sample preparation • Environmental • Food • Pharmaceuticals

R.A. Lorenzo Ferreira (✉) • A.M. Carro Díaz
Department of Analytical Chemistry, Nutrition and Bromatology, Faculty of Chemistry and Health Research Institute of Santiago de Compostela (IDIS), Campus Vida, University of Santiago de Compostela, Santiago de Compostela, Spain
e-mail: rosaantonia.lorenzo@usc.es; tuchi.carro@usc.es

M.A. Lage Yusty
Department of Analytical Chemistry, Nutrition and Bromatology, Faculty of Pharmacy, Campus Vida, University of Santiago de Compostela, Santiago de Compostela, Spain
e-mail: maria.lage.yusty@usc.es

J.-M. Mérillon, K.G. Ramawat (eds.), *Sweeteners*, Reference Series in Phytochemistry,
https://doi.org/10.1007/978-3-319-27027-2_19

Abbreviations

ACS-K	Acesulfame
ADA	American Diabetes Association
ADI	Acceptable daily intake
AHA	American Heart Association
ALI	Alitame
ANSES	Agency for Food Environmental and Occupational Health & Safety
APCI	Atmospheric pressure chemical ionization
API	Atmospheric pressure ionization
ASP	Aspartame
C18	Octadecylsilane
CE	Capillary electrophoresis
CFR	Code of Federal Regulations
CID	Collision induced dissociation
CP-ANN	Counter-propagation artificial neural networks
CYC	Cyclamate
DAD	Diode array detector
DKP	Diketopiperazine
DUL	Dulcin
EDCs	Endocrine disrupting chemicals
EFSA	European Food Safety Authority
ELSD	Evaporative light scattering detector
EOF	Electroosmotic flow
ESI	Electrospray ionization
FAO	Food and Agriculture Organization of the United Nations
FASI	Field-amplified sample injection
FT	Fourier Transform
GA	Glycyrrhizic acid
GC-ECD	Gas chromatography-electron capture detector
GRAS	Generally recognized as safe
HILIC	Hydrophilic interaction liquid chromatography
HPTLC	High-performance thin-layer chromatography
HRMS	High-resolution mass spectrometry
HS-SDME	Headspace single-drop microextraction
IC	Ion chromatography
ISA	International Society of Automation
JECFA	Joint Expert Committee on Food Additives
JRC	Joint Research Centre
LODs	Limits of detection
LOQs	Limits of quantification
LVI	Large-volume injection

MDL	Method detection limit
MEKC	Micellar electrokinetic chromatographic
MeOH	Methanol
MIP	Molecularly imprinted polymer
MRM	Multiple reaction monitoring
MSTFA	*N*-methyl-*N*-trimethylsilytrifluoroacetamide
NEO	Neotame
NHDC	Neohesperidin dihydrochalcone
NMR	Nuclear magnetic resonance
PCR	Principal component regression
PLE	Pressurized liquid extraction
PLS	Partial least squares
PPCPs	Pharmaceuticals and personal care products
Q	Single quadrupole
QqQ	Triple quadrupole
QTOF	Quadrupole time-of-flight
RP	Reversed phase
RRHD	Rapid resolution high definition
SAC	Saccharin
SGFE	*Siraitia grosvenorii* Swingle fruit extract
SIM	Selected ion monitoring
SIR	Selective ionization recording
SPE	Solid phase extraction
SPM	Suspended particulate matter
SPME	Solid phase microextraction
STV	Stevioside
SUC	Sucralose
SWTs	Sweeteners
TMCS	Trimethylchlorosilane
TRIS	Tris (hydroxymethyl) amino methane
UHPLC-MS/MS	Ultra-high performance liquid chromatography-tandem mass spectrometry
WHO	World Health Organization
WWTPs	Wastewater treatment plants

Contents

1 Introduction 442
2 Artificial Sweeteners Trace Analysis Techniques 448
2.1 Sample Preparation Strategies 448
2.2 Liquid Chromatography Mass-Spectrometry Methods for Quantification 454
2.3 Other Determination Techniques 467
3 Conclusions and Futures Perspectives 470
References 471

1 Introduction

Sweetener is a food additive used to impart a sweet taste to foods or in table-top sweeteners, included in the Regulation 1333/2008 [1]; it serves for one or more of the following purposes: replacing sugars for the production of energy-reduced food, noncariogenic food, or food with no added sugars and replacing sugars where this permits an increase in the shelf-life of the food or producing food intended for particular nutritional uses.

These additives are known as low-calories or nonnutritive sweeteners. Each provides little or no energy, in most cases because it activates sweet taste receptors at very low concentrations relative to sugar, with estimates of the potency of artificial sweeteners (SWTs) ranging from about 200 times to up to 20,000 times the sweetness of sucrose. They are of plant origin or obtained by chemical synthesis and are used in the food industry for their sweetening power [2].

It is important to balance the calories you consume with the calories you burn by maintaining a sensible, balanced diet combined with regular physical activity. Energy density (kcal g^{-1}) of foods is an important determinant of energy intake in a meal or over the course of the day. By substituting sugar for low calorie sweeteners, it is possible to lower the energy density of foods and drinks offering an easy method of reducing calories while maintaining the palatability of the diet. As such, low calories sweeteners can play a helpful role in assisting the achievement of weight maintenance or weight loss, as part of a balanced diet [3].

The low-calorie sweeteners should have the following characteristics: have a similar taste profile as sucrose, noncaloric at normal usage levels, noncariogenic, safe, natural in origin, commercially available at a competitive price, easy to use, stable under a range of processing and usage conditions, inert and compatible with a wide range of food ingredients, stable on storage, provide some bulking effect and mouthfeel, biodegradable, etc.

The intense sweeteners currently authorized in Europe comprise 11 compounds of various chemical natures [4–7], while nine high-intensity sweeteners are FDA-approved as food additives in the United States [8]. They are used in the formulation of foods, beverages, and as excipients in pharmaceutical industry, essentially for their sweetening role but also for their technological properties (stabilizers, texturizers) (Tables 1 and 2).

Regulation 1129/2011 [4–6] includes the Union list of food additives approved for use in foods and conditions of use:

- The name of the food additive and its E number. The "E number" refers to Europe and shows that the additive is regarded as safe in Europe. In effect, the E is a guarantee of safety. Food additives must be included either by name or by an E number in the ingredient list.
- The foods to which the food additive may be added (at the maximum intended use level or the maximum permitted level): dairy products, edible ice, fruit and vegetables, confectionery, cereals, meat, fish, table-top sweeteners, beverages, etc.

Table 1 Discovery date, start date use, ADI (Acceptable Daily Intake), and regulatory status of sweeteners

Sweetener	N° EFSA	Start date-use EU	Start date-use FDA	Times sweeter than sucrose	ADI EFSA (mg kg^{-1}) BW	[a]ADI FDA (mg kg^{-1}) BW	Regulate status FDA
AcesulfameK	E 950	1983	1988	130–200 ×	0–9	0–15	21 CFR 172.800
Aspartame	E 951	1983	1981	200 ×	0–40	0–50	21 CFR 172.804
Cyclamates	E 952	1954	NA	30–50 ×	0–7	0–11	
Saccharins	E 954	1887	1879	300–500 ×	0–5	0–15	21 CFR 180.37
Sucralose	E 955	2000	1998	600 ×	0–15	0–5	21 CFR 172.831
Thaumatin	E 957	1985	1988	2000–3000 ×		NS	GRAS
Neohesperidine DC	E 958	1988	NA	400–600 ×	0–5		
Steviol glycosides	E 960	2011	2008	200–400 ×	0–4	0–4	GRAS
Neotame	E 961	2000	1998	8000 ×	0–3	0–3	21 CFR 172.829
Salt of aspartame-acesulfame	E 962	1994	NA	350 ×	0–15		
Advantame	E969	2014	2014	20,000 ×	0–5	0–5	21 CFR 172.803
Luo Han Guo SGFE		NA	2009	100–250 ×	–	–	GRAS
Alitame	E 956	NA	NA	2000 ×	–	0–1	

NS means not specified, *NA* not approved, *BW* body weight, *CFR* Code of Federal Regulations, *GRAS* generally recognized as safe
[a]ADI established by the Joint FAO/WHO Expert Committee on Food Additives (JECFA)

Table 2 Metabolic, physiological properties, and stability of low-calorie sweeteners

Sweetener	Metabolic and physiological properties	Stability
Acesulfame K	Not metabolized by the human body and excreted unchanged	Heat stable, suitable for cooking and baking. Readily soluble
Aspartame	Digested like other proteins to its components all of which occur in the diet in greater quantities Metabolized into aspartate, phenylalanine, and methanol	Loses sweetening properties when exposed to high temperature therefore not recommended for baking. Can be added to foods at the end of cooking cycle
Cyclamates	Generally not metabolized and excreted unchanged	Good stability at high and low temperature can be used in cooking and baking. Good solubility
Saccharins	Not metabolized by the human body and excreted unchanged	Can be used in cooking and baking Heat stable
Sucralose	Not metabolized by the human body and excreted unchanged	Good stability in very high temperature can be used in cooking and baking. Good solubility
Thaumatin	Although it is caloric, its contribution is negligible at concentration used	Heat stable, soluble in water
Neohesperidine DC	Metabolisms carried out by intestinal microflora. 90% of material is excreted in the first 24 h, primarily in urine	Low solubility in water at room temperature. Solubility increases with temperature. Stable at room temperature
Steviol glycosides	Are broken down to steviol in the gut. Steviol is excreted in the urine as steviol glucuronide	Steviol glucosides are heat stable
Neotame	Rapidly but not completely absorbed. Absorbed neotame is completely excreted in urine and feces	Better stability than aspartame Solubility is high in ethanol and sufficient in water and increases with increasing temperature. Stable in dry form and an liquid form over a wide range of pHs and temperatures
Salt of aspartame-acesulfame	As it dissociates into its component parts on dissolution without the presence of potassium, exposure is actually to aspartame or acesulfame and it presents no new toxicological issues	Stability on dissolution is the same as for individual components
Alitame	7–22% is excreted unchanged in feces. The remainder is hydrolyzed to aspartic and alanine amide	Heat stable at temperatures normally used for food additives
Advantsame	Rapidly but poorly absorbed and the main excretion route is via feces	Stable under normal storage conditions

– The conditions under which the food additive may be used.
– Restrictions on the sale of the food additive directly to the final consumer.

Each low sweeteners used in food and drink production has its own unique taste profile, technical characteristics, and benefits. Low sweeteners can be used alone or in combination with each other as a blend. It is possible to set the taste of sweetness to the demands of a products and consumer taste, while taking into account factors such as stability and cost.

The Acceptable Daily Intake (ADI) is a guideline quantity that represents the amount of low calorie sweetener that can be safely consumed on a daily basis throughout a person's lifetime without any health problems. Joint FAO (Food and Agricultural Organization, on de UN) WHO (World Health Organization) Expert Committee of Food Additives (JECFA) introduced the concept of the ADI for the safety regulation of all food and drink additives in 1961. Other international scientific authorities such as US-FDA (Drug and Food Administration of United States), EFSA (European Food Safety Authority), and ISA (International Society of Automation) use the same method of deriving the ADI independently, guaranteeing consistency of food safety worldwide.

ADI "not specified" is applicable to a food substance of very low toxicity in which, on the basis of the available data (chemical, biochemical, toxicological, and other), the total dietary intake of the substance arising from its use at the levels necessary to achieve the desired effect and from its acceptable background in food does not represent a hazard to health.

GRAS (**G**enerally **R**ecognized **A**s **S**afe) notices have been submitted to FDA for three types of high-intensity sweeteners, thaumatin (extracted from the arils of the fruit of *Thaumatococcus daniellii*), certain steviol glycosides obtained from the leaves of the stevia plant (*Stevia rebaudiana* Bertoni), and extracts obtained from *Siraitia grosvenorii* (Swingle fruit), also known as Luo Han Guo or monk fruit [8, 9], on the priority list of substances proposed for evaluation.

Furthermore, EFSA is reevaluating all food additives by 2020. Over the past few decades the European Union (EU) has put in place a robust food safety system that helps to ensure that consumers are protected from possible food-related risks. Many sweeteners were approved more recently and are scheduled for review after 2015. EFSA can also reprioritize a food additive in light of new information; for example, safety of advantame [10], the deadline for the artificial sweetener aspartame was brought forward from 2013 to 2020 due to concerns raised regarding recent studies [11]. The reevaluation of all approved sweeteners listed in Directive 94/35/EC [12] shall be completed by 31 December 2020 [13].

The metabolic and physiological properties and stability of low-calorie sweeteners are shown in Table 2 [3, 9, 14, 15].

The influence of low-calorie sweeteners on hunger, satiety, and energy intake has been addressed in many laboratory studies and reviews [3, 16]. While the use of low calorie sweeteners does not, in itself, result in a rapid weight loss, it may promote long-term dietary compliance by improving the diversity, variety, and the overall palatability of a reduced energy diet. Several studies have examined the acute effects of low sweeteners on hunger and food intakes. They concluded that replacing sucrose by low calorie sweeteners in foods or drinks does not increase food intake or hunger in children nor has shown to increase food intake in normal weight or overweight men and women. Studies with adults have reported either unchanged or reduced motivation to eat, regardless or whether the low calorie sweetener was delivered in a solid or liquid medium. Mattes and Popkin [17] indicate that the substitution of low-calorie sweeteners for a nutritive sweetener generally elicits incomplete energy compensation, but evidence of long-term efficacy for weight management is not available. The addition of low-calorie sweeteners to diets poses no benefit for weight loss or reduced weight gain without energy restriction. They have concluded that the available evidence either refuted or was insufficient to refute or support each of these potential mechanisms or hypotheses for low-calorie sweeteners increasing appetite, hunger, or energy intake. The Academy of Nutrition and Dietetics updates its position low calorie sweeteners allow a more versatile approach to weight management and may even encourage compliance with a diet [18]. Even modest amounts of weight loss have been shown to contribute significantly to a reduction in risk associated with obesity and overweight, such as diabetes and heart disease. In 2012 the American Heart Association (AHA) and the American Diabetes Association (ADA) [19] issued a joint scientific statement on low calorie sweeteners and their potential usefulness in helping people achieve and maintain a healthy body weight and help people with diabetes to control their glucose level. Other authors [20] indicate that they do not raise blood glucose levels and can be used to control weight and to treat hypoglycemia. If they are used excessively, they can increase weight, promote obesity, and can cause impairment of normal metabolic responses. However, conclude that artificial sweeteners should be used in a limited amount. Moreover, use of natural sweeteners should be increased.

The opinion of the French Agency for Food, Environmental and Occupational Health & Safety (ANSES) [21] on the assessment of the nutritional benefits and risks related to intense sweeteners indicates there is sufficient scientific information to support the claims that intense sweeteners as all sugar replacers lead to a lower rise in blood sugar levels after meals and maintain tooth mineralization by decreasing tooth demineralization again if consumed instead of sugars.

However, EFSA's experts [22] could find no clear cause and effect relationship to justify the claims that intense sweeteners when replacing sugars maintain normal blood sugar levels or maintain/achieve a normal body weight. After an analysis of all of the scientific literature, it appears that, despite a large number of studies, the data are insufficient to determine any long-term nutritional benefits related to the consumption of products containing low-calorie sweeteners as sugar substitutes. ANSES and Bellisle [16, 21] particularly stress the lack of relevant conclusive studies on the supposed benefits of artificial sweeteners, in the context of their broad, long-standing use in nutrition, although the available data do not show any risks related with their consumption.

On the other hand, it seems that the abuse of artificial sweeteners is not convenient, especially in children, because it could persistently alter sweet preferences, leading to enhanced intake of sugars throughout adulthood. Also it could interfere with learning of basic relations between sweet tastes and the delivery of calories, which in turn could negatively affect regulation of metabolic processes and could alter the composition of the gut microbiota, which in turn can contribute to metabolic dysregulation [23–25]. Artificial sweeteners should be used in a limited amount; their excessive use can increase weight, promote obesity, and cause impairment of normal metabolic responses [20].

The determination of low-calorie sweeteners is of great importance not only to identify and quantify their content in foods but also because due to increase consumption; they can be found in the environment and are considered emerging pollutants. They are predominately used in the food industry for the production of sugar-free low calorie foodstuffs, to replace sugar, and to enhance the taste of some personal care products, such as toothpaste as well as drugs and sanitary products. Thus, there are a variety of beverages, foods, and food supplements in which artificial sweeteners are present, so it is necessary to provide validated analytical methodology for their detection and quantification. The need to monitor the use and consumption in different countries is required to know if intake levels frequently exceed the advised acceptable daily intake (ADI).

Artificial sweeteners are highly consumed with increasing trends in consumption. Excretion after human consumption (not metabolized by the human body and excreted unchanged (Table 2)) is undoubtedly a major source of artificial sweeteners in the environment, but it is surely not the only one. Sweeteners have been detected worldwide in a variety of environmental media; nevertheless, monitoring of their presence is still not required by any existing regulations [26]. Its current presence in wastewater has made that some sweeteners are regarded as high-priority emerging contaminants.

Acesulfame, saccharin, cyclamic acid, and sucralose are artificial sweeteners with very limited metabolism in the human digestive system [3, 14, 27]. Therefore, they pass through the human body virtually unchanged, ending up in wastewaters. Artificial sweeteners were ubiquitously present in wastewater-contaminated surface waters. As very water-soluble and relatively persistent compounds they can also persist through conventional water treatment, thus ending up in drinking water.

In recent years, several reviews were published in relation to presence of artificial sweeteners in foodstuffs [28] and as recognized class of emerging environmental contaminants [26, 29]. To avoid major overlap with previously published reviews, we focus our attention on the analytical strategies from the point of view of their presence in foods, pharmaceuticals, and environmental matrices published in the last years, with special emphasis on discussion of the different aspects about their determination. The main objective of this chapter is to present the advances in sample preparation methods used for liquid chromatography-mass spectrometry (LC-MS) analysis of artificial sweeteners in foodstuffs, environmental matrices, and pharmaceuticals. A schematic diagram of sample preparation steps and analytical techniques used in artificial sweeteners analysis is shown in Fig. 1.

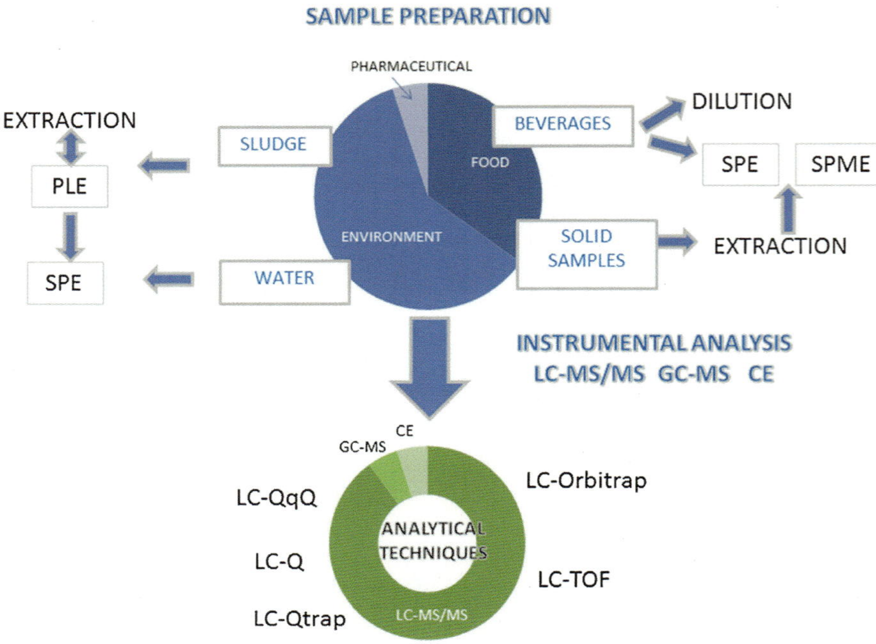

Fig. 1 Main sample preparation steps and analytical techniques used in artificial sweeteners analysis

2 Artificial Sweeteners Trace Analysis Techniques

Environmental samples and foods are complex mixtures of different components contained in varying amounts, making analysis a challenging task. As a consequence, the analysis of sweeteners is important in the evaluation of food safety [12, 30], including the potential detrimental effects on human health. Therefore, reliable analytical methods are needed to control the levels of sweetener in several foodstuff matrices, pharmaceuticals, and aquatic environment samples [28, 29]. Due to the complexity of the food-based or sewage matrices, sample preparation is a previous step before the final analysis. Solid phase extraction (SPE) has proven to be a useful and common sample preparation technique for the determination of artificial sweeteners. It is compatible with many chromatographic techniques, allowing preconcentration and clean-up of samples. Liquid chromatography is the most widely used technique for the determination, combined with mass or tandem mass spectrometry detection (MS or MS/MS).

2.1 Sample Preparation Strategies

The presence of sweeteners in foods, environment matrices, and pharmaceuticals at different concentrations levels led to the development of numerous analytical procedures for their determination. Sample preparation technique depends on the matrix and must include the compounds extraction, clean-up, and concentration steps. In simple matrices, as drinks, the dilution of the sample and direct analysis are suitable. In matrices more complex as food or environmental, extraction and additional sample clean-up to eliminate matrix constitutes is usually necessary to obtain appropriate liquid chromatographic separation and tandem MS detection. Solid phase extraction (SPE) is a well-established technique for the preconcentration and clean-up of target compounds from aqueous samples. The target analytes are transferred from the matrix to a solid sorbent to be retained by different mechanisms in the retention stage and after selecting the suitable solvent they are eluted from the sorbent in the elution step.

The applicability of various SPE sorbents for isolation of sweeteners was investigated by Zygler et al. [31] using several types of LC–MS compatible buffers. Furthermore, in SPE selective highly extractions when molecularly imprinted polymers are used as sorbents can be achieved [32]. Critical evaluation of the procedures described in this chapter involves the matrix type and the more appropriate sample preparation method for the corresponding instrumental technique.

2.1.1 Isolation and Preconcentration of Artificial Sweeteners from Beverages and Processed Foods

Sweeteners are commonly used in various types of beverages (soft drinks, wines, fruit beverages, fermented milk drinks) and processed food as candies, chewing gum, yogurt products, jams, pickles, canned fruits, dried fruits, various sauces, dehydrated soups, jellies, and bakery products [28] for maintaining characteristics of aliment low-calorie and food quality.

Generally, the beverages contain considerable amount of different sweeteners which have good solubility in water. The carbonated drinks are degassed in an ultrasonic bath, in order to remove the carbon dioxide gas that might be present in the beverage, and then the samples are simply diluted 50- to 100-fold with deionized water [33–35], Milli-Q water [36, 37], methanol: water (8:92, v/v) [38] or (1: 1 v/v) [39], mobile phase [40] or 500 times with 0.1% aqueous formic acid [41]. Instant products as nectars, juice, instant pudding, and sauces were solubilized in water using an ultrasonic bath for 15 min and centrifuged to separate solids [36] or add trichloroacetic acid for the precipitation of proteins, before the centrifugation and dilution [41].

In matrices more complex containing milk or solids, the sample was purified by SPE using an Oasis HLB cartridge, after dilution [42]. A variety of methods involving pre-treatment by solvent extraction with diethyl ether and methanol to remove oil components were described [43]. Shah et al. [44] used 0.075% formic acid adjusted to pH 4.5 with *N*,*N*-diisopropylethylamine to dissolve yogurt samples that were shaken using a digital Vortex for 30 min. The resultant supernatant was cleaned-up using a 3 mL C18 cartridge conditioned with MeOH and extraction buffer. The isolated analytes were eluted with methanol. A liquid-liquid extraction method combined with SPE was developed for the determination of neotame in nonalcoholic beverages [45]. The sample was extracted with formic acid/

triethylamine/ultrapure water (4:125:5000 v/v/v) in an ultrasonic bath and the supernatant was loaded into C18-SPE cartridge and NEO was eluted with methanol. Zygler et al. [46] determined nine high-intensity sweeteners in a variety of food samples (i.e., beverages, dairy, and fish products). The samples were extracted using a buffer composed of formic acid and N,N-diisopropylethylamine at pH 4.5 in ultrasonic bath during 10 min. The obtained extracts were cleaned-up using Strata-X 33 μm Polymeric SPE column.

The isolation of sweeteners from beverages was also performed using microextraction methods. Hashemi et al. [47] used a headspace single-drop microextraction (HS-SDME) method to determine cyclamate by gas chromatography. The procedure is based on the reaction of cyclamate with nitrite in slight acidic medium and extraction of cyclohexene formed in a microdrop for direct injection into the gas chromatograph. Several extracting solvents such as toluene, benzene, propanol, 1-butanol, benzylalcohol, xylene, and n-dodecane were investigated for extraction of cyclohexene. The use of 2.5 μL of n-dodecane gave the best extraction efficiency, integrating sample clean-up, preconcentration, and sample introduction into one step.

Solid phase microextraction (SPME) was used for the selective isolation of acesulfame. Moein et al. [48] developed a sol–gel based molecularly imprinted polymer nanofiber by electro-spinning technique on the surface of a stainless steel bar. The fiber was applied for online selective SPME and determination of acesulfame coupled with HPLC.

2.1.2 Isolation and Preconcentration of Artificial Sweeteners from Environmental Matrices

Since the first reports documenting the high production, consumption, and widespread occurrence of artificial sweeteners in the aquatic environment, they are considered as priority emerging contaminants [49, 50]. Analytical methods currently used for the determination of SWTs in food and environmental matrices have been reported, and several critical reviews have recently been published [28, 29]. The following three topics were focused: (1) overview of analytical methods for trace analysis, (2) occurrence in the aquatic environment, and (3) advanced treatment processes for the removal of artificial sweeteners. In addition to these reviews, other reviews have appeared describing certain aspects related to the presence of some particular sweetener such as sucralose [51] detected in municipal effluents and surface waters in the United States and Europe, epidemiological studies concerning the use aspartame [52] as low-calorie sweetener, and stevia [53] highlighting its remarkable potential as an intense high-potency sweetener.

Environmental Water and Wastewater

The concentrations of SWTS in environmental water and wastewater were reported in the order of μg L^{-1} which requires a method detection limit (MDL) at the ng L^{-1} level [28, 29]. Several methods for the analysis of environmental contaminants based con LC-MS/MS were developed by direct analysis using large-volume injection (LVI) or concentration by SPE [54]. LVI is an analytical technique that is performed by the direct-injection of a large sample volume onto the liquid chromatographic column and only requires centrifugation or filtration of water samples. Wu et al. [55] integrated the large volume injection approach (500 μL injection) with ultra-high performance liquid chromatography-tandem mass spectrometry (UHPLC-MS/MS) for quantification of acesulfame and sucralose, with MDL of 0.2 and 5 ng L^{-1}, respectively. Similarly, Berset et al. [56] developed a DI-HPLC–MS/MS method using an injected volume of 100 μL for quantitative determination of acesulfame, cyclamate, saccharin, sucralose, neohesperidine, neotame, aspartame, and its metabolite diketopiperazine in wastewater effluents, surface waters, groundwater, and tap water samples.

SPE is widely accepted as extraction and clean-up method to isolate sweeteners from waste water and surface water. The typical SPE phases used were evaluated in different publications [26, 57–61]. Scheurer et al. [57] tested different cartridges: BakerbondTM styrol-divinylbenzene (SDB 1), Isolute® ENV+, C18, Bond Elut® PPL, Strata™ X, Strata™ X-AW and and Oasis® HLB, WAX, MAX, and MCX. This work concluded that the styrol-divinylbenzene phase SPE columns were the most effective for removing the matrix effect. For five of seven analytes under investigation, method recoveries >75% were obtained for tap water (50 mL), adjusting the sample at pH 3 prior to loading, with polymeric sorbent SDB 1 cartridges, which were eluted using methanol. Zygler et al. [26] also investigated the applicability of several commercially available C18-bonded silica, phenyl-bonded silica, and polymeric SPE sorbents for isolation of nine intense sweeteners in relation to the composition and pH value of buffers used. In this study, very high recoveries (better than 92%) for all studied compounds were obtained using formic acid–*N*, *N*-diisopropylethylamine buffer adjusted to pH 4.5 and C18-bonded silica sorbents. Other polymeric sorbents as Strata-X polymeric RP and Oasis HLB were found to be suitable for quantitative extraction of sweeteners from aqueous solutions. But, in this work real aqueous matrices were not evaluated. Three different approaches of polymeric SPE sorbents (reversed-phase, weak and strong anion-exchange mixed-mode) have been compared for the determination of six artificial sweeteners in environmental waters [58]. The reversed-phase sorbents (Oasis® HLB and Strata™ X) showed the best performance. Oasis®HLB provided recoveries between 73% and 112% and limits of quantification of 0.01–0.5 μg L^{-1} when river water and effluent and influent wastewater were tested. Oasis® HLB cartridges (500 mg) also were used to extract seven SWTs from influent waters, both secondary and tertiary effluent and river samples [59, 60]. The SPE conditions were optimized to enable the direct injection of the organic extract into the hydrophilic interaction liquid chromatography (HILIC) system. The samples were adjusted to pH 3, loaded, and the phase was subjected to a washing step with H_2O (pH = 3) to remove salts and highly polar compounds. Finally, the analytes were eluted with a mixture of NH_4OH:MeOH:ACN (1:4:15). Contrary, when ten commercial SPE cartridges were tested by Gan et al. [62], satisfactory recoveries (77–99%) for seven SWTs were obtained using a Poly-Sery PWAX cartridge with 25 mM sodium acetate solution (pH 4) as wash buffer and methanol containing 1 mM tris (hydroxymethyl) amino methane (TRIS) as elution solvent versus relative lower recoveries obtainer with

Waters Oasis® WAX. The reason maybe the difference in the structure of the sorbents, CNW® Poly-Sery PWAX is packed with amino functionalized styrene/divinylbenzene copolymer, a packing different to Oasis® WAX. This method was successfully applied to the analysis of wastewater, tap water, surface water, and groundwater [61]. Artificial SWTs were simultaneous determined with 24 pharmaceuticals and personal care products (PPCPs) and endocrine disrupting chemicals (EDCs) by SPE [63]. Five cartridges were tested for simultaneous extraction of all target analytes. Good recoveries ($\geq$70%) were observed for all compounds when extraction was performed using Chromabonds® HR-X (500 mg, 6 mL) cartridges under acidic condition (pH 2).

Numerous studies have demonstrated the presence of artificial sweeteners in domestic wastewater, surface water, and groundwater [28, 64–66]; therefore, sweeteners are considered as ideal indicators for the investigation of wastewater contamination in water supplies. Oasis®WAX was the phase selected to investigate the use of two artificial sweeteners, cyclamate and acesulfame, as an indicator set for contamination by wastewater within the rural catchment of a karst spring [67]. The combination of rather persistent acesulfame with more degradable cyclamate can provide valuable additional information on the origin and/or the age of contamination. Acesulfame, saccharin, and cyclamate were selected for Roy et al. [68], and they were detected in leachate or leachate-impacted groundwater being saccharin the dominant sweetener and cyclamate was detected less frequently. Also Oasis®WAX was selected to extract sucralose, acesulfame, saccharin, and cyclamate from boreal lakes and rivers [27] finding concentrations similar to previously reported in European surface waters.

Alternatively, SPE is a technique that has the possibility to operate in mode *online* coupling to liquid chromatography. An *online* SPE system was used for the determination of sucralose in reclaimed and drinking waters [69, 70]. 10.0 mL of sample was injected into a 10.0 mL loop and then loaded onto a SPE column (HyperSep Retain PEP) by the loading LC pump, followed by a wash step to remove interferences. The sucralosa was retained in the SPE column and the matrix that is not retained during the extraction process was directed to waste. After 5.3 min, when the valve was switched to Inject Position, the solvent flows through the SPE column to chromatographic system. The use of an orbitrap HRMS detector in combination with a fast and robust online SPE preconcentration methodology that does not require sample pretreatment is proposed as an alternative for ultra-trace quantitation of sucralose in environmental aqueous samples [71]. The *online* SPE column was a Hypersep Retain PEP® (20 mm $\times$ 3 mm, 1.2 μm) presenting high sensitivity and selectivity with method detection limits (MDL, 1.4 ng L^{-1}) which are lower than any MDL reported in the literature.

Sludges

Sludge originates from the process of treatment of waste water. Due to the physical-chemical processes involved in the treatment, the sludge tends to concentrate heavy metals and poorly biodegradable trace organic compounds as well as potentially pathogenic organisms present in waste waters [72]. The Sewage Sludge Directive

86/278/EEC seeks to encourage the use of sewage sludge in agriculture and to regulate its use in such a way as to prevent harmful effects on soil, vegetation, animals, and man [73]. The artificial SWTs are trace organic which are present in sewage sludge as was reported for sucralose in the Swedish Screening Program 2007 [74, 75]. Sewage sludge samples were treated with diluted HCl, the suspension was centrifuged and the liquid phase collected to be processed by solid phase extraction on Oasis® HLB and the extract was cleaned by passing through a mixed-mode ion exchange SPE-cartridge (Isolute-MM). From the first data obtained, these studies concluded that sucralose was not significantly accumulated in sewage sludge. Pressurized liquid extraction (PLE) followed by liquid chromatography–tandem mass spectrometry was proposed for the simultaneous determination of sweeteners in sewage sludge [76, 77]. A procedure based on PLE using water followed by SPE was optimized for the determination of six artificial SWTs in sewage sludge [76]. The extraction cell was filled with the sample and sand and two cellulose filters placed on the bottom and the top. The extraction was carried out with aqueous formate buffer (pH 3.5) at 80 °C during a single static cycle of 21 min. Finally, the water extract was concentrated by SPE following a procedure for wastewater samples, previously described [58]. Acesulfame, cyclamate, saccharin, and sucralose were found in the samples at concentrations ranging from 17 to 628 ng g^{-1} dw. The PLE optimized conditions by Arbeláez et al. [77] were: MeOH:water (1:1 v/v) as the extraction solvent, 5 min preheating period, 40 °C, extraction pressure of 1500 psi, a static period of 5 min in one cycle, flush volume of 40% of the cell volume, and nitrogen purge time of 90 s. Due to the high matrix effect present in the sample, a clean-up with C18 in-cell was proposed and the extract obtained was subjected to another SPE with Oasis® HLB. The method was successfully applied and of the eight compounds, five were determined in all of the samples analyzed, with acesulfame and saccharin being recorded at the highest concentrations of up to 481 and 591 μg kg^{-1} (dw), respectively.

A study to evaluate the mass loadings, removal efficiencies, and environmental emission of sucralose, saccharin, aspartame, and acesulfame was based on the concentrations measured in wastewater influent, primary effluent, effluent, suspended particulate matter (SPM), and sludge collected from two wastewater treatment plants (WWTPs) in the Albany area of New York State [78]. The SPM and sludge samples were extracted with 6 mL of methanol/water mixture (5:3 v/v) using an ultrasonic bath. Extracts were centrifuged and purified by passage through Oasis®HLB cartridges. Aspartame and saccharin were significantly removed from wastewater treatment plants; however, sucralose and acesulfame were removed at <2.0%. This procedure to extract sludge samples was adapted for Subedi et al. [79] with the modification of the used cartridges, Sep-pak® Vac C18 to purify the extract. The media concentrations of SWTs in sludge from domestic WWTPD and mixed WWTPM (domestic plus industrial) were on the order of aspartame > saccharin > acesulfame > sucralose. These results were compared with those reported in the USA [78], which were sucralose > aspartame ≈ saccharin > acesulfame. The differences in the patterns of SWTs in sludge in the USA and Korea suggest differences in per-capita consumption of individual artificial SWTs and the removal in WWTP treatment processes. Soil samples and dust samples were extracted with 25 mL of Milli-Q water buffered at pH 4 [80] and then supernatant was concentrated using a SPE protocol as described above for water samples [62]. Saccharin, cyclamate, and acesulfame were the dominant artificial SWTs in both gas and particulate phase, with concentrations varying from 0.02 to 1940 pg m^{-3}.

2.1.3 Isolation and Preconcentration of SWTs from Pharmaceuticals

Artificial sweeteners are used as excipients in the pharmaceutical industry in different pharmaceutical formulations [81]. They are added to chewable tablets and liquid preparations to mask the unpleasant taste of the medicament at concentrations regulated by FDA and by the Joint Food and Agriculture Organization of the United Nations (FAO)/World Health Organization (WHO) Expert Committee on Food Additives (JECFA) that attributed an Acceptable Daily Intake (ADI) for each SWT [82]. The development of analytical methods for the simultaneous quantification of active ingredient and excipients in pharmaceutical formulation is required for quality control purpose and also counterfeit drug control.

The presence of sweeteners and other substances in 73 pharmaceutical preparations of 35 medicines for oral administration were evaluated according to drug labeling information of the excipients [83]. No data were reported in relation to analytical methodology for sample dissolution. The artificial sweetener sodium saccharin was found in 38.3% of samples. Aspartame, cyclamate, acesulfame-K, and saccharin were determined in food products and pharmaceutical samples by CZE [84, 85] and diet supplements [86]. The liquid samples were appropriately diluted with water and the solid samples were weighed, solubilized, and diluted appropriately in water. The validated method was suitable to quantification of all analytes for the routine analysis.

Molecularly imprinted polymer (MIP) for aspartame has been prepared to selectively and specifically separate aspartame from aqueous solution and various pharmaceutical samples which contain aspartame as the major constituent [87]. The MIP was prepared using *N*-(2-ammonium-ethylpiperazinium) maleimidopropane sulfonate copolymer bearing zwitterionic centers along the backbone via a surface confined grafting procedure. The quantitative aspartame recovered from the sample was 88%.

2.2 Liquid Chromatography Mass-Spectrometry Methods for Quantification

There are analytical methodologies available to determine sweeteners in different matrices such as food, drinks, dietary products, drugs and sanitary products, and environmental samples. Analytical methods are faster, highly efficient, accurate, and sensitive for the identification and quantification of sweeteners in foodstuffs due to advances in the development of chromatographic instrumentation and mass spectrometric analyzers. High performance liquid chromatography (HPLC)–MS is a unique tool for reliable characterization of complex mixtures. Its excellent figures of merit are a consequence of the combination of the separative power of HPLC to the power of MS to identify molecular structure. HPLC is the most popular method using reversed phase (RP) columns based on C18 or C8 stationary phases [26]. Usually, the mobile phase combines water and (methanol and acetonitrile) using buffers such as ammonium acetate, acetic acid, or formic acid to search a compromise between chromatographic separation and electrospray ionization (ESI) sensitivity [26].

Chromatographic separations may be faster achieved by increasing the flow rate of the mobile phase, by decreasing the length of the column, reducing the stationary phase particle diameter, or increasing the temperature of analysis. The emergence of ultra high performance liquid chromatography (UHPLC) in 2004 (termed UPLC by Waters Corp., Milford, MA) makes possible to use columns packed with sub 2 μm particle size that allow good resolution and short analysis time [88]. Recent developments in special phases that promote polar compound retention are being evaluated, namely bridged ethylsiloxane/silica hybrid (BEH) C18 (1.7 μm) [42] and high strength silica (HSS) T3 C18 (1.8 μm) [27], both columns incorporated trifunctional ligand bonding chemistries on the particles, Kinetex® C18 (1.7 μm) based on core-shell silica support [38] and Hypersil GOLD™ C18 (1.9 μm) based on high purity silica technology with a proprietary bonding and endcapping procedure [36], were evaluated. The BEH chemistry utilizes new endcapping processes that ensure good peak shape for basic analytes. The HSS T3 C18 bonded phase was fabricated to retain and separate small water-soluble polar organic compounds. By nano-structuring technology, a durable, homogeneous porous shell is grown on a solid silica core to create a core-shell particle that produces less band broadening and high efficiencies compared to fully porous particles in RP separations. Hypersil GOLD columns provide outstanding peak symmetry.

The column temperature increase results in enhanced efficiency, mass transfer, and linear velocity [88]. However, high temperatures are not used routinely in the analysis of sweeteners. As an example, the analysis of sweeteners by LC-MS/MS in the range of 110–150 °C was proposed using a Shodex ETRP1 column (4 μm) [37].

Various traditional detectors are used in combination with liquid chromatography for sweeteners analysis but with complex matrices such as food and environmental samples, mass spectrometry has become the technique of choice in order to ensure selectivity and confirmation of target analytes. Several papers have demonstrated the utility of HPLC in combination with ultraviolet (UV) detection for acesulfame-K (ACS-K), aspartame (ASP), and saccharin (SAC) in commercial soft drinks [89] and mixtures of ASP and ACS-K in artificial sweeteners using partial least square (PLS-2) multivariate calibration [90]. Fast and high resolution LC methods were achieved by recent developments in UHPLC techniques with Diode Array Detector (DAD) and Rapid Resolution High Definition (RRHD) columns, such as the screening of food additives, including three sweeteners (ACS-K, ASP, and SAC) [91, 92]. Kailasam describes an application using a ZORBAX Eclipse Plus C18 (1.8 μm) column to separate nine additives in soft drinks and colas within a run time of 1 min even with 1 μL injection volume [91]. The separation of five additives in soft drinks and coffee sweeteners, described by Pedjie, was carried out using a

Restek® Pinnacle® DB C18 (3 μm) column with a run time of 3.5 min and 4 μL injection volume [92]. Furthermore, the separation by UPLC-DAD of ACS-K, ASP, SAC, cyclamate (CYC), and neotame (NEO) from food was affected by the variables such as pH of the buffer solution, proportion of solvents in the mobile phase, flow rate, and column temperature during the optimization of chromatographic method [36]. For this reason, multivariate central composite design was used for the simultaneous optimization of 13 responses applying the Derringer and Suich desirability function. After optimization, the method was applied with good resolution (except for CYC) and low analysis time (11 min) for 25 samples from 9 food matrices (ready to drink tea, soft drink, nectar, juice, instant juice, instant pudding, jam, tomato sauce, and barbecue sauce). However, UV detection is not suitable for cyclamate (CYC) and sucralose (SCL) because they do not absorb in the UV/visible range due to a lack of chromophore group [28]. Another detector, which has been successfully combined with HPLC, was evaporative light scattering detector (ELSD). The eluent stream passes through a nebulizer into an evaporation chamber, where the solvent is evaporated to leave a mist of tiny sample particles. These scatter a light beam, and the extent of the light scattering is proportional to the amount of sample present. A simultaneous identification and quantification of nine sweeteners, i.e., ACS-K, ASP, CYC, SAC, sucralose (SCL), neohesperidine dihydrochalcone (NHDC), NEO, alitame (ALI), and dulcin (DUL), in beverages, canned or bottled fruits, and yogurts was performed by HPLC-ELSD [93]. ELSD can detect virtually any analyte without the need for chromophores or fluorophores but the analyte should be less volatile than the mobile phase. Separation of all sweeteners was obtained in less than 25 min using an aqueous buffer solution composed of formic acid and trimethylamine ($pH = 4.5$), with an injection volume of 8 μL.

All these detectors qualify substances based on retention time, and they are quantified based on the height or the area of the chromatographic signal. The resolution obtained is high, but the determination and precise quantification of substances can be difficult if multiple components elute at approximately the same time during simultaneous multianalyte analysis. However, LC-MS systems combine outstanding resolution liquid chromatography separation with outstanding capabilities qualitative mass spectrometry. Comparatively, mass spectrometry (MS) is a highly sensitive detection technique that ionizes sample components by various methods, separating the resulting ions in vacuum according to their mass-to-charge ratios and measures the intensity of each ion. The mass spectra obtained can indicate the level of concentration of ions that have a given mass, being very useful for qualitative analysis. When changing from single MS to tandem mass spectrometry (MS/MS), selectivity is greatly enhanced, and therefore improved detection and quantitation limits can be achieved.

It provides a wealth of structural information, and at the same time increases selectivity, which allows identification and quantitation of even co-eluting compounds. But the belief that the importance of chromatographic separation and the sample clean-up is less important because the mass spectrometry provides high selectivity and sensitivity is an error and should be considered during the validation process [94].

The most common soft ionization sources or atmospheric pressure ionization (API) are ESI and atmospheric pressure chemical ionization (APCI). In most of the published studies, LC-MS and LC-MS/MS methods are used for the determination of artificial sweeteners (Table 3). The instruments used were equipped with ESI source which uses electrical energy to assist the transfer of ions from solution into the gaseous phase before they are subjected to mass spectrometric analysis. Most of the studies used negative ESI mode for the mass-spectrometric detection [29]. Neutral and relatively nonpolar molecules, lacking a functional group capable of carrying charge, do not ionize well using ESI. The constant consumption of sample and the susceptibility to ion suppression effects are two important drawbacks that ESI presents. Some MS instruments are equipped with both ESI and APCI sources. The sensitivity of LC-MS analysis can be improved when APCI is used. APCI presents some advantages over ESI, in the ionization of thermally stable polar and nonpolar compounds and decreasing signal suppression for several compounds in different matrices. However, thermal decompositions of labile compounds in the heated nebulization or high demands on solvent purity are some of the weakness of APCI.

2.2.1 Methods Based on LC-MS

Single quadrupole (Q) mass analyzers were used in several applications for sweeteners analysis. A disadvantage of single quadrupole instruments is the high intensity of background signals produced from sample matrix. The matrix effect may either reduce (ionization suppression) or increase (ionization enhancement) the analyte signal. Different ways for ionization suppression/enhancement evaluation were reported due to the need to adequately compensate these effects and internal standard/matrix-matched/standard addition calibration methods were considered [95].

A HPLC/ESI-MS method was developed for the simultaneous determination of seven artificial sweeteners (ACS-K, ASP, CYC, SAC, SCL, NEO, ALI) and one natural sweetener, stevioside, in different food samples [96]. The compounds were quantified using selective ionization recording (SIR) or selected ion monitoring (SIM), where only one of the ions produced in the ion source is focused and detected. The cycle time for a given ratio m/z is very short and many cycles can be performed for a certain time. The *m/z* were 178, 397, 377, 293, 641, 312, 162, and 182 to CYC, SCL, NEO, ASP, stevioside (STV), ALI, ACS-K, and SAC, respectively, using warfarin sodium (SIR *m/z* 307) as an internal standard (IS). ESI was operated in negative ion mode to generate quasimolecular ions $[M - H]^-$. The quantification was based on internal standard calibration to avoid matrix effect. It was observed, the ion suppression by the sample matrix increased significantly for the analytes eluted near the dead time of the column, implying that the strong polar components in the sample matrix intensively suppressed the ionization of the analytes. The limits of detection (LODs) were below 0.10 μg mL^{-1}, whereas the limits of quantification (LOQs) were below 0.30 μg mL^{-1}. The recoveries were in the range of 95.4–104.3%, with relative standard deviation of $<10\%$ for the analytes. Compared to the HPLC-ELSD method [94], the present method had a higher sensitivity and partial structural information.

Table 3 Applications of liquid chromatography coupled with mass spectrometry for the analysis of most popular artificial sweeteners

Analyte	Samples	Analytical technique	Column/temperature	Mobile phase	LOD/LOQ	Recovery	Ref.
ACS-K, ASP, CYC, SAC, SCL, NEO, GA, ALI, DUL, DKP	Beverages	UHPLC-MS/MS (MRM mode) ESI^- and ESI^+	ACQUITY BEH C18 (100 × 2.1 mm, 1.7 μm)/40 °C	Gradient elution with 0.1% formic acid in water and ACN	LODs: 0.002–0.56 μg mL^{-1} LOQs: 0.007–1.9 μg mL^{-1}	94.3–106.1%	[42]
ACS, SAC, CYC, SCL	River and lake water	UPLC–MS/MS (MRM mode) ESI^-	Acquity UPLC HSS T3 (50 × 2.1 mm, 1.8 μm)/40 °C	Gradient elution with 2 mM NH_4OAc in water and MeOH	LOD 1–20 ng L^{-1} LOQ 13–200 ng L^{-1}	52–83%	[27]
ACS-K, ASP, CYC, SAC, SCL, NEO, NHDC	Beverages	UPLC-PDA and LC-MS/MS (MRM mode) ESI^-	UPLC: Kinetex C18 (50 × 2.1 mm, 1.7 μm)/30 °C LC: ZORBAX ECLIPSE C8 (150 × 4.6 mm, 5 μm)	UPLC: Gradient elution with 1 mM phosphate buffer in water/ACN LC: Gradient elution with 20 mM NH_4OAc in water and MeOH	LOQs UPLC-PDA: 10–100 ng mL^{-1} LOQs LC-ESI-MS/MS: 0.05–5 ng mL^{-1}	UPLC-PDA: 95.7–108.4% LC-ESI-MS/MS:90–108.4%	[38]
ACS-K, ALI, ASP, CYC, NEO, NHDC, SAC, SCL STV	Beverages	HTLC-MS/MS (MRM mode) ESI^-	Shodex ETRP1 (150 × 3.0 mm, 4 μm)	Gradient elution with 5 mM NH_4OAc in water/MeOH	LODs: 0.05–10 mg L^{-1} LOQs: 0.17–33.3 mg L^{-1}	86–113%	[37]
ACS-K, ASP, CYC, SAC,	Food (candied and preserved	HPLC/MS (SIR mode) ESI^-	Spherigel C18 (250 × 4.5 mm, 5 μm)/25 °C	Gradient elution with (0.8 mL of formic acid and	LODs < 0.10 μg mL^{-1} LOQs < 0.30 μg mL^{-1}	95.4–104.3%	[96]

SCL, NEO, ALI, STV	fruit, cake, beverages)			1.5 mL of triethylamine in 1 L of water) in MeOH and acetone			
ACS-K, ASP, CYC, SAC, SCL, NHDC, NEO, ALI, DUL	Food (beverage, yogurt, fish product)	HPLC/DAD-MS (SIM mode) ESI^-	Nucleodur C18 Pyramid (250 × 3 mm, 5 μm)/22 °C	Gradient elution with (1.5 mL (20 mmol L^{-1}) of formic acid in 2 L of water) in MeOH and acetone	LODs < 0.25 μg mL^{-1} LOQs < 2.5 μg mL^{-1}	84.2–106.7%	[46]
SCL	Surface water	LC-MS/MS (MRM mode) ESI^-	Restek Ultra aqueous C18 (100 × 2.1 mm, 3 μm), or Hypersil Gold (100 × 2.1 mm, 3 μm)	Gradient elution with 0.1% acetic acid in water and ACN	LOD 10 ng L^{-1}	26–62%	[98]
ACS-K, CYC, SAC, SCL	Wastewater, groundwater	LC-MS/MS (SRM mode) ESI^-	Gemini C18 (150 × 2 mm, 5 μm)	Gradient elution with 1 mM NH_4OAc in MeOH	LODs 2–65 μg L^{-1}	–	[64]
SCL	Water and wastewater	HPLC-MS/MS (MRM mode) ESI^- and ESI^+	XBridge-C18 (150 × 2.1 mm, 3.5 μm) ESI^- Luna C18-HST (100 × 3 mm, 2.5 μm) ESI^+	Gradient elution with 0.2% NH_4OH and ACN	LOQ 100 ng L^{-1}	91%	[69]
ACS-K, ASP, CYC, SAC,	Wastewater	HPLC-MS/MS (MRM	Zorbax Eclipse XDB-C8	Gradient elution with 20 mM	–	>75% except ASP, 41% and NHDC 59%	[57]

(*continued*)

Table 3 (continued)

Analyte	Samples	Analytical technique	Column/temperature	Mobile phase	LOD/LOQ	Recovery	Ref.
SCL, NHDC, NEO		mode) ESI^-	(150 × 4.6 mm, 5 μm)	NH_4OAc in water and MeOH			
ACS-K, ASP, CYC, SAC, SCL, ALI, NHDC, NEO and five common steviol glycosides	ACS-K, ASP, CYC, SAC, SCL, ALI, NHDC, NEO and five common steviol glycosides	HPLC-MS/MS (MRM mode) ESI^- and ESI^+	Ascentis Express C18 (100 mm × 4.6 mm, 2.7 μm)/40 °C	Gradient elution with 0.1% acetic acid in water and MeOH/acetone	LODs 1.08–4.52 ng mL^{-1} LOQs 3.23–13.56 ng mL^{-1}	97.0–105.7%	[40]
SAC, CYC, ASP-K, ACE, NHDC, SCL, STV, GA	River water, wastewater	UHPLC-MS/MS (SRM mode) ESI^-	Ascentis Express RP-Amide (100 × 2.1 mm, 2.7 μm)/25 °C	Gradient elution with acetic acid in water/ACN	LOD 8–40 ng L^{-1} river water LOD 10–500 ng L^{-1} wastewater	71–104% river water 69–108% wastewater	[60]
ACS-K, ASP, CYC, SAC, SCL, NHDC	Wastewater	LC-MS/MS (MRM mode) ESI^-	Luna C18 and Luna HILIC (100 × 2.0 mm, 3 μm)/40 °C	C18 column: Gradient elution with 5 mM NH_4OAc in water and MeOH HILIC column: Gradient elution with 5 mM NH_4OAc in water and ACN/water	C18 LOQs: 0.01–0.5 μg L^{-1}	73–106% river water 75–112% wastewater	[58]
ACS-K, ASP, CYC, DUL, NEO, NHDC,	Food	LC-MS/MS (MRM mode)	Luna Phenyl-Hexyl (150 × 4.6 mm, 5 μm)/35 °C	Gradient elution with 10 mM	LOQs 0.1–0.5 μg g^{-1}	85–120%	[34]

SAC, SCL STV, GA		ESI^- and ESI^+		NH_4OAc in water and MeOH			
SCL	Water	LC-MS/MS LC-QTOF-MS (MRM mode) ESI^- and ESI^+	Zorbax Eclipse Plus C18 (50 × 2.1 mm, 1.8 μm)	Gradient elution with 0.1% formic acid in water/ ACN	$LOD_{LC\text{-}MS/MS}$ 15 ng L^{-1} $LOD_{LC\text{-}QTOF\text{-}MS}$ 400 ng L^{-1}	–	[101]
ACE, SCL	Well water and wastewater	LVI-UPLC-TOF-MS/ MS (MRM mode) ESI^-	Dikma C18 (100 × 3 mm, 3 μm)	Gradient elution with 0.1% formic acid in water and MeOH	LODs: ACE (0.2 ng L^{-1}) and SUC (5 ng L^{-1})	–	[55]
SCL	Drinking and reclaimed water	LC-MS/MS (SRM mode) $APCI^-$	Hypersil Gold PFP (100 × 2.1 mm, 1.9 μm)	Gradient elution with 0.1% formic acid in water/ ACN	LODs: 8.5 ng L^{-1} and 2.7 μg L^{-1} in drinking and reclaimed waters	85–113%	[71]

Zygler et al. [46] proposed a HPLC/MS method for simultaneous determination of nine sweeteners, ACS-K, ASP, CYC, SAC, SCL, NHDC, NEO, ALI, and DUL, in food products (i.e., beverage, yogurt, fish product), including a clean-up step. Methyl derivative of cyclamic acid was used as IS. Quantification of all compounds was performed using SIM mode. ESI in negative ion mode was employed. The most intense signal is obtained from $[M - H]^-$ except for DUL which came from a formic acid–DUL adduct $[DUL + HCOO^-]^-$. ACS-K, SAC, and SCL presented nonlinear response in the considered concentration range. The LODs were below 0.25 μg mL^{-1}, whereas the LOQs were below 2.5 μg mL^{-1}. No matrix effect was observed and external calibration approach has been employed. The recoveries for all sweeteners were in the range of 84.2–106.7%, with relative standard deviation of <10%. Possibly, the decomposition of ASP and NEO dipeptides by proteolytic enzymes released from fish cells during homogenization of the sample, caused losses observed for these two sweeteners.

2.2.2 Methods Based on LC-MS/MS

Often more structured information is needed than has been generated by the ionization method used. This information can be obtained by coupling two analyzers separated by a collision cell (tandem mass spectrometers) [97]. Triple quadrupole (QqQ) are among the most common MS/MS systems operating as tandem in space analyzers. The first analyzer is used to select the compound of interest. Then, this ion goes to the collision cell normally pressurized with an inert gas such as argon. Fragmentation of this ion in the cell is produced by collision induced dissociation (CID). MS/MS acts as a mass filter to selectively monitor a specific molecular ion (selected reaction monitoring, SRM). For quantification, the acquisition of two states selected ion is performed. The first spectrometer is programmed to transmit the first ion of interest to the collision cell, and then one of the product ions is monitored after the second spectrometer. With the selection of a suitable internal standard, the mass spectrometer can be changed rapidly between the four ions (two precursor ions and two associated product ions) and the relative intensities of the product ions monitored. This technique is known as multiple reaction monitoring (MRM) and is able to achieve high levels of specificity. Analytical methodologies based on mass spectrometry for the determination of artificial sweeteners in environmental media were reviewed by Kokotou et al. [26]. Beam-type analyzers are used in tadem-in-space instruments, whereas trapping instruments are classified in tandem-in-time. In a QTrap system the third quadrupole can be switched between ion trap mode and quadrupole mode, so the instrument combines useful features of both triple quadrupole and ion trap analyzers. Ion trap mode allows to enhance sensitivity in product ion scanning and to induce additional fragmentation.

Most MS/MS systems are tandem-in-space instruments such as QqQ and quadrupole time-of-flight (QTOF). The third quadrupole of QqQ can be replaced by a TOF analyzer to produce a QTOF mass spectrometer. This system passes all ions in a pulse and separates them in time but does not scan, unlike QqQ. However, the advantage of the QTOF is the speed in which an MS/MS spectrum can be obtained. High-resolution mass spectrometry (HRMS) is based on full scan data giving the possibility of retrospective analysis of data based on an a posteriori hypothesis, unlike MS/MS. HRMS allows the elucidation of the elemental composition of analytes on the basis of accurate mass and isotopic standards. Recently, the increasing interest in the use of HRMS in the environmental field was observed using hybrid tandem mass instruments, for example, QTOF, QTrap, and the Orbitrap mass analyzer, which offer a combination of the characteristics of the two analyzers. The Orbitrap device consists of a small electrostatic device into which ion packets are injected at high energies to orbit around a central, spindle-shaped electrode. The image current of the axial motion of the ions is picked up by the detector and this signal is Fourier transformed to yield high resolution mass spectra.

Triple Quadrupole (QqQ) System

Recently, artificial sweeteners have been shown to play an important role as emerging contaminants in the aquatic environment, especially SCL. Many of the studies used QqQ MS and negative ESI mode and were summarized in a review [29]. One of them, organized by the European Commission's Joint Research Centre (JRC) evaluated the occurrence of sucralose in European surface waters by LC–MS/MS using negative ESI and triple quadrupole [98]. The sucralose molecule shows the characteristic MRM transition $[M - H]^- \rightarrow [M - H\text{–}Cl]^-$ 395 → 359 under chlorine loss. To avoid strong ion suppression quantification was performed by isotope dilution with deuterated sucralose d6 internal standard. The recoveries obtained are not completely satisfactory varying from $62 \pm 9\%$ in tap water, ~55% for the river water, to ~26% for the wastewater. Since the sweeteners can be quite persistent in WWTPs, the suitability of ACS-K [64] or SCL [69] as chemical markers for domestic wastewater was investigated by HPLC–ESI-MS/MS. Quantification was made using peak area ratios relative to the internal standard and with standard addition to account for matrix effects [64]. SCL proved to be an indicator for the presence of conventional biologically treated municipal and domestic wastewater, septic system sources to water bodies [69], and finished drinking water [99] in the United States. Occurrence data for 85 trace organic compounds was examined using HPLC–ESI-MS/MS with negative and positive ESI modes and two based C18 columns. The sucralose ionization suppression could be largely corrected by dilution, using sucralose d6 isotope dilution, but the recovery for sucralose-d6 was greater than 50%. For this reason, gemfibrozil were used as IS and diluted tenfold before analysis to minimize matrix effects [69]. Minten et al. developed a LC-MS/MS method to assess the efficacy of treatment plants for removing SCL. A special MRM technique where the two quadrupoles monitor the same m/z was applied. The sodium adduct of SCL was used for quantification, because lower detection limits were obtained when it was compared to the sucralose quasi-molecular ion in negative ion mode [100].

APCI represents an alternative ionization source for species in LC analysis, such as SCL, that are difficult to ionize or tend to show low sensitivity in LC-MS/MS analysis. To this end, a (SPE)-LC-APCI/MS/MS was proposed and validated for the determination of SCL at low ng L^{-1} levels in drinking and reclaimed waters from South Florida, United States. [70]. APCI operated in the negative mode showed better sensitivity than ESI, where adducts ions with methanol and inorganic ions are found. The two SRM transitions monitored for SCL were 397 → 361 and 397 → 359 for quantitation and confirmation, respectively. For sucralose-d6 the transitions were 403 → 367 and 403 → 365 for quantitation and confirmation, respectively. Sucralose is frequently detected at significant levels in the aquatic environment, possibly due to its high resistance to photodegradation, minimum sorption, and high solubility. Experiments of the potential of sucralose photodegradation using multiple light sources and water matrices were tested. The findings are in agreement with previous results indicating that sucralose is a good tracer of anthropogenic pollution of waters [69, 99].

Many authors enhanced sweeteners analysis using stationary phases that reduce analysis time or alternatives to offset the matrix effect. Thus, the occurrence of four sweeteners in the aquatic environment from Switzerland was investigated by Buerge et al. [64] using a Gemini C18 (5 μm). ACS-K, CYC, SAC, SCL were determined in ~17 min and Aspartame was not selected in this study because was assumed it is quickly biodegraded in WWTPs [64]. An increase in temperature can have beneficial effects. As the temperature increases, shorter run times can be achieved; the viscosity of the mobile phase decreased, resulting in lower pressures and lower mobile-phase viscosity also improves diffusion in the chromatographic system, giving narrower peaks. A development based on Zorbax Eclipse XDB-C8 (5 μm) at 40 °C with a flow rate of 0.8 mL min^{-1} was carried out for the separation of ACS-K, ASP, CYC, SAC, SCL, NHDC, and NEO [57]. However, no retention time data were provided. Also, Kubica et al. [40] presented a HPLC-ESI-MS/MS for the determination of ACS-K, ASP, CYC, SAC, SCL, ALI, NHDC, NEO and five common steviol glycosides in soft and alcoholic beverages. An Ascentis Express C18 column (2.7 μm) set at 40 °C allows the complete separation of analytes in 16 min. Sensitivity of ASP, ALI, and NEO was increased using positive ESI mode. Using considerable dilution of the analyzed samples, no matrix effects were observed. The demands to process hundreds of samples in a short period of time have resulted LC-MS/MS equipment using reverse phase columns with a particle size smaller. Different chromatographic columns were tested, i.e., Zorbax Eclipse XDB-C18 (1.8 and 5 μm), Kinetex C18 (2.6 μm), Ascentis Express C18 (2.7 μm), Ascentis Express HILIC (2.6 μm), and Ascentis Express RP-Amide (2.7 μm) in the determination of SAC, CYC, ASP-K, ACE, NHDC, SCL, STV, and GA in river water and wastewater [60]. The last column provided the best resolution for the eight sweeteners analyzed with a column temperature of 25 °C in less than 12 min. In another case where two types of chromatographic separation were compared, the reversed-phase provided better performance than hydrophilic interaction (HILIC) in sweeteners analysis of wastewater [58]. The MS/MS analysis showed that predominant precursor ions were $[M - H]^-$ when ESI^- was applied, except for SCL and STV, since their adduct with $Cl^-$$[M + Cl35]^-$was the predominant precursor ion [60]. The matrix effect was evaluated with different clean-up solvents, elution sol-vents, sample volumes, and commercial SPE cartridges. Moreover, ion suppression could not be completely eliminated and two isotopically labeled standards (ASP-d3 and SCL-d6) as surrogates also eight-point matrix-matched calibration curves were used to the compensate the matrix effect and recoveries for each analyte.

The maximum permissible amount of sweeteners in food varies significantly due to their frequent synergistic use therein. Hence the analytical methodology, based on HPLC or UHPLC-MS/MS, has great interest for the control of food quality and application of the regulation. Based on green chromatography approach, high temperature liquid chromatography HTLC–MS/MS method which included a dual temperature and organic modifier gradient was used in the determination of nine sweeteners in drink samples [37]. Temperature in the range of 110–150 °C was applied, whereas the mobile phase is formed by water and percentages of ethanol varying from 3% to 20%. The separation was achieved in less than 20 min (23 min, including column re-equilibration). Also, the advantages of two different instrumental systems (PDA and MS/MS detection) using two different LC systems (UPLC and HPLC) were developed and compared for the control of the content of ACS-K, ASP, CYC, SAC, SCL, NEO, and NHDC in 66 beverage products available on the Spanish market from national and international industries [38]. A reduced analysis cost was obtained with UPLC-PDA that showed other advantages such as good resolution and high sensitivity in a run time of 3 min for five target sweeteners when a Kinetex C18 column (1.7 μm) was used. Also, no matrix effects are reported. LC-ESI-MS/MS method guaranteed the unequivocal determination of all the sweeteners in the same run and confirmed the results obtained by UPLC-PDA. Matrix matched with IS calibration curves were applied to correct matrix effect in MS/MS method. The validation parameters of UPLC-PDA method were successfully compared with other methods reported.

UPLC–MS/MS has demonstrated to be a powerful tool with the ability to transfer existing HPLC conditions directly provided fast chromatographic separations by increasing temperature of column such as in the investigation of the occurrence of ACS, SAC, CYC, and SCL in boreal rivers and lakes [27]. An Acquity UPLC HSS T3 (1.8 μm) set at 40 °C provided a separation in less than 2 min. The analytes were identified by retention times and MRM with specific transitions for each compound. SCL was quantified by means of a chlorine adduct precursor $[M + Cl]^-$, *m/z* 433, which had a higher peak area than $[M - H]^-$ ion, m/z 395. Transition *m/z* 395 > 359 was used as a qualifier. Despite the matrix effect was corrected by the response of the mass-labeled surrogates in the IS calibration, low values of recoveries were obtained.

UHPLC-MS/MS using both negative and positive ESI modes was applied for the simultaneous analysis of ten compounds, namely ACS-K, ASP, CYC, SAC, SCL, NEO, GA, ALI, DUL, rebaudioside A (REB), and diketopiperazines (DKP) which is the main decomposition product of ASP, in various types of beverages sold in Japan [42]. All the compounds were ionized in negative ESI mode, except ALI and DUL, which were ionized in positive ESI mode. Several C18 columns were tested but ACQUITY UPLC BEH C18 column set at 40 °C showed better separation of all the components from highly polar ACS-K to weakly polar GA and excellent peak shapes in less than 7 min of elution time. The matrix effect was negligible because the samples were diluted 500 times. SCL showed the highest LOD and LOQ (0.56 and 1.9 μg mL^{-1}, respectively). Multiple sweeteners (ACS-K, SCL and ASP) were detected simultaneously in two carbonated drinks. ACS-K and SCL were detected in two sport drinks. It is noted that four sweeteners (ACS-K, SCL, ASP, and REB) were added as sugar substituents to low-sugar and low-calorie products.

Hybrid Analyzers (QTrap and QTOF) and Orbitrap HRMS System

Several studies proposed analytical methodologies based on the use of QTrap. A development based on Zorbax Eclipse XDB-C8 (5 μm) at 40 °C with a flow rate of 0.8 mL min^{-1} was carried out for the separation of ACS-K, ASP, CYC, SAC, SCL, NHDC, and NEO [57]. However, no retention time data were provided. All sweeteners were quantified by external standard calibration, except sucralose, which was quantified by internal standard calibration (sucralose-d6) to correct matrix effect. The recoveries for six of the seven tested artificial sweeteners are lower due to ion suppression and not to losses during sample extraction. In order to increase the ionization yield 20 mM Tris (hydroxymethyl) amino methane was added post-column. This strong base facilitates deprotonation of the weakly acidic analytes. Furthermore, reliable information about sweeteners content in foodstuffs is a concern to both consumers and quality control agencies. In this regard, LC-MS/MS method was used for the simultaneous separation of ACS-K, ASP, CYC, DUL, NEO, NHDC, SAC, SCL STV, and glycyrrhizic acid (GA) on an analytical column, Luna Phenyl-Hexyl at 35 °C [34]. Various food matrices including 27 beverages (16 alcoholic and 11 nonalcoholic beverages) and 15 pickled foods (1 pickled pepper, 3 candies, and 11 candied fruits) were analyzed. Negative ESI mode was applied for all sweeteners except for DUL and NEO for which was used positive ESI mode. Matrix effect was corrected by matrix-matched calibration.

Studies of some sweeteners such as SCL or ACS in water used LC-MS/MS based on QTrap operating with negative electrospray ionization in MRM mode. For example, Mawhinney et al. quantified the amount of SCL present in select US drinking water systems [99]. Sucralose was quantified using the $(M - H)^{-}$ precursor ion at *m/z* 395.0 and the Cl product ion at *m/z* 35.0 and confirmed using the $(M - H)^{-}$ precursor ion at *m/z* 397.0 and the Cl^{-} product ion at *m/z* 35.0. The sample integrity during long-term storage was performed on a set of sample extracts stored for 36–48 months at −80 °C. SCL showed to be a relatively inert and stable molecule, very soluble in MeOH, the storage solvent.

Another discussion was the fact that SCL has a complicated chlorine isotopic distribution and a difference in sensitivity about 400 times less than ACE. Therefore, different approaches were used with the aim of enhancing the SCL sensitivity. A study compared a LC-MS/MS with QqQ operating in MRM mode (in negative and positive ion mode) with an LC-QTOF-MS operating in MS mode (negative and positive ion mode) [101]. In the first system, SCL exhibited the precursor $[M - H]^{-}$ ion at *m/z* 395 using negative ESI mode. Under positive ESI, SCL ionized as the sodium adduct, at m/z 239 and m/z 221 MRM transitions, as expected from the results obtained in TOF analyses. In positive ion mode, SCL forms a strong sodium adduct that can be easily fragmented. In negative ion mode, the fragmentation is less favored, resulting in less sensitivity and selectivity when using MS-MS conditions. This is because they only include the loss and detection of a chlorine atom. LC-QTOF-MS provides high-resolution accurate mass determination. The most sensitive analytical methodology for the analysis of SCL in water samples was LC/MS-MS under positive mode of ionization providing low limits of detection of 15 ng L^{-1}. This same problem was retaken by Wu et al. who used integrated large volume injection (500 μL) approach with UHPLC-MS/MS sensitive method to identify MRM transitions that can produce higher sensitivity signals for SCL [55]. First, the QTrap-MS analysis was at negative ESI MRM transition *m/z* 395–359 allowed to obtain sensitivity in the range of μg L^{-1}. Using positive ESI, sodium adducts $[M + Na]^+$ were adopted as the precursor ion but salt effects reduce ionization efficiency, and the sodium adducts of SCL do not provide reproducible signals for quantification [100]. For this reason, the study of Wu et al. delved about the chlorine adducts of SCL in negative ESI. The same was confirmed for SUC-d6. The source of Cl^- for the formation of $[M + Cl]^-$ is probably due to the fragmentation of SCL in the ion source or the impurities in the SCL standards since no HCl is added in the mobile phase. Different precursor ions, i.e., $[M - H]^-$, $[M + Cl]^-$, and $[M + HCOO]^-$ were compared using the injection volume of 20 μL. The lower LOD of SCL was 0.22 μg L^{-1}, for the ion transition of $[M + HCOO]^-$ *m/z* 441.0–395.0. Moreover, the LODs for ACE and SCL were reduced to 0.2 and 5 ng L^{-1}, respectively, with the UHPLC-TOF-MS/MS method including an injection volume of 500 μL.

The orbitrap HRMS is an alternative that avoids detection by MS/MS but allows simultaneous quantitation of analytes and confirmation, through the use of high-to-power resolution mass spectrometry to detect the intact anion. This system combine very low maintenance costs (relative to other HRMS machines such as FT-ICRs which require cryogenic gases) with the high resolving power and virtually background-free detection associated with Fourier-transform mass spectrometers. Very high sensitivity and selectivity were obtained by adding a basic buffer to enhance negative-mode ionization of the complete SCL isotopic signature and its detection and quantitation at ultra-trace levels (LOD 1.4 ng L^{-1} and LOQ 5.7 ng L^{-1}) lower than any limits reported before in the literature [71]. In a recent work, Wu et al. [55] recommended 0.1% formic acid as modifier for LC-MS/MS analysis of SCL. However, Batchu et al. [71] proposed the use of a basic mobile phase which enhances the deprotonation product minimizing the production of chloride and forming adducts that would decrease the yield of the analytical signal. The proposed LC-HRMS method was applied to a set of seawater and estuarine samples and WWTP.

2.3 Other Determination Techniques

The LC-MS and LC-MS/MS methods proved to be the best choice for environmental and food analysis of artificial sweeteners, as seen above, since low LODs can be achieved without the need of the time-consuming derivatization step. However, some chromatographic methods, which include derivatization, were used, as shown in the following developments and applications. Three chromatographic techniques were applied for the determination of SCL [102, 103] or CYC [104] using derivatization step. GC methods show high resolution, but they require derivatization steps prior to analysis. 70 μL of MSTFA–1% TMCS (*N*-methyl-*N*-trimethylsilytrifluoroacetamide w/1% trimethylchlorosilane) with 30 μL of pyridine was added after extraction of SCL and heated for 30 min at 70 °C, before GC with detection by ion trap mass spectrometer operated in scan mode from 60 to 650 *m/z* [102]. This study presented the first concentration data for SCL for North American coastal and open ocean waters. A method based on Gas Chromatography-Electron Capture Detector (GC-ECD) was developed, including the CYC conversion to *N*, *N*-dichlorocyclohexylamine with sodium hypochlorite for 5 min at room temperature [104]. The method presented several advantages such as the simplicity of the sample preparation steps, the high stability of the derivatization products, the high speed of the chromatographic separation, and the high selectivity of the detection. The LOD and LOQ for yellow wine and fruit juice were 0.05 and 0.2 mg L^{-1}, respectively. The LOD and LOQ for cake and preserved fruit were 0.25 and 0.8 mg kg^{-1}, respectively. A high-performance thin-layer chromatography (HPTLC) method was developed to analyze SCL in water [103]. Separation was performed in parallel on a HPTLC plate silica gel 60 F254 with a mixture of isopropyl acetate, MeOH, and water (15:3:1, v/v/v) within 15 min. Various postchromatographic derivatization reactions (with *p*-aminobenzoic, β-naphthol, and aniline diphenylamine o-phosphoric acid) were compared to selectively detect SCL in effluent and surface water matrices, due to the weak native UV absorption of SCL ($\leq$200 nm). Thereby the last derivatization reagent was slightly preferred. The limit of quantification (LOQ) of SCL in drinking and surface water was 100 ng L^{-1} and the recovery was 80%. The comparison of the means obtained by HPTLC and the respective means of six laboratories, analyzed by HPLC–MS/MS or HPLC-TOF-MS with the use of mostly isotopically labeled standards, was not statistically significant.

The ecotoxicological impact of artificial sweeteners and its transformation products requires further research, as evidenced by the detection of SAC and CYC in leachate or leachate-impacted groundwater at levels comparable to those of untreated wastewater at 14 of 15 municipal landfill sites tested, including several closed for >50 years [68]. For this study Ion chromatography (IC) system coupled to QTrap-MS/MS operating in negative ESI mode was used. Two MRM transitions were monitored for both sweeteners.

Other spectroscopic methods were also described in the literature. Raman spectroscopy is a photonic high-resolution technique that provides, in few seconds, chemical and structural information. The Fourier Transform (FT) technique can detect all wavelengths at the same time improving the resolution, acquisition time spectrum, and signal-to-noise (S/R) ratio of conventional Raman spectroscopy. All this combined with chemometric tools, i.e., partial least squares (PLS), principal component regression (PCR), and counter-propagation artificial neural networks (CP-ANN) methods, offer the possibility of reliable quantification of ASP in commercial sweeteners [105]. ASP concentrations in the range of 17.46–35.93% were obtained in excipient formulations which contained, by weight, 55.8–76.9% lactose, 2.7–5.5% leucine, 2.9–4.7% CMCNa, and up to 0.6% magnesium stearate.

Another spectroscopic technique used in quality control is based on Nuclear Magnetic Resonance (NMR) which is concerned with the magnetic properties of certain nuclei. High resolution 1H-NMR combined to principal component analysis (PCA) was used for the identification and the quantification of different soft drinks additives that included three sweeteners (ACS-K, CYC, and ASP) [106]. Moreover, the combination between NMR and chemometric data analysis was a suitable tool for cola drink authentication.

HPLC methods demand a large amount of organic solvents which is harmful to the operator. Additionally, both HPLC and GC require complicated and extensive sample preparation, which consumes time and effort. In this context, Capillary Electrophoresis (CE) is an analytical separation technique that generally offers shorter analysis time and relatively low consumption of consumables. Thus, several studies demonstrated the utility of CE for the analysis of artificial sweeteners. Capacitively coupled contactless conductivity detection (C^4D) was applied for detection on CE of ASP, CYC, SAC, and ACS-K in commercial samples of soft drinks and tabletop sweetener formulations [35]. The pH of the separation buffer controls both the analyte charge and the level of electroosmotic flow (EOF). A relatively high pH value of 9.4 was necessary to use in order to render all sweeteners in the anionic charged form required for separation and detection. However, CE separation with EOF reversion demonstrated problems, especially with broadening of the SAC peak, due to interactions of the SAC with EOF inverters tested. Separations with good resolution can be obtained in less than 6 min. This analysis time is smaller than half of that required in the previous isotachophoresis method also using conductivity detection for determination of sweeteners in chewing gums and candies [107]. The LODs obtained with the CE-C^4D method are lower (1.4–4.2 mg L^{-1}) than those usually attained by CE with photometric detection. Several approaches were developed to enhance CE-C^4D technique. The use of superimposed hydrodynamic pumping was found to be of great benefit in the determination of the same sweeteners in low calorie soft drinks, sweets, and a tabletop sweetener formulation by CE-C^4D with analysis times of less than 1 min [108]. Band broadening was avoided by using capillaries of a narrow 10 μm internal diameter. The use of surface modification to eliminate or reverse the EOF was not necessary due to the superimposed bulk flow. The conditions for rapid separations not only led to higher limits of detection, but also a narrower dynamic range. On-line preconcentration such as field-amplified sample injection (FASI) was used to overcome the sensitivity CE-C^4D limitations because it has enrichment factors of several 100-fold or higher [33]. The FASI-CE-C^4D method demonstrated to be applicable for simultaneous determination of ACS-K, SAC, and CYC in beverages. The separation was achieved within 10 min using 20 mmol L^{-1} HAc (pH 3.3.) as running buffer. No EOF modifier is required to be added into the background electrolyte (BGE) so the problems of broadening of the analyte peaks are avoided.

On the other hand, a simple capillary zone electrophoresis (CZE) method with UV detection was developed for the determination of NEO in nonalcoholic beverage [45]. Both borate and phosphate were tested as the background electrolyte. The results showed that the borate buffer gave a better peak shape and shorter migration time (less than 5 min) than the phosphate buffer. Buffer pH 8 proved to be ideal on migration behavior. LOD was 0.118 μg mL^{-1}. Excellent LODs (4.4 μg L^{-1} for ACS-K, 6.7 μg L^{-1} for SAC and 8.8 μg L^{-1} for CYC) were achieved. In most of the previous papers, separation time until 6–10 min was required to separate four sweeteners. A new sub-minute method based on CZE using indirect UV detection and short-end injection procedure (SEIP) was developed for the analysis of ASP, CYC, SAC, and ACE-K [86]. The BGE used in these measurements were buffered at pH around 9, to guarantee total ionization of the sweeteners in the anionic form, and ASP was also buffered at pH 1.4 to measure its mobility in the cationic form. The LOD and LOQ were lower than 6.5 and 21.5 mg L^{-1}, respectively.

Finally, a modification of CE, micellar electrokinetic chromatographic (MEKC) method which is suitable for neutral and charged analytes to be separated in a single injection was able to separate 11 food additives simultaneously within 30 min [109]. Six sweeteners, i.e., ALI, ASP, SAC, NEO, STV, and ACS-K, were determined in beverage, yogurt, and candied fruit samples using MEKC. The separation buffer consisted of 20 mmol L^{-1} sodium tetraborate, 42 mmol L^{-1} boric acid (pH 8.83), and 100 mmol L^{-1} sodium deoxycholate. The detection wavelength was 214 nm and the LODs were in the ranges of 0.25–2.5 mg L^{-1}. To demonstrate the accuracy of the proposed MEKC method, a FAPAS® proficiency test sample containing caffeine and SAC was successfully analyzed.

3 Conclusions and Futures Perspectives

This chapter has focused on the analytical strategies for the determination of artificial sweeteners. Monitoring foodstuffs for additives is an area of increasing concern and importance. First, depending the selected matrix, different strategies for sample preparation have been considered. Many of the methods of liquid environmental matrices are based on the SPE and microextraction techniques coupled with powerful analytical techniques (LC-MS and LC-MS/MS) which have obtained lower limits of detection and high selectivity in their determination. The trends in sample preparation include miniaturization, automation, high throughput performance, online coupling with analytical instruments, and low-cost operation using little or no solvent consumption [110]. Last years, high selective materials as MIPs have become an alternative to existing commercial sorbents for SPE. The investigation of smart materials [111] based on carbon nanotubes, nano inorganic oxides, ionic liquids, and MIP sol-gels will be explored in extraction and microextraction techniques (SPME and MEPS) of artificial sweeteners from complex matrices. To date, there are very few studies related to the presence of artificial sweeteners in sludge and sample preparation methods for their extraction. Further investigation is required to identify and quantify these compounds and the degradation products formed in wastewater and sludge.

After a detailed overview of the literature on the use of LC-MS methods in foodstuffs and environmental field, it seems clear that tandem MS hyphenated to UHPLC is still the reference technique for determination of sweeteners in complex matrices, mainly because of the resolution and sensitivity attainable by use of these instruments working in MRM mode. Often identification and quantification are both required. Most current LC based methods rely on QqQ-MS/MS. LC-MS/MS data on precursor and product ions under negative and positive ionization mode are often used. ASP, SAC, and SCL can be determined (individually or as sodium adducts) in both ionization modes, although higher sensitivity and more fragments were obtained only through the positive ionization mode. However, the only possible way to achieve simultaneous determination of artificial sweeteners by LC-MS and LC-MS/MS techniques is by running analysis under negative ESI. Throughout literature a large majority of methods referred to the determination of SCL but an advantage of LC-MS/MS is the capacity to the simultaneous analysis of a wide variety of sweeteners within a single analytical run with minimal incremental cost. This has the potential to simplify laboratory set and provide additional useful information (e.g., metabolite profiles). The internal standardization method was mainly used in quantification due to the need to adequately compensate the matrix effects. QTrap and TOF mass analyzers have enabled popularization of HRMS coupled to LC, leading to new perspectives in food quality control and regulation enforcement of these compounds considered as environmental pollutants. LC-HRMS provide significant advantage in the possibility of discrimination of ions due to high mass resolving power; however, the main drawback of LC-HRMS methods is in quantification. On the other hand, different stationary phases [88], some of them based on stimuli-responsive polymers, can offer advantages for chromatographic development related to higher specificity for retention and sequential elution of artificial sweeteners at cost-effective prices. These polymers can be integrated in the stationary phase in the form of cross-linked networks or grafted on to solid beads or inert surfaces.

In summary, the application of extraction approaches using specific sorbents, green solvents, and miniaturized analytical systems reducing labor consumption will contribute to the development of environment-friendly analytical techniques to determination of trace emerging pollutants such as artificial sweeteners. Furthermore, the advantages of chromatographic methodologies, i.e., the use of monolithic columns, the use of column temperature, and the use of sub-2 μm particle size column or fused-core column technologies, can be good alternatives for high-efficiency and fast LC separations.

Acknowledgments The authors would like to thank the Spanish Ministry of Science and Innovation (Project AGL-2014-53647-R) and the Dirección General de Tráfico (Ministry of Interior, Spain), Project SPIP2015-01838, and FEDER for their financial support.

References

1. Regulation (EC) No 1333/2008 of the European Parliament and of the Council of 16 December 2008 on food additives. Off J Eur Union L 354:16–33
2. US FDA (United States Food and Drug Administration) (2014) http://www.fda.gov/food/ingredientspackaginglabeling/foodadditivesingredients/ucm397716.htm. Accessed 22 Jan 2016

3. ISA (International Sweeteners Association) (2014) Low calories sweeteners and you. http://www.sweeteners.org/en. Accessed 25 Jan 2016
4. Commission Regulation (EU) No 1129/2011 of 11 November 2011 amending Annex II to Regulation (EC) No 1333/2008 of the European Parliament and of the Council by establishing a Union list of food additives. Off J Eur Union L 295:1–177
5. Commission Regulation (EU) No 1131/2011 of 11 November 2011 amending Annex II to Regulation (EC) No 1333/2008 of the European Parliament and of the Council with regard to steviol glycosides. Off J Eur Union L 295:205–211
6. Commission Regulation (EU) No 497/2014 of 14 May 2014 amending Annex II to Regulation (EC) No 1333/2008 of the European Parliament and of the Council and the Annex to Commission Regulation (EU) No 231/2012 as regards the use of advantame as a sweetener text with EEA relevance. Off J Eur Communities 143:6–13
7. EFSA (European Food SafetyAuthority) (2014) http://www.efsa.europa.eu/en/topics/topic/sweeteners. Accessed 22 Jan 2016
8. FDA (Food and Drug Administration) http://www.fda.gov/Food/IngredientsPackagingLabeling/FoodAdditivesIngredients/ucm397725.htm. Accessed 22 Jan 2016
9. Grembecka M (2015) Natural sweeteners in a human diet. Rocz Panstw Zakl Hig 66:195–202
10. EFSA (European Food Safety Authority) (2013) Scientific opinion on the safety of advantame for the proposed uses as a food additive. EFSA J 11:3301–3369. doi:10.2903/j.efsa.2013.3301
11. EFSA (European Food Safety Authority) (2013) Scientific opinion on the re-evaluation of aspartame (E 951) as a food additive. EFSA J 11:3496–3759. doi:10.2903/j.efsa.2013.3496
12. European Commission (2006) Directive 2006/52/EC of European Parliament and Council of 5 July 2006 amending Directive 95/2/EC on food additives other than colours and sweeteners and Directive 94/35/EC on sweeteners for use in foodstuffs. Off J Eur Union L 204:10–22. Brussels
13. Commission Regulation (EU) No 257/2010 of 25 March 2010 setting up a programme for the re-evaluation of approved food additives in accordance with Regulation (EC) No 1333/2008 of the European Parliament and of the Council on food additives. Off J Eur Communities 80:19–27
14. Spillane WJ (2006) Optimising sweet taste in food. CRC Press, Boca Raton
15. Kaneko R, Kitabatake N (2001) Sweetness of sweet protein thaumatin is more thermoresistant under acid conditions than under neutral or alkaline conditions. Biosci Biotech Bioch 65:409–413. doi:10.1271/bbb.65.409
16. Bellisle F (2015) Intense sweeteners, appetite for the sweeteners taste, and relationship to weight management. Curr Obes Rep 4:106–110. doi:10.1007/s13679-014-0133-08
17. Mattes RD, Popkin BM (2009) Nonnutritive sweetener consumption in humans: effects on appetite and food intake and their putative mechanisms. Am J Clin Nutr 89:1–14. doi:10.3945/ajcn.2008.26792
18. Ficth C, Keim KS (2012) Position of the Academy of Nutrition and Dietetic: use of nutritive and nonnutritive sweeteners. J Acad Nutr Diet 112:739–758
19. AHA (American Heart Association), ADA (American Diabetes Association) (2012) Nonnutritive sweeteners: current use and health perspectives: a scientific statement. Circulation 126:509–519. doi:10.1161/CIR0b013e31825c42ee
20. Ain Q, Khan SA (2015) Artificial sweeteners: safe or unsafe? Recent Adv Endocr 65:225–227
21. ANSES (French Agency for Food, Environmental and Occupational Health & Safety) (2015) OPINION of the French Agency for Food, Environmental and Occupational Health & Safety of 19 November 2014, revised on 9 January 2015 on the assessment of the nutritional benefits and risks related to intense sweeteners. www.anses.fr. ANSES Opinion Request no. 2011-SA-0161
22. EFSA (2011) Scientific opinion on the substantiation of health claims related to intense sweeteners and contribution to the maintenance or achievement of a normal body weight (ID 1136, 1444, 4299), reduction of post-prandial glycaemic responses (ID 4298), maintenance of normal blood glucose concentrations (ID 1221, 4298), and maintenance of tooth mineralisation by decreasing tooth demineralisation (ID 1134, 1167, 1283) pursuant to Article 13(1) of Regulation (EC) No 1924/20061. EFSA J 9:2229–2255

23. Pepino MY (2015) Metabolic effects of non-nutritive sweeteners. Physiol Behav 152:450–455. doi:10.1016/j.physbeh.2015.06.024
24. Swithers SE (2015) Artificial sweeteners are not the answer to childhood obesity. Appetite 93:85–90. doi:10.1016/j.appet.2015.03.027
25. Shankar P, Ahuja S, Sriram K (2013) Non-nutritive sweeteners: review and update. Nutrition 29:1293–1299. doi:10.1016/j.nut.2013.03.024
26. Kokotou MG, Asimakopoulos AG, Thomaidis NS (2012) Artificial sweeteners as emerging pollutants in the environment: analytical methodologies and environmental impact. Anal Methods 4:3057–3070. doi:10.1039/c2ay05950a
27. Perkola N, Sainio P (2014) Quantification of four artificial sweeteners in Finnish surface waters with isotope-dilution mass spectrometry. Environ Pollut 184:391–396. doi:10.1016/j.envpol.2013.09.017
28. Zygler A, Wasik A, Namieśnik J (2009) Analytical methodologies for determination of artificial sweeteners in foodstuffs. Trends Anal Chem 28:1082–1102. doi:10.1016/j.trac.2009.06.008
29. Lange FT, Scheurer M, Brauch HJ (2012) Artificial sweeteners a recently recognized class of emerging environmental contaminants: a review. Anal Bioanal Chem 403:2503–2518. doi:10.1007/s00216-012-5892-z
30. GRAS (Generally Recognized as Safe) (2013) Code of Federal Regulations Title 21, Parts 172 and 189. Food and Drug Administration (FDA). Available from: http://www.accessdata.fda.gov/scripts/cdrh/cfdocs/cfcfr/CFRSearch.cfm?CFRPart=172 http://www.accessdata.fda.gov/scripts/cdrh/cfdocs/cfcfr/CFRSearch.cfm?CFRPart=189. Accessed 20 Mar 2016
31. Zygler A, Wasik A, Namiésnik J (2010) Retention behaviour of some high-intensity sweeteners on different SPE sorbents. Talanta 82:1742–1748. doi:10.1016/j.talanta.2010.07.070
32. Carro-Díaz AM, Lorenzo-Ferreira RA (2013) Molecularly imprinted polymers for sample preparation. In: Alvarez-Lorenzo C, Concheiro A (eds) Handbook of molecularly imprinted polymers. Smithers Rapra Technology, Shropshire. ISBN 978-1-84735-959-9
33. Yang L, Zhou S, Xiao Y, Tang Y, Xie T (2015) Sensitive simultaneous determination of three sulfanilamide artificial sweeters by capillary electrophoresis with on-line preconcentrationand contactless conductivity detection. Food Chem 188:446–451. doi:10.1016/j.foodchem.2015.04.060
34. Chang CS, Yeh TS (2014) Detection of 10 sweeteners in various foods by liquid chromatography/tandem mass spectrometry. J Food Drug Anal 22:318–328. doi:10.1016/j.jfda.2014.01.024
35. Bergamo AB, Fracassi da Silva JA, Pereira de Jesus D (2011) Simultaneous determination of aspartame, cyclamate, saccharin and acesulfame-Kin soft drinks and tabletop sweetener formulations by capillary electrophoresis with capacitively coupled contactless conductivity detection. Food Chem 124:1714–1717. doi:10.1016/j.foodchem.2010.07.107
36. Botelho Dias C, Dillenburg Meinhart A, Queiroz Pane D, Augusto Ballus C, Teixeira Godoy H (2015) Multivariate optimisation and validation of a method for the separation of five artificial sweeteners by UPLC-DAD in nine food matrices. Food Anal Methods 8:1824–1835. doi:10.1007/s12161-014-0056-8
37. Ordoñez EY, Rodil R, Quintana JB, Cela R (2015) Determination of artificial sweeteners in beverages with green mobile phases and high temperature liquid chromatography–tandem mass spectrometry. Food Chem 169:162–168. doi:10.1016/j.foodchem.2014.07.132
38. Lorenzo RA, Pena MT, Fernández P, González P, Carro AM (2015) Artificial sweeteners in beverages by ultra performance liquid chromatography with photodiode array and liquid chromatography tandem mass spectrometry. Food Control 47:43–52. doi:10.1016/j.foodcont.2014.06.035
39. Ma K, Li X-J, Wang H-F, Zhao M (2015) Rapid and sensitive method for the determination of eight food additives in red wine by ultra-performance liquid chromatography tandem mass spectrometry. Food Anal Methods 8:203–212. doi:10.1007/s12161-014-9893-8
40. Kubica P, Namieśnik J, Wasik A (2015) Determination of eight artificial sweeteners and common *Stevia rebaudiana* glycosides in non-alcoholic and alcoholic beverages by

reversed-phase liquid chromatography coupled with tandem mass spectrometry. Anal Bioanal Chem 407:1505–1512. doi:10.1007/s00216-014-8355-x

41. de Queiroz Pane D, Botelho Dias C, Dillenburg Meinhart A, Augusto Ballus C, Teixeira Godoy H (2015) Evaluation of the sweetener content in diet/light/zero foods and drinks by HPLC-DAD. J Food Sci Technol 52:6900–6913. doi:10.1007/s13197-015-1816-1
42. Sakai H, Yamashita A, Tamura M, Uyama A, Mochizuki N (2015) Simultaneous determination of sweeteners in beverages by LC-MS/MS. Food Addit Contam Part A 32:808–816. doi:10.1080/19440049.2015.1018341
43. Ohtsuki T, Sato K, Abe Y, Sugimoto N, Akiyama H (2015) Quantification of acesulfame potassium in processed foods by quantitative 1H NMR. Talanta 131:712–718. doi:10.1016/j.talanta.2014.08.002
44. Shah R, Farris S, De Jager LS, Begley TH (2015) A novel method for the simultaneous determination of 14 sweeteners of regulatory interest using UHPLC-MS/MS. Food Addit Contam Part A 32:141–151. doi:10.1080/19440049.2014.994111
45. Hu F, Xu L, Luan F, Liu H, Gao Y (2013) Determination of neotame in non-alcoholic beverage by capillary zone electrophoresis. J Sci Food Agric 93:3334–3338. doi:10.1002/jsfa.6181
46. Zygler A, Wasik A, Kot-Wasik A, Namieśnik J (2011) Determination of nine high-intensity sweeteners in various foods by high-performance liquid chromatography with mass spectrometric detection. Anal Bioanal Chem 400:2159–2172. doi:10.1007/s00216-011-4937-z
47. Hashemi M, Habibi A, Jahanshahi N (2011) Determination of cyclamate in artificial sweeteners and beverages using headspace single-drop microextraction and gas chromatography flame-ionisation detection. Food Chem 124:1258–1263. doi:10.1016/j.foodchem.2010.07.057
48. Moein MM, Javanbakht M, Karimi M, Akbari-adergani B (2015) Molecularly imprinted sol–gel nanofibers based solid phase microextraction coupled on-line with high performance liquid chromatography for selective determination of acesulfame. Talanta 134:340–347. doi:10.1016/j.talanta.2014.11.011
49. Richardson SD (2010) Environmental mass spectrometry: emerging contaminants and current issues. Anal Chem 82:4742–4774. doi:10.1021/ac101102d
50. Richardson SD, Ternes TA (2014) Water analysis: emerging contaminants and current issues. Anal Chem 86:2813–2848. doi:10.1021/ac500508t
51. Tollefsen KE, Nizzetto L, Huggett DB (2012) Presence, fate and effects of the intense sweetener sucralose in the aquatic environment. Sci Total Environ 438:510–516. doi:10.1016/j.scitotenv.2012.08.060
52. Marinovich M, Galli CL, Bosetti C, Gallus S, La Vecchia C (2013) Aspartame, low-calorie sweeteners and disease: regulatory safety and epidemiological issues. Food Chem Toxicol 60:109–115. doi:10.1016/j.fct.2013.07.040
53. Lemus-Mondaca R, Vega-Gálvez A, Zura-Bravo L, Ah-Hen K (2012) *Stevia rebaudiana* Bertoni, source of a high-potency natural sweetener: a comprehensive review on the biochemical, nutritional and functional aspects. Food Chem 132:1121–1132. doi:10.1016/j.foodchem.2011.11.140
54. Backe WJ, Fieldls JA (2012) SPE necessary for environmental analysis? A quantitative comparison of matrix effects from large-volume injection and solid-phase extraction based methods. Environ Sci Technol 46:6750–6758. doi:10.1021/es300235z
55. Wu M, Qian Y, Boyd JM, Hrudey SE, Le XC, Li XF (2014) Direct large volume injection ultra-high performance liquid chromatography-tandem mass spectrometry determination of artificial sweeteners sucralose and acesulfame in well water. J Chromatogr A 1359:156–161. doi:10.1016/j.chroma.2014.07.035
56. Berset JD, Ochsenbein N (2012) Stability considerations of aspartame in the direct analysis of artificial sweeteners in water samples using high-performance liquid chromatography–tandem mass spectrometry (HPLC–MS/MS). Chemosphere 88:563–569. doi:10.1016/j.chemosphere.2012.03.030

57. Scheurer M, Brauch HJ, Lange FT (2009) Analysis and occurrence of seven artificial sweetenersnin German waste water and surface water and in soil aquifer treatment (SAT). Anal Bioanal Chem 394:1585–1594. doi:10.1007/s00216-009-2881-y
58. Ordoñez EY, Quintana JB, Rodil R, Cela R (2012) Determination of artificial sweeteners in water samples by solid-phase extraction and liquid chromatography–tandem mass spectrometry. J Chromatogr A 1256:197–205. doi:10.1016/j.chroma.2012.07.073
59. Salas D, Borrull F, Fontanals N, Marcé RM (2015) Hydrophilic interaction liquid chromatography coupled to high-resolution mass spectrometry to determine artificial sweeteners in environmental waters. Anal Bioanal Chem 407:4277–4285. doi:10.1007/s00216-014-8330-6
60. Arbeláez P, Borrull F, Pocurull E, Marcé RM (2015) Determination of high-intensity sweeteners in river water and wastewater by solid-phase extraction and liquid chromatography–tandem mass spectrometry. J Chromatogr A 1393:106–114. doi:10.1016/j.chroma.2015.03.035
61. Gan Z, Sun H, Feng B, Wang R, Zhang Y (2013) Occurrence of seven artificial sweeteners in the aquatic environment and precipitation of Tianjin, China. Water Res 47:4928–4937. doi:10.1016/j.watres.2013.05.038
62. Gan Z, Sun H, Wang R, Feng B (2013) A novel solid-phase extraction for the concentration of sweeteners in water and analysis by ion-pair liquid chromatography–triple quadrupole mass spectrometry. J Chromatogr A 1274:87–96. doi:10.1016/j.chroma.2012.11.081
63. Tran NH, Hu J, Ong SL (2013) Simultaneous determination of PPCPs, EDCs, and artificial sweeteners in environmental water samples using a single-step SPE coupled with HPLC–MS/MS and isotope dilution. Talanta 113:82–92. doi:10.1016/j.talanta.2013.03.072
64. Buerge IJ, Buser HR, Kahle M, Muller MD, Poiger T (2009) Ubiquitous occurrence of the artificial sweetener acesulfame in the aquatic environment: an ideal chemical marker of domestic wastewater in groundwater. Environ Sci Technol 43:4381–4385. doi:10.1021/es900126x
65. Van Stempvoort RD, Roy JW, Brown SJ, Bickerton G (2011) Artificial sweeteners as potential tracers in groundwater in urban environments. J Hydrol 401:126–133. doi:10.1016/j.jhydrol.2011.02.013
66. Hillebrand O, Nödler K, Sauter M, Licha T (2015) Multitracer experiment to evaluate the attenuation of selected organic micropollutants in a karst aquifer. Sci Total Environ 506–507:338–343. doi:10.1016/j.scitotenv.2014.10.102
67. Zirlewagen J, Licha T, Schiperski F, Nödler K, Scheytt T (2016) Use of two artificial sweeteners, cyclamate and acesulfame, to identify and quantify wastewater contributions in a karst spring. Sci Total Environ 547:356–365. doi:10.1016/j.scitotenv.2015.12.112
68. Roy JW, Van Stempvoort DR, Bickerton G (2014) Artificial sweeteners as potential tracers of municipal landfill leachate. Environ Pollut 184:89–93. doi:10.1016/j.envpol.2013.08.021
69. Oppenheimer J, Eaton A, Badruzzaman M, Haghani AW, Jacangelo JG (2011) Occurrence and suitability of sucralose as an indicator compound of wastewater loading to surface waters in urbanized regions. Water Res 45:4019–4027. doi:10.1016/j.watres.2011.05.014
70. Batchu SR, Quinete N, Panditi VR, Gardinali PR (2013) Online solid phase extraction liquid chromatography tandem mass spectrometry (SPE-LC-MS/MS) method for the determination of sucralose in reclaimed and drinking waters and its photo degradation in natural waters from South Florida. Chem Cent J 7:141–149. doi:10.1186/1752-153X-7-141
71. Batchu SR, Ramirez CE, Gardinali PR (2015) Rapid ultra-trace analysis of sucralose in multiple-origin aqueous samples by online solid-phase extraction coupled to high-resolution mass spectrometry. Anal Bioanal Chem 407:3717–3725. doi:10.1007/s00216-015-8593-6
72. http://ec.europa.eu/environment/waste/sludge/index.htm. Accessed 15 Mar 2016
73. Council Directive (1986) Council Directive of 12 June 1986 on the protection of the environment, and in particular of the soil, when sewage sludge is used in agriculture. Off J Eur Communities 181:0006–0012
74. Brorström-Lundén E, Svenson A, Viktor T, Woldegiorgis A, Remberger M, Kaj L, Dye C, Bjerke A, Schlabach M (2008) Measurements of sucralose in the Swedish Screening Program

2007. PART I. Sucralose in surface waters and STP samples. Swedish Environmental Research Institute, Stockholm

75. Brorström-Lundén E, Svenson A, Viktor T, Woldegiorgis A, Remberger M, Kaj L, Dye C, Bjerke A, Schlabach M (2008) Measurements of sucralose in the Swedish Screening Program 2007. PART II. Sucralose in biota samples and regional STP samples. Swedish Environmental Research Institute, Stockholm
76. Ordoñez EY, Quintana JB, Rodil R, Cela R (2013) Determination of artificial sweeteners in sewage sludge samples using pressurised liquid extraction and liquid chromatography–tandem mass spectrometry. J Chromatogr A 1320:10–16. doi:10.1016/j.chroma.2013.10.049
77. Arbeláez P, Borrull F, Marcé RM, Pocurull E (2015) Trace-level determination of sweeteners in sewage sludge using selective pressurized liquid extraction and liquid chromatography–tandem mass spectrometry. J Chromatogr A 1408:15–21. doi:10.1016/j.chroma.2015.07.001
78. Subedi B, Kannan K (2014) Fate of artificial sweeteners in wastewater treatment plants in New York State, U.S.A. Environ Sci Technol 48:13668–13674. doi:10.1021/es504769c
79. Subedi B, Lee S, Moon H-B, Kannan K (2014) Emission of artificial sweeteners, select pharmaceuticals, and personal care products through sewage sludge from wastewater treatment plants in Korea. Environ Int 68:33–40. doi:10.1016/j.envint.2014.03.006
80. Gan Z, Sun H, Yao Y, Zhao Y, Li Y, Zhang Y, Hu H, Wang R (2014) Distribution of artificial sweeteners in dust and soil in China and their seasonal variations in the environment of Tianjin. Sci Total Environ 488–489:168–175. doi:10.1016/j.scitotenv.2014.04.084
81. Rowe RC, Sheskey PJ, Cook WG, Fenton ME (2012) Handbook of pharmaceutical excipients, 7th edn. Pharmaceutical Press/American Pharmacists Association, Philadelphia. ISBN 978 0 85711 027 5
82. Pifferi G, Restani P (2003) The safety of pharmaceutical excipients. Il Farmaco 58:541–550. doi:10.1016/S0014-827X(03)00079-X
83. Balbani APS, Stelzer LB, Montovani JC (2006) Pharmaceutical excipients and the information on drug labels. Rev Bras Otorrinolaringol 72:400–406
84. Corrêa de Carvalho R, Pereira Netto AD, Ferreira de Carvalho Marques F (2014) Simultaneous determination of strontium ranelate and aspartame in pharmaceutical formulation for the treatment of postmenopausal osteoporosis by capillary zone electrophoresis. Microchem J 117:214–219. doi:10.1016/j.microc.2014.06.024
85. Pereira Vistuba J, Danielli Dolzana M, Vitali L, Leal de Oliveira MA, Micke GA (2015) Sub-minute method for simultaneous determination of aspartame, cyclamate, acesulfame-K and saccharin in food and pharmaceutical samples by capillary zone electrophoresis. J Chromatogr A 1396:148–152. doi:10.1016/j.chroma.2015.03.070
86. Grembecka M, Szefer P (2012) Simultaneous determination of caffeine and aspartame in diet supplements and non-alcoholic beverages using liquid-chromatography coupled to corona CAD and UV-DAD detectors. Food Anal Methods 5:1010–1017. doi:10.1007/s12161-011-9334-x
87. Singh M, Kumar A, Tarannum N (2013) Water-compatible 'aspartame'-imprinted polymer grafted on silica surface for selective recognition in aqueous solution. Anal Bioanal Chem 405:4245–4252. doi:10.1007/s00216-013-6812-6
88. Núñez O, Gallart-Ayala H, Martins CPB, Lucci P (2012) New trends in fast liquid chromatography for food and environmental analysis. J Chromatogr A 1228:298–323. doi:10.1016/j.chroma.2011.10.091
89. Dossi N, Toniolo R, Susmel S, Pizzariello A, Bontempelli G (2006) Simultaneous RP-LC determination of additives in soft drinks. Chromatographia 63:557–562. doi:10.1365/s10337-006-0793-y
90. Cantarelli MA, Pellerano RG, Marchevsky EJ, Camiña JM (2009) Simultaneous determination of aspartame and acesulfame-K by molecular absorption spectrophotometry using multivariate calibration and validation by high performance liquid chromatography. Food Chem 115:1128–1132. doi:10.1016/j.foodchem.2008.12.101

91. Kailasam S (2010) Ultrafast and sensitive analysis of sweeteners, preservatives and flavorants in nonalcoholic beverages using the Agilent 1290 Infinity LC system Agilent Technologies, Inc. Publication Number 5990-5590EN. www.agilent.com/chem/1290
92. Pedjie N (2010) Analysis of common sweeteners and additives in beverages with the PerkinElmer Flexar FX-15 System equipped with a PDA Detector PerkinElmer, Inc. 009473_01. www.perkinelmer.com
93. Wasik A, McCourt J, Buchgraber M (2007) Simultaneous determination of nine intense sweeteners in foodstuffs by high performance liquid chromatography and evaporative light scattering detection – development and single-laboratory validation. J Chromatogr A 1157:187–196. doi:10.1016/j.chroma.2007.04.068
94. Kruve A, Rebane R, Kipper K, Oldekop ML, Evard H, Herodes K, Ravio P, Leito I (2015) Tutorial review on validation of liquid chromatography–mass spectrometry methods: part I. Anal Chim Acta 870:29–44. doi:10.1016/j.aca.2015.02.017
95. Kruve A, Rebane R, Kipper K, Oldekop ML, Evard H, Herodes K, Ravio P, Leito I (2015) Tutorial review on validation of liquid chromatography–mass spectrometry methods: part II. Anal Chim Acta 870:8–28. doi:10.1016/j.aca.2015.02.016
96. Yang DJ, Chen B (2009) Simultaneous determination of nonnutritive sweeteners in foods by HPLC/ESI-MS. J Agric Food Chem 57:3022–3027. doi:10.1021/jf803988u
97. Hird SJ, Lau BPY, Schuhmacher R, Krska R (2014) Liquid chromatography-mass spectrometry for the determination of chemical contaminants in food. Trends Anal Chem 59:59–72. doi:10.1016/j.trac.2014.04.005
98. Loos R, Gawlik BM, Boettcher K, Locoro G, Contini S, Bidoglio G (2009) Sucralose screening in European surface waters using a solid-phase extraction-liquid chromatography–triple quadrupole mass spectrometry method. J Chromatogr A 1216:1126–1131. doi:10.1016/j.chroma.2008.12.048
99. Mawhinney DB, Young RB, Vanderford BJ, Borch T, Snyder SA (2011) Artificial sweetener sucralose in U.S. drinking water systems. Environ Sci Technol 45:8716–8722. doi:10.1021/es202404c
100. Minten J, Adolfsson-Erici M, Björlenius B, Alsberg T (2011) A method for the analysis of sucralose with electrospray LC/MS in recipient waters and in sewage effluent subjected to tertiary treatment technologies. Int J Environ Anal Chem 91:357–366. doi:10.1080/03067310903582333
101. Ferrer I, Zweigenbaum JA, Thurman EM (2013) Analytical methodologies for the detection of sucralose in water. Anal Chem 85:9581–9587. doi:10.1021/ac4016984
102. Mead RN, Morgan JB, Avery GB Jr, Kieber RJ, Kirk AM, Skrabal SA, Willey JD (2009) Occurrence of the artificial sweetener sucralose in coastal and marine waters of the United States. Mar Chem 116:13–17. doi:10.1016/j.marchem.2009.09.005
103. Morlock GE, Schuele L, Grashorn S (2011) Development of a quantitative high-performance thin-layer chromatographic method for sucralose in sewage effluent, surface water, and drinking water. J Chromatogr A 1218:2745–2753. doi:10.1016/j.chroma.2010.11.063
104. Yu S, Zhu B, Lv F, Li S, Huang W (2012) Rapid analysis of cyclamate in foods and beverages by gas chromatography-electron capture detector (GC-ECD). Food Chem 134:2424–2429. doi:10.1016/j.foodchem.2012.04.028
105. Mazurek S, Szostak R (2011) Quantification of aspartame in commercial sweeteners by FT-Raman spectroscopy. Food Chem 125:1051–1057. doi:10.1016/j.foodchem.2010.09.075
106. Maes P, Monakhova YB, Kuballa T, Reusch H, Lachenmeier DW (2012) Qualitative and quantitative control of carbonated cola beverages using 1H NMR spectroscopy. J Agric Food Chem 60:2778–2784. doi:10.1021/jf204777m
107. Herrmannová M, Krivánková L, Bartos M, Vytras K (2006) Direct simultaneous determination of eight sweeteners in foods by capillary isotachophoresis. J Sep Sci 29:1132–1137. doi:10.1016/j.foodchem.2010.07.107

108. Stojkovic M, Mai TD, Hauser PC (2013) Determination of artificial sweeteners by capillary electrophoresis with contactless conductivity detection optimized by hydrodynamic pumping. Anal Chim Acta 787:254–259. doi:10.1016/j.aca.2013.05.039
109. Wu Y, Li J, Zhao S, Ding X (2016) Simultaneous determination of 11 food additives by micellar electrokinetic capillary chromatography. Food Anal Methods 9:589–595. doi:10.1007/s12161-015-0225-4
110. Moein MM, Abdel-Rehim A, Abdel-Rehim M (2015) Microextraction by packed sorbent (MEPS). Trends Anal Chem 67:34–44. doi:10.1016/j.trac.2014.12.003
111. Lorenzo RA, Carro AM, Concheiro A, Alvarez-Lorenzo C (2015) Stimuli-responsive materials in analytical separation. Anal Bioanal Chem 407:4927–4948. doi:10.1007/s00216-015-8679-1

Characterization of Artificial Sweeteners Using Raman Spectroscopy

21

Anna Grazia Mignani, Leonardo Ciaccheri, Gabriele Manca, Andrea Azelio Mencaglia, Tom Verschooten, Heidi Ottevaere, and Hugo Thienpont

Abstract

A selection of fourteen common and commercially available table-top artificial sweeteners was considered. The samples contained aspartame and saccharin as high-intensity sweeteners and dextrose, sorbitol, sucrose, and maltodextrin as low-intensity sweeteners. These were all examined both in powder form and as aqueous solutions. Raman spectra, excited at 1064 nm, were acquired using a compact dispersive scheme. These spectra provided fluorescence-free Raman signatures from which to identify the most significant peaks of the various sweeteners. These peaks were also compared with ones obtained by means of computational analysis, in order to show the effect of the entire sweetener matrix. The spectroscopic data were then processed by means of chemometric analysis for distinguishing what kind of sweetener was present in a given sample. First, Principal Component Analysis was applied for the purpose of data dimensionality reduction and explorative investigation and provided good clustering depending on the type of sweetener. Next, the K-nearest neighbors method was used in order to assign the samples to predefined classes. An excellent identification in accordance with the type of high- or low-power sweetener was thus obtained. These

A.G. Mignani (✉) • L. Ciaccheri (✉) • A.A. Mencaglia
CNR-Istituto di Fisica Applicata "Nello Carrara", Sesto Fiorentino (FI), Italy
e-mail: a.g.mignani@ifac.cnr.it; l.ciaccheri@ifac.cnr.it; a.mencaglia@ifac.cnr.it

G. Manca
CNR-Istituto di Chimica dei Composti Organometallici and CNR-Centro Ricerca&Impresa, Sesto Fiorentino (FI), Italy
e-mail: gabriele.manca@iccom.cnr.it

T. Verschooten • H. Ottevaere • H. Thienpont
Department of Applied Physics and Photonics, Brussels Photonics Team, Vrije Universiteit Brussel (VUB), Brussel, Belgium
e-mail: t.verschooten@gmail.com; hottevae@b-phot.org; hthienpo@vub.ac.be

J.-M. Mérillon, K.G. Ramawat (eds.), *Sweeteners*, Reference Series in Phytochemistry,
https://doi.org/10.1007/978-3-319-27027-2_18

results confirm the success of Raman spectroscopy in attaining a straightforward analysis of intact food, with high potentials for its use as a non-destructive and "green" analytical method for quality control in the food industry.

Keywords
Artificial sweetener • Raman spectroscopy • Chemometrics • Pattern analysis • Classification

Contents

1 Artificial Sweeteners: Modern Substitutes for Sugar 480
2 The Table-Top Artificial Sweeteners Analyzed 482
3 Raman Spectroscopy: The Instrumentation 483
4 Experimental Results and Data Processing 485
4.1 Raman Spectra 485
4.2 Chemometric Processing 486
4.3 Distinguishing the Artificial Sweeteners in Powders 488
4.4 Distinguishing the Artificial Sweeteners in Aqueous Solutions 488
5 Prospects 489
References 490

1 Artificial Sweeteners: Modern Substitutes for Sugar

Artificial sweeteners have been in use since 1879, that is, since the discovery of saccharin. Many others have been created since then and are currently being used by increasing numbers of people as an alternative to sugar in food and drink. These sugar substitutes have a sweetening capacity that is hundreds of times greater than that of sugar while providing a pleasant sweet taste. Their very low energy content, i.e., a few kcal per gram, can be regarded as negligible and makes them particularly suitable for weight maintenance or reduction as well as for other dietary needs. They also help in preventing dental decay, since they do not promote the growth of bacteria that cause cavities. Indeed, artificial sweeteners can be used with complete confidence, thanks to scientific studies which have demonstrated their safety and to numerous worldwide laws that dictate the amounts that can be used and the required labeling [1, 2].

The confectionery section of the food industry frequently needs to identify the type of artificial sweetener in a mixture prior to the introduction of the latter into the production chain. Several analytical methods based on different techniques are available for determining artificial sweeteners both qualitatively and quantitatively. A comprehensive review of conventional analytical techniques and their comparison, together with an extensive list of related references, can be found in Ref. [3]. These traditional methodologies, namely, high-performance liquid chromatography, ion chromatography, gas chromatography, capillary electrophoresis, flow-injection analysis, and electrochemical techniques, require the preparation of sample, are both destructive and time consuming, and involve the use of expensive instrumentation.

Optical spectroscopy is currently emerging as a modern and "green" analytical technique for intact food analyses, thanks to the non-destructive nature of light measurements which enable rapid checks without making use of reagents or chemical treatments, thus avoiding the problem of waste disposal [4, 5]. Moreover, spectroscopy always earns the highest scores of green effectiveness in the eco-scale for assessing the analytical procedures [6, 7]. While absorption and fluorescence spectra shows broad peaks resulting from the convolution of the many overlapping bands, which are poorly resolved for the purposes of multicomponent analysis, vibrational spectroscopy show sharp bands that identify the molecular composition. Indeed, infrared and Raman spectroscopy can immediately lead to the detection of multiple components and to their quantification [8–10]. Ultraviolet spectroscopy has been employed for determining aspartame and acesulfame-k in solutions [11], while ultraviolet combined with visible spectroscopy has been exploited for measuring the concentration of saccharin and acesulfame-k in different sweeteners and fruit juice powders [12]. In addition, attenuated total reflectance Fourier transform infrared spectroscopy has been used to determine the concentration of aspartame in soft drinks [13], acesulfame-k in diet foods [14], and aspartame and acesulfame-k in table-top sweeteners [15]. Raman spectroscopy carried out with Fourier transform-based detection units has demonstrated its effectiveness in quantifying aspartame [16, 17], as well as sodium saccharin and sodium cyclamate [18]. Recently, dispersive Raman spectroscopy was used for the detection of counterfeit stevia products [19].

The objective of this paper is to assess how Raman spectroscopy can be used for qualitatively distinguishing different types of artificial sweeteners in a mixture. The results will be of interest for industrial applications in the field of industrial quality control. A selection of fourteen common and commercially available table-top artificial sweeteners was considered. The samples contained aspartame and saccharin as high-intensity sweeteners and dextrose, sorbitol, sucrose, and maltodextrin as low-intensity sweeteners. They were examined in the form of both powder and aqueous solutions. Raman spectra were measured using an innovative compact instrument that provided excitation at 1064 nm and made use of a dispersive detection unit. Thanks to the long excitation wavelength, the spectra provided fluorescence-free Raman signatures from which to identify the most significant peaks of the various sweeteners. These peaks were also compared with those obtained by means of computational analysis in order to show the effect of the entire sweetener matrix. The spectroscopic data were then processed by means of chemometric analysis for identifying which kind of sweetener was present in a given sample. First, Principal Component Analysis was applied for the purpose of data dimensionality reduction and explorative investigation and provided a good clustering depending on the type of sweetener. The K-nearest neighbors method was then used for the assignment of the samples to predefined classes. An excellent allocation in accordance with the type of sweetener was thus obtained. Indeed, Raman spectroscopy in combination with chemometric data processing made it possible to create identification maps that could distinguish the type of both high-power sweeteners (saccharin or aspartame) and low-power sweeteners (sorbitol, dextrose, sucrose, maltodextrin). These results confirm the ability of Raman spectroscopy in obtaining a straightforward analysis of intact food, with high potentials for its use as a non-destructive and "green" analytical method for quality control purposes within the food industry.

2 The Table-Top Artificial Sweeteners Analyzed

Table 1 summarizes the table-top artificial sweeteners examined in this study, together with their origin, main ingredients, and the codes provided for their rapid identification. These fourteen selected samples were collected in Europe, USA, and Asia, and represent a picture of the most popular sweeteners among those available commercially. They consist of mixtures of sweeteners that have a sweetening capacity ranging from high to low. Saccharin and aspartame, which are used in low concentrations, are high-power sweeteners. Sorbitol, dextrose, sucrose, and maltodextrins, which are used in higher concentrations, are those with a lower sweetening capacity. Other additives, which are used as stabilizers, are typically based on corn, cellulose, and calcium silicate.

Before proceeding with experimental spectroscopy, the vibrational frequencies of the high- and low-power sweeteners identified were calculated analytically, for their optimized molecular geometries. The geometrical optimizations of saccharin, aspartame, sorbitol, dextrose, sucrose, and maltodextrin were carried out within a Gaussian 09 package [20]. The calculations were performed at a B3LYP-DFT [21] level of theory with the 6-31 + G(d,p) basis set for all the atomic species. Raman frequencies

Table 1 Artificial sweeteners examined and their characteristics

Code	Brand	Country	Ingredients
DM	Diet Magnelli	Italy	Dextrose, Sodium Cyclamate, Sodium saccharin
PF	Peso Forma	Italy	Aspartame, Dextrose
DI	Dietor	Italy	Mannitol, Sorbitol, Fructose, Sodium saccharin
TS	Tropicana Slim	Indonesia	Sorbitolo, Aspartame, Corn powder
SL	Sweet'n Low	USA	Dextrose, Sodium saccharin, Tartar cream, Calcium silicate
EQ	Equal Original	USA	Dextrose, Aspartame, Acesulfame K, Maltodextrins
AC	American Coffee Services	USA	Dextrose, Saccharin, Maltodextrins
SW	Sweet Light	Italy	Sucrose, Aspartame, Acesulfame K
CA	Canderel	Belgium	Dextrose, Aspartame, Acesulfame K, Silicon dioxide
NZ	Natrena	Belgium	Aspartame, Maltodextrins
SU	Surari	Japan	Sucrose, Aspartame, Acesulfame K
DR	Darmar	Italy	Dextrose, Sodium cyclamate, Sodium saccharin
VI	Viander	Italy	Dextrose, Aspartame
CV	Café Valet	USA	Dextrose, Saccharin, Maltodextrins

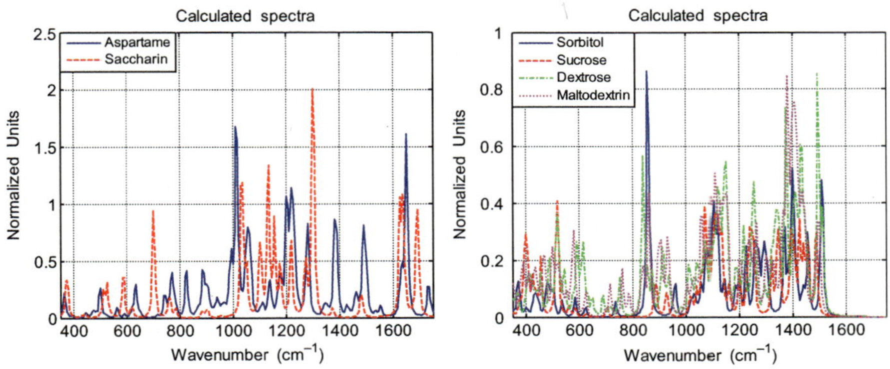

Fig. 1 Calculated Raman spectra. *Left*: aspartame and saccharin; *Right*: sorbitol, dextrose, sucrose, and maltodextrin

were calculated for the fully optimized molecular geometries. The real wavenumbers confirmed the minima features of the molecules. Maltodextrins, which are chains of glucose units, were represented in our calculation by maltose, which is a chain of only two glucose units. Figure 1 shows the calculated Raman spectra. Table 2 summarizes the main frequencies of the Raman spectra and the vibrational assignments.

3 Raman Spectroscopy: The Instrumentation

The Raman effect can be observed when monochromatic light impinges on a molecule and interacts with its electron cloud. In addition to the incident light scattered with an unchanged wavelength (elastic or Rayleigh scattering), a small fraction of the order of 10^{-6} of the incident light is scattered inelastically and undergoes a wavelength shift. This wavelength shift does not depend on the illumination wavelength but only on the molecular vibrational levels. Hence, since it is a unique expression of the molecular structure, the Raman spectrum is considered to be a molecular fingerprint – a signature [22, 23].

The instrument for Raman spectroscopy used in this experiment provided laser excitation at 1064 nm, which is not the most popular wavelength for Raman experiments because the Raman signal is inversely proportional to the fourth power of the illuminating wavelength. However, the use of this long excitation wavelength makes it possible to avoid fluorescence effects that are common in sweeteners and that could mask the weak Raman signal. The detection unit was based on a dispersive scheme that provided a more compact unit as compared to the Fourier transform configurations frequently used in other Raman instruments. Three spectrometers made it possible to operate over a wide wavenumber range (300–3200 cm^{-1}), and a thermoelectric-cooled InGaAs array that was set at -55 °C served as detector. A resolution of 4 cm^{-1} was obtained. As shown in Fig. 2, an optical fiber was used to guide the laser light to a microoptic unit. This unit

Table 2 Calculated and measured wavenumbers of the artificial sweeteners and vibrational assignments

Structure	Experimental (cm^{-1})	Calculated (cm^{-1})	Main character assignments
Aspartame 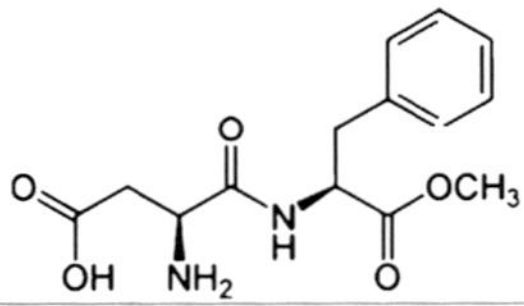	622.0	634.0	Aryl ring deformation and C–H bending
	818.0	824.0	Aryl C–H wagging
	1005.0	1013.6	Aryl ring deformation
Saccharin anion 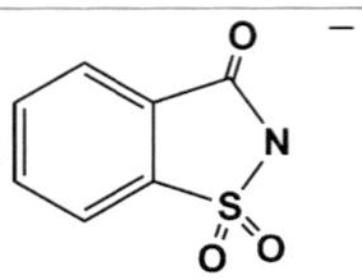	711.0	701.5	Skeleton deformation
	1023.0	1032.0	Aryl ring deformation
	1143.0	1132.2	Aryl ring deformation and C–H bending
	1154.0	1155.0	Aryl ring deformation and C–H bending
	1298.0	1300.0	C–N stretching and C–H bending
Sorbitol 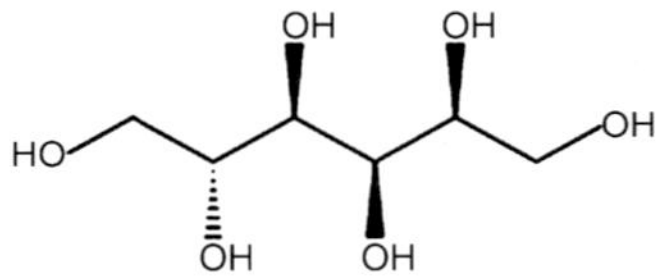	878.0	857.0	Skeletal vibration and C–H twisting
	1056.0	1075.7	C–O stretching and O–H bending
	1093.0	1107.0	C–C stretching and O–H bending
	1134.0	1119.6	C–O and C–C stretching, and O–H bending
Dextrose 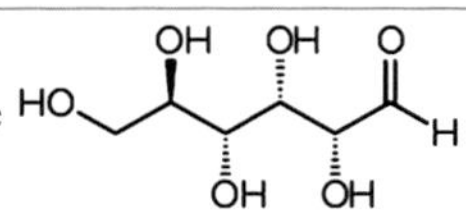	416.0	400.0	O–H bending
	515.0	516.8	Skeletal deformation
	563.0	562.0	Skeletal
	857.0	885.0	C–H bending and C–C stretching
	919.0	926.0	C–H bending and C–C stretching
	1075.0	1073.3	Skeletal deformation
	1125.0	1114.3	C–O stretching and C–H bending
Sucrose 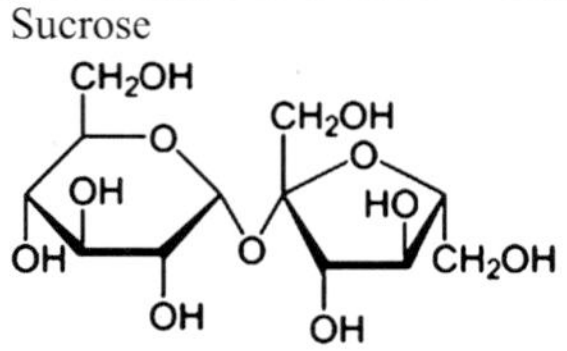	527.0	517.0	Skeletal 6-ring deformation
	546.0	523.0	Skeletal 6-ring deformation
	640.0	613.1	Skeletal 5-ring deformation
	850.0	838.0	Molecular vibration
	921.0	914.0	C–H bending and C–C stretching
	1040.0	1042.0	C–C and C–O stretching and O–H bending
	1126.0	1123.0	C–C and C–O stretching

(*continued*)

Table 2 (continued)

Structure	Experimental (cm^{-1})	Calculated (cm^{-1})	Main character assignments
Maltodextrin 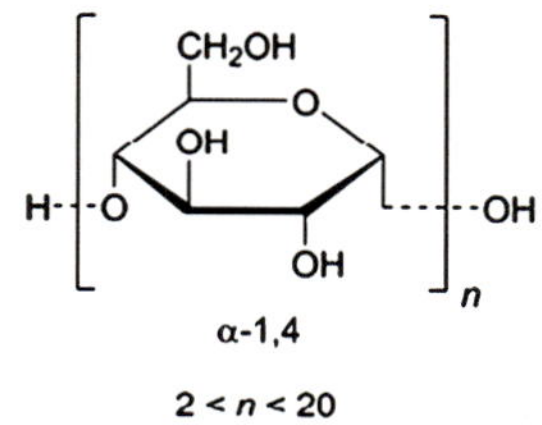	481.0	467.0	O–H bending and ring deformation
	852.0	859.0	C–H bending and C–O stretching
	933.0	906.4	C–H bending and C–C stretching
	1084.0	1098.3	C–H bending and C–C/C–O stretching
	1125.0	1115.0	6-ring deformation

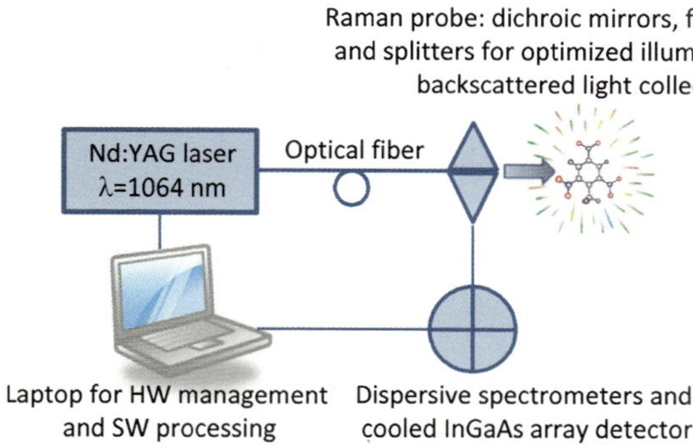

Fig. 2 Setup for Raman spectroscopy

was a compact device capable of optimizing the sample illumination and the backscattered light collection. The samples were analyzed inside a 4 ml vial. This vial was inserted in a suitable holder that was butt-coupled to the microoptic unit. A thorough description of this instrument is given in Ref. [24]. The entire instrument was portable; it was interfaced to a laptop PC which included software for the management of hardware options and for spectra acquisition, display, and first processing [25].

The sweeteners of Table 1 were measured both in powder form and as aqueous solutions. The solutions were prepared using a 10% concentration of the powder.

4 Experimental Results and Data Processing

4.1 Raman Spectra

Raman measurements were performed over the entire operational range of the instrument, that is, 300–3200 cm^{-1}, for both powders and solutions. Powders were measured using 300 mW laser power, 25 s. integration time, and two acquisitions; 4 replicas were measured for each sample and the spectrum of that sample was calculated as the average of the replica measurements. Solutions were measured with 400 mW laser power, 30 s. integration time, and three acquisitions; also in this case, 4 replicas were measured for each sample and the spectrum of that sample was calculated as the average of the replica measurements.

A different pre-processing was applied for the spectra of powders and solutions because of the different spectral behaviors. The spectra of powders underwent a baseline correction by subtracting the baseline estimated in the 1700–2000 cm^{-1}; the spectra were then normalized to the unit area in the 350–1700 cm^{-1} range. The baseline correction for the solutions was performed by considering the baseline in the 2300–2600 cm^{-1} band. The influence of water to the spectra was estimated in the 1600–2300 cm^{-1} band, which was almost free from interference by sweeteners and was then subtracted. Lastly, the spectra were normalized to the unit area in the 300–1600 cm^{-1} band.

Figures 3 and 4 show the resulting Raman spectra of powders and solutions, respectively, depicted in the ranges showing the most significant differences: namely, 400–1250 cm^{-1} for the powders and 350–1600 cm^{-1} for the solutions. Each figure consisted of two identical figures: the left figure groups the spectra depending on whether saccharin or aspartame is present, while the right figure distinguishes the spectra of the samples made up of sorbitol, dextrose, sucrose, or maltodextrin. Table 2 shows a comparison of theoretically calculated and experimentally measured Raman frequencies.

4.2 Chemometric Processing

The spectroscopic data were first processed by means of Principal Component Analysis (PCA) for the purpose of distinguishing the different types of sweeteners. PCA, which is a convenient and popular method for explorative analysis and data dimensionality reduction, is capable of providing the coordinates for identifying the samples

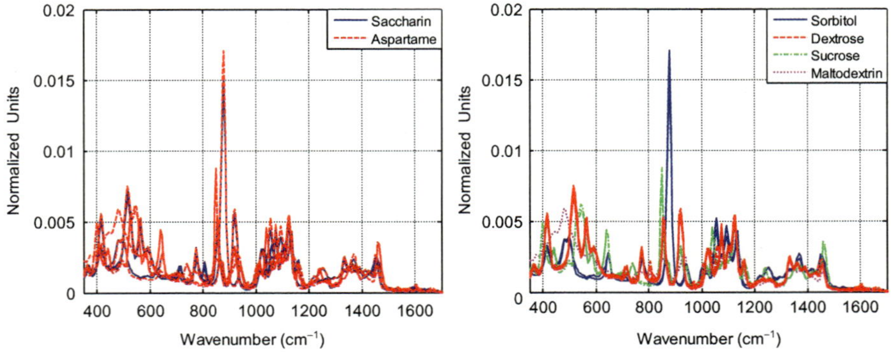

Fig. 3 Measured Raman spectra of all samples in powder form. *Left*: spectra grouped depending on whether saccharin or aspartame is present. *Right*: spectra grouped depending on whether sorbitol, dextrose, sucrose, or maltodextrin is present

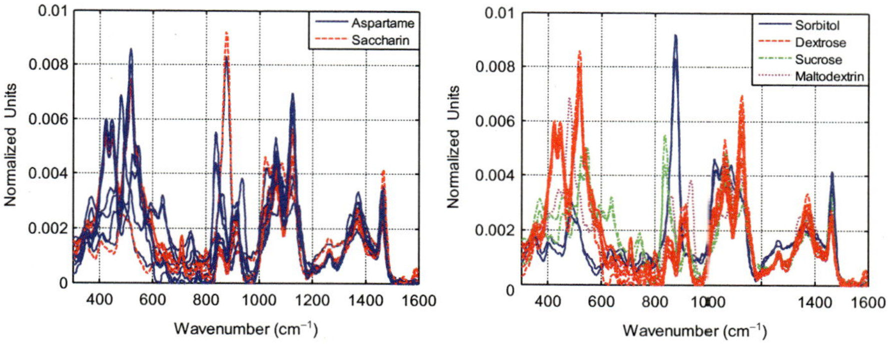

Fig. 4 Measured Raman spectra of all aqueous solutions. *Left*: spectra grouped depending on whether saccharin or aspartame is present. *Right*: spectra grouped depending on whether sorbitol, dextrose, sucrose, or maltodextrin is present

in a 2D or 3D map [26]. This method combines the spectroscopic data that characterize each sample in linear manner in order to produce new variables. The coefficients that provide the weight of each variable in the linear combination are called *loadings*. The new variables, which are mutually uncorrelated, are called Principal Components (PCs). The first PC (PC1) has the largest variance among all possible linear combinations of the starting variables, while the PCn has the largest variance among all linear combinations of the starting variables that are orthogonal to PC1 … PC(n − 1). This means that high-order PCs carry little weight in characterizing the samples and can be disregarded with a minimal loss of information.

A discriminant analysis was then achieved by means of the K-nearest neighbors (KNN) method [27]. KNN automatically classifies unknown objects within the set of predefined classes, without having to hypothesize any data density distribution. KNN simply measures the distance of the unclassified pattern from all patterns of the training set; it then selects the K-nearest patterns (the K nearest neighbors) and examines their classes. The unclassified object is assigned to the class most greatly represented within the set of K neighbors. In our analysis, K = 1 was chosen for the number of neighbors. Each sample was classified using the other 13 as a training set, and the resulting class was compared with the true one. The percentage of correct classifications was then calculated. In order to visualize the "influence" zone of each class, that is, the zone in which an unknown sample should lie in order to be assigned to a given class, a uniformly spaced grid of points was generated in the PCA space, and these points were then classified using all 14 samples as a training set for the KNN algorithm. The decision borders between the different zones were then added to the score plots.

In practice, PCA and KNN were used sequentially: the dimensionality of the pattern space was firstly reduced by means of PCA, and KNN was then applied for the classification. PCA processing was carried out by means of the CAMO-Unscrambler® commercial software [28], while KNN calculations were carried out by means of programs implemented in Matlab® code.

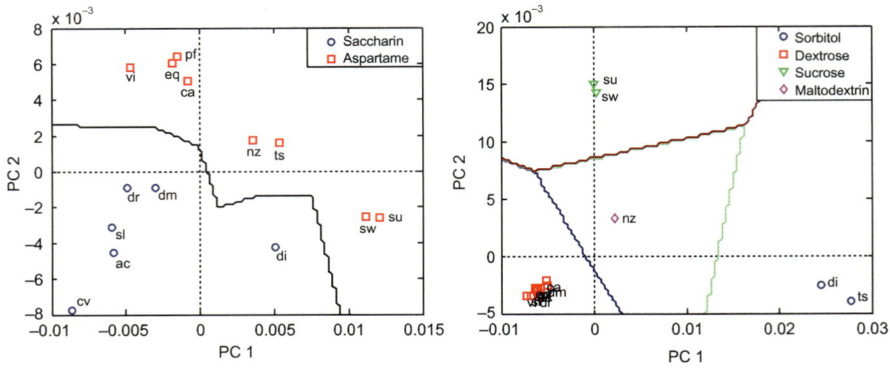

Fig. 5 Artificial sweeteners in powder form: 2D plot of PCA scores and decision borders obtained by means of KNN classification. *Left*: distinguishing the samples according to the type of high-power sweetener; *Right*: distinguishing the samples according to the type of low-power sweetener

4.3 Distinguishing the Artificial Sweeteners in Powders

The distinction between samples containing sorbitol, dextrose, sucrose, or maltodextrin was achieved by processing the entire preprocessed spectra with the use of PCA. Figure 5-right shows the 2D plot of the PCA scores that made a clear distinction between the samples depending on the type of low-power sweetener. The decision borders obtained by means of the KNN classification are also shown: they were achieved by considering the PC1, PC2, and PC3 as predictors, and an accuracy of 92.9% was obtained. The Natrena sample (nz in the graph) was obviously misclassified, because it was alone in its class and, consequently, had no neighbors of the same class around it.

The distinction between the samples containing saccharin and aspartame was achieved by processing the second derivative of the preprocessed spectra in the 670–730 cm^{-1} band using PCA. In fact, restricting the processing to this limited band removed almost all interference on the part of dextrose, sorbitol, and maltodextrin, with the exception of the weak dextrose band at 718 cm^{-1}, which was managed using the second derivative processing. Figure 5-left shows the 2D plot of PCA scores which clearly distinguishes between the samples according to the type of high-power sweetener. The decision borders obtained by means of KNN classification are also shown; these were attained by considering PC1 and PC2 as predictors, and an accuracy of 92.9% was obtained.

4.4 Distinguishing the Artificial Sweeteners in Aqueous Solutions

Just as in the case of sweeteners in powder form, also for aqueous solutions the distinction between samples containing sorbitol, dextrose, sucrose, or maltodextrin was achieved by processing the full preprocessed spectra by means of PCA. Figure 6-right shows the 2D plot of PCA scores which clearly distinguishes between samples

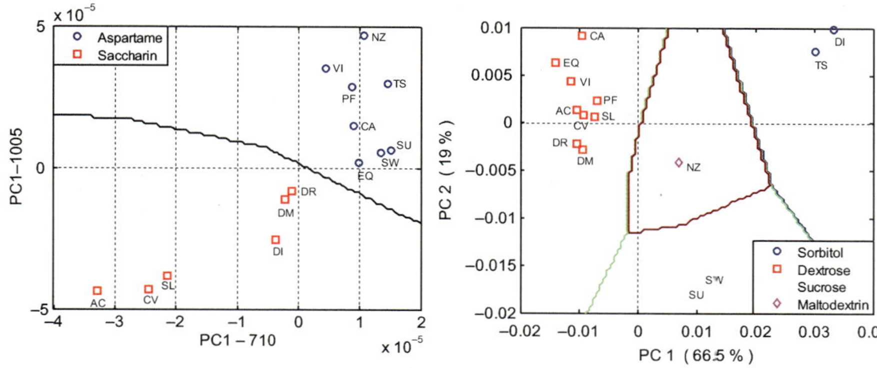

Fig. 6 Artificial sweeteners in aqueous solutions: 2D plot of PCA scores and decision borders obtained by means of KNN classification. *Left*: distinguishing the samples according to the type of high-power sweetener; *Right*: distinguishing the samples according to the type of low-power sweetener

depending on the type of low-power sweetener. The decision borders obtained by means of KNN classification are also shown: they were attained by considering PC1, PC2, and PC3 as predictors and also in this case an accuracy of 92.9% was obtained. As in the case of the sweeteners in powder form, the Natrena sample (NZ in the graph) was misclassified due to the fact that it was a single sample.

The distinction between samples containing saccharin and aspartame was achieved by using PCA to process the second derivative of the preprocessed spectra in two bands: B1 = 670–730 cm^{-1}, which contained a strong saccharin peak at 710 cm^{-1}, and B2 = 970–1030 cm^{-1}, which contained a strong aspartame peak at 1005 cm^{-1}. The use of two bands would not be necessary, as the B1 band provides sufficient discrimination capability. However, the use of both bands greatly improved the resolution of the two groups. The B2 band did not give satisfactory results for the powder samples because the interference from the low-power sweeteners was too great. However, in the water solutions, some of their peaks were weakened, and thus the use of the B2 band for discrimination purposes was relevant. Figure 6-left shows the 2D plot of PCA scores which clearly distinguishes the samples depending on the type of high-power sweetener. PC1 calculated in the B1 and B2 bands were considered. The decision borders obtained by means of KNN classification are also shown: these were achieved by considering as predictors PC1 calculated in the B1 band and PC1 calculated in the B2 band, and an accuracy of 100% was obtained.

5 Prospects

Raman spectroscopy excited at 1064 nm and measured by means of a compact dispersive detector demonstrated good potentials for distinguishing the composition of different types of popular table-top artificial sweeteners, both in powder form and as aqueous solutions. The fluorescence effects that are typically present in organic components were avoided by using a long excitation wavelength, while an effective cooling system enabled us to acquire low-noise Raman signals, thus making a qualitative analysis possible. This study is only the first step in a broader experiment that will consider other artificial sweeteners worldwide, as well as mixtures of natural and artificial sweeteners. Moreover, a model building for predicting the concentration of the sweeteners, as well as the degrees Brix, is in progress.

Acknowledgements The authors wish to thank the Ente Cassa di Risparmio di Firenze, the MIUR-PON contract #00636 "Fingerimball," FWO (G008413N), IWT, the MP1205 COST Action, the Methusalem and Hercules foundations, and the OZR of the Vrije Universiteit Brussel for partial funding. Project HP10CHEVJ8 (CINECA) and CREA Center (Centro Ricerche Energia e Ambiente) are also acknowledged for the computational resources.

References

1. Online (2016) The Sugar Association. http://www.sugar.org/
2. Online (2016) International Sweeteners Association. http://www.sweeteners.org
3. Zygler A, Wasik A, Namieśnik J (2009) Analytical methodologies for determination of artificial sweeteners in foodstuffs. TRAC Trends Anal Chem 28:1082–1102. doi:10.1016/j.trac.2009.06.008
4. de la Guardia M, Garrigues S (eds) (2011) Challenges in green analytical chemistry. RSC Publishing, Cambridge. ISBN 978-1-84973-132-4
5. Armenta S, Garrigues S, de la Guardia M (2008) Green analytical chemistry. TRAC Trends Anal Chem 27:497–511. doi:10.1016/j.trac.2008.05.003
6. Gałuszka A, Migaszewski ZM, Namieśnik J (2013) The 12 principles of green analytical chemistry and the SIGNIFICANCE mnemonic of green analytical practices. TRAC Trends Anal Chem 50:78–84. doi:10.1016/j.trac.2013.04.010
7. Gałuszka A, Konieczka P, Migaszewski ZM, Namieśnik J (2012) Analytical eco-scale for assessing the greenness of analytical procedures. TRAC Trends Anal Chem 37:61–72. doi:10.1016/j.trac.2012.03.013
8. Moros J, Garrigues S, de la Guardia M (2010) Vibrational spectroscopy provides a green tool for multi-component analysis. TRAC Trends Anal Chem 29:578–591. doi:10.1016/j.trac.2009.12.012
9. Chalmers JM, Griffiths PR (eds) (2001) Handbook of vibrational spectroscopy. Selected chapters on food science. Applications in life, pharmaceutical and natural sciences, Vol 5. Wiley, Chichester. ISBN: 978-0-471-98847-2
10. Li-Chan ECY, Chalmers J, Griffiths P (eds) (2010) Applications of vibrational spectroscopy in food science (2 volume set). Wiley, Chichester. ISBN 978-0-470-74299-0
11. Cantarelli MA, Pellerano RG, Marchevsky EJ, Camiña JM (2009) Simultaneous determination of aspartame and acesulfame-k by molecular absorption spectrophotometry using multivariate calibration and validation by high performance liquid chromatography. Food Chem 115:1128–1132. doi:10.1016/j.foodchem.2008.12.101
12. Llamas NE, Di Nezio MS, Palomeque ME, Fernández Band BS (2008) Direct determination of saccharin and acesulfame-k in sweeteners and fruit juices powders. Food Anal Methods 1:43–48. doi:10.1007/s12161-007-9006-z
13. Khurana HK, Cho IK, Shim JY, Li QX, Jun S (2008) Application of multibounce attenuated total reflectance Fourier transform infrared spectroscopy and chemometrics for determination of aspartame in soft drinks. J Agric Food Chem 56:778–783. doi:10.1021/jf0724116

14. Shim JY, Khurana IK, Li QX, Jun S (2008) Attenuated total reflectance-Fourier transform infrared spectroscopy coupled with multivariate analysis for measurement of acesulfame-k in diet foods. J Food Sci 73:C426–C431. doi:10.1111/j.1750-3841.2008.00751.x
15. Armenta S, Garrigues S, de la Guardia M (2004) FTIR determination of aspartame and acesulfame-k in tabletop sweeteners. J Agric Food Chem 52:7798–7803. doi:10.1021/jf0492181
16. Mazurek S, Szostak R (2011) Quantification of aspartame in commercial sweeteners by FT-Raman spectroscopy. Food Chem 125:1051–1057. doi:10.1016/j.foodchem.2010.09.075
17. Peica N (2009) Identification and characterization of the E951 artificial food sweetener by vibrational spectroscopy and theoretical modelling. J Raman Spectrosc 40:2144–2154. doi:10.1002/jrs.2384
18. Armenta S, Garrigues S, de la Guardia M (2004) Sweeteners determination in table top formulations using FT-Raman spectrometry and chemometric analysis. Anal Chim Acta 521:149–155. doi:10.1016/j.aca.2004.05.077
19. Vargas Jentzsch P, Torrico-Vallejos S, Mendieta-Brito S, Ramos LA, Ciobotă V (2016) Detection of counterfeit stevia products using a handheld Raman spectrometer. Vib Spectrosc 83:126–131. doi:10.1016/j.vibspec.2016.01.015
20. Frisch MJ, Trucks GW, Schlegel HB, Scuseria GE, Robb MA, Cheeseman JR, Scalmani G, Barone V, Mennucci B, Petersson GA, Nakatsuji H, Caricato M, Li X, Hratchian HP, Izmaylov AF, Bloino J, Zheng G, Sonnenberg JL, Hada M, Ehara M, Toyota K, Fukuda R, Hasegawa J, Ishida M, Nakajima T, Honda Y, Kitao O, Nakai H, Vreven T, Montgomery JA Jr, Peralta JE, Ogliaro F, Bearpark M, Heyd JJ, Brothers E, Kudin KN, Staroverov VN, Kobayashi R, Normand J, Raghavachari K, Rendell A, Burant JC, Iyengar SS, Tomasi J, Cossi M, Rega N, Millam JM, Klene M, Knox JE, Cross JB, Bakken V, Adamo C, Jaramillo J, Gomperts R, Stratmann RE, Yazyev O, Austin AJ, Cammi R, Pomelli C, Ochterski JW, Martin RL, Morokuma K, Zakrzewski VG, Voth GA, Salvador P, Dannenberg JJ, Dapprich S, Daniels AD, Farkas Ö, Foresman JB, Ortiz JV, Cioslowski J, Fox DJ (2009) Gaussian09. Gaussian, Inc., Wallingford
21. Becke AD (1993) Density-functional thermochemistry. III. The role of exact exchange. J Chem Phys 98:5648–5652
22. Ferraro JR, Nakamoto K (eds) (2003) Introductory Raman spectroscopy, 2nd edn. Elsevier Science. ISBN: 978-0-12-254105-6. http://www.fulviofrisone.com/attachments/article/406/introductory%20raman%20spectroscopy.pdf
23. Larkin PJ (2011) IR and Raman spectroscopy: principles and spectral interpretation. Elsevier, Amsterdam. ISBN 978-0-12-386984-5
24. Mignani AG, Ciaccheri L, Mencaglia AA, Di Sanzo R, Carabetta S, Russo M (2016) Dispersive Raman spectroscopy for the nondestructive and rapid assessment of the quality of Southern Italian honey types. J Lightwave Technol 34:4479–4485. doi:10.1109/JLT.2016.2539550
25. Online (2016) Model BRAM-1064-HR. Bayspec Inc., San Jose www.bayspec.com
26. Jackson JE (2003) A user's guide to principal components. Wiley, Hoboken. ISBN 978-0-471-47134-9
27. Cover TM, Hart PE (1967) Nearest neighbor pattern classification. IEEE Trans Inf Theory 13:21–27. doi:10.1109/TIT.1967.1053964
28. Online (2016) http://www.camo.com/rt/Products/Unscrambler/unscrambler.html, Version 10.3

Sweet-Tasting Protein Thaumatin: Physical and Chemical Properties

22

Tetsuya Masuda

Abstract

Thaumatin is a sweet-tasting protein isolated from the fruits of *Thaumatococcus daniellii* Benth, a plant native to tropical West Africa. Thaumatin consists of a single-chain of 207 amino acid residues and elicits sweet taste at only a concentration of 50 nM, a value 100,000 times larger than that of sucrose on a molar basis. The intensely sweet taste of thaumatin has potential as a low-calorie sweetener as well as substitute for sucrose for industrial applications, and it may be useful tool in clarifying the mechanisms how we perceive of sweet taste. Nowadays, thaumatin has widely been used not only a natural sweetener but also as a substance for flavor enhancers as well as masking unpleasant taste in the food and pharmaceutical industries. This chapter describes recent progress as well as historical backgrounds on thaumatin: features of the protein, the recombinant production, the sweetness-determinants, the docking simulation with sweet taste receptors, and the atomic resolution structure with its applications.

Keywords

Thaumatin • Sweet-tasting protein • Structure-sweetness relationship • Chemical modification • Site-directed mutagenesis • Sweet taste receptors • Atomic structural analysis

Contents

1 Introduction 494
2 History of Thaumatin 495
 2.1 Thaumatin Is a Proteinaceous Sweetener 495
 2.2 Characteristics of Thaumatin 496

T. Masuda (✉)
Division of Food Science and Biotechnology, Graduate School of Agriculture, Kyoto University, Uji, Japan
e-mail: t2masuda@kais.kyoto-u.ac.jp

J.-M. Mérillon, K.G. Ramawat (eds.), *Sweeteners*, Reference Series in Phytochemistry,
https://doi.org/10.1007/978-3-319-27027-2_10

2.3 Primary Structure of Thaumatin 497
3 Expression of the Recombinant Thaumatin 498
3.1 Production of Thaumatin by Bacteria 498
3.2 Production of Thaumatin by Fungi 498
3.3 Production of Thaumatin by Yeast *S. cerevisiae* 499
3.4 Production of Thaumatin by Yeast *Pichia pastoris* 499
3.5 Production of Thaumatin by Transgenic Plant 501
4 Elucidation of the Sweetness-Determinants of Thaumatin 501
4.1 Chemical Modification Approaches 501
4.2 Site-Directed Mutagenesis Approaches 503
4.3 The Sweetness-Determinants Were Surveyed by Monoclonal Antibody Mapping 507
5 Prediction of the Mechanism of Interaction Between Sweet-Tasting Proteins and Their Receptors 507
5.1 Sweet Taste Receptor 507
5.2 Interpretations of Binding Sites of Sweeteners by Cell-Based Assay 508
5.3 Wedge Model 509
5.4 The Complex Model Between the T1R2-T1R3 Sweet Receptor and Thaumatin ... 510
6 Three-Dimensional Structure of Thaumatin 510
6.1 Early Three-Dimensional Structure Analysis of Thaumatin 511
6.2 Atomic Resolution Structural Analysis of Thaumatin 511
6.3 Structure of Thaumatin II 514
6.4 Neutron Crystallographic Analysis 515
6.5 Thaumatin Crystals as a Tool in Checking Various Effects 515
6.6 Atomic Structure of the Thaumatin at pH 8.0 516
7 Conclusions 517
References 517

1 Introduction

Many low-molecular weight molecules including amino acids, saccharides, polyols, peptides, and other synthetic compounds are well known to elicit the sensation of sweetness [1], whereas most proteins are tasteless and flavorless. However, some proteins are known to elicit a sweet taste response on the human palate. To date, six proteins, thaumatin [2], monellin [3, 4], mabinlin [5], curculin (neoculin) [6–8], brazzein [9], and egg white lysozyme [10, 11] were identified as eliciting a sweet taste. Among sweet-tasting proteins, thaumatin and monellin are derived from plants in tropical Africa and elicit a sweet taste at a low concentration approximately 100,000-fold less than that of sucrose on a molar basis. In this review, the author focuses on one of the potent sweet-tasting proteins, thaumatin, and its physical and chemical properties including recent insights as well as the early studies.

In the second section, the history and characteristic of thaumatin are reviewed. Thaumatin is a protein and nontoxic, and it has been used as a sweetener and for flavor enhancement and synergistic properties in food products [12]. Because of the difficulty and limitation of obtaining the natural source of thaumatin, numerous attempts to produce thaumatin in different microorganisms and in transgenic plants have been performed [13, 14]. The production of an active sweet form is still difficult since thaumatin contains eight disulfide bonds in a molecule [15]. These topics will be mentioned in Sect. 3.

The potent sweetness features of sweet-tasting protein would be useful as sugar-substitutes and would serve to clarify the perception of sweetness. A number of investigations have been performed to assess sweetness-structural relationships in sweet-tasting proteins. The effects of charged residues on the elicitation of sweetness have been thoroughly investigated by site-direct mutagenesis as well as chemical modification approaches, suggesting that sweet-tasting proteins interact with sweet receptors through a multipoint interaction involving both positive and negative charges but mainly basic residues [16–24]. The elucidation of the sweetness-determinants of thaumatin will be described in Sect. 4.

Assessment of the structure-sweetness relationships in sweet-tasting proteins may provide valuable information on the mechanism for the elicitation of sweetness in proteins as well as for the interaction between sweet-tasting proteins and sweet taste receptors. The properties of the sweet taste receptors and docking simulation between thaumatin and sweet taste receptors also will be described in Sect. 5.

The three-dimensional structures of thaumatin have been determined, and recently atomic resolution structural analyses for thaumatin [25–28] as well as monellin [29] have been reported. However, no common feature among sweet-tasting proteins has been identified in the tertiary structure or in the amino acid sequence. Most sweet proteins, thaumatin and monellin in particular, are basic proteins and have high isoelectric points. The detailed structural features of thaumatin and the applications for crystallographic studies will be explained in Sect. 6. Finally, the properties of the structure of thaumatin from the early insights to the recent progress will be mentioned with future prospects.

2 History of Thaumatin

2.1 Thaumatin Is a Proteinaceous Sweetener

The fruits of *Thaumatococcus daniellii* Benth contain a very sweet-tasting substance. In 1852, this miraculous berry distributed in West Africa was first described by Daniell [30]. It is called "Katemfe" by the native habitants. Its principles of sweetness are a proteinaceous substance, thaumatin; the extraordinary impression on palate was caused by the ingestion of a berry, so that the sour substances become intensely sweet, and unpleasant qualities, such as citric acid, tartaric acid, lime juice, and vinegar, were also lost. It has also been reported that the people of Sierra Leone commonly have recourse to them, with the view of rendering more sweet and palatable and acidulated kind of bread, sour fruits, and bad palm wine [31]. When a sufficient quantity of berry has been taken, these unique features are commonly perceptible through the day. A single fruit is composed of two or three triangular, freshly pericarps, and each triangular fruit contains a large black seed covered by a thin layer of transparent gel, with a light-yellow *aril* at the base. In Ghana, there are two distinct varieties of fruits differing in shape, size, texture, and chemical composition. The one found in the east of the Volta River is relatively large, and the other found in the Ashanti region of Ghana is typically small [32].

Inglett and May found that the sweet principle existed in aril and is soluble in water [33]. The isolation of proteinaceous sweet principles was first described by van der Wel and Loeve, researchers in Unilever Research [2].The light-yellow arils from each seed were collected and homogenized in water. The solid materials were removed by filtration, and extraction of the sweet principle was concentrated by ultrafiltration, and the concentrate was freeze-dried. Eight hundred milligram of freeze-dried extract was dissolved in 15 ml of water and subsequently purified by a Sephadex G-50 gel filtration column. The fractions with a sweet taste were pooled, concentrated, and freeze-dried. Fifty milligram of the fractions dissolved in 1 ml of 20 mM phosphate buffer, pH 7.0 were further purified by a cation-exchange column (SE-Sephadex G-25) with a linear gradient of sodium chloride and yielded two peaks containing strongly sweet-tasting materials. They named the two sweet-tasting basic proteins, thaumatin I and thaumatin II.

2.2 Characteristics of Thaumatin

Thaumatin is isolated from the arils of *Thaumatococcus daniellii* Benth, as mentioned above [2]. Van der Wel and Loeve first identified two major components (*thaumatin I* and *thaumatin II*) that have a molecular mass of around 21 kDa and have an isoelectric point of 12. Thaumatins are nearly 100,000 times sweeter than sucrose on a molar basis and about 1,600 times sweeter on a weight basis, and the threshold value of sweetness of thaumatin is around 50 nM [2, 32]. A third small peak eluted just before thaumatin I is designated as thaumatin 0 which seems not to elicit a sweet taste. By carefully fractionated with a shallow salt gradient using a Whatman CM32 column, at least three minor variants (thaumatin a, thaumatin b, and thaumatin c) were obtained besides the two major variants (thaumatin I and thaumatin II) [32]. These minor variants have similar amino acid contents and molecular weight but are less sweet than thaumatin I and thaumatin II. As thaumatin a, thaumatin b, and thaumatin c were eluted just before thaumatin I as was thaumatin 0, these three minor components seem to correspond to thaumatin 0. The contents of *thaumatin variants* in fruits at different stages of maturation and different regions of Ghana were investigated [34]. A significant increase of total sweet protein levels was observed over the whole ripening periods. Two major components of thaumatin I and thaumatin II and a minor component of thaumatin 0 were observed in Ashanti region fruits, whereas only thaumatin I and a small amount of thaumatin 0 were included in the fruit from the Kadjebe region. Thus, thaumatin I and thaumatin II are different gene products, and thaumatin is regulated during development. Analysis of the metal ion content of an extract from aril showed the existence of aluminum ions, up to six residues per molecule of thaumatin [35]. The yield of extraction by aluminum ions is increased compared with other mono-, di-, or tri-valent ions. These results suggested that aluminum ions tightly bind to thaumatin molecules serving to increase the stability and solubility of the proteins. The specificity and binding sites of aluminum ions are still obscure and should be investigated in the future.

2.3 Primary Structure of Thaumatin

At least five different variants of thaumatin existed in the fruit. In 1979, Iyengar et al. reported the *amino acid sequence* of thaumatin I, which consists of a single-chain of 207 amino acid residues, and neither carbohydrate nor unusual/modified amino acids are contained in the molecule [36]. The molecular cloning and the nucleotide sequence of thaumatin II was also reported by Edens et al. who found that thaumatin contains the amino-terminal extension of 22 amino acids, which is hydrophobic and resembles an excretion-related signal sequence [37]. Beside the amino-terminal 22 amino acids extensions, thaumatin contains the acidic 6 amino acids-long carboxyl terminal. Comparison to the amino acid sequence of thaumatin I determined by Iyengar et al. the deduced amino acid sequence of thaumatin II differed at five positions (Lys46 instead of Asn, Arg63 instead of Ser, Arg67 instead of Lys, Gln76 instead of Arg, and Asp113 instead of Asp) (Table 1).

Lee et al. reexamined the sequence of two major variants of plant thaumatin and designated them as *thaumatin A* and *thaumatin B* [38]. The sequence of thaumatin A and thaumatin B differed from that of thaumatin I determined by Iyengar et al. by one amino acid (Asp113 instead of Asn) or two amino acids (Lys46 instead of Asn and Asp113 instead of Asn), respectively. Our group also determined the *nucleotide sequence* of thaumatin I as well as thaumatin II [39]. The deduced amino acid sequence of the nucleotide sequence of thaumatin I is identical to that of thaumatin A but differed from that of thaumatin I reported by Iyengar et al. at residue 113 (Asp113 instead of Asn). The nucleotide sequence of thaumatin II we cloned was the same as the sequence reported by Edens et al.

Table 1 Amino acid sequences of thaumatin variants

	Amino acid position (nucleotide)					Structure
Thaumatin	46	63	67	76	113	PDB ID, highest resolution
I[a]	Asn	Ser	Lys	Arg	Asn	Asherie et al. 2vhk, 0.94 Å [25]
II[b]	Lys[c](AAG)	Arg[c] (CGC)	Arg[c] (CGG)	Gln[c] (CAG)	Asp[c] (GAC)	Masuda et al. 3wou, 0.99 Å [28]
A[a]	Asn	Ser	Lys	Arg	Asp	Ko et al. 1thv, 1.75 Å [108]
B[a]	Lys	Ser	Lys	Arg	Asp	Ma et al. 1rqw, 1.05 Å [http://www.rcsb.org/pdb/explore/explore.do?structureId=1RQW]
I[b]	Asn[c] (AAC)	Ser[c] (AGC)	Lys[c] (AAG)	Arg[c] (CGG)	Asp[c] (GAC)	Masuda et al., 3al7, 1.10 Å [26]

[a]Directly determined protein sequence by the Edman degradation method [36, 38]
[b]The nucleotide sequences were from GeneBank of thaumatin II (**J01209**) [37] and thaumatin I (**AF355098**) [39]
[c]Amino acid residues deduced from thaumatin nucleotides

3 Expression of the Recombinant Thaumatin

As mentioned in the introduction, it is difficult to obtain a sufficient amount of the natural source of thaumatin since the yields are highly dependent on the climates and the harvesting time in the rainforest of West Africa. In this section, the recent developments in the production of the recombinant thaumatin by different microorganisms and by transgenic plants are reviewed.

3.1 Production of Thaumatin by Bacteria

Production of thaumatin was first attempted by Edens et al. using *Escherichia coli* K-12 under control of *lac* and *trp* promoter/operator systems [37]. Faus et al. constructed the *synthetic gene* encoding the amino acid sequence of thaumatin II and attempted to express recombinant thaumatin in *E. coli* strains DH5 [40]. However, these studies did not assess the sweetness of recombinant thaumatin. Daniell et al. attempted to produce recombinant thaumatin II and successfully obtained 40 mg/L of recombinant sweet thaumatin using the reduced/oxidized glutathione renaturation system from inclusion bodies [41].

Besides *E. coli* expression systems, the expression of thaumatin was attempted using *Bacillus subtillis* [42] and *Streptomyces lividans* [43].

3.2 Production of Thaumatin by Fungi

An early study by Hahm and Batt successfully secreted the sweet-form thaumatin II in the culture medium of *Aspergillus oryzae* using the *S. cerevisiae* glyceraldehydes 3-phosphate dehydrogenase (GPD) promoter [44]. However, the yields were low, 50 ng/L. To increase the yields of products, a gene encoding the thaumatin II with the optimized for its expression in filamentous fungi was synthesized, [40] and expression of thaumatin was attempted in *Penicillium roquefortii* [45]. The recombinant thaumatin elicited a sweet taste, but the yield was not high. The expression of thaumatin in *Aspergillus nigar var. awamori* was attempted by the same group; however, yields were low at concentrations of 5–7 mg/L [46].

In order to increase the yields, Moralejo et al. examined the efficiency of various promoters [47]. The *gdhA* promoter of *A. awamori* and the *gpdA* promoter of *A. nidulans* are suitable for expression, and the highest yields were at the concentrations of 9–11 mg/L for approximately 72 h of incubation. The double transformant, which contains both the B2 promoter of *A. chrysogenum* and the *gdhA* promoter of *A. awamori,* extended the secretion of thaumatin until 96 h with a slight increase of yields of 14 mg/L. The secreted recombinant thaumatin II was purified to homogeneity, yielding one major component and two minor components. The three forms of recombinant thaumatin II differed in amino acids at the N-terminal, that is, the major component was attached to the Lys-Arg residue at the N-terminal, and the minor components were attached to the Arg residue or the

Arg-Lys-Arg residues at the N-terminal. All three forms elicited a sweet taste. Although no significant effect on the sweetness was induced by the insufficient processing of N-terminal end, heterogeneous components might interfere with obtaining the high quality of crystals for high-resolution structural analysis. Reduction of proteolytic degradation by gene silencing as well as by direct gene deletion of the aspergillopepsin B resulted in increase of secretion yield of recombinant at a concentration of around 10 mg/L [48]. The optimization of the carbon and nitrogen source of culture medium resulted in the increase of the yields at a concentration of 105 mg/L [49]. Introduction of protein disulfide isomerase encoding gene *pdiA* under the control of the glucoamylase promoter resulted in a high-yields expression (150 mg/L) of the recombinant thaumatin II [50]. Introduction of the chaperone *bipA* gene also resulted in higher levels of secreted thaumatin than the control strain by 2.5-fold [51].

3.3 Production of Thaumatin by Yeast *S. cerevisiae*

Expression of thaumatin by yeast was first performed using *S. cerevisiae* under the control of glyceraldehyde-3P-dehydrogenase (GAPDH) promoter [52]. The plant origin preprothaumatin II gene that contains the 22 amino acids signal sequence at N-terminal and 6 amino acids at C-terminal besides mature thaumatin was used for expression. The results showed that both 22 amino acids presequence and 6 amino acids prosequence were clipped off just as during processing by plants. Although Edens and van der Wel used *S. cerevisiae* and *Kluyveromyces lactis*, secretion of thaumatin was very low, and sweetness could not be assessed [53]. In order to increase the production level, Lee et al. attempted to synthesize the genes encoding thaumatin with yeast-preferred codons. The synthetic thaumatin genes were inserted into an expression vector that contains yeast 3-phosphoglycerate kinase (PGK) promoter and terminator. The recombinant thaumatin was located in the insoluble fractions of yeast at a high level; however, none elicit a sweet taste [38]. Large-scale fermentation was performed, and a total 160 mg of recombinant thaumatin was obtained from 12 L of SD^{-}leu medium [54]. By the introduction of yeast invertase secretion signal sequence between the PGK promoter region and the thaumatin gene, approximately 800 mg of the purified recombinant thaumatin that elicit a sweet taste was obtained from 5.7 L of fermentation broth [54].

3.4 Production of Thaumatin by Yeast *Pichia pastoris*

The methylotrophic yeast *Pichia pastoris* has been developed as a host for high-level production of recombinant proteins [55–57], including sweet-tasting proteins [24, 39, 58–63]. The first attempt was performed using plant thaumatin II cDNA. The gene was cloned into the yeast-shuttle vector, pPIC9K, which contains the highly efficient and inducible AOX1 promoter. The mature thaumatin gene was inserted under the *S. cerevisiae* prepro α-mating factor secretion signal sequence. Since transformant contains a kanamycin resistance gene, positive colonies were easily selected by G418 antibiotics [64, 65]. Approximately 25 mg/L of the recombinant thaumatin II was obtained, but additional amino acid residues at both N- and C-terminal ends were attached besides the mature protein [58]. Since this recombinant thaumatin II elicited a sweet taste as did plant thaumatin, N- and C-terminal regions of the thaumatin II molecule may not play an important role in eliciting the sweet taste of thaumatin.

To date, the thaumatin II gene was preferably used to express the recombinant protein by microorganisms. This is partly because the information on the nucleotide sequence of thaumatin was limited to only thaumatin II. We successfully found the thaumatin I gene from the cDNA of *Thaumatococcus daniellii* Benth and attempted to express it by *P. pastoris* [39]. This thaumatin I gene was cloned to the yeast shuttle vector, pPIC6α. This shuttle vector contains AOX1 promoter and α-factor secretion signal as does a pPIC9K vector, and notably, it contains blasticidin resistance gene. After transformation into *P. pastoris* X-33, the stable transformants were easily obtained with the selection of blasticidin in a week. Approximately 30 mg/L of recombinant thaumatin I was obtained. Unexpected amino acid residues were attached at the N-terminal end. This is partly due to the processing deficiency of Kex2 protease from yeast and the hydrophobic environment of the N-terminal region of thaumatin. Sensory analysis revealed that recombinant thaumatin I containing the additional amino acid residues at N-terminal elicited a sweet-taste sensation as did the thaumatin I from plants.

A bottleneck for the secretion of thaumatin is the processing deficiency of Kex2 protease and the attachment of unexpected amino acid residues at the N-terminal end. To overcome these problems, the roles of the *signal sequences* as well as *prosequences* were investigated [60]. The expression plasmids with different presequences and prosequences were constructed and transformed into *P. pastoris*. The transformants containing the prethaumatin gene that has the native plant signal secreted thaumatin molecules in the medium. The N-terminal amino acid sequence of the secreted thaumatin molecule was Ala, suggesting it was processed correctly. The secretion level of the recombinant thaumatin I increased at a concentration of about 60 mg/L, which is twofold higher than that of α-factor secretion signal. The importance of the five prosequences for production of the thaumatin molecule has been suggested [37, 52]. To assess these effects, the expression vectors containing five prosequences were made to analyze the production level and the processing efficiency. The results showed that the secretion yield was slightly decreased (46 mg/L) by the introduction of the prosequence. In *Pichia* systems, the production yield of thaumatin was not positively affected by the C-terminal prosequence, and the prosequence was not processed. The recombinant thaumatin I, of course, elicited a sweet taste as did plant thaumatin I. Recently, Masuda et al. successfully obtained approximately 100 mg/L of recombinant thaumatin using an expression vector that possesses three copies of the thaumatin gene containing the 22-amino acid presequence [62].

3.5 Production of Thaumatin by Transgenic Plant

The recombinant thaumatin II produced by transgenic *S. tuberosum* is in the order of 100 nM and elicited a sweet taste [66]. Western blot analysis suggested that the mobility of the recombinant thaumatin II is the same as that of plant thaumatin, suggesting both its signal sequence and prosequence were correctly processed. Recently, transgenic plants carrying thaumatin II gene have been developed to improve and modify the taste of plants. These include cucumber, pear, tomato, and strawberry [67–70]. Total soluble protein from transgenic cucumber leaves was extracted. Western blot analysis showed that the presence of thaumatin II was detected in all cucumber fruits evaluated as sweet in taste. Total soluble proteins from tomato and pear were also analyzed by western blot, and a 22 kDa band was detected. Recently, Firsov et al. reported that the yields of thaumatin in transgenic tomato fruits was approximately 50 μg/mg of total soluble protein [71], and more recently, recombinant thaumatin from tomato fruits yields 50 mg/kg fresh fruit [72].

4 Elucidation of the Sweetness-Determinants of Thaumatin

The potent sweetness features of sweet-tasting protein would be useful as sugar-substitutes, and a number of studies have been performed so far. Furthermore, the elucidation of sweetness-determinants of protein would provide valuable insights on the perception of sweetness. In this chapter, the author introduces the results of chemical modification study at first, and then the recent results obtained by the site-directed mutagenesis studies are reviewed.

4.1 Chemical Modification Approaches

In the early studies, chemical modification approaches had been extensively employed to elucidate the *sweetness-structure relationships* in elicitation sweetness of thaumatin. However, these results were fragmentary and not conclusive. Van der Wel described the results and problems of chemical modification studies in his review [73].

Here, the results of chemical modification of lysine, arginine, and glutamic/aspartic acid residues are reviewed, since the information on these residues are relatively abundant and conclusive, and deeply influences the subsequent site-directed mutagenesis studies.

4.1.1 Chemical Modification of Lysine Residue

The effects of acetylation and *methylation* of lysine residues on the sweetness of thaumatin were first reported by van der Wel and Bel. The results showed that the sweetness intensity of the acetylated thaumatins decreases with increase in the number of amino groups acetylated, and finally, sweetness is abolished when four lysine residues were acetylated. On the other hand, reductive methylation of seven lysine residues of thaumatin resulted in no reduction of sweetness. Although the *acetylation* decreases net charge and lowers the isoelectric point of the protein, methylation hardly changes the isoelectric point of the protein. These results suggested that the total net charge of thaumatin might play a role in elicitation of sweetness [74]. The modification of lysine residues by succinic acid anhydride resulted in a significant reduction of sweetness of thaumatin [75]. Introduction of one succinyl group reduced the sweetness by 50%, and the sweetness was nearly quenched by introduction of two succinyl groups. *Modification of amino groups using pyridoxal 5′-phosphate (PLP)* resulted in a reduction of sweetness [73, 76]. Removal of the phosphate group of PLP by alkaline phosphatase had no influence on sweetness [76]. For the moment, the conclusion might be justified that the net charge of thaumatin as well as thaumatin's basicity are not the dominating factor in determining the intensity of its sweetness, and some of the lysine residues are involved in elicitation of sweetness of thaumatin. Thus, characteristic features of lysine residues are attractive to identify the sweetness-determinants of thaumatin; however, these previous chemical modification studies did not identify the labeled amino acid residues. Kaneko and Kitabatake prepared a variety of lysine-modified thaumatin derivatives by chemical modification with PLP. After chromatographic purification, they successfully obtained five specific lysine-modified thaumatin derivatives (PLP-Lys78, PLP-Lys97, PLP-Lys106, PLP-Lys137, and PLP-Lys187) [77]. The sweetness of these derivatives was markedly reduced (Table 2). In contrast, the intensity of sweetness returned to that of unmodified thaumatin when these phosphopyridoxylated lysine residues were dephosphorylated by alkaline phosphatase, except Lys106. The authors suggested that Lys106 itself plays a critical role in sweetness, or an amino acid residue that is critical for sweetness is located near Lys106, or there may be other reasons. Since a significant structural change in secondary structure was not observed in all lysine derivatives, the loss of sweetness was not a result of major disruption in protein structure.

Table 2 Sensory analysis of PLP, PL and PM variants

PLP- or PM-variants	Sweetness threshold (nM)	Dephosphorylated-PLP variants (PL)	Sweetness threshold (nM)
PLP-Lys78	140 ± 50	PL-Lys78	50
PLP-Lys97	320 ± 140	PL-Lys97	50
PLP-Lys106	260 ± 120	PL-Lys106	260 ± 120
PLP-Lys137	200	PL-Lys137	50
PLP-Lys187	260 ± 120	PL-Lys187	50
PM-Asp60	50		
PM-Ala207	50		
PM-Asp21/Glu42/Asp129	44 ± 12		

Kaneko and Kitabatake [77]

4.1.2 Chemical Modification of Arginine Residue

Modification of the guanidine group of the arginine residue was performed with cyclohexane-1,2-dione in the presence of borate buffer [74].

Despite the drastic decrease of the isoelectric point of thaumatin, the sweetness of thaumatin was not changed even when 6 out of the 11 arginine residues were modified. Kitabatake and Kusunoki also performed chemical modification of arginine residues by cyclohexane-1,2-dione in borate buffer at pH 9.0 [78]. Modification of one or two arginine residues resulted in the loss of sweetness. By treatment with hydroxylamine, nonsweet derivatives restored the sweetness as like the original thaumatin. Thus, the results were controversial, further investigations would be required to clarify the role of arginine residues on the elicitation of sweetness of thaumatin.

4.1.3 Chemical Modification of Glutamic/Aspartic Acid Residues

The carboxylic group was modified to acid amide group by *N*-ethyl-*N*′-(3-dimethylaminopropyl)-carbodiimide [74]. *Amidation of the carboxylic groups* increased the isoelectric point of thaumatin and caused some increase of sweetness. In contrast, incorporation of NH-methyl or NH-ethyl group decreased the sweetness. Kaneko and Kitabatake prepared carboxyl-group-modified thaumatin derivatives by treatment with pyridoxamine (PM) and further purified by SP Sephadex C-25 column chromatography followed by ion-exchange HPLC [77]. Modification of the carboxyl group of Asp21, Glu42, Asp60, Asp129, or Ala207 (C-terminal) did not significantly affect the sweetness of thaumatin I, whereas PM-Asp21/Glu42/Asp129-thaumatin I was a little sweeter (Table 2). Elimination of the negative charge or a decrease in the number of negative charges of thaumatin molecule seems to increase sweetness in specific conditions. However, the details of properties are still unknown and need to be clarified.

4.2 Site-Directed Mutagenesis Approaches

4.2.1 Early Site-Directed Mutagenesis Investigation

Chemical modification results should be taken into the consideration based on the following disadvantages, that is, a number of efforts are required to obtain homogeneous modified product, conformational changes might be induced by a number of modifications, and steric hindrance introduced by modification sometimes might interfere with the adjacent active residues. To date, the site-directed mutagenesis approaches have been extensively employed to elucidate the functions, mechanisms, and behavior of proteins, such as enzymes. Due to the difficulty to obtain an active (sweet) form of thaumatin by expression from microorganisms, the progress of site-direct mutagenesis studies on thaumatin seems to be relatively slow compared with other sweet-tasting proteins, such as monellin and brazzein. This might be partly because thaumatin contains eight disulfide bonds in a molecule.

The site-directed mutagenesis studies of thaumatin were first reported by Kim and Weickmann [79]. Mutation at Lys67, Lys137, or Tyr169 reduced sweetness by fivefold. The authors concluded that these residues contribute to determining sweet taste.

4.2.2 Mutation of Basic Amino Acid Residue (Lys or Arg)

Since our group successfully has expressed the recombinant thaumatin at high yields using the *P. pastoris* system, it is important to prepare various types of mutant proteins to examine structure-sweetness relationships in thaumatin in depth. Furthermore, chemical modification studies suggested the importance of lysine residues on the face of the cleft-containing side. Taken together, we first focused on the basic amino acid residue such as *lysine residue* and *arginine residue*. On the *cleft-containing side*, four lysine residues (K49, K67, K106, and K163) and three arginine residues (R76, R79, and R82) are present, so we attempted to prepare the mutant in which alanine residues were substituted for lysine or arginine residues, and the sweetness of each mutant protein was evaluated by sensory analysis in humans [61] (Fig. 1 and Table 3). The results showed that all four lysine residues and three arginine residues played significant roles in thaumatin sweetness. Of these amino acid residues, *Lys67* and *Arg82* were particularly important for eliciting the sweetness. Since the sweetness of K67A and R82A decreased by approximately 19-fold and 24-fold, respectively, we also prepared two further mutant thaumatin I proteins, R82K and K67R. The sweetness of R82K was 240 nM and reduced approximately fivefold, suggesting that not only the positive charge but also the structure of the side chain of the arginine residue at position 82 influences the sweetness of thaumatin. As to K67R, the threshold value of sweetness is 44 nM, which is almost the same as recombinant thaumatin I (45 nM), suggesting that at the position of 67, only the positive charge of the Lys67 side chain affects sweetness. The locations of the two critical residues for sweetness, *Arg82* and *Lys67*, are shown in Fig. 1.

Previous chemical modification studies by phosphopyridoxylation of lysine residues resulted in a dramatic decrease of the sweetness intensity of thaumatin as mentioned above. Selectively modified five positively charged lysine residues are all

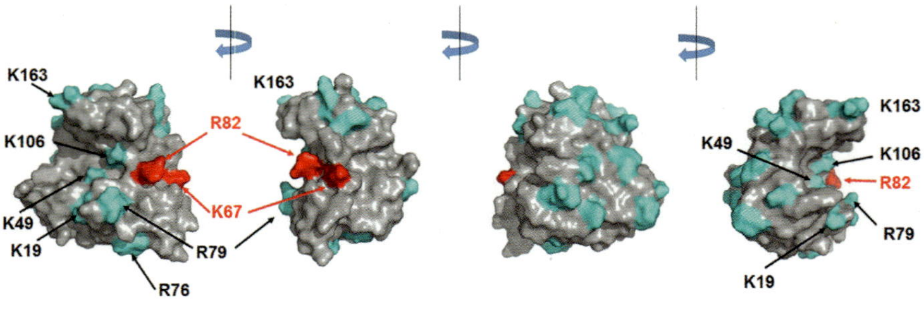

Fig. 1 Four views of thaumatin showing the relative positions of basic amino acid residues. All the lysine and arginine residues are shown in *cyan*. The residues chosen for mutagenesis are indicated in an *arrow*. Two critical residues, R82 and K67, are shown in *red*. Each figure is related by a 90° rotation. Molecular models were generated with *PyMOL* (DeLano, W. L. (2002). The pyMOL/Molecular Graphics System. DeLano Scientific, San Carlos, CA, USA)

Table 3 Effects of mutation of Lys or Arg residues on sweetness of thaumatin

	Sweetness threshold (nM)
K19A	70 ± 0
K49A	140 ± 10
K67A	870 ± 92
K106A	140 ± 19
K163A	180 ± 24
R76A	95 ± 12
R79A	170 ± 24
R82A	1100 ± 58
K67R	44 ± 6
K67E	1500 ± 61
R82K	240 ± 36
R82Q	1010 ± 135
R82E	8800 ± 1600
Plant thaumatin I	45 ± 9
Recombinant thaumatin I	45 ± 12

Ohta et al. [61, 80]

located on the same side of the thaumatin molecule, affecting sweetness, suggesting that these residues might play significant roles in the elicitation of the sweetness of thaumatin. Although one lysine residue, Lys106, in particular, seemed to be highly important for eliciting the sweetness, the threshold value of sweetness of K106A was 140 nM, which is about a 3.1-fold reduction of sweetness. Results of site-directed mutagenesis of K106A differ from those of chemical modification, since the threshold values of PLP-K106 as well as dephosphorylated-PLP-K106 are around 300 nM. This discrepancy might be partly due to the introduction of the relatively bulky PLP molecule at the position of ϵ-amino group of Lys106. The bulky PLP molecule might interfere with the adjacent other amino acid residues. Thus, we reanalyzed the structure of the surface of thaumatin molecule and the basic amino acid residues located to the adjacent of Lys106. The distance between Lys106 and Lys49 is about 9 Å, and that of Lys106 and Arg82 is 10–11 Å. The threshold value of K49A was 140 nM, and the sweetness was reduced only by 3.1-fold as like K106A, whereas the threshold value of R82A was 1,100 nM and a substantial reduction of sweetness. Taken together, we concluded that bulky PLP molecule as well as dephosphorylated PLP molecule at the position of Lys106 might affect the interaction sites of thaumatin, particularly against Arg82, causing an unexpected reduction of sweetness. Site-directed mutagenesis results of R82K might support this conclusion, since the subtle conformational change resulted in a substantial reduction of sweetness.

In order to clarify the environments around two vital residues, Lys67 and Arg82, for sweetness, various mutants of thaumatin altered at Arg82 as well as Lys67 were prepared, and their sweetness levels were quantitatively evaluated by cell-based assays using HEK293 cells expressing human sweet receptors besides sensory analysis by humans [80]. Mutations at Arg82 had a more deteriorative effect on sweetness than mutations at Lys67. Particularly, a charge inversion at Arg82 (R82E) resulted in the abolishment of the response to sweet receptors, and the threshold value was around 10 μM. These results indicate that Arg82 plays a central role in determining the sweetness of thaumatin. A strict spatial charge location at residue 82 appears to be required for interaction with sweet receptors.

4.2.3 Mutation of Acidic Amino Acid Residues (Asp or Glu)

The importance of basic amino acid residues has long been known for all sweet-tasting proteins, including thaumatin. As mentioned above, selective chemical modification and site-directed mutagenesis of thaumatin revealed that most lysine residues (Lys19, Lys49, Lys67, Lys78, Lys97, Lys106, Lys137, Lys163, and Lys187) located on the *cleft-containing side* are essential for sweetness, and particularly Lys67 is important. On the contrary, the possible effects of acidic amino acid residues on the structure-activity relationship of thaumatin are not fully understood. In the structure, thaumatin hosts 19 acidic residues, but it is possible to restrict the choice using, as a guide, previous knowledge on the most likely region of the protein surface that interacts with the receptor with aid of the preliminary results of docking studies, based on the so-called *wedge model* for the interaction of sweet proteins with the sweet taste receptor [81]. In the case of thaumatin, the two most important basic residues, namely Lys67 and Arg82, are at the center of this region and are surrounded by six acidic residues, Asp21, Glu42, Asp55, Asp59, Asp60, and Glu89. We prepared six mutant thaumatin constructs (D21N, E42Q, D55N, D59A, D60A, and E89Q) and had their sweetness evaluated by a human panel [82] (Fig. 2, Table 4). Although most of the acidic residues did not play a significant role in sweetness, removal of the negatively charged residue *Asp21* produced the sweetest thaumatin mutant, D21N, with a threshold value 31 nM, which was much sweeter

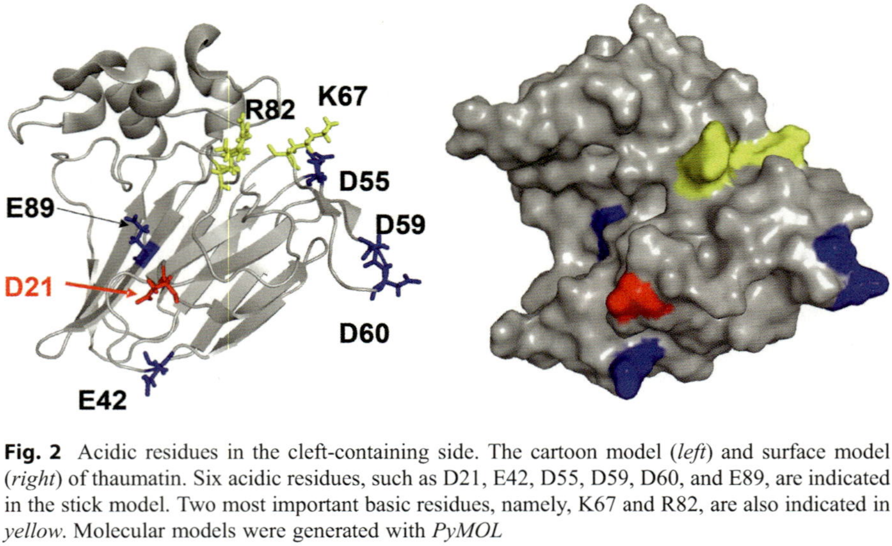

Fig. 2 Acidic residues in the cleft-containing side. The cartoon model (*left*) and surface model (*right*) of thaumatin. Six acidic residues, such as D21, E42, D55, D59, D60, and E89, are indicated in the stick model. Two most important basic residues, namely, K67 and R82, are also indicated in *yellow*. Molecular models were generated with *PyMOL*

Table 4 Effects of mutation of Asp or Glu residues on sweetness of thaumatin

	Sweetness threshold (nM)
D21N	31 ± 4
D55N	57 ± 4
E42Q	65 ± 13
D59A	87 ± 13
D60A	63 ± 7
E89Q	68 ± 2
Plant thaumatin I	51 ± 4

Masuda et al. [82]

than wild type thaumatin and, together with the Y65R mutant of single-chain monellin, one of the two sweetest proteins known so far. Models derived from tethered docking provide a deeper insight in our understanding of how thaumatin and its D21N mutant interact with the sweet receptor. Detailed explanation of docking simulation between thaumatin and sweet receptor will be described in a later section.

4.3 The Sweetness-Determinants Were Surveyed by Monoclonal Antibody Mapping

It is quite interesting that *monoclonal antibodies* to thaumatin as an antigen cross-reacted with another sweet-tasting protein, monellin [83]. Slootstra et al. identified epitopes for monoclonal antibodies and suggested that two regions of thaumatin, amino acids 19–29 and 77–84, were the important determinants of sweet taste (Fig. 3). Interestingly, an aspartame-like site that is formed by Asp21 and Phe80 was found, and the sequence of 19-KGDAALDAGGR-29 loop is similar to peptide-sweeteners, such as L-Asp-D-Ala-L-Ala-methyl ester and L-Asp-D-Ala-Gly-methyl ester. Interestingly, the results obtained by mapping with monoclonal antibodies partly well coincided with the results from site-directed mutagenesis; indeed, the residue of Arg82, one of the two critical residues for sweetness of thaumatin, and the residue of Asp21, which plays a critical role in enhancement of sweetness, are included in the epitope of monoclonal antibody.

5 Prediction of the Mechanism of Interaction Between Sweet-Tasting Proteins and Their Receptors

5.1 Sweet Taste Receptor

Current knowledge is that a single receptor accounts for the sweet taste of all sweet molecules, from sugars to sweet proteins. Sweet taste receptor is a metabotropic *G-protein-coupled receptor (GPCR)* composed of two GPCR named T1R2 and

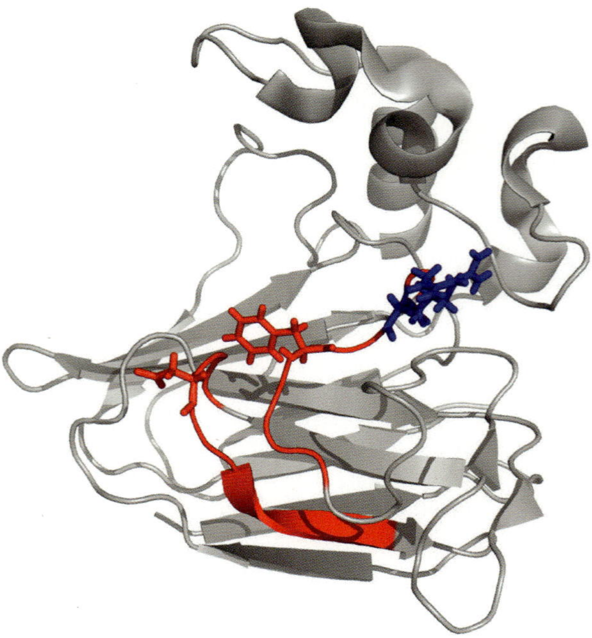

Fig. 3 Sweetness-determinants are surveyed by antibody mapping. The epitope of monoclonal antibodies are shown in *red*. An aspartame-like residues (Asp21-Phe80) are shown in the stick model. The critical Arg82 is also indicated in *blue* with stick model. Molecular models were generated with *PyMOL*

T1R3 [84–89]. *T1R2* and *T1R3* belong to the class C GPCR subunits and possess a large N-terminal extracellular domain, which is composed of two parts: a *Venus flytrap module (VFTM)* and *cysteine-rich domain (CRD)*. CRD links the VFTM to the following heptahelical *transmembrane domain (TMD)*.

5.2 Interpretations of Binding Sites of Sweeteners by Cell-Based Assay

Functional expression of sweet taste receptor (T1R2/T1R3) in human embryonic kidney HEK293 cells revealed that diverse natural and synthetic sweeteners, including carbohydrates, amino acids and derivatives, and sweet-tasting proteins, were detected by this receptor. Previous studies have shown that sweet-tasting proteins as well as aspartame can be perceived by humans, apes, and Old World monkeys but not New World monkeys and rodents [90, 91]. *Species differences in the response to sweeteners* would provide valuable information on the molecular mechanism by which sweet receptors function as well as aid the identification of interaction sites in receptors [91–95]. Studies with mouse/human chimeras of T1R2 and T1R3 suggested that the CRD of human T1R3 is essential for binding to brazzein and thaumatin, and the N-terminal VFTM of T1R3 is required for neoculin. The N-terminal VFTM of T1R2 is required for response to monellin and brazzein and small size sweeteners, such as aspartame and neotame. The TMD of T1R3 is essential for response to cyclamate and lactisole [93, 96]. As to thaumatin, CRD within hT1R3 is important for the response [95]. Detailed investigations were further attempted by the point mutation in CRD, suggesting that five amino acid residues (Gln504, Ala537, Arg556, Ser559, and Arg560) in the CRD of human T1R3 are important for the response to thaumatin [97]. These five residues involved in the response to thaumatin were dispersed in the CRD of hT1R3 and widely distributed when compared to brazzein, suggesting the unique intense sweet-taste of thaumatin might be attributed to the different receptor activation mechanism compared to the small molecule sweetener.

The *cell-based assay* using HEK293 cells stably expressing the human T1R2/T1R3 sweet taste receptor also suggested that a decrease in the intracellular cAMP accumulation of thaumatin was observed as like that of lysozyme as well as aspartame. The effect of sweetness suppression on thaumatin and lysozyme by inhibitor lactisole was apparently detected both in vitro and in vivo experiments [98].

Recently, the functions of sweet taste receptor (T1R2/T1R3) from a species of New World primate, squirrel monkey, has been reported and found that the sweet taste receptor of squirrel monkey did not respond to artificial sweeteners aspartame, neotame, cyclamate, saccharin, and sweet-tasting protein monellin, but it did respond to thaumatin at high concentrations (>18 μM) [99]. Molecular models indicated that electrostatic properties of the receptors probably mediate the species-dependent response to sweet-tasting proteins.

The mice lacking T1R3 subunit showed no preference for artificial sweeteners and had diminished but not abolished, behavioral and nerve responses to sugars [100]. These results indicate that T1R3-independent sweet-responsive receptors and/or pathways exist in taste cells, and the mechanisms should be clarified.

Thus as these cell-based assays did not directly detect the interaction of sweet proteins with the receptor, more attention should be paid to the environment around interaction sites. In the case of sweet-tasting proteins, particularly for thaumatin, the dimension of a typical CRD is not consistent with the large surface spun by key residues of thaumatin, and an alternative explanation should be formulated. It is likely that even a slight tilt of the VFTD towards the membrane surface, induced by altering the architecture of the CRD, would prevent binding.

5.3 Wedge Model

Most small sweet molecules such as sugars bind to the cavities of the VFTD, which proved more difficult to account for the interaction of the relatively large sweet proteins, mainly because of their dimensions. Considering that the molecular volumes of thaumatin and aspartame are estimated to be 27,000 $Å^3$ and 270 $Å^3$ respectively, suggesting thaumatin is substantially larger than small molecular weight sweeteners [82]. Thus it seems to be difficult to imagine that sweet proteins can bind to the same sites of the sweet receptor that bind small ligands. The first interpretation of the mechanism of interaction of sweet proteins with the sweet receptor was proposed by Temussi [101]. He named this model as “wedge model,” as large molecules cannot be hosted by the orthosteric binding sites of the VFTD of either T1R2 or T1R3 protomers. The basis of the wedge model is the homology between the sweet receptor and the *mGluR1 glutamate receptor*; the template structure of the mGluR1 receptor was used to build the homology model of T1R2-T1R3 [102]. As like mGluR1, the sweet taste receptor exists as an equilibrium mixture between a resting conformation and the active form, even when the ligands are not bound. Small molecular weight sweet compounds shift the equilibrium by entering one or two of the orthosteric sites inside the VFTD. Larger molecules like sweet proteins achieve the same results by binding to a secondary, external site of the active form.

5.4 The Complex Model Between the T1R2-T1R3 Sweet Receptor and Thaumatin

By applying up-to-date site-directed mutagenesis results, we successfully obtained a reasonable model to explain the results from site-directed as well as chemical modification studies [82].

The complex model between the T1R2-T1R3 sweet receptor and thaumatin, derived from tethered docking in the framework of the *wedge model*, confirmed that each of the positively charged residues critical for sweetness, including two important residues, Arg82 and Lys67, is close to a receptor residue of opposite charge to yield optimal electrostatic interaction. As to the model for sweeter D21N mutant, it should be noted that the distance between D21 and its possible counterpart D433 (located on the T1R2 protomer of the receptor) is safely large to avoid electrostatic repulsion but, at the same time, amenable to a closer approach if D21 is mutated into the corresponding asparagine. The interaction regions for D21N mutant is relatively larger compared to the recombinant thaumatin and might serve to enhance the sweetness as a result.

6 Three-Dimensional Structure of Thaumatin

The three-dimensional structure of thaumatin consists of mainly three domains (Fig. 2). The core domain consists of an 11-stranded, flattened β-sandwich folded into 2 Greek-key motifs. All β-strands are antiparallel except the parallel N-terminal and the C-terminal strands. The second domain consists of two β-strands forming a β-ribbon and connected by an Ω-loop. The third domain consists of a disulfide-rich region. It contains one α-helix and three short helical fragments. Two of the helical segments are connected by an unusually sharp turn, stabilized by a disulfide bridge.

The three-dimensional structure of thaumatin is the remarkable homology to *pathogenesis-related (PR) proteins* and t*haumatin-like proteins* [103]. Despite their structural similarity, none of the PR-5 proteins has been reported to have a sweet taste [104, 105].

At first, the author explains the history of the determination of the structure of thaumatin and preparations of the crystals suitable for structural analysis. Then, recent progress as well as topics will be described.

6.1 Early Three-Dimensional Structure Analysis of Thaumatin

Thaumatin has been known to crystallize rapidly in the presence of tartrate ions and has frequently been used in crystallization studies. To date, in 2016, more than hundreds of structures of thaumatin have been deposited in the Protein Data Bank (PDB). Most of them are derived from plant sources, and no further purification tends to be performed for crystallization.

The three-dimensional structure of thaumatin was first reported at the resolution of 3.1 Å by de Vos et al. using the crystal of the orthorhombic form [106] (Table 5). Ogata et al. also reported the crystal structure of thaumatin I and refined it at a resolution better than 1.65 Å using a combination of energy minimization and restrained least-squares methods [107]. The orthorhombic forms of crystal were used for the structure determination, and the final model was refined with a crystallographic *R*-factor of 0.167 at 1.65 Å. The model has good stereochemistry, with root-mean-square deviations from ideal values for bond and angle distances of 0.014 Å and 0.029 Å, respectively. The structure of the thaumatin consists of all 207 amino acids with 28 alternative conformations and 236 water molecules.`

Ko et al. reported the structure of thaumatin in three crystal forms (*monoclinic*, *orthorhombic*, and *tetragonal*) [108]. The crystals are of space groups *C2* with a = 117.7 Å, b = 44.9 Å, c = 38.0 Å, and $\beta = 94.0°$, $P2_12_12_1$ with a = 44.3 Å, b = 63.7 Å and c = 72.7 Å, and a tetragonal form $P4_12_12$ with a = b = 58.6 Å and c = 151.8 Å. The structures of all three crystals have been solved by molecular replacement. In a monoclinic crystal, the structure was refined to *R*-factors of 0.184 at 2.6 Å, and the orthorhombic and tetragonal crystals were refined at 1.75 Å to *R*-factors of 0.165 and 0.181, respectively. No solvent was included in the monoclinic crystal, whereas 123 and 105 water molecules were included in orthorhombic and tetragonal structures, respectively. The tartrate molecule was also included in the tetragonal structure.

The structure of thaumatin in *hexagonal* crystal form was determined at the resolution of 1.6 Å by Charron et al. [109]. The structure was refined to a final *R*-factor of 0.211 and R_{free} of 0.223. Although the protein fold is identical to that found in other crystal forms, the proportions of lattice interactions involving hydrogen bonds and hydrophobic or ionic groups were greatly different, and the distribution of acidic and basic amino acid residues involved in contacts were different from each other.

6.2 Atomic Resolution Structural Analysis of Thaumatin

Using purified thaumatin I, Asherie et al. found that the crystal habit and solubility largely depended on the chirality of the precipitant of the tartrate ion. The solubility

Table 5 High resolution structures of thaumatin

Thaumatin variants	Crystal systems	Cells dimensions (Å)	Resolution, PDB ID
Thaumatin I	$P4_12_12$	57.88, 57.88, 149.95	0.94 Å, 2vhk [25]
Thaumatin I	$P4_12_12$	57.80, 57.80, 149.91	0.95 Å, 2vhr [25]
Thaumatin I	$P2_12_12_1$	50.89, 54.52, 70.87	0.95 Å, 2vu6 [25]
Thaumatin I	$P4_12_12$	57.84, 57.84, 149.55	0.95 Å, 5avg [128]
Thaumatin I	$P4_12_12$	57.85, 57.85, 149.98	0.95 Å, 2vi3 [25]
Recombinant thaumatin II	$P4_12_12$	57.69, 57.69, 150.011	0.99 Å, 2vi3 [28]
Recombinant thaumatin I	$P2_12_12_1$	43.64, 63.53, 71.77	1.00 Å, 3vhg [27]
Recombinant thaumatin I	$P2_12_12_1$	43.64, 63.53, 71.77	1.00 Å, 3vjq [27]
Thaumatin I	$P2_12_12_1$	50.94, 54.30, 70.81	1.04 Å, 2vil [25]
Thaumatin B	$P4_12_12$	57.85, 57.85, 150.13	1.05 Å, 1rqw, Ma et al. 2009 [http://www.rcsb.org/pdb/explore/explore.do?structureId=1RQW]
Thaumatin I	$P4_12_12$	57.91, 57.91, 150.88	1.05 Å, 2vi2 [25]
Thaumatin I	$P2_12_12_1$	43.81, 63.91, 71.68	1.08 Å, 2vu7 [25]
Thaumatin I	$P4_12_12$	57.92, 57.92, 149.95	1.10 Å, 2vi4 [25]
Recombinant thaumatin I	$P4_12_12$	57.70, 57.70, 149.78	1.10 Å, 3al7 [26]
Thaumatin I	$P4_12_12$	57.70, 57.70, 149.85	1.10 Å, 3ald [26]
Thaumatin	$P4_12_12$	58.53, 58.53, 151.35 (293K)	1.20 Å, 1kwn [129]
Thaumatin	$P4_12_12$	57.84, 57.84, 149.58	1.20 Å, 4tvt [130]
Thaumatin	$P4_12_12$	57.84, 57.84, 150.37	1.20 Å, 4zg3 [131]
Thaumatin	$P4_12_12$	57.93, 57.93, 150.64	1.20 Å, 5a47 [132]
Thaumatin	$P4_12_12$	57.80, 57.80, 149.96	1.25 Å, 1lxz [115]
Thaumatin	$P4_12_12$	57.91, 57.91, 150.13	1.25 Å, 3qy5, Stojanoff et al. 2011 http://www.rcsb.org/pdb/explore/explore.do?structureId=3QY5

(*continued*)

Table 5 (continued)

Thaumatin variants	Crystal systems	Cells dimensions (Å)	Resolution, PDB ID
Thaumatin II	$P4_12_12$	57.77, 57.77, 150.13	1.27 Å, 3aok, [26]
Thaumatin	$P4_12_12$	57.87, 57.87, 149.96	1.36 Å, 1ly0, [115]
Thaumatin	$P4_12_12$	57.98, 57.98, 150.38	1.38 Å, 4axr [133]
Thaumatin	$P4_12_12$	58.00, 58.00, 150.52	1.38 Å, 4axu [133]
Thaumatin I	$P2_12_12_1$	43.69, 63.65, 71.74	1.39 Å, 3vhf, [27]
Thaumatin	$P4_12_12$	57.78, 57.78, 150.09	1.40 Å, 2blr [134]
Thaumatin	$P4_12_12$	57.78, 57.78, 150.09	1.40 Å, 2blu [134]
Recombinant thaumatin I	$P4_12_12$	58.51, 58.51, 151.71 (293 K)	1.46 Å, 3x3s, Masuda et al. 2015 http://www.rcsb.org/pdb/explore/explore.do?structureId=3X3S
Recombinant thaumatin I	$P4_12_12$	58.60, 58.60, 151.55 (298 K)	1.48 Å, 3x3o, Masuda et al. 2015 http://www.rcsb.org/pdb/explore/explore.do?structureId=3X3O
Recombinant thaumatin I	$P4_12_12$	58.53, 58.52, 151.59 (293 K)	1.48 Å, 3x3p, Masuda et al. 2015 http://www.rcsb.org/pdb/explore/explore.do?structureId=3X3P
Thaumatin	$P4_12_12$	57.91, 57.91, 150.32	1.48 Å, 4dc5 [135]
Thaumatin	$P6_1$	144.83, 144.83, 47.74	1.60 Å, 1pp3 [109]
Thaumatin A	$P2_12_12_1$	44.30, 63.70, 72.70	1.75 Å, 1thv [108]
Thaumatin A	$P4_12_12$	58.60, 58.60, 151.80	1.75 Å, 1thw [108]
Thaumatin B	C2	117.70, 44.90, 38.00	2.60 Å, 1thu [108]
Thaumatin	$P2_12_12_1$	52.25, 53.32, 74.42	3.20 Å, 1thi [106]

of thaumatin crystals (bipyramidal crystals form) increased with temperature in L-tartrate, whereas it (prismatic and stubby crystals form) decreased with temperature in D-tartrate [110]. Subsequently, their study revealed the importance of using pure protein and stereochemically pure precipitants in the crystallization of thaumatin [111]. Asherie et al. also attempted to crystallize thaumatin with the three stereoisomers (L-, D-, and meso-) of tartrate [25]. All three precipitants produced high-quality crystals that diffracted at an atomic resolution of the highest of 0.94 Å–1.00 Å. The structure was refined to a final R-factor ranging from 0.117 to 0.128 and R_{free} ranging from 0.126 to 0.143. The root-mean-square deviations from ideal values for bond lengths ranged from 0.010 Å to 0.021 Å, and bond angles ranged from 1.524° to 1.961°.

The homogeneous preparation might affect the quality and nucleation of the crystals. Indeed, high resolution analysis has been reported as mentioned above. However, it is still unknown how the purified plant source thaumatin I influenced the discrimination the electron densities on the residues different between thaumatin variants. Thus, we attempted to determine the structure of the recombinant thaumatin I at atomic resolution [26]. The recombinant form of thaumatin I was produced in the yeast *P. pastoris* like mentioned in the Sects. 3 and 4, and the purified protein samples were further crystalized in the presence of tartrate ions. The crystal was diffracted to a resolution of 1.1 Å, and the model was refined with anisotropic B-parameters and riding hydrogen atoms. The R-values for recombinant thaumatin I was 9.11%, indicating the final model to be of good quality. The root-mean-square deviations from ideal values for bond length and bond angle were 0.017 Å and 0.034 Å, respectively. The final model of recombinant thaumatin I consisted of 207 residues with a total of 3,386 protein atoms, including 1,633 hydrogen atoms, two tartrate ions, four glycerol molecules, and 476 water molecules. The electron-density maps around Asn46 and Ser63, which differ among thaumatin variants, were significantly improved, and a number of hydrogen atoms became visible in an OMIT map. Although it has been quite difficult to determine whether the residue at 113 is Asp or Asn in plant thaumatin directly from the density maps, the OMIT map of Asp113 in recombinant thaumatin is quite different from that of Asn104, suggesting the usefulness for the direct discrimination on the residue at position 113.

Thus, the high-quality structure of recombinant thaumatin with hydrogen atoms should provide details about sweetness-determinants in thaumatin and valuable insights into the mechanism of its interaction with taste receptors.

6.3 Structure of Thaumatin II

As to thaumatin I, atomic resolution structural analysis of recombinant thaumatin I as well as purified plant thaumatin I has been extensively examined, whereas little attention has been paid to thaumatin II. In the early studies, most researchers used the gene for thaumatin II, a comparison of the structures between thaumatin I and thaumatin II would provide an important insight for further research on production of the recombinant thaumatin in microorganisms as well as transgenic plants. Thaumatin I differed from thaumatin II at four positions (N46K, S63R, K67R, and R76Q) (Table 1). Despite four amino acid differences between thaumatin I and II, the crystals of plant thaumatin II are very small, and their reflections are limited to the resolution of 1.27 Å [112]. Overall structure of thaumatin II is similar to thaumatin I, but a slight shift of the Cα atom of Gly96 in thaumatin II was observed. Later, the author attempted to produce a recombinant thaumatin II in the yeast *P. pastoris*, and the structure of the recombinant thaumatin II including hydrogen atoms was refined to an *R*-factor of 9.74% at a resolution of 0.99 Å [28]. Atomic resolution structural analysis with riding hydrogen atoms illustrated the differences in the direction of the side chains more precisely, and the electron density maps of the C-terminal regions were markedly improved. Though it had been suggested that the three consecutive glycine residues (Gly142-Gly143-Gly144) have highly flexible conformations, Gly143, the central glycine residue was successfully modeled in two conformations for the first time. The side-chain root-mean-square deviation values for two residues (Arg67 and Arg82) critical for sweetness exhibited substantially higher values, suggesting that these residues are highly disordered. Taken together, these results provided the important insight that the flexible conformations in two critical residues favoring their interaction with sweet taste receptors are prominent features of the intensely sweet taste of thaumatin.

6.4 Neutron Crystallographic Analysis

Structural information on the charge distribution and protonation state of hydrogen atoms would be useful for understanding the elicitation of the sweetness of thaumatin as well as the interaction with sweet receptors. Besides high-resolution x-ray crystal analysis, neutron crystallographic studies are thought to provide a number of insights for hydrogen atoms as well as water molecules involved in the perception of sweetness. This attempt was first performed by Teixeira et al. [113, 114]. Although the largest drawback of the neutron crystallographic studies is to obtain a substantially large size of crystals, large hydrogenated crystals were prepared in deuterated crystallization buffer using the gel-acupuncture method. Data were collected up to a resolution of 2 Å and the feasibility of a full neutron crystallographic analysis aimed at providing relevant information on the location of hydrogen atoms, the distribution of charge on the protein surface and localized water in the structure. The results obtained by neutron crystallographic studies will contribute to further understanding of the molecular mechanisms underlying the perception of taste in detail and shed light on the effect of pH on the structural stability of thaumatin-like proteins.

6.5 Thaumatin Crystals as a Tool in Checking Various Effects

Effects of the *glycerol molecule on the quality of crystal* were investigated [115]. Although only minor changes were detected in the overall structure, the number of water molecules was reduced by approximately 20% in the presence of 25% (v/v) glycerol. Notably, the overall quality of the crystals prepared in the presence of glycerol as a cryoprotectant was enhanced.

Thaumatin was also used as a model for crystallization in microgravity-environment experiments conducted on two US Space Shuttle missions (USML-2 and LMS) [116]. The quality of *space-grown thaumatin crystals*. The quality of the thaumatin crystals grown in a microgravity (space-grown thaumatin crystals) environment were superior to crystals grown on earth. The diffraction limit and diffraction properties were significantly improved, and the mosaicity of space-grown crystals was significantly less than that of crystals grown on earth.

Although atomic resolution structural analysis provide valuable information including hydrogen atoms, alternative motion of side chain of amino acid, the exposure of high-brilliance synchrotron sources induced radiation damage of crystals. Using thaumatin crystals as a model, *the temperature-dependence of radiation damage* to thaumatin crystals was examined and found that the amount of damage for a given dose decreased sharply as the temperature decreased from 300 to 220 K and then decreased more gradually on further cooling below the protein-solvent glass transition [117]. Furthermore, radiation damage was dominated by diffusive motions in the protein and solvent above 200 K, and large-scale conformational and molecular motions ceased below 200 K. These results provided the important insight that data collection at around 220 K may provide a viable alternative for structure determination when cooling-induced disorder at 100 K was excessive.

Warkentin et al. also suggested that the nature of the radiation damage changed dramatically at approximately 180 K, and the radiation fingerprint is more prominent in tertiary structures such as local packing, solvent accessibility, and crystal contacts than in primary structure [118]. Above 180 K, the role of solvent diffusion was apparent, and solvent-exposed turns and loops were especially sensitive to radiation damage. In contrast, below 180 K sensitivity is correlated with poor local packing.

Attempt to obtain the high resolution at room temperature is difficult since the crystals are susceptible to radiation damage. To prevent and reduce these side effects, cryo-cooling techniques are often applied; however, unexpected artifacts of cryo-protectants such as glycerol may be introduced in the molecules.

Serial femtosecond crystallography (SFX) using ultrashort pulses from X-ray free-electron lasers (XFELs) has attracted keen interest, since it inherits the advantage of acquiring minimal-radiation damaged protein structures at ambient temperature using the “diffraction-before-destruction” approach [119–124]. SFX analysis of thaumatin has been successfully performed to collect data from small protein crystals in micron to submicron scales [125, 126].

6.6 Atomic Structure of the Thaumatin at pH 8.0

The sweetness remains when thaumatin is heated at 80 °C for 4 h under acid conditions; it rapidly declines when heating at a pH above 6.5. The reduction in sweetness caused by heating at above pH 7.0 is attributed to the formation of aggregates, and the disulfide interchange reaction was catalyzed by cysteine, that is, a free sulfhydryl residue was formed via the β-elimination of a disulfide bond [127]. We attempted to crystalize the recombinant thaumatin I at pH 8.0, and its structure was determined at a resolution of 1.0 Å [27]. The root-mean-square deviation value of a Cα atom was substantially greater in the large disulfide-rich region of domain II, especially residues 154–164 compared to the crystal structure of thaumatin at pH 7.3 and 7.0. These results suggested that a loop region in domain II is affected by changes in pH. Furthermore, *B*-factors of Lys137, Lys163, and Lys187 were significantly affected by changes in pH, suggesting a striking increase in the mobility of three lysine residues, which could facilitate a reaction with a free sulfhydryl residue produced via the β-elimination of disulfide bonds by heating at a pH above 7.0. The increase in mobility of lysine residues as well as a loop region in domain II might play an important role in the heat-induced aggregation of thaumatin above pH 7.0. The investigations of the structure of thaumatin at acidic pH are now in progress, and it will provide important insights for the heat-resistance at acidic conditions.

7 Conclusions

Life-style related diseases such as hypertension, hyperlipidemia, diabetes, and obesity have become major and serious global problems. These diseases seem to be linked to the staggering increases in obesity. However, as sweet taste is an important taste modality, the use of the low-calorie sucrose substitutes in foods, beverages, and medicines should be considered. Sweet-tasting proteins are potential low-calorie substitutes for sugars. Since thaumatin elicits a sweet taste at a low concentration, approximately 100,000-fold less than that of sucrose on a molar basis, the structure-sweetness relationships in sweet-tasting proteins would help our understanding of the mechanism of how thaumatin interacts with sweet receptors. Furthermore, biotechnological production of sweet-tasting proteins and its sweeter forms in sufficient amounts would play a significant role in the future, considering the limitation of the natural source of thaumatin. In this point, the sweetest mutant of D21N thaumatin, which we recently have found, could provide good clue not only for designing more effective sweeteners but for our understanding of the mechanism of interaction of thaumatin with sweet receptors.

Acknowledgments This work was supported by a Grant-in-Aid for Young Scientists (B) (T.M., no. 19780074) and Scientific Research (C) (T.M., no. 22580105, 25450167) from The Japan Society for the Promotion of Science. The structure of thaumatin was determined at BL26B1, BL38B1, and BL44XU in SPring-8 (proposal number 2009A1379, 2009B1096, 2010B1064, 2011B1073, 2012A1048, 2012B1067, 2013A1053, 2013B1069, 2014A1063, 2014B1181, 2014B1339, 2014B2020, 2015A1037, 2015B2037, 2016A2548, 2016A2552).

References

1. Nabors LO'B, Gelardi RC (1991) Alternative sweeteners. Marcel Dekker, New York
2. van der Wel H, Loeve K (1972) Isolation and characterization of thaumatin I and II, the sweet-tasting proteins from *Thaumatococcus daniellii* Benth. Eur J Biochem 31:221–225
3. van der Wel H (1972) Isolation and characterization of the sweet principal for *Dioscoreophyllum cumminsii* (Stapf) Diels. FEBS Lett 21:88–90
4. Moris JA, Cagan RH (1972) Purification of monellin, the sweet principal for *Dioscoreophyllum cumminsii*. Biochim Biophys Acta 261:114–122

5. Liu X et al (1993) Purification, complete amino acid sequence and structure characterization of the heat stable sweet protein, mabinlin II. Eur J Biochem 211:281–287
6. Yamashita H et al (1990) Purification and complete amino acid sequence of a new type of sweet protein with taste-modifying activity, curculin. J Biol Chem 265:15770–15775
7. Shirasuka Y et al (2004) Neoculin as a new taste-modifying protein occurring in the fruit of *Curculigo latifolia*. Biosci Biotechnol Biochem 68:1403–1407
8. Suzuki M et al (2004) Recombinant curculin heterodimer exhibits taste-modifying and sweet-tasting activities. FEBS Lett 573:135–138
9. Ming D, Hellekant G (1994) Brazzein, a new high-potency thermostable sweet protein from *Pentadiplandra brazzeana* B. FEBS Lett 355:106–108
10. Masuda T et al (2001) Sweetness and enzymatic activity of lysozyme. J Agric Food Chem 49:4937–4941
11. Maehashi K, Udaka S (1998) Sweetness of lysozyme. Biosci Biotechnol Biochem 53:605–606
12. Etheridge K (1994) The sales and marketing of talin. In: Witty M, Higginbotham JD (eds) Thaumatin. CRC Press, Boca Raton
13. Faus I (2000) Recent developments in the characterization and biotechnological production of sweet-tasting proteins. Appl Microbiol Biotechnol 53:145–151
14. Masuda T, Kitabatake N (2006) Developments in biotechnological production of sweet proteins. J Biosci Bioeng 102:375–389
15. van der Wel H et al (1984) Assignment of the disulphide bonds in the sweet-tasting protein thaumatin I. Eur J Biochem 144:41–45
16. Esposito V et al (2006) The importance of electrostatic potential in the interaction of sweet proteins with the sweet taste receptor. J Mol Biol 360:448–456
17. Assadi-Porter FM et al (2010) Key amino acid residues involved in multi-point binding interactions between brazzein, a sweet protein, and the T1R2-T1R3 human sweet receptor. J Mol Biol 398:584–599
18. Jin Z et al (2003) Critical regions for the sweetness of brazzein. FEBS Lett 544:33–37
19. Somoza JR, Cho JM, Kim SH (1995) The taste-active regions of monellin, a potently sweet protein. Chem Senses 20:61–68
20. Xue WF et al (2009) Role of protein surface charge in monellin sweetness. Biochim Biophys Acta 1794:410–420
21. Kurimoto E et al (2007) Curculin exhibits sweet-tasting and taste-modifying activities through its distinct molecular surface. J Biol Chem 282:33252–33256
22. Nakajima K et al (2008) Acid-induced sweetness of neoculin is ascribed to its pH-dependent agonistic-antagonistic interaction with human sweet taste receptor. FASEB J 22:2323–2330
23. Masuda T, Ide N, Kitabatake N (2005) Effects of chemical modification of lysine residues on the sweetness of lysozyme. Chem Senses 30:253–264
24. Masuda T, Ide N, Kitabatake N (2005) Structure-sweetness relationship in egg white lysozyme: role of lysine and arginine residues on the elicitation of lysozyme sweetness. Chem Senses 30:667–681
25. Asherie N et al (2009) Tartrate chirality determines thaumatin crystal habit. Cryst Growth Des 9:4189–4198
26. Masuda T et al (2011) High-resolution structure of the recombinant sweet-tasting protein thaumatin I. Acta Crystallogr Sect F Struct Biol Cryst Commun 67:652–658
27. Masuda T et al (2012) Atomic structure of the sweet-tasting protein thaumatin I at pH 8.0 reveals the large disulfide-rich region in domain II to be sensitive to a pH change. Biochem Biophys Res Commun 419:72–76
28. Masuda T, Mikami B, Tani F (2014) Atomic structure of recombinant thaumatin II reveals flexible conformations in two residues critical for sweetness and three consecutive glycine residues. Biochimie 106:33–38
29. Hobbs JR, Munger SD, Conn GL (2007) Monellin (MNEI) at 1.15 Å resolution. Acta Crystallogr Sect F Struct Biol Cryst Commun 63:162–167

30. Danieli WF (1852) On the *Synsepalum dulcificum*, De Cand; or miraculous berry of Western Africa. Pharmaceutical J 11:445–448
31. Danieli WF (1855) Katemfe, or the miraculous fruit of Soudan. Pharmaceutical J 14:158–159
32. Higginbotham JD (1979) Protein sweeteners. In: Hough CAM, Parker KJ, Viltos AJ (eds) Developments in sweeteners I. Applied Science, London
33. Inglett GE, May JF (1968) Tropical plants with unusual taste properties. Econ Bot 22:326–331
34. Mackenzie A et al (1985) Changes in the sweet proteins (thaumatins) in *Thaumatococcus Danielli* fruits during development. Phytochemistry 11:2503–2506
35. Higginbotham JD (1977) Useful taste properties of amino acids and proteins. In: Birch GC, Brennan JG, Parker KJ (eds) Sensory properties of foods. Applied Science, London
36. Iyengar RB et al (1979) The complete amino-acid sequence of the sweet protein thaumatin I. Eur J Biochem 96:193–204
37. Edens L et al (1982) Cloning of cDNA encoding the sweet-tasting plant protein thaumatin and its expression in *Escherichia coli*. Gene 18:1–12
38. Lee JH et al (1988) Expression of synthetic thaumatin genes in yeast. Biochemistry 27:5101–5107
39. Ide N et al (2007) Cloning of the thaumatin I cDNA and characterization of recombinant thaumatin I secreted by *Pichia pastoris*. Biotechnol Prog 23:1023–1030
40. Faus I et al (1996) Expression of a synthetic gene encoding the sweet-tasting protein thaumatin in *Escherichia coli*. Biochem Biophys Res Commun 229:121–127
41. Daniell S et al (2000) Refolding the sweet-tasting protein thaumatin II from insoluble inclusion bodies synthesised in *Escherichia coli*. Food Chem 71:105–110
42. Illingworth C, Larson G, Hellekant G (1988) Secretion of the sweet-tasting plant protein thaumatin by *Bacillus subtilis*. Biotechnol Lett 10:587–592
43. Illingworth C, Larson G, Hellekant G (1989) Secretion of the sweet-tasting plant protein thaumatin by *Streptomyces lividans*. J Ind Microbiol 4:37–42
44. Hahm YT, Batt CA (1990) Expression and secretion of thaumatin from *Aspergillus oryzae*. Agric Biol Chem 54:2513–2520
45. Faus I et al (1997) Expression of a synthetic gene encoding the sweet-tasting protein thaumatin in the filamentous fungus *Penicillium roquefortii*. Biotechnol Lett 19:1185–1191
46. Faus I et al (1998) Secretion of the sweet-tasting protein thaumatin by recombinant strains of *Aspergillus niger* var. *awamori*. Appl Microbiol Biotechnol 49:393–398
47. Moralejo FJ et al (1999) Thaumatin production in *Aspergillus awamori* by use of expression cassettes with strong fungal promoters and high gene dosage. Appl Environ Microbiol 65:1168–1174
48. Moralejo FJ et al (2002) Silencing of the Aspergillopepsin B (*pepB*) gene of *Aspergillus awamori* by antisense RNA expression or protease removal by gene disruption results in a large increase in thaumatin production. Appl Environ Microbiol 68:3550–3559
49. Moralejo FJ et al (2000) Overexpression and lack of degradation of thaumatin in an aspergillopepsin A-defective mutant of *Aspergillus awamori* containing an insertion in the *pep*A gene. Appl Microbiol Bioetechnol 54:772–777
50. Moralejo FJ et al (2001) A defined level of protein disulfide isomerase expression is required for optimal secretion of thaumatin by *Aspergillus awamori*. Mol Genet Genomics 266:246–253
51. Lombraña M et al (2004) Modulation of *Aspergillus awamori* thaumatin secretion by modification of *bipA* gene expression. Appl Environ Microbiol 70:5145–5152
52. Edens L et al (1984) Synthesis and processing of the plant protein thaumatin in yeast. Cell 37:629–633
53. Edens L, van der Wel H (1985) Microbial synthesis of the sweet-tasting plant protein thaumatin. Trends Biotechnol 3:61–64
54. Weickmann JL, Blair LC, Wilcox GL (1994) High level expression of thaumatin in *Saccharomyces cerevisiae*. In: Witty M, Higginbotham JD (eds) Thaumatin. CRC Press, Boca Raton

55. Cregg JM et al (2000) Recombinant protein expression in *Pichia pastoris*. Mol Biotechnol 16:23–52
56. Cereghino JL, Cregg JM (2000) Heterologous protein production in the methylotrophic yeast *Pichia pastoris*. FEMS Microbiol Rev 24:45–66
57. Macauley-Patrick S et al (2005) Heterologous protein production using the *Pichia pastoris* expression system. Yeast 22:249–270
58. Masuda T et al (2004) Cloning, expression, and characterization of recombinant sweet-protein thaumatin II using the methylotrophic yeast *Pichia pastoris*. Biotechnol Bioeng 85:761–769
59. Masuda T, Ueno Y, Kitabatake N (2005) High yield secretion of the sweet-tasting protein lysozyme from the yeast *Pichia pastoris*. Protein Expr Purif 39:35–42
60. Ide N, Masuda T, Kitabatake N (2007) Effects of pre- and pro-sequence of thaumatin on the secretion by *Pichia pastoris*. Biochem Biophys Res Commun 363:708–714
61. Ohta K et al (2008) Critical molecular regions for elicitation of the sweetness of the sweet-tasting protein, thaumatin I. FEBS J 275:3644–3652
62. Masuda T et al (2010) High-yield secretion of the recombinant sweet-tasting protein thaumatin I. Food Sci Technol Res 16:585–592
63. Poirier N et al (2012) Efficient production and characterization of the sweet-tasting brazzein secreted by the yeast *Pichia pastoris*. J Agric Food Chem 60:9807–9814
64. Scorer CA et al (1994) Rapid selection using G418 of high copy number transformants of *Pichia pastoris* for high-level foreign gene expression. Bio/Technology 12:181–184
65. Tuite MF, Clare JJ, Romanos MA (1999) Expressing cloned genes in the yeasts *Saccharomyces cerevisiae* and *Pichia pastoris*. In: Higgins SJ, Hames BD (eds) Protein expression. Oxford University Press, New York
66. Witty M (1990) Preprothaumatin II is processed to biological activity in *Solanum tuberosum*. Biotechnol Lett 12:131–136
67. Szwacka M et al (2002) Variable properties of transgenic cucumber plants containing the thaumatin II gene from *Thaumatococcus daniellii*. Acta Physiol Plant 24:173–185
68. Lebedev VG et al (2002) Pear transformation with the gene for supersweet protein thaumatin II. Acta Hortic 596:199–202
69. Bartoszewski G et al (2003) Modification of tomato taste in transgenic plants carrying a thaumatin gene from *Thaumatococcus daniellii* Benth. Plant Breed 122:347–351
70. Schestibratov KA, Dolgov SV (2005) Transgenic strawberry plants expressing a thaumatin II gene demonstrate enhanced resistance to *Botrytis cinerea*. Sci Hortic 106:177–189
71. Firsov A et al (2012) Transgenic tomato plants as supersweet protein thaumatin II producers. Appl Biochem Microbiol 48:746–751
72. Firsov A et al (2016) Purification and characterization of recombinant supersweet protein thaumatin II from tomato fruit. Protein Expr Purif 123:1–5
73. Van der Wel H (1994) Structure-activity relationship in the thaumatin molecule. In: Witty M, Higginbotham JD (eds) Thaumatin. CRC Press, Boca Raton
74. Van der Wel H, Bel WJ (1976) Effect of acetylation and methylation on the sweetness intensity of thaumatin I. Chem Senses Flavor 2:211–218
75. Van der Wel H (1986) The sweet proteins. In: Hudson BJF (ed) Developments in food proteins. Applied Science, London
76. Shamil S, Beynon RJ (1990) A structure-activity study of thaumatin using pyridoxal 5-′-phosphate (PLP) as a prove. Chem Senses 15:457–469
77. Kaneko R, Kitabatake N (2001) Structure-sweetness relationship study of sweet protein thaumatin: importance of lysine residues. Chem Senses 26:167–177
78. Kitabatake N, Kusunoki M (1994) Role of arginine and lysine residues in sweetness expression of thaumatin. In: Kurihara K, Suzuki N, Ogawa H (eds) Olfaction and taste XI. Springer, Japan
79. Kim S-H, Weickmann JL (1994) Crystal structure of thaumatin I and its correlation to biochemical and mutational studies. In: Witty M, Higginbotham JD (eds) Thaumatin. CRC Press, Boca Raton

80. Ohta K et al (2011) Introduction of a negative charge at Arg82 in thaumatin abolished responses to human T1R2-T1R3 sweet receptors. Biochem Biophys Res Commun 413:41–45
81. Temussi PA (2011) Determinants of sweetness in proteins: a topological approach. J Mol Recognit 24:1033–1042
82. Masuda T et al (2016) A hypersweet protein: removal of the specific negative charge at Asp21 enhances thaumatin sweetness. Sci Rep 6:20255
83. Slootstra JW et al (1995) Possible active site of the sweet-tasting protein thaumatin. Chem Senses 20:535–543
84. Hoon MA et al (1999) Putative mammalian taste receptors: a class of taste-specific GPCRs with distinct topographic selectivity. Cell 96:541–551
85. Kitagawa M et al (2001) Molecular genetic identification of a candidate receptor gene for sweet taste. Biochem Biophys Res Commun 283:236–242
86. Max M et al (2001) Tas1r3, encoding a new candidate taste receptor, is allelic to the sweet responsiveness locus *Sac*. Nat Genet 28:58–63
87. Montmayeur JP et al (2001) A candidate taste receptor gene near a sweet taste locus. Nat Neurosci 4:492–498
88. Sainz E et al (2001) Identification of a novel member of the T1R family of putative taste receptors. J Neurochem 77:896–903
89. Nelson G et al (2001) Mammalian sweet taste receptors. Cell 106:381–390
90. Hellekant G, Danilova V (1996) Species differences toward sweeteners. Food Chem 56:323–328
91. Liu B et al (2011) Molecular mechanism of species-dependent sweet taste toward artificial sweeteners. J Neurosci 31:11070–11076
92. Jiang P et al (2004) The cystein-rich region of T1R3 determines responses to intensely sweet proteins. J Biol Chem 279:45068–45075
93. Xu H et al (2004) Different functional roles of T1R subunits in the heteromeric taste receptors. Proc Natl Acad Sci U S A 101:14258–14263
94. Koizumi A et al (2007) Taste-modifying sweet protein, neoculin, is received at human T1R3 amino terminal domain. Biochem Biophys Res Commun 358:585–589
95. Ohta K et al (2011) The cysteine-rich domain of human T1R3 is necessary for the interaction between human T1R2-T1R3 sweet receptors and a sweet-tasting protein, thaumatin. Biochem Biophys Res Commun 406:435–438
96. Jiang P et al (2005) Lactisole interacts with the transmembrane domains of human T1R3 to inhibit sweet taste. J Biol Chem 280:15238–15246
97. Masuda T et al (2013) Five amino acid residues in cysteine-rich domain of human T1R3 were involved in the response for sweet-tasting protein, thaumatin. Biochimie 95:1502–1505
98. Ide N et al (2009) Interactions of the sweet-tasting proteins thaumatin and lysozyme with the human sweet-taste receptor. J Agric Food Chem 57:5884–5890
99. Liu B et al (2015) Functional characterization of the heterodimeric sweet taste receptor T1R2 and T1R3 from a New World monkey species (squirrel monkey) and its response to sweet-tasting proteins. Biochem Biophys Res Commun 427:431–437
100. Damak S et al (2003) Detection of sweet and umami taste in the absence of taste receptor T1r3. Science 301:850–853
101. Temussi PA (2002) Why are sweet proteins sweet? Interaction of brazzein, monellin and thaumatin with the T1R2-T1R3 receptor. FEBS Lett 526:1–4
102. Kunishima N et al (2000) Structural basis of glutamate recognition by a dimeric metabotropic glutamate receptor. Nature 407:971–977
103. Koiwa H et al (1999) Crystal structure of tobacco PR-5d protein at 1.8 Å resolution reveals a conserved acidic cleft structure in antifungal thaumatin-like proteins. J Mol Biol 286:1137–1145
104. Dudler R, Mauch F, Reimmann C (1994) Thaumatin-like proteins. In: Witty M, Higginbotham JD (eds) Thaumatin. CRC Press, Boca Raton

105. Peng Z et al (1997) Taste properties of grape (*Vitis vinifera*) pathogenesis-related proteins isolated from wine. J Agric Food Chem 45:4639–4643
106. de Vos AM et al (1985) Three-dimensional structure of thaumatin I, an intensely sweet protein. Proc Natl Acad Sci U S A 82:1406–1409
107. Ogata CM et al (1992) Crystal structure of a sweet tasting protein thaumatin I, at 1.65 Å resolution. J Mol Biol 228:893–908
108. Ko TP et al (1994) Structures of three crystal forms of the sweet protein thaumatin. Acta Crystallogr Sect D Struct Biol Cryst Commun 50:813–825
109. Charron C, Giegé R, Lorber B (2004) Structure of thaumatin in a hexagonal space group: comparison of packing contacts in four crystal lattices. Acta Crystallogr Sect D Struct Biol Cryst Commun 60:83–89
110. Asherie N et al (2008) Solubility of thaumatin. Crystal Growth Des 8:1815–1817
111. Asherie N et al (2008) Effects of protein purity and precipitant stereochemistry on the crystallization of thaumatin. Crystal Growth Des 12:4200–4207
112. Masuda T et al (2011) Crystal structure of the sweet-tasting protein thaumatin II at 1.27 Å. Biochem Biophys Res Commun 410:457–460
113. Teixeira SC et al (2008) A preliminary neutron crystallographic study of thaumatin. Acta Crystallogr Sect F Struct Biol Cryst Commun 64:378–381
114. Teixeira SC et al (2010) Sweet neutron crystallography. Acta Crystallogr Sect D Struct Biol Cryst Commun 66:1139–1143
115. Charron C et al (2002) Crystallization in the presence of glycerol displaces water molecules in the structure of thaumatin. Acta Crystallogr Sect D Struct Biol Cryst Commun 58:2060–2065
116. Ng JD et al (1997) Comparative analysis of thaumatin crystals grown on earth and in microgravity. Acta Crystallogr Sect D Struct Biol Cryst Commun 53:724–733
117. Warkentin M (2010) Glass transition in thaumatin crystals revealed through temperature-dependent radiation-sensitivity measurements. Acta Crystallogr Sect D Struct Biol Cryst Commun 66:1092–1100
118. Warkentin M (2012) Spatial distribution of radiation damage to crystalline proteins at 25-300 K. Acta Crystallogr Sect D Struct Biol Cryst Commun 68:1108–1117
119. Chapman HN et al (2011) Femtosecond X-ray protein nanocrystallography. Nature 470:73–77
120. Neutze R et al (2000) Potential for biomolecular imaging with femtosecond X-ray pulses. Nature 406:752–757
121. Emma P et al (2010) First lasing and operation of an ångstrom-wavelength free-electron laser. Nat Photonics 4:641–647
122. Schlichting I, Miao J (2012) Emerging opportunities in structural biology with X-ray free-electron lasers. Curr Opin Struct Biol 22:613–626
123. Barty A et al (2012) Self-terminating diffraction gates femtosecond X-ray nanocrystallography measurements. Nat Photonics 6:35–40
124. Ishikawa T et al (2012) A compact X-ray free-electron laser emitting in the sub-ångström region. Nat Photonics 6:540–544
125. Sugahara M et al (2015) Grease matrix as a versatile carrier of proteins for serial crystallography. Nat Methods 12:61–63
126. Nass K et al (2016) Protein structure determination by single-wavelength anomalous diffraction phasing of X-ray free-electron laser data. IUCrJ 3:180–191
127. Kaneko R, Kitabatake N (1999) Heat-induced formation of intermolecular disulfide linkages between thaumatin molecules that do not contain cysteine residues. J Agric Food Chem 47:4950–4955
128. Sugiyama et al (2012) Growth of protein crystals in hydrogels prevents osmotic shock. J Am Chem Soc 134:5786–5789
129. Sauter et al (2002) Towards Atomic Resolution with Crystals Grown in Gel: The Case of Thaumatin Seen at Room Temperature. Proteins 48:146–150

130. Teplitsky et al (2015) High throughput screening using acoustic droplet ejection to combine protein crystals and chemical libraries on crystallization plates at high density. J Struct Biol 191:49–58
131. Wagner et al (2016) In-vacuum long-wavelength macromolecular crystallography. Acta Crystallogr D Struct Biol 72:430–439
132. Zander et al (2015) Meshandcollect: An Automated Multi-Crystal Data-Collection Workflow for Synchrotron Macromolecular Crystallography Beamlines. Acta Crystallogr D Biol Crystallogr 71:2328–2343
133. Cipriani et al (2012) Crystal Direct: A New Method for Automated Crystal Harvesting Based on Laser-Induced Photoablation of Thin Films. Acta Crystallogr D Biol Crystallogr 68:1393–1399
134. Nanao et al (2005) Improving Radiation-Damage Substructures for Rip. Acta Crystallogr D Biol Crystallogr 61:1227–1237
135. Kissick et al (2013) Towards protein-crystal centering using second-harmonic generation (SHG) microscopy. Acta Crystallogr D Biol Crystallogr 69:843–851

Part V

Industrial Applications

The Recent Development of a Sweet-Tasting Brazzein and its Potential Industrial Applications

23

Fabrice Neiers, Christian Naumer, Michael Krohn, and Loïc Briand

Abstract

Brazzein is a small heat- and pH-stable sweet-tasting protein isolated from the West African plant, *Pentadiplandra brazzeana*. Brazzein combines a highly sweet potency, a long history of human consumption, and a remarkable stability, giving it great potential as a natural sweetener. Due to the difficulties of obtaining brazzein from its natural source, several efforts have been made to express brazzein using various heterologous expression systems. This chapter describes the biochemical, structural, sensory, and physiological properties of brazzein. We will summarize the current knowledge of the structure-activity relationship of brazzein. The biotechnological production of brazzein using various expression systems will also be reviewed. Furthermore, the emerging application of brazzein in the food industry to replace traditional sugars by acting as a natural, good, low-calorie sweetener will be discussed.

Keywords

Brazzein • High-potency sweeteners • Sweet-tasting protein • Sweet-taste receptor • Structure-function relationship

Abbreviations

CRD	Cysteine-rich domain
GPCR	G-protein-coupled receptor
GST	Glutathione-S-transferase

F. Neiers (✉) • L. Briand (✉)
Centre des Sciences du Goût et de l'Alimentation, INRA UMR1324, CNRS UMR6265, Université de Bourgogne – Franche-Comté, Dijon, France
e-mail: Fabrice.Neiers@u-bourgogne.fr; loic.briand@inra.fr

C. Naumer (✉) • M. Krohn (✉)
BRAIN AG, Zwingenberg, Germany
e-mail: cn@brain-biotech.de; mk@brain-biotech.de

J.-M. Mérillon, K.G. Ramawat (eds.), *Sweeteners*, Reference Series in Phytochemistry,
https://doi.org/10.1007/978-3-319-27027-2_2

HPLC	High performance liquid chromatography
NMR	Nuclear magnetic resonance
NSB-H	Non-Specific Binding-Hypothesis
NTD	N-terminal domain
pyrE	Pyroglutamic acid
SDS-PAGE	Sodium dodecyl sulfate-polyacrylamide gel Electrophoresis
SUMO	Small Ubiquitin-like Modifier
TMD	Transmembrane domain

Contents

1 Introduction 528
2 Sweet-Tasting Proteins 529
3 Discovery of Brazzein 532
4 General Properties of Brazzein 533
4.1 Amino Acid Sequence and Biochemical Properties of Brazzein 533
4.2 Secondary and Tertiary Structures of Brazzein 533
4.3 Structure-Function Relationships of Brazzein 534
4.4 Sensory Properties of Brazzein 535
4.5 Physiological Properties of Brazzein 537
4.6 Proposed Mode of Sweet Taste Receptor Activation 537
5 Biotechnological Expression of Brazzein 539
5.1 Bacterial Expression of Brazzein 539
5.2 Yeast Expression of Brazzein 541
5.3 Brazzein Expression using Transgenic Plants and Animals 541
6 Brazzein and Food Application 542
7 Patent Analysis on Brazzein 543
8 Conclusion 543
References 544

1 Introduction

The global increase of obesity and diabetes mellitus has led to an explosive interest for healthy and low-caloric natural sweeteners with favorable tasting properties. Consumed by indigenous people for centuries, sweet-tasting proteins may constitute an interesting group of natural sweeteners. Six sweet-tasting proteins have been identified, which are able to induce an intense sweetness in humans, including thaumatin, monellin, mabinlin, brazzein, pentadin, and neoculin. Discovered in the fruits of the tropical plant *Pentadiplandra brazzeana* Baillon, brazzein is the smallest (6.5 kDa) sweet-tasting protein with a taste close to that of natural sugars. Its high sweet potency coupled with an absence of bitterness makes it a good alternative to natural sweeteners such as steviosides for instance. In addition, brazzein presents physicochemical characteristics such as high water-solubility and extreme thermostability, which are essential for food applications. This chapter will describe the biophysical, structural, and sensory properties of brazzein. In the oral cavity, sweet taste recognition is achieved by a heterodimeric receptor made of the assembly of two subunits, the T1R2 (taste receptor type 1, member 2) and T1R3 (taste receptor type 1, member 3).

Remarkably, this single heterodimeric receptor is able to detect all of the sweet-tasting compounds including natural sugars, artificial and natural sweeteners including sweet-tasting proteins. The three-dimensional structure of brazzein has been solved by nuclear magnetic resonance (NMR) and X-ray crystallography. Numerous site-directed mutagenesis experiments have allowed identification of amino acid residues on brazzein that are important for its sweetness. However, the exact mechanism of T1R2/T1R3 receptor activation is still unclear and needs to be clarified. This chapter will summarize the structure-activity relationship of brazzein and will describe the current knowledge on the putative molecular mechanism of receptor activation. Because its natural source is limited, brazzein has been expressed using various recombinant expression systems including bacteria, yeast, and transgenic plants and animals. The expression system most suitable for efficient industrial production will also be reviewed.

2 Sweet-Tasting Proteins

Around the world, five sweet-tasting proteins have been identified: brazzein, neoculin (previously named curculin), mabinlin, monellin, and thaumatin [1]. A sixth sweet-tasting protein, pentadin, has been discovered but is still not characterized. A seventh protein, miraculin, which is a taste-modifying protein, can be added to the list. Miraculin is not sweet by itself but is able to convert sourness into sweetness. These seven proteins all originate from tropical forests from south Asia or Africa. Brazzein and pentadin originate from the same plant found in Congo, Gabon, and Cameroon tropical forests. Miraculin is found in the same geographical area. Monellin is also found in African tropical forests but in different geographical areas including south Mozambique and Angola. Thaumatin was discovered in a rainforest plant located from Sierra Leone to Togo. The two Asian sweet-tasting proteins, neoculin and mabinlin, are found in south Malaysia and Yunnan Chinese provinces, respectively. Primates eating fruits contribute to seed dispersal in their feces [2]. In the tropical forest, the evolutionary pressure due to primate preference for the sweetest fruits contributes to the increase of sugar content. In this context, the apparition of sweet-tasting proteins in some fruits appears to be a selective advantage for the plants in using primates as a seed disperser. This evolutionary point of view explains the more favorable context for the apparition of sweet-tasting proteins in tropical forests where primates contribute to seed dispersal. Additionally, this theory is well supported by the differences observed among the sweet-tasting proteins. Indeed, they all present a different amino acid sequence, proving a non-related origin and supporting a selective pressure that contributed to the apparition at least six times during the plant evolution. Interestingly, among the five well-characterized sweet-tasting proteins, all present a different three-dimensional structure, which was not a surprise due to the sequence differences resulting in large molecular mass differences (from 6.4 kDa for brazzein to 24 kDa for neoculin and probably much more for miraculin, Table 1). Not only do they

Table 1 Presentation of the main characteristics of the sweet-tasting and taste-modifying proteins

	Brazzein	Neoculin	Mabinlin	Monellin	Pentadin	Thaumatin	Miraculin
Plant name	*Pentadiplandra brazzeana*	*Curculingo latifolia*	*Capparis masakai*	*Dioscoreophyllum cumminsii*	*Pentadiplandra brazzeana*	*Thaumatococcus danielli*	*Richadella dulcifica*
Sweetness factor of the sweetest variant (Weight basis)	1,900	550	375	3,000	500	3,000	Not applicable
Structure, first PDB deposit (PDB code)	1BRZ	2D04	2DS2	1MON	Unknown	1THI	Unknown
Oligomeric state (molecular mass of the native form)	Monomer (6.5 kDa)	Homo-dimer (24 kDa)	Hetero-dimer (12.4 kDa)	Hetero-dimer (11.4 kDa)	Unknown (12 kDa)	Monomer (22.2 kDa)	Homo-dimer or homo-tetramer (49.2 or 98.4 kDa)

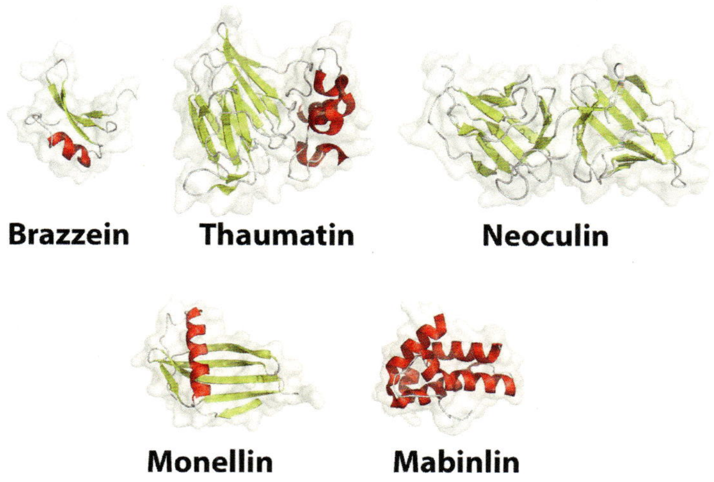

Fig. 1 Three-dimensional structure of sweet-tasting proteins. The five known sweet-tasting protein structures are represented in cartoon mode. The α-helices and β-sheet are represented in *red* and *yellow* colors, respectively. The protein surfaces are represented in *white*. The PDB codes used to build up the figure are 1BRZ for brazzein, 2VHR for thaumatin, 2DO4 for neoculin, 3MON for monellin, and 2DS2 for mabinlin

present different folding but the amount of secondary structure is totally different, from folding composed of only alpha helices to folding made of only beta sheets for mabinlin and neoculin, respectively (Fig. 1). Interestingly, the quaternary structures of the seven sweet-tasting and sweet-modifying proteins, when they are known, are also highly diverse including monomeric, homodimeric, and also heterodimeric states (Table 1). These structural differences may explain the sweetness, the off-taste, and the thermal stability differences observed between these proteins. The most intense sweet-tasting proteins compared to sucrose on a weight basis are brazzein, monellin, and thaumatin, with sweetness potencies of 1,900 and 3,000 for the last two, respectively. The least intense sweet-tasting protein is mabinlin (375-fold for the form II, the sweetest isoform) (Table 1). As observed for curculin (which is also sweet by itself in contrast to miraculin), miraculin is able to modify the sour taste into an intense sweet taste [3]. Other than their sweetness, one of the common features of the different sweet-tasting proteins is their basic isoelectrical point, except that of brazzein, which is acidic. Among all the known sweet-tasting proteins, brazzein is the most stable with the lowest lingering off-taste, making it one of the most interesting if we also consider its intense sweetness.

3 Discovery of Brazzein

Pentadiplendra brazzeana Baillon is a plant growing in west tropical African countries such as Gabon, Cameroon, and Nigeria. This climbing shrub living on the edge of the forest was described in 1868 by Henri Ernest Baillon, a French botanist and physician [4]. The plant was rediscovered by Marcel and Anette Hladik, two French primatologists studying apes habits [5]. *Pentadiplendra brazzeana* bears red globular fruits of approximately 5 cm in diameter (Fig. 2). These berries present a thick epicarp and contain five reniform seeds surrounded by a soft layer of red pulp. This well-known fruit is named "oubli" by the local population. This French word, meaning "forgot," refers to the fruit's sweetness, which is so intense and attractive that the children who eat them forget to go back with their mothers [6]. The search for a sweet-tasting protein in the fruit of *Pentadiplendra brazzeana* started at the end of the 1980s. From a water extract of the pulp fruit collected in Gabon, a protein named pentadin was first isolated by van der Wel et al. [7]. The sweetness intensity of pentadin was estimated to be approximately 500 times higher than a 10% sucrose solution on a weight basis. SDS-PAGE analysis in the presence and absence of a reducing agent indicated the presence of subunits of approximately 12 kDa linked by disulfide bonds. Unfortunately no further characterization of pentadin was reported in the literature since. Further investigations would be necessary to determine the identity of the pentadin. The discovery of brazzein was reported in 1994. Ming and collaborators isolated the brazzein protein from the red pulp fruit and determined its amino acid sequence by Edman degradation of isolated peptide fragments [8]. The sweetness and physicochemical properties of brazzein were also reported.

Fig. 2 Picture of the *Pentadiplendra brazzeana* fruits

4 General Properties of Brazzein

4.1 Amino Acid Sequence and Biochemical Properties of Brazzein

Brazzein has been isolated in the pulp of the fruit between the pericarp and the seeds with a content of approximately 0.2–0.05% by weight. Brazzein is a highly soluble protein (more than 50 g/L) with an isoelectric point of 5.4 [8]. Brazzein is a single chain protein with a molecular mass of 6.49 kDa. Brazzein has been observed in two forms from its natural source [8]. The major form (~80%), composed of 54 amino acid residues, is called pyrE-bra. This form contains a pyroglutamic acid (pyrE) at its N-terminal position. The minor form (representing approximately 20%), made of 53 amino acid residues, is called des-pyrE-bra and lacks the N-terminal pyrE (Fig. 3). The pyrE residue at the N-terminus results from the chemical cyclization of the N-terminal glutamine residue. Interestingly, sensory studies have shown that the des-pyrE-bra form is twice as sweet as pyrE-bra [9, 10]. Due to the presence of four disulfide bridges, brazzein is remarkably pH- (from 2.5 to 8) and heat-stable. Ming and Hellekant reported that brazzein remains sweet after an incubation of 2 hours at 98 °C or 4.5 h at 80 °C [8].

4.2 Secondary and Tertiary Structures of Brazzein

The first brazzein three-dimensional structure was solved in 1998 using proton NMR spectroscopy [11]. No surface similarity or any three-dimensional fold similarity was found with the other sweet-tasting proteins. The brazzein fold consists of a "cysteine-stabilized alpha-beta" motif (CSαβ). In this motif, the main characteristic is the stabilization of the unique α-helix by two disulfide bonds with the closest β-strand. Additionally, the brazzein structure presents two other disulfide bonds. These four disulfide bonds explain the high thermal stability of the protein. It has been proposed that brazzein evolved from a plant protease inhibitor from deletion mutations, indeed the most related amino acid sequence corresponds to a trypsin inhibitor [11]. Concerning the brazzein fold homologues, it appears that the proteins presenting the same CSαβ fold mainly encompass a defense function and are present

```
                     1          10         20         30         40         50
pyrE-bra      pyrEDKCKKVYEN YPVSKCQLAN QCNYDCKLDK HARSGECFYD EKRNLQCICD YCEY
des-pyrE-bra     -DKCKKVYEN YPVSKCQLAN QCNYDCKLDK HARSGECFYD EKRNLQCICD YCEY
Q1-bra           QDKCKKVYEN YPVSKCQLAN QCNYDCKLDK HARSGECFYD EKRNLQCICD YCEY
```

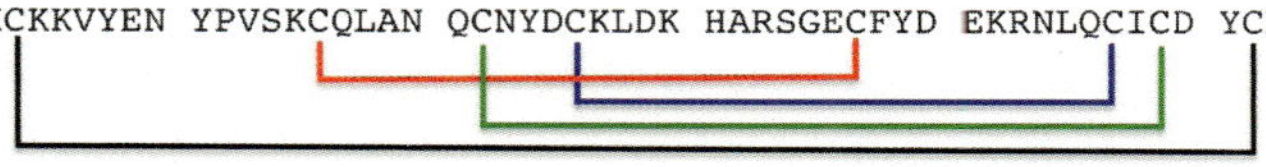

Fig. 3 Amino acid sequences of the two sweet-tasting constituents isolated from the fruit of *Pentadiplandra brazzeana* Baillon: pyrE-bra and the recombinant Q1-bra. pyrE-bra form contains a pyrE residue at its N-terminus while the des-pyrE-bra lacks this N-terminal residue. The recombinant Q1-bra form possesses a N-terminal glutamine residue (Q1) instead of a pyrE residue. The four disulfide bridges are presented

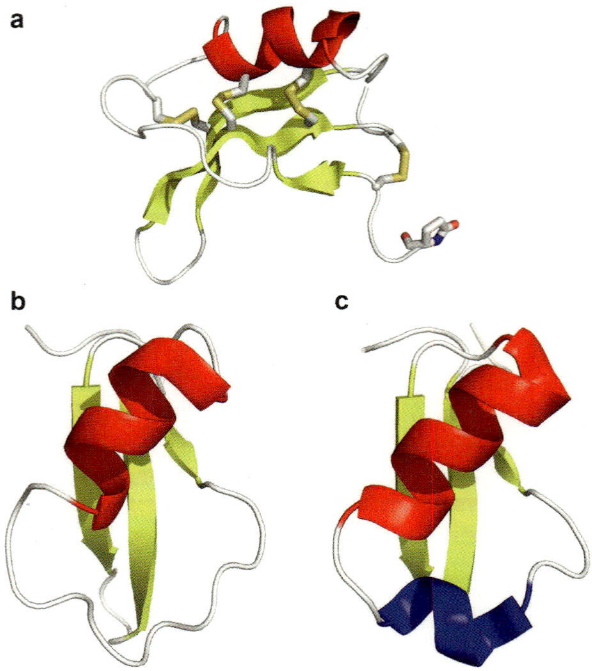

Fig. 4 Three-dimensional structure of brazzein. In the panel (**b**) and (**c**), NMR and X-ray solved brazzein structures are compared. Structures obtained using NMR (PDB entry 1BRZ) (**a** and **b**) or X-ray diffraction (PDB entry 4HE7) (**c**) for the resolution are represented. In panel (**a**), the four disulfide bonds are represented as well as the N-terminal des-pyro Glu in stick representation. The alpha helix is indicated in *red*; the beta-strands in *yellow*. An additional 3_{10} helix found in the X-ray structure is indicated in *blue* (**c**)

in many plants, but also fungi and insects [12]. Among this family the main structural difference found in brazzein structure is the presence of an additional 3_{10} helix in the first loop located between the first beta strand and the alpha helix (indicated in blue, Fig. 4). Interestingly, this 3_{10} helix is observed in the structures obtained at low pH as shown for the X-ray structure presented in the Fig. 4, for which the protein crystals were obtained at pH 4.0. Likewise, the NMR structure shown here obtained at pH 5.2 does not present this 3_{10} helix [13]. At pH 5.2, this helix already appears less stable because it is not observed in another NMR structure obtained at the same pH (PDB entry 2KGQ). However, this loop formation is not a major determinant for the sweet taste of brazzein because brazzein presents sweetness at both of these pH values.

4.3 Structure-Function Relationships of Brazzein

Extensive site-directed mutagenesis experiments [14–21] combined to sensory analysis or cellular assays have been conducted to elucidate the structure-function relationships of brazzein (Fig. 5). Three main sites have been identified as responsible for the sweetness of brazzein: site 1 (Loop residue R43), site 2 (N- and C-terminal regions/residue E36 and Loop 33), and site 3 (Loop residues 9–19). Numerous mutations in these regions led to reduced or profoundly reduced sweetness. Remarkably, deletion of the C-terminal residue (Y54, site 2) completely

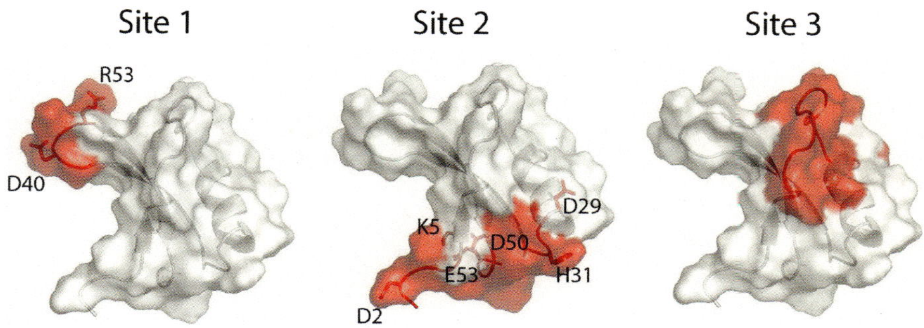

Fig. 5 The three main sites responsible for the sweetness of brazzein. Site 1 (Loop residue R43), site 2 (N- and C-terminal regions/ residue E36 and Loop 33), and site 3 (Loop residues 9–19) are represented in *red*

abolished the sweetness of brazzein [14]. The substitution of this residue by a tryptophan significantly increased the sweetness of brazzein, suggesting that this residue is involved in the interaction with the sweet taste receptor [14]. In contrast, the natural minor form, des-pyrE-bra, which lacks the N-terminal pyrE, is sweeter than the major form [9, 10]. Interestingly, some mutations located in the three major sites have been shown to increase the sweetness of brazzein [14–21]. These mutations include D2E, K5R, D29A, D29N, H31R, D40K, E41A, D50K, and E53R (Fig. 5). Mutations that alter the disulfide bonds lead to non-sweet protein variants [14, 20]. NMR studies revealed that these modifications led to large conformational changes of brazzein [14]. Recently, Lee and collaborators reported that triple mutations of critical amino acid residues (H31R/E36D/E41A) have an additive effect on brazzein sweetness in agreement with the hypothesis of multiple binding sites on sweet taste receptors [17].

4.4 Sensory Properties of Brazzein

For the food industry, a novel high-intensity sweetener should have the same techno-functional properties as carbohydrate sweeteners but preferably without their calorie burden. To date, neither artificial nor natural high-intensity sweeteners have been identified that are able to mimic the organoleptic or application profile of carbohydrate sweeteners. As main disadvantages, they often exhibit off-taste (bitterness, metallic, or licorice) and a non-sugar-like temporal profile (late sweetness onset and/or sweetness lingering). Ethnobotanical history of human brazzein consumption and its status as a natural compound are probable explanations for only minor toxicological concerns, allowing several laboratories during recent years to do initial human sensory evaluations (e.g., sip and spit tests). Nevertheless, publicly available information from descriptive sensory evaluations on brazzein is still limited, often lacks quantitative data, and is probably strongly affected by differences in protein purity. The latter is important because contaminating small peptides can have a large

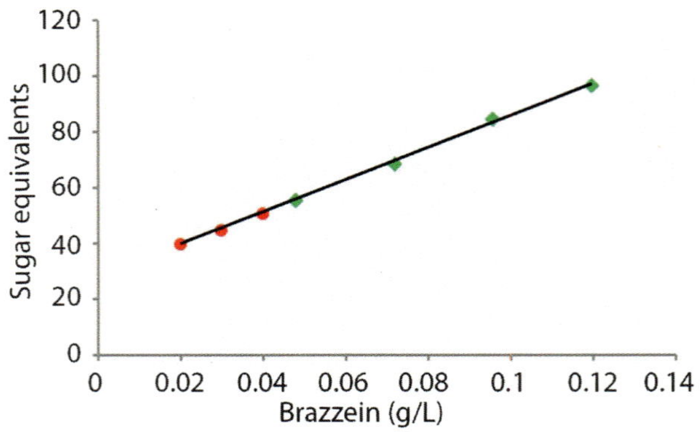

Fig. 6 Acidified brazzein solutions (0.5 g/l citric acid) were ranked in comparison to sugar solutions (isosweetness) by a panel of trained tasters (filled diamonds). *Filled dots* represent calculated data

contribution to the taste profile. In an industrial environment, sensory evaluation (BRAIN AG, own data) by a human taste panel with wild-type brazzein (des-pyrE-brazzein; 98% purity according to HPLC analysis) produced from a recombinant yeast bioprocess was conducted. The sweetness intensity of brazzein was described as 1,900 times that of sucrose at a sugar equivalent level of 40 g/L [10]. At a sugar equivalent level of 100 g/L, it declines to 800 times (Fig. 6). These values are important for any type of food application because typical single sweetener concentrations in beverages are in a range of 50–600 mg/L. To get an impression of whether a sweetener is suited for industrial food applications, an examination of its basic sensory attributes is helpful. Such an evaluation of valid sensory attributes is given in the spider plot of Fig. 7. Brazzein reveals nearly no bitter or metallic off-notes and has a good isosweetness intensity with high sugar similarity. However, as often seen for noncarbohydrate sweeteners, brazzein has a delayed sweetness onset, a licorice off-note, and shows concise sweetness lingering. A good model for lingering effects and, probably, also for the delayed onset of noncarbohydrated sweeteners is the so-called Non-Specific Binding Hypothesis (NSB-H) [22]. NSB-H explains lingering by the rapid, weak, and non-specific binding of a sweetener like brazzein to non-receptor sites and cell membranes. Subsequently, NSB-H could lead to a delayed maximum activation of the sweet receptor and potentially a longer term and ongoing activation from the nonspecifically bound brazzein reservoir in the oral cavity.

Based on the available sensory data, the application potential of brazzein in nutritional products should be summarized as most likely suited for calorie-reduced products. Brazzein also shows synergistic effects with other sweeteners as described by Hellekant and Danilova [23]. Therefore, blends and formulation development for defined food product matrices should have the potential to optimize a brazzein-based product taste profile. Compared to existing high-intensity sweeteners, brazzein has some advantageous properties, and an approved brazzein would most likely already be part of existing products. However, high costs for approval and production combined with a limited product-range potential are probably the main reasons that brazzein is still not on the market.

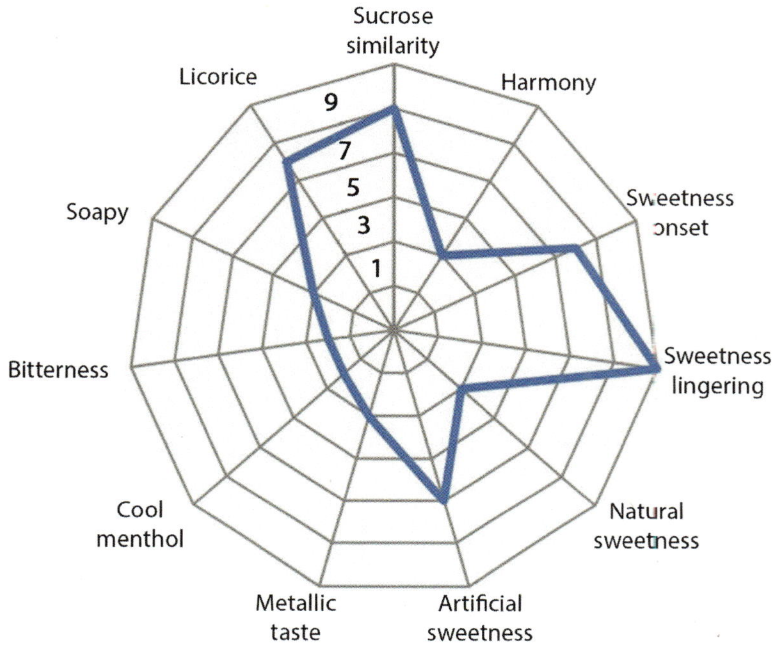

Fig. 7 A descriptive sensory evaluation of a 0.035 % brazzein solution (0.5 g/l citric acid) was conducted by a panel of 10 trained tasters. The *blue line* is a graphical representation of the overall taste profile of brazzein regarding the depicted sensory attributes (number of tasters confirming an attribute)

4.5 Physiological Properties of Brazzein

Brazzein has been used by indigenous people for centuries as sweetener in food or drinks [6]. *Chorda tympani* nerve recordings have shown that brazzein is perceived as sweet by primates such as chimpanzees or rhesus monkeys [23]. Contrariwise, marmosets and New World monkeys are not sensitive to brazzein. The New and Old world monkeys share an ancestry approximately 38 million years ago, maybe explaining the sweet taste receptor differences as the origin of the New World monkey insensitivity for brazzein [24]. Cellular-based assays have shown that mouse T1R2/T1R3 is insensitive to brazzein [25]. Surprisingly, the fruit fly *Drosophila melanogaster* has been shown to be responsive to brazzein and responded appetitively to and ingested brazzein [26].

4.6 Proposed Mode of Sweet Taste Receptor Activation

In the mouth, the detection of sweet compounds is achieved by a heterodimeric receptor made of the obligate assembly of two G-protein-coupled receptor (GPCR) subunits (Fig. 8). These subunits are named T1R2 (taste receptor type 1, member 2)

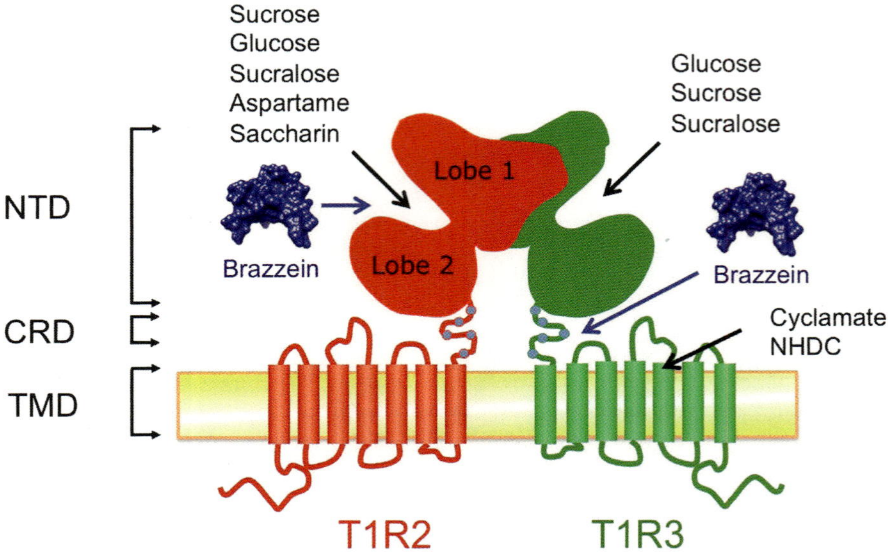

Fig. 8 Schematic representation of the sweet taste T1R2/T1R3 receptor. The two subunits possess a large N-terminal domain (*NTD*), connected to the transmembrane domain (*TMD*) by a short cysteine-rich domain (*CRD*). T1R2-NTD contains the primary binding site of sweet compounds where natural sugars, sucralose, aspartame, and saccharin bind. Cyclamate and neohesperidin dihydrochalcone (*NHDC*) interact with T1R3-TMD, whereas natural sugars and sucralose bind into T1R3-NTD. It has been proposed that brazzein binds into both T1R2-NTD and T1R3-CRD

and T1R3 (taste receptor type 1, member 3). T1R2 and T1R3 subunits are members of subclass 3 of GPCRs. The members of this family share structural features including a large N-terminal domain (NTD) connected to the transmembrane domain (TMD) by a short cysteine-rich domain (CRD). The T1R2/T1R3 sweet taste receptor [27, 28] is able to detect a wide chemical variety of sweet tasting compounds, including natural sugars (such as sucrose, fructose, and glucose), sugar alcohols (such as xylitol, erythritol, and sorbitol), and artificial (such as aspartame, sucralose, saccharin, acesulfame-K, and cyclamate) and natural sweeteners (such as stevioside, glycyrrhizin, and mogrosides). Interestingly, the human T1R2/T1R3 receptor is also activated by all the sweet-tasting proteins [10, 27, 29–31]. Mouse-human chimera, molecular modeling, and site-directed mutagenesis studies have revealed that the NTD of T1R2 contains the primary binding site of sweet compounds. This NTD of T1R2 contains two lobes, forming a cavity in which sweeteners bind. However, two additional binding sites have been identified in the NTD and TMD of T1R3 (Fig. 8). Indeed, cyclamate and the sweetener neohesperidin dihydrochalcone (NHDC) have been shown to bind into the TMD of the T1R3 subunit [32, 33], whereas natural sugars and sucralose interact with the NTD of the T1R3 subunit [34, 35]. As with the other sweet-tasting proteins, the mechanism for receptor activation by brazzein is still unknown. Molecular modeling combined to docking studies has proposed a wedge model in which brazzein docks between the open cleft of the T1R2 or

T1R3-NTDs [36, 37]. These proposed models of activation have not been further confirmed by site-directed mutagenesis. Interspecies mixed sweet taste receptors and NMR saturation transfer revealed the importance of T1R2-NTD and T1R3-CRD for brazzein sensitivity [25, 38]. Although the functional role of the CRD of T1R3 remains to be elucidated, it has been shown that this domain is also involved in the response to thaumatin [31]. Using site-directed mutagenesis, it has been shown that both T1R2- and T1R3-NTDs are important for brazzein activation, suggesting multipoint binding interactions between brazzein and the sweet taste receptor [14]. Further studies will be required to elucidate the molecular mechanism of receptor activation by brazzein.

5 Biotechnological Expression of Brazzein

Access to *Pentadiplandra brazzeana* Baillon and producing the fruits containing brazzein is difficult; its culture is also complex, and finally the brazzein extraction is expensive. The only way to easily obtain brazzein is the use of recombinant DNA technology. In the context of the evident commercial interest of brazzein, the recombinant DNA biotechnology provides an alternative solution for brazzein production. Heterologous production of brazzein is complicated by the fact that the protein contains four disulfide bridges and requires a specific N-terminal sequence to be fully active. The various levels of production are summarized in Table 2.

5.1 Bacterial Expression of Brazzein

Among the different organisms used for recombinant protein production, the bacterium *Escherichia coli* is the most commonly used. The first success concerning recombinant brazzein production was made in 2000 with this organism. Assadi-Porter and collaborators successfully produced and purified a sweet tasting brazzein using *Escherichia coli*. However, this expression presented two main limitations. First the production of a nuclease fused protein required an additional step to remove the fused protein. Secondly more than 70% of the protein is produced within inclusion bodies, requiring chemical refolding steps [9]. A brazzein without tags was produced under soluble form in a second study at a yield of 30–35 mg/L of protein, comparable to the yield of soluble protein found in the Assadi-Porter study. Unfortunately, this brazzein did not present the same sweetness compared to the plant brazzein [39]. Importantly, this brazzein is not produced with the N-terminal pyrE, which can explain the lower sweetness. In the same study, the authors produced and purified a C-terminal His-tagged brazzein, which is not sweet. In addition, the producing yield of this form was three times lower compared to the non-tagged form. The most promising study using *Escherichia coli* for recombinant brazzein production used a Small Ubiquitin-like Modifier (SUMO) protein with a His-tag at the N-terminal extremity of the brazzein [29]. The producing yield is not

Table 2 Organisms and main characteristics of recombinant brazzein production

Organism	Protein characteristics	Purified yield	Sweetness
Escherichia coli [9]	N-terminal pyroglutamate (pyEDK-)/C-terminal staphylococcal nuclease fusion	130–150 mg/L of culture after purification and tag cleavage. Purified from the inclusion bodies	Comparable to the wild type
Escherichia coli [39]	Modified N-terminal sequence (MAQDK-)	Unknown, a producing yield of 30–35 mg/L of soluble protein was used for the purification	Less sweet compared to the wild type
Escherichia coli [39]	Modified N-terminal sequence (MAQDK-), His-tag in C-terminal	Unknown, a producing yield of 5–10 mg/L of soluble protein was used for the purification	Not sweet
Escherichia coli [29]	N-terminal His tag + SUMO fusion	50 mg/10 g of wet producing cells after purification and tag cleavage. Purified from the soluble protein fraction.	Comparable to wild-type brazzein, but no pyroglutamate at the N-terminal after tag cleavage
Lactococcus lactis [39–41]	Modified N-terminal sequence (MAQDK-)	Only detectable on western blot analysis	Less sweet compared to the wild-type brazzein
Pichia pastoris [10]	N-terminal pyroglutamate/des-pyrE-bra/Q1-bra	90, 30, and 90 mg/L	Comparable to wild-type brazzein
Kluyveromyces lactis [43]	des-pyrE-bra	104 mg/L	Comparable to wild-type brazzein
Zea mays [44]	N-terminal without pyroglutamate (DK-)	400 μg/g of corn seed	1,700 times that of sucrose on a per weight basis at 2.0 % sucrose
Mus musculus [45]	Not indicated in the study	Detectable on western blot analysis	Sweet

comparable with the previous study, due to the fact it is indicated as 50 mg/10 g of wet producing cells after purification. However, this production appears as mainly soluble with an efficient tag cleavage (more than 90% compared to 50% for the previous nuclease tag) and a quick purification protocol (2 days).

Lactic acid bacteria such as *Lactococcus lactis* have the main advantage of being recognized as safe (GRAS status), allowing the agro-industry to use it easily to produce proteins such as brazzein. One group made several efforts to produce and increase brazzein production using this microorganism [39, 40]. In the first study, they demonstrated that *Lactococcus lactis* allows production of brazzein at a very low scale because the protein is not detectable by Coomassie staining but only by using western blot analysis [39]. This production yield was first improved 17-fold by optimizing the growth conditions [41] and then more than 800-fold using another expressing plasmid in another cell type. Unfortunately, the produced brazzein was not sweet.

5.2 Yeast Expression of Brazzein

Attempts at recombinant expression of brazzein have been made in yeast. The first expression of brazzein using yeast was achieved using *Saccharomyces cerevisiae* [42]. Brazzein was intracellularly expressed as a fusion with Glutathione-S-transferase (GST). The identity of the expressed protein was confirmed using antibodies, but the recombinant brazzein was not further characterized. In 2012, we reported that the methylotrophic yeast *Pichia pastoris* expression system is suitable for high-level expression of active brazzein [10]. Recombinant brazzein was efficiently secreted into the buffered minimal medium using the yeast prepropeptide signal from the *Saccharomyces cerevisiae* α-mating factor under the control of a methanol-inducible alcohol oxidase promoter. We expressed the two natural forms of brazzein, pyrE-bra and des-pyrE-bra, and an additional isoform, named Q1-bra, which is not naturally present in the fruit. The Q1-bra form differs from the pyrE-bra form in having a glutamine residue instead of pyrE at its N-terminus. Over a secretion period of 6 days, *Pichia* cells secreted approximately 120 mg/L of brazzein. We also described the purification and the biochemical characterization of the recombinant brazzeins. Mass spectrometry and $^{(1)}$H NMR spectroscopy indicated that the recombinant pyrE-bra and des-pyrE-bra brazzein forms were properly folded and identical to the natural proteins. In addition, we demonstrated that recombinant brazzein is functional and able to activate the human heterodimeric sweet taste T1R2/T1R3 receptor using cellular assays. Finally, sensory studies revealed that recombinant brazzein has sensory properties similar to those of the two natural brazzein forms. Brazzein was also expressed using the yeast *Kluyveromyces lactis* [43]. Brazzein was secreted into the culture medium using the yeast prepropeptide signal from the *Saccharomyces cerevisiae* α-mating factor. After 96 h of expression, the amount of purified recombinant des-pyrE-bra was approximately 104 mg per liter of culture. N-terminal amino acid sequencing of secreted brazzein confirmed the identity of the purified protein and the proper cleavage of the signal peptide. The authors checked the conformational state of the recombinant brazzein using circular dichroism. Sensory analysis revealed an intrinsic sweetness of purified brazzein produced by yeast *Kluyveromyces lactis*, which was identical to the natural brazzein.

5.3 Brazzein Expression using Transgenic Plants and Animals

Maize plants were genetically engineered to efficiently express and produce brazzein [44]. Interestingly, brazzein was produced in the corn's seed allowing an easy and economical production. The brazzein production rate of 400 μg/g of corn seed is the highest level obtained for a sweet-tasting protein produced in a plant. Two different forms of brazzein were produced in the maize with differences in N-terminal (QDKCKK- and DKCKK-), and the second one appeared twice as sweet compared to the first one [44]. Importantly, the brazzein produced in maize was described as sweet within the corn seed and was still sweet after processing conducive to the flour, easily allowing industrial production. The second expressing system using a complex organism is transgenic mice modified to produce brazzein in their milk. Even if the production level is low, the produced brazzein tasted sweet, demonstrating that it is possible to produce brazzein using transgenic animals, even if applied on cattle or goats which, as suggested by the author, will be more appropriate for mass production [45].

6 Brazzein and Food Application

Endemic consumption of brazzein either by raw fruits or as a sweetening agent is a long-known ethnobotanical heritage. Despite this long period, fundamental information on the brazzein content of *P. brazzeana* fruits is still only disclosed as 0.2–0.05% (w/w) by Ming et al. [46]. Being a protein, brazzein can be regarded as a low calorie sweetener (4 cal/mg protein) with a low potential for gastrointestinal distress. Due to current knowledge and the safe history of use, allergenicity of brazzein is not an obvious issue. However, in the case of industrial applications and due to its stability at 98 °C for 2 h or at 80 °C for 4.5 h and pH ranging from 2.5 to 8 [9], we need further evaluation of its potential as a food allergen. In addition to the low performance of proteinogenic sweeteners in certain bakery processes due to denaturation at high temperatures, industrial scale up either by plant farming or fermentation is still one of the hurdles to overcome to raise the probability of market acceptance of brazzein as an economically viable sweetener. Up to date thaumatin is the only approved proteinogenic low-calorie sweetener, which is, however, more often applied as a flavor modifier.

Industrial scale up for food and beverage applications is a prerequisite, and expression of the brazzein gene has been described several times in literature. Shake flask expression has been performed in bacteria [9, 39] as well as eukaryotes [10, 43]. The yields described here range from as low as 1.6 mg/L to ~120 mg/L [10, 43]. Higher yields were all achieved using eukaryotic expression systems. With yields in the 100 mg/L range, scale up into a fermenter appears to be feasible. However, production in a fermenter has, to our knowledge, not been described in a peer-reviewed journal. Nevertheless, several patents [47–49] describe successful expressions from the 1 to 1,000 L scale.

The patent covering the expression in a gram-positive expression host (*Bacillus subtilis*) [47] also gives an idea of the modifications to the host that are necessary to achieve a commercially viable expression level. In the patent, several strains carrying mutations beneficial for the expression of brazzein as well as numerous optimizations of the gene/promoter structure are evaluated. A standard fed-batch strategy is applied and the yield achieved is described as "industrially relevant." The same company (Novozymes) filed another patent [49] describing the production of brazzein in 10 L-fermenters using a filamentous fungi (*Aspergillus oryzae*). Here also the yield is described as industrially relevant.

Another patent describes the use of *Pichia pastoris* as a production host and goes into great detail concerning the host and the process. In the host used, the brazzein gene is transcribed from a constitutive GAP promoter. To increase the transcript level, several copies of the expression cassette have been integrated into the genome of the host organism. The fermentation is done in a fed-batch mode reaching high cell densities. A scale-up to the 1,000 L scale was also done. The yield reached was 0.5 g/L in the supernatant, which is dependent on the cell density of 0.2–0.3 g/L of culture broth. The purification is also described using common process unit operations such as ultrafiltration and ion exchange chromatography. Using these techniques, a purification yield above 90% is reached. The described process would be commercially feasible if the yield per liter of culture broth were higher. Considering other examples for bioprocess costs [50], to be commercially viable a yield greater than at least 5 g/L of culture broth would probably be needed.

7 Patent Analysis on Brazzein

In addition to the original patent on brazzein [46], several patent applications were published which disclose brazzein variants [51, 52]. Some introduce single point mutations [51], and some combine multiple point mutations to increase the sweetness of the molecule [52]. Since the expiration of the original patent on brazzein [46] in 2014, there have been significantly more patent applications published on the subject [47–49, 53]. This shows a continued commercial interest in brazzein. Nevertheless, since the early days of brazzein discovery in the late 1980s, there is still no industrial upscale or any known technical solution published with the potential to meet the market needs.

8 Conclusion

Due to its high sweetness potency, good sensory properties, and a long history of human consumption, brazzein can be viewed as a promising natural sweetener. In addition, brazzein presents physicochemical properties such as high water-solubility and extreme stability to the pH and the temperature, which are essential for food applications. One of the major problems hindering the use of brazzein for industrial application is the difficulty of obtaining the sweet-tasting protein from its natural source. Recently, many efforts have been made to overexpress brazzein using various heterologous systems including bacteria, yeast, animals, and plants. Yeast expression systems including *Pichia pastoris* have been proved to be a very efficient and economical expression system to secrete brazzein. The secreted protein has been shown to be easy to purify with characteristics identical to the natural protein.

Finally, the large amount of brazzein obtained by biotechnology paves the way for food applications using brazzein as an alternative low-calorie, high-intensity sweetener.

References

1. Faus I (2000) Recent developments in the characterization and biotechnological production of sweet-tasting proteins. Appl Microbiol Biotechnol 53(2):145–151
2. Hladik A, Hladik CM (1969) Rapports trophiques entre végétation et Primates dans la forêt de Barro Colorado (Panama). LaTerre et la Vie 23:25–117
3. Theerasilp S, Kurihara Y (1988) Complete purification and characterization of the taste-modifying protein, miraculin, from miracle fruit. J Biol Chem 263(23):11536–11539
4. Baillon H (1868) Adansonia, recueil périodique d'observations botaniques. Bull Soc Linn Paris 1:611
5. Hladik A, Bahuchet S, Ducatillion C, Hladik CM (1984) Les plantes a tubercules de la foret dense d'afrique centrale. Rev Ecol (Terre Vie) 39:249–290
6. Hladik CM, Hladik A (1988) Sucres et 'Faux Sucres' de la foret equatoriale: evolution et perception des produits sucres par les populations forestieres d'Afrique. J Agric Trad Bota Appl XXXV:35–50
7. Van der Wel H, Larson G, Hladik A, Hladik CM, Hellekant G, Glaser D (1988) Isolation and characterization of pentadin, the sweet principle of *Pentadiplandra brazzeana* Baillon. Chem Senses 14(1):75–79. doi:10.1093/chemse/14.1.75
8. Ming D, Hellekant G (1994) Brazzein, a new high-potency thermostable sweet protein from *Pentadiplandra brazzeana* B. FEBS Lett 355(1):106–108
9. Assadi-Porter FM, Aceti DJ, Cheng H, Markley JL (2000) Efficient production of recombinant brazzein, a small, heat-stable, sweet-tasting protein of plant origin. Arch Biochem Biophys 376 (2):252–258. doi:10.1006/abbi.2000.1725
10. Poirier N, Roudnitzky N, Brockhoff A, Belloir C, Maison M, Thomas-Danguin T, Meyerhof W, Briand L (2012) Efficient production and characterization of the sweet-tasting brazzein secreted by the yeast *Pichia pastoris*. J Agric Food Chem 60(39):9807–9814. doi:10.1021/jf301600m
11. Caldwell JE, Abildgaard F, Dzakula Z, Ming D, Hellekant G, Markley JL (1998) Solution structure of the thermostable sweet-tasting protein brazzein. Nat Struct Biol 5(6):427–431
12. Dias RO, Franco OL (2015) Cysteine-stabilized alphabeta defensins: from a common fold to antibacterial activity. Peptides 72:64–72. doi:10.1016/j.peptides.2015.04.017
13. Nagata K, Hongo N, Kameda Y, Yamamura A, Sasaki H, Lee WC, Ishikawa K, Suzuki E, Tanokura M (2013) The structure of brazzein, a sweet-tasting protein from the wild African plant *Pentadiplandra brazzeana*. Acta Crystallogr D Biol Crystallogr 69(Pt 4):642–647. doi:10.1107/S0907444913001005
14. Assadi-Porter FM, Maillet EL, Radek JT, Quijada J, Markley JL, Max M (2010) Key amino acid residues involved in multi-point binding interactions between brazzein, a sweet protein, and the T1R2-T1R3 human sweet receptor. J Mol Biol 398(4):584–599. doi:10.1016/j.jmb.2010.03.017
15. Jin Z, Danilova V, Assadi-Porter FM, Aceti DJ, Markley JL, Hellekant G (2003) Critical regions for the sweetness of brazzein. FEBS Lett 544(1–3):33–37
16. Jin Z, Danilova V, Assadi-Porter FM, Markley JL, Hellekant G (2003) Monkey electrophysiological and human psychophysical responses to mutants of the sweet protein brazzein: delineating brazzein sweetness. Chem Senses 28(6):491–498
17. Lee JW, Cha JE, Jo HJ, Kong KH (2013) Multiple mutations of the critical amino acid residues for the sweetness of the sweet-tasting protein, brazzein. Food Chem 138(2–3):1370–1373. doi:10.1016/j.foodchem.2012.10.140

18. Lim JK, Jang JC, Kong JN, Kim MC, Kong KH (2015) Importance of Glu53 in the C-terminal region of brazzein, a sweet-tasting protein. J Sci Food Agric. doi:10.1002/jsfa.7501
19. Singarapu KK, Tonelli M, Markley JL, Assadi-Porter FM (2016) Structure-function relationships of brazzein variants with altered interactions with the human sweet taste receptor. Protein Sci 25(3):711–719. doi:10.1002/pro.2870
20. Walters DE, Cragin T, Jin Z, Rumbley JN, Hellekant G (2009) Design and evaluation of new analogs of the sweet protein brazzein. Chem Senses 34(8):679–683. doi:10.1093/chemse/bjp048
21. Yoon SY, Kong JN, Jo DH, Kong KH (2011) Residue mutations in the sweetness loops for the sweet-tasting protein brazzein. Food Chem 129:1327–1330
22. DuBois GE (2011) Validity of early indirect models of taste active sites and advances in new taste technologies enabled by improved models. Fragr J 26(4). doi:10.1002/ffj.2042
23. Hellekant G, Danilova V, Ninomiya Y (1997) Primate sense of taste: behavioral and single chorda tympani and glossopharyngeal nerve fiber recordings in the rhesus monkey, *Macaca mulatta*. J Neurophysiol 77(2):978–993
24. Danilova V, Danilov Y, Roberts T, Tinti JM, Nofre C, Hellekant G (2002) Sense of taste in a new world monkey, the common marmoset: recordings from the chorda tympani and glossopharyngeal nerves. J Neurophysiol 88(2):579–594
25. Jiang P, Ji Q, Liu Z, Snyder LA, Benard LM, Margolskee RF, Max M (2004) The cysteine-rich region of T1R3 determines responses to intensely sweet proteins. J Biol Chem 279(43): 45068–45075. doi:10.1074/jbc.M406779200
26. Gordesky-Gold B, Rivers N, Ahmed OM, Breslin PA (2008) Drosophila melanogaster prefers compounds perceived sweet by humans. Chem Senses 33(3):301–309. doi:10.1093/chemse/bjm088
27. Li X, Staszewski L, Xu H, Durick K, Zoller M, Adler E (2002) Human receptors for sweet and umami taste. Proc Natl Acad Sci U S A 99(7):4692–4696. doi:10.1073/pnas.072090199
28. Nelson G, Chandrashekar J, Hoon MA, Feng L, Zhao G, Ryba NJ, Zuker CS (2002) An amino-acid taste receptor. Nature 416(6877):199–202. doi:10.1038/nature726
29. Assadi-Porter FM, Patry S, Markley JL (2008) Efficient and rapid protein expression and purification of small high disulfide containing sweet protein brazzein in *E. coli*. Protein Expr Purif 58(2):263–268. doi:10.1016/j.pep.2007.11.009
30. Li DF, Jiang P, Zhu DY, Hu Y, Max M, Wang DC (2008) Crystal structure of Mabinlin II: a novel structural type of sweet proteins and the main structural basis for its sweetness. J Struct Biol 162(1):50–62. doi:10.1016/j.jsb.2007.12.007
31. Masuda T, Taguchi W, Sano A, Ohta K, Kitabatake N, Tani F (2013) Five amino acid residues in cysteine-rich domain of human T1R3 were involved in the response for sweet-tasting protein, thaumatin. Biochimie 95(7):1502–1505. doi:10.1016/j.biochi.2013.01.010
32. Jiang P, Cui M, Zhao B, Snyder LA, Benard LM, Osman R, Max M, Margolskee RF (2005) Identification of the cyclamate interaction site within the transmembrane domain of the human sweet taste receptor subunit T1R3. J Biol Chem 280(40):34296–34305. doi:10.1074/jbc.M505255200
33. Winnig M, Bufe B, Kratochwil NA, Slack JP, Meyerhof W (2007) The binding site for neohesperidin dihydrochalcone at the human sweet taste receptor. BMC Struct Biol 7:66. doi:10.1186/1472-6807-7-66
34. Maitrepierre E, Sigoillot M, Le Pessot L, Briand L (2012) Recombinant expression, in vitro refolding, and biophysical characterization of the N-terminal domain of T1R3 taste receptor. Protein Expr Purif 83(1):75–83. doi:10.1016/j.pep.2012.03.006
35. Nie Y, Vigues S, Hobbs JR, Conn GL, Munger SD (2005) Distinct contributions of T1R2 and T1R3 taste receptor subunits to the detection of sweet stimuli. Curr Biol 15(21):1948–1952. doi:10.1016/j.cub.2005.09.037
36. Temussi PA (2002) Why are sweet proteins sweet? Interaction of brazzein, monellin and thaumatin with the T1R2-T1R3 receptor. FEBS Lett 526(1–3):1–4

37. Walters DE, Hellekant G (2006) Interactions of the sweet protein brazzein with the sweet taste receptor. J Agric Food Chem 54(26):10129–10133. doi:10.1021/jf062359y
38. Cornilescu CC, Cornilescu G, Rao H, Porter SF, Tonelli M, DeRider ML, Markley JL, Assadi-Porter FM (2013) Temperature-dependent conformational change affecting Tyr11 and sweetness loops of brazzein. Proteins 81(6):919–925. doi:10.1002/prot.24259
39. Berlec A, Jevnikar Z, Majhenic AC, Rogelj I, Strukelj B (2006) Expression of the sweet-tasting plant protein brazzein in *Escherichia coli* and *Lactococcus lactis*: a path toward sweet lactic acid bacteria. Appl Microbiol Biotechnol 73(1):158–165. doi:10.1007/s00253-006-0438-y
40. Berlec A, Strukelj B (2009) Large increase in brazzein expression achieved by changing the plasmid /strain combination of the NICE system in *Lactococcus lactis*. Lett Appl Microbiol 48 (6):750–755. doi:10.1111/j.1472-765X.2009.02608.x
41. Berlec A, Tompa G, Slapar N, Fonovic UP, Rogelj I, Strukelj B (2008) Optimization of fermentation conditions for the expression of sweet-tasting protein brazzein in *Lactococcus lactis*. Lett Appl Microbiol 46(2):227–231. doi:10.1111/j.1472-765X.2007.02297.x
42. Guan Z, Hellekant G, Yan W (1995) Expression of sweet protein brazzein by *Saccharomyces cerevisiae*. Chem Senses 20:701
43. Jo HJ, Noh JS, Kong KH (2013) Efficient secretory expression of the sweet-tasting protein brazzein in the yeast *Kluyveromyces lactis*. Protein Expr Purif 90(2):84–89. doi:10.1016/j.pep.2013.05.001
44. Lamphear BJ, Barker DK, Brooks CA, Delaney DE, Lane JR, Beifuss K, Love R, Thompson K, Mayor J, Clough R, Harkey R, Poage M, Drees C, Horn ME, Streatfield SJ, Nikolov Z, Woodard SL, Hood EE, Jilka JM, Howard JA (2005) Expression of the sweet protein brazzein in maize for production of a new commercial sweetener. Plant Biotechnol J 3(1):103–114. doi:10.1111/j.1467-7652.2004.00105.x
45. Yan S, Song H, Pang D, Zou Q, Li L, Yan Q, Fan N, Zhao X, Yu H, Li Z, Wang H, Gao F, Ouyang H, Lai L (2013) Expression of plant sweet protein brazzein in the milk of transgenic mice. PLoS One 8(10):e76769. doi:10.1371/journal.pone.0076769
46. Hellekant BG, Ming D (1994) Brazzein sweetener. PCT/US1994/001493
47. Breuner A (2012) Sweet-tasting polypeptide from gram-positive bacteria. PCT/EP2012/055123
48. Carlson A, Armentrout RW, Ellis TP (2010) Enhanced production and purification of a natural high intensity sweetener PCT/US2009/058996
49. Vind J, Tams JW, Beier L, Olsen CL (2011) Method of producing a sweet protein. PCT/EP2010/061435
50. Tufvesson P, Lima-Ramos J, Jensen JS, Al-Haque N, Neto W, Woodley JM (2011) Process considerations for the asymmetric synthesis of chiral amines using transaminases. Biotechnol Bioeng 108(7):1479–1493. doi:10.1002/bit.23154
51. Assadi-Porter FM (2009) Protein sweetener PCT/US2008/055913
52. Kong KH (2013) Brazzein variant having higher sweetness and method for preparing multi-variant PCT/KR2009/004855
53. Miles L, Louie M, Subramanian M (2010) Sweetener preparations and methods of use. PCT/US2009/056852

Sugar Alcohols as Sugar Substitutes in Food Industry

24

Małgorzata Grembecka

Abstract

Among nutritive sweeteners, there can be distinguished polyhydric alcohols (polyols), also known as sugar alcohols, because they are derived from simple carbohydrates, obtained by the substitution of the aldehyde group by the hydroxy one. They are natural sugar alternatives but are also referred to as semisynthetic sweeteners. There are many advantages of sugar alcohols, so they are becoming more and more popular among both consumers and producers. They are characterized by a lower caloric value and glycemic index than sugars and exhibit prebiotic and anticaries effects. All sugar alcohols can be used as bulking agents, which can substitute sugar or corn syrups 1:1 ratio. However, their sweetness varies from 25% to 100% as compared with sucrose, so they are usually combined with intense sweeteners or sugar in order to obtain the required flavor and level of sweetness. Additionally, they promote mouthfeel and eliminate improper taste. Therefore, they can be used as reduced-calorie sugar alternatives.

Keywords

Sugar alcohol • Polyol • Bulk sweeteners • Sugar-free products • Cooling effect • Anticariogenic • Sugar substitutes

Contents

1 Introduction 548
2 Nutritive and Nonnutritive Sugar Substitutes 549
3 Sugar Alcohols 551
3.1 General Characteristics of Sugar Alcohols 551
3.2 Metabolism of Sugar Alcohols 554
3.3 Production of Sugar Alcohols 556

M. Grembecka (✉)
Department of Food Sciences, Faculty of Pharmacy with Subfaculty of Laboratory Medicine, Medical University of Gdansk, Gdansk, Poland
e-mail: mgrembecka@gumed.edu.pl

J.-M. Mérillon, K.G. Ramawat (eds.), *Sweeteners*, Reference Series in Phytochemistry,
https://doi.org/10.1007/978-3-319-27027-2_23

3.4 Biological Activities of Sugar Alcohols 559
3.5 Properties and Applications 560
4 Polyols in the Food Industry 565
5 Summary 566
References 567

1 Introduction

Since prehistoric times, a man exhibits the evolutionally shaped preference for a sweet taste. It is related to the fact that people, similarly to animals, learned to use sweet taste to identify foods with high caloric and nutritional values. As indicated by the scientific literature, the increased preference for sweet taste of foods and beverages has already been confirmed worldwide for both infants and children [1]. However, today we encounter a great variety of readily available sugary foods, which are often of very low nutritional value, but with high energy value; thus, their preference over others can cause many health problems. Among all the environmental factors, nutrition has the greatest impact on the development and well-being of a human body. It determines the harmonious growth of the organism, health, resistance to infection, well-being, and ability to physical and mental work. Metabolic processes run without interruption in the human body, and therefore, it needs nutrients, such as proteins, carbohydrates, and fats, and energy, of which a source is food. Human demand on the particular energy components is individual, dependent mainly on age, sex, physical activity, metabolism, or illnesses. According to the latest recommendations, carbohydrates should provide 50-70% of energy, fats 20–35% (including polyunsaturated fatty acids 3%), and proteins 10–15% [2].

Epidemiological studies, carried out throughout the world, indicate a relationship between energy imbalance of the human body and the incidence of metabolic disorders such as overweight, obesity, insulin resistance, and cardiovascular disease [3–5]. Overweight and obesity constitute major problems of the modern world [6]. Currently, there are over 1.9 billion adults overweight and over 600 million of them are obese [7]. What is more frightening, there is an increasing number of children, even under the age of 5, who are becoming overweight or obese. According to World Health Organization (WHO), the number of such children amounted to 42 million in 2013 [7].

As the obesity prevalence has doubled since 1980, the number of people who care about their weight is dramatically rising. Simultaneously, the enormous trade pressure and incredible competitiveness are forcing producers to continually reduce costs of food production while maintaining its attractiveness in the eyes of the consumer, who is usually looking for sweet products. Therefore, producers are forced to replace sugar with its natural substitutes or synthetic sweeteners, which are classified as food additives. According to the Regulation of the European Parliament and the European Community Council in 2008, food additive is any substance, which under normal circumstances is not consumed alone as food and is not used as a characteristic ingredient of food, regardless of its nutritive value, the intentional addition for technological reasons to food during its manufacture, processing, preparation, treatment, packaging, transport, or storage results, or may be reasonably expected to result in it or its by-products becoming, directly or indirectly, a component of the food [8].

2 Nutritive and Nonnutritive Sugar Substitutes

Sweeteners, which are responsible for the development of sweet taste, can be divided according to different criteria. Among the most important are the origin (natural or synthetic agents), the technological function (semisynthetic fillers and sweeteners), consistency (powders and syrups), and nutritional value (nutritive or energy-free) [9]. Among sweeteners which were authorized to use by the Joint FAO/WHO Expert Committee on Food Additives (JECFA) can be mentioned 27 compounds (Table 1), and 10 of them belong to sugar alcohols [10].

Natural substances that are widely considered responsible for the occurrence of sweet taste are carbohydrates. It is a diverse group of organic compounds widely distributed in nature, which main task is to provide energy, but also to perform

Table 1 Acceptable daily intake (ADI) of sweetening food additives [10]

Sweetening food additives	E number according to the EU	ADI mg/kg b.w./day
Acesulfame-K	E950	15
Advantame	E969	5
Alitame		1
Aspartame	E951	40
Aspartame–acesulfame salt	E962	As for components
Calcium cyclamate/cyclamic acid/sodium cyclamate	E952	11
Calcium saccharin/potassium saccharin/saccharin/ sodium saccharin	E954	5
Erythritol	E968	Not specified
Isomalt	E953	Not specified
Lactitol	E966	Not specified
Maltitol/maltitol syrup	E965	Not specified
Mannitol	E421	Not specified
Neohesperidin DC	E959	5
Neotame	E961	2
Polyglycitol syrup	E964	Not specified
Sorbitol/sorbitol syrup	E420	Not specified
Steviol glycosides	E960	4
Sucralose	E955	15
Thaumatin	E957	Not specified
Xylitol	E967	Not specified

structural functions and be a source of dietary fiber. They also determine the proper economy of nutrients and participate in the regulation of satiety and hunger [2]. Carbohydrates found in food differ significantly in view of physiological, nutritional, and chemical properties and content in food products. According to the FAO/WHO experts, carbohydrates are classified into three main groups, i.e., sugars (monosaccharides and disaccharides), oligosaccharides, and polysaccharides. Polyols (sugar alcohols) and modified starch also belong to carbohydrates [2, 11].

Polyhydric alcohols (polyols), also known as sugar alcohols, are carbohydrates and natural sugar substitutes as well as food additives. They are becoming increasingly popular among consumers, mainly due to their lower caloric content, the glycemic index, and the beneficial anticaries effects. Sugar alcohols are often used in combination with other sweeteners in order to increase the sweetening effect of food products [12–14].

Natural sugar substitutes are generally regarded as safe, but it is recommended to control the presence of these substances in food products (Table 1). There are seven sugar alcohols approved for use in the European Union and Poland, e.g., sorbitol (E420), mannitol (E421), isomalt (E953), maltitol (E965), lactitol (E966), xylitol (E967), and erythritol (E968) [8]. Their addition must be marked on the package, as consumed in excess they have a laxative effect due to slower and incomplete digestion [15–17].

Unlike natural sugar substitutes, artificial additives do not deliver calories and does not affect blood glucose levels. They improve the organoleptic characteristics of many products, affect the food stability, increase its attractiveness, and meet the need of human perception of sweet taste [18–20]. They are characterized by a minimum caloric value at the used doses, so they give dietetic character to foodstuffs which they are added to. They are usually used together, since products in which aspartame is used in combination with other sweeteners exhibit a greater sweetness than would result from the sum of sweetening powers of the individually used substances. Acceptable daily intake (ADI) of aspartame, set by the Joint FAO/WHO Expert Committee on Food Additives (JECFA) in 1980, amounts to 40 mg/kg body weight/day [21], which is the highest value noted for this group of additives (Table 1).

Studies have shown that artificial sweeteners' use, in order to decrease the correlation between sweet taste and food energy value in rats, caused an increase in total caloric intake. These results indicate that the consumption of products containing artificial sweeteners can lead to weight gain as well as homeostasis and physiological processes disorders [22–24]. In addition, removing sugar from food as unnecessary and replacing it with artificial sweeteners may lead to unintended metabolic disorders, affecting the satiety center [23, 25] or lead to preferences of very sweet product.

The ideal sweetener for food industry should characterize with chemical and biological stability and the same properties as sucrose or glucose. Consumers demand food products with added sweeteners to have identical or very similar taste and appearance as with traditional sugars [26].

3 Sugar Alcohols

As a group, polyols are similar to carbohydrates but they have a hydroxyl group instead of aldehyde or ketone one. These are white crystalline powders without odor. Although they are similar, they differ significantly in their characteristics.

3.1 General Characteristics of Sugar Alcohols

3.1.1 Energy Value

Due to incomplete digestion, sugar alcohols characterize with lower caloric value which varies according to different legislations (Table 2). In Europe, they are given 2.4 kcal/g except for erythritol which energy value is set at 0 kcal/g. However, American and Japanese legislation set caloric value of polyols within the range of 0.0 (erythritol) and 3.0 kcal/g (sorbitol, xylitol) [8, 14, 17, 27, 28]

3.1.2 Heat of Solution

Every sugar alcohol is characterized by negative heat of solution, which is defined as energy needed to dissolve crystals. This energy is absorbed from the surrounding environment; thus, dissolution of polyols creates cooling sensation in the mouth. The lowest heat of solution is attributed to erythritol, whereas the highest is to isomalt (Table 3). According to de Cock [29], such cooling sensation is perceived when heat of solution is lower than −20 cal/g. This feature is of great importance to pharmaceutical products which characterize with soothing effects, i.e., lozenges, cough drops, throat medication, and breath mints [29].

3.1.3 Solubility

This property is especially important to food manufacturers. Among all polyols, erythritol, isomalt, and mannitol characterize with the lowest solubility and fast crystallization. However, maltitol and sorbitol have these parameters relatively high, i.e., 175 g and 235 g/100 g of water at 25 °C [31]. Food producers mix maltitol syrups with maltitol powder in order to control its crystallization [31].

Table 2 Sweetness and caloric values of particular sugar alcohols [8, 14, 17, 27, 28]

		Caloric value [kcal/g]		
Compound	Sweetness[a]	EU	USA	Japan
Erythritol	0.6–0.8	0	0	0
Isomalt	0.45–0.65	2.4	2	2
Lactitol	0.3–0.4	2.4	2	2
Maltitol	0.9	2.4	2.1	2
Mannitol	0.5–0.7	2.4	1.6	2
Sorbitol	0.5–0.7	2.4	2.6	3
Xylitol	1.0	2.4	2.4	3

[a]sucrose sweetness = 1 [13]

Table 3 Physical properties of sugar alcohols [28, 30]

Compound	Molecular weight	Heat of solution	Viscosity at 25 °C	Hygroscopicity
Erythritol	122	−43	Very low	Very low
Isomalt	344	−9	High	Low
Lactitol	344	−14	Very low	Low
Maltitol	344	−19	Medium	Medium
Mannitol	182	−29	Low	Low
Sorbitol	182	−26	Medium	High
Xylitol	152	−36	Very low	Medium

3.1.4 Toxicology

All polyols have been intensively tested in both animals and humans, and no health risk was found; thus, no ADI have been specified for whole group [10, 32–34] (Table 1). Currently their use is approved in most countries, except for maltitol, which awaits the approval of Food and Drug Administration (FDA), but it can be used as the petition regarding this compound has been already accepted [35].

3.1.5 Erythritol (E968)

It is a linear four-carbon compound with a molecular formula $C_4H_{10}O_4$ and systematic name (2R,3S)-1,2,3,4-butanetetrol. It is also available on the global market under names erythrite, meso-erythritol, or tetrahydroxybutane [14]. It is a sugar alcohol which can be found in natural sources such as beverages, fruits, and seaweeds and which was discovered in 1848 by John Stenhouse, a Scottish chemist [29].

Consumption of this sugar alcohol from natural sources has been estimated at the level of 25 mg/person/day in the USA, whereas in Japan at 106 mg/person/day [29]. Its safety has been confirmed, and it is authorized to use as a food additive in the European Union (EU) under Annex II of Regulation (EC) 1333/2008 [36], which can be used at *quantum satis*. Owing to its properties, it can be used as flavor enhancer, humectant, and sweetener [14].

3.1.6 Isomalt

Isomalt constitute an equimolar mixture of two disaccharides, each composed of two sugars: first of glucose and mannitol (α-D-glucopyranosido-1,6-mannitol) and the second of glucose and sorbitol (α-D-glucopyranosido-1,6-sorbitol). Its hydrolysis results in mixture of glucose (50%), sorbitol (25%), and mannitol (25%) [37]. It can be found under such names as hydrogenated isomaltulose and isomaltitol. Isomalt is an odorless, crystalline, and nonhygroscopic substance, which can be used as an anticaking, bulking, and glazing agent as well as a stabilizer, sweetener, and thickener [14]. Isomalt, which is a white crystalline powder without odor, has an E number E953. It decomposes in temperatures higher than 160 °C.

3.1.7 Lactitol

Lactitol, having the formula $C_{12}H_{24}O_{11}$, has a systematic name 4-*O*-β-L-galactopyranosyl-L-glucitol and a molecular weight 344. This sugar alcohol can be applied as an emulsifier, sweetener, and thickener [14]. According to the European legislation, it is given number E966. It is available in anhydrous and monohydrate forms, which appear as crystalline powder. It is a stable compound even in temperatures above 160 °C.

3.1.8 Maltitol

Maltitol has a formula $C_{12}H_{24}O_{11}$ and a systematic name 4-*O*-α-D-glucopyranosyl-D-glucitol [14]. The molecular weight of pure maltitol powder amounts to 344 [31]. It can be used as a bulking agent, an emulsifier, a humectant, a stabilizer, a sweetener, and a thickener [14]. Although it resembles sugars, it does not undergo caramelization and the Maillard reaction during processing. In the European Union, it has an E number E965.

3.1.9 Mannitol

Mannitol has a number E421, according to the European legislation, and a molecular formula $C_6H_{14}O_6$. It can be found under such names as D-mannitol and mannite [14]. It is a nonhygroscopic substance, which is poorly soluble in water (20 g 100 g^{-1} water) and stable in temperatures above 160 °C. It is an isomer of sorbitol, which is used as an anticaking and bulking agent, humectant, stabilizer, sweetener, and thickener [14].

3.1.10 Sorbitol

Sorbitol is a poorly digestible sugar alcohol, which is labeled in the European Union as E number E420. It has a formula $C_6H_{14}O_6$ and a systematic name D-glucitol. Sorbitol is well soluble (235 g 100 g^{-1}), hygroscopic, and stable in high temperatures. Due to its properties, it can be used as a bulking agent, a humectant, a sequestrant, a stabilizer, a sweetener, and a thickener. Although it is an isomer of mannitol, these two sugar alcohols characterize with different properties and applications [14].

3.1.11 Xylitol

Xylitol, which can be utilized as an emulsifier, a humectant, a stabilizer, a sweetener, and a thickener, has a molecular formula $C_5H_{12}O_5$ and a molecular weight of 152 (Table 3). According to the IUPAC, it is D-erythro-pentitol In Europe it is known as a food additive (E967) safe for use by both adults and children. However, it is also often called birch sugar, as initially birchwood was used for its production. It is quite soluble in water (169 g 100 g^{-1}), heat stable, and hygroscopic [14, 38]. It has been used as a sweetener since the 1960s [38].

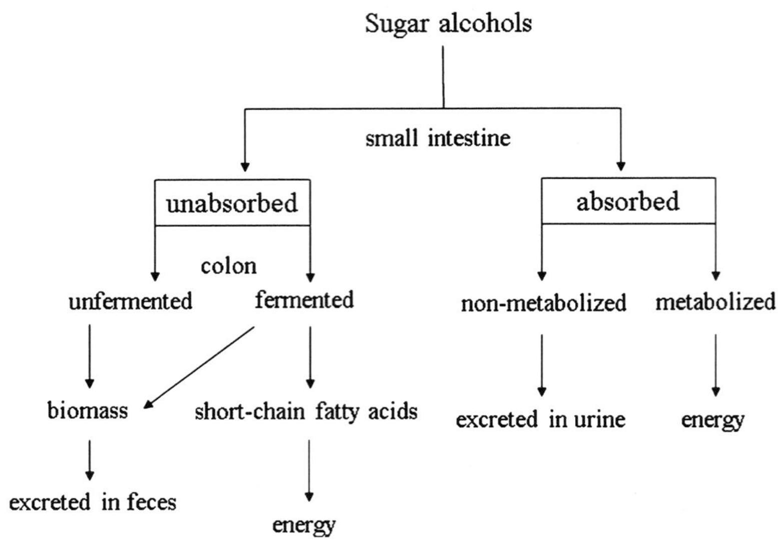

Fig. 1 Metabolism of sugar alcohols in the human organism

3.2 Metabolism of Sugar Alcohols

Lower caloric value of these compounds results from their slower and incomplete absorption. They are absorbed through a passive diffusion, and the rate of their uptake in the small intestine diminishes with the increase of their molecular mass. There are two pathways of polyols digestion, i.e., they can be metabolized or excreted in urine or feces (Fig. 1). It was observed that when the dose of polyols increases, the part absorbed lowers. Due to polyols fermentation in the colon, there can be observed gastrointestinal effects such as noisy bowels, flatulence, or diarrhea [15–17]. However, such side effects are dependent on one's tolerance which can be improved by regular consumption. Polyols are also characterized by a laxative threshold which is defined as the onset of diarrhea in half of consumers.

3.2.1 Erythritol

Erythritol characterizes with a high rate of absorption in the small intestine amounting up to 60–90%. However, it does not undergo fermentation process in the intestines and is excreted intact in urine within 24 h [14, 17, 29, 39–42]. Bornet et al. [43] have found, in studies conducted on healthy volunteers, that majority (approx. 80%) of the consumed erythritol (1 g/kg in 250 mL of beverage) was rapidly absorbed and excreted intact in the urine within a 24-h period. According to Roberts and Renwick [44], erythritol absorption is of first order, and its rate is greater than that of elimination.

Laxative effects, which are usually associated with consumption of polyols, are unlikely to be observed in case of erythritol as it is not fermented; thus, its laxative threshold is about zero per day [17, 29, 39–42]. According to Oku and Okazaki [45], women are less susceptible to gastrointestinal effects (diarrhea) due to consumption of erythritol than men. What is more, such effects were noted only at very high doses consumed, up to 1000 mg/kg body [39]. However, it was also found that erythritol is characterized by similar tolerability between children and adults in terms of laxation on a body weight basis [36]. Therefore, it can be concluded that this polyol's fate in the human organism is identical independently of the age of consumer [36]. According to de Cock [29], the maximum tolerated dose of erythritol amounts to 40 g when consumed as a bolus liquid or 80 g when consumed spread over the day.

3.2.2 Isomalt

Among polyols, isomalt characterizes with a very low absorption percentage, which amounts to 10% [17, 46]. Moreover, 90% of the absorbed fraction undergoes fermentation in the intestines, producing short-chain fatty acids, CO_2, CH_4, and H_2 [17, 46]. Isomalt is a source of butyrate, which increases the growth of bifidobacteria leading to a probiotic effect [26, 47]. Excess consumption of this polyol results in a laxative effect, which depends on individual sensitivity, the moment, and frequency of consumption [26]. The outcome of this intolerance is greater when isomalt is consumed in liquid food.

3.2.3 Lactitol

It is absorbed in about 2% by passive diffusion in the small intestine. The remaining part is undigested and passes to the distal part of the large intestine, where it is fermented by gut microbiota converting lactitol into biomass, short-chain fatty acids, carbon dioxide, a small amount of hydrogen, and organic acids [30, 32, 38]. It can be also a source of energy for such intestinal microflora in the colon as *Bifidobacterium* and *Lactobacillus* spp. [38, 48]. These two species are beneficial to human health; thus, lactitol can be used as a prebiotic. Lactitol fermentation, led by these two species, lowers the pH due to the production of butyrate [30]. This polyol can be also slowly hydrolyzed by galactosidase-containing enzymes into galactose and sorbitol [30].

According to the Scientific Committee on Food of the European Community, laxative symptoms are unlikely to observe when 20 g of lactitol is consumed per day [38]. When more than 20 g is consumed in a single dose, bloating and flatulence can occur [30]. Higher doses consumed might also result in a laxative effect, which depends on the person's diet, age, general gut health, and the mode and frequency of digestion [14, 30].

3.2.4 Maltitol

After uptake, maltitol must be hydrolyzed into glucose and sorbitol in order to be absorbed [17]. Its absorption percentage ranges from 5% to 80% [17]. However, the maltase enzyme usually does not have enough time to hydrolyze all the maltitol as the process goes very slowly, so the remaining part is moved to the lower gut where is being fermented [16, 35]. There is produced gas and the water balance is disrupted resulting in osmotic laxation. However, maltitol is well tolerated and laxative effects were observed for daily doses exceeding 25–30 g/kg body weight [33].

3.2.5 Mannitol

Among all polyols, mannitol is the least well tolerated as its daily laxative threshold amounts to 20 g, but one serving should supply less than 10 g in order to prevent side effects [42]. It is only partially absorbed, in about 25%, in the small intestine, whereas the unabsorbed fraction is slowly metabolized by colonic bacteria in the gastrointestinal tract [17, 49, 50]. Main products of bacterial fermentation include organic acids which can be used by human organism [51].

3.2.6 Sorbitol

Sorbitol, similarly to mannitol, is slowly absorbed in the gastrointestinal tract and its metabolism takes place in the liver. Its absorption rate ranges from 25 to 80% of the ingested dose as an oral solution [17]. The non-absorbed fraction is metabolized by colonic bacteria [17, 49, 50, 52, 53]. Therefore, consumption of doses exceeding 50 g per day or 10 g per single serving can result in laxation resulting from an osmotic imbalance in the gut [53].

3.2.7 Xylitol

Xylitol is either metabolized directly in the liver via the glucuronic acid–pentose phosphate shunt of the pentose phosphate pathway or indirectly by fermentation conducted by intestinal flora [38, 54]. Almost 50% of the ingested xylitol is absorbed in the small intestine, whereas 50–70% of the remaining part is fermented in the large bowel. The fermentation products include minor amounts of gas (H_2, CH_4, CO_2) as well as short-chain volatile fatty acids, which usually constitute bacterial biomass. It is a well-tolerated compound up to 100 g per day [55].

3.3 Production of Sugar Alcohols

3.3.1 Erythritol

It is widely distributed in nature, mainly in seaweeds and fruits as well as other products [56]. Its higher levels were determined in sake (1150 mg L^{-1}), miso bean paste (1310 mg L^{-1}), or soy sauce (910 mg L^{-1}). Among fruits, melons contain 22–47 mg of erythritol per 1 kg, whereas pears 0–40 mg kg^{-1} [29]. It was first isolated from the algae *Protococcus vulgaris* (1852) and then from *Trentepohlia jolithus* (1900). Its synthesis procedure from 2-butene-1,4-diol was patented in 1943 by German scientists Reppe and Schnabel [29]. Although its chemical synthesis is possible, it is rarely used due to high cost of production and complexity [29]. Erythritol is produced almost uniquely by fermentation conducted by various microorganisms, including some species of lactic acid bacteria, i.e., *Oenococcus oeni*, *Leuconostoc mesenteroides*, and *Lactobacillus sanfranciscencis* [57–59], and osmophilic yeasts such as *Moniliella pollinis* [12, 18, 27, 29, 60, 61]. However, the highest yield and productivity is obtained for two species, i.e., *Trichosporonoides megachiliensis* and *Pseudozyma tsukubaensis* [56]. New reports have shown that it can be also produced by *Yarrowia lipolytica* MK1 [62–64] from glycerol. The improvement of production yield can be obtained by applying two strategies, i.e., optimization of culture conditions for microbial fermentation and random mutagenesis by UV and chemical mutagen [65–67].

3.3.2 Isomalt

This polyol was developed and is manufactured and marketed by BENEO-Palatinit GmbH, which is owned by Südzucker AG (Germany) [26]. Depending on applications, there were developed different types of isomalt such as isomalt ST, isomalt HC, isomalt GS, isomalt DC, and isomalt LM [26]. The only sugar which is used as a substrate for its production in a two-step process is sucrose. Firstly, sugar is transformed by enzymatic transglucosidation into isomaltulose (6-*O*-α-D-glucopyranosyl-D-fructose), which is then hydrogenated into isomalt, a compound being composed of two disaccharide alcohols, i.e., 6-*O*-α-D-glucopyranosyl-D-sorbitol (1,6-GPS) and 1- *O*-α-D-glucopyranosyl-D-mannitol dihydrate (1,1-GPM) [26, 46]. The isomalt type is characterized by varied ratio of GPS/GPM, which in isomalt ST is around 1 1, whereas in GS 3:1 [26].

3.3.3 Lactitol

It has not been found in nature and was synthesized for the first time in 1920 [38]. It is produced similarly to other polyols, by a catalytic hydrogenation using Raney nickel as the catalyst [30, 32, 38]. In general, 30–40% lactose solution is hydrogenated at 100 °C and hydrogen pressure of 40 bar or more. Much stricter conditions (130 °C, 90 bar) would result in partial epimerization to lactulose and even hydrolyzation to galactose and glucose, which could be subsequently hydrogenated to such sugar alcohols as lactitol, lactulitol, sorbitol, and galactitol [38].

3.3.4 Maltitol

There are several routes of obtaining maltitol available for manufacturers. It can be produced from maltose or very high maltose glucose syrup of high purity by a catalytic hydrogenation [31]. Maltose can be obtained by crystallization from the maltose syrup and then is hydrogenated to maltitol, which is obtained by melt crystallization. High-purity maltitol can be also recovered using liquid chromatography. The hydrogenation reaction take place at high temperature (usually 100–150 °C) and high pressure (100–150 bar) and is catalyzed by a suitable catalyst, usually Raney nickel. Maltitol was first produced by Hayashibara in Japan and then licensed to Towa (Japan), Cerestar/Cargill (EU), and Roquette (EU and the USA) [31].

According to the European legislation, maltitol powder (on a dry basis) must contain at least 98% of this polyol. Products with lower concentrations are defined as dried maltitol syrups [31]. However, Food Chemicals Codex in the USA demands maltitol powder to contain at least 92% and not more than 100.5% of this polyol on a dry basis [31].

3.3.5 Mannitol

Mannitol can be produced by fermentation, extraction from seaweed, or through a catalytic hydrogenation of fructose, sucrose, or glucose/fructose (1:1) mixture [49, 68, 69]. During hydrogenation, both sorbitol and mannitol are produced, and they are separated based on their solubility. Mannitol solubility amounts to 22 g 100 g^{-1} water, whereas sorbitol equals 235 g 100 g^{-1} water. Hydrogenation process has low efficiency, and there is obtained mixture with only 25% of mannitol, which need to be purified. That is why, there are now studied biological ways of mannitol production. It was found that heterofermentative lactic acid bacteria are able to convert D-fructose into D-mannitol in mild conditions [50, 70–72]. Moreover, cyanobacteria are also considered as a mannitol producers [73]. As mannitol can be found in several natural sources, including the exudates of certain trees, especially from manna ash (*Fraxinus ornus*), figs, olives, larches, edible fungi, yeasts, and seaweed, its extraction from these sources is also considered [49, 52, 69, 74–76].

3.3.6 Sorbitol

It can be found in nature, especially in stone and berries from trees of the genus *Sorbus*. There are several routes of sorbitol manufacturing, which can be obtained in both liquid and crystalline form [52, 77]. The most common is the production which is based on a catalytic hydrogenation of glucose or sucrose at high temperatures, using hydrogen gas and nickel catalyst [12, 14, 27, 74, 78]. Sorbitol in crystalline form is obtained through evaporation of its solution and then crystallization. Besides hydrogenation, electrochemical reduction of dextrose in alkaline conditions can be also used [12, 14, 27]. Moreover, a few microorganisms can be found, such as *Zymomonas mobilis* and *Candida boidinii* that are able to produce sorbitol [72, 74, 79–82].

3.3.7 Xylitol

It was synthesized for the first time by E. Fischer and his associate in 1891 [38]. Similarly to other polyols, xylitol can be obtained by a catalytic hydrogenation of the corresponding sugar, i.e., xylose, which can be obtained from birch trees, hardwood, or lignocellulosic biomass [18, 34, 56, 83–86]. First, xylose is obtained by xylan hydrolysis, and then chromatographically purified. Such mixture is then hydrogenated using nickel catalyst. However, commercial production can begin with xylose solution hydrogenation, and then xylitol is purified and crystallized in orthorhombic form [27, 30]. As this process is very expensive, there are researched new ways of xylitol manufacturing, including biotechnological ones. There are microorganisms such as *Candida tropicalis* and *C. guilliermondii* that can naturally produce xylitol during xylose assimilation [57, 74. 76]. Although *Candida* genus is the best source of xylitol, it cannot be used in food industry as it is a pathogenic one [14, 72, 74]. It is also possible to engineer metabolically other species, such as microalgae or fungi that do not contain xylose reductase to produce xylitol efficiently [56, 87, 88]. Moreover, bacterial production of xylitol would have overwhelming advantages such as rapid cell growth, easy genetic manipulation, and usage of inexpensive growth medium [56, 89]. There were also developed other biotechnological xylitol production systems from corncobs, the waste of sugarcane, and other fibers, but they were not introduced on a commercial scale up to date [30, 90–94].

3.4 Biological Activities of Sugar Alcohols

3.4.1 Sugar Alcohols Versus Blood Glucose

The Food and Agriculture Organization definition of glycemic index (GI) is as follows: "the incremental area under the blood glucose response curve of a 50 g carbohydrate portion of a test food expressed as a percent of the response to the same amount of carbohydrate from a standard food taken by the same subject" [25]. Insulinemic index is similar to GI, but it concerns measurement of insulin instead of glucose [17]. Control of glycemic and insulinemic responses is very important to diabetics. Polyols are characterized by low glycemic and insulinemic responses; thus, the intake results in a lower increase in postprandial glucose (Table 4). They are also associated with lipogenesis inhibition as well as lower insulin production. Therefore, products containing polyols are recommended for people with diabetes [14–16, 95, 96]. According to the study conducted by Holub et al. [97], 30 g of isomalt significantly improves the metabolic control of diabetes. What is more, it was found that mixtures of polyols, fats, and starch- and protein-based foods yield lower GI than cumulative GI of the components [17].

3.4.2 Colon Health

Polyols, especially those that characterize with low uptake in the intestines as well as low fecal excretion, have a beneficial effect on colon health. It is due to positive influence on the growth of gastrointestinal flora, which consists of glucidolytic and acid-forming organisms [17]. One of the products of polyols fermentation is butyric acid, which serves as a source of energy for the intestinal mucosa. It is also suspected of slowing cancer cell's growth [98]. Lactitol is known for its ability to decrease NH_3 and toxic microbial substances levels, which is especially important during hepatic encephalopathy treatment [17, 99].

Table 4 Glycemic and insulinemic indexes of sugar alcohols [17]

Sugar alcohol	Glycemic index	Insulinemic index
Erythritol	0	2
Isomalt	9	6
Lactitol	6	4
Maltitol	35	27
Mannitol	0	0
Sorbitol	9	11
Xylitol	13	11

3.4.3 Oral Cavity Health

All sugar alcohols are non-cariogenic due to oral microorganisms' inability to ferment these compounds. They are also non-acidogenic as the pH of interdental plaque does not fall below 5.7; thus, there is no favorable environment for tooth demineralization. Moreover, according to the Scientific Opinion from the European Food Safety Agency (EFSA), polyols help in tooth mineralization as well as neutralize plaque acids and reduce tooth demineralization [15].

It was found that chewing gum containing any of polyols three or more times daily for longer periods has beneficial anticaries effect. What is more, in vitro studies have suggested that xylitol and maltitol chewing gums help in remineralization process. According to Mäkinen et al.'s [100] study of 6-month usage of erythritol, xylitol, and sorbitol, the use of the two first in the form of chewable tablets was associated with a significant reduction of mutans streptococci both in the plaque and saliva. It was concluded that both erythritol and xylitol exhibit similar anticariogenic properties, though their mechanism may differ [100]. Clinical studies have shown that consumption of xylitol by mothers significantly reduces the incidence of dental caries in their children aged 19–31 months, who are usually infected by *S. mutans* through contacts with mother [101, 102]. Among all polyols, xylitol and erythritol have proven to have the biggest bacteriostatic effect. Milgrom et al. [103] have conducted clinical trial concerning 15-month-old children, who were given xylitol or sorbitol syrup with a syringe for 12 months, and it was found that only the xylitol syrup exhibited effective anticaries action. However, according to couple literature reviews, xylitol is ineffective at low doses; thus, to obtain anticariogenic effects, it should be consumed at least three times per day, with a daily dose of 5–6 g [34, 104, 105]. It is recommended to chew gum or tablets coated with xylitol after meals, especially those containing sugar [14]. According to the US FDA and European Commission, all products which contain polyols can have a health claim on the labeling stating "does not promote tooth decay" [15].

3.5 Properties and Applications

3.5.1 Erythritol

This polyol characterizes with sweetness corresponding to around 60% that of sucrose (Table 2), but it can be increased, even by about 30%, through blending with other polyols, i.e., sorbitol and xylitol, and intense sweeteners such as aspartame, sucralose, or rebaudioside A (rebiana) [12, 29, 106]. Erythritol characterizes with no aftertaste and is able to mask the unwanted taste of intense sweeteners [12, 29, 106]. It gives a strong cooling effect in the mouth and is a nonhygroscopic substance that crystallizes quickly [12, 18, 21, 42, 45]. It does not take part in the Maillard-type browning reactions and is stable in both acidic and alkaline environments as well as high temperatures up to 160 °C [29].

Besides being non-glycemic, it is also a non-cariogenic substance as it inhibits the growth of mutans streptococci because it cannot be metabolized by them [27, 107–109]. According to Runnel et al. [109], the consumption of candies with erythritol by 7–8-year-old children resulted in reduced plaque growth and lower levels of plaque acetic acid and propionic acid.

What is more, in vitro studies have shown that erythritol is an excellent radical scavenger with membrane-protecting properties [29, 40]. It is also able to inhibit chemical-induced hemolysis [29] and protect endothelial cells under hyperglycemic conditions [40]. Consuming it simultaneously with fructose in equimolar amounts causes its impaired absorption [110].

Erythritol, which is a low-calorie, tooth-friendly, bulk sweetener [14], can be used to provide volume, texture, and microbiological stability in tooth-friendly chewing gums, candy products, ice creams, and also hypocaloric beverages [12, 29]. Due to its properties and resemblance to sugar, it is often used in tabletop sweeteners and calorie-reduced beverages [29]. Moreover, mixture of erythritol and maltitol can be applied in bakery products as it improves baking stability and result in higher sweetness and humectancy properties [29].

3.5.2 Isomalt

It can be used in food products, which are heated due to isomalt resistance to the loss of sweetness at high temperatures [111]. It has 45–65% of the sweetness of sucrose, which depends on concentration, temperature, and the type of isomalt [46, 111]. Its sweetness is also increased when isomalt is mixed with intense sweeteners. This polyol is very similar to sugar physically (white, crystalline, and odorless) but does not crystallize so quickly, so it can be used for decorative purposes [12]. It is characterized by low hygroscopicity and no aftertaste [26]. What is more, it is frequently used in combination with nonnutritive sweeteners and bulking agents due to its ability to mask their unwanted aftertaste [46]. Isomalt is also able to enhance flavor transfer in foods [46, 111] and is characterized by the highest heat of solution among all polyols (Table 3). Therefore, it does not give cooling effect in the mouth.

Similarly to other sugar alcohols, products with isomalt can be labeled as safe for teeth as it is a non-cariogenic substance [15, 112]. Moreover, it was found that toothpastes with isomalt enhance teeth remineralization [112].

Products with this sugar alcohol can be recommended for people with diabetes as it does not increase blood glucose or insulin levels (Table 4) [14]. Isomalt in food industry can be used as a sweetener, bulking agent, anticaking agent, and glazing agent [113]. Due to low hygroscopicity, it can be stored for long periods [26]. Nowadays, it finds its application in such products as hard candies, toffees, chewing gum, chocolates, baked goods, nutritional supplements, cough drops, and throat lozenges [113]. Isomalt coatings are resistant to cracking and chipping and can have various colors [26].

3.5.3 Lactitol

This polyol characterizes with 40% of sucrose sweetness and no aftertaste [38]. Its mild sweetness increases with concentration, though the required taste is usually obtained by addition of intense sweeteners [38]. Due to the fact that it dissolves at lower temperatures than sucrose, there are lower costs of production of food with lactitol [30, 32]; thus, it is frequently used as a sucrose 1:1 substitute in calorie-controlled foods. What is more, it can be stored for long periods due to its high stability in various conditions and low hygroscopicity [30]. Similarly to other polyols, lactitol does not take part in the Maillard reaction and enhances perception of product flavor [32]. Therefore, it is usually used in products with such sweeteners as acesulfame-K aspartame, cyclamate, neotame, saccharin, stevia sweeteners, and sucralose. It has the second lowest heat of solution next to isomalt; thus, it gives a very small cooling effect (Table 3).

Besides being non-cariogenic substance, it also functions as a prebiotic, which positively influences the growth of colonic beneficial bacteria such as *bifidobacteria* and *lactobacilli* [15, 30, 32, 48, 114, 115]. What is more, lactitol decreases the population of intestinal putrefactive bacteria, the pH in the intestines, and production and absorption of ammonia [30, 32, 48, 114, 115]. There were also studies on this polyol application in encephalopathy and constipation treatment [116–120]. Food industry uses lactitol as a sweetener, a thickener, and an emulsifier [121]. Lactitol does not influence blood glucose effect; therefore, it is used in food for diabetics. Owing to its features, it finds application in low-calorie, low-fat, and/or sugar-free foods such as ice cream, chocolate, hard and soft candies, baked goods, sugar-reduced preserves, chewing gums, and sugar substitutes [15, 32, 116, 121–123].

3.5.4 Maltitol

It is a polyol that highly reminds sucrose in view of its properties [31, 35]. It characterizes with 90% of sucrose sweetness as well as clean and pleasant taste (Table 2). Maltitol has a very small cooling effect but can be used as a fat substitute, which provides creamy texture to food [31, 33, 35, 124]. It does not caramelize and is stable in high temperatures. Due to its slow absorption, the insulin response is significantly reduced [15, 17]. When applied with short-chain fructooligosaccharides in sugar-free food product formulations, it decreases postprandial glycemic responses [125]. It does not increase the risk of caries, because bacteria in the oral cavity do not decompose it into acids, which cause the disintegration of the enamel and the formation of cavities. Therefore, its non-cariogenic properties allow its usage in many sugar-free foods. Moreover, its physical and chemical properties allow its application in baked products as well as a variety of reduced-calorie and reduced-fat products [27, 31, 33]. Tableted gums owe their crunchy texture and a glossy surface to maltitol. It is also used as the best replacement for sugar in baked goods and chocolates [31], though the range of products with this sugar alcohols is still limited. It is mainly due to its cost, market demand, and difficulty of production and availability of substrates [31].

3.5.5 Mannitol

This sugar alcohol is characterized by a cooling effect that positively influences the food products' taste [17, 49, 74]. Although it is only 50% as sweet as sucrose, its combination with other ingredients and sweeteners results in synergistic effect of sweetness and better taste [14]. Due to its properties including a pleasant taste, thermal stability and a high melting point (165–169 ° C), and non-cariogenicity, it is frequently used in food and pharmaceutical industry in nutritional tablets, chocolate-flavored coating agents for ice cream and confections, or "breath-freshening" and "sugar-free" products [14]. Its very low hygroscopicity allows its application as a bulking agent for sugar-free coatings and a dusting powder for chewing gum, which prevents the gum from sticking to industrial equipment and wrappers [14, 49, 53, 126]. Mannitol is characterized by glycemic and insulinemic indexes amounting to 0; thus, it can be recommended for diabetics (Table 4) [17, 50, 69, 108]. It was also found that mannitol can constitute a feedstock for bioethanol production [127].

Besides food and pharmaceutical industry, mannitol finds its application in medical practice [128]. It exhibits antioxidant effects and can be used as scavenger of hydroxyl radicals, which can decrease the risks associated with free radicals during ischemia–reperfusion injury [40, 49, 69, 74, 128–130]. Moreover, it is alleged of being a health-promoting agent that can add extra nutritional value to food products and protect from colon cancer development [49, 50, 131]. It was found that inhaled mannitol helps in mucus and cough clearance in asthmatics and other hypersecretory diseases [132, 133]. What is more, it might have positive influence on renal function during endovascular aortic aneurysm repair when combined with hydration [134]. As it has a low molecular weight 182 (Table 3), it can be freely filtered through the renal tubules [128]. Therefore, it can be used as an osmotic diuretic which quickly reduces blood pressure within the skull (intracranial pressure) and the eye (glaucoma). It was also found that mannitol influences the release of renal prostaglandins which result in renal vasodilation and in increase in tubular urine flow [128]. It finds also its application in kidney failure with low urine output and during cardiopulmonary bypass in the circuit prime of a heart–lung machine. It helps protecting renal system during cardiac, vascular, and renal transplantation surgeries. Mannitol can aid in the management of rhabdomyolysis as well as in bowel preparation before colorectal surgery [128]. It also promotes excretion of toxic materials in the urine [128].

3.5.6 Sorbitol

It is characterized by similar properties to mannitol, except for solubility in water which is 20-fold higher than the mannitol one [74, 77, 79]. Its sweetness equals 60% that of sucrose but can be increased when sorbitol is mixed with other sweeteners. The low heat of solution results in a cooling sensation in the mouth (Table 3). It characterizes with a pleasant taste which can mask the unwanted one of other substances as it combines easily with other food components including sugars, gelling agents, proteins, and fats. Similarly to other polyols, it does not take part in the Maillard reactions, which is an important feature for food products of delicate flavor and the intrinsic qualities that must be maintained during production process [52, 135, 136].

Sorbitol has several technological functions as it can be used as a sweetener, an excellent humectant, a softener, and a texturizing and anti-crystallizing agent [74, 80]. It is also an important substance for fruits as it take part in carbon metabolism and influence the quality of starch accumulation and sugar–acid balance [12].

Besides having several advantages such as thermal stability, chemical inertness, and easy compression, it has one great disadvantage from the technological point of view, which is high hygroscopicity [52, 53, 126, 135, 136]. On the other hand, due to this feature, sorbitol can be used by food producers as a perfect humectant in the production of confectionery, baked goods, and chocolate [14].

Products with this sugar alcohol can be labeled as safe to teeth as sorbitol is a non-cariogenic substance [15]. As it is characterized by a low glycemic index, it can be used in products for diabetics and other pharmaceuticals (Table 4). Sorbitol also constitutes an important precursor of the vitamin C production, sorbose, and surfactants [3, 79, 82]. Approximately 25% of the total sorbitol produced is used in syrups [27], whereas another 25% is used for the synthesis of vitamin C [72, 79, 137]. Moreover, it can be found as an additive in many products including sugar-free candies, chewing gums, and sugar-free foods such as frozen desserts and baked goods as well as cosmetics [14].

3.5.7 Xylitol

Among all polyols, xylitol is characterized by the greatest sweetness, the same as sucrose, pleasant taste, and the lowest heat of solution [14, 27]. It gives the greatest cooling sensation and quickly dissolves [30, 85, 107]. It is mainly used as a bulk sweetener with non-insulin stimulant properties, which is safe to use by people with diabetes [14, 18, 83].

Xylitol exhibits several biological activities [107]. Adequate doses of this sugar alcohol result in significant anticariogenic effects as well as in a decrease of plaque formation [18, 34, 83, 103, 107, 138]. Results of several clinical trials have proved that xylitol inhibits the growth and metabolism of *Streptococcus mutans* and *Streptococcus sobrinus*, which are responsible for caries and dental plaque acid production, respectively [55, 139–142]. It is due to the fact that xylitol is not fermented in the mouth; therefore, it cannot be converted to acids by bacteria inhabiting in the oral cavity, including *Streptococcus mutans*. What is more, xylitol consumption by mothers resulted in their children lower occurrence of teeth decay [101]. Besides being an anticariogenic agent, it also helps in teeth remineralization through oral pH increase and formation of complexes with Ca(II) [107, 143, 144]. Xylitol also resembles a polypeptide statherin, which provides an appropriate protective conditions for teeth [107]. Its anticaries activity increases with concentration but also in the presence of other polyols [34, 107]. According to Ly et al. [34], xylitol works most efficiently when applied 3–5 times a day, and the total daily dose ranges from 6 to 10 g. However, its consumption exceeding 10 g per day does not result in better efficiency [34]. Moreover, xylitol eliminates unpleasant mouth odor and has a refreshing abilities [14]. In addition, xylitol stimulates the production of saliva, which is an important feature for the elderly as well as for the treatment of xerostomia [107, 145].

This compound also exhibits antifungal and antibacterial effects. There are also studies confirming the xylitol effectiveness in reduction of ear infections incidence and pneumococcal nasal colonization [55, 74, 107, 146–148]. It has been shown in clinical studies that a 5% xylitol solution inhibits the growth of *Streptococcus pneumoniae*, whereas addition of this sugar alcohol to the nasal aerosol results in antimicrobial activity against other bacteria including *Haemophilus influenzae* [55]. Preventive use of xylitol can decrease incidence of children middle ear infections by 25–40% [55]. At the same time, it inhibits the growth of *Candida albicans* and *Helicobacter pylori* [14, 30]. Xylitol also stimulates absorption of calcium from the intestines, aiding in bone mineralization. Studies on rats have shown that diet supplemented in 10% with xylitol improves bone metabolism and has a beneficial effect in the initial phase of arthritis, type II [149].

This sugar alcohol can be also used as an energy source in infusion therapy [150] as well as a preventive factor of adrenocortical suppression during steroid therapy [150] and phenylenediamine-induced hepatotoxicity [151]. Moreover, it can be applied as an antioxidant agent of fish oil [152] or a sanitizer [153].

4 Polyols in the Food Industry

Each food product must be safe, should have proper aesthetic values and good taste, and fulfill the consumers' wishes, who prefer to buy sweet products. Food producers are facing the new demand from consumers who have increased interest in healthy lifestyles; thus, a range of new sweeteners that provide alternatives to sugar is being increasingly used. There are many advantages of sugar alcohols, so they are becoming more and more popular among both consumers and producers. They are characterized by a lower caloric value and glycemic index than sugars and exhibit prebiotic and anticaries effects [12, 17, 27, 108]. Similarly to carbohydrates, they are just not assigning an appropriate sweet taste, but also play role of texture-, filling, preservative and retaining moisture substances, which also give a cooling feeling in the mouth [12, 27].

Polyols can be obtained in two forms, i.e., solid and liquid. Their technological application greatly depends on molecular weight. In general, with its decrease, polyol osmolality increases, freezing point decreases, viscosity decreases, and boiling point increases [154]. All of these properties can influence spread, volume, texture, shelf life, and mouthfeel of various food products [154].

All sugar alcohols can be used as bulking agents, which can substitute sugar or corn syrups 1:1 ratio. Additionally, they promote mouthfeel and eliminate improper taste. Therefore, they can be used as reduced-calorie sugar alternatives. Moreover, their application results in products that have lower glycemic index and are suitable for diabetics [14].

One of the most important properties for food industry is polyols' ability to control the moisture of the final products. This feature allows on varied application of particular polyhydric alcohols. Polyols such as sorbitol or xylitol can help to retain moisture, whereas others including mannitol, which are characterized with very low hygroscopicity, allow its application as a dusting powder for chewing gum, which prevents the gum from sticking to industrial equipment and wrappers [14, 49, 53, 126]. Moreover, polyols with low hygroscopicity can provide some products with the required by a consumer crispness [154].

In general, solubility of polyols increases with temperature. Sorbitol, xylitol, maltitol, and lactitol belong to polyols well soluble in water, whereas isomalt, mannitol, and erythritol are characterized by significantly lower solubility.

A feature which is related to solubility is crystallization, as less soluble compounds are characterized by a higher tendency to form crystals. Similarly to sugars, heat and supersaturation with carefully controlled cooling can prevent crystallization [154]. What is more, addition of sorbitol can inhibit crystallization, thus influencing the smoothness and creaminess of the final product; thus, this polyol is frequently used in syrups and jams [31, 33, 35, 124]. On the other hand, hard-boiled candies and lozenges require addition of sugar alcohols of low hygroscopicity, especially isomalt. The rate of crystallization can be modified through stirring, agitation, and addition of other compounds [35, 154]. Due to variations in crystallization rate, sugar alcohols also find application in cold and frozen foods. Nguyen and Ulrich [155] have concluded that polyols (isomalt and xylitol) are potential multifunctional agents in the freeze-casting process.

Although sugar alcohols are usually characterized by lower sweetness as compared to sucrose, except for xylitol, they can be used in combination with other sweeteners in order to increase the sweetening effect. Therefore, it is easy for manufacturers to change the sweetness pattern of a food product and the scope of products available for consumers expands. What is more, polyol sweetness is not influenced by temperature. Most of polyols exhibit distinctive cooling effect (the strongest, erythritol; the lowest, isomalt), which can be desirable for peppermint and menthol products but unwanted in case of baked products and chocolate (Table 3). However, this problem can be solved by mixing sugar alcohols with sugar which reduces the cooling effect.

Contrary to sugar, polyols do not caramelize or undergo the Maillard reaction. This property might be of great value for products, in which change of color is undesirable. As these compounds are stable within wide range of temperatures and pH, thus, they can improve the shelf life of the product.

There are only two disadvantages of polyols, i.e., laxative effect and their price. When consumed in excess, they have a laxative effect due to their slower and incomplete digestion; thus, their presence must be mentioned on the label [15–17].

5 Summary

Due to the constantly evolving processing and globalization, the current market of processed foods is rich in added sugars and saturated fats that results in energy oversupply that is usually stored in the body as fat. Simultaneously, consumers are becoming more and more aware of the nutrition role in the healthy lifestyles. Therefore, manufacturers are searching for a range of new sweeteners that provide alternatives to sugar. Polyols are bulk sweeteners that constitute excellent substitutes to sugar in food industry due to various physical, chemical, and biological properties. There are developed new food products with sugar alcohols that have positive effect on human health, including prebiotic and anticariogenic ones. What is more, such products can be recommended not only to people with diabetes but also to others who suffer from civilizational diseases.

References

1. Ventura AK, Mennella JA (2011) Innate and learned preferences for sweet taste during childhood. Curr Opin Clin Nutr Metab Care 14(4):379–384
2. Jarosz M (2012) Normy żywienia dla populacji polskiej – nowelizacja. Instytut Żywności i Żywienia Warszawa (in Polish)
3. ADA Reports (2004) Position of the American Dietetic Association: use of nutritive and nonnutritive sweeteners. J Am Diet Assoc 104:255–275
4. Lê L, Tappy L (2015) Contributions of sugars to metabolic disorders in human models. In: Goran MI, Tappy L, Lê K-A (eds) Dietary sugars and health. CRC Press Taylor & Francis Group, Boca Raton
5. Lustig RH, Schmidt L, Brindis CD (2012) Public health: the toxic truth about sugar. Nature 482:27–29
6. WHO. Health topics: Obesity http://www.who.int/topics/obesity/en/. Accessed 10 Dec 2015
7. WHO http://www.who.int/mediacentre/factsheets/fs311/en/. Accessed 10 Dec 2015
8. Regulation of the European Parliament and Council Regulation (EC) No 1333/2008 of 16 December 2008 on food additives
9. Grembecka M (2015) Natural sweeteners in a human diet. Rocz Panstw Zakl Hig 66(3):195–202
10. WHO. Evaluations of the Joint FAO/WHO Expert Committee on Food Additives (JECFA) http://apps.who.int/food-additives-contaminants-jecfa-database/search.aspx?fc=66. Accessed 15 Dec
11. Cummings JH, Stephen AM (2007) Carbohydrate terminology and classification. Eur J Clin Nutr 61(Suppl 1):S5–S18
12. Barbieri G, Barone C, Bhagat A, Caruso G, Conley ZR, Parisi S (2014) Sweet compounds in foods: sugar alcohols. In: The influence of chemistry on new foods and traditional products. Springer International Publishing, Cham
13. Health Canada (2005) http://www.hc-sc.gc.ca/fn-an/securit/addit/sweeten-edulcor/polyols_polydextose_factsheet-polyols_polydextose_fiche-eng.php. Accessed 1 Nov 2015
14. Grembecka M (2015) Sugar alcohols: their role in the modern world of sweeteners: a review. Eur Food Res Technol 241(1):1–14
15. EFSA (2011) Scientific opinion on the substantiation of health claims related to the sugar replacers xylitol, D-tagatose, xylitol, sorbitol, mannitol, maltitol, lactitol, isomalt, erythritol, D-tagatose, isomaltulose, sucralose and polydextrose and maintenance of tooth mineralisation by decreasing tooth demineralisation. EFSA J 9(4): 2076 http://www.efsa.europa.eu/en/efsajournal/doc/2076.pdf. Accessed 29 Nov 2015
16. Grabitske HA, Slavin JL (2008) Perspectives in practice low-digestible carbohydrates in practice. J Am Diet Assoc 108:1677–1681
17. Livesey G (2003) Health potential of polyols as sugar replacers, with emphasis on low-glycaemic properties. Nutr Res Rev 16:163–191
18. Chattopadhyay S, Raychaudhuri U, Chakraborty R (2014) Artificial sweeteners – a review. J Food Sci Technol 51:611–621
19. Shankar P, Ahuja S, Sriram K (2013) Non-nutritive sweeteners: review and update. Nutrition 29:1293–1299
20. Sylvetsky AC, Brown RJ, Rother KI (2015) Biological and health effects of nonnutritive sweeteners. In: Goran MI, Tappy L, Lê K-A (eds) Dietary sugars and health. CRC Press Taylor & Francis Group, Boca Raton

21. INCHEM. Aspartame. http://www.inchem.org/documents/jecfa/jecmono/v16je03.htm. Accessed 16 Dec 2015
22. Hampton T (2008) Sugar substitutes linked to weight gain. JAMA 299:2137–2138
23. Swithers SE, Davidson TL (2008) A role for sweet taste: calorie predictive relations in energy regulation by rats. Behav Neurosci 122:161–173
24. Swithers SE, Sample CH, Davidson TL (2013) Adverse effects of high-intensity sweeteners on energy intake and weight control in male and obesity-prone female rats. Behav Neurosci 127(2):262–274
25. Report of a Joint FAO/WHO Expert Consultation 14–18 April 1997. Carbohydrates in human nutrition. http://www.fao.org/docrep/w8079e/w8079e00.htm, Rome. Accessed 16 Dec 2015
26. Sentko A, Bernard J (2012) Isomalt. In: O'Brien Nabors L (ed) Alternative sweeteners. CRC Press, Taylor & Francis Group, Boca Raton
27. Evrendilek GA (2012) Sugar alcohols (Polyols). In: Varzakas T, Labropoulos A, Anestis S (eds) Sweeteners: nutritional aspects, applications, and production technology. CRC Press, Boca Raton
28. de Cock P (2012) Erythritol. In: O'Brien Nabors L (ed) Alternative sweeteners. CRC Pres, Boca Raton
29. de Cock P (2012) Erythritol. In: O'Donnell K, Kearsley MW (eds) Sweeteners and sugar alternatives in food technology. Wiley-Blackwell, West Sussex
30. Zacharis C (2012) Lactitol. In: O'Donnell K, Kearsley MW (eds) Sweeteners and sugar alternatives in food technology. Wiley-Blackwell, West Sussex
31. Kearsley MW, Boghani N (2012) Maltitol. In: O'Brien Nabors L (ed) Alternative sweeteners. CRC Press, Taylor & Francis Group, Boca Raton
32. Koivistoinen M (2007) Lactitol. In: Wilson R (ed) Sweeteners. Blackwell Publishing/Leatherhead Publishing, Oxford/Surrey
33. Lawson P (2007) Erythritol. In: Wilson R (ed) Sweeteners. Blackwell Publishing/Leatherhead Publishing, Oxford/Surrey
34. Ly KA, Milgrom P, Rothen M (2006) Xylitol, sweeteners, and dental caries. Pediatr Dent 28:154–163
35. Kearsley MW, Deis RC (2012) Maltitol powder. In: O'Donnell K, Kearsley MW (eds) Sweeteners and sugar alternatives in food technology. Wiley-Blackwell, West Sussex
36. EFSA (2015) Scientific opinion on the safety of the proposed extension of use of erythritol (E 968) as a food additive. EFSA J 13(3):4033
37. Joint Expert Committee on Food Additives (JECFA) "Isomalt". International Programme on Chemical Safety (IPCS) http://inchem.org/documents/jecfa/jecmono/v20je14.htm. Accessed 18 Dec 2015
38. Zacharis C, Stowell J (2012) Lactitol. In: O'Brien Nabors L (ed) Alternative sweeteners. CRC Press, Taylor & Francis Group, Boca Raton
39. Bernt WO, Borzelleca JF, Flamm G, Munro IC (1996) Erythritol: a review of biological and toxicological studies. Regul Toxicol Pharmacol 24:S191–S197
40. den Hartog GJ, Boots AW, Adam-Perrot A, Brouns F, Verkooijen IW, Weseler AR, Haenen GR, Bast A (2010) Erythritol is a sweet antioxidant. Nutrition 26:449–458
41. Munro IC, Bernt WO, Borzelleca JF, Flamm G, Lynch BS, Kennepohl E, Bär EA, Modderman J (1998) Erythritol: an interpretive summary of biochemical, metabolic, toxicological and clinical data. Food Chem Toxicol 36:1139–1174
42. Friedman RB (2008) Monosaccharides and polyols in foods. In: Fraser-Reid B, Tatsuta K, Thiem J (eds) Glycoscience. Springer, Berlin/Heidelberg
43. Bornet FR, Blayo A, Dauchy F, Slama G (1996) Plasma and urine kinetics of erythritol after oral ingestion by healthy humans. Regul Toxicol Pharm 24:S280–S285
44. Roberts A, Renwick A (2009) Toxicokinetics. In: Ballantyne B, Marrs TC, Syversen T (eds) General and applied toxicology, vol 1, 3rd edn. Wiley, Chichester
45. Oku T, Okazaki M (1996) Laxative threshold of sugar alcohol erythritol in human subjects. Nutr Res 16:577–589

46. Sentko A, Willibald-Ettle I (2012) Isomalt. In: O'Donnell K, Kearsley MW (eds) Sweeteners and sugar alternatives in food technology. Wiley-Blackwell, West Sussex
47. Gostner A, Blaut M, Schäffer V, Kozianowski G, Theis S, Klingeberg M, Dombrowski Y, Martin D, Ehrhardt S, Taras D, Schwiertz A, Kleessen B, Lührs H, Schauber J, Dorbath D, Menzel T, Scheppach W (2006) Effect of isomalt consumption on faecal microflora and colonic metabolism in healthy volunteers. Br J Nutr 95(1):40–50
48. Ballongue J, Schumann C, Quignon P (1997) Effects of lactulose and lactitol on colonic microflora and enzymatic activity. Scand J Gastroenterol 222:41–44
49. Ghoreishi SM, Shahrestani RG (2009) Innovative strategies for engineering mannitol production. Trends Food Sci Technol 20:263–270
50. Wisselink HW, Weusthuis RA, Eggink G, Hugenholtz J, Grobben GJ (2002) Mannitol production by lactic acid bacteria: a review. Int Dairy J 12:151–161
51. Xiao J, Li X, Min X, Sakaguchi E (2013) Mannitol improves absorption and retention of calcium and magnesium in growing rats. Nutrition 29:325–331
52. Deis RC, Kearsley MW (2012) Sorbitol and mannitol. In: O'Donnell K, Kearsley MW (eds) Sweeteners and sugar alternatives in food technology. Wiley-Blackwell, West Sussex
53. Jamieson PR (2012) Sorbitol and mannitol. In: O'Brien Nabors L (ed) Alternative sweeteners. CRC Press, Taylor & Francis Group, Boca Racton
54. Sheet BS, Artık N, Ayed MA, Abdulaziz OF (2014) Some alternative sweeteners (xylitol, sorbitol, sucralose and stevia): review. Karaelmas Sci Eng J 4(1):63–70
55. Lee BD, Park MK (2014) Effects and safety of xylitol on middle ear epithelial cells. Int Adv Otol 10:19–24
56. Park Y-C, Oh EJ, Jo J-H, Jin Y-S, Seo J-H (2016) Recent advances in biological production of sugar alcohols. Curr Opin Biotechnol 37:105–113
57. Veiga-da-Cunha M, Santos H, van Schaftingen E (1993) Pathway and regulation of erythritol formation in *Leuconostoc oenos*. J Bacteriol 175:3941–3948
58. Stolz P, Biicker G, Hammes WP, Vogel RF (1995) Utilization of electron acceptors by lactobacilli isolated from sourdough. Z Lebensm Unters Forsch 201:91–96
59. Richter H, Vlad D, Unden G (2001) Significance of pantothenate for glucose fermentation by *Oenococcus oeni* and for suppression of the erythritol and acetate production. Arch Microbiol 175:26–31
60. Embuscado ME, Patil SK (2001) Erythritol. In: Dekker M (ed) Food science and technology, vol 17, Alternative sweeteners, 3rd edn. Marcel Dekker, New York
61. Lin S-J, Wen C-Y, Wang P-M, Huang J-C, Wei C-L, Chang J-W, Chu W-S (2010) High-level production of erythritol by mutants of osmophilic *Moniliella* sp. Process Biochem 45:973–979
62. Mirończuk AM, Dobrowolski A, Rakicka M, Rywińska A, Rymowicz W (2015) Newly isolated mutant of *Yarrowia lipolytica* MK1 as a proper host for efficient erythritol biosynthesis from glycerol. Process Biochem 50:61–68
63. Tomaszewska L, Rywińska A, Rymowicz W (2014) High selectivity of erythritol production from glycerol by *Yarrowia lipolytica*. Biomass Bioenerg 64:309–320
64. Yang L-B, Zhan X-B, Zheng Z-Y, Wu J-R, Gao M-J, Lin C-C (2014) A novel osmotic pressure control fed-batch fermentation strategy for improvement of erythritol production by *Yarrowia lipolytica* from glycerol. Bioresour Technol 151:120–127
65. Sawada K, Taki A, Yamakawa T, Seki M (2009) Key role for transketolase activity in erythritol production by *Trichosporonoides megachiliensis* SN-G42. J Biosci Bioeng 108:385–390
66. Ghezelbash GR, Nahvi I, Emamzadeh R (2014) Improvement of erythrose reductase activity, deletion of by-products and statistical media optimization for enhanced erythritol production from *Yarrowia lipolytica* Mutant 49. Curr Microbiol 69:149–157
67. Ghezelbash G, Nahvi I, Malekpour A (2014) Erythritol production with minimum by-product using *Candida magnoliae* mutant. Appl Biochem Microbiol 50:292–296
68. Makkee M, Kieboom APG, Van Bekkum H (1985) Production methods of D-mannitol. Starch-Starke 37:136–141

69. Song SH, Vieille C (2009) Recent advances in the biological production of mannitol. Appl Microbiol Biotechnol 84:55–62
70. Soetaert W, Buchholz K, Vandamme EJ (1995) Production of D-mannitol and D-lactic acid by fermentation with Leuconostoc mesenteroides. Agro Food Ind Hi Tec 6:41–44
71. von Weymarn N, Hujanen M, Leisola M (2002) Production of D-mannitol by heterofermentative lactic acid bacteria. Process Biochem 37:1207–1213
72. Akinterinwa O, Khankal R, Cirino PC (2008) Metabolic engineering for bioproduction of sugar alcohols. Curr Opin Biotech 19:461–467
73. Jacobsen JH, Frigaard N-U (2014) Engineering of photosynthetic mannitol biosynthesis from CO_2 in a cyanobacterium. Metab Eng 21:60–70
74. Ortiz ME, Bleckwedel J, Raya RR, Mozzi F (2013) Biotechnological and in situ food production of polyols by lactic acid bacteria. Appl Microbiol Biotechnol 97:4713–4726
75. Ruperez P, Toledano G (2003) Celery by-products as a source of mannitol. Eur Food Res Technol 216:224–226
76. Wilson P (2007) Mannitol. In: Wilson R (ed) Sweeteners. Blackwell Publishing/Leatherhead Publishing, Oxford/Surrey, pp 219–225
77. Lawson P (2007) Sorbitol and sorbitol syrup. In: Wilson R (ed) Sweeteners. Blackwell Publishing/Leatherhead Publishing, Oxford/Surrey, pp 227–238
78. Kusserow B, Schimpf S, Claus P (2003) Hydrogenation of glucose to sorbitol over nickel and ruthenium catalysts. Adv Synth Catal 345:289–299
79. Silveira MM, Jonas R (2002) The biotechnological production of sorbitol. Appl Microbiol Biotechnol 59:400–408
80. Jonas R, Silveira MM (2004) Sorbitol can be produced not only chemically but also biotechnologically. Appl Biochem Biotechnol 118:321–336
81. Silveira MM, Wisbeck E, Lemmel C, Erzinger G, da Costa JP, Bertasso M, Jonas R (1999) Bioconversion of glucose and fructose to sorbitol and gluconic acid by untreated cells of *Zymomonas mobilis*. J Biotechnol 75:99–103
82. Ladero V, Ramos A, Wiersma A, Goffin P, Schanck A, Kleerebezem M, Hugenholtz J, Smid EJ, Hols P (2007) High-level production of the low-calorie sugar sorbitol by Lactobacillus plantarum through metabolic engineering. Appl Environ Microbiol 73:1864–1872
83. Nigam P, Singh D (1995) Processes for fermentative production of xylitol – a sugar substitute. Process Biochem 30:117–124
84. Zhang J, Geng A, Yao C, Lu Y, Li Q (2012) Xylitol production from D-xylose and horticultural waste hemicellulosic hydrolysate by a new isolate of *Candida athensensis* SB18. Bioresour Technol 105:134–141
85. Granström TB, Izumori K, Leisola M (2007) A rare sugar xylitol. Part I: the biochemistry and biosynthesis of xylitol. Appl Microbiol Biotechnol 74:277–281
86. Zakaria A (2001) Production of natural and rare pentoses using microorganisms and their enzymes. Electron J Biotechnol 4:103–111
87. Pourmir A, Noor-Mohammadi S, Johannes TW (2013) Production of xylitol by recombinant microalgae. J Biotechnol 165(3–4):178–183
88. Hirabayashi S, Wang J, Kawagishi H, Hirai H (2015) Improving xylitol production through recombinant expression of xylose reductase in the white-rot fungus *Phanerochaete sordida* YK-624. J Biosci Bioeng 120:6–8
89. Kim SM, Choi BY, Ryu YS, Jung SH, Park JM, Kim G-H, Lee SK (2015) Simultaneous utilization of glucose and xylose via novel mechanisms in engineered *Escherichia coli*. Metab Eng 30:141–148
90. Tada K, Horiuchi J, Kanno T, Kobayashi M (2004) Microbial xylitol production from corn cobs using *Candida magnoliae*. J Biosci Bioeng 98:228–230
91. Latif F, Rajoka MI (2001) Production of ethanol and xylitol from corn cobs by yeasts. Bioresour Technol 77:57–63
92. Buhner J, Agblevor FA (2004) Effect of detoxification of dilute- acid corn fiber hydrolysate on xylitol production. Appl Biochem Biotechnol 119:13–30

93. Dominguez JM, Gong CS, Tsao GT (1996) Pretreatment of sugar cane bagasse hemicellulose hydrolysate for xylitol production by yeast. Appl Biochem Biotechnol 57–58:49–56
94. Santos JC, Pinto IR, Carvalho W, Mancilha IM, Felipe MG, Silva SS (2005) Sugarcane bagasse as raw material and immobilization support for xylitol production. Appl Biochem Biotechnol 121–124:673–683
95. Wheeler ML, Pi-Sunyer X (2008) Carbohydrate issues. Type and amount. J Am Diet Assoc 108:S34–S39
96. Nakamura S (2005) Bioavailability of cellobiose and other non-digestible and/or nonabsorbable sugar substitutes and related topics. Nutrition 21:1158–1159
97. Holub I, Gostner A, Hessdorfer S, This S, Bender G, Willinger B, Schauber J, Melcher R, Allolio B, Scheppach W (2009) Improved metabolic control after 12-week dietary intervention with low glycaemic isomalt in patients with type 2 diabetes mellitus. Horm Metab Res 41 (12):886–892
98. Hamer HM, Jonkers D, Venema K, Vanhoutvin S, Troost FJ, Brummer RJ (2008) Review article: the role of butyrate on colonic function. Aliment Pharmacol Ther 27:104–119
99. Blanc P, Daures JP, Rouillon JM, Peray P, Pierrugues R, Larrey D, Gremy F, Michel H (1992) Lactitol or lactulose in the treatment of chronic hepatic encephalopathy: results of a meta-analysis. Hepatology (Baltimore) 15:222–228
100. Mäkinen KK, Saag M, Isotupa KP, Olak J, Nõmmela R, Söderling E, Mäkinen PL (2005) Similarity of the effects of erythritol and xylitol on some risk factors of dental caries. Caries Res 39(3):207–215
101. Söderling E, Isokangas P, Pienihakkinen K, Tenovuo J, Alanen P (2001) Influence of maternal xylitol consumption on mother-child transmission of mutans streptococci: 6-year-follow-up. Caries Res 35:173–177
102. Nakai Y, Shinga-Ishihara C, Kaji M, Moriya K, Murakami-Yamanaka K, Takimura M (2010) Xylitol gum and maternal transmission of mutans streptococci. J Dent Res 89(1):56–60
103. Milgrom P, Zero DT, Tanzer JM (2009) An examination of the advances in science technology of prevention of tooth decay in young children since the Surgeon General's Report on Oral Health. Acad Pediatr 9:404–409
104. Söderling E (2009) Controversies around Xylitol. Eur J Dent 3(2):81–82
105. Söderling E (2009) Xylitol, mutans streptococci, and dental plaque. Adv Dent 21:74–78
106. Boileau A, Fry JC, Murray R (2012) A new calorie-free sugar substitute from the leaf of the stevia plant arrives in the UK. Nutr Bull 37(1):47–50
107. Mäkinen KK (2011) Sugar alcohol sweeteners as alternatives to sugar with special consideration of xylitol. Med Princ Pract 20:303–320
108. Livesey G (2012) Glycaemic responses and toleration. In: O'Donnell K, Kearsley MW (eds) Sweeteners and sugar alternatives in food technology. Wiley-Blackwell, West Sussex
109. Runnel R, Mäkinen KK, Honkala S, Olak J, Mäkinen P-L, Nõmmela R, Vahlberg T, Honkala E, Saag M (2013) Effect of three-year consumption of erythritol, xylitol and sorbitol candies on various plaque and salivary caries-related variables. J Dent 41:1236–1244
110. Kim Y, Park SC, Wolf BW, Hertzler SR (2011) Combination of erythritol and fructose increases gastrointestinal symptoms in healthy adults. Nutr Res 31:836–841
111. Ferguson T, Sentko A, Willibald-Ettle I (2007) Isomalt. In: Wilson R (ed) Sweeteners. Blackwell Publishing/Leatherhead Publishing, Oxford/Surrey, pp 167–177
112. Takatsuka T, Exterkate RAM, ten Cate JM (2008) Effects of isomalt on enamel de-and remineralization, a combined in vitro pH-cycling model and in situ study. Clin Oral Invest 12:173–177
113. FAO-WHO Food Standards. Codex Alimentarius. GSFA Online. Isomalt http://www.codexalimentarius.net/gsfaonline/additives/details.html?id=180. Accessed 7 Nov 2015
114. Finney M, Smullen J, Foster HA, Brokx S, Storey DM (2007) Effects of low doses of lactitol on faecal microflora, pH, short chain fatty acids and gastrointestinal symptomology. Eur J Nutr 46:307–314

115. Ouwenhand AC, Tiihonen K, Saarinen M, Putaala H, Rautonen N (2009) Influence of a combination of *Lactobacillus acidophilus* NCFM and lactitol on healthy elderly: intestinal and immune parameters. Br J Nutr 101:367–375
116. Chen C, Li L, Wu Z, Chen H, Fu S (2007) Effects of lactitol on intestinal microflora and plasma endotoxin in patients with chronic viral hepatitis. J Infect 54:98–102
117. Mas A, Rodes J, Sunyer L, Rodrigo L, Planas R, Vargas V, Castells L, Rodríguez-Martínez D, Fernández-Rodríguez C, Coll I, Pardo A (2003) Comparison of rifaximin and lactitol in the treatment of acute hepatic encephalopathy: results of a randomized, double-blind, double-dummy, controlled clinical trial. J Hepatol 38:51–58
118. Salerno F, Moser P, Maggi A, Vitaliani G, Benetti G (1994) Effects of long-term administration of low-dose lactitol in patients with cirrhosis but without overt encephalopathy. J Hepatol 21:1092–1096
119. Cammà C, Fiorello F, Tinè F, Marchesini G, Fabbri A, Pagliaro L (1993) Lactitol in treatment of chronic hepatic encephalopathy. A meta-analysis. Dig Dis Sci 38:916–922
120. Morillas RM, Sala M, Planas R (2014) Prevention of hepatic encephalopathy. Med Clin 142:512–514
121. FAO-WHO Food Standards. Codex Alimentarius. GSFA Online. Lactitol http://www.codexalimentarius.net/gsfaonline/additives/details.html?id=156. Accessed 7 Nov 2015
122. Kummel KF, Brokx S (2001) Lactitol as a functional prebiotic. Cereal Food World 46:424–429
123. Gibson GR, Roberfroid MB (1995) Dietary modulation of the human colonic microbiota: introducing the concept of prebiotics. J Nutr 125:1401–1412
124. FAO-WHO Food Standards. Codex Alimentarius. GSFA Online. Maltitol. http://www.codexalimentarius.net/gsfaonline/additives/details.html?id=159. Accessed 7 Nov 2015
125. Respondek F, Hilpipre C, Chauveau P, Cazaubiel M, Gendre D, Maudet C, Wagner A (2014) Digestive tolerance and postprandial glycaemic and insulinaemic responses after consumption of dairy desserts containing maltitol and fructo-oligosaccharides in adults. Eur J Clin Nutr 68 (5):575–580
126. Gombás Á, Szabó-Révész P, Regdon G, Erös I (2003) Study of thermal behaviour of sugar alcohols. J Therm Anal Calorim 73:615–621
127. Wang J, Kim YM, Rhee HS, Lee MW, Park JM (2013) Bioethanol production from mannitol by a newly isolated bacterium, *Enterobacter* sp. JMP3. Bioresour Technol 135:199–206
128. Shawkat H, Westwood M-M, Mortimer A (2012) Mannitol: a review of its clinical uses. CEACCP 12:82–85
129. Monedero V, Pérez-Martínez G, Yebra MJ (2010) Perspectives of engineering lactic acid bacteria for biotechnological polyol production. Appl Microbiol Biotechnol 86:1003–1015
130. Saha BC, Racine FM (2010) Effects of pH and corn steep liquor variability on mannitol production by Lactobacillus intermedius NRRL B-3693. Appl Microbiol Biotechnol 87:553–560
131. Gaspar P, Neves AR, Ramos A, Gasson MJ, Shearman CA, Santos H (2004) Engineering *Lactococcus lactis* for production of mannitol: high yields from food-grade strains deficient in lactate dehydrogenase and the mannitol transport system. Appl Environ Microbiol 70:1466–1474
132. Daviskas E, Anderson SD, Eberl S, Young IH (2010) Beneficial effect of inhaled mannitol and cough in asthmatics with mucociliary dysfunction. Respir Med 104:1645–1653
133. Daviskas E, Anderson SD, Young IH (2010) Effect of mannitol and repetitive coughing on the sputum properties in bronchiectasis. Respir Med 104:371–377
134. Kalimeris K, Nikolakopoulos N, Riga M, Christodoulaki K, Moulakakis KG, Dima C, Papasideris C, Sidiropoulou T, Kostopanagiotou G, Pandazi A (2014) Mannitol and renal dysfunction after endovascular aortic aneurysm repair procedures: a randomized trial. J Cardiothorac Vasc Anesth 28(4):966–971
135. Le AS, Mulderrig KB (2001) Sorbitol and mannitol. In: Nabors LO (ed) Alternative sweeteners, 3rd edn. Marcel Dekker, New York

136. Basu S, Shivhare US (2013) Rheological, textural, microstructural, and sensory properties of sorbitol-substituted mango jam. Food Bioprocess Technol 6:1401–1413
137. Budavari S, O'Neil M, Smith A, Heckelman PE, Kinneary JF (1996) The Merck index. An encyclopedia of chemicals, drugs, and biologicals. Merck, Whitehouse Station, pp 1490–1491
138. Söderling E, Hirvonen A, Karjalainen S, Fontana M, Catt D, Seppä L (2011) The effect of xylitol on the composition of the oral flora: a pilot study. Eur J Dent 5:24–31
139. Söderling EM, Ekman TC, Taipale TJ (2008) Growth inhibition of *Streptococcus mutans* with low xylitol concentrations. Curr Microbiol 56:382–385
140. Lee SH, Choi BK, Kim YJ (2012) The cariogenic characters of xylitol-resistant and xylitol-sensitive *Streptococcus mutans* in biofilm formation with salivary bacteria. Arch Oral Biol 57:697–703
141. Misra S, Raghuwanshi S, Gupta cP, Saxena RK (2012) Examine growth inhibition pattern and lactic acid production in *Streptococcus mutans* using different concentrations of xylitol produced from *Candida tropicalis* by fermentation. Anaerobe 18:273–279
142. ElSalhy M, Sayed Zahid I, Honkala E (2012) Effects of xylitol mouthrinse on *Streptococcus mutans*. J Dent 40:1151–1154
143. Tanzer JM (1995) Xylitol chewing gum and dental caries. Int Dent J 45:65–76
144. Bahador A, Lesan S, Kashi N (2012) Effect of xylitol on cariogenic and beneficial oral streptococci: a randomized, double-blind crossover trial. Iran J Microbiol 4:75–81
145. Ship JA, McCutcheon JA, Spivakovsky S, Kerr AR (2007) Safety and effectiveness of topical dry mouth products containing olive oil, betaine, and xylitol in reducing xerostomia for polypharmacy-induced dry mouth. J Oral Rehabil 34(10):724–732
146. Vernacchio L, Vezina RM, Mitchell AA (2007) Tolerability of oral xylitol solution in young children: implications for otitis media prophylaxis. Int J Pediatr Otorhinolaryngol 71:89–94
147. Nyyssölä A, Pihlajaniemi A, Palva A, von Weymarn N, Leisola M (2005) Production of xylitol from D-xylose by recombinant Lactococcus lactis. J Biotechnol 118:55–66
148. Uhari M, Tapiainen T, Kontiokari T (2000) Xylitol is preventing acute otitis media. Vaccine 19:144–147
149. Kaivosoja SM, Mattila PT, Knuuttila MLE (2008) Dietary xylitol protects against the imbalance in bone metabolism during the early phase of collages type II- induced arthritis in dark agouti rats. Metab Clin Exp 57(8):1052–1055
150. Georgieff M, Moldawer LL, Bistrian BR, Blackburn GL (1985) Xylitol, an energy source for intra-venous nutrition after trauma. J Parenter Enteral Nutr 9:199–209
151. Sood C, Khan S, O'Brien PJ (1997) Phenylenediamine induced hepatocytes cytotoxicity redox. Cycling mediated oxidative stress without oxygen activation. Biochim Biophys Acta 1335:343–352
152. Faraji H, Lindsay RC (2004) Characterization of antioxidant activity of sugars and polyhydric alcohols in fish oil emulsions. J Agric Food Chem 52:7164–7171
153. Kwon NH, Kim SH, Kim JY, Lim JY, Kim JM, Jung WK, Park KT, Bae WK, Noh KM, Choi JW, Hur J, Park YH (2003) Antimicrobial performance of alkaline ionic fluid (GC-100X) and its ability to remove *Escherichia coli* O157:H7 from the surface of tomatoes. J Food Prot 66:1604–1610
154. Nabors L, Hedrick T (2012) Sugar reduction with polyols. IFT 66(9) http://www.ift.org/Food-Technology/Past-Issues/2012/September/Features/Sugar-Reduction-with-Polyols.aspx?view=print&page=viewall
155. Nguyen PTN, Ulrich J (2015) Sugar alcohols – multifunctional agents in the freeze casting process of foods. J Food Eng 153:1–7

Tagatose Stability Issues in Food Systems 25

Leonard N. Bell

Abstract

Tagatose is a monosaccharide that may be used in food and beverage products as a bulk low-calorie sweetener, a flavor enhancer, or a prebiotic. Several studies have shown that tagatose, as a powder or dissolved in solution can break down and cause discoloration. The extent of degradation depends upon product composition and the storage environment. Tagatose powder is susceptible to sticking, deliquescence, discoloration, and degradation that are enhanced by higher relative humidities and temperatures. In solution, citrate and phosphate buffers catalyze tagatose degradation, especially at higher pH levels. Amino acids and other solutes can react with tagatose to promote browning reactions. Degradation and browning are both accelerated by increases in temperature. During pasteurization, tagatose losses would be less than 1%. Refrigeration would enable 98% of the original tagatose concentration to remain in beverages stored for 6 months. Food product formulations and proper storage conditions can be used to optimize tagatose stability while controlling discoloration.

Keywords

Tagatose • Prebiotic • Stability • Degradation • Shelf life • Browning

Contents

1 Introduction 576
2 Stability Background 577
  2.1 Degradation Pathways 577
  2.2 Kinetic Modeling 578
3 Tagatose Stability in the Solid State 579
4 Tagatose Stability in Solution 581
  4.1 pH Effect 581

L.N. Bell (✉)
Department of Poultry Science, Food Science Program, Auburn University, Auburn, AL, USA
e-mail: BELLLEO@auburn.edu

J.-M. Mérillon, K.G. Ramawat (eds.), *Sweeteners*, Reference Series in Phytochemistry,
https://doi.org/10.1007/978-3-319-27027-2_32

4.2 Buffer Type and Concentration 581
4.3 Nonenzymatic Browning Reactions 583
4.4 Temperature Effects 585
5 Conclusions 587
References 588

1 Introduction

The two most commonly recognized monosaccharides are glucose (**1**) and fructose (**2**). However, another monosaccharide that displays interesting properties is tagatose (**3, 4**). Tagatose is chemically classified as a ketose, differing from fructose by the reversed orientation of the hydroxyl group on the fourth carbon. As with all monosaccharides, tagatose dissolved in water undergoes structural rearrangement called mutarotation, resulting in five possible structures: α-D-tagatofuranose, β-D-tagatofuranose, α-D-tagatopyranose (**4**), β-D-tagatopyranose (**3**), and acyclic D-tagatose [1]. Researchers have determined that the most predominant structure in solution is α-D-tagatopyranose (**4**), which makes up 70–80% of the possible structures [2, 3]. The acyclic keto form is the least predominant at approximately 0.3% [3].

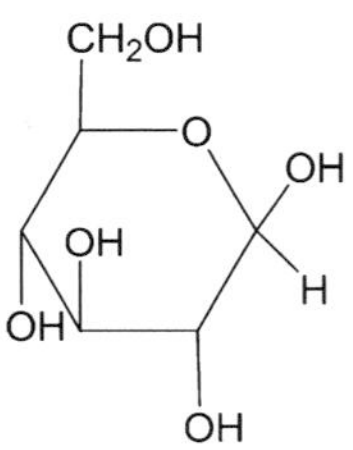

1

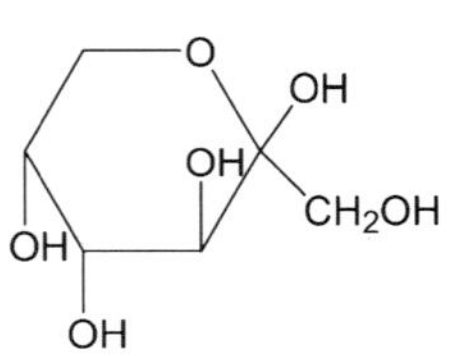

2

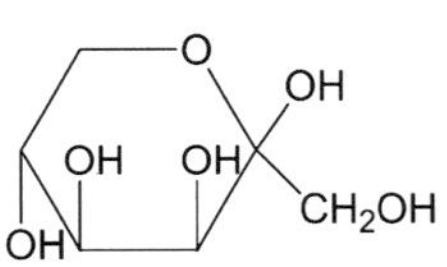

3

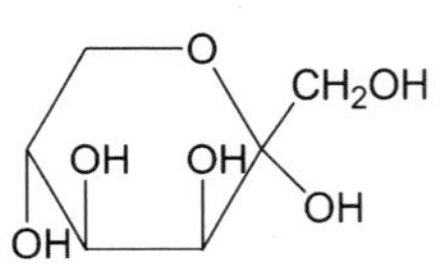

4

Tagatose is naturally found in limited quantities in dairy products and some fruits [4]. The quantity of tagatose in cow's milk was reported to be 2–3 g kg^{-1}, while in raisins, oranges, and apples, the amounts were 0.7 g kg^{-1}, 1.5 g kg^{-1}, and 3.5 g kg^{-1}, respectively [4]. Extracting the low amounts of tagatose from natural sources is cost prohibitive. Therefore, tagatose to be used as a food ingredient must be synthesized; it is manufactured either by chemical processes or biocatalytic production using enzymes [5, 6].

Interest in tagatose as a food ingredient has stemmed primarily from numerous studies that suggest it has prebiotic properties. A prebiotic substance is defined as a "nondigestible food ingredient that beneficially affects the host by selectively stimulating the growth and/or activity of one or a limited number of bacteria in the colon, and thus improves host health" [7]. In the upper gastrointestinal tract, the absorption of tagatose is relatively low [8–10]. Once in the large intestines, the microflora ferments tagatose into short-chain fatty acids, including butyrate and propionate [9–13]. Researchers demonstrated that tagatose fed to humans lowered the number of coliform bacteria while promoting higher numbers of lactic acid bacteria [9].

Because tagatose is not directly metabolized, its caloric value is less than 1.5 cal g^{-1} compared to 4 cal g^{-1} for glucose and fructose [8, 14, 15]. In addition, tagatose has a sweetness value approximately 92% that of sucrose [15]. Thus, this prebiotic has the potential to function as a reduced-calorie bulk sweetener.

Another function of tagatose in some foods is that of a flavor enhancer. Tagatose lessens the residual sweetness associated with high-intensity sweeteners in beverages [4]. Tagatose also improves the sensory characteristics of chewing gum and enhances the flavor of mint tablets [4]. However, a flavor-enhancing effect was not observed in bakery products; consumer sensory panels were unable to differentiate between control bakery products and those containing 1 or 2% tagatose [16]. However, low amounts of tagatose did not adversely affect the flavor either, enabling tagatose to be incorporated into bakery products for the prebiotic benefit [16].

Tagatose as a food ingredient has been approved by the European Union, Australia, New Zealand, South Africa, Korea, and Brazil [4]. Although considered generally recognized as safe (GRAS) in the United States, tagatose incorporation must not exceed specified limits depending upon the food product [17]. For example, tagatose may only be used up to 3% in non-diet soft drinks, 7.5% in ice cream, 10% in cookies, and 33% in ready-to-eat breakfast cereals [17].

The utilization of tagatose as a food ingredient depends upon its stability. The ingredient itself must remain physically and chemically stable prior to its incorporation into the food. Once in the food system, tagatose must remain stable to the stresses of food processing and for the duration of product storage. Aspects related to tagatose stability are addressed subsequently.

2 Stability Background

2.1 Degradation Pathways

Tagatose and other monosaccharides can proceed through a variety of degradation reaction pathways. The types and rates of these reactions are affected by moisture content and temperature exposure. If dissolved in solution, the solution's composition and pH also impact the degradation reactions.

Monosaccharide degradation, in the absence of amine-containing compounds, initiates through ionization and mutarotation [18, 19]. Structural rearrangement of the acyclic monosaccharide produces an enediol anion. This enediol progresses through additional degradative pathways, including cleavage, to ultimately produce short-chain carboxylic acids [18, 20]. Elevated pH levels [19] and temperatures [21] favor the above degradation pathway. The degradation is also accompanied by the formation of brown colors [20].

The presence of amino acids or other amine-containing compounds enables monosaccharides to participate in the Maillard reaction [18, 22, 23]. The initial step of this reaction involves an unprotonated amine, as a nucleophile, attacking the acyclic monosaccharide to produce a glycosylamine [18]. Through a series of structural rearrangements and other reactions, brown nitrogenous polymers called melanoidins are ultimately produced [22, 23]. The monosaccharides are structurally degraded during the Maillard reaction.

The literature does not contain detailed mechanistic studies specifically involving tagatose degradation. However, researchers observed a drop in solution pH during the course of the reaction, formation of brown discoloration, and generation of caramel-like aromas during tagatose stability studies [24, 25]. Based on these observations, they hypothesized that tagatose degradation follows mechanisms similar to those of other monosaccharides [24, 25].

2.2 Kinetic Modeling

Previous studies have used first-order kinetics to model monosaccharide degradation [21, 26]. The first-order kinetic model, shown in Eq. 1, has also been applied to tagatose degradation [24, 25]:

$$[\text{Tagatose}] = [\text{Tagatose}]_o e^{-kt} \tag{1}$$

In Eq. 1, [Tagatose] is the tagatose concentration at some time, t, and $[\text{Tagatose}]_o$ is the initial tagatose concentration at time zero. By regressing the degradation data to the first-order kinetic equation, the first-order rate constant, k, can be determined. Figure 1 shows an example of a first-order kinetic plot for tagatose degradation in 0.1 M phosphate buffer at pH 7 and 60 °C; the rate constant would be calculated using the slope of the regression line. Rate constants indicate the speed of tagatose loss and enable predicting the extent of its loss. Additional details regarding kinetic modeling, stability testing, and shelf life prediction appear in the literature [28].

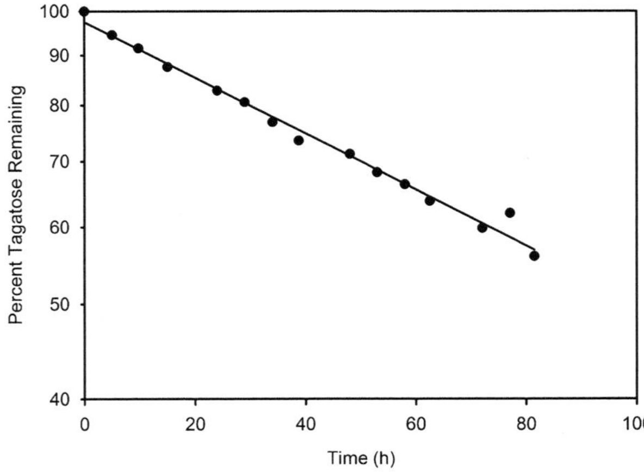

Fig. 1 First-order kinetic plot of tagatose degradation in 0.1 M phosphate buffer at pH 7 and 60 °C (Data from [27])

3 Tagatose Stability in the Solid State

Tagatose exists as a white to off-white, crystalline powder [15, 29]. Three stability concerns with this product are (1) changes to its physical properties, (2) potential chemical degradation of the tagatose, and (3) discoloration of the powder.

Powdered ingredients, especially those containing sugars, may be susceptible to caking and sticking upon exposure to humidity and heat [30]. The resulting loss of product flowability, dispersion, and dissolution hamper its utilization as an ingredient. Some ingredients may even experience a liquefaction process known as deliquescence when exposed to extreme humidities and/or temperatures. This critical relative humidity is referred to as the deliquescence relative humidity [31].

To evaluate potential physical stability issues associated with long-term storage of tagatose powder, samples were stored at various relative humidities and temperatures [29]. When tagatose was stored at 33% relative humidity and 20 °C, the powder remained free flowing for a year. However, an increase in either temperature or relative humidity resulted in the powder clumping to various extents, with some samples completely caking in as little as 1 month. At relative humidities above 85%, deliquescence of tagatose was observed. Table 1 summarizes the combined effect of temperature and relative humidity on the physical stability of tagatose powder [29].

When tagatose maintained its solid properties, no chemical degradation was observed during a year of storage [29]. However, approximately 20% of the tagatose was lost in samples held 6 months at 40 °C and 85% relative humidity due to deliquescence enhancing the molecular mobility necessary for degradation and the elevated temperature providing the required energy. Although only one sample displayed an actual loss of tagatose, measureable browning was observed in two-thirds of the samples [29]. A visualization of this browning is shown in Fig. 2.

Table 1 Physical characteristics of tagatose as affected by relative humidity (RH) and temperature after 12 months of storage and the time[a] at which final physical state was observed (Grant and Bell, Journal of Food Science, Copyright © 2012 Institute of Food Technologists. Reprinted with permission from John Wiley and Sons)

	Temperature		
Relative humidity (%)	20 °C	30 °C	40 °C
33	Free flowing (13 months)[a]	Partially caked (1 month)	Fully caked (1 month)
54	Fully caked (10 months)	Fully caked (4 months)	Fully caked (4 months)
75	Fully caked (1 month)	Fully caked (1 month)	Fully caked (1 month)
85	Fully deliquesced (6 months)	Fully deliquesced (2 months)	Fully deliquesced (1 month)

[a]Times reported include the 1-month equilibration time and the 12-month experimental storage time

Fig. 2 The effect of time and relative humidity (*RH*) on browning of tagatose during storage at 40 °C (Grant and Bell, Journal of Food Science, Copyright © 2012 Institute of Food Technologists. Reprinted with permission from John Wiley and Sons)

Depending upon the reaction conditions, browning can occur with minimal loss of tagatose.

Considering both physical and chemical stabilities, Grant and Bell concluded that optimal storage conditions for tagatose powder would be below 33% relative humidity and 20 °C [29]. Proper packaging and handling procedures to prevent moisture gain into the powder would maintain the desirable physical properties of tagatose.

4 Tagatose Stability in Solution

Ingredients incorporated into foods may break down during processing or storage. Tagatose as a powder was deemed to be fairly stable with respect to its degradation; however, upon moisture gain leading to its dissolution (i.e., deliquescence), chemical degradation was observed [29]. In solutions, the stability of tagatose to thermal processing and long-term storage depends upon the solution properties, such as the pH and interactions with other dissolved substances.

4.1 pH Effect

Tagatose in buffer solutions degraded faster at pH 7 than pH 3 [24, 25]. In 0.1 M citrate buffer at 20–40 °C, degradation occurred 5–11 times faster at pH 7 than pH 3 [24]. The results were more dramatic in 0.1 M phosphate buffer, with tagatose degradation occurring 31–58 times faster at pH 7 than pH 3 [24, 25]. The first step leading to monosaccharide degradation requires removing a proton to create an anion [18]. Because this proton removal happens more efficiently at pH 7 than pH 3, ionization and subsequent tagatose loss also occur faster in higher pH solutions.

Tagatose stability in two beverages, reduced-fat milk and diet lemonade, was evaluated from 20–81 °C [32]. Degradation rates were faster in the milk (pH 6.6) than in the lemonade (pH 2.8). For example, the time required for a 50% tagatose loss (i.e., the half-life) at 81 °C was 3.6 days in milk compared to 26 days in lemonade [32]. The pH was not the only difference between the two beverages; the milk also contained proteins which could enhance tagatose degradation due to the Maillard reaction. Both pH and compositional differences would affect tagatose degradation rates in the beverages.

The three studies all demonstrated that tagatose loss in solutions occurs more rapidly in higher pH solutions [24, 25, 32]. However, the effects of other ingredients, such as buffer salts and proteins, must also be considered.

4.2 Buffer Type and Concentration

Two studies specifically evaluated the effect of buffer type on tagatose stability [24, 25]. At pH 7, tagatose degradation proceeded faster in phosphate buffer than citrate buffer [24, 25]. For example, tagatose loss in 0.1 M buffer occurred about 6 times faster in phosphate buffer solutions as compared to citrate buffer solutions at 20–40 °C [24]. Similar observations were noted for tagatose degradation at high temperatures [25]. At pH 3, differences between tagatose degradation rates in phosphate and citrate buffers were minimal. These studies suggested a buffer species in the pH 7 phosphate buffer was responsible for its enhanced catalytic effect as compared to citrate buffer.

Buffer concentration also affects tagatose stability. Ryu et al. observed no tagatose loss in unbuffered solutions at pH 3–5 after heating for 5 h at 100 °C [33]. Two later studies noted that at pH 7 the tagatose degradation rates increased as citrate or phosphate buffer concentrations increased [24, 25]. At 20–40 °C, tagatose was, on average, 3.7 times more stable in 0.02 M phosphate buffer than 0.1 M phosphate buffer [24]. The presence of buffer salts catalyzed the tagatose degradation pathways [24, 25], whereas degradation was not detected in the absence of the catalytic buffer salts [33]. The trends were less clear at pH 3 due to extremely slow degradation rates.

Several kinetic studies have pointed to the dibasic phosphate anion (HPO_4^{-2}) as the highly catalytic buffer species that causes enhanced reactivity [34–36]. The degradation pathway of monosaccharides involves proton transfers. The small dibasic phosphate anion has proton donating and accepting groups in close proximity, which can facilitate these proton transfers more effectively than the much larger citrate anion. As buffer concentration increases, the concentration of HPO_4^{-2} also increases, causing faster tagatose degradation rates. The concentration of HPO_4^{-2} is also much higher at pH 7 than pH 3, where HPO_4^{-2} is practically nonexistent. Thus, the effect of buffer concentration on tagatose degradation is more difficult to observe at the lower pH. The stability of tagatose therefore depends upon the interaction between pH, buffer type, and buffer concentration.

Figure 3 summarizes the combined effects of pH, buffer type, and buffer concentration on tagatose stability. A larger half-life indicates greater stability. Stability is generally better at pH 3 than 7, in citrate buffer than phosphate buffer, and at lower buffer concentrations.

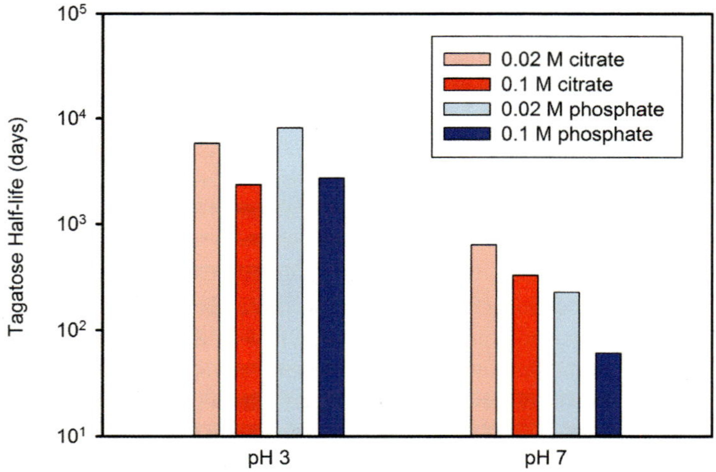

Fig. 3 Half-life of tagatose in buffer solutions at 40 °C (Data from [24])

4.3 Nonenzymatic Browning Reactions

Several types of nonenzymatic browning reactions exist that involve monosaccharides, including caramelization reactions and the Maillard reaction. Potential concerns associated with these reactions are the loss of the monosaccharide, undesirable discoloration, and the formation of undesirable aromas and flavors.

As mentioned previously, tagatose degrades in buffer solutions by intramolecular rearrangement and cleavage. A brown discoloration accompanies the degradation, which is believed to be a caramelization-type reaction [25]. Another potential tagatose degradation mechanism is through its reaction with amine-containing compounds via the Maillard reaction. This reaction is initiated by an unprotonated amine "attacking" the carbonyl group of a monosaccharide. As the reaction proceeds, the monosaccharide is lost and a brown discoloration appears.

Heating various tagatose–glycine solutions at 70–100 °C resulted in the formation of brown discoloration [33]. Tagatose–glycine solutions became increasingly dark as either pH or temperature increased. Without glycine present, tagatose loss was not observed from pH 3–5 solutions heated at 100 °C for 5 h. The researchers concluded this browning came from tagatose and glycine reacting through the Maillard reaction [33]. Unfortunately, tagatose concentrations were not measured in the glycine-containing solutions; therefore, the extent of tagatose loss is unknown.

Tagatose degradation and brown pigment formation were evaluated in buffer solutions stored at 20–40 °C with and without glycine [37]. Data collected were consistent with that presented by Ryu et al. [33] where the glycine-containing solutions showed enhanced browning. However, tagatose degradation rates were not enhanced, but were similar between the solutions with and without glycine [37]. In another study, tagatose solutions with and without glycine were stored at 60–80 °C [27]. Glycine again promoted enhanced browning, but like Dobbs [37], the presence of glycine again had only minimal effects on the degradation rates of tagatose. The larger effect of glycine on browning rates than tagatose degradation rates in pH 7 buffer solutions is shown in Fig. 4.

The presence of glycine in a tagatose solution would have been expected to increase tagatose loss due to an additional reaction pathway (i.e., the Maillard reaction) beyond that encountered in the absence of glycine. However, two stability studies conducted over different temperature ranges failed to show enhanced tagatose loss associated with the Maillard reaction [27, 37]. Similarly, fructose in a pH 7 phosphate buffer solution also did not show enhanced degradation in the presence of lysine [38].

The two potential degradation mechanisms, the intramolecular rearrangement of tagatose and the Maillard reaction, are both initiated by mutarotation producing very low concentrations of acyclic tagatose, which rearranges into an enediol. The enediol can then proceed through a series of reactions to form small organic acids, isomerize into an aldose (which can then react reversibly with glycine through the Maillard reaction), or revert back to the ketose (which may also react reversibly with glycine). An undetermined rate-limiting step may exist in the early transformation pathways

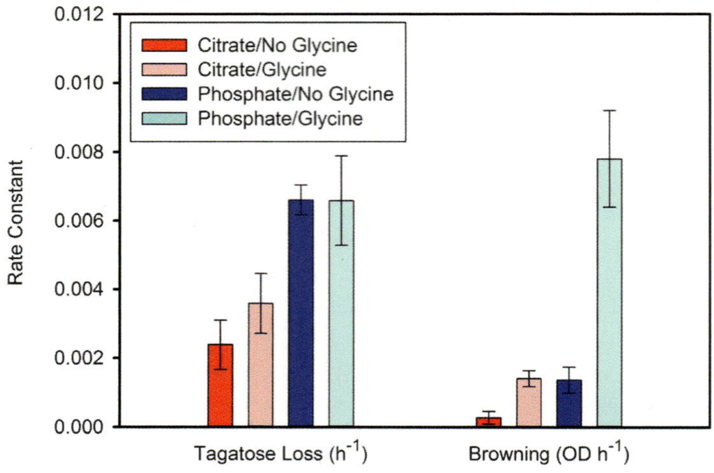

Fig. 4 Effect of glycine on the rate constants for tagatose loss and browning in buffer solutions at pH 7 and 60 °C (Data from [27])

of acyclic tagatose such that the tagatose degradation rate does not depend upon faster reactions occurring subsequently, such as the Maillard reaction. The data in Fig. 4, where rates of tagatose loss are similar with and without the amino acid, but browning is more pronounced, support this hypothesis.

As noted previously, tagatose was lost faster in reduced-fat milk than diet lemonade at 20–80 °C [32]. The tagatose-containing milk also experienced major browning during the experiments which was not observed in the lemonade [1]. Thus, the Maillard reaction was likely occurring between tagatose and the dairy proteins. However, whether the Maillard reaction contributed to tagatose degradation in milk cannot be definitively concluded.

A more recent study evaluated the discoloration of tagatose solutions at 100 °C as affected by various added solutes [39]. In unbuffered solutions, tagatose browning increased as pH increased from pH 3 to 9, which is consistent with previously published data that also showed browning increased as pH increased [25, 33]. This nonenzymatic browning reaction was not attributed to the Maillard reaction because amino acids were not present in the solution. Browning of tagatose solutions was enhanced by the presence of citric acid and lactic acid [39]. Luecke and Bell [25] speculated that tagatose browning reactions were acid–base catalyzed, which is supported by the faster browning in the presence of the organic acids in comparison to a blank [39]. Sodium bisulfite and potassium metabisulfite displayed anti-browning effects [39]. Sulfites are well known for their food preservation properties, including the inhibition of browning reactions [40]. Figure 5 compares the effects of some of the solutes on the development of browning from tagatose. Unfortunately, the researchers did not measure tagatose concentrations along with the browning.

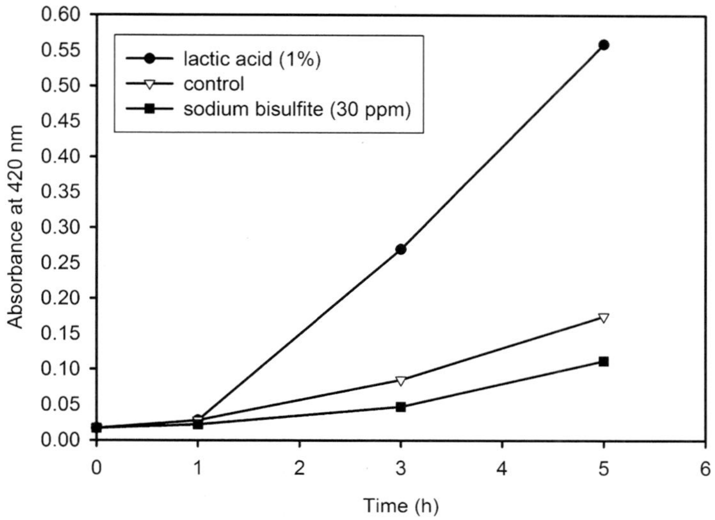

Fig. 5 Effect of added solutes on the browning of 1 M tagatose aqueous solutions at 100 °C (Adapted from [39])

The extent of tagatose loss as affected by the added solutes remains unknown. However, it is clear that tagatose browning in solution was impacted by other solutes.

4.4 Temperature Effects

In addition to the pH, buffer type and concentration, amino acids, and other solutes, tagatose degradation in solution is significantly affected by processing and storage temperatures. The activation energy is the key parameter that describes the temperature sensitivity of the tagatose degradation rates [28]. Larger activation energies indicate the reaction has a greater sensitivity to changes in temperature. The activation energy is also vital for enabling shelf life predictions [28]. In order to predict the extent of tagatose loss and consequently the product shelf life, tagatose degradation data must be collected over several temperatures. The data is then modeled using the Arrhenius equation:

$$k = Ae^{-Ea/RT} \quad (2)$$

where k is the rate constant, A is the pre-exponential factor, Ea is the activation energy, R is the ideal gas constant, and T is temperature in Kelvin.

The effects of temperature on tagatose degradation kinetics were presented by three studies [24, 25, 32]. The tagatose degradation rates were extremely low in acidic buffer solutions, preventing the determination of reliable activation energies. Using tagatose stability data collected from pH 7 buffer solutions at 20, 30, and 40 °C, activation energies for tagatose degradation ranged from 12.6 to 16.4 kcal mol^{-1} [24]. A second study evaluated the thermal stability of tagatose in pH 7 buffer solutions held at 60, 70, and 80 °C and found activation energies ranging from 31.4 to 38.7 kcal mol^{-1} [25]. By combining the two related data sets, the reliability of the regressed data can be improved. Bell presented the combined data for tagatose degradation in 0.1 M phosphate buffer and citrate buffer at pH 7 in the form of Arrhenius plots and calculated the activation energies over the 20–80 °C temperature range [1]. In citrate buffer, activation energies were 26.2–28.3 kcal mol^{-1}, while in phosphate buffer they were 23.3–24.4 kcal mol^{-1} [1]. Tagatose degradation in phosphate buffer has a lower activation energy than in citrate buffer because of the catalytic ability of the dibasic phosphate anion, as mentioned earlier. Catalysts function to lower the activation energies of a reaction. The activation energy for tagatose degradation in reduced-fat milk at pH 6.6 was 24.6 kcal mol^{-1} [32], which is similar to that from the combined data presented by Bell [1].

Using the degradation data and activation energies, the extent of tagatose degradation can be predicted under various time/temperature combinations. Table 2 lists the amount of tagatose lost in various solutions and beverages during various types of processing and storage conditions.

Pasteurization (basic, HTST, UHT) of tagatose-containing solutions would result in losses of less than 1%. If sterilization was required for a beverage, a larger amount

Table 2 Percent tagatose loss from various solutions during processing and storage. Values are calculated based on previously published data [24, 25, 32]

	Processing			Storage	
Sample type	63 °C 30 min[a]	72 °C 15 s[b]	135 °C 5 s[c]	4 °C 180 day	25 °C 180 day
Phosphate buffer					
0.02 M pH 3	<1	<1	n.d.[d]	<1	1–2
0.1 M pH 3	<1	<1	n.d.	<1	1–2
0.02 M pH 7	<1	<1	<1	<1	12
0.1 M pH 7	<1	<1	<1	2	36
Citrate buffer					
0.02 M pH 3	<1	<1	n.d.	<1	1–2
0.1 M pH 3	<1	<1	n.d.	<1	1–2
0.02 M pH 7	<1	<1	<1	<1	4
0.1 M pH 7	<1	<1	<1	<1	7
Beverages					
Milk	<1	<1	<1	<1	4
Diet lemonade	<1	<1	n.d.	<1	<1

[a]Basic pasteurization
[b]High-temperature short time (HTST) pasteurization
[c]Ultrahigh-temperature (UHT) pasteurization
[d]*n.d.* not determined

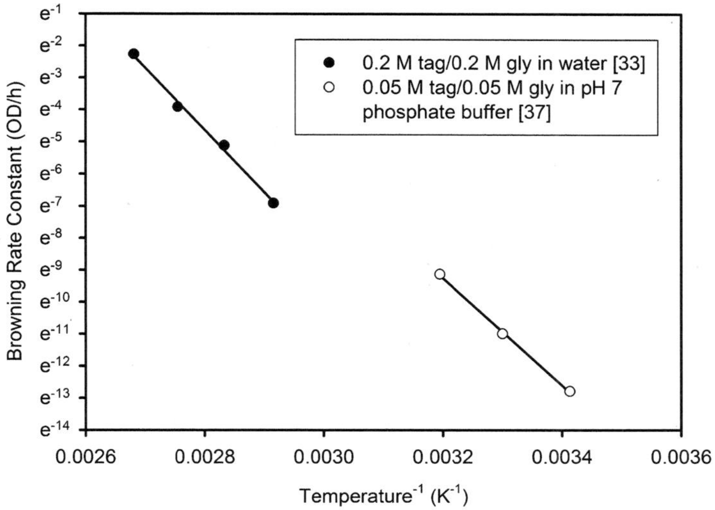

Fig. 6 Rate constants for Maillard browning between tagatose and glycine as affected by temperature depicted as an Arrhenius plot (Data from [33, 38])

of tagatose would be lost. During 6 months of refrigerated storage, beverages would lose very little tagatose. However, in a shelf-stable beverage held 6 months at 25 °C, larger amounts of tagatose would be lost, especially at pH 7. Table 2 also shows the improvement in tagatose retention by lowering pH, lowering buffer concentration, and using citrate buffer instead of phosphate buffer.

Researchers also noted that the extent of browning (both Maillard and non-Maillard) increased with increasing temperature [24, 25, 33, 37, 39]. Figure 6 shows the effect of temperature on the rate of Maillard browning between glycine and tagatose in the form of an Arrhenius plot [33, 37]. The activation energy for brown pigment formation, based on data from [33], was determined to be 38.5 kcal mol^{-1}. Similar results were noted at lower temperatures as well [37], where the activation energy was 33.5 kcal mol^{-1}. In the absence of glycine, non-Maillard browning involving tagatose also increased as temperature increased [24, 25, 39]. Maillard browning reactions involving tagatose, based on the higher activation energy, are more temperature sensitive than the tagatose degradation reactions.

5 Conclusions

Tagatose has great potential as a bulk sweetening agent with beneficial prebiotic properties. Sticking, clumping, and discoloration of tagatose powder are possible during storage at abusive temperatures and relative humidities. In most food systems, tagatose degradation is not expected to be significant, either during storage or processing. Both pH and ingredient interactions would affect the extent of tagatose loss. Although tagatose loss may be minimal, noticeable browning has been observed in tagatose-containing beverages and solutions. Food discoloration may be the limiting factor related to loss of shelf life.

References

1. Bell LN (2015) Tagatose stability in beverages as impacted by composition and thermal processing. In: Preedy VR (ed) Processing and impact on active components in food. Academic, London
2. Angyal SJ, Bethell GS (1976) Conformational analysis in carbohydrate chemistry. III the 13-C N.M.R. spectra of the hexuloses. Aust J Chem 29:1249–1265
3. Wolff GJ, Breitmaier E (1979) 13-C-NMR-spektroskopische bestimmung der keto-form in wäßrigen lösungen der D-fructose, L-sorbose und D-tagatose. Chem Z 103:232–233
4. Skytte UP (2006) Tagatose. In: Mitchell H (ed) Sweeteners and sugar alternatives in food technology. Blackwell Publishing, Ames
5. Jayamuthunagai J, Gautan P, Srisowmeya G, Chakravarthy M (2016) Biocatalytic production of D-tagatose: a potential rare sugar with versatile applications. Crit Rev Food Sci Nutr. doi:10.1080/10408398.2015.1126550
6. Vastenavond CM, Bertelsen H, Hansen SJ, Laursen RS, Saunders J, Eriknauer K (2011) Tagatose (D-tagatose). In: O'Brien-Nabors L (ed) Alternative sweeteners, 4th edn. CRC, Baton Rouge
7. Gibson GF, Roberfroid MB (1995) Dietary modulation of the human colonic microbiota: introducing the concept of prebiotics. J Nutr 125:1401–1412
8. Livesey G, Brown JC (1996) D-tagatose is a bulk sweetener with zero energy determined in rats. J Nutr 126:1601–1609
9. Bertelsen H, Jensen BB, Buemann B (1999) D-tagatose – a novel low-calorie bulk sweetener with prebiotic properties. World Rev Nutr Diet 85:98–109
10. Laerke HN, Jensen BB (1999) D-tagatose has low small intestinal digestibility but high large intestinal fermentability in pigs. J Nutr 129:1002–1009
11. Laerke HN, Jensen BB, Hojsgaard S (2000) In vitro fermentation pattern of D-tagatose is affected by adaptation of the microbiota from the gastrointestinal tract of pigs. J Nutr 130:1772–1779
12. Bertelsen H, Andersen H, Tvede M (2001) Fermentation of D-tagatose by human intestinal bacteria and dairy lactic acid bacteria. Microb Ecol Health Dis 13:87–95
13. Venema K, Vermunt SHF, Brink EJ (2005) D-tagatose increases butyrate production by the colonic microbiota in healthy men and women. Microb Ecol Health Dis 17:47–57
14. Levin GV, Zehner LR, Saunders JP, Beadle JR (1995) Sugar substitutes: their energy values, bulk characteristics, and potential health effects. Am J Clin Nutr 62(Suppl):1161S–1168S
15. Levin GV (2002) Tagatose, the new GRAS sweetener and health product. J Med Food 5:23–36
16. Armstrong LM, Luecke KJ, Bell LN (2009) Consumer evaluation of bakery product flavor as affected by incorporating the prebiotic tagatose. Int J Food Sci Techol 44:815–819
17. Cheeseman MA (2011) Agency response letter GRAS Notice No. GRN 000352. U.S. Food and Drug Administration. http://www.fda.gov/food/ingredientspackaginglabeling/gras/noticeinventory/ucm245241.htm. Accessed 28 Mar 2016
18. Wong DWS (1989) Mechanism and theory in food chemistry. Van Nostrand Reinhold, New York
19. Robyt JF (1998) Essentials of carbohydrate chemistry. Springer, New York
20. De Bruijn JM, Kieboom APG, van Bekkum H, van der Poel PW (1986) Reactions of monosaccharides in aqueous alkaline solutions. Sugar Technol Rev 13:21–52
21. Wolfenden R, Yuan Y (2008) Rates of spontaneous cleavage of glucose, fructose, sucrose, and trehalose in water, and the catalytic proficiencies of invertase and trehalas. J Am Chem Soc

130:7548–7549. Supplemental information available at http://pubs.acs.org/doi/suppl/ 10.1021/ja802206s. Accessed 28 Mar 2016
22. Hodge JE (1953) Chemistry of browning reactions in model systems. J Agric Food Chem 1:928–943
23. Friedman M (1996) Food browning and its prevention: an overview. J Agric Food Chem 44:631–653
24. Dobbs CM, Bell LN (2010) Storage stability of tagatose in buffer solutions of various composition. Food Res Int 43:382–386
25. Luecke KJ, Bell LN (2010) Thermal stability of tagatose in solution. J Food Sci 75(4): C346–C351
26. Bamford CH, Bamford D, Collins JR (1950) Kinetic studies on carbohydrates in alkaline conditions. II The kinetics of the rearrangements of glucose and fructose in alkaline solution. Proc Royal Soc London Ser A Math Phys Sci 204(1076):85–98
27. Luecke KJ (2009) Thermal stability of tagatose in solution. Thesis, Auburn University
28. Bell LN (2007) Nutraceutical stability concerns and shelf life testing. In: Wildman REC (ed) Handbook of nutraceuticals and functional foods, 2nd edn. CRC, Boca Raton
29. Grant LD, Bell LN (2012) Physical and chemical stability of tagatose powder. J Food Sci 77(3): C308–C313
30. Adhikari B, Howes T, Bhandari BR, Truong V (2001) Stickiness in foods: a review of mechanisms and test methods. Int J Food Prop 4:1–33
31. Mauer LJ, Taylor LS (2010) Water-solids interactions: deliquescence. Annu Rev Food Sci Technol 1:41–63
32. Bell LN, Luecke KJ (2012) Tagatose stability in milk and diet lemonade. J Food Sci 71(1): H36–H39
33. Ryu SY, Roh HJ, Noh BS, Kim SY, Oh DK, Lee WJ, Yoon JR, Kim SS (2003) Effects of temperature and pH on the non-enzymatic browning reaction of tagatose-glycine model system. Food Sci Biotechnol 12:675–679
34. Bell LN, Wetzel CR (1995) Aspartame degradation in solution as impacted by buffer type and concentration. J Agric Food Chem 43:2608–2612
35. Bell LN (1997) Maillard reaction as influenced by buffer type and concentration. Food Chem 59:143–147
36. Pachapurkar D, Bell LN (2005) Kinetics of thiamin degradation in solutions under ambient storage conditions. J Food Sci 70:C423–C426
37. Dobbs CM (2008) Storage stability of tagatose in buffer solutions of various composition. Thesis, Auburn University
38. Ajandouz EH, Tchiakpe LS, Ore FD, Benajiba A, Puigserver A (2001) Effects of pH on caramelization and Maillard reaction kinetics in fructose-lysine model systems. J Food Sci 66:926–931
39. Kwon SY, Baek HH (2014) Effects of temperature, pH, organic acids, and sulfites on tagatose browning in solutions during processing and storage. Food Sci Biotechnol 23:677–684
40. Lindsay RC (2008) Food additives. In: Damodaran S, Parkin KL, Fennema OR (eds) Fennema's food chemistry, 4th edn. CRC, Boca Raton

Low Calorie Intense Sweeteners: Safety Aspects

26

Qurrat ul-Ain, Madiha Sikander, Sohaib Ahmed Khan, Muhammad Sikander Ghayas Khan, and Rabia Ghayas

Abstract

The safety of low calorie sweeteners (LCS) is always the subject of question. Artificial sweeteners or intense sweeteners are sugar substitutes that are utilized as a different option for common sugar. They are commonly sweeter than normal sugar and as they contain no calories, they can be utilized aptly to control weight and obesity. Because of this, they are presently utilized by a large number of individuals worldwide without knowing their potential hazards on the body. Broad investigative exploration has shown the safety of the six low-calorie sweeteners at present endorsed for use in foods in the U.S and Europe (stevia, acesulfame-K, aspartame, neotame, saccharin and sucralose) each with a satisfactory status. This chapter means to cover the health concerns, highlighting dangers of consuming artificial sweetener. Artificial sweeteners are added in food, beverages, medications and hygiene items. The principle explanation

Q. ul-Ain (✉)
Shalamar Hospital, Lahore, Pakistan

Dow Medical College DUHS, Civil Hospital Karachi, Karachi, Pakistan
e-mail: qurrat.dumhs@gmail.com

M. Sikander
University of Lahore, Lahore, Pakistan
e-mail: dr.madihakhan83@gmail.com

S.A. Khan
Civil Hospital, Karachi, Pakistan
e-mail: drsohaib136@gmail.com

M.S.G. Khan
Riphah International University, Lahore campus, Lahore, Pakistan
e-mail: sikandar.khandr@riphah.edu.pk

R. Ghayas
Arif Memorial Teaching Hospital, Lahore, Pakistan
e-mail: dr.ghayasrabia@gmail.com

J.-M. Mérillon, K.G. Ramawat (eds.), *Sweeteners*, Reference Series in Phytochemistry,
https://doi.org/10.1007/978-3-319-27027-2_28

behind their overall use is that they don't raise blood glucose levels, used to control weight and to treat hypoglycemia. Apart from these advantages, artificial sweetener can affect the body in several ways for instance leukemia, lymphoma and others which is supported by researches. Use of artificial sweeteners also contributes in obesity as well as in diabetes mellitus type 2. To control the above mentioned complications usage of artificial sweetener must be implied to certain limitation and use of natural sweeteners must be encouraged.

Keywords
Artificial sweetener • Low calorie sweetener • Aspartame • Saccharin • Non nutritional sweetener • Intense sweeteners

Abbreviations

Ace-K	Asculfame potassium
ADI	Acceptable daily intake
AS	Artificial sweetener
ATP	Adenosine tri phosphate
DNA	Deoxy Ribo Nucleic acid
E-Number	European number
EU	European Union
FDA	Food and Drug Administration
JECFA	Joint Expert Committee on Food Additives
LCS	Low calorie sweetener
MUD	Maximum usable dose
NNs	Non Nutritional sweeteners
NOAEL	No observed adverse effect level
pKa	Acid dissociation constant
PKU	Phenyl ketone urea
SCF	Scientific committee on food
US	United States

Contents

1 Introduction 593
2 History of Low Calorie Intense Sweeteners Discovery 594
3 Criteria for Safety Evaluation 595
3.1 Toxicological Testing 595
3.2 Risk Assessment and Acceptable Daily Intake 596
3.3 Re-Evaluation of Safety of Sweeteners 596
3.4 Risk Assessment of LCS 597
4 Chemical and Physical Properties of FDA Approved Intense Sweeteners 597
4.1 Acesulfame Potassium (E950) 597
4.2 Aspartame (951) 599
4.3 Cyclamate E952 601
4.4 Saccharin (E954) 602
4.5 Sucralose (E955) 602
4.6 Neotame E961 604

5 Sweetener Metabolism 605
5.1 Metabolism of Artificial Sweeteners 605
6 Safety Aspects of Intense Sweeteners 606
6.1 Gastrointestinal Tract Disturbance 606
6.2 Obesity 607
6.3 Diabetes Mellitus Type 2 607
6.4 Coronary Heart Disease 607
6.5 Brain Damage 608
6.6 Phenylketonuria 608
6.7 Early Menarche 608
6.8 Cancer 608
6.9 Preterm Delivery 609
7 Conclusion 609
8 Cross-References 610
References 610

1 Introduction

Sweeteners are food additives that are used to improve the taste of everyday foods. Natural sweeteners are sweet-tasting compounds with some nutritional value the major ingredient of natural sweeteners is either mono- or disaccharides. Artificial sweeteners, on the other hand, are compounds that have very little or no nutritional value. This is possible because artificial sweeteners are synthesized compounds that have high-intensities of sweetness, meaning less of the compound is necessary to achieve the same amount of sweetness.

High intensity, low calorie sweeteners is an important class of sweeteners. This class comprise of compounds which are sweeter than common table sugar. So in result, small quantities are required to produce the same intensity as that of sugar and energy input is often low or negligible. Sensation of sweet taste comes after artificial sweetener is comparatively different than that of table sugar. Therefore to achieve the natural essence of sugar various blends of ASs are used in food industries. The beverages containing AS which are labeled as light or diet have different mouth feel. In Europe and United States six AS have been approved for use yet [1] (Fig. 1).

Intense sweeteners, as their name infers, are commonly sweeter than sugar. For instance saccharin, aspartame and acesulfame K (pro K) which are up to 200 times sweeter than sucrose. Sucralose and neotame have been more as of latest endorsed and are around 600 and 7,000 times sweeter than sugar correspondingly. Artificial sweeteners are regularly used to give sweetness in an extensive variety of food items including refreshments, dairy items, confectioneries, pastries and ice cream. Some artificial sweeteners are metabolized during digestion e.g., aspartame while others are not metabolized and are discharged unchanged e.g., sucralose.

Low calorie sweeteners are utilized in variety of food and beverage items, especially in the manufacturing of low calorie forms of foods and beverages. Though getting good taste by keeping its shelf life long is not an easy and straight away job. Supplanting sugar with low calorie sweeteners in foods shows some specific challenges, as sugar has various practical attributes providing sweet taste. Frequently,

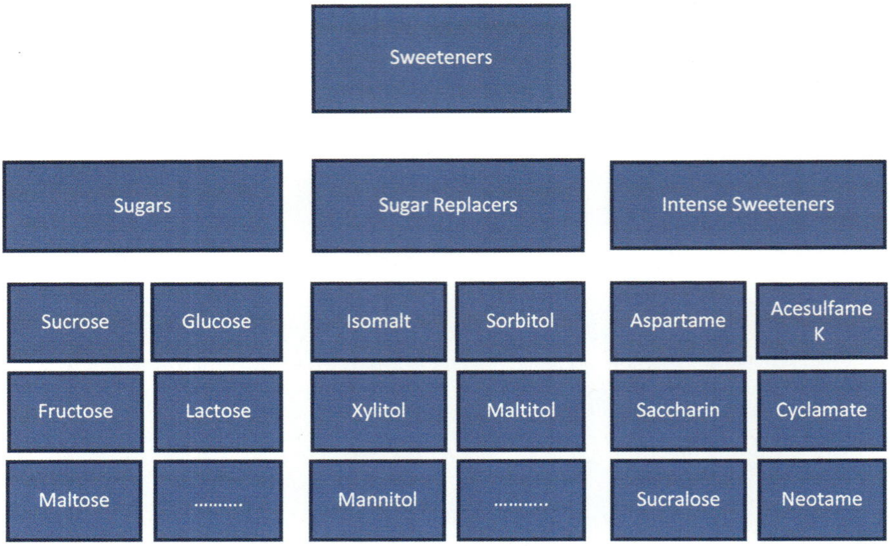

Fig. 1 Brief classification of sweeteners

blends of bulk and intense sweeteners are required to coordinate the taste and texture of sugar-sweetened items. With a comprehension of the specialized qualities of various sweeteners, it is conceivable to create lower calorie items that still taste great and give more customer choice [2].

2 History of Low Calorie Intense Sweeteners Discovery

The novel artificial sweetener, saccharin, was at first created to address sugar deficiencies in the Second World War. Saccharin was found at first by the physicist Ira Remsen in 1878 at Johns Hopkins University. Though, it was inadvertently re-found in 1879 by his post-doctoral student, Constantine Fahlberg, while dealing with derivatives of coal tar items. Fahlberg's native origin is itself not as much as clear, being referenced as American, Russian and German-American in various sources. Remsen was not inspired by commercializing saccharin, but rather Fahlberg filed a patent and guaranteed to be the sole pioneer, much to the dismay of Remsen. Through the end of the 1800s and start of the 1900s, utilization of saccharin started to develop raising concerns by Harvey Wiley, the principal driver of the 1906 Pure Food and Drug Act and organizer of the Food and Drug Administration (FDA). Wiley scrutinized the security of saccharin and in 1908 proposed banning its utilization in the food supply. He spoke to then president Theodore Roosevelt to bolster a boycott. Though, Roosevelt was not persuaded of the detrimental effects and is notably reported as expressing, "Any individual who says saccharin is damaging to health is an idiot. Dr. Rixey offers it to me consistently" (supposedly as a way to deal with moderate his weight). In spite of broad audit by different administrative bodies and leeway by every one of them, inquiries concerning the security of LCS continue [3, 4].

3 Criteria for Safety Evaluation

Food added substances that have been assessed and thought to be safe for utilization by people under the proposed states of utilization are assigned an "E-number" (E = European). In spite of the fact that this is intended to bring a surety for safe use, E-numbers are once in a while wrongly seen as speaking to non-natural constituents or even a danger to human health despite the fact that it is feasible for a food added substance to be of common starting point or be natural origin or identical to nature [5].

Regarding LCS manufacturers, LCS use, maximum usable dose (MUD) levels for every sweetener have been determined for important food and refreshment items which should not be surpassed. The factual levels utilized in items, nonetheless, won't generally equate to the MUD as this will likewise be impacted by extra variables, for example, the inborn sweetening properties of the sweetener and the coveted taste of the final product. Moreover, it is feasible for a few LCS to be utilized as a part of blend as they have been appeared to work synergistically to upgrade the sweetness power or veil undesirable persistent flavors. All the time, this will bring about a lower level of LCS being utilized as a part of items than would be required in the event that they were being utilized separately [6].

Wide range database is required to lead a safety assessment for administrative purposes of any substance as a food added substance. It ought to incorporate results from studies on ingestion, circulation and digestion system in testing animals and human subjects, in vitro and in vivo toxicological testing, regulatory information, specialized information in connection to identity, purity, strength and potential breakdown items, production process, technological requirement, worth to customers, proposed applications, levels of utilization in various food classifications and evaluated contact in result of the proposed use [7, 8].

3.1 Toxicological Testing

The point of toxicological testing is to figure out if the substance, when utilized in various ways and amounts proposed, would represent any apparent danger to the health of the normal consumer and to those whose food utilization pattern, physiological or health status might make them susceptible, e.g., pregnancy, young age or diabetes. The toxicological testing project of food added substances incorporates fundamental and supplementary studies. The annex to direction on compliance for food added substance assessments by the SCF gives contemplations basic the central toxicological prerequisites and examinations on the degree and utilization of different studies and relevance [9].

3.2 Risk Assessment and Acceptable Daily Intake

The hazard evaluation of sweeteners is performed taking after a general system for risk appraisal of chemicals in food, an investigative procedure that requires aptitude in toxicology and nutrition (for the intake evaluation). The strategy comprises of four stages: hazard identification, hazard characterization, exposure assessment and risk characterization. As the consequence of risk interpretation acceptable daily intake (ADI) has been built up for every sweetener. The ADI is the measure of the food added substance, represented as mg/kg body weight, that can be ingested day by day over a lifetime without causing any calculable health hazard. Artificial sweeteners that are as of now all owed for food use in the EU have been allotted a numerical ADI, with one special case. Bulk sweeteners that are presently allowed for food use in the EU were observed to be "acceptable" by the SCF. This shows the normal presentation to the substance, emerging from its utilization or utilizations in food at the level important to accomplish the preferred technological impact as known at the period of assessment, does not express to a hazard to human health [9].

3.3 Re-Evaluation of Safety of Sweeteners

There is at present no procurement for occasional audits of the safety of allowed food added substances. Nonetheless, the safety evaluation of sweeteners (and other food added substances) depends on the learning and information accessible at the season of evaluation. At the point when new toxicology information get to be accessible in the documented literature, national specialists and worldwide expert boards of trustees consider them with alert and might embrace a re-assessment of their safety. Reliant upon the result, three situations are conceivable: the ADI can be managed or changed, or the utilization of a sweetener as a food added substance can be observed to be inadmissible. The SCF distributed its first opinion on sweeteners in 1985. The main re-assessment of the security of a percentage of the sweeteners was directed in. From that point forward, the safety of a few sweeteners has been re-assessed on a few events. Nordic specialists talked about the security features of sweeteners in 1989 [10].

In 1999 the Nordic Working Group on Food Toxicology and Risk Assessment chose to inspect whether the safety assessments, which shaped the premise for states of utilization, were still substantial and sufficient in the light of standards for security evaluation sat that time, and whether critical new toxicological studies had been documented subsequent to past assessments. Thus, a report covering all food added substances allowed in EU by April 2000 was submitted to the Nordic Council of Ministers. This report contains monographs of 12 sweeteners. The second part of every monograph is a concentrate of the background information as reported by the SCF and JECFA, supplemented by a short depiction of the central studies for their evaluation of the sweetener and also important studies on that compound documented after the assessments by the SCF and JECFA [11].

3.4 Risk Assessment of LCS

The danger evaluation of LCS contains four stages:

1. Hazard identification: recognizes the antagonistic health impacts connected to the substance being referred to. For this reason, investigative information from experiences from human exposures, concentrates on in vitro studies are required.
2. Hazard characterization: prompts determination of the basic information set in which the critical adverse impact is recognized. This information set is utilized to build up the dosage response effect for the substance. On the other hand, if the information exhibit that the substance is non-genotoxic the "no observed adverse effect level" (NOAEL) is resolved from the sensitive study in the most delicate species tried. The acceptable daily intake (ADI) is built up from the NOAEL by partitioning it by a security component, which considers species contrasts in between of people and test animals, and variety within human subjects.
3. Exposure evaluation: in view of data in regards to the levels of a substance proposed for use in various foodstuff sand estimations of the intake of the applicable foodstuffs in the nation or area being referred to. The objective is to decide exposure to the substance (by means of intake of food items) in the population as such considered and in special population groups, and in people (maximum/minimum, daily/over time). Data on food utilization might be gotten from food supply information, family unit reviews, singular dietary overviews, total diet studies and/or biomarkers. The figures are made accessible for the hazard characterization process.
4. Risk assessment: coordinates data from exposure appraisal and risk assessment into advice suitable for use in choice making or hazard administration. The conclusions might be that the anticipated/present exposure is secure by set up ADI or that declines in exposure are required to meet the ADI standards [12, 13].

4 Chemical and Physical Properties of FDA Approved Intense Sweeteners

Under this heading various properties of artificial sweeteners which are approved by FDA will be discussed (Table 1).

4.1 Acesulfame Potassium (E950)

Acesulfame potassium is a calorie-free artificial sweetener usually known as Sunett and Sweet One. It was found by a German scientific expert, Karl Clauss, in 1967. It has a white crystalline structure and is around 180–200 times sweeter than sucrose. At high dosage, it has a tendency to have a persistent bitter flavor however less so as lower quantity. Kraft Foods licensed the utilization of sodium ferulate to cover this bitter sensation. The US FDA affirmed the utilization of Ace-K alongside the Kraft

Table 1 Characteristics of 5 FDA Approved Artificial Sweeteners

Artificial sweetener	Year of FDA approval	Year of discovery	Times sweeter than sugar	Acceptable daily intake	Calories KCAL/KG	Pregnancy status	Association with cancer	Side effects	Used for baking	Brand name
Aspertame	1981	1965	200	50 mg/kg	4	Approved	No	Headache, dizziness, mood changes, skin reactions	No	Nutra sweet, equal, sugar twin
Saccharin	2000	1879	200–700	5 mg/kg	0	Not approved	Listed as anticipated carcinogen in 1981, removed in 2000	Irritability, Muscle Dysfunction, Crossess Placenta	Yes	Sweet N low, sweet twin, necta sweet
Acesulfame-K	1998	1967	200	15 mg/kg	0	Approved	No	Bitter taste	Yes	Ack, sunett, sweet N safe, sweet one
Neotame	2002		7,000–13,000	18 mg/kg	0	Approved	No	As toxic as aspertame. Both substances break to methanol	Yes	Neotame
Sucralose	1998	1976	500	5 mg/kg	0	Approved	No	Skin rashes, diarrhea, muscle aches, abdominal cramping, headaches, bladder issues	Yes	Splenda

FDA Food and drug administration

Food protected rendition in 2003. Ace-K, not at all like its well-known opponent sweetener aspartame, is stable in high-warm circumstances and is in this manner frequently utilized as a part of prepared items. Presently it can be found in numerous tabletop sweeteners, pastries, puddings, bakery products, soda pops, confections (counting breath mints, hack drops and lozenges), dairy items, canned foods and alcoholic beverages. Ace K has unique long time shelf life and due to this is perfect for use in confections, canned foods and alcoholic drinks.

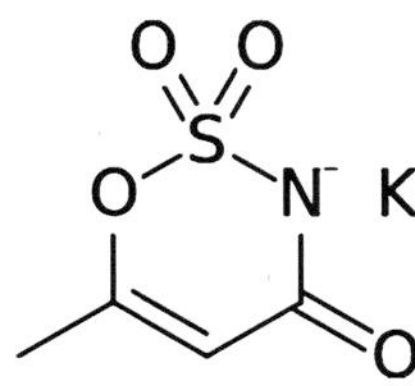

IUPAC name: potassium 6-methyl-2,2-dioxo-2H-1,2λ6,3-oxathiazin-4-olate.

Common name: Acesulfame potassium (E950)

Ace-K is a white crystalline powder with molecular formula C4H4KNO4S, molecular weight of 201.24 g/mol, and a density of 1.81 g/cm^3. It has a structural resemblance to that of saccharin. Acesulfame K is obtained by chemical synthesis and purified through recrystallization. The compound is freely soluble in water and very slightly soluble in ethanol, and stable in the high temperature, pH and time ranges. No by-products can be found in beverages under normal usage and storage conditions. Its stability and solubility in water facilitate its use in foodstuffs. Another critical part of Ace-K is its capacity to stay stable and hold its sweetness under purifying conditions which frequently opens dairy items to a wide variety of temperatures and pH values. Ace- K does not participate in tooth rot making therefore it is a perfect possibility for "sugar free" confections and light or diet drinks. Considering Ace-K's intense bitter prolong flavor impression, it is regularly utilized along with another sweetener as a part of general items. Ordinarily aspartame or sucralose are utilized to adjust the bitter aftertaste impression. Regular items which contain exclusively Ace-K or a mix of Ace-K and another sweetener are: Diet Rite Cola, Pepsi One/Pepsi Max, Coca-Cola Zero, Diet Coke with Splenda, Trident gum, and without sugar Jell-O [14, 15].

4.2 Aspartame (951)

Aspartame was found unintentionally by scientific expert James M. Schlatter in 1965 when he licked his finger which happened to have become sullied by a compound he composed while attempting to produce anti-ulcer medication. Aspartame was not promptly affirmed by the US FDA at the point when an extensive study demonstrated an immediate association among aspartame and bladder cancer in rats. In 1980 there were no further certain studies connecting aspartame to brain damage or cancer, furthermore, it was then affirmed as a general sweetener. In 1983 it was further affirmed for use in carbonated drinks then further again in 1996 when it was took into account use in different drinks, bakery products and desserts. In 1996, the FDA expelled all confinements from aspartame use. Between its disclosure and today, aspartame has gotten to be a standout amongst the most studied artificial sweeteners worldwide. Aspartame has basically no bitter prolonged flavor impression permitting it to be utilized as a part of numerous items as a sugar elective. It is a non-nutritive sweetener which makes it exceptionally prominent among individuals hoping to watch calories, stay in better general health or basically enjoy a hefty portion of the low-or lessened calories items accessible today. A few parts of aspartame make it an exceptionally alluring sweetener in natural product seasoned items, particularly gum, since aspartame has a capacity to "amplify" flavors making them appear to be sweeter and given them above all full-bodied taste. Non-significant part of aspartame is its low working temperature. Since it loses sweetness in high-heat or high-pressure circumstances, it is fairly constrained for use in bakery products. This issue can be resolved by utilizing extra sweeteners as a part of the initial preparing process of bakery products, for example, Ace- K or sucralose then including aspartame at a later stage. Likewise, aspartame can be utilized for heating for its flavor while another sweetener can be utilized for its sweetness, (for example, Ace-K).

Aspartame is frequently used to protect or strengthen taste in an item however normally depends on another source of sweetness if the item is to boiled or baked. Aspartame can be found in more than 6,000 items, including carbonated soda pops, powdered soda pops, chewing gum, sugary confectionaries, gelatins, dessert blends, puddings and fillings, frozen desserts, yogurt, tabletop sweeteners, and a few pharmaceuticals, for example, vitamins and sugar free cough syrups..

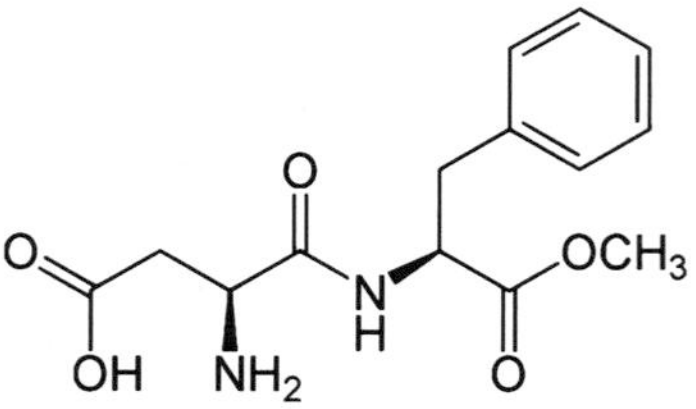

IUPAC name: Methyl L-α-aspartyl-L-phenylalaninate

Common name: Aspartame (E951)

Aspartame (*N*-L-alpha-Aspartyl-L-phenylalanine 1-methyl ester) is the methyl ester of the dipeptide of the normal amino acids L-aspartic acid and L-phenylalanine. It hydrolyzes, or breaks down, into its amino acids when warmed to high temperatures, subsequently losing its sweetness. Aspartame is made out of 57.1% carbon, 6.2% hydrogen, 9.5% nitrogen, and 27.2% oxygen [5]. It has the concoction equation C14H18N2O, a molar mass of 294.3 g/mol, and a thickness of 1.3 g/cm^3.

Aspartames' segments, aspartic acid, phenylalanine, and methanol, happen normally in food items, yet aspartame itself does not present and should be made. Aspartame is made through fermentation and synthesis procedures. Foods and drinks that contain aspartame must carry a label demonstrating that the item contains phenylalanine. This announcement is for the advantage of people with the inherited sickness phenylketonuria, who should entirely confine their consumption of this amino acid. Normal and healthy people don't have to limit their phenylalanine consumption [16–19].

4.3 Cyclamate E952

Cyclamate, similar to aspartame, was found incidentally. In 1937, Michael Sveda recklessly set a cigarette into a white powder and when he set the cigarette once again into his mouth, he found a sweet and charming taste holding up. Cyclamate was endorsed by the FDA in 1958 however then banned in 1969 after various studies connecting cyclamate to malignancy. As of date, a wild battle is being driven by cyclamate makers to reapprove the sweetener in the United States. A few people find that cyclamate has a bitter aftertaste sensation yet not as much as that of Ace-K. Every now and again cyclamate is utilized alongside different sweeteners, especially sucralose in a proportion of 10 sections cyclamate to 1 section sucralose, to cover its "off-taste" and lingering flavor. This synergistic impact frequently shocks specialists in that the resultant sweetness is fundamentally higher than what might be anticipated from two sweeteners joined. Cyclamate has a long time span of usability and a wide working temperature taking into consideration it to be heated and frozen with no impact on sweetness or stability. It is utilized as a tabletop sweetener, in diet drinks, and in other low-calorie foods. Additionally, cyclamate is helpful as a flavor enhancer and in addition a good seasoning ingredient for some pharmaceuticals and toiletries. Cyclamate (Cyclohexylsulfamic acid) is the sodium or calcium salt of cyclamic acid.

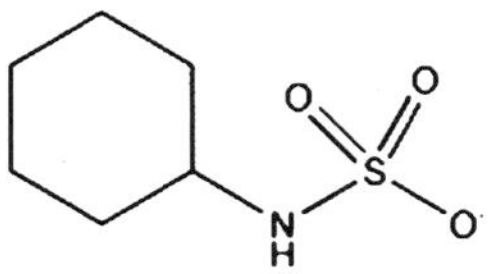

Cyclamate

IUPAC name: *N*-cyclohexylsulfamate

Common name: Cyclamate (E952)

Cyclamate is 30–50 times sweeter than sugar making it the slightest strong of the industrially utilized artificial sweeteners. It is frequently utilized synergistically with other simulated sweeteners, particularly saccharin; the blend of 10 sections cyclamate to 1 section saccharin is basic and veils the off-tastes of both sweeteners. Cyclamate is stable under high temperature, making it perfect for baking. Cyclamic acid is sparingly solvent in water, and is gradually hydrolyzed in boiling hot water. Sodium cyclamate and calcium cyclamate are both unreservedly solvent in water. The allowed levels of utilization shift from 250 to 1,500 mg/kg relying upon food classification. It takes around 1.5 l of soda containing 250 mg/l cyclamates to accomplish the ADI built up by the SCF [20–22].

4.4 Saccharin (E954)

Saccharin is one of the oldest artificial sweeteners. It was created by a John Hopkins University graduate understudy in 1879. It was initially utilized as an additive and germ-free yet turned into an extremely prevalent sweetener amid the first and second World Wars. It found an extremely tumultuous association with the FDA yet categorically was banned in 1977. Saccharin is temperamental when warmed yet it doesn't respond artificially with different foods. This limits it altogether in cooking applications however some sweetness is held and no health hazards have been seen upon saccharin's breakdown. Saccharin has an extremely observable metallic lingering flavor which numerous individuals find not pleasant. Saccharin had innumerable regular uses in drinks and foods. In times of sugar deficiencies, saccharin was regularly apportioned in tablet structure thinking of it as is 200–700 times sweeter than sugar. Today it is vigorously controlled because of its disputable health impacts. In the United States, the FDA requires that any items delivered with or using saccharin as a sweetener have unclearly obvious cautioning name. Saccharin (benzoic sulfimide) is an extremely stable natural acid with a pKa of 1.6 and concoction recipe C7H5NO3S. Its' synthetic arrangement is 45.9% carbon, 2.7% hydrogen, 7.7% nitrogen, 26.2% oxygen, and 17.5% sulfur. It has a molar mass of 183.2 g/mol and a thickness of 0.83 g/cm^3. In acid structure saccharin is not water solvent. Along these lines, the structure utilized as a artificial sweetener is its sodium salt. Saccharin can be produced utilizing the Remsen-Fahlberg and Maumee or Sherwin-Williams technique

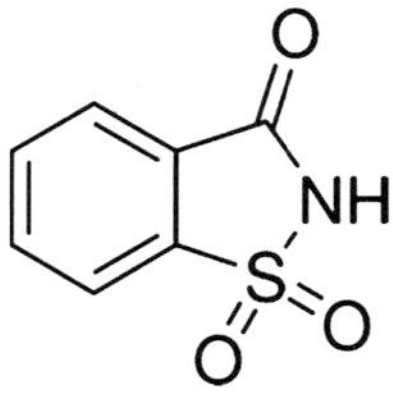

IUPAC name: 2H-1λ6,2-Benzothiazol-1,1,3-trione

Common name: Saccharin (E954)

Saccharin is allowed in an extensive variety of food items and drinks, including table-top sweeteners, pastries, yogurt, frozen yogurt, prepared bakery products, jam, jelly, sodas, desserts, mustard and sauces. The allowed levels of utilization fluctuate from 100 to 500 mg/kg relying upon the food classification. It takes around 0.5 kg of desserts containing 500 mg/kg or 3.5 l of soda containing 500 mg/l saccharin to accomplish the ADI built up by the SCF [23–25].

4.5 Sucralose (E955)

Sucralose was found in 1976 and got FDA endorsement in 1998. It is one of the fresher sweeteners accessible in today's business sector which makers are discovering hard to satisfy. The demand for sucralose, generally known as Splenda, is high while the assembling of sucralose is moderately tedious and costly creating costs of sucralose to be essentially higher than those of its rivals. Sucralose is a standout amongst the most stable artificial sweeteners accessible in today's business sector and can be utilized as a part of about each application sugar is utilized. Sucralose is gotten from sugar and is thusly fundamentally the same to it in its synthetic structure and reactivity. Sucralose is altogether more stable than aspartame accordingly permitting it an essentially more time span of usability without loss of sweetness.

Sucralose is otherwise called trichlorosucrose. Sucralose is gotten from sucrose by the particular substitution of three hydroxyl bunches by chlorine particles. Sucralose is promptly dissolvable in water, lower alcohols and other polar solvents, giving arrangements of impartial ph. In corrosive arrangements, e g., in some soda pops, sucralose hydrolyses gradually to its constituent monosaccharides, 4-chloro-4-deoxygalactose (4-CG) and 1,6-dichloro-1,6-dideoxyfructose (1,6DCF). Sucralose is not metabolized or put away in the body. After it is ingested, it is immediately retained and after that quickly discharged unaltered. The SCF communicated its first feeling on sucralose in 1989. Around then the SCF considered sucralose to be toxicologically unsuitable as a few exceptional inquiries rose up out of the assessment of the accessible information. In 2000, the SCF considered further studies and built up the ADI of 015 mg/kg body weight. Sucralose is allowed in an extensive variety of food items and drinks. Among these are soda pops, pastries, frozen yogurt, ice cream, jam and sandwich spreads. The allowed levels of utilization shift from 10 mg/l to 1,000 mg/kg relying upon the food classification. It takes around 2 kg of a pastry containing 400 mg/kg or 3 l of soda pop containing 300 mg/l sucralose to accomplish the ADI built up by the SCF.

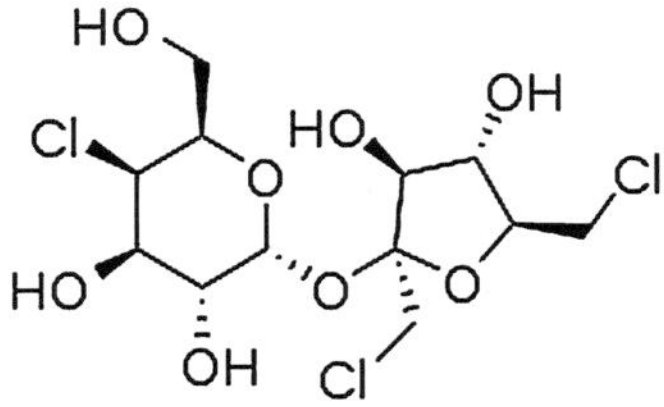

IUPAC name: 1,6-Dichloro-1,6-dideoxy-β-D-fructofuranosyl-4-chloro-4-deoxy-α-D-galactopyranoside

Common name: Sucralose (E955)

Basically, sucralose is utilized all alone on account of its similitude to sugar yet in a few events is utilized to enhance the taste or properties of a specific food or drink containing other artificial sweeteners. Sucralose can likewise be utilized alongside sugar alcohols which are frequently in fluid structure. Sucralose is exceptionally solvent in a wide range of sorts of fluids making it a standout amongst the most adaptable synthetic sweeteners today. Sucralose (1′, 4, 6′- Trichloro-galactosucrose) is a chlorinated sugar with substance equation C12H19C13O8 and a molar mass of 397.64 g/mol. Sucralose is a steady particle that keeps up its sweetness property when presented to high temperatures, making it suitable for use in baking [26–28].

4.6 Neotame E961

Neotame is chemically the same to aspartame however is considerably more stable and more powerful. Aspartame is around 200 times sweeter than sucrose sugar while Neotame is somewhere around 7,000 and 13,000 times sweeter than sucrose sugar. Neotame is utilized as a part of foods and drinks, including chewing gum, carbonated soda pops, ready to-drink drinks, tabletop sweeteners, frozen pastries and novelties, puddings and fillings, dairy items, (for example, yogurt), bakery products and confections. It can likewise be utilized as a part of both cooking and preparing applications as a result of its great heat resistance. Neotame, alongside numerous other artificial sweeteners, is frequently utilized as a part of mix with different sweeteners. It doesn't have an especially solid persistent flavor and in light of its high strength, is frequently utilized alone or with sweeteners, for example, aspartame or Ace-K. Likewise, due to its strength, it is extremely alluring by mass makers of food items since extensive sums can be created, inexpensively and not substantial sums are required for adequate sweetening purposes. Its' compound equation is C20H24N2O5 and it has a molar mass of 378.46 g/mol. Neotame is delivered by including a 6-carbon (neohexyl) gathering to the amine nitrogen of aspartame. Peptidases, which would regularly break the peptide bond between the aspartic corrosive and phenylalanine moieties, are adequately obstructed by the vicinity of the 3, 3dimethylbutyl moiety, in this manner diminishing the creation of phenylalanine, along these lines making its utilization by the individuals who experience the ill effects of phenylketonuria safe.

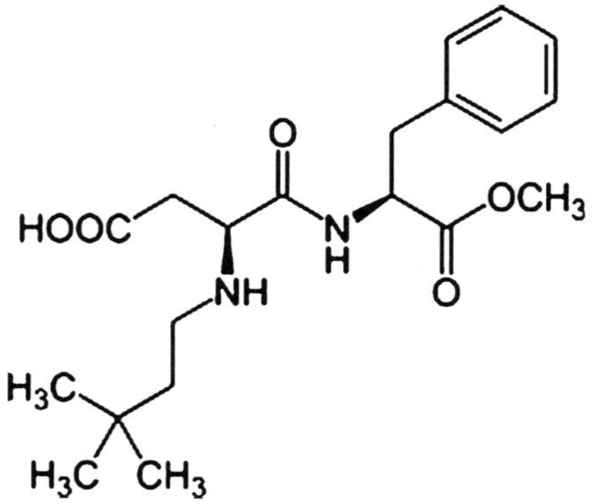

IUPAC name: (3S)-3-(3,3-Dimethylbutylamino)-4-[[(2S)-1-methoxy-1-oxo-3-phenylpropan-2-yl]amino]-4-oxobutanoic acid

Common name: Neotame (E961)

Neotame has a perfect, sweet have an aftertaste like sucrose, and flavor-improving properties. It is modestly heat stable. Neotame is quickly metabolized and totally dispensed with, and does not gather in the body. The major metabolic pathway is hydrolysis of the methyl ester, which yields de-esterified neotame and methanol in equimolar sums. Since just little measures of neotame are expected to

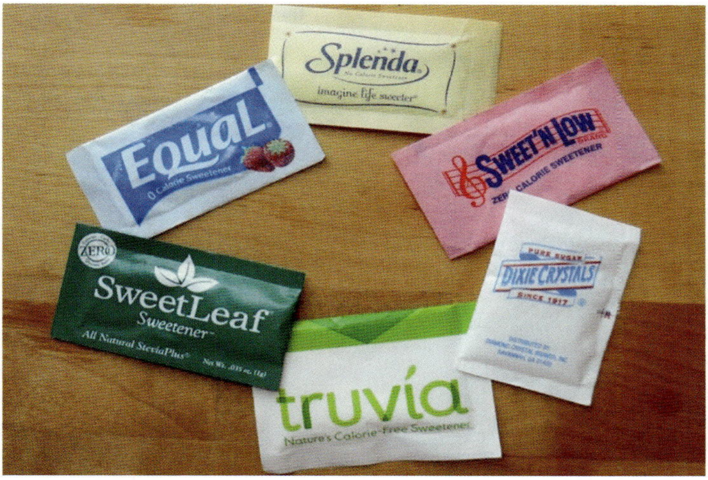

Fig. 2 Available brands of various artificial sweeteners

sweeten nourishments, the exposure of methanol that might come about because of ingestion of neotame-containing food and refreshments is viewed as irrelevant contrasted and that from other dietary sources [29–31] (Fig. 2).

5 Sweetener Metabolism

The digestion system of sweeteners can be isolated into two primary classes – the digestion system of natural sweeteners and the digestion system of artificial sweeteners. The real distinction that isolates these two classes originates from the way that natural sweeteners contain some type of starch (sugar) while artificial sweeteners don't. Thus, natural sweeteners, for example, table sugar (sucrose), nectar, and organic product sugar (fructose) prompt pathways that outcome in the generation of ATP. Then again, artificial sweeteners have almost no healthful worth to the human body. Research demonstrates that for most artificial sweeteners, more than 90% of the starting compound can be found in excrement and pee natural. This area gives data on the two noteworthy metabolic pathways that regular sweeteners are prepared by alongside information from exploratory trials that demonstrate the destiny of various artificial sweeteners once they are ingested [32].

5.1 Metabolism of Artificial Sweeteners

Generally, artificial sweeteners go through the body naturally. Numerous tests have been performed attempting to recoup ingested artificial sweeteners from excrement and pee of animals and people and found that the greater part of expended artificial sweetener can be recuperated [33].

6 Safety Aspects of Intense Sweeteners

Numerous medical concerns are connected with artificial sweeteners. This is the fundamental explanation behind their worldwide prevalence. By, Phyllis Roxland, "it empowers individuals that are carb, sugar or calorie conscious to take in a wider variety of food that they would either not be permitted to eat or could just eat in such modest sums that they were not satisfying." They are utilized as a part of diabetes mellitus, as a different option for sugar since they don't raise blood glucose levels. As they contain no calories, they can likewise be appropriately utilized for weight control. Various diabetics face trouble in adjusting to non-sugary food items. Artificial sweeteners make these troublesome move simpler, permitting individuals with diabetes to eat their most loved foods. They can likewise be utilized as a part of receptive hypoglycemia, dental consideration to stay away from caries and help improving flavors. Articulation from American Heart Association and American Diabetes Association in 2011 spelled out those non-nutritive sweeteners might be useful for health by decreasing or controlling weight and can have other useful metabolic impacts too [34].

6.1 Gastrointestinal Tract Disturbance

Episodic proof of bloating and unpredictable bowel movement is common in the utilization of artificial sweeteners. The impact that artificial sweeteners have on the gastrointestinal tract for the most part needs to do with their association with the microbial verdure. The human gut harbors an extremely various group of microbial living beings which assume imperative parts in the assimilation of food. The human digestive system, for instance, harbors a rough of 100 trillion micro flora forms of no less than 1,000 distinctive species [35]. Sucralose is a high-intensity sweetening exacerbate that is basically a chlorinated disaccharide. In a trial done by the Duke University Medical Center [36], Sprague–Dawley rats were given the artificial sweetener Splenda (principle component sucralose) at various time periods. In conclusion the aftereffects of this study demonstrate that sucralose oppresses some valuable micro flora. Additionally, sucralose affects raised expression of P-gp and CYP proteins which basically assume the part of letting medications and toxicants bypass the body's digestion system framework and render them less bioavailable. Aspartame hydrolyzes into its part atoms (L-aspartic acid, L-phenylalanine, and methanol) inside of the gut and the expansion of these segments inside of the gut were viewed as a probability for gastrointestinal issues created by aspartame. Manufactured sweeteners, while they are low caloric and are not metabolized much inside of the human body; do appear to appreciably affect the micro flora inside of the gastrointestinal tract.

6.2 Obesity

Obesity is one of the most serious problems that Americans face today. An obese lifestyle leads to many other health complications such as diabetes, high blood pressure, cancer, gallbladder disease, metabolic syndrome, osteoarthritis, heart disease, and depression [37]. Larger part of the studies surveying the relationship of NNSs with the body weight include the substitution of caloric sweeteners with NNSs in drinks, yet meager information is accessible with respect to substitution in food. San Antonio Heart Study, reported huge weight pick up and danger of stoutness in ordinary weight or overweight persons at benchmark who expending misleadingly sweetened drinks as contrasted with nonconsumers [38]. Sweetness decoupled from caloric substance offers incomplete, however not finish, actuation of the food reward pathways. Actuation of the hedonic part might add to expanded craving. Animals look for food to fulfill the inborn longing for sweetness, even without vitality need. Absence of complete fulfillment, likely in light of the inability to enact the postingestive part, development energizes the nourishment looking for manner. Decrease in reward reaction might add to heftiness. Disabled actuation of the mesolimbic pathways taking after milkshake ingestion was seen in hefty pre-adult young ladies [39]. Ordinarily, an orosensory boost is noiselessly created by sweet food items, educating the body concerning exposure of calories. This is trailed by a few GI reflexes that set up the body for intake. However, when sugar substitutes are utilized alone, the association between sweet sensation and caloric admission is lost and capacity of the body in controlling intake of food changes. This prompts disarray of the body and ultimately weight pick up [34]. The utilization of simulated sweeteners decreases the caloric intake of a consumer, yet it impedes the mind's capacity to evaluate caloric intake in light of sweetness [37].

6.3 Diabetes Mellitus Type 2

In spite of the fact that artificially sweetened drinks and organic product too showed positive relationship with frequency of type 2 diabetes [40]. Utilization of artificial sweeteners can expand the risk of type 2 diabetes mellitus. In 2014, an Israeli research introduced test prove that AS might irritate, as opposed to put a stop to, metabolic scatters, for example, type 2 diabetes. Another exploration led in 2013 showed that weight control plans sweetened with either common or artificial sugars are connected with an expansion in type 2 diabetes. However, more exploration is required [34].

6.4 Coronary Heart Disease

Numerous researchers attempt to discover out the relationship between the utilization of NNSs and the cardio-metabolic and glycemic variables. Utilization of ≥ 2 diet soft drink day by day was significantly connected with the improvement of coronary heart disease and chronic kidney problem when contrasted with diet soda utilization of <1 serving month to month when tracked for 11–12 years [38].

6.5 Brain Damage

Anecdotal evidence suggests that artificial sweeteners have negative effects on the central nervous system, causing difficulty to concentrate and carry out mental operations. Aspartame is most broadly utilized NNS however it is likewise most dubious among the NNSs. It is connected with intense symptoms, for example, queasiness, spewing, migraine particularly headache, and dry mouth. Mukhopadhyay et al., built up that aspartame is connected with a huge expansion of up to 2.5–4.2-fold in chromosomal aberration. Long-term utilization of aspartame can cause hepatocellular harm, adjust the hepatic cell reinforcement equalization. Aspartame present in food jars put away at high temperatures and at pH >6, it can break into its metabolite deketopiperazine which is a central nervous system cancer-causing agent and is under dynamic research [38].

6.6 Phenylketonuria

Phenylalanine, a vital amino acid, is shaped from aspartame digestion system. However, patients with phenylketonuria (PKU), an inalienable error of digestion system. Elevated amounts of phenylalanine in youngsters can bring about a number of conceivably exceptionally negative health impacts, for example, obstructed mental health. Thus, the ADI is not applicable to PKU patients and as such, items containing aspartame must carry a label that they are a source of phenylalanine. This notice, however, does not identify with any potential threat of aspartame to those in normal community [41].

6.7 Early Menarche

In African American and Caucasian young ladies from the United States, more prominent utilization of energized and artificially sweetened diet drinks at 9–10 years of age is tentatively connected with a higher danger of ahead of early menarche, independent of adiposity. Utilization of artificial sweetened beverages was likewise connected with a higher risk of early menarche [42].

6.8 Cancer

Artificial sweeteners might affect the body effectively. They can bring about variety of risks including cancer. In 1970, a study was done which demonstrated the relationship of saccharin with bladder disease in research center rats. They are additionally connected with malignancies such as leukemia, lymphoma and various myelomas (in men). In another study which surveyed the conceivable impacts of five nonnutritive sweeteners on cell aberration, morphology and cells' DNA by using Caco2, HT-29 (colon) and HEK-293 (renal) cell lines. DNA harm, if any, prompted by AS, was concentrated on. Results demonstrated that cells turned out to be less which are very much characterized and compliment at higher AS dosage. Colon cells were observed to be more influenced than renal cells. It was likewise seen that sodium saccharin and sucralose created more DNA aberrations in all cell lines than other ASs [34].

6.9 Preterm Delivery

Although many AS are considered safe during pregnancy, women with any form of diabetes (gestational or diabetes mellitus) and insulin resistance must limit their use of these substitutes. Saccharin has been shown to cause effects as anemia, iron and vitamin A deficiency, depressed growth and elevated vitamin E in rats. They are also linked with premature deliveries prompted by two observational studies published in 2010 and 2012. Therefore pregnant women must be advised to avoid these sweeteners [34].

7 Conclusion

The health impacts of sugar sweetened drinks, artificially sweetened drinks, and natural product juice have gotten impressive consideration from experimental and open groups. Artificial sweeteners are generally new and their utilizations are being investigated and augmented each day. Much debate encompasses artificial sweeteners and their health impacts as synthetic sweeteners might separate into unsafe and safe blends. New artificial sweeteners are continually being investigated and because of their minimal effort and simplicity of creation, they will probably turn into the essential sweetening mixes later on. It can be for the most part presumed that the sweeteners themselves are not made out of any unsafe chemicals and are unrealistic to represent a noteworthy danger to human health. In any case, the assembling and generation forms that these sweeteners experience should be precisely directed. As artificial sweeteners are utilized as a part of a huge number of regular items, the security of people in general is at danger to conceivable health risks that might happen as an aftereffect of free assessments of the assembling process. Last items should be completely tried all the time and creation strategies should be assessed and kept up to regulation benchmarks.

Aspartame, has been thought to bring about cerebrum harm as a result of one of its segment atoms, phenylalanine. Phenylalanine causes brain damage in individuals who have a hereditary disorder (homozygous phenylketonuria) that renders them not able to metabolize phenylalanine. Aspartame's joining into the overall population's eating routine raised the issue that the populace would be presented to expending superfluously high dosages of phenylalanine, bringing about conceivable brain harm even in people who don't have homozygous phenylketonuria. While the normal individual will most likely be unable to expend unreasonably high measures of aspartame – enough to bring about cerebrum harm, a few people with a more inactive, diet drink subordinate way of life, or people who depend vigorously on simulated sweeteners in any capacity might expend overabundance phenylalanine and experience neurological disorientation. Through the examination of overview results with distributed writing, any reasonable person would agree that the open need more intensive instruction with respect to artificial sweeteners and their health impacts.

The consistent media buildup must be taken incredulously and concerned people must attempt to take in the genuine certainties behind sweeteners safety through substantial experimental examination. Also, the intake of sweeteners must be precisely checked on an individual premise as was prescribed by the specialists met. While attempting new items, consumers must be careful about any unexpected responses and must ensure that they are not expending sweeteners or sweetener containing items in abundance. As a rule, artificial sweeteners, when utilized as a part of balance are a satisfactory substitute to normal sweeteners. They have all the earmarks of being a solid and charming approach to advance food sweetness, taste, and composition while keeping significant safety chances low.

8 Cross-References

▸ Analytical Strategies to Determine Artificial Sweeteners By Liquid Chromatography-Mass Spectrometry
▸ Characterization of Artificial Sweeteners Using Raman Spectroscopy
▸ Health Implications of Fructose Consumption in Humans
▸ Nonnutritive Sweeteners and Their Role in the Gastrointestinal Tract
▸ Sugar Alcohols as Sugar Substitutes in Food Industry
▸ Sweet-Taste Receptor Signaling Network and Low-Calorie Sweeteners
▸ Sweeteners: Regulatory Aspects
▸ Tagatose Stability Issues in Food Systems
▸ The Role of Dietary Sugars and Sweeteners in Metabolic Disorders and Diabetes
▸ Xylitol: One Name, Numerous Benefits
▸ Xylitol as Sweetener

References

1. Walters ED (2009) All about sweeteners. All About Sweeteners. Web, 23 Sept. 2009
2. Suez J, Korem T, Zeevi D, Zilberman-Schapira G, Thaiss CA, Maza O, Weinberger A (2015) Artificial sweeteners induce glucose intolerance by altering the gut microbiota. Obstet Gynecol Surv 70(1):31–32
3. Kauffman GB, Priebe PM (1978) The discovery of saccharin: a centennial retrospect. Ambix 25(3):191–207
4. Yang Q (2010) Gain weight by "going diet?" Artificial sweeteners and the neurobiology of sugar cravings: neuroscience 2010. Yale J Biol Med 83(2):101

5. Joint, F, Organization, W. H, Additives, W. E. C. o. F (1987) Principles for the safety assessment of food additives and contaminants in food. World Health Organization, Geneva
6. Mortensen A, Larsen JC (2012) Risk assessment of sweeteners used as food additives. In: Sweeteners: nutritional aspects, applications, and production technology, vol 419. CRC Press, Hoboken
7. Mortensen A (2006) Sweeteners permitted in the European Union, safety aspects. Scand J Food Nutr 50(3):104–116
8. Capar SG, Mindak WR, Cheng J (2007) Analysis of food for toxic elements. Anal Bioanal Chem 389(1):159–169
9. Renwick A, Barlow S, Hertz-Picciotto I, Boobis A, Dybing E, Edler L,. . . Lambe J (2003) Risk characterisation of chemicals in food and diet. Food Chem Toxicol 41(9):1211–1271
10. Mulla SJ, Ansari ARM (2015) Review on artificial sweeteners used in formulation of sugar free syrups. Int J Adv Pharm 4(2):05–09
11. Ibrahim OO (2015) High intensity sweeteners chemicals structure, properties and applications. Nat Sci Discovery 1(4):88–94
12. Breinholt V, Dragsted L, Hossaini A, Lam HR, Mortensen A, Rasmussen E,. . . Penttilä P (2002) Food additives in Europe 2000, status of safety assesments of food additives presently permitted in the EU. Nordiccouncil of ministers, Denmark
13. Renwick A (1993) Data-derived safety factors for the evaluation of food additives and environmental contaminants. Food Addit Contam 10(3):275–305
14. Renwick A, Lazarus N (1998) Human variability and noncancer risk assessment – an analysis of the default uncertainty factor. Regul Toxicol Pharmacol 27(1):3–20
15. Lebedev I (2010) Popular sweeteners and their health effects. Worcester Polytechnic Institute, Worcester, Massachusetts, U.S.A.
16. O'Donnell K, Kearsley M (2012) Sweeteners and sugar alternatives in food technology. Wiley, Chichester
17. Rycerz K, Jaworska-Adamu JE (2013) Effects of aspartame metabolites on astrocytes and neurons. Folia Neuropathol 51(1):10–17
18. Magnuson B, Burdock G, Doull J, Kroes R, Marsh G, Pariza M,. . . Williams G (2007) Aspartame: a safety evaluation based on current use levels, regulations, and toxicological and epidemiological studies. Crit Rev Toxicol 37(8):629–727
19. Efsa A (2013) Panel (EFSA panel on food additives and nutrient sources added to food). Scientific opinion on the re-evaluation of aspartame (E 951) as a food additive. EFSA J 11(12):3496
20. Militana H, Ferguson N (2010) Carbohydrates, sugars and artificial sweeteners. Webinar, Newyork state department of Health, U.S.A.
21. Renwick A, Thompson J, O'Shaughnessy M, Walter E (2004) The metabolism of cyclamate to cyclohexylamine in humans during long-term administration. Toxicol Appl Pharmacol 196 (3):367–380
22. Smith J, Hong-Shum L (2011) Food additives data book. Wiley, Oxford
23. Gardner C, Wylie-Rosett J, Gidding SS, Steffen LM, Johnson RK, Reader D, Lichtenstein AH (2012) Nonnutritive sweeteners: current use and health perspectives a scientific statement from the American Heart Association and the American Diabetes Association. Diabetes Care 35(8):1798–1808
24. Whitehouse CR, Boullata J, McCauley LA (2008) The potential toxicity of artificial sweeteners. AAOHN J 56(6):251–259
25. Association AD (2010) Standards of medical care in diabetes – 2010. Diabetes Care 33(Suppl 1): S11–S61
26. Kim J-W, Baek H-H (2011) Safety of saccharin and its current status of regulation in the World. Korean J Food Sci Technol 43(6):659–674
27. Mukherjee A, Chakrabarti J (1997) In vivo cytogenetic studies on mice exposed to acesulfame-K – a non-nutritive sweetener. Food Chem Toxicol 35(12):1177–1179
28. Grotz VL, Munro IC (2009) An overview of the safety of sucralose. Regul Toxicol Pharmacol 55:1

29. Martínez C, González E, García RS, Salas G, Constantino-Casas F, Macías L,... Durán-de-Bazúa C (2010) Effects on body mass of laboratory rats after ingestion of drinking water with sucrose, fructose, aspartame, and sucralose additives. Open Obes J 2:116–124
30. Mayhew DA, Comer CP, Stargel WW (2003) Food consumption and body weight changes with neotame, a new sweetener with intense taste: differentiating effects of palatability from toxicity in dietary safety studies. Regul Toxicol Pharmacol 38(2):124–143
31. Li X, Servant G, Tachdjian C (2011) The discovery and mechanism of sweet taste enhancers. Biomol Concepts 2(4):327–332
32. Zhu L, Wang G, Dong B, Peng C, Tian Y, Gong L (2016) Effects of sweetener neotame on diet preference, performance and hematological and biochemical parameters of weaned piglets. Anim Feed Sci Technol 214:86
33. Brown RJ, De Banate MA, Rother KI (2010) Artificial sweeteners: a systematic review of metabolic effects in youth. Int J Pediatr Obes 5(4):305–312
34. Ain Q, Khan S (2015) Artificial sweeteners: safe or unsafe? JPMA. J Pak Med Assoc 65(2):225–227
35. Krishnan R, James HM, Bais R, Rofe AM, Edwards JB, Conyers RA (1980) Some biochemical studies on the adaptation associated with xylitol ingestion in rats. Aust J Exp Biol Med Sci 58:627–638
36. Abou-Donia MB, El-Masry EM, Abdel-Rahman AA, McLendon RE, Schiffman SS (2008) Splenda alters gut microflora and increases intestinal p-glycoprotein and cytochrome p-450 in male rats. J Toxicol Environ Health A 71(21):1415–1429
37. Fowler SP, Williams K, Resendez RG, Hunt KJ, Hazuda HP, Stern MP (2008) Fueling the obesity epidemic? Artificially sweetened beverage use and long-term weight gain. Obesity 16(8):1894–1900
38. Yarmolinsky J, Duncan BB, Chambless LE, Bensenor IM, Barreto SM, Goulart A C,... Schmidt MI (2015) Artificially sweetened beverage consumption is positively associated with newly diagnosed diabetes in normal-weight but not in overweight or obese Brazilian adults. J Nutr 146(2):290–297
39. Stice E, Spoor S, Bohon C, Veldhuizen MG, Small DM (2008) Relation of reward from food intake and anticipated food intake to obesity: a functional magnetic resonance imaging study. J Abnorm Psychol 117(4):924
40. Imamura F, O'Connor L, Ye Z, Mursu J, Hayashino Y, Bhupathiraju SN, Forouhi NG (2015) Consumption of sugar sweetened beverages, artificially sweetened beverages, and fruit juice and incidence of type 2 diabetes: systematic review, meta-analysis, and estimation of population attributable fraction. BMJ 351:h3576
41. Logue C, Peters SJ, Gallagher AM, Verhagen H (2015) Perspectives on low calorie intense sweeteners with a focus on aspartame and stevia. Eur J Food Res Rev 5(2):104
42. Mueller NT, Jacobs DR, MacLehose RF, Demerath EW, Kelly SP, Dreyfus JG, Pereira MA (2015) Consumption of caffeinated and artificially sweetened soft drinks is associated with risk of early menarche. Am J Clin Nutr 102(3):648–654

Sweeteners: Regulatory Aspects

27

Jose Manuel García-Almeida, Isabel M. Cornejo-Pareja, Araceli Muñoz-Garach, Ana Gómez-Pérez, and Jorge García-Alemán

Abstract

Evidence suggests that excessive intake of added sugars has adverse effects on cardiometabolic health. In this chapter we review the role and potential benefits of noncaloric sweeteners, as a part of the diet. After their emergence and because of their supposed beneficial effects in different conditions as diabetes or obesity, an increasing number of studies show controversial results. Some of them even suggest that they are ineffective. In conclusion, further research and results are needed to provide convincing evidence of their long-term effectiveness and the absence of negative effects secondary to their use. It is interesting to examine the distinctive properties of sweeteners compared with sugar, the gold standard. We will focus on other substances that are commonly used to sweeten foods instead of sugar.

Keywords

Artificial sweeteners • Sugar and sugar substitutes • Nonnutritive sweetener • Low and noncaloric sweetener • Glycemic response

Abbreviations

ACS	The American Cancer Society
ADA	The American Diabetes Association
ADI	Acceptable daily intake
AHA	The American Heart Association
APM	Aspartame

J.M. García-Almeida (✉) • I.M. Cornejo-Pareja • A. Muñoz-Garach • A. Gómez-Pérez • J. García-Alemán
Endocrinology and Nutrition Department, Universitary Hospital Virgen de la Victoria, Malaga, Andalucía, Spain
e-mail: jgarciaalmeida@yahoo.com; isabelmaria_cornejo@hotmail.com; aracelimugar@gmail.com; anamgp86@gmail.com; jgarcialeman@hotmail.com

J.-M. Mérillon, K.G. Ramawat (eds.), *Sweeteners*, Reference Series in Phytochemistry,
https://doi.org/10.1007/978-3-319-27027-2_12

BMI	Body mass index
DM	Diabetes mellitus
EDI	Estimated daily intake
EFSA	European Food Safety Authority
EU	European Union
FFQ	Food frequency questionnaire
FOS	Fructooligosaccharides
GI	Glycemic index
GIP	Glucose-dependent insulinotropic peptide
GLP	Glucagon-like peptides
GRAS	Generally recognized as safe
HIS	High-intensity sweeteners
JECFA	Joint FAO/WHO Expert Committee on Food Additives
LNCS	Low and noncaloric sweeteners
mRNA	Messenger RNA
MS	Member states
NNS	Nonnutritive sweeteners
SCF	Scientific Committee on Food
WHO	World Health Organization

Contents

1 Introduction 614
2 Recommendations and Policy Responses 617
3 Legal Aspects 618
4 Trends in Caloric and Low-Calorie Sweeteners, Food, and Beverage Sales 620
5 Sweeteners: Types and Key Characteristics 621
6 The Health Effects of Sweetener Consumption 628
6.1 Sweeteners and Energy Intake 629
6.2 Sweeteners and Appetite Regulation 631
6.3 Sweeteners and Their Effect on Body Weight 632
6.4 Sweeteners and Diabetes 634
6.5 Sweeteners and Tooth Decay 635
6.6 Sweeteners and Cancer 635
7 A Global View of Sweetener Use: SWOT Analysis 636
8 Recommendations 637
9 Conclusion 638
References 639

1 Introduction

The term "sweetener" refers to those food additives which are able to mimic the sweetness of sugar and provides less than 3.8 kilocalories per gram and/or are used in low quantities so the amount of calories provided is negligible. Some of them are natural extracts and others are synthetic. In the case of the latter, they are also known as artificial sweeteners. The use of low and noncaloric sweeteners (LNCS), as a substitute of the sugar content in food and drink, has experienced a big growth over the past 35 years [1]. Analysis of consumption trends of sugar-sweetened beverages around the world, in terms of calories sold per person-day and volume per person-day, shows that the four regions with the highest consumption are North America, Latin America, Australasia, and Western Europe; and the fastest absolute growth in sales of sugar-sweetened beverages by country between 2009 and 14 was seen in Chile [2, 3]. It is therefore essential to clarify the effectiveness and safety of these substances for the consumers to have clear information.

New eating patterns, characterized by dense calorie diets with high consumption of processed foods, changing their fat and sugar content, mark a notable shift away from the traditional Mediterranean diet and leads to an imbalance in energy intake and expenditure. Sugar-sweetened beverages and processed food are major sources of added sugars in most countries and are void of much nutritional value.

There has been an exponential growth both in energy-rich foods and those that are supposedly low calorie. Industry pressure plays a pivotal role in these food consumption patterns. In our opinion, we need to stop these growing food sales worldwide. In the past few years, many governments have initiated actions to reduce consumption of sugar-sweetened beverages.

A nationally representative sample of US households in 2013 showed that 68% of all calories consumed contained caloric sweeteners and 2% contained LNCS, and now is a trend toward consuming LNCS. It is vital to conduct extensive research and to take a strict regulatory approach on these issues [2].

There is currently no conclusive data on the effects of sweeteners on crucial factors such as energy intake, appetite, and their relationship with sweet taste. Furthermore, the exact quantities of these sweeteners that foods contain are unknown. For this reason, it would be of great interest to accurately quantify the prevalence of use of products containing LNCS. This chapter tries to summarize the current principal scientific and legislative findings on this issue with an eye on improving the rational use of these substances in our diet [4].

The health concept is very broad and its determinants encompass biological aspects, such as genetic characteristics, and other socioeconomic and cultural aspects which, as a whole, determine an individual's health status (Fig. 1). Over time, changes in disease patterns, probably associated with lifestyle changes in the general population, have led to an increase in the incidence of many chronic diseases (such as obesity, type 2 diabetes, and metabolic syndrome), which ultimately result in an increase in cardiovascular morbidity and mortality. Recent studies have shown a strong association between sales of sugar-sweetened beverages and the prevalence of diabetes [5]. Interest in the potential role of sweeteners has grown, due to the need to find alternatives to prevent disease and maintain good health by following a healthy diet.

Likewise, a recent meta-analysis informs [6] that consumption of sugar-sweetened beverages over 10 years may be related to a substantial number of cases of new onset diabetes, independently of adiposity (linked to 4–13% of type 2 diabetes incidence in the United States and 2–6% in the United Kingdom over 10 years). Artificially sweetened beverages and fruit juice are candidate alternatives

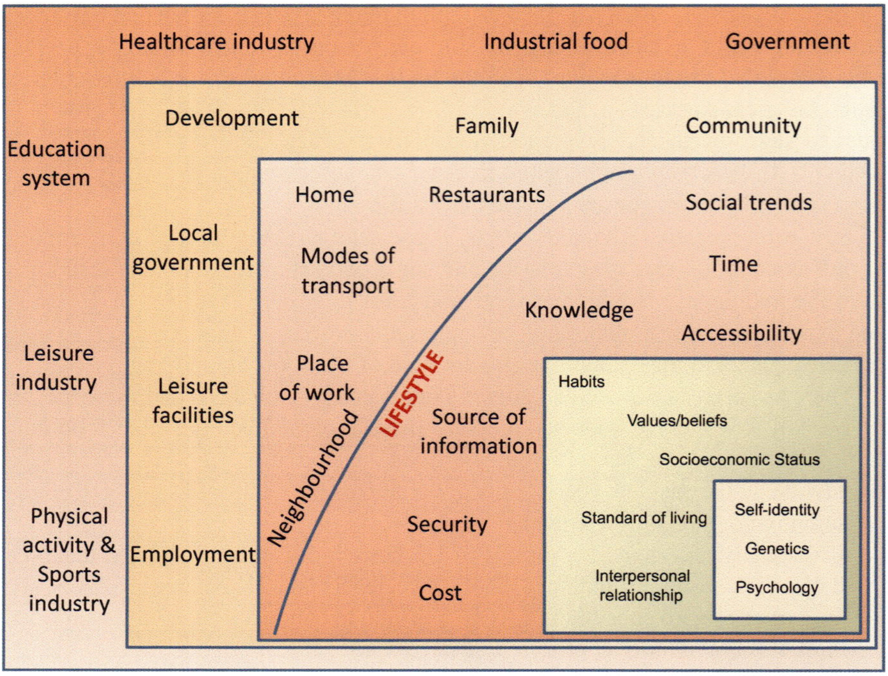

Fig. 1 Global determinants of health: the multifactorial relationship between individuals and environmental factors, in relation to food choices and healthy behavior (Amended by Ref. [58])

to sugar-sweetened beverages, but their prospective associations with type 2 diabetes have not yet been well established because only a few studies have examined these associations [7–9]. Reverse causality could also exist in these studies because of unmeasured comorbid conditions and existing public interest over the health effects that might alter consumption of beverages, particularly artificially sweetened beverages, in adults at high risk of type 2 diabetes or higher weight. Nonetheless, both artificially sweetened beverages and fruit juice were unlikely to be healthy alternatives to sugar-sweetened beverages and do not provide substantial benefit for the prevention of type 2 diabetes [10].

Obesity has become one of the biggest global health challenges of the twenty-first century. Because overweight and obesity are major causes of comorbidities, including cardiovascular disease, hypertension, type 2 diabetes, certain cancers, and other health conditions [11]. In the last years, the increase in childhood and adolescent obesity is particularly alarming; this problem is especially evident in high-income countries such as the United States, the United Kingdom, and Australia, where people are experiencing rapid changes in their eating habits and increases in childhood obesity.

The considerable raise in sugary drink consumption among adults and children in the United States and other countries is regarded as a potential contributor to the obesity pandemia. Recent evidence shows that sucrose consumption in drinks is approaching 15% of the American public's daily caloric intake, accounting for up to 357 kcal per drink. Identifying strategies that help regulating body weight is imperative. So, regulatory strategies to limit the sale and, consequently, the consumption of these drinks are being developed.

2 Recommendations and Policy Responses

Extensive meta-analyses show that the risk of cardiometabolic problems, diabetes, or obesity, resulting from added sugars in beverages, is substantial and the risk increases with the amount consumed [12–16]. The exceptions are studies funded by the sugar and beverage industries, which the authors of some reviews suggest are biased [17]. In the past 3 years, two major meta-analyses of the effect of added sugars in food show that the risk of adverse cardiometabolic outcomes from caloric sweeteners in food is lower than the risk of caloric sweeteners in beverages [15, 16]. However, the World Health Organization (WHO) and other health organizations recommend the reduction of added sugar intake not only in beverages but also in food.

On the basis of this risk, WHO published a revised guideline in 2015, in which it urged the reduction of added sugar intake to less than 10% of total energy intake (strong recommendation) and a further reduction to below 5% of total energy intake (conditional recommendation). On the other hand, medical associations have promoted the lower limit of 5% [18]. These studies and guidelines help to provide a basis to promote the development of health recommendations and government/political decisions aimed to limit sugary drink consumption, particularly those served at a low cost and in oversized portions, with a view to reverse the growing childhood obesity. These kinds of intervention could prevent young people from developing type 2 diabetes and its complications.

In view of the adverse health effects associated with the widespread consumption of sugar-sweetened beverages, many national governments have taken action to reduce consumption [19, 20]. Although it has identified 72 policy actions that have been implemented in 49 countries around the world (the policies vary substantially from place to place), the most common actions implemented so far include taxation, reduction of availability in schools and improvement in the quality of foods they provide, and restrictions on children marketing of sugar-sweetened beverages and foods high in fat, sugar, and salt content.

In 2004, the American Academy of Pediatrics recommended sweetened drinks to be replaced in school by water, white and flavored milks, or 100% fruit and vegetable beverages. Since then, school nutrition has undergone a significant transformation. Federal, state, and local regulations and policies, along with alternative products developed by industry, have helped decrease the availability of nutrient-poor foods and beverages in school.

Optimal nutrition for individuals older than 2 years is described by the 2010 Dietary Guidelines for Americans (DGAs) that recommended consumption of a diverse, nutrient dense diet based on five food groups: vegetables, fruits, grains and whole grains, low-fat or no-fat milk and dairy, and quality protein sources, but does not exceed caloric needs. The Academy of Nutrition and Dietetics has long espoused the perspective that "all foods fit"; a balanced, nutrient-rich dietary pattern still can accommodate all types of foods, when foods with added sugars and fats are consumed judiciously. Data from three School Nutrition Dietary Assessment (SNDA) surveys between 1995 and 2009 showed fast improvements in nutrition quality, increasingly aligned with DGA recommendations [21].

Other common policy actions are public awareness campaigns that tend to promote drinks that are low in calories as alternatives, and positive and negative front-of-pack labeling. But, practically none of these actions is implemented in low-income countries.

This finding suggests that more policy actions are needed beyond high-income countries where rates of consumption are rising. If these actions are implemented, we could expect long-term declines in the purchase of processed foods and increase in the purchase of unprocessed food after the tax was introduced. However, this assumption does not take into account the changes in industry behavior in response to taxation policies, so the long-term effects are difficult to predict.

Therefore, and from the perspective of sweetener consumption, survey data confirms that they are currently looking to use LNCS to reduce total caloric intake, promote weight loss, and/or prevent the development of diseases like diabetes.

Besides the interest generated by their potential preventative role in the development of chronic metabolic disease, we could also highlight their effect on oral cavity diseases like tooth decay. In particular, polyalcohols can reduce the risk of tooth decay. For example, xylitol is considered to be cariostatic and helps to prevent tooth decay [22]. However consumers are also concerned about the risk associated with their use, such as "artificial or natural" elements and whether they pose a risk to health.

The estimated sweetener consumption is complex to identify, but there are approximately more than 6,000 products which contain them in the United States, mainly soft drinks. Information on nutrition labels is often incomplete, without any details on the exact amount. Data from NHANES 2007–2008 24-h recalls and food frequency questionnaires show an increase in the American population's consumption of sweeteners which, interestingly, is not associated with a reduction in sugary foods.

3 Legal Aspects

With regard to the legal aspects of the use of sweeteners at the European level, the European Parliament and Council Directive 94/35/CE of 30 June 1994 [23] on all sweeteners for use in foodstuffs arose as the initial governing regulation. It is a specific directive from the Framework Directive on food additives used as sweeteners. The articles of this law contain explanations and specific provisions for the use of sweeteners in food and drinks. The maximum usage levels for each of the LNCS are set out in specific food categories in the Directive's annex. Over the years, this Directive has been amended three times to accommodate technological advances in the field of sweeteners. Later on, the European Parliament and Council adopted a regulatory framework (Regulation No.1333/2008) which, from January 2011, consolidated all the existing authorizations for sweeteners and food additives into a single legal text. At present, the following LNCS are authorized in the European Union (EU): acesulfame K (E950), aspartame (E951), aspartame-acesulfame salt (E962), cyclamate (E952), neohesperidin dihydrochalcone (E959), saccharin (E954), sucralose (E955), thaumatin (E957), and neotame (E961) [24]. Other sweeteners such as polyalcohols (sorbitols (E420), mannitol (E421), xylitol (E-957), and erythritol (E-968)) have also been approved. (International Sweeteners Association at http://www.info-edulco-rants.org/es/recursos-profesionales/folleto-isa).

Following the European Food Safety Authority (EFSA)'s favorable opinion, the use of stevia derivatives, steviol glycosides, were finally approved as a natural noncaloric sweetener throughout the European market. They can be used as food additives and thereby provide a healthy and natural alternative for sweetening foods, especially for diabetic patients or those who wish to be fit, for example, flavored drinks or diet foods designed for weight control.

Annex II to Regulation (EC) No. 1333/2008 of the European Parliament and Council was amended with the introduction of the Commission's Regulation (EU) No. 1131/2011 of the 11 November 2011, with regard to steviol glycosides (E960), and limits on the use of sweeteners in different foods and drinks were established (soft drinks, fermented dairy products, flavored ice creams, table sweeteners, diet products for weight control).

The safety of sweeteners is evaluated by the national authorities, the EU Scientific Committee on Food (SCF), and the Joint FAO/WHO Expert Committee on Food Additives (JECFA). The SCF was responsible for it from 1974 until 2003, the year in which it became the responsibility of the European Food Safety Authority (EFSA) (http://efsa.europa.eu/). Within the EFSA, the Scientific Panel on Food Additives and Nutrient Sources (ANS) is currently responsible for the regulation of these substances [25].

Legal aspects need to be reviewed on a continual basis to update new scientific developments published on the safety or effective use of sweeteners. As there are very diverse molecules, numerous potential risk sources appear: interference with absorption, metabolism, or the excretion of any intermediate metabolite, as well as any allergic reaction, tissue accumulation, effects on normal intestinal flora, changes in blood sugar regulation, or interaction with other pharmaceuticals or drugs.

The European Food Safety Authority (EFSA) has recently produced a scientific evaluation of the safety of aspartame (APM). In order to carry out this comprehensive risk assessment, the EFSA (http://www.efsa.europa.eu/en/press/news/130108.htm) has carried out a thorough review of the scientific literature on APM and its breakdown products, including new studies on humans. In this safety reevaluation by the EFSA, it was concluded that APM does not present any risk of toxicity to consumers at current exposure levels. The current acceptable daily intake (ADI) is considered safe for the general population and consumer exposure is normally below the ADI. When establishing the ADI for APM, the ANS commission also considered the results of long-term studies on phenylalanine, an APM metabolite, both in toxicity and carcinogenesis tests on animals and humans, specifically the fetal development in mothers who consume this sweetener.

At the same time, in North America the US Food and Drug Administration (FDA) has evaluated its safety since 1958, and seven sweeteners have been approved for use in the United States under the category of generally regarded as "safe for consumption": acesulfame K, aspartame, neotame, saccharin, stevia, sucralose and luo han guo.

The American FDA regulations also refer to the concept of estimated daily intake (EDI), which is a conservative estimate of the probable daily intake over a lifetime and the concentration of food additives in commonly eaten foods. Another important concept concerning consumer safety is GRAS (generally recognized as safe), which implies that, although the potential risks are not yet completely understood, experience through common use has not raised any problems. This is the accepted recognition to market stevia currently in the United States, pending further information in the future.

Information on the correct use of these substances comes from knowing the differences on the nutrition fact labels of commonly consumed products that contain sweeteners. The presence of sweeteners should be listed in the food's ingredients, along with calorie, fat, or carbohydrate content, in the nutritional labeling information.

However, with the exception of warnings about phenylalanine from APM or the amount of saccharin, this information is usually missed or incomplete. It opens up a major area for improvement in the field of sweetener use, to provide consumers with the best information in the future scientific research, although limited in humans according to the Evidence Analysis Library of the Academy of Dietetics and Nutrition (http://www.adaevidencelibrary.com/files/Docs/NNSResourceDraft3.pdf), showing that artificial sweeteners are safe to use in general population, including pregnant women and children. Most studies have not found any adverse effects related to the consumption of sweeteners, even when they are consumed in large quantities. Special population groups, such as pregnant women, should limit their use even though they have been approved by the FDA.

In this chapter we will review the main sweeteners and their metabolic effects and analyze their potential strengths, weaknesses, opportunities, and threats (SWOT system).

4 Trends in Caloric and Low-Calorie Sweeteners, Food, and Beverage Sales

A recent study of the US food illustrates global trends and help to ascertain the extent of the use of caloric sweeteners and LNCS in packaged foods and beverages [2]. This study shows a significant increase in the proportion of purchased products, especially beverages, with both caloric sweeteners and LNCS, and also a significant increase in purchases, especially beverages, with no added sweeteners between 2000 and 2013. In 2013, 30% of all food calories and 37% of all beverage calories purchased were in products that did not have added sweeteners, which lends support to earlier work showing that intake of total added sugars in the United States has declined since 2000 and that much of this shift is caused by a reduction in the consumption of caloric beverages. Sixty-eight percent of packaged foods and beverages available in the United States contain caloric sweeteners, 74% include both caloric and LNCS, and 5% have only LNCS.

Another study using data for global trends in beverage sales from the Euromonitor Passport International database [2] shows North America and Latin America are the largest consumers of sugar-sweetened beverages, while sales in Asia Pacific are substantially lower. In 2009–2014 sales fell in North America, Australia, and Western Europe, but increased in all other regions. It is curious that worldwide, sales of caloric soft drinks remained fairly unchanged, whereas sales of fruit drinks and energy drinks have increased. Three of the six countries with the highest per person daily calories sold from sugar-sweetened beverages are in Latin America: Chile, Mexico, and Argentina. The United States and Saudi Arabia are also in the top six. This situation is different from year 2000, when the United States was the largest consumer, before the substantial decline in consumption of sugar-sweetened beverages.

The United States is one example in which the sales have shifted to beverages with low caloric content, often by replacing caloric sweeteners with low-calorie sweeteners. So, there is enormous heterogeneity in sales volumes and trends between regions, but also differ between countries in the same region.

5 Sweeteners: Types and Key Characteristics

Sweeteners can be grouped according to their calorie content (caloric or noncaloric), their source (natural or artificial), or even their chemical structure (Fig. 2). Naturally sourced sweeteners are not necessarily safer or more effective and, in this respect, there is a great deal of consumer missing information about them. In this chapter we will focus on the most common and those with existing scientific studies of interest. The current classification of the main sweeteners is presented in Table 1.

Sugars are carbohydrates and therefore contain 4 cal per gram They are found naturally in many foods such as fruit, vegetables, cereals, and milk. It is important to have good oral hygiene to prevent harm to teeth, although the most recent scientific literature also points out that the stickiness of food and the frequency with which these foods are consumed could have an impact on tooth decay.

Sucrose has a moderately high glycemic index (GI). Other natural caloric sweeteners, like honey and maple syrup, contain sugar, but they also have other nutritional qualities. Their GI is somewhat lower than sugar. Saccharose is included in the group of natural sweeteners; the most commonly used are saccharose or sucrose, fructose, glucose (GI of 100 and a sweetness relative to sucrose of 0.5–1), and maltose (GI 105 and a sweetness relative to sucrose of 0.5). Fructose is typically used as a substitute for sucrose in diabetic patients and as a sweetener in the manufacture of many products labeled as "suitable for diabetics." However, it has been recently

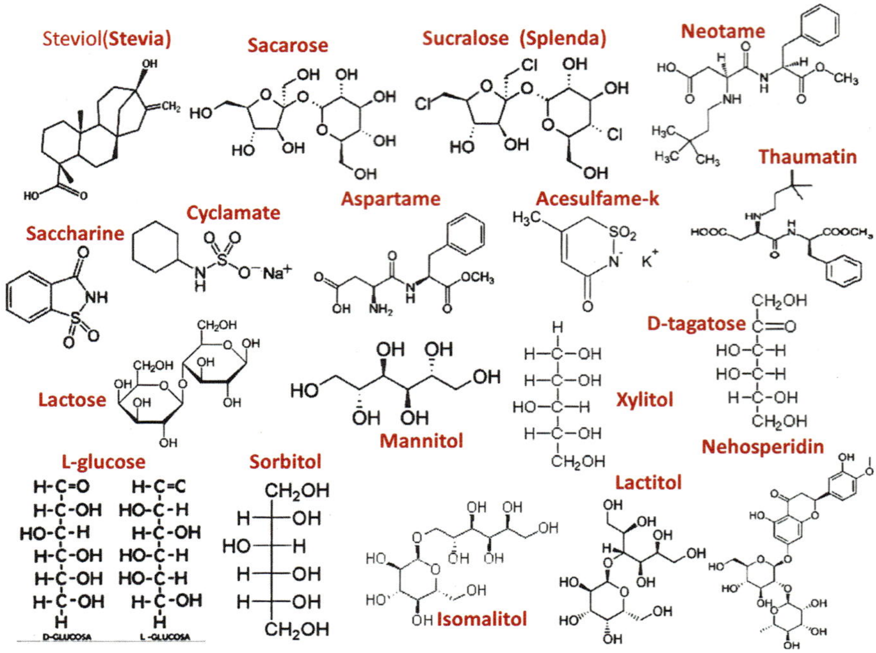

Fig. 2 The chemical structure of sweeteners

Table 1 Classification of sweeteners

Caloric	Natural	Sugars	Sucrose, glucose, dextrose, fructose, lactose, maltose, Natural galactose and trehalose, tagatose, Sucromalt[a]
		Natural caloric sweeteners	Honey, maple syrup, palm or coco sugar, and sorghum syrup
	Artificial	Modified sugars	High fructose corn syrup, caramel, inverted sugar
		Sugar alcohols	Sorbitol, xylitol, mannitol, erythritol maltitol, isomaltulose, lactitol, glycerol
Noncaloric	Natural	Noncaloric sweeteners	Luo han guo, stevia, thaumatin, pentadin, monellin, brazzein
	Artificial	Sweeteners	Aspartame, sucralose, saccharin, neotame, acesulfame K, cyclamate, neohesperidin DC, alitame, advantame

[a]Caloric value similar to fructose, although it really is an artificial oligosaccharide

proven that diets high in fructose, especially if added to manufactured foods, can cause hyperinsulinemia, hypertriglyceridemia, and insulin resistance, which was a determining factor in the recommendation that diabetics should limit its use. Its properties include a calorie content of 4 kcal/g, a GI of 23, and a sweetness relative to sugar of between 1 and 2. On the other hand, tagatose and trehalose have different calorie contents, 1.5 and 3.6 and a sweetness of 0.9 and 0.45 respectively.

Fructooligosaccharides (FOS) have half the calories per gram than sucrose or glucose with a sweetness relative to sucrose of 0.3–0.6. Inulin is a fructan with a degree of polymerization of 20–60 fructose monomers and a documented prebiotic effect, which is found naturally in a native Andean tuber, the yacon (12.5 g/100 g), and had historically grown in various Latin American countries. This tuber is mainly used as a sweetener and the possibility of nutraceutical properties due to its high content of various minerals and vitamin C and B group. Coco sugar is another traditional product which can be used as an alternative to sugar in diabetic patients as it is considered to be low GI. It consists of sucrose and amino acids such as glutamine and stands out for its high mineral and group B vitamin content.

Alcohols derived from sugar or polyalcohols are also carbohydrates which are produced artificially, although we can find them naturally in plants and cereals, in small amounts. They generally contain less calories per gram than sugar and are not associated with tooth decay. Despite the fact that they are carbohydrates, the body cannot fully metabolize them and, consequently, they tend to have less than 4 cal per gram and a very low GI.

Some of the carbohydrates used as sweeteners (i.e., polydextrose or xylitol) have been proposed as ingredients for functional foods useful to control intake because of their low energy content, due to their partial metabolism (1.5–3 kcal/g) and the possible effects of some of them on appetite suppression. A large number of these products are increasingly used as sweeteners in "sugar-free" products. The chemical structure of these substances (Table 2) causes them to have greater sweetening powers when they interact with taste receptors and a lower absorption by the digestive tract; consequently, they have a lower usable calorie content than sugar. Limits on the amount consumed are related to their secondary gastrointestinal effects.

The manufacture of sugars modified by enzymatic starch conversion, which are frequently used in industrial cooking or in processed foods, gives rise to a blend of carbohydrates which are usually high in calories and have an elevated GI. One product regularly used in the industry which belongs to this group and has a nutritional value close to 4 kcal/g, typical in carbohydrates, is high-fructose corn syrup with a relative sweetness of 1.

Sucromalt is an artificial oligosaccharide (http://www.aesan.msc.es/AESAN/docs/docs/cadenaalimentaria/tabla_decisiones_2013.pdf) which is converted from sucrose and maltose to fructose and a glucose oligosaccharide with links at 1–3 and 1–6 alternatively. It has been extensively used in the design of low GI foods. Its nutritional value is similar to fructose and its sweetness relative to sucrose is 0.7.

There are also natural sweeteners (stevia, luo han guo, thaumatin, and brazzein) whose calories are insignificant compared to the quantities usually used for sweetening purposes. These are not carbohydrates; therefore, they do not have a GI. They are considered high-intensity sweeteners (HIS).

Over the past few years, stevia is probably one of the sweeteners which have generated most interest in scientific and informative forums. It is used as a sugar

Table 2 Description of sugar alcohols

Sugar alcohols	Nomenclature	Nutritional value (kcal/g)	Sweetness, relative to sucrose	Maximum tolerable quantity without gastrointestinal symptoms	Presence	GI
Erythritol	E968	0.2	0.75	At higher doses	In fruits and other fermented foods	1
Hydrolyzed hydrogenated starch (Lycasin) polyglycitol syrup	E964	≤3	0.4–0.9	–	Sports drinks (e.g., Powerade), ice cream	–
Lactitol	E966	2	0.5	≥20	Sweets, biscuits, ice cream	3
Maltitol	E965	2.1	1	30–50	Chewing gum, sweets jelly sweets	35
Mannitol	E421	1.6	0.7	10–20	Chewing gum[a]	2
Sorbitol	E420	2.6	0.5–1	>80	Chewing gum[a]	4
Xylitol	E967	2.4	1	>50	Chewing gum, breath mints, toothpaste, and mouthwash	12

[a]Furthermore it also contains isomalt, aspartame, and acesulfame K. They amount to 61.7 g of polyalcohols/100 g

substitute and it has a slow taste onset at the beginning and is longer lasting, although in high concentrations, it can have a bitter taste similar to "liquorice." Although the word "stevia" refers to the whole plant, only certain parts of the stevia leaf are sweet. These sweet components are known as steviol glycosides (an alcohol which can be naturally found in the plant). Furthermore, the term "stevia" typically refers to a crude preparation (whether it's powder or liquid) made from the plant's leaves, and these preparations contain a mixture of various components, not just those that give the leaf its sweet flavor.

Steviol glycosides are the sweet components of the stevia leaf and there are several types, although the most common are stevioside and rebaudioside A. Stevioside is the most common steviol glycoside in the stevia leaf and has been widely studied. On the other hand, rebaudioside A is a better tasting steviol glycoside and is metabolized in the same way as a stevioside. These sweeteners are up to 480 times sweeter than sugar, because their leaves are naturally 15–30 times sweeter than sugar. It is a natural product which has a GI of zero and so it is suitable for diabetics. It is heat stable and suitable for cooking as well as suitable for use in processed foods. It has been used for centuries by the indigenous people of Paraguay, South America, and also in Asia (Japan) since the 1970s. Its standardization in the American market from 2008 onwards with GRAS recognition has proven complicated. The native plant contains proteins, fiber, iron, phosphorus, calcium, potassium, zinc, and vitamin A, and its derivatives often provide varying amounts of the active compounds which is a major limitation of their use. Various studies have been conducted to examine their effects on weight, appetite, or gut flora with results that are not entirely conclusive. More studies are needed to clarify this matter.

The EFSA recommended that ADI for stevia, or steviol glycosides, is consistent with the level adopted in the past by the Joint FAO/WHO Expert Committee on Food Additives, JECFA.

Luo han guo is a noncaloric high-intensity natural sweetener (300 times sweeter than sugar cane). It is extracted from monk fruit, in China, where it has been used for hundreds of years. Its sweetness comes from a substance called mogroside in the pulp of the fruit. One of the advantages over stevia is the lack of a bitter aftertaste characteristic of stevia. Its GI is zero and it has been approved by the FDA for use in GRAS recognition for the additives of certain foods from 2010. In Europe, it's still under evaluation [22].

One subset of natural sweeteners, which is still completely outside of the commercial market, is "sweet proteins." Despite of the fact that seven sweet proteins have been identified (thaumatin, monellin, mabinlin, pentadin, brazzein, curculin, and miraculin), only two have been commercialized: thaumatin and brazzein. All these proteins have been extracted from plants which grow in rain forests. Sweet proteins tend to have lingering aftertastes, a characteristic which clearly distinguishes them from sugar. Thaumatin is the most advanced sweet protein regarding product development and its situation with the regulatory authorities.

"Artificial sweeteners" per se usually refer to the various existing compounds on the market which are characterized by the fact that they are noncaloric, have no glycemic effect whatsoever, and are HIS. This group is of interest in research, with the aim to prove their safety and provide firm data on their possible therapeutic effects on patients with diabetes or other specific health problems. The consumer interest in these products has increased significantly looking for low-calorie products (Table 3).

Saccharin continues to dominate the global market of HIS for levels of consumption, with millions of tonnes in 2010. Asia continues to be the world's biggest consumer of saccharin. Saccharin was the first artificial sweetener discovered more than 120 years ago. Like most artificial sweeteners, it was discovered by accident looking for other unrelated substances. It is 300 times sweeter than sugar, but it has a slightly unpleasant metallic aftertaste. It has a GI of zero; it contains no calories and is suitable for diabetics. It does not tolerate high temperatures so it is not suitable for cooking. It blends well with other sweeteners or even with a small amount of sugar as in some "diet" or "zero" drinks.

Cyclamate is the second oldest artificial sweetener in use today. It is the least powerful of this group, only 40 times stronger than sugar. For this reason, it is often blended with other sweeteners like saccharin. It is heat stable and has a long shelf life which makes it suitable for cooking and food processing. It has a GI of zero and contains no calories. Like saccharin, it is also widely used in Asia. Its use is authorized in Europe and 50 other countries, but it has been prohibited in the United States since 1969 due to a reported association with developing bladder tumors in animal models and has not been reviewed since then.

Sucralose is a modified form of common sugar (sucrose) which has no calories and is 600 times sweeter than sugar. Its flavor is considerably different to table sugar and does not decompose when heated. It is commonly used all over the world, alone or with other sweeteners, and can be found in more than 4,500 foods and drinks.

Neohesperidin dihydrochalcone is a sweetener derived from the chemical modification of a substance found in bitter oranges. It is between 250 and 1,800 times sweeter than sucrose and has a long-lasting sweet flavor with a liquorice aftertaste. It has been approved in Europe but not by the FDA.

Aspartame (APM) is an artificial sweetener which is almost 200 times sweeter than sugar. It is a protein and as such, contains 4 cal per gram. However, it is so sweet that only a small amount is needed and so its calorific value is insignificant. It continues to be one of the most commonly used and well-known intense sweeteners, thanks largely to its strong market position in the United States, its main producer, which consumes 60% of the global demand for this substance. It decomposes when heated and is therefore not suitable for cooking. It has almost completely replaced saccharin as the most commonly used sweetener in "diet" drinks. There have been huge controversies over its safety although agency reports claim that it is safe for consumption. It is the biggest source of complaints to the FDA, more than any other product or medication.

Acesulfame K is another compound 130–200 times sweeter than sucrose. It is not metabolized and is eliminated and unchanged. It's frequently used in soft drinks, fruit nectars, table sweeteners, dairy products, oven-baked goods, toothpaste, and pharmaceutical products. There is a combination of APM and acesulfame whose

Table 3 Description of sugar alcohols

Sweetener	Nomenclature	Nutritional value (kcal/g)	Sweetness relative to sucrose	ADI[a] (mg/kg weight/day)		Maximum amount of sweetener (mg/day) in a 70 kg subject	No. of drinks/Over = ADI for a 70Kg subject
				EU	FDA		
Acesulfame K	E950	0	200	0–9	15	630	16/13
Aspartame	E951	4	160–220	0.40	50	2,800	15/70
Cyclamate: Cyclamic acid and sodium and calcium salts	E-952	0	30	0–7	Not permitted	490	–
Lu han guo or concentrated fruit extracts sweetener	Natural sweetener	0	150–250	Not permitted	Unspecified. Included in GRAS status	–	–
Neohesperidin DC	E959	0	1,500	0–5	–	350	–
Neotame	E961	0	8,000	0–2	18	140	Absent in carbonated drinks and not consumed in products
Saccharin and its sodium, potassium and calcium salts	E954	0	300	0–5	Unspecified	350	44/9
Stevia (steviol glycoside)	E960	0	300	0–4	4	280	16.5/31
Sucrose (Splenda)[c]	E955	0	600	0–15	5	1,050	15/95.5
Thaumatin	E957	approx. 0	2,000–3,000	Unspecified or by JECFA	Not specified. Included in GRAS status	–	–

[a]ADI: acceptable daily intake.
[c]Splenda®.

composition is 64–36%, respectively. It is known by the E number E-962 and has an immeasurable nutritional value, and its sweetness relative to sucrose is 350.

Neotame is a dipeptide derived from APM and has sweetness 8,000 times higher than sugar. Unlike APM it doesn't decompose when heated and therefore is suitable for cooking and for use in processed food. It has zero calories per portion and a GI of zero, which makes it suitable in diabetic diet. It is not metabolized to phenylalanine and so it is safe for phenylketonuria patients. It is mainly used by food manufacturers, in blends with sucrose, and other HIS. Since it was introduced to Europe in 2010, its use has grown considerably. Alitame is 2,000 times sweeter than sugar. It is a dipeptide made from aspartic acid and alanine. It is stable, have 0 cal and a GI of zero. It has been approved in Europe (E956) but not in the United States.

There are many new sweeteners like advantame, a derivative of the same amino acids as APM with vanillin, a component of vanilla. Compared to APM (about 200 times sweeter than sugar), advantame is between 20,000 and 40,000 times sweeter than sugar. It has been authorized in Australia and New Zealand and is considered GRAS as flavoring for nonalcoholic drinks, chewing gum, and dairy products.

6 The Health Effects of Sweetener Consumption

Current literature [26] and a systematical review [27] provide a rigorous evaluation of the available scientific literature to date and indicates that substituting LNCS for sugar modestly reduces body weight, body mass index (BMI), fat mass, and waist circumference. Another study in normal-weight Dutch children [28] also concluded that masked replacement of sugary beverages with artificially sweetened beverages resulted in less weight gain, but long-term data are lacking. Replacement of caloric sweeteners with lower- or no-calorie alternatives is one strategy that may help reduce energy intake, thereby facilitating weight loss or weight maintenance and prevention of weight gain [22]. But, general and metabolic impact of using these substances, mainly added to food and drinks, can affect the quality of the final product (nutritional and organoleptic properties), energy consumption, and body weight. Existing data are insufficient to clearly support or refute the effectiveness of substitution with LNCS as a mean of reducing added sugar intake. It is important not to lose sight of the impact of incorporating LNCS containing beverages and foods on overall diet quality when assessing potential health benefits versus risks [10].

Before choosing one of these substances for its supposed metabolic effects it should be compared with sugar as a reference standard. However, the current lack of knowledge on their possible effects is significant, and therefore it is difficult to support their use based on scientific evidence full of contrasts [3].

In theoretical models, these lower-calorie substances, which have less effect on blood sugar, could have a beneficial impact on diabetes; however, this correlation is unlikely. Recent results obtained through short-term intervention models show that artificial sweeteners, especially in drinks, may be useful in reducing energy intake as well as body weight and reducing the risk of type 2 diabetes and cardiovascular disease, if it's compared with sugar consumption. Long-term research is needed to confirm these hypotheses [29].

A consensus among the organizations (The American Diabetes Association, ADA, and the American Heart Association, AHA) has recently been published in order to clarify certain aspects of the effects on appetite and components of cardiometabolic syndrome. There are significant limitations on the interpretation of research data due to inherent difficulties in the design to change diet carbohydrate content without changing fat or protein, because to maintain the calorie content, proteins or fats need to increase, and this change can affect appetite.

The majority of data involving humans comes from observational studies and certain randomized controlled trials on changing sweeteners in soft drinks. In many of these studies, the data of sweetener consumption in food frequency questionnaires (FFQ) on their exact composition in the products consumed is not well documented due to incomplete information on labels or industry references on the quantities contained in processed foods. On the other hand, experimental research on animals provides important data on the potential adverse effects or toxicity of sweeteners. However, extrapolation of these results to general population has major limitations [8]. Although the body of evidence on LNCS and body weight has grown in recent years, several research questions remain without answer.

6.1 Sweeteners and Energy Intake

A priori, it would be logical to think, from an energy intake point of view, that substitution of sugar with lower-calorie sweeteners should reduce the total energy consumption.

To date, research examining LNCS and body weight has produced mixed results about the use of LNCS. A recent review of randomized controlled trials [30] and new controlled trial [31] provides suggestive evidence that LNCS do not increase the desire or inclination to consume more sweet foods, contrary to past hypotheses [32, 33]. Furthermore, LNCS could help improve adherence to weight loss or maintenance plans by helping to reduce energy intake and maintaining the palatability of foods and beverages with fewer calories than sugar [34]. However, leading nutrition and health authorities recommend a multifaceted approach to weight loss and weight maintenance – one that includes an overall healthy dietary pattern, physical activity, and other lifestyle behavior changes [35, 36].

However, the research to date about LNCS and body weight is controversial. Because, it should also be taken into account that in processed products, not only the sweetness provided by sugar needs replacing but also the physical or other technological properties of sugar. The result is that a product reformulated with less sugar is often higher in calories than the "full sugar" version because the sugar has been replaced with other higher-calorie nutrients, such as fat.

Some studies on humans have shown a short-term reduction in calorie intake resulting from only a partial compensation of the calories that are not ingested when compared with sucrose, mainly in soft drinks. However, there is also epidemiological data which connects the use of sweeteners with weight gain. It would appear that the dissociation between the sensation of sweetness and the reduced calorie content produced by sweeteners could cause an increase in appetite, giving rise to a higher energy consumption and weight gain. This operative conditioning theory (Pavlov's Model) has been demonstrated in animal models [37, 38].

On the other hand, dietary intake is also influenced by the mechanisms and behaviors involved in food selection. From the behavioral point of view, persons might consumpt LNCS as a way to consume an unhealthy diet or even larger portion sizes because of the common belief that "diet" products (which typically contain LNCS) have fewer calories.

Several observational studies have also described the association of caloric and LNCS beverages use with an overall poorer quality diet due to the loss of healthy eating patterns which include fruit and vegetables. There are many associated confounding factors which make it difficult to draw clear conclusions, such as the fact that these low-calorie foods are frequently associated with other higher-calorie foods, and individuals choose them precisely in order to reduce their overall calorie intake. A recent study has examined this association using data on yearly purchases by individuals included in Homescan from 2000 to 2010 (Nielsen Consumer Panel and Retail Measurement) [39]. Findings suggest that consumption of any type of sweetened beverage might negatively affect diet. Another study on the long-term associations between different profiles of beverage consumers and dietary patterns [40] found that households in which either caloric or LNCS beverages were consumed were significantly less likely to follow healthier dietary patterns compared with nonconsumers. However, LNCS beverage consumers also had a higher probability of following a prudent dietary pattern, which is characterized by eating fruits/vegetables augmented with snacks and LNCS desserts. Other study reported that consumers of LNCS beverages had a lower cardiometabolic risk [41].

Furthermore, it is known that, repeated exposure to LNCS uncoupled with energy was hypothesized to modify the natural relationship between sweet taste and energy, which could affect appetite and energy intake by disrupting hormonal and neurobehavioral pathways that control satiety [32, 42]. Both in humans and animals, food consumption causes a thermogenic response in the cephalic phase of digestion. This response prepares the gastrointestinal tract for the arrival of nutrients. There is evidence in rodent models that the chronic use of sweeteners, such as saccharin, causes a reduction in this stimulus and slows down the thermic effect of food and perhaps other aspects of metabolic equilibrium [37]. As opposed a recent randomized controlled trial showed that children randomized to LNCS or caloric sweetener beverages had no differences in satiety [43].

It has been proven that sweeteners can play an active role in the gastrointestinal tract by reacting with the sweet taste receptors (T1R family of receptors and a-gustducin) and mediating changes in peptide hormone responses such as glucagon-like peptides (GLP) in intestinal L cells. Therefore it has been hypothesized that the concomitant intake of artificial sweeteners together with food or drinks containing sugar could enable faster absorption of sugar, as well as enhancing GLP-1 and insulin secretion, affecting weight, appetite, and blood sugar [44].

6.2 Sweeteners and Appetite Regulation

The mechanisms that sweeteners use to modulate appetite include:

(a) Cephalic phase stimulation. In this respect, some studies maintain the hypothesis that failure to stimulate the cephalic phase response can increase the risk of obesity; conversely, others claim that stimulation of cephalic phase responses, from ingesting or simply being exposed to sweet foods, can be problematic because it stimulates both appetite and food intake. Another proposed mechanism could be mediated by the direct effect of LNCS on insulin secretion and glucose metabolism [30].

(b) Nutritional and osmotic effects. It is known that the stomach produces appetite signals, mainly based on the volume it can or cannot cover, whereas the bowel is more sensitive to signs of the presence of nutrients. This hypothesis does not appear to be fulfilled as strictly as the presence of osmoreceptors, at an intestinal level, and chemoreceptors, at a gastric level. It has been proven that with gastric distension, whether it's due to the presence of nutrients or for another reason (gastric balloon), the feeling of fullness increases. Drinks which contain caloric sweeteners have more energy in terms of osmotic load which can be the same or even less than the load produced by LNCS, which means that with the same osmolarity, the calorie content of LNCS is less; therefore gastric emptying doesn't just depend on osmolarity (chemoreceptors/osmoreceptors).
However, caloric sweeteners cause slower emptying regardless of the osmotic effects.

Activation of signals both in the gut and in the stomach, from the presence of nutrients, has a synergistic effect on satiety. It has been hypothesized that drinks containing LNCS can weaken this effect present in those which contain nutritive sweeteners, even though no clear data really exists on this matter [30].

(c) Responses of gastrointestinal peptides. Each macronutrient stimulates the release of peptides in the digestive tract with varying degrees of effectiveness. So, it has been proven that carbohydrates stimulate the secretion of GLP-1, which plays a significant role both as a satiety and incretin factor.
It is thought that LNCS do not permit such a release of peptides, and therefore, in theory, this would mean a lesser feeling of satiety and would cause an increase in energy intake.

Some more recent evidence shows that there are receptors, with properties similar to the sweet receptors located on the tongue, in the gastrointestinal tract that stimulate the release of GLP-1, which could give LNCS a role in regulating these incretin systems [45].

(d) Palatability. Another of the major benefits of using LNCS, as part of the diet, is to improve the organoleptic properties of the food in question, thus enabling improved acceptance both of the foods themselves and reduced calorie meals in which any food of this type is used, compared with its original higher-calorie version, which may contain sugar as such and which undoubtedly contributes to optimal organoleptic properties. This might be a huge advantage in overweight, obese, or diabetic patients to improve adherence to treatment regimes and changes in nutritional habits. The hypothesis is whether the degree of food palatability affects appetite sensation, but following numerous studies, there is still no conclusive evidence on this matter.

(e) Changes in gut microbiota. It has been proven that changes in bacterial populations, which make up the gut microbiota, can contribute to the low-grade chronic inflammatory process which is observed in some obese patients and which seems to promote weight gain at the expense of fat mass, as well as actively contributing to the development of the comorbidities typically associated with obesity, such as insulin resistance [46].
APM releases a methanol molecule, which is metabolized into a formaldehyde molecule, a highly reactive substance, classified as carcinogenic. However, the amount of these dangerous substances that are ingested is usually below the levels of risk. Therefore, it is not unusual for very small amounts of sweeteners to alter intestinal flora, as they act as the first line of defense in the gut and are therefore in direct contact with the sweetener and its metabolic compounds. During low-calorie diets for weight control, the use of sweeteners like APM can change the optimal functioning of gut microbiota [47].
(f) Overcompensation. Studies show that saving/withdrawing energy by substituting foods with LNCS could subsequently lead to overcompensation of food intake later on, which may even exceed the energy deficit induced by the sweetener and therefore cause a positive energy balance.
(g) Loss of signal fidelity. Certain sensory properties of food influence the metabolic response required for each product. So, if the sensory input of sweetness by LNCS leads to an inaccurate or inconsistent prediction, energy regulation could be affected and lead to a positive energy balance due to excessive intake caused by these signals.
(h) Activation of reward systems. It's possible that the improved palatability of sweetened products could play a role stimulating food reward.
(i) Learning through the positive reinforcement of sweet flavors. It refers to the possibility that repeated exposure to LNCS can perpetuate a preference for sweet products in the diet, including those sweetened with caloric sweeteners [8].

6.3 Sweeteners and Their Effect on Body Weight

For many years, weight management has been one of the main reasons for the extensive use of sweeteners as part of a regular diet. However, from 1986 onwards doubts surfaced over the possible effects on weight gain, according to the results of surveys conducted by the American Cancer Society (ACS) [30].

Furthermore, in many instances an increase in its use has not been accompanied by a reduction in nutritive sweeteners, which they are intended to replace with the aim of reducing calorie intake; consequently, intake remained unchanged. The extent to which foods are chosen where nutritive sweeteners have been replaced by other LNCS should also not be overlooked, as, in the majority of cases, this leads to an increase in fat and protein content which may be trying to compensate for the calorie deficit caused by the food containing LNCS.

Changes in appetite regulation are responsible for changes in energy intake and therefore managing body weight. So, all the mechanisms of appetite regulation mentioned in the previous section may be involved in weight gain.

Changes in neural response mechanisms have been proposed as a possible explanation for the weight gain associated with sweetener use. It is known that the act of eating and the satisfaction derived from it is the result of sensory stimulation from foods after ingestion. In humans, when a food is swallowed, the taste, which is detected by receptors found in the oral cavity, ascends via the thalamus and reaches the area of the anterior insula/frontal operculum as well as the frontal orbital cortex. Similarly, the amygdalae also make connections, via the taste pathway, at every level. Last but not least, the role played by the mesolimbic dopaminergic system is discussed, as it is responsible for recognizing the stimuli and the pleasure/satisfaction sensation following the ingestion of food.

Following studies on rats, it has demonstrated that the hypothalamus mediates the postprandial effect on food reward system, given its different functions in the secretion of various peptides which regulate energy, osmotic equilibrium, and behavior in the presence of food. We have more evidence which shows that artificial sweeteners do not activate the food reward cascade in the same way as natural sweeteners, as it appears that the lack of calories suppresses the post-ingestive component. Moreover, the gustatory branch activation mechanism also differs in each case.

The sweet taste of LNCS may boost the appetite and dependency on such flavors, and there is a high correlation between the repeated exposure to a flavor and the degree of preference for it. Research in this field, but on the reduction of fat and salt in the diet, showed how with reduced exposure, the group's preference for these products diminished. A possible theory has been put forward that the presentation of unsweetened diets could be one of the keys to reduce sugar consumption and consequently reverse the obesity epidemic [48].

These aforementioned hypotheses, which have already been brought to light in previous studies, are also reflected in the research carried out by the American Academy of Nutrition and Dietetics [4], where it was observed that the sweeter a product is, the higher the consumption of food or drink will be. For the test subjects, the effect on appetite, caused by the repeated exposure to sweeteners, is due to an interruption of the hormonal and neurobehavioral pathways responsible for controlling hunger and satiety.

With regard to the risks of cardiovascular disease associated with weight gain, the prospective observational studies which exist to date only allow the identification of casual associations, but they are not in any way determinants; in many cases reverse causality is very plausible for some of the significant associations observed between obesity or type 2 diabetes and LNCS [8].

6.4 Sweeteners and Diabetes

The potential benefit attributed to nonnutritive sweeteners (NNS) for diabetics is the reduction in calories and carbohydrates which improves weight control and blood sugar, respectively.

Various studies have shown that the use of LNCS does not seem to affect blood sugar or plasma lipid levels in adults with diabetes; it has not been researched sufficiently in children [49]. Diabetics should take into account total carbohydrate intake in order to improve blood sugar control. It has been suggested that blood sugar and weight control can be improved by using LNCS better than with sugary foods.

Two transporters are involved in the control of glucose absorption through the gut wall, the sodium-glucose linked transporter (SGLT1) which has an active role as a transporter in the apical membrane, and the facilitated glucose transporter (GLUT2). Both are present in the basolateral and apical membrane. The cells responsible for their absorption are enterocytes. Sugars, as well as LNCS that may be present in the diet, increase SGLT1 mRNA, protein expression, and the absorption capacity for glucose; furthermore, given the relationship between SGLT1 activity and the insertion of GLUT2 in the apical membrane, T1R3 stimulation (sweet taste receptor subunit) also promotes a greater insertion of GLUT2.

Enteroendocrine cells communicate with enterocytes by producing signals which are detected by the latter, increasing SGLT1 expression. These incretin signals include glucose-dependent insulinotropic polypeptides (GIP) and GLP-1, which have numerous effects on glucose metabolism, including the stimulation of insulin release, inhibition of glucagon secretion, reduced gastric emptying, and an increase in the feeling of fullness. As for the other mechanisms that have been described, the date available comes from in vitro and other short-term studies, as well as studies on animals, which implies significant limitations on the extrapolation of results to human studies [50].

The effects of specific sweeteners on postprandial glycemia, insulin, and blood lipids have also been studied. In the comparison of a diet rich in sucrose versus another which, in contrast, contains LNCS, there is a significant increase in both postprandial blood glucose and postprandial insulinemia and blood lipid levels in a slightly overweight but healthy population, for the group with a diet rich in sucrose [51].

If we look at the latest American Diabetes Association (ADA 2015) recommendations [52], with a B level of evidence, it states that in the case of type 2 diabetes mellitus and those at risk, intake of sugar-sweetened beverages should be limited or avoided to reduce risk for weight gain and worsening of cardiometabolic risk profile. There are no specific ADA recommendations that include other sources of sweeteners, apart from in soft drinks, where decreased consumption is recommended. We can also add that there are no recommendations on limiting the use of other sources of sweeteners.

Information on sweetener use should be transmitted clearly during diabetic education sessions. Accurate proven information, based on the best scientific evidence available, is necessary, so that relevant decisions and recommendations concerning their consumption can be made. In this respect, it is essential to disprove the myths, which frequently surrounds this issue, as well as combat the disinformation/misinformation that come across on the Internet and in the media on a daily basis. Research on LNCS gathered by regulatory agencies (FDA) contributes to their safe use and their potential benefits in controlling blood sugar.

6.5 Sweeteners and Tooth Decay

A cavity is formed by the localized destruction of hard dental tissue by acidic material which comes from the fermentation processes carried out by certain pathogenic bacteria, cariogenic bacteria, from fermentable carbohydrates present in the diet. Other factors which contribute to the development of tooth decay are microbiological changes in bacterial flora, saliva composition and its pH buffering capacity, the type of sugary foods consumed and the frequency of their consumption, and the quality and regularity of oral hygiene. Among the large group of sweeteners and according to the health claims that they help prevent cavities, the sugar alcohols erythritol, D-tagatose, sucralose, and isomaltulose, have been approved for consumption [22].

6.6 Sweeteners and Cancer

The role of low-calorie sweeteners on cancer risk has been widely debated since the 1970s, when animal studies found an excess bladder cancer risk in more than one generation of rodents treated with extremely high doses of saccharin. This was however not confirmed in subsequent studies, and mechanistic data showed different saccharin metabolism in rodents and humans. However, others subsequent case–control studies of bladder or low urinary tract cancers found no significant association with consumption of sweeteners [53].

The Italian network of cases and control study, performed between 1991 and 2004, showed a lack of association between LNCS and the risk of several common neoplasms (oral cavity, pharynx, esophagus, colon, rectum, renal cell carcinoma, etc.) [54]. Very recent studies done with humans have analyzed the possible epidemiological relationship between the consumption of LNCS and different kinds of cancer (hematopoietic or brain malignancies, gastric, pancreatic, or endometrial carcinoma, etc.), without finding any type of relationship or trend [55, 56].

Consensus concerning the effect of diet and nutritional supplement usage is an evolving area. When studying the relation between sweeteners and cancers, the role of obesity and sugar on carcinogenesis should be taken into account, since the use of sweeteners is inversely correlated with sugar. And obesity and sugar has been directly associated to the risk of certain cancers [57].

7 A Global View of Sweetener Use: SWOT Analysis

Once all the matters relating to the general aspects of sweetener use and their possible metabolic effects on the body have been addressed, an overview of their use will be given, in accordance with the SWOT analysis system (Fig. 3).

Strengths

- Growing interest and the discovery of nutraceutical products with sweetening properties.
- The supposed beneficial effects of sweetener use as a whole, on metabolism, in different diseases (obesity, diabetes, tooth decay) carving out a niche for them in the food industry.
- LNCS provide sweetness without the extra calories or in the worst case, never as high as sugar.
- Large potential market demand associating them with a more balanced diet, although according to international organization recommendations such as the FAO/WHO, the consumption of simple carbohydrates (sugars) below 10% of the total dietary energy is acceptable as part of a healthy balanced diet.
- They appear to help limit refined sugar consumption in the diet

Weaknesses

- The general population, including many health professionals, often lack correct knowledge on the particular characteristics of the different sweeteners available in the market, such as advising on and/or choosing a particular sweetener based on its properties.

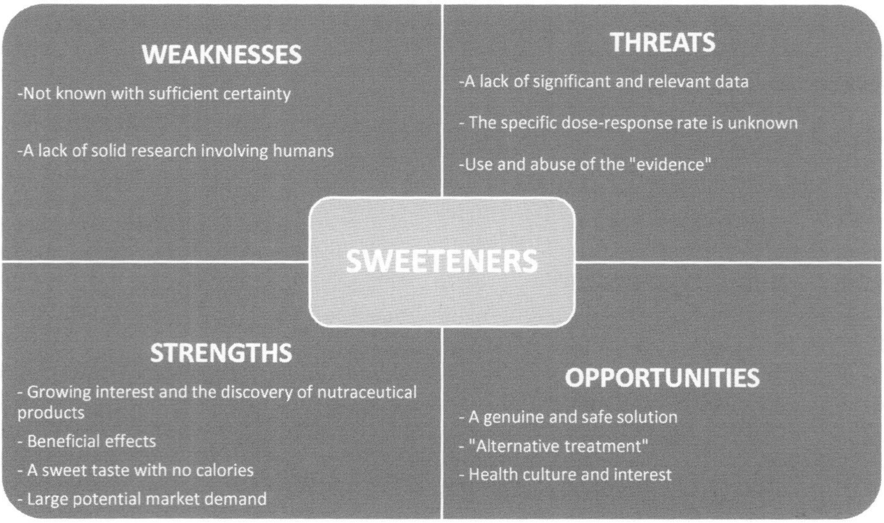

Fig. 3 Global View of Sweetener Use: SWOT Analysis

- Even though the production-demand binomial is gradually becoming more evident in society, in the search for "possible solutions" to improve health, there is actually a real lack of research on humans which confirms their potential benefits.

Opportunities

- They could eventually be a genuine and safe solution, taking into account an increase in chronic illnesses in society today (diabetes, obesity).
- They could become an "alternative treatment," for the prevention and favorable evolution/management of certain illnesses.
- As the years go by, the culture and interest in looking after the body and achieving optimum health grows. Included here are any criteria which could be a way of achieving this (i.e., physical activity, specific diets, consumption of diet foods).

Threats

- A lack of significant and relevant scientific data.
- The specific dose–response rate, which clarifies the metabolic effects of their use, is unknown.
- Derived from previous ideas, use and abuse of "evidence" to date in order to recommend for use.

8 Recommendations

Acording to Existing evidence shows that sugar is not really the problem, but its excess. Increased sugar intake amplifies the risk of weight gain, diabetes, and many other cardiometabolic problems. Further research is needed to investigate the health effects of beverages containing LNCS and 100% fruit juices.

Considering the existing controversy over the potential benefits and the value and cost of sweeteners for public health nowadays because of the high incidence of chronic diseases (in particular obesity and its long-term consequences), sweeteners may offer to people who need to lose weight "a little help," as well as both a primary and secondary preventative measure in the treatment of obesity and its associated conditions, but it is not going to "solve the problem." Healthy structured eating patterns which tend toward calorie balance through a variety of natural, healthy foods and are adapted to our customs are essential, without forgetting to combine this with a healthy active lifestyle. It is important not to lose sight of the impact of incorporating LNCS containing beverages and foods on overall diet quality when assessing potential health benefits versus risks.

We need to avoid the indiscriminate use of them [30], as their potential interest as a tool for preventing excess weight or diabetes, even in the healthy population who want to take care of their health, has not been proven with evidence that supports their beneficial effects over the alternative standard caloric sweeteners [3].

In any case, the American Academy of Nutrition and Dietetics' position that any individual can use them safely, but only if they form part of a diet based on dietary recommendations and reference intakes for the general population, without forgetting personal preferences and health objectives. Irrespective of LNCS use in the diet, it is essential to control the total energy intake and increase the degree of physical activity in order to maintain body weight.

Recommendations from scientific societies (ADA, AHA) confirm that sugar alcohols and NNS are safe if daily dietary intake is within the levels established by the regulatory agencies (FDA, AESAN) [22].

9 Conclusion

WHO, major scientific bodies, and most countries recognize the importance of reducing consumption of sugar-sweetened beverages to improve public health. However, the absence of a consensus of evidence on beverages containing LNCS and fruit juices creates a practical conundrum for policy formulation.

The study of the potential benefits and risks of substituting LNCS for added sugars is fraught with important limitations and challenges. So, existing data are insufficient to clearly support or refute the effectiveness of substitution with LNCS as a mean to reduce added sugar intake.

In addition, the existing evidence on using LNCS as part of the population's regular diet and nutrition is lacking in long-term results which are of significant scientific relevance, and the majority are epidemiological studies.

There are plenty of results on their effects and benefits from studies on animals, but not so many on humans due to bias and limitations on the interpretation and extrapolation of population data. On the other hand, it is necessary to determine the exact dose–response rate, which explains the metabolic effects of their use.

Likewise, and even though sugar consumption can be limited in patients with metabolic disorders, there is no evidence that recommendations on the use of sweeteners has been sufficiently scientifically proven to recommend the supposed long-term benefits of their use.

In 2009 the AHA concluded that limiting added sugars is a core strategy for maintaining optimum nutrition and a healthy weight. ADA has included monitoring carbohydrate consumption (which includes limiting added sugars) in their clinical practice recommendations, as a key strategy [8].

Finally, it should be noted that all LNCS approved for use are deemed to be safe, within permitted usage levels. Intake estimates are difficult to assess, if you also take into consideration that, in the majority of cases, food products contain a mixture of them which makes them even more difficult to calculate. It needs more studies to make practical decisions about how to treat drinks with LNCS and new designs of policies to tackle consumption of sweetened beverages. Until further studies are available to help address the stated limitations and theses challenges of investigating, it would be wise to continue to strive for optimizing overall diet quality, including limiting added sugars and avoiding excessive energy intake.

It is essential that future studies on their consumption include a sufficient number of subjects, consumers in the 95th percentile, and even other groups that may have an above average intake (e.g., diabetics) or groups with particular issues (pregnant women or children).

On the other hand, a key challenge for policy makers and researchers is the absence of a consensus on the relation of beverages containing LNCS with cardiometabolic outcomes, since decisions about whether these are healthy substitutes for sugar-sweetened beverages are an integral part of policy design. Ongoing randomized controlled trials should hopefully help to reach a consensus on the effects of LNCS on cardiometabolic health outcomes.

The evaluation of not only sugar taxes but also new marketing controls and front-of-pack labeling is important and represents one of the next frontiers – namely, these policies can effectively reduce consumption of sugar.

References

1. Anderson GH, Foreyt J, Sigman-Grant M, Allison DB (2012) The use of low-calorie sweeteners by adults: impact on weight management. J Nutr 142(6):1163S–1169S
2. Popkin BM, Hawkes C (2015) Sweetening of the global diet, particularly beverages: patterns, trends, and policy responses. Lancet Diab Endocrinol [Internet]. [citado 25 de enero de 2016]; Recuperado a partir de: http://linkinghub.elsevier.com/retrieve/pii/S2213858715004192
3. Wiebe N, Padwal R, Field C, Marks S, Jacobs R, Tonelli M (2011) A systematic review on the effect of sweeteners on glycemic response and clinically relevant outcomes. BMC Med 9:123
4. Ng SW, Slining MM, Popkin BM (2012) Use of caloric and noncaloric sweeteners in US consumer packaged foods, 2005–2009. J Acad Nutr Diet 112(11):1828–1834.e1-6
5. Basu S, Yoffe P, Hills N, Lustig RH (2013) The relationship of sugar to population-level diabetes prevalence: an econometric analysis of repeated cross-sectional data. PLoS One 8(2): e57873
6. Imamura F, O'Connor L, Ye Z, Mursu J, Hayashino Y, Bhupathiraju SN, et al (2015) Consumption of sugar sweetened beverages, artificially sweetened beverages, and fruit juice and incidence of type 2 diabetes: systematic review, meta-analysis, and estimation of population attributable fraction. [citado 18 de enero de 2016]; Recuperado a partir de: http://www.bmj.com/content/351/bmj.h3576.full
7. Xi B, Li S, Liu Z, Tian H, Yin X, Huai P et al (2014) Intake of fruit juice and incidence of type 2 diabetes: a systematic review and meta-analysis. PLoS One 9(3):e93471
8. Gardner C, Wylie-Rosett J, Gidding SS, Steffen LM, Johnson RK, Reader D et al (2012) Nonnutritive sweeteners: current use and health perspectives: a scientific statement from the American Heart Association and the American Diabetes Association. Diabetes Care 35 (8):1798–1808
9. Greenwood DC, Threapleton DE, Evans CEL, Cleghorn CL, Nykjaer C, Woodhead C et al (2014) Association between sugar-sweetened and artificially sweetened soft drinks and type 2 diabetes: systematic review and dose–response meta-analysis of prospective studies. Br J Nutr 112(5):725–734
10. Gardner C (2014) Non-nutritive sweeteners: evidence for benefit vs. risk. Curr Opin Lipidol 25 (1):80–84
11. Office of the Surgeon General (US), Office of Disease Prevention and Health Promotion (US), Centers for Disease Control and Prevention (US), National Institutes of Health (US) (2001) The surgeon general's call to action to prevent and decrease overweight and obesity [internet]. Office

of the Surgeon General (US), Rockville. [citado 31 de enero de 2016]. Recuperado a partir de: http://www.ncbi.nlm.nih.gov/books/NBK44206/

12. Hu FB (2013) Resolved: there is sufficient scientific evidence that decreasing sugar-sweetened beverage consumption will reduce the prevalence of obesity and obesity-related diseases. Obes Rev 14(8):606–619
13. Malik VS, Popkin BM, Bray GA, Després J-P, Willett WC, Hu FB (2010) Sugar-sweetened beverages and risk of metabolic syndrome and type 2 diabetes: a meta-analysis. Diabetes Care 33(11):2477–2483
14. Anand SS, Hawkes C, de Souza RJ, Mente A, Dehghan M, Nugent R et al (2015) Food consumption and its impact on cardiovascular disease: importance of solutions focused on the globalized food system: a report from the workshop convened by the World Heart Federation. J Am Coll Cardiol 66(14):1590–1614
15. Te Morenga LA, Howatson AJ, Jones RM, Mann J (2014) Dietary sugars and cardiometabolic risk: systematic review and meta-analyses of randomized controlled trials of the effects on blood pressure and lipids. Am J Clin Nutr 100(1):65–79
16. Te Morenga L, Mallard S, Mann J (2013) Dietary sugars and body weight: systematic review and meta-analyses of randomised controlled trials and cohort studies. BMJ 346:e7492
17. Bes-Rastrollo M, Schulze MB, Ruiz-Canela M, Martinez-Gonzalez MA (2013) Financial conflicts of interest and reporting bias regarding the association between sugar-sweetened beverages and weight gain: a systematic review of systematic reviews. PLoS Med 10(12): e1001578; dicsussion e1001578
18. Guideline: Sugars Intake for adults and children. Geneva: World Health Organization. (2015) ISBN-13: 978-92-4-154902-8
19. Thow AM, Hawkes C (2014) Global sugar guidelines: an opportunity to strengthen nutrition policy. Public Health Nutr 17(10):2151–2155
20. WCRF (2015) Curbing global sugar consumption: effective food policy actions to help promote healthy diets and tackle obesity. World Cancer Research Fund International, London
21. Council on School Health, Committee on Nutrition (2015) Snacks, sweetened beverages, added sugars, and schools. Pediatrics 135(3):575–583
22. Fitch C, Keim KS, Academy of Nutrition and Dietetics (2012) Position of the Academy of Nutrition and Dietetics: use of nutritive and nonnutritive sweeteners. J Acad Nutr Diet 112(5):739–758
23. Directiva 94/35/CE del Parlamento Europeo y del Consejo de 30 de Junio de 1994. Diario Oficial de las Comunidades Europeas,10.09.94, N° L 237/3
24. Renwick A et al (2000) Edulcorantes bajos en calorías: funciones y beneficios (monografía). International Sweeteners Association. En: Anales de la Universidad de Chile [Internet]. [citado 18 de enero de 2016]. pp 1–31. Recuperado a partir de: http://www.revistas.uchile.cl/index.php/ANUC/article/viewArticle/2524/2436
25. Mortensen A (2006) Sweeteners permitted in the European Union: safety aspects. Scand J Food Nutr 50(3):104–116
26. Roberts JR (2015) The paradox of artificial sweeteners in managing obesity. Curr Gastroenterol Rep [Internet]. [citado 18 de enero de 2016];17(1). Recuperado a partir de: http://link.springer.com/ 10.1007/s11894-014-0423-z
27. Miller PE, Perez V (2014) Low-calorie sweeteners and body weight and composition: a meta-analysis of randomized controlled trials and prospective cohort studies. Am J Clin Nutr 100(3): 765–777
28. De Ruyter JC, Olthof MR, Seidell JC, Katan MB (2012) A trial of sugar-free or sugar-sweetened beverages and body weight in children. N Engl J Med 367(15):1397–1406
29. Raben A, Richelsen B (2012) Artificial sweeteners: a place in the field of functional foods? Focus on obesity and related metabolic disorders. Curr Opin Clin Nutr Metab Care 15(6):597–604
30. Mattes RD, Popkin BM (2009) Nonnutritive sweetener consumption in humans: effects on appetite and food intake and their putative mechanisms. Am J Clin Nutr 89(1):1–14

31. Piernas C, Tate DF, Wang X, Popkin BM (2013) Does diet-beverage intake affect dietary consumption patterns? Results from the Choose Healthy Options Consciously Everyday (CHOICE) randomized clinical trial. Am J Clin Nutr 97(3):604–611
32. Blundell JE, Hill AJ (1986) Paradoxical effects of an intense sweetener (aspartame) on appetite. Lancet Lond Engl 1(8489):1092–1093
33. Stellman SD, Garfinkel L (1986) Artificial sweetener use and one-year weight change among women. Prev Med 15(2):195–202
34. Foreyt J, Kleinman R, Brown RJ, Lindstrom R (2012) The use of low-calorie sweeteners by children: implications for weight management. J Nutr 142(6):1155S–1162S
35. McGuire S (2011) U.S. Department of Agriculture and U.S. Department of Health and Human Services, Dietary Guidelines for Americans, 2010. 7th Edition, Washington, DC: U.S. Government Printing Office, January 2011. Adv Nutr Bethesda Md 2(3):293–294
36. Seagle HM, Strain GW, Makris A, Reeves RS, American Dietetic Association (2009) Position of the American Dietetic Association: weight management. J Am Diet Assoc 109 (2):330–346
37. Swithers SE, Martin AA, Davidson TL (2010) High-intensity sweeteners and energy balance. Physiol Behav 100(1):55–62
38. Davidson TL, Swithers SE (2004) A Pavlovian approach to the problem of obesity. Int J Obes Relat Metab Disord J Int Assoc Study Obes 28(7):933–935
39. Piernas C, Ng SW, Mendez MA, Gordon-Larsen P, Popkin BM (2015) A dynamic panel model of the associations of sweetened beverage purchases with dietary quality and food-purchasing patterns. Am J Epidemiol 181(9):661–671
40. Piernas C, Mendez MA, Ng SW, Gordon-Larsen P, Popkin BM (2014) Low-calorie- and calorie-sweetened beverages: diet quality, food intake, and purchase patterns of US household consumers. Am J Clin Nutr 99(3):567–577
41. Duffey KJ, Steffen LM, Van Horn L, Jacobs DR, Popkin BM (2012) Dietary patterns matter: diet beverages and cardiometabolic risks in the longitudinal Coronary Artery Risk Development in Young Adults (CARDIA) Study. Am J Clin Nutr 95(4):909–915
42. Swithers SE, Davidson TL (2008) A role for sweet taste: calorie predictive relations in energy regulation by rats. Behav Neurosci 122(1):161–173
43. De Ruyter JC, Katan MB, Kuijper LDJ, Liem DG, Olthof MR (2013) The effect of sugar-free versus sugar-sweetened beverages on satiety, liking and wanting: an 18 month randomized double-blind trial in children. PLoS One 8(10):e78039
44. Brown RJ, de Banate MA, Rother KI (2010) Artificial sweeteners: a systematic review of metabolic effects in youth. Int J Pediatr Obes 5(4):305–312
45. Brown RJ, Walter M, Rother KI (2009) Ingestion of diet soda before a glucose load augments glucagon-like peptide-1 secretion. Diabetes Care 32(12):2184–2186
46. Pepino MY, Bourne C (2011) Non-nutritive sweeteners, energy balance, and glucose homeostasis. Curr Opin Clin Nutr Metab Care 14(4):391–395
47. Wu GD, Chen J, Hoffmann C, Bittinger K, Chen Y-Y, Keilbaugh SA et al (2011) Linking long-term dietary patterns with gut microbial enterotypes. Science 334(6052):105–108
48. Yang Q (2010) Gain weight by "going diet?" Artificial sweeteners and the neurobiology of sugar cravings: neuroscience 2010. Yale J Biol Med 83(2):101–108
49. Scientific Opinion on the substantiation of health claims related to intense sweeteners and contribution to the maintenance or achievement of a normal body weight (ID 1136, 1444, 4299), reduction of post-prandial glycaemic responses (ID 4298), maintenance of normal blood glucose concentrations (ID 1221, 4298), and maintenance of tooth mineralisation by decreasing tooth demineralisation (ID 1134, 1167, 1283) pursuant to Article 13(1) of Regulation (EC) No 1924/2006. (2011) EFSA J 9(6): 2229 [26 pp.]
50. Renwick AG, Molinary SV (2010) Sweet-taste receptors, low-energy sweeteners, glucose absorption and insulin release. Br J Nutr 104(10):1415–1420

51. Raben A, Møller BK, Flint A, Vasilaris TH, Christina Møller A, Juul Holst J, et al (2011) Increased postprandial glycaemia, insulinemia, and lipidemia after 10 weeks' sucrose-rich diet compared to an artificially sweetened diet: a randomised controlled trial. Food Nutr Res 2011. 55 5961 - DOI: 10.3402/fnr.v55i0.5961
52. American Diabetes Association (2015) Standards of medical care in diabetes. Diabetes Care 38(Suppl 1):S1–S2
53. Morrison AS, Buring JE (1980) Artificial sweeteners and cancer of the lower urinary tract. N Engl J Med 302(10):537–541
54. Gallus S, Scotti L, Negri E, Talamini R, Franceschi S, Montella M et al (2007) Artificial sweeteners and cancer risk in a network of case–control studies. Ann Oncol Off J Eur Soc Med Oncol ESMO 18(1):40–44
55. Lim U, Subar AF, Mouw T, Hartge P, Morton LM, Stolzenberg-Solomon R et al (2006) Consumption of aspartame-containing beverages and incidence of hematopoietic and brain malignancies. Cancer Epidemiol Biomark Prev Publ Am Assoc Cancer Res Cosponsored Am Soc Prev Oncol septiembre de 15(9):1654–1659
56. Bosetti C, Gallus S, Talamini R, Montella M, Franceschi S, Negri E et al (2009) Artificial sweeteners and the risk of gastric, pancreatic, and endometrial cancers in Italy. Cancer Epidemiol Biomark Prev Publ Am Assoc Cancer Res Cosponsored Am Soc Prev Oncol 18(8): 2235–2238
57. Blackadar CB (2016) Historical review of the causes of cancer. World J Clin Oncol 7(1):54–86
58. Anderseon GH, Foreyt J, Sigman-Grant M, Allison DB (2012) The use of low calorie sweeteners by adults: impact on weight management. J Nutr 142(6):1163S–1169S

Index

A
alcohol/aqueous alcohol, 107
Abiotic stresses, 197
Absorption, 367, 412, 413, 418, 419, 425
Acceptable daily intake (ADI), 103, 123, 443, 454, 550, 619
Accumulation, thaumatin, 199, 201
Acesulfame, 447
Acesulfame-potassium, 5, 212, 312, 383, 626
Acetic acid, 137
Achenes, 52
Acute otitis media (AOM), 425, 426
Added sugar, 227
Adduct, 464
Adherence, 412, 414, 426, 427
ADI. *See* Acceptable daily intake (ADI)
Adiponectin, 237
Advantame, 379, 445
Aftertaste, 191, 196, 204, 560, 561
Agency for Food Environmental and Occupational Health & Safety (ANSES), 446
Agence Française de Sécurité Sanitaire des Aliments (AFSSA), 124
Aglycone, 154, 155, 246
  of kaempferol, 360
Agrobacterium, 28, 172
  transformation by, 171
Agronomic researches, 73
Agro-techniques, 38
Aldolase B, 292
Aldosterone, 20
Alien DNA, 203
Alitame, 456
Allergenicity, 542
Allergens, 187
Allogamous species, 40
Alpha (α) glucosidase, 419
18α-glycyrrhizinic acid, 251
Alveolar/bronchial adenoma, 333
Alveolar carcinomas, 333
Amendments, 57
American Diabetes Association (ADA), 629, 634
American Heart Association (AHA), 629
Amino acids, 7, 20–30, 169, 188–189, 356
Amino-based/reversed-phase columns, 120
AMP-activated protein kinase (AMPK), 402
AMP deaminase 2, 236
Animal models, 367, 370
Annual crop, 67
Anoxerigenic properties, 401
Anti-caries action, 560
Anticonvulsant effects, 277
Antidiabetic effects, 159
Anti-filarial agent, 254
Antifungal activity, 187
Anti-hyperglycaemic effects, 267
Anti-hyperglycemic activity, 354
Anti-hypertensive, 276, 354
Anti-inflammatory compound, 94
Antimicrobial activity, 93, 273
Antimicrobial effects, 264, 411, 425
Antimicrobial properties, 354
Antimutagenic activity, 95, 97
Antioxidant(s), 94, 357, 413, 423, 424
  activity, 360
  compounds, 356
  defense, 367
  effects, 264
  enzymes, 365, 368, 369
  potential in vitro, 368
  systems, 70
Antiparasitic activity, 254
Antiradical activity, 360
Antitumor activity, 274

J.-M. Mérillon, K.G. Ramawat (eds.), *Sweeteners*, Reference Series in Phytochemistry,
https://doi.org/10.1007/978-3-319-27027-2

Antitussive activity, 93
Antiulcer activity, 94–95
Antoni type A pattern, 328, 329
Antoni type B pattern, 327, 329
AOM. *See* Acute otitis media (AOM)
Apigenin, 360
Apoplast, 200
Apoptosis, 95, 275
Appearance time (AT), 115
Appetite, 630
  regulation, 633
Aqueous extract, 108
Aqueous solutions, 481, 488–489
Arbuscular mycorrhizal fungi (AMF), 58, 59
Arcuate nucleus, 402
Arginine, 503, 504
Artificial sweetener(s), 168, 212, 229, 480, 593, 596, 597, 601, 604–606, 625
Ascorbic acid, 108, 362
Aspartame (APM), 4, 7, 19, 21, 212, 229, 379, 445, 481, 482, 488, 489, 593, 599–601, 626
  absorption, distribution, metabolism of, 315–316
  acute effects of, 316
  carcinogenicity bioassays on, 323
  chronic toxicity, 317–323
  decomposition of, 315
  epidemiological studies on cancer risks and exposure, 338
  genotoxicity of, 317
  sub-acute tests, 316
  technical characteristics, production, uses and stability of, 313–315
Aspergillus nigar var. *awamori*, 498
Aspergillus oryzae, 161, 498, 543
Asteraceae, 107, 264
Atherosclerosis, 125
Atherosclerotic plaque, 355
Atopic dermatitis, 426, 427
2,2'-azino-bis(3-ethylbenzthiazoline-6-sulphonic acid (ABTS), 361

B
Bacillus subtilis, 215
Bacteria, 274, 411, 412, 414, 415, 426, 427
  biomass, 556
  expression, of brazzein, 539–541
  fermentation, 556
Bacteriostatic effect, 560
Baillon, Henri Ernest, 532
Baked goods, 561, 562, 564
Barley, 197
Beneficial consortium, 60
5-Benzyl-3, 6-dioxo-2 piperazine acetic acid (DKP), 313, 314
β-1,3-glucanase, 193
Beta-cell, 412, 420, 421
β-D-glucosyl moieties, 105
β-glucosidase, 160
Beverage(s), 125, 229, 442, 449
Bifidobacteria, 555, 562
Biliary cholesterol, 142
Bioactivation, 279
Bioavailability, 159, 278, 367
Bioconversion, 161–162
Biodistribution, 267
Biofertilizers, 59
Biofilm, 412, 414, 426, 427
Biological aspects, sweeteners, 615
Biological models, 368
Biological significance, 369
Biological system, 364, 370
Biophase, 323, 333
Biotransformation, 273
  mogrosides (*see* Mogrosides)
Birch sugar, 553
Bitterness, 115
Black licorice, 115
Blood glucose, 412, 418, 421, 550, 559, 561, 562
  control, 5
  level, 354
Blood pressure, 275, 276
Blood triglyceride concentration, 296
Body weight, 316, 318, 323, 329, 333, 396
Brainstem, 402
Brain tumors, 318, 338
Brazzein, 23, 30, 192, 528
  bacterial expression of, 539–541
  food application, 542–543
  patent analysis on, 543
  recombinant production, 540
  structure/function relationships, 534
  transgenic plants and animals, 541
  yeast expression of, 541
Breeding program, 48
Browning, 579, 580, 583, 584, 587
Bulk density, 131
Bulking agent(s), 547, 553, 561, 563, 565
Bulk sweetener, 561, 566

C

C-13 and C-19 position, 105
Cakes, 139
Caking, 579
Calcium, 275, 412, 413, 415, 416, 424, 425
Calorically dense, 2
Caloric value, 411, 423, 547, 550, 551, 554, 565
Calorie content, 290, 629
Calorie contribution, 353
Calreticulin, 197
Camellia sinensis teas, 362, 369
Cancer, 248–249, 274, 275, 635
Candida, 136
  C. guilliermondii, 132
  C. tropicalis, 135
  C. utilis, 217
Candy, 115
Capillary electrophoresis, 469
Capitulum inflorescences, 40
Caramelize, 562, 566
Carbohydrate(s), 226, 414, 419, 422, 547, 548, 550, 551, 565
  derived low-calorie sweeteners, 395
  intake, 634
Carbohydrate-free protein, 219
Carbohydrate response element binding protein (ChREBP), 422
Carcinogenesis, 635
Carcinogenicity bioassays, 313, 317, 318, 323–336
Carcinogenic risks, 396
Cardiometabolic risk factors, 239, 301–304, 634
Caries, 412, 414, 415, 417
Cariogenic agent, 414, 415, 417
Carotenoids, 363, 369
Cartridges, 451
Catalase, 365, 367
Catalytic hydrogenation, 557, 558
35S Cauliflower mosaic virus promoter, 198
Cellular antioxidant activity (CAA)
  assay, 366, 370
  values, 366
Cellular based assays, 537
Cellulase, 160
Cellulose, 133
Cephalic phase stimulation, 631
Cesare Maltoni Cancer Research Center, 323
Chemical assays, 360
Chemical conversion, 160
Chemical detection, 121
Chemical inertness, 564
Chemical methods, 363, 368
Chemical modification, 159
Chemical structure, sweeteners, 621
Chemometric, 468
  analysis, 481
  data processing, 481
  processing, 486–487
Chemopreventive effects, 275
Chemoprotective agents, 355
Chewing gum, 414, 415, 417, 426, 427
Chickens, metabolic studies of steviol glycosides in, 118
Childhood, sweeteners usage during, 9
Children, 313, 316, 338, 342, 344
Chlorogenic acids, 358, 359, 369
Cholesterol, 413, 421
Chorda tympani nerve, 537
Chromatography, 178
Chromosome aberrations, 317
Chronic lifestyle diseases, 168, 179
Chronic toxicity, 317
Circular dichroism analysis, 214
Cisgenic plants, 203, 204
Clinical changes, 316
Clinical evidence, 272
Closed cultivation system, 176
Codon modifications, 173
Codon optimization, 27, 173
Cola soft drinks, 111, 315
Collagen, 94, 413, 424, 426, 428
Colonic bacteria, 556
Columns, 455
Commercial development, 31
Commercialization, 179
Commercial product, 29
Compartmentalization, 200
Compression, 564
Conservation tillage, 50
Control samples, 110
Conventional tillage, 49
Cookies, 139
Cooling effect, 560, 562, 566
Corncob, 135
Corn flour, 29
Cough syrups, 139
C-reactive protein, 237
Crispness, 566
Crop management, 73
Crop yields, 67
Crude extracts, 368
Crystalline form, 558
Crystallization, 551, 557, 558

CSαβ fold, 533
CSαβ motifs, 25
Cucumbers, 193
Cucurbitaceae, 154
Cucurbitadienol synthase, 162
Cultivation space utilization, 176
Cumulative prevalence, 342
Cyclamate, 19, 212, 312, 452, 538, 626
Cyclamic acid, 447
Cyclooxygenase-2, 249
Cystallization, 566
Cytochrome P450 enzymes (CYP3A), 251
Cytokines, 271
Cytoplasm, 137

D

Defensin(s), 24
Defensin rs-afr2, 201
Deglucosylation, 161
Degradation, 108, 577, 579, 581, 582, 585, 587
  compounds, 112
  products, 109, 314
Deliquescence, 579, 581
Demineralization, 414, 415, 417
Demulcent, 93
De novo lipogenesis, 292, 293, 296, 297, 301, 303
De novo lipogenic enzymes, 294
Dental, 412, 414, 415, 417
  care, 417
  caries, 9, 290, 355
  cavities, 142
  health, 140
  protection, 411
Deoxyribose assay, 361
Derivatization, 468
Dextranases, 160
Dextrose, 481, 482, 488
Diabetes, 10, 227, 271, 353, 418, 420, 421, 423, 424, 428, 446, 559, 561, 564, 567
Diabetic(s), 4, 117, 316, 412, 413, 418, 419, 421, 424, 634
  patients, 267
Diet(s), 205
  soda, 314, 338, 340, 341
Dietary carbohydrates and sugars, 287
Dietary-flavonoids, 360
Dietary Guidelines for Americans (DGAs), 617
Dietary pattern, 629
Dietary polyphenols, 369
Dietary recommendations, 548
Dietary sugar, 237
Dietary supplement(s), 103, 123, 352
Dietary Supplement Health and Education Act (DSHEA), 124
7,12-dimethylbenz(a)anthracene (DMBA), 253
Dioscoreophyllum cumminsii, 212
1,1-diphenyl-2-picrylhydrazyl (DPPH) radical, 361, 363
Directly quench free radicals, 364
Disaccharides, 239, 288
Disulfide bridges, 533
Diterpene glycosides, 356
Diuresis, 275
DNA oxidative damage, 365, 370
DNA protection, 365
Dosage forms, 280
Dose-dependent, 318, 362
Dried seafood, 116
Drinks, 115
Drought, 70
Drug discovery, 98
Drying, 68
Dry leaf yield, 67
Dulcin (DUL), 456
Dulcoside A, 104
Dyslipidemia, triglycerides. *See* Triglycerides

E

Ear, 426, 429
  infections, 425, 564
Eating pattern, sweetener, 615
EFSA. *See* European Food Safety Authority (EFSA)
Electricity costs, 177
Electroosmotic flow, 469
Electrostatic potentials, 192
Enamel, 412, 415, 417
Endogenous antioxidative systems, 358
Energy drinks, 138
Energy expenditure, 301
Energy intake, 629–630
Energy output, 304
Energy value, 548, 550, 551
Enhance taste, 196
Enoxolone, 246
Enteroendocrine cells, 398
E-number, 442
Environmental water, 450
Enzymatic conversion, 160
Enzymatic hydrolysis, 121
Enzyme immobilization, 161

Epidemiological case–control studies, 336–341
Epidermal growth factor receptor, 248
ER-retention signals, 200
Erythritol, 549, 552, 554, 556, 559, 561, 566
Erythrocytes, 140
Escherichia coli, 215, 498, 539
Esterases, 315
Estimated daily intake (EDI), 620
Estrogens, 237
European Food Safety Authority (EFSA), 103, 124, 313, 314, 343, 344, 619
Excipients, 442
Exercise, 300–301
Exg1, 162
Exogenous antioxidants, 358
Expectorant, 93
Expression cassette, 170
Expression systems, 215
Extinction time (ET), 115
Extraction, of *S. rebaudiana*, 107
Extra-hepatic insulin sensitivity, 299–300
(E,Z)-2,6 nonadienal, 196

F

Fat, 413, 421, 423
Fat and protein content, 633
Fermentation, 161, 411, 414, 554, 559
Fermenter, 542
Ferric reducing/antioxidant power (FRAP) assay, 361
Ferric reducing power, 362
Ferrous ion chelating ability, 362
First-pass metabolism, 267
F2-isoprostanes, 370
Flavones, 360
Flavonoid(s), 89, 360, 369
  concentration, 362
  glycosides, 359, 360
Flavonols, 360
Flavour(s), 217, 633
Flavoured milk, 117
Flavour enhancer, 204, 552, 577
Flowering, 39, 66
5-Fluorocytosine, 203
Folin–Ciocalteu method, 361
Food(s), 442
  additive(s), 230, 278, 352, 369, 548–550, 552, 553
  industry, 535
  matrix, 367
  reward system, 633
  supplement, 220
Food and Agriculture Organization (FAO), 124, 445
Food Drug Administration (FDA), 123, 143
Food Safety and Standards Authority of India (FSSAI), 124
Food Standards Australia New Zealand (FSANZ), 124
Formaldehyde, 315, 323
Free fatty acid (FFA), 420, 421
Free radical(s), 252, 357
Freezing point, 565
Fructokinase, 234, 292
Fructolysis, 293
Fructolytic enzymes, 294
Fructooligosaccharides (FOS), 623
Fructose, 227, 621
  malabsorption, 304
Fruit, 201
  beverage products, 369
  cream, 315
  ripening, 172
  structure, 173
  yield, 177
  yogurt, 315
Full production chain, 73
Functional components, 355
Functional T1R2-T1R3 receptor, 397
Fungal diseases, 65
Fungi, 274

G

Galactosidase, 160
Gamma-thionins, 24
Gas chromatography (GC), 468
Gas chromatography-electron capture detector, 468
Gastric emptying, 413, 423
Gastrointestinal effects, 554, 555
Gastrointestinal flora, 559
Gastrointestinal peptid, 631
Gastrointestinal problems, 143
Gastrointestinal tract, 120, 315, 379
Gavage, 315, 317
gdhA promoter, 498
GD-Searle, 313, 318
Gender(s), 304, 318, 341, 342
Gene dosage, 174
Generally recognised as safe (GRAS), 4, 20, 27, 123, 620
Gene silencing, 172
Genetic backgrounds, 173
Genetic engineering, 170

Genotoxicity, 317
Germ flour, 28
Ghrelin, 401
Gibberellins (GAs), 71
Glabridin, 92
Global trends, sweeteners, 620
Glucagon, 269
Glucagon-like peptide-1 (GLP-1), 230, 630
Glucanase, 160
Gluconeogenesis, 299, 354
Glucose, 140, 227, 269, 411, 412, 418, 420, 422, 423
  absorption, 387, 412, 419
  metabolism, 233
  production, 296
  tolerance, 354
  uptake, 270
  utilization, 354
Glucosensors, 398
Glucose-6-phosphatase, 294
Glucose transporters, 292
  GLUT1, 292
  GLUT2, 404, 634
  GLUT3, 292
  GLUT4, 271, 292
  GLUT5, 293
Glucuronic acid, 132
Glucuronidation, 266
Glucuronides, 278
Glutamic/aspartic acid, 503
Glutathione (GSH), 316
Glutathione reductase (GR), 316
Glycemia, 419
Glycemic control, 411, 412, 418, 419, 429
Glycemic index (GI), 131, 228, 237, 411, 419, 423, 547, 550, 559, 564, 565, 621
Glycemic load (GL), 233, 237
Glycemic response, 10, 235
Glycocalyx, 427
Glycolytic pathway, 18, 141
Glycoprotein, 169
Glycoside hydrolases, 160, 161
Glycoside sweeteners, 20
Glycosidic bonds, 155, 161
Glycosidic losses, 122
Glycosylation capacity, 170
Glycyrrhetinic acid, 92, 246
Glycyrrhiza glabra, 89, 246
Glycyrrhiza glandulifera, 90
Glycyrrhizin, 91, 246
Glyoxalase I, 256
Gold standard, 364
Good Agricultural and Collection Practices of Medicinal and Aromatic Plants (GACP-MAP), 44
gpdA promoter, 498
G-protein-linked taste receptors, 18
“Green” analytical technique, 481
Green coffee, 369
Green tea, 366
Grouping, steviol glycosides, 105
Growing market demands, 38
Growth, 412, 414, 415, 426, 427
  inhibiton, 412
  regulators, 54
Guaraní indigenous, 352
Guaraní people, 38
Gustatory, 26
Gut malabsorption, 304
Gut microbiota, 388, 632
Gymnemic acid, 169

H

Haematological neoplasias into multiple organs, 343
Hairy root, 197
Hamsters, metabolic studies of steviol glycosidesin, 118
Hansenula polymorpha, 216
Harvard study, 340
Harvesting regimes, 66
Harvest time, 66, 72
Hausner ratio, 132
Health Canada, 124
Health food industry, 351
Health recommendations, 617
Health status, 234, 615
Heat of solution, 551, 552, 562, 564
Heat shock protein 18.2 (HSP) terminator, 172
Heat stability, 30
Hemangiosarcomas, 336
Hemicellulosic xylan, 143
Hepatic insulin sensitivity, 299
Hepatic steatosis, 297
Hepatitis, 95, 252
Hepatitis B virus, 274
Hepatocarcinoma, 333
Hepatocellular carcinoma, 95
Herbaceous shrub, 39
Herbicide applications, 63
Heterologously monellin, 218
Heterologous production, 170, 539
High caloric food, 212

Higher temperature, 112
High fat diet, 421, 423
High fructose corn syrup (HFCS), 233, 234
High-intensity sweeteners (HIS), 623
High-Mobility Group Box 1, 249
High performance liquid chromatography (HPLC), 121
High-performance thin-layer chromatography, 468
High potency sweeteners (HPS), 3, 113
High-resolution mass spectrometry (HRMS), 111, 462
HILIC-MS/MS method, 121
Histiocytic sarcoma, 328, 332
HIV, 251
^{1}H NMR spectroscopy, 122
Homozygous transgenes, 174
Hot-water extraction, 107
hT1R2–hT1R3 receptor, 170
Human cells, 364, 370
Human consumption, 125
Human studies, 367
Human subjects, 421
Humectant, 21, 552, 553, 563, 564
Hyacinth, 196
Hybrid analyzers, 466
Hydrochloric acid (HCl), 160
Hydrogen peroxide, 133
Hydrolyzation, 557
Hydrolysates, 132
Hydroxyl radical(s), 362
  scavenger activity, 363
Hydroxymethyle furfural, 133
11β-Hydroxysteroid dehydrogenase, 255
Hygroscopicity, 552, 561, 566
Hypercholesterolemia, 141
Hyperglycemia, 141, 420, 421
Hyperlipidemia, 212, 231
Hypertension, 125, 232, 277
Hypertensive, 275
Hyperuricemia, 235
Hyphae, 201
Hypoglycemia, 269
Hypokalemia, 97
Hypothalamus, 403

I

Ice-tea, 117
Immune-modulator, 96
Incretin signals, 634
Industry pressure, 615
Inflammation, 237
Influenza-A H1N1 virus, 96
Infusion system, 419
Inosine monophosphate, 235
Insect damage, 65
Insect pollinators, 52
Insect-repellant qualities, 64
Insulin, 96, 354, 412, 418, 422, 429
  resistance, 231, 298, 299, 412, 420, 421
  sensitivity, 270
Insulinemic index(es), 559, 563
Insulinotropic effect, 267, 269
Integrated pest management (IPM), 65
Integrated weed management, 63
Intense sweeteners, 593, 594
  chemical and physical properties, FDA approved, 597–605
  safety aspects, 606–609
Intensive sweeteners, 41
Intercellular layer space, 171
Intercropping, 49
Interim sacrifices, 318
Interleukin 1β, 249
Internal standard, 466
International Agency for Research on Cancer (IARC), 344
International Conference on Harmonization (ICH), 109
Inter-species, 273, 278
Intestinal microbiota, 230
Intracellular antioxidant activity, 366
Intrahepatic fat, 295, 297–299, 301
In vitro biological, 364
In vitro proliferation, 95
In vitro screening, 364
In vitro studies, 317
In vivo absorption, 370
In vivo antioxidant effects, 369
In vivo proliferation, 95
In vivo studies, 317
Ion exchange resins, 138
Irrigation scheduling, 60
Isodityrosine bridges, 133
Isomalt, 549, 552, 555, 557, 559, 561, 562, 566
Isosteviol, 274
Isotachophoresis, 469

J

Joint FAO/WHO Expert Committee on Food Additives (JECFA), 103
Juices, 117

K
Katemfe, 186, 495
Ketohexokinase, 236
Kinetics, 578, 585
Kluyveromyces lactis, 216, 541
K-nearest neighbor (KNN), 481, 487
Kre6, 162

L
Laccases, 135
Lactate production, 296
Lactitol, 549, 553, 555, 557, 559, 561, 562, 566
Lactobacilli, 562
Lactococcus lactis, 540
L-amino acids, 24
Large-volume injection, 450
L-aspartic acid, 313
Laxative effect(s), 550, 554–556, 566
Laxative threshold, 554, 556
LC-MS/MS, 464
LC-TOF method, 122
Leguminosae, 90
Leishmaniasis, 254
Lemon-lime soft drinks, 111
Leptin, 401
Lettuce. *See* Transgenic lettuce
Licochalcones, 94
Licocoumarone, 95
Licorice, 89
Life-span exposure, 330
Lifestyle changes, 615
Light exposure, 73
Lignin, 133
Lignocellulosic materials, 132
Limits of quantification (LOQ), 465
Lipid(s), 413, 421
  metabolism, 420–422
  peroxidation, 94, 365, 368, 370
Lipogenesis, 234, 421
Lipogenic pathway, 235
Lipoprotein metabolism, 237
Liquid chromatography, 454–458
Liquiritin, 92
Liquorice, 246
Listeria monocytogenes, 363
Liver cells, 366
Liver toxicity, 317
Long-day conditions, 46
Low-calorie, 561, 562
  materials, 2
  soft drinks, 116
  value, 3
Low calorie sweeteners (LCSs), 395, 445, 593
  risk assessment, 597
  types, 395
  uses, 395
Low pH, 112
Low-potency sweeteners, 3
L-phenylalanine methyl ester, 313, 314–315
Luo Han Guo, 625
Luo-Han-Kuo (LHK), 154, 155, 157, 159, 160, 163
Luteolin, 360
Lymphocytic lymphoma, 328, 330
Lymphoimmunoblastic lymphoma, 328, 331
Lysine, 501, 502

M
Mabinlin, 528
Macronutrient deficiency, 57
Magnesium isoglicyrrhizinate, 251
Maillard reaction, 553, 562, 563, 566, 577, 581, 583, 587
Maize plants, 541
MALDI-MS, 203
Malignant schwannomas, 327, 329
Malignant tumors, 318, 326, 330, 333, 342
Maltase, 160
Maltitol, 549, 553, 555, 557, 559, 562, 566
Maltitol powder, 551, 553, 557
Maltodextrin, 289, 481, 482, 488
Maltose, 557
Mammary adenocarcinomas, 330
Mannitol, 549, 553, 556, 558, 559, 562–563, 565, 566
Mass balance, 109, 110
Mass spectrometry, 122, 454–457
Matrix-matched calibration, 466
Matrix metalloproteinase, 254
Mature age, 343
Mature thaumatin, 198
Maximum permitted levels (MPL), 314
Medicinal plants, 89
Meta-analysis, 277
Metabolic disorders, 550
Metabolic dysfunctions, 231
Metabolic effects, 628
Metabolic syndrome, 125, 143, 232, 264
Metabolism, 118, 378, 548, 554, 556, 563, 565
Metabolization, 366
Methicilin-resistant *Staphylococcus aureus*, 255

Micellar electrokinetic chromatographic method, 470
Microbial activities, 417
Microbial load, 363, 369
Microbial production, 215
Micronuclei, 317
Microorganisms, 93, 142, 273
Micro-propagation, 51
Milk chocolate, 315
Mineral fertilizers, 56
Minor degradation compounds, 111
Minor steviol glycosides, 356
Miraculin, 529
 accumulation, 173
 concentration, 177
 definition, 168
 dimerization, 169
 heterologous production of, 170
 sweetening effect of, 169
 sweetness receptor, 169
 taste-modifying effect of, 169
 taste-receptor membrane, 169
 transgenic lettuce, 175
 transgenic tomatoes, 171–174, 176–178
Mäkinen, 417
MNEI-mutants, 214
Mock beverages, 109
Moderately salt-tolerant plant, 62
Modulate antioxidant enzymes and inhibit oxidases, 360
Mogrol, 154, 155, 157, 159, 161
Mogroside(s), 154
 bioconversion, 161
 biosynthesis, 162
 chemical conversion, 160
 extraction and contents, 155–158
 functional properties, 159
 in vitro enzymatic conversion, 160–161
 mogroside III, 157, 160
 mogroside V, 157, 159, 161
 structure, 154–155, 158
 sweetness, 157–158
Molecularly imprinted polymer, 454
Molecular weight, 552, 553, 563, 565
Monellin, 192, 212, 528
Monoculture, 49
Monocytic leukemia, 239, 288, 328, 333
Monosaccharides, 288
MS medium, 54
Mulaithi, 90
Multinational companies, 43
Multiple harvests, 67
Multiple malignant tumors, 342
Multiple myeloma, 341
Multiple reaction monitoring (MRM), 462, 463
Multiple transgene copies, 174
Mutagenesis, 26
Mutagenic, 317
Mutans streptococci, 560
Mutarotation, 576, 577, 583
Mycobacterium tuberculosis, 274
Mycorrhizal infection, 59
Myeloid leukemias, 328, 332

N

N-acetylmuramoyl peptide, 96
NADPH, 136
Nagoya protocol, 49
Native to Paraguay, 107
Natural antioxidants, 370
Natural caloric sweeteners, 231, 352
Natural Health Products (NHPs), 124
Natural raw material, 351
Natural substitutes, 548
Natural sugar substitutes, 550
Natural sweetener, 219, 621
Near infrared (NIR) spectroscopy, 121
Negative ion mode, 463
Neoculin, 528
Neohesperidin dihydrochalcone (NHDC), 456, 538, 626
Neoplasms, 635
Neotame, 7, 312, 383, 628
Neuro-astroglial system, 404
New generation sweeteners, 396
N fertilization, 69
N form, 70
Nickel-immobilized metal-affinity chromatography (IMAC), 178
Nitric oxide, 362
Non-alcoholic fatty liver disease, 286
Non-caloric sweetener, 19, 105
Non-calorific values, 130
Non-cariogenic substance, 560, 562, 564
Non-covalent interactions, 219
Non esterified fatty acid (NEFA), 412, 413, 420, 421
Non-fasting blood glucose (NFBG), 418
Non-glycemic sweeteners, 19
Non-Hodgkin lymphoma, 341
Non-neoplastic lesions, 318
Non-nutritive sweeteners (NNS), 228, 378–388, 607, 608

Non-specific binding hypothesis, 536
Normoglycemic, 412, 413, 419, 420, 423
North-eastern Paraguay, 38
Not-sweet constituents, 356
Novel food, 43
NPY/AgRP/GABA-neurons, 402
Nuclear factor-kB, 249
Nuclear magnetic resonance (NMR), 111
  folding, 26
  structure, 534
Nucleus accumbens, 403
Number of animals, 313, 323
Nurses' Health Study, 340
Nutrient requirement, 56
Nutritional and osmotic effects, 631
Nutritional composition, 355
Nutritional doses, 56
Nutritional value, 548, 549, 563
Nutrition labels, 618
Nutritive sweeteners, 228

O
Oat(s), 131
OAT1, 267
OAT3, 267
Obese individuals, 316
Obesity, 9–10, 21, 212, 412, 413, 420, 423, 616
Off-taste, 117, 535
Online SPE, 452
Open reading frame (ORF), 162
Operating costs, 177
Opioid receptors, 277
Optical spectroscopy, 481
Oral cavity, 560, 564
Oral ingestion, 315
Orbitrap, 463, 466
Orexigenic properties, 401
Orexins, 401
Organic acids, 108
Organic fertilizers, 57
Organoleptic analysis, 24, 203
Osmolality, 565
Osmotic adjustment, 70
Otitis media, 142
Overcompensation, 632
Overweight, 232, 548
Oxidants, 357
Oxidative damage, 358
  to lipids, 370
  markers, 370
Oxidative phosphorylation, 364, 369
Oxidative stress, 237, 271, 358, 413, 423
  in vivo, 367
  protective agents against, 365
Oxidized low density lipoprotein (ox-LDL), 367, 370
Oxygen radical absorbance capacity (ORAC) assays, 362, 364, 366

P
Palatability, 631
Palmitic acids, 134
Parenteral nutrition, 140
Pasteurization, 586
Patent analysis, on brazzein, 543
Patent applications, 48
Pectin, 138
Pentadin, 528
Pentadiplandra brazzeana, 532, 533
Peptidases, 315
Perennial crop, 67
Pericarp, 218
Peroxidase enzymes, 363, 369
Peroxisome, 137
Peroxisome proliferation activated receptors, 96, 249
Peroxyl radical scavengers, 363, 364
Pests and diseases, 64
pH range, 111
Pharmaceutical(s), 447, 454
Pharmaceutical industries, 130
Pharmacokinetic(s), 279
  of steviol, 119
  studies, 265–267, 316
Pharmacophors, 97
Pharmacopoeias, 97
Phase I metabolites, 273
Phenolic compounds, 89
Phenolic content, 358
Phenolic metabolome, 358, 360
Phenylketonuria, 21
Phosphatidylcholine hydroperoxide, 370
Phosphoric acid, 160
Phosphorus nutrition, 59
Photodegradation, 464
Photoperiodism, 46
Photostability, 109
Photosynthesis, 177
Physical activity, 304, 629
Physical properties, 552
Physiological properties, 445

Phytopathogenic fungi, 197, 204
Phytotherapeutics, 97
Pichia pastoris, 216, 499–500, 541
Pigs, metabolic studies of steviol glycosidesin, 118
Pioglitazone, 96
Plant-based expression systems, 187
Plant density, 55
Plant factory, 175–177
Plant growth retardants, 72
Plant growth promoting rizhobacteria (PGPRs), 59
Planting time, stevia, 55
Plaque, 412, 414, 416, 417, 560
Plasma levels, 119
Plastic mulch, 50
Pneumococci, 426
Policy actions, 617
Polyalcohol, 623
Polyhydric alcohols, 550, 565
Polyol(s), 131, 424, 428, 429, 550, 552, 554, 556–558, 566
Polypeptide chains, 214
Polyphenol(s), 91, 239, 271, 288, 358, 361, 369
  oxidase, 363, 369
Polysaccharides, 289
POMC/CART neurons, 402
Positive reinforcement, 632
Postprandial, 412, 419, 423
Posttranslational modifications, 173
PR5 protein, 187
Prebiotic(s), 555, 562, 565, 567, 576, 577, 587
Precursor ions, 467
Pregnancy, sweeteners usage during, 8
Pregnane X receptor, 251
Pre-harvest factors, 69
Prenatal life, 318, 323, 328–336
Preprothaumatin II, 188, 189
Pressurized liquid extraction, 453
Primary antioxidant properties, 362
Primary energy substrates, 291
Principal component analysis (PCA), 481, 486
Probiotic effect, 555
Processed foods, 449
Processed fruits, 363
Product reformulation, 629
Professional Follow-up Study, 340
Proliferators activated receptor gamma (PPARγ), 422
Promoter(s), 28, 171, 175
Proof-of-concept, 28
Proper machinery, 68
Protein accumulation, 28
Proteinase degradation, 191
Proteinase inhibitor, 191
Protein-derived low calorie sweeteners, 395
Protein sweeteners, 22, 30
Purification, 177, 543
Purified recombinant thaumatin, 203
Purified steviol glycosides, 368
Pustulanases, 160
Pyridoxal, 5′-phosphate (PLP), 502
Pyroglutamic acid, 24, 533

Q

QqQ MS, 463
Quadrupole time-of-flight, 462
Quality control, of stevia, 44
Quantification, 454–457
Quantum satis, 552
Quaternary structures, 531
Quercetin, 360

R

Radioactivity, 119
Raman spectroscopy, 468, 481, 483, 486, 489
Ramazzini Institute, 323, 342
Randomized controlled trials (RCTs), 396
Randomized trial, 272
Raney nickel, 132, 557
Rat feces, 119
Rats, metabolic studies of steviol glycosidesin, 118
Raw material, 38
Reactive oxygen species (ROS), 358
  scavengers, 368
  scavenging capacity, 363
Reb A, 48, 72, 351, 363
Rebaudioside A, 104, 266, 356, 383
Rebaudioside C, 104
Rebaudioside E, 103
Recombinant brazzein, 26
Recombinant miraculin. *See* Miraculin
Recombinant protein, 215
Recombinant thaumatin, 197, 203
Recombinant thaumatin II, 188
Recombinant yeast strain, 216
Regulatory framework, 619
Relative sweetness, 157, 192
Remineralization, 412, 413, 415, 417, 424
Respiratory tract infections, 425

Reverse causality, 616
Reward systems, 632
Rhinosinusitis, 425, 427
Rhizobium etli, 160
Rhizosecretion, 198
Rice straw, 135
Richadella dulcifica, 168
Root(s), 91
Root beer soft drinks, 111
Rotation, 49

S
Saccharides, 104
Saccharin, 7, 19, 212, 229, 383, 447, 481, 482, 488, 489, 593, 594, 602, 626
Saccharomyces cerevisiae, 135, 162, 216, 499, 541
Saccharomyces genome database (SGD), 162
Saccharose/sucrose, 621
Safe sweetener, 130
Safety, 352, 427, 429
  assessment, 177
  consideration, 4
  thaumatin, 204
  xylitol, 429
Salt adaptation, 61
Saponins, 20, 246
Satiety, 630, 631
Screening, 370
Secondary antioxidant ability, 362
Secretion of lipoproteins, 294
Sedative effects, 278
Seed(s), 40
  formation, 40
  germination, 52
  propagation, 51
Selected ion monitoring, 457
Semi-synthetic fillers, 549
Sensitive experimental model, 342
Sensory evaluations, 535
Sensory properties, 632
Sensory stimulation, 633
Serine protein inhibitors, 24
SGD. *See* Saccharomyces genome database (SGD)
Shelf life, 139, 578, 585, 588
Short-chain volatile fatty acids, 556
Short-day plant, 46
Siamenoside I, 157, 161
Single-copy gene, 174
Single quadrupole, 457
Single-step purification, 178
Siraitia grosvenori. *See* Luo-Han-Kuo (LHK)
Site-directed mutagenesis, 534
Site-specific agronomic techniques, 44
Skin lesions, 427
Sludges, 452–454
SOD mRNAs, 367
Sodium-glucose linked transporter (SGLT1), 404, 634
Soft drinks, 111, 116, 231, 233, 237, 455
Soil water consumption, 61
Solid phase extraction (SPE), 448, 453
Solid phase microextraction, 450
Solubility, 551, 558, 563, 566
Sorbents, 451
Sorbitol, 139, 481, 482, 488, 549, 553, 555, 560, 563, 566
Soy sauce, 116
Sporophytic self-incompatibility, 52
Sprague Dawley rats, 318, 323–330
Spray dried, 108
SRM transitions, 464
SSB consumption, 233
Stability, 577, 579, 583
Stability of steviol glycosides, 108
Stable expression, 176
Standardized method, 364
Staphylococcus aureus, 427
Staphylococcus epidermidis, 427
Starch, 289
Starting materials, 51
S-taste fibers, 26
Steatosis, 272
Stem cuttings, 51
Sterol response element binding protein 1c (SREBP-1c), 422
Stevia, 125, 264, 623
  crude extracts, 363, 368
  feeding, 368
  leaves, 42, 270
Stevia plants
  area selection, 47
  on film, 50
  global cultivation, 42
  growth rate, 43
  in vitro culture, 54
  organic production, 45
  production costs, 43
  varietal diversity in, 47
  varieties, 47

Stevia rebaudiana, 102, 350, 351
anti-cariogenic activity, 355
antimicrobial properties, 354
antioxidant(s), 357–358
antioxidant activity, 360, 362
atherosclerosis, 355
Bertoni, 279
crude extracts, 363
diabetes and blood glucose maintenance, 353
flavonoids, 360
functional components, 355
hypertension, 354
nutritional composition, 355
nutritional implications, 353
obesity, 353
phenolic composition, 358, 359
Stevia's crude extracts, 351
worldwide demand, 41
Steviol, 104, 118, 264, 279, 280
derivatives, 363
equivalents, 352
glucuronide, 118, 265, 363
Steviol glycosides, 38, 125, 265, 351, 352, 445
analysis, 120
growing market for, 41
Stevioside, 104, 265, 351, 356, 363, 383
Sticking, 579
Strawberries, 195
Streptococcus, 140
S. mutans, 412, 414, 415, 564
S. pneumoniae, 426
Stress response, 69
Strong sweetness, 218
Study population, 341
Sub-acute tests, 316–317
Subcellular targeting, 174
Subsidiary substrates, 291
Sucralose, 8, 192, 229, 312, 383, 384, 447, 452, 538, 626
Sucromalt, 623
Sucrose, 18, 113, 229, 246, 312, 313, 411, 414, 417, 419, 423, 427, 428, 446, 481, 482, 488
Sugar(s), 226, 363
alcohol(s), 130, 288, 550, 552–554, 557, 559, 563–566
consumption, 291
free alternatives, 2
substitute(s), 204, 353
Sugarcane bagasse, 135
Sugar-free food, 562, 564
Sugar-like taste, 204
Sugar-sweetened
beverages, 231–233, 621
drinks, 227
foods, 227
Sugary drink, 616
Sulfuric acid (H_2SO_4), 160, 161
Superoxide anion, 362
Superoxide dismutase, 365
Supportive evidence, 370
Survival, 318, 323, 329
Sustainability, 38
Sustainable production, 57
Sweetener(s), 18, 168, 170, 179, 384, 411, 424, 428, 442, 548, 550, 560–563, 565, 566, 614
consumption, 618
opportunities, 637
potency exceeding, 3
strengths, 636
threats, 637
weaknesses, 636
Sweetening effect, 169
Sweeter, 353
Sweet herb, 350
Sweet herb of Paraguay, 102
Sweetness, 196, 385, 411, 428, 550, 551, 560, 564, 566
intensity, 536
receptor, 170
Sweet proteins, 215, 625
Sweet taste, 191, 384
preference, 228, 289, 305, 397
receptors, 290
signaling, 398
Sweet-tasting proteins, 168, 528
Sweet wood, 90
Swine flu, 96
SWOT analysis, 636
Synergistic and antagonistic interactions, 367
Synthesis, 556, 564
Synthetic sweeteners, 548

T
Table-top artificial sweeteners, 481
Table-top sweeteners, 115
Tagatose, 576, 587
Tan seed, 41
Targeting sequences, 28

Taste, 385
  modifier, 204
  receptor, 22, 385
  receptor type 1, member 2, 537
Taste-modifying protein, miraculin. *See* Miraculin
Taxation policies, 618
Teeth remineralization, 561, 564
Temperature, 111, 218
Terminator, 172
Test model, 364
Tetramer, 169
Thaumatin, 186, 189, 212, 445, 528
  characteristics of, 496
  expression, 199
  primary structure of, 497
  proteinaceous sweetener, 495–496
  recombinant production, 498
  site-directed mutagenesis, 503–507
  sweetness-determinants, 501–503, 507
  sweet-tasting proteins and receptors, 507, 510
  thaumatin I, 496
  thaumatin II, 496
  thaumatin A, 188
  thaumatin B, 188
  three-dimensional structure, 189, 510–517
Thaumatin-like proteins, 193
Thaumatococcus daniellii Benth, 186, 495, 496
Thermal stability, 179, 562, 564
Thermic effect, 630
Thermogenic response, 141
Thermoresistance, 191
Thermostable, 24, 218
Thin layer chromatography (TLC), 120
Thiobarbituric acid reactive substances (TBARS), 365
Tissue-specific E8 promoter, 172
Toll-like receptor-4, 249
Tomato, 193, 217. *See also* Transgenic tomatoes
Tooth, 414, 417
  decay, 635
  demineralization, 412, 560
  mineralization, 560
Tooth-friendly, 561
Total malignant tumors, 326
Toxicokinetics, 119
Toxicological studies, 143
T7 promoter, 215
T1R family, 22
T1R2/T1R3, 18
  receptor, 192, 538
  receptor heterodimer, 404
  sweet taste receptor, 538
T2R receptors, 22
Traceability, 73
Trace analysis, 448
Traditional foods, 125
Transcript expression, 218
Transgene, 218
  copies, 200
  silencing, 171, 175, 176
Transgenic fruits, 217
Transgenic lettuce, 175–176, 218
Transgenic lines, 218
Transgenic mice, 317, 542
Transgenic plant, miraculin. *See* Miraculin
Transgenic tomatoes, 171, 218
  production of, 176–177
  purification, 177–178
  recombinant miraculin protein, expression of, 171–174
Transgenic vegetables, 217
Transitional cell carcinomas, of renal pelvis, 326
Translational efficiency, 173
Tridentate, 142
Trifluoroacetic acid (TFA), 160
Triglyceride(s), 287
Triglyceride synthesis, 234
Triokinase, 292
Triple quadrupole, 462
Triterpenoid, 154, 159, 160
Tromboxane synthase, 249
Tumor, 275, 342
Tumor Necrosis Factor-α, 249
Turku, 417
Twin rows arrangement, 71
Type 2 diabetes (T2D), 21, 125, 212, 232, 413, 418, 420, 423
Type 2 diabetes mellitus, 634

U

Ubiquitin, 175
Ultra-HPLC methods with electrospray ionization mass spectrometry (UHPLC-MS), 122
Ultrasonic bath, 449
Uncouplers, 364
Uric acid, 236

Urinary excretion, 120
US Food and Drug Administration, 620
US National Cancer Institute (NCI), 317, 340
US National Toxicology Program (NTP), 318

V

Vagus, 402
Vasodilation, 275, 276
Vegetative portions, 53
Verocay bodies, 328, 329
Visceral adiposity, 234
Viscosity, 552, 565
Vitamin C, 369

W

Wastewater, 450
Water-deficit stress, 61
Wedge model, 192, 506
Weed competition, 62
Weight, 411, 413, 422, 423, 628
- gain, 378
- loss, 232
- sweetener effect on, 632–633

Wister rats, 316
World Health Organisation (WHO), 124, 246, 344, 445, 617

X

Xylanase, 193
Xylitol, 410–429, 549, 553, 556, 558, 560, 564, 566
Xylitol dehydrogenase, 136
Xylose assimilation, 558
Xylose reductase, 137

Y

Yeast expression, of brazzein, 541
Yeast expression systems, 216
Yerba mate, 369
Yogurt, 116

Printed by Printforce, the Netherlands